# Proceedings

Today, a steadily growing store of information is called for in order to understand the increasingly complex technologies used in modern automobiles. Functions, modes of operation, components and systems are rapidly evolving, while at the same time the latest expertise is disseminated directly from conferences, congresses and symposia to the professional world in ever-faster cycles. This series of proceedings offers rapid access to this information, gathering the specific knowledge needed to keep up with cutting-edge advances in automotive technologies, employing the same systematic approach used at conferences and congresses and presenting it in print (available at Springer.com) and electronic (at SpringerLink and Springer für Professionals) formats.

The series addresses the needs of automotive engineers, motor design engineers and students looking for the latest expertise in connection with key questions in their field, while professors and instructors working in the areas of automotive and motor design engineering will also find summaries of industry events they weren't able to attend. The proceedings also offer valuable answers to the topical questions that concern assessors, researchers and developmental engineers in the automotive and supplier industry, as well as service providers.

Peter E. Pfeffer
Editor

# 6th International Munich Chassis Symposium 2015

chassis.tech plus

Springer Vieweg

*Editor*

Prof. Dr. Peter E. Pfeffer
Munich University of Applied Sciences
Munich, Germany

ISSN 2198-7432
ISSN 2198-7440 (electronic)
ISBN 978-3-658-09710-3
ISBN 978-3-658-09711-0 (eBook)
DOI 10.1007/978-3-658-09711-0

Library of Congress Control Number: 2015937195

Springer Vieweg


Printed on acid-free paper

Springer Vieweg is part of Springer Science+Business Media
(www.springer.com)

# WELCOME

Connectivity has arrived in the vehicle – whether it is in-car internet or car-to-car communication. For the chassis too, the connected car is increasingly becoming a driver of innovation. Predictive and intelligent chassis systems and automated driving are just some of the topics being addressed. In addition to enhancing driving comfort and safety, interconnecting the powertrain with the chassis can also provide new functions, not only in cars but also in commercial vehicles. What is more, modularization, electrification of the powertrain, intelligent development methods and efforts to reduce fuel consumption are also driving innovations in chassis systems.

On 16 and 17 June 2015, the 6th International Munich Chassis Symposium chassis.tech plus will bring together experts in the field of the chassis, steering systems, brakes, and wheels / tires for an exchange of ideas and experience and constructive discussions on the latest topics. They will give us an insight into their research and development activities. On behalf of the Scientific Advisory Board, we would like to cordially invite you to take part in this year's symposium.

Renowned keynote speakers will give a comprehensive overview of challenges and solutions for chassis technology. In the well-established manner, four parallel sections will address issues relating to the chassis, steering systems, brakes, and wheels / tires. The program of lectures will be rounded off by two plenary sections at the beginning and end of the symposium.

We look forward to welcoming you at the Hotel Bayerischer Hof in the heart of Munich and we hope you enjoy an exciting symposium.

Prof. Dr. Peter E. Pfeffer
Munich University of Applied Sciences
Scientific Director of the Symposium

# INDEX

## CHASSIS.TECH PLUS SECTION

### KEYNOTE LECTURES

**The all new BMW 7 Series** 1
Peter Langen, M. Wachinger, Dr. C. Dorrer, W. Nixel, M. Schwarz, BMW Group

**Highly automated driving for commercial vehicles** 5
Markus Kirschbaum, Daimler AG

**Automated driving, electrification and connectivity – the evolution of vehicle motion control** 17
Alexander Häußler, Robert Bosch GmbH

### CONSUMER PROTECTION AND METHODS

**Future consumer protection demands on vehicle safety** 33
Andreas Rigling, ADAC e. V. Technik Zentrum

**Model-based development methods – What can chassis and powertrain development learn from each other?** 35
Bernhard Schick, M. Paulweber, AVL List GmbH, Austria

# PARALLEL STRAND I

## NEW CHASSIS SYSTEMS

**The chassis of the all new Audi Q7** 37
Carsten Jablonowski, V. Underberg, M. Paefgen, AUDI AG

**Network topology for chassis – potential of ethernet-based systems** 51
Kristian Trenkel, P. Wunner, iSyst Intelligente Systeme GmbH

**Suspension design of the Visio.M electric research vehicle** 79
Andreas Schultze, T. Helfrich, Prof. Dr. M. Lienkamp,
Institute of Automotive Technology (FTM), TU Munich

## ACTIVE CHASSIS SYSTEMS

**Development of an active motion system of tire contact point control** 95
Hiroshi Shibuya, H. Iida, H. Kanayama, D. Fujii, X. Carrera Akutain,
K. Shima, Toyota Motor Corporation, Japan

**The influence of the modeling depth of active chassis systems with respect to the development stage and their interaction with driving characteristics** 103
Kilian Dettlaff, Prof. Dr. J. Wiedemann,
Institute for Internal Combustion Engines
and Automotive Engineering (IVK), University of Stuttgart;
U. Schaaf, I. Scharfenbaum, Dr. A. Wagner, AUDI AG

**Smart electromechanical system to improve vehicle handling and stability by toe and camber control on the rear wheel** 123
Isabel Ramirez Ruiz, Ferrari S.p.A., Italy;
Dr. M. Alirand, N. Kieny, Siemens Industry Software SAS, France;
Prof. F. Cheli, Politecnico di Milano, Italy

# PARALLEL STRAND II

## ENERGY EFFICIENCY, SAFETY AND RESOURCES

**Tire use and road safety – background to policy recommendations for new EU measures** 129
Sven Jansen, Dr. A. Schmeitz,
TNO Technical Sciences/Automotive, The Netherlands;
L. Akkermans, Transport & Mobility Leuven, Belgium

**The influence of wheel and tire aerodynamics in WLTP** 149
Dr. Timo Kuthada, F. Wittmeier, Institute of
Automotive Engineering and Vehicle Engines Stuttgart (FKFS)

**Towards a comprehensive approach for the sustainability assessment of a product: product social impact assessment** 161
Dr. Marzia Traverso, P. Tarne, Dr. V. Wagner, BMW Group

## DRIVER ASSISTANCE SYSTEMS

**A vehicle lateral control approach for collision avoidance by emergency steering maneuvers** 175
Martin Keller, Prof. Dr. Dr. T. Bertram, Institute of
Control Theory and Systems Engineering (RST), TU Dortmund;
Dr. C. Haß, Dr. A. Seewald, TRW Automotive GmbH

**Collision avoidance with combined braking and steering** 199
Carlo Ackermann, J. Bechtloff, Prof. Dr. Dr. R. Isermann,
Institute of Automatic Control and Mechatronics (IAT),
TU Darmstadt

**Driver assistance for trucks – from lane keeping assistance to smart truck maneuvering** 215
Alexander Gaedke, R. Greul, S. Kanngießer, N. Boos,
Robert Bosch Automotive Steering GmbH

# CHASSIS.TECH SECTION

## DEVELOPMENT PROCESS

**Development of a driving dynamics-oriented suspension design during the early concept phase** 233
Karthik Vemireddy, T. Dittmar, Prof. Dr. L. Eckstein,
Institute for Automotive Engineering (ika), RWTH Aachen University;
L. Hesse, P. Rettweiler, fka Forschungsgesellschaft
Kraftfahrwesen mbH Aachen

**Development of a chassis model including elastic behavior for real-time applications** 257
Frédéric Etienne Kracht, Prof. Dr. D. Schramm,
Dr. B. Hesse, Chair of Mechatronics, Y. Zhao, Institute for
Mechatronics and System Dynamics, University of Duisburg-Essen;
Dr. M. Unterreiner, Dr. Ing. h.c. F. Porsche AG

**Lightweight design in subassemblies with changing design spaces to find an overall weight optimum for series-produced cars** 283
Gerhard Steber, BMW Group; Prof. Dr. R. Lachmayer,
Institute of Product Development (IPeG),
Leibniz University Hannover

## VEHICLE LATERAL DYNAMICS

**Objektive Ratingmethode für Handling- und Komfortkriterien für den Einsatz im Fahrversuch und in der Simulation (Objective method for rating ride and handling criteria in simulation and vehicle testing)** 305
Joachim Ecker, Dr. P. Schöggl, E. Bogner, M. Oswald,
AVL List GmbH, Austria

**Virtual chassis tuning with emphasis on the damper characteristics – a method for optimal integrative damper adjustment by means of vertical and lateral dynamics simulation and evaluation criteria** 325
Florian Klinger, Dr. J. Edelmann, Prof. Dr. M. Plöchl,
Institute of Mechanics and Mechatronics, TU Vienna, Austria;
S. Jeindl, B. Angrosch, MAGNA Steyr Engineering, Austria

**Importance of body rigidity in the transient stage of the maneuver** 347
Charlie Gagliano, Honda R&D Americas, Inc., USA;
T. Geluk, Siemens Industry Software NV, Belgium

## RIDE COMFORT AND TESTING

**Contribution to the objective evaluation of roll dynamics** 359
Andreas Apfelbeck, M. Schwarz, S. Wegner, BMW Group;
Dr. R. Henze, Prof. Dr. F. Küçükay,
Institute of Automotive Engineering (IfF), TU Braunschweig

**Improved prediction of ride comfort characteristics by considering suspension friction in the automotive development process** 377
Christian Angrick, Prof. Dr. G. Prokop,
Institute for Automotive Technologies Dresden (IAD),
Dresden University of Technology;
Dr. P. Knauer, Dr. A. Wagner, AUDI AG

**Endurance tests of electronic suspension for motorcycles – a system approach** 405
Frederik Harnischmacher, KTM AG, Austria;
Prof. Dr. T. Kuttner, Department of Mechanical Engineering,
University of Federal Armed Forces Munich

# STEERING.TECH SECTION

## STEERING TECHNOLOGY

**Availability and fail-safety approaches for electric power steering systems – trends and requirements** 421
Eberhard Kübler, T. Pötzl, Dr. T. Frenz, J. Sauler, Robert Bosch Automotive Steering GmbH

**Innovative software functions to operate electric power steering systems in sports cars – Unterstützungskraftregelung (UKR)** 423
Dr. Christoph Bittner, A. Uselmann, K. M. Krüger, G. Rivera, Dr. Ing. h.c. F. Porsche AG

**Steering System Fingerprint – a tool for steering system performance check and overview** 443
Frank Esser, T. Vercammen, Ford-Werke GmbH

## TEST BENCH METHODS

**Model-based steering ECU calibration on a steering-in-the-loop test bench** 455
Dr. Hans-Michael Koegeler, B. Schick, AVL List GmbH, Austria; Alessandro Contini, Prof. Dr. P. E. Pfeffer, Munich University of Applied Sciences; M. Lugert, T. Schöning, Hyundai Motor Europe Technical Center GmbH

**Test infrastructure for EPS steering systems – balancing between requirement-based, experience-based and free testing** 467
Thomas Maur, TRW Automotive GmbH

**Realistic dynamic testing of EPS motors and ECUs by means of a hardware-in-the-loop test bench** 489
Hermann Briese, E. Farshizadeh, S. Oedekoven, DMecS Development of Mechatronic Systems GmbH & Co. KG; T. Schubert, Prof. Dr. H. Henrichfreise, Cologne Laboratory of Mechatronics (CLM), Cologne University of Applied Sciences

## DRIVING SIMULATION AND TESTING

**A simulator study on the controllability of steering systems with reduced maximum steering wheel angle** 507
Christian Dreßler, S. Eßers, TAKATA AG

**Implementation and testing of different control strategies on a steer-by-wire research platform** 519
Michele Sigilló, M. Dold, C. Delmarco, K. Polmans, ThyssenKrupp Presta AG, Liechtenstein

**Driving quality optimization based on cross-linked cause and effect chain models using the example of energy-efficient steering assistance** 541
Marinette Iwanicki, M. El-Haji, Institute of Vehicle Systems Technology (FAST), T. Freudenmann, Institute of Product Engineering (IPEK), Karlsruhe Institute of Technology (KIT)

# BRAKE.TECH SECTION


## ENVIRONMENTAL ASPECTS AND FUTURE TECHNOLOGIES

**The consequences of a closed rim design for the brakes of a high-efficiency vehicle** 567
Dr. Ralf Stroph, S. Gielisch, Dr. A. Pruckner, BMW Group

**The contribution of brake emissions to the total vehicle emissions** 585
Jürgen Lange, R. Steege, D. Welp, TMD Friction Holdings GmbH

**An innovative production method for a C/C-SiC brake disc, suitable for a large-scale production** 605
Dr. Daisy Julia Nestler, N. Nier, K. Roder, A. Todt, Prof. Dr. B. Wielage, Prof. Dr. G. Wagner, Institute of Materials Science and Engineering (IWW), Prof. Dr. L. Kroll, E. Päßler, Institute of Lightweight Structures (IST), Prof. Dr. S. Spange, J. Weißhuhn, Dr. H. Würfel, Institute of Chemistry, TU Chemnitz

## CONTROL AND SIMULATION

**Combined control strategy for the combustion engine and brake system to enhance the driving dynamics and traction of front-wheel-drive vehicles** 629
Daniel Killian, Prof. Dr. M. Lienkamp, Institute of Automotive Technology (FTM), TU Munich; S. Fischer, Elektronische Fahrwerksysteme GmbH; S. Poltersdorf, Dr. R. Schwarz, AUDI AG

**Real-time simulation of braking interventions in heavy commercial vehicles** 647
Dr. Philipp Wagner, T. Ille, Dr. C. Kohrs, MAN Truck & Bus AG; F. Bauer, Institute of Mechanics, University of Federal Armed Forces Munich

**Efficient digital development of brake components with multiple requirements** 663
Konrad Meister, Dr. Tobias Rößler, Dr. V. Fäßler, Dr. S. Staudacher, TWT GmbH

## BRAKE TECHNOLOGY

**Electrically controlled parking brake (EPB) for heavy commercial vehicles** 681
Dr. Falk Hecker, T. Weinhold,
Knorr-Bremse Systeme für Nutzfahrzeuge GmbH

**Electric parking brake meets drum brake – synergy or contradiction** 683
Christian Breiten, B. Schmittner, Continental Teves AG & Co. oHG

**Large aircraft landing gears – a brief overview** 685
Hans-Martin Besch, formerly Airbus Operations GmbH

# TIRE.WHEEL.TECH SECTION

## ROAD SURFACE, TIRE AND WHEEL INTERACTION

**Continuous wheel force measurement for passenger vehicles and commercial vehicles** 717
Dr. Manfred Kraus, Dr. M. Bäuml,
Schaeffler Technologies AG & Co. KG

**Adaptive state observers for driving dynamics – online estimation of tire parameters under real conditions** 719
Jakob Bechtloff, C. Ackermann, Prof. Dr. Dr. R. Isermann,
Institute of Automatic Control and Mechatronics (IAT),
TU Darmstadt

**Damage on alloy wheels for motor vehicles – limits of rework** 735
Thomas Kollmeier, TÜV SÜD Product Service GmbH

## SIMULATION

**Evaluation of different modeling approaches for the tire handling simulations – analysis and results** 749
Francesco Calabrese, Dr. M. Bäcker, A. Gallrein,
Fraunhofer Institute for Industrial Mathematics (ITWM)

**Rolling resistance modeling for electric vehicle consumption** 775
Andrea Ficht, Prof. Dr. M. Lienkamp,
Institute of Automotive Technology (FTM), TU Munich

**The non-steady-state tire model as a set of physical submodels for driver assistance systems analysis** 799
Pavel Sarkisov, Prof. Dr. G. Prokop,
Institute for Automotive Technologies Dresden (IAD),
Dresden University of Technology;
Dr. S. Popov, Bauman Moscow State Technical University, Russia

## DESIGN AND TESTING

**Evolution of the requirements on vehicle tires and insights from 15 years of test operation** 815
Lars Netsch, M. Staude, TÜV SÜD Product Service GmbH

**Concurrent design of vehicle tires and axles** 839
Jens Wimmler, Dr. M. Wahle, Dr. M. Zimmermann, BMW Group; Prof. Dr. D. Schramm, Chair of Mechatronics, University of Duisburg-Essen

**New testing technology to evaluate lateral ice performance of tires** 853
Klaus Wiese, J. Heyne, Prof. Dr. B. Wies, T. Neddenriep, Continental Reifen Deutschland GmbH

# CHASSIS.TECH PLUS SECTION

## PLENARY LECTURES

**Sense-Plan-Act – the role of chassis systems** 867
Ralph Lauxmann, Continental Teves AG & Co. oHG

**Consumer acceptance ensured through safety** 869
Klemens Schmiederer, TÜV SÜD AG

# Speakers

**Carlo Ackermann**
Institute of Automatic Control
and Mechatronics (IAT),
TU Darmstadt

**Christian Angrick**
Institute for
Automotive Technologies
Dresden (IAD),
Dresden University
of Technology

**Andreas Apfelbeck**
BMW Group

**Jakob Bechtloff**
Institute of Automatic Control
and Mechatronics (IAT),
TU Darmstadt

**Hans-Martin Besch**
formerly Airbus Operations GmbH

**Dr. Christoph Bittner**
Dr. Ing. h.c. F. Porsche AG

**Christian Breiten**
Continental Teves AG & Co. oHG

**Hermann Briese**
DMecS Development
of Mechatronic Systems
GmbH & Co. KG

**Francesco Calabrese**
Fraunhofer Institute
for Industrial Mathematics (ITWM)

**Alessandro Contini**
Munich University
of Applied Sciences

**Kilian Dettlaff**
Institute for Internal
Combustion Engines
and Automotive Engineering (IVK),
University of Stuttgart

**Christian Dreßler**
TAKATA AG

**Joachim Ecker**
AVL List GmbH, Austria

**Frank Esser**
Ford-Werke GmbH

**Andrea Ficht**
Institute of
Automotive Technology (FTM),
TU Munich

**Alexander Gaedke**
Robert Bosch
Automotive Steering GmbH

**Charlie Gagliano**
Honda R&D Americas, Inc., USA

**Alexander Häußler**
Robert Bosch GmbH

**Frederik Harnischmacher**
KTM AG, Austria

**Dr. Falk Hecker**
Knorr-Bremse Systeme
für Nutzfahrzeuge GmbH

**Marinette Iwanicki**
Institute of Vehicle
Systems Technology (FAST),
Karlsruhe Institute
of Technology (KIT)

**Carsten Jablonowski**
AUDI AG

**Sven Jansen**
TNO Technical Sciences/Automotive, The Netherlands

**Martin Keller**
Institute of Control Theory and Systems Engineering (RST), TU Dortmund

**Daniel Killian**
Institute of Automotive Technology (FTM), TU Munich

**Markus Kirschbaum**
Daimler AG

**Florian Klinger**
Institute of Mechanics and Mechatronics, TU Vienna, Austria

**Dr. Hans-Michael Koegeler**
AVL List GmbH, Austria

**Thomas Kollmeier**
TÜV SÜD Product Service GmbH

**Frédéric Etienne Kracht**
Chair of Mechatronics, University of Duisburg-Essen

**Dr. Manfred Kraus**
Schaeffler Technologies AG & Co. KG

**Eberhard Kübler**
Robert Bosch Automotive Steering GmbH

**Dr. Timo Kuthada**
Institute of Automotive Engineering and Vehicle Engines Stuttgart (FKFS)

**Jürgen Lange**
TMD Friction Holdings GmbH

**Peter Langen**
BMW Group

**Ralph Lauxmann**
Continental Teves AG & Co. oHG

**Thomas Maur**
TRW Automotive GmbH

**Konrad Meister**
TWT GmbH

**Dr. Daisy Julia Nestler**
Institute of Materials Science and Engineering (IWW), TU Chemnitz

**Lars Netsch**
TÜV SÜD Product Service GmbH

**Isabel Ramirez Ruiz**
Ferrari S.p.A., Italy

**Andreas Rigling**
ADAC e. V. Technik Zentrum

**Dr. Tobias Rößler**
TWT GmbH

**Pavel Sarkisov**
Institute for Automotive Technologies Dresden (IAD), Dresden University of Technology

**Bernhard Schick**
AVL List GmbH, Austria

**Klemens Schmiederer**
TÜV SÜD AG

**Andreas Schultze**
Institute of
Automotive Technology (FTM),
TU Munich

**Hiroshi Shibuya**
Toyota Motor Corporation, Japan

**Michele Sigilló**
ThyssenKrupp Presta AG,
Liechtenstein

**Gerhard Steber**
BMW Group

**Dr. Ralf Stroph**
BMW Group

**Dr. Marzia Traverso**
BMW Group

**Kristian Trenkel**
iSyst Intelligente Systeme GmbH

**Karthik Vemireddy**
Institute for
Automotive Engineering (ika),
RWTH Aachen University

**Dr. Philipp Wagner**
MAN Truck & Bus AG

**Klaus Wiese**
Continental Reifen
Deutschland GmbH

**Jens Wimmler**
BMW Group

# The all new BMW 7 series

Peter Langen, Head of the Vehicle Dynamics Division at BMW

## Abstract

Improvements in driving characteristics can be seen with each vehicle generation throughout the automotive sector. A high level of quality and constant improvements in handling, comfort, and safety combined with reductions in the levels of CO2 emissions are expected by the customers.

In order to fulfill these expectations, BMW has turned to completely electrified control- and servo-systems, whose advantages include reduced fuel consumption, independence from the combustion engine, improved quality, as well as a programmable, interlinked functionality. An example of this is the simultaneous integration of the electronic damper control, air spring system, active roll stabilization, and electric power steering at the front and rear axles. The new BMW architecture combines this integrated control approach with a stiff, lightweight body structure (the "Carbon Core") to provide the customer with an excellent mix of vehicle dynamics and ride comfort.

BMW's lateral and vertical dynamic management system ensures that all control functions are working at their optimum and in harmony with each other, even in the face of complex situations. This system will be used to improve the performance of the new 7-series, and will receive information from periphery sensors and from navigation data, so that it can adapt to both the driver's intentions and the current environmental conditions.

In the future, solely improving the traditional core attributes of a car will no longer suffice to meet customer expectations. Due to increasing traffic density and growing urbanization, a social change is occurring where the act of driving is perceived more as a burden than as a pleasure. Even when an optimal route can be calculated for the driver, it is not always possible to avoid heavy traffic. Many would prefer to use this time more productively and avoid experiencing the associated levels of stress.

The rising demand for a new way of driving is being addressed through such automated functions as Active Assist, which will be available at the push of a button in the new 7-series. Using Active Assist, the customer can decide whether they want assistance with lane keeping and velocity control, thereby unloading some of the burden in traffic and facilitating a more enjoyable driving experience. A key aspect of this new technology is its homologation, which requires close collaboration with the associated authorities. We are pleased to rise to this exciting challenge.

What's next? BMW will be raising the maximum vehicle speeds available with our Lane Keeping and Traffic Jam Assistants. The Adaptive Cruise Control will also be linked with our Stop&Go system. Furthermore, the customer will soon be able to experience high levels of vehicle automation in the form of automated parking in longitudinal and lateral parking spaces, as well as remote parking without a driver behind the wheel.

Whether driving dynamically on long winding roads or comfortably gliding along with Active Assist, the future offers both sides of sheer driving pleasure in a single BMW – safely and efficiently!

# Highly automated driving for commercial vehicles

Dipl. Ing Markus Kirschbaum

# Summary

Highly automated driving is a further decisive step towards accident-free driving based on intelligent interaction of driving assistant systems, which in turn will result in increased road safety. In addition, this concept will reduce the stress to which drivers are subject and increase their comfort level in monotonous traffic situations (e.g. long stretches of highway driving or stop-and-go situations), provide greater efficiency and economy and contribute to ecological sustainability. The Future Truck 2025 has now demonstrated for the first time how much freedom for new tasks can be created in the logistics field, even while the vehicle is in motion, to improve the working environment of truck drivers. This shift of emphasis from being "just" a driver to being a transport manager offers enormous potential, but is accompanied by a new kind of driver/vehicle interaction.

The first phase describes the overall functionality and the components required. By adding connectivity through V2V, V2I and the Internet, all functionality is enhanced still further, making additional opportunities available to both the customer and driver.

Tomorrow's driver's cab shows very clearly that providing a corresponding interior and actively involving the truck driver in the vehicle's functionality generates understanding and, as a result, a higher level of user acceptance. To this end, it is essential to provide reproducible driving behavior that is transparent and fully understood by the driver.

# 1 Highly automated Driving in the Future Truck 2025

The Future Truck 2025 was presented to the public at the IAA Commercial Vehicles Trade Show in 2014. Technical demonstrations followed on a limited section of the A14 highway in Magdeburg. The vehicle convincingly demonstrates how efficiency and safety can be maximized by the use of highly automated driving.

## 1.1 Motivation

The considerable increase in goods traffic anticipated in coming years in conjunction with a slowly growing infrastructure shows how crucial it is to make optimum use of existing resources. [1] In addition, forwarders are subject to increasing pressure from costs, in particular fuel costs, highway toll charges and a lack of suitable drivers. There is a suitable solution to these issues:

We must make better use of the existing infrastructure, apply assistance and telematics systems to a greater extent and interlink them more efficiently and make the truck driver's job much more attractive.

## 1.2 Functionality

The core function of this concept is highly automated driving on highways without the need for the driver to grasp the steering wheel and thus allowing him to perform other activities that are not directly related to driving the vehicle.

After entering the highway, the driver of the Mercedes-Benz Future Truck 2025 takes up a position in the traffic flow in the appropriate lane. The system then offers the driver the "Highway Pilot". If the driver activates the "Highway Pilot functionality" , the system switches to highly automated mode. Fig. 1

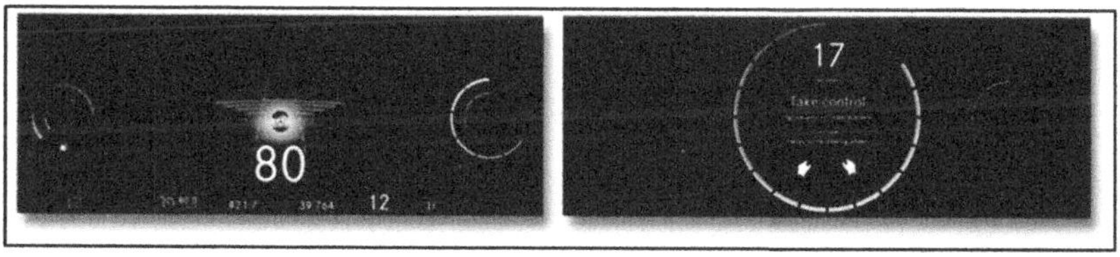


Fig. 1: Instrument display in the Highway Pilot mode and the instruction to the driver to take over control later on

Depending on the current traffic situation, the Future Truck 2025 operates highly automated and does not require another vehicle in front to guide it through traffic. Within its traffic lane, it runs highly automated. If another vehicle is driving in front of the truck, it can use that vehicle's speed as a guide within the permitted speed limits and maintain the required safe distance. As a result, there is no inherent danger if another vehicle moves in in front of the truck. Safe distances are automatically maintained – the Mercedes-Benz Future Truck 2025 adapts perfectly to its surroundings. The Future Truck 2025 operates independently of other vehicles. Through its additional networked systems, it is fully aware of its infrastructure and is thus able to improve its logistical efficiency. Open standards and communication protocols are a prerequisite for this functionality.

## 1.3 Challenges facing highly automated Driving

The technical preconditions have been demonstrated for the first time in the Mercedes-Benz Future Truck 2025, but adjustments to the legal framework will be required to permit this new driving standard to come public roads. Further adaptation of the Vienna Convention on Road Traffic (1968) [2], which has been signed and ratified by almost all European States, is in progress. This is an international treaty designed to increase road safety by establishing standard traffic rules among the contracting parties and states that the driver must at all times and in all circumstances be in control of his vehicle.

UN/ECE Regulation R 79 for steering systems, which is based on the Vienna Convention on Road Traffic, permits corrective steering functions but does not allow automated steering at speeds in excess of 10 km/h. This is a crucial aspect in permitting automated parking assistance and stop-and-go control systems.

Technically and legally it is a further requirement that data shall be transmitted safely and surely. This applies both to actions from outside affecting an individual vehicle, and the transmission of data in V2V and V2I communications and the Internet.

Besides vehicle homologation issues, there are also matters relating to road traffic law, such as liability for traffic violations and in the event of accidents, which it will not be possible to prevent all together. The same applies to such matters as insurance and product liability. Fig. 2

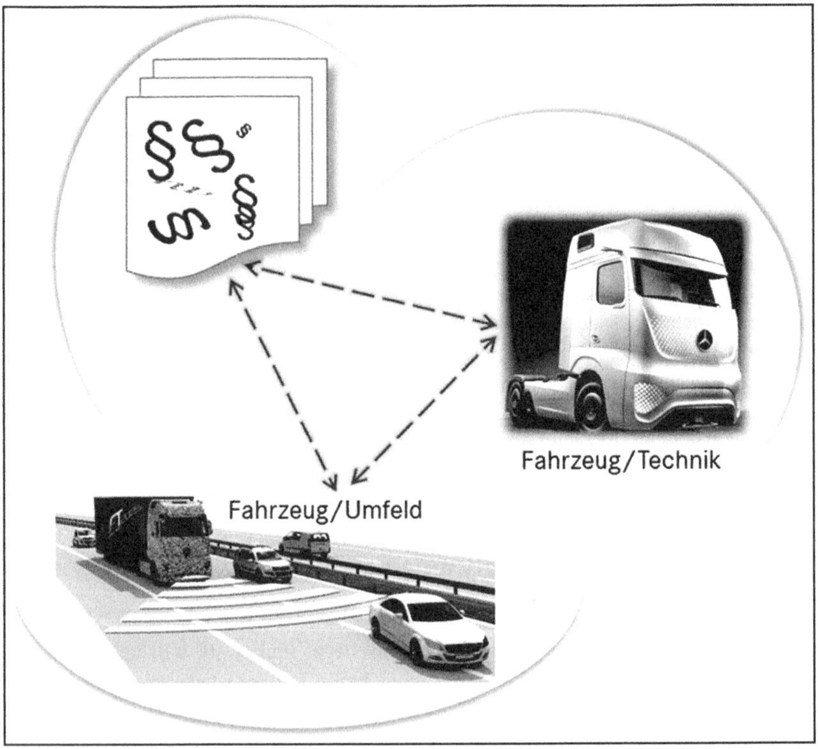


Fig. 2: The legal framework and data protection will also influence highly automated driving

# 2 The Technical Background of the Future Truck 2025

To enable highly automated driving, a system has been installed on-board that provides basic functionality even when connectivity is interrupted. V2X and Internet connectivity cannot be reliably guaranteed at all times because they are dependent on the strength of the network and the density of usable V2V vehicles.

## 2.1 Radar Sensors and a Camera Scan the Surroundings

Highly automated driving in the Mercedes-Benz Future Truck 2025 is enabled by means of two independent kinds of sensors. Camera and radar systems detect and monitor the environment around the vehicle for highly automated driving. Fig. 3

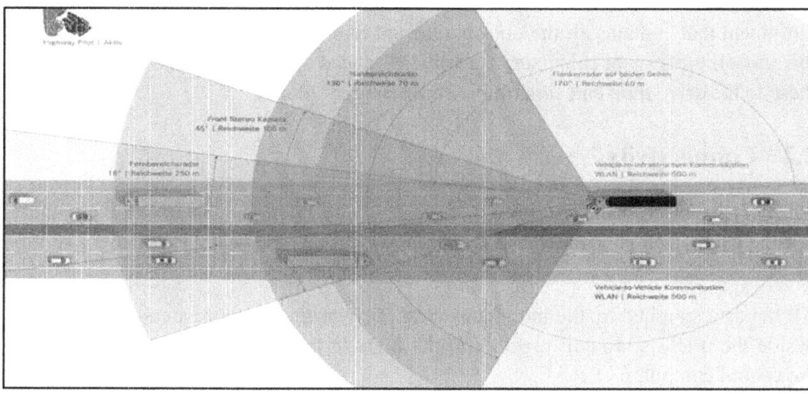

Fig. 3: Detecting the surroundings using radar and camera sensors

In the lower part of the vehicle front, a radar sensor scans the near and far ranges to the front. The front radar sensor covers a range of 250 m and a spread of 18 degrees. The close-range sensor covers 70 m and has a spread of 130 degrees. The radar sensor forms the basis for such safety systems as adaptive cruise control and emergency braking assistance, which are already in production today.

Additionally, there is a stereo camera system installed above the dashboard behind the windscreen that monitors the space in front of the vehicle. In today's optional lane keeping assistance systems, there is a mono-camera in this position. The stereo camera has a range of 100 m and covers a horizontal spread of 45 degrees and a vertical spread of 27 degrees.

The stereo camera fitted to the Mercedes-Benz Future Truck 2025 identifies single- and two-lane roads, detects pedestrians, moving and stationary objects, and all other objects within the area monitored. The camera detects everything that stands out in some way from its background and can, as a result, accurately detect empty space. The front stereo camera also detects the information presented by road signs.

As well as detecting objects and free space, this stereo camera system can detect lanes, which is a key function for highly automated driving.

The road areas to the left and right of the truck are monitored by radar systems mounted on the sides of the vehicle. They are mounted to the left and right in front of the rear axle of the vehicle. Their range is 60 m and they cover a spread of 170 degrees in a longitudinal direction.

Drawing together all this information, the system generates a model of the truck's environment that contains all the objects relevant to highly automated driving. Based on this model, the system calculates the vehicle's speed and steering angle and transmits these to the drive train and superimposed steering system.

## 2.2 Connectivity

All the sensors on-board the Future Truck 2025 are interlinked to each other (multi-sensor fusion) and contribute to a comprehensive view of the environment. The sensors detect every stationary or moving object in the area around the truck.

All the data supplied by the individual sensors covering the entire area in front of and beside the truck is brought together in the high-performance multi-core processor in the central computer.

The sensor and camera technology is designed to work efficiently up to the permitted maximum speed for heavy goods vehicles (HGV). The system ensures the truck retains its position in the center of its lane by means of fully-automatic steering adjustments. The system also has a digital three-dimensional map on board as already available as Predictive Powertrain Control (PPC) systems. The truck is, therefore, always fully aware of the intended route and the surrounding topography.

### 2.2.1 V2V and V2I for highly automated Driving

The Highway Pilot is ideally supplemented by V2V and V2I communications. The messages exchanged should contain the following information: Vehicle position and type, vehicle dimensions, direction of travel and speed, any acceleration or braking activities and the currently steered curve.

The frequency of the transmission of that information will depend on the speed of the vehicle and the intensity of its changes of direction. The transmission frequency ranges from one telegram per second during uneventful driving to ten-times as often where noticeable changes occur.

Transmission uses WLAN technology over the pan European uniform G5 frequency at 5.9 gigahertz. The basis for this is the ITS Vehicle Station (Intelligent Transport Systems and Services) on-board the vehicle.

Communication between the Future Truck 2025 and other vehicles (V2V) covers a range of approx. 500 m in all directions. Using this medium, vehicles can mutually communicate details of their motions and can respond in good time where necessary. This includes, for example, response to vehicles pulling in in front of the truck on the highway or the approaching end of a tailback. The more vehicles communicate in this way with each other, the more dynamically and flexibly they can respond to each other and the current situation.

Ideally, this will result in an uninterrupted chain of communication along the road which will notify the drivers and vehicles as required about the exact traffic situation even far ahead of the vehicle.

Vehicle-to-infrastructure (V2I) communication means that all these notifications and signals are also sent to external recipients, such as traffic management and control points. These organizations can respond flexibly, for example, by adjusting the maximum permitted speed or opening up additional overflow lanes. In return, messages can be transmitted to vehicles to warn about short-term roadworks, for example. Fig. 4

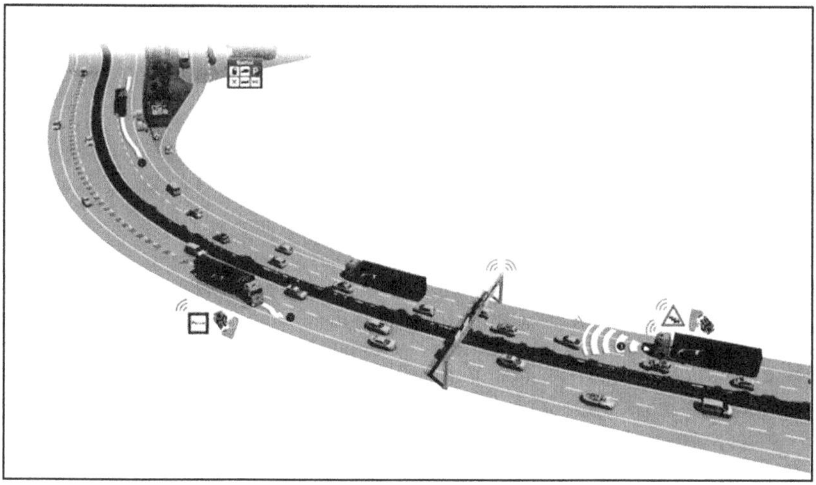

Fig. 4: V2V and V2I can optimize the traffic flow and increase efficiency

All this data informs the driver and the vehicle computer right away about events that are beyond their respective visual ranges. In this way, the driver and vehicle become aware of obstacles ahead long before a hazard situation can occur.

### 2.2.2 For the Vehicle Owner and the Driver

However, in order to take full advantage in terms of logistical efficiency and to optimize the benefits provided by the infrastructure, interconnection between all the various systems would be beneficial. This development will, no doubt, progress step-by-step. Open standards and communication protocols are a prerequisite for this functionality.

Highly automated driving can take the pressure off the driver so he does not have to concentrate on the driving at all time, in particular on stretches that are monotonous and prone to induce drowsiness. Since the truck can regulate its speed automatically and can use the navigation system independently to find the optimal route, and since the freight forwarder, sender and recipient of the freight are all aware in real-time of the current location, progress of the journey and anticipated arrival time, there is far less pressure on the driver to keep to the schedule. It is this pressure that accounts to a large extent for stress in the driver's cab.

At the same time, the driver gains time for other tasks and can communicate safely with his environment. It is conceivable that the driver might take on tasks previously dealt with by scheduling clerks or keep up with social contacts. In particular, self-employed drivers will be able to pursue office activities in comfort while traveling.

Taking on additional tasks will, in this way, make marked changes to the truck driver's profession by offering the opportunity to become more of a transport manager rather than just a truck driver. The truck driver's profession will become more attractive – highly automated driving is, therefore, a clear response to the problem of a lack of drivers. The truck and its driver team up through highly automated driving.

# 3 Tomorrow's Workplace

The interior of the Future Truck 2025 stands out because of the reduction of functions to only what is necessary and the purist design. The traveling workplace is designed to be unfussy and tidy, as one would expect of a modern, paperless office. Modern technology is in a fascinating contrast here to natural materials, warmth and comfort.

The dashboard is tidy and minimalist, displays take the place of instruments and outside mirrors. The displays appear to hover in front of the piano-finish surfaces in the cockpit. Instead of intrusive air jets, the air-conditioning system uses indirect air flows, the touchpad takes the place of traditional rows of switches. Fig. 5

Fig. 5: The workplace in the Future Truck 2025

When the truck is traveling in the highly automated mode, the driver can, if he so wishes, move his seat backwards and, at the same time, turn it through 45 degrees towards the open space – allowing him to take up a more relaxed, comfortable working position. The indirect lighting also brightens the interior of the cab without creating dazzle.

In the workplace of the future, the driver communicates while on the road via the tablet computer which is slotted into the new design of central console, ready to be taken out for other purposes. Using this tablet, the driver can process various documents, arrange schedules for new destinations, accept and plan new orders and organize his next break. The computer screen can be freely configured, allowing the driver to call up any of the journey data he may require.

Instead of the steering wheel and pedals, the tablet computer becomes the main work medium on long, highly automated journeys. The Mercedes-Benz Future Truck 2025 will result in a new form of work in the driver's cab of long-distance trucks. Fig. 6

Fig. 6: The driver works with the tablet PC and the driver's seat is swiveled through 45°

# 4 Summary and Outlook

Highly automated driving in commercial vehicles is a necessary move in view of continued increases in goods traffic to ensure the existing infrastructure is used to the optimum. At the same time, the Highway Pilot represents a further step towards improved road safety because interlinked communications will allow hazardous situations to be communicated much faster and appropriate action taken.

Highly automated driving on highways relies to a large extent on sensors integrated in the vehicle. Connectivity allows these driving functions to be optimized still further and opens up new opportunities. In addition to the technical challenges, there are issues to be discussed concerning data security, homologation and road traffic law.

State-of-the-art workplaces for truck drivers are a key challenge for the future involving harmonizing design and functionality with a feel-good factor. Confidence in highly automated driving must be quickly established, while ensuring that drivers are presented with sufficient challenges. Many HMI functions will be transferred from the steering wheel or dashboard to positions much closer to the driver.

# Bibliography

[1] BMVI Verkehrsprognose 2030 (2014) (Road-Traffic Forecast for 2030 by German Ministry of Transport (BMVI) (issued 2014)).

[2] 0.741.10 Vienna Convention on Road Traffic dated 8 November 1968, updated 15 Feb. 2013

# Automated driving, electrification and connectivity – the evolution of vehicle motion control

Alexander Häußler, Robert Bosch GmbH

# 1 Introduction

Looking back at the last 50 years, driving a car was a completely different experience than today. In the 60's and 70's power steering was limited to expensive hydraulic systems on only the most expensive cars. Through much of the 80's, smaller cars like the Fiat Panda, VW Golf and Renault Rapid were manufactured without power brakes. In other words, the driver had no assistance in powering and controlling the brake and the steering system (see Figure 1).

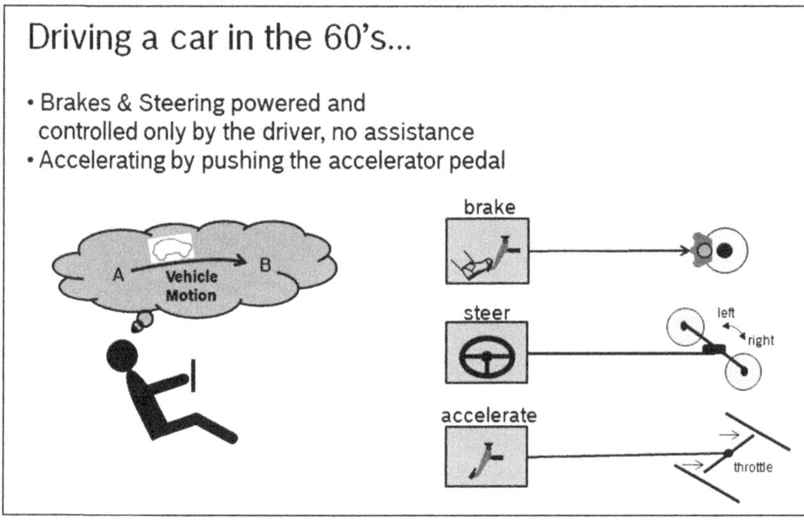


Figure 1: Driving a vehicle with direct access to brake & steering actuators.

When passive safety systems such as the safety belt, were launched in the 70's, many drivers initially refused to use them for fear of tying themselves to the car in the case of an accident.

Since then, there have been significant changes on both the demands of drivers on driving and the technical capabilities regarding comfort and safety on the road. No matter what the future brings, one thing remains the same…when a driver wants to get from A to B there is a requirement for "vehicle motion control".

# 2 Market Trends & Drivers

Vehicle Motion Control, meaning how the driver and his or her passengers get from A to B, will be strongly influenced by 3 major market trends and drivers (see Figure 2).

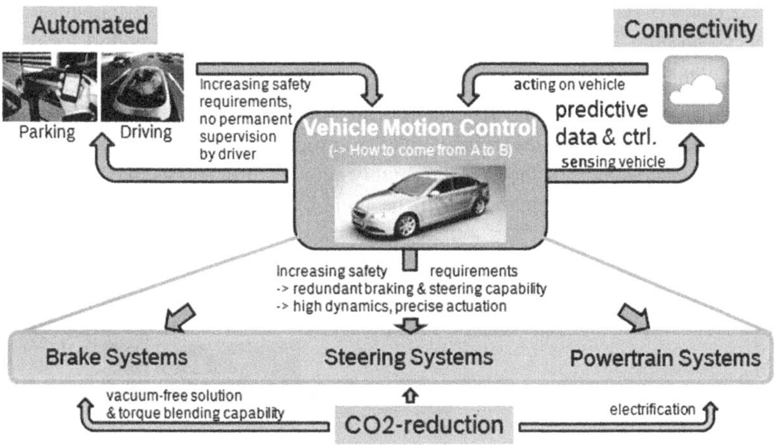


Figure 2: Vehicle Motion Control - Market Trends & Drivers

First, we have the trend of increasing automation from driver assistance up to fully automated driving. Through the step by step introduction of functions with increasing degrees of automation, increasing safety requirements are emerging for vehicle motion control. The longitudinal and lateral guidance of the vehicle must be done in a safe and comfortable way without permanent supervision by the driver.

These increasing safety requirements for the system as well as for vehicle motion control will lead to increasing safety requirements, especially concerning availability of the actuator systems (brake system, steering system and powertrain system).

In addition to these new safety requirements, the actuator systems are driven by a second trend... that is $CO_2$-reduction. The combination of automated driving and increased safety is leading to new actuator systems with new constraints. At the same time the new requirements offer new potential in the field of vehicle motion control.

A third trend affecting the future of Vehicle Motion Control is connectivity. Connectivity effectively increases the field of view of the vehicle via data from the cloud and through this allows new driving strategies and a predictive Vehicle Motion Control.

## Evolution toward Automated Driving

The step by step introduction of automated driving functions is shown in figure 3.

Today's Driver Assistance functions like Lane Keeping Support or Adaptive Cruise Control (ACC) support the driver through lateral **or** longitudinal guidance in a given driving lane via vehicle motion control. The next generation of partially automated assistance functions like Integrated Cruise Assist or Highway Assist support the driver through longitudinal **and** lateral guidance in a driving lane.

For both driver assistance functions and partially automated functions, the overall system must be supervised by the driver. With the introduction of highly automated functions this will no longer be the case. For highly automated functions like Highway Pilot, vehicle guidance on highways will be done by the system without permanent supervision by the driver.

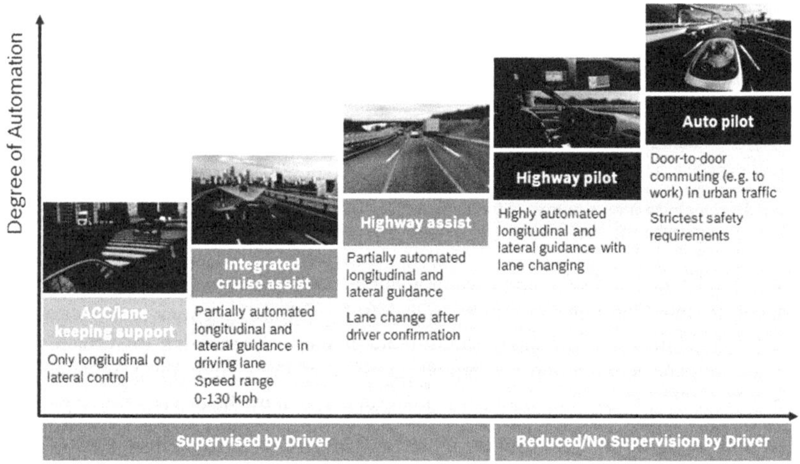


Figure 3: Evolution toward Automated Driving

This means in the case of a primary system failure the system must provide a backup and have the capability to bring the vehicle to a safe state on its own. For Highway Pilot this use case requires a safe brake and steering to stop capability. Therefore, especially the brake system, as part of the actuator network (steering system, brake system, powertrain system), must be designed with redundant braking capabilities (fail-operational).

In addition to continuous operational availability of actuator systems themselves, an increasing number of functions influencing vehicle motion require an abstraction layer "Vehicle Motion Control" that handles complexity by coordinating and controlling access to different actuator systems (brake, steering, powertrain systems).

For the development of this SW-System "Vehicle Motion Control" the derivation of functional system architecture, based on existing and upcoming use cases, plays a decisive role.

Figure 4 shows selected uses cases for different degrees of assisted and automated functions. Today's driver assistance functions like ACC or Lane Keeping Support are comfort-orientated functions for driving in a given lane which have direct functional access to single actuator systems like steering, brake and powertrain. The focus of the use case "driving in the lane" is on haptics for the driver.

For the use case "lane change", as part of partial automation, Vehicle Motion Control has to coordinate the parallel use of different actuators for trajectory control and account for driver/vehicle interaction during lane change.

In high automation, safety requirements become a dominant key success factor for Vehicle Motion Control. Amongst others this leads to the planning and control of safety trajectories and fail-operational motion control within a highly available actuator network.

In addition, for highly automated functions a precise trajectory planning and control is necessary.

For the use case "automated cut in" in a Highway Pilot a trajectory that is realizable and collision-free while remaining comfortable for the driver must be determined and implemented.

For this purpose, available driving torque must be considered in addition to the dynamics of the braking and steering systems. For Vehicle Motion Control this means the potential (limits) of the actuator systems must be considered predictively and as part of an entire integrated system.

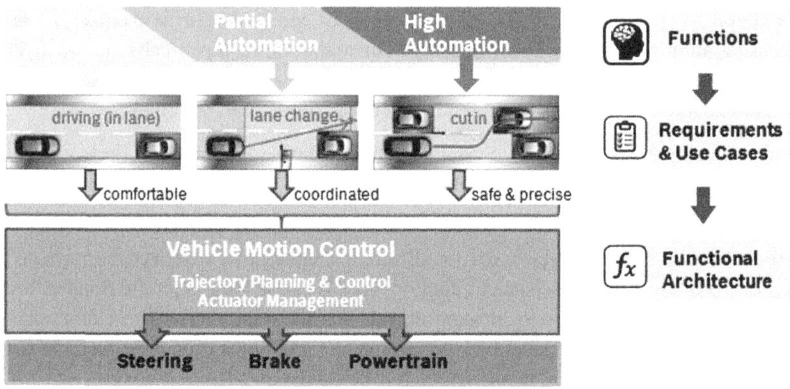


Figure 4: Degree of automation – functional architecture

The abstraction layer for motion control and actuator management "Vehicle Motion Control" at Bosch is in its introductory phase for partially automated functions and will be further developed for series production for highly automated driving.

A Highway Pilot which drives at approximately 130 km/h on highways has been realized at Bosch in the form of a prototype and has been undergoing field testing since the beginning of 2013 on public highways.

Commercial introduction of Highway Pilot is expected for the end of this decade.

## Evolution of the Brake System

The brake system for passenger cars can be divided into the brake boost element, the modulation element and the foundation element.

The brake boost element is responsible for amplifying the brake force of the driver when the driver is pressing the brake pedal.

The foundation elements are the components at the wheel such as the brake disc which provides the actual braking effect.

The modulation element was introduced with systems like ABS and ESP® to modulate the brake force for each wheel. This modulation also allows for active braking without the driver actuating the brake pedal.

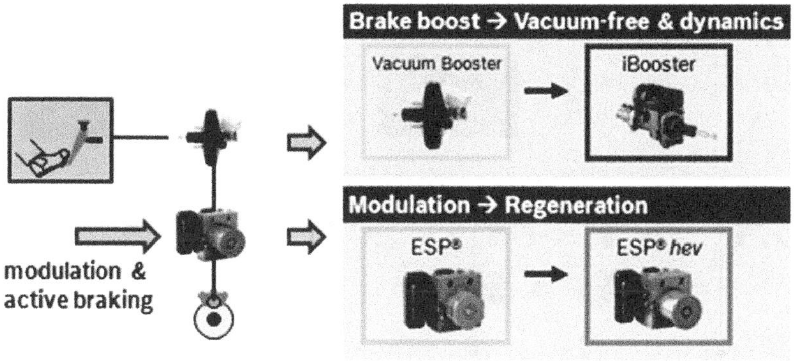


Figure 5: Bosch-portfolio for boosting and modulation

For the design of upcoming brake systems an important market driver is the trend toward CO2-reduction. This leads to the need for vacuum-free solutions as well as a torque blending capability for the brake system. Figure 5 shows the Bosch portfolio designed to meet these requirements. With the iBooster a brake boost system was introduced in the market which meets the requirements for vacuum-free boosting. The system detects the pedal movement via a pedal travel sensor and activates the iBooster motor instead of an air valve from a vacuum booster. This system offers the highest dynamics for automated emergency functions.

In addition to the trend toward CO2-reduction the trend toward increasing degrees of automation up to highly automated driving leads to new safety requirements for Vehicle Motion Control and derived availability requirements for employed actuators like the brake system.

Figure 6 shows the step-by-step evolution of the brake system with focus on the step-by-step introduction of automated driving functions.

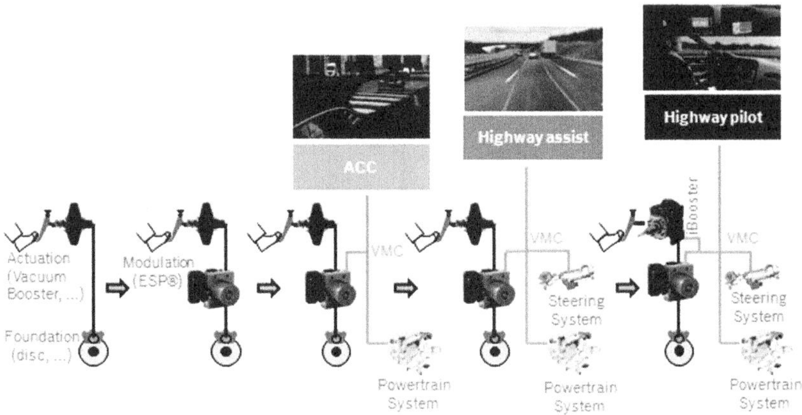


Figure 6: Evolution of the Brake System based on Bosch Portfolio

With the expansion of the brake system through a modulation unit capable of braking independently from the driver, the basic prerequisite for the introduction of driver assistance systems like Adaptive Cruise Control (ACC) was met. This intelligent cruise control system slows down and speeds up the car automatically to keep pace with the car in front of you. The system employs Vehicle Motion Control (VMC) to govern the deceleration of the vehicle with the brake system and the acceleration of the vehicle through the powertrain system. For partially automated systems like Highway Assist the coordination and control task of VMC is extended by inclusion of the steering system.

For highly automated systems like Highway Pilot, a redundant braking and steering capability is required in the network of actuators coordinated by VMC.

The redundant braking capability, necessary for automated driving is accounted for in the Bosch braking system portfolio through the combination of iBooster and ESP®.

## Evolution Steps of Powertrain – Electrification

For the development of future powertrain systems the main market driver will be CO2-reduction. There is still some potential for CO2-reduction in conventional power trains with combustion engines but the main potential is seen in electrification of the powertrain. To achieve the goal of highest CO2-reduction potential only, the best alternative would be the introduction of pure e-vehicles. For cost reasons there will be a step-by step introduction of electrification (see Fig.7).

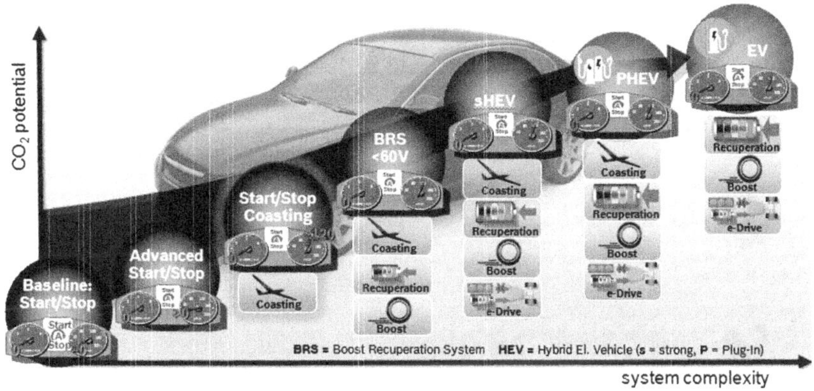


Figure 7: Evolution Steps of Powertrain - Electrification

The step-by-step introduction of increasing electrification begins with Start/Stop at standstill and low speed, coasting at higher speeds, followed by hybrid electric vehicles initially equipped with the Boost Recuperation System as a mild hybrid and in the end moving to the pure electric vehicles.

This electrification of the powertrain leads to new actuators influencing vehicle motion. Therefore for Vehicle Motion Control new driving strategies with respect to CO2-reduction must be considered as well as new actuator potential such as the dynamic response characteristics of e-drives.

# 3 Vehicle Motion Control – System of Systems

As we move from manual driving, via existing driver assistance systems up to highly automated driving, the number and complexity of cross domain functions influencing vehicle motion control increases significantly

To master these functional chains and the variety of driving tasks an abstraction layer "Vehicle Motion Control" is needed (see Fig. 8), which handles complexity by coordinating and controlling access to different actuators; braking, steering, and powertrain.

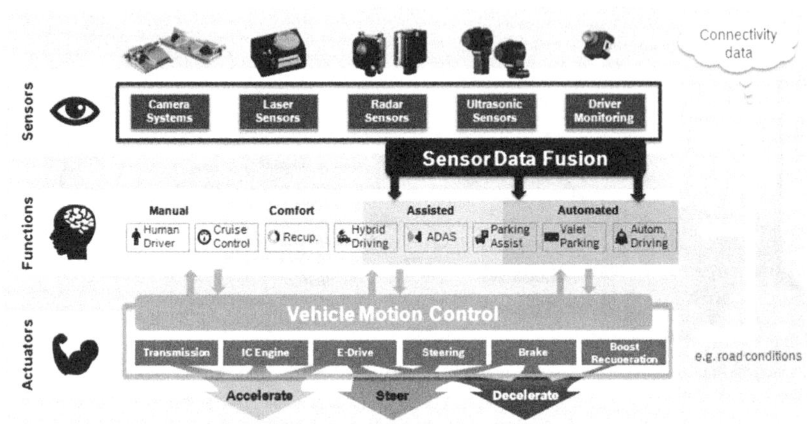


Figure 8: Vehicle Motion Control as system of systems

For assisted and automated functions it is also important to have an appropriate environmental model, based on several surround sensors, for example camera systems, laser sensors, radar sensors, ultrasonic sensors and map data, in order to derive an adequate driving decision for Vehicle Motion Control.

Looking at the respective potential of different actuators shown in figure 8, it shows that different actuator systems can be used to realize the same driving decision. This means, for example, that a driving task "decelerate" can not only be realized through the brake system, but also through a boost recuperation system or an E-drive. In the same way, a driving task "steering" can not only be realized through the steering system itself, but with some limitations also through the brake system or an e-powertrain.

In this context "Vehicle Motion Control" as system of systems is the SW-system that enables optimal combinations of different actuator capabilities concerning vehicle dynamics for each use case. It takes over the coordination of longitudinal and lateral motion of a vehicle in a central way. It can be allocated over different ECUs and it is the key enabler for future cross domain functions with respect to agility, comfort and safety.

# 4 Future Cross Domain Functions - Examples

In this chapter three examples for future cross domain functions will be discussed.

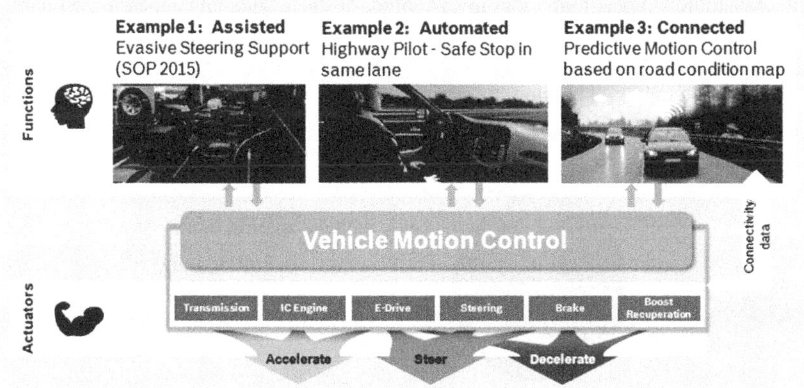


Figure 9: Vehicle Motion Control for assisted, automated and connected use cases

The first example is the advanced driver assistance function, Evasive Steering Support, which will enter series production in 2015. This example focuses on Vehicle Motion Control using the potential of different actuators in the same use case.

The second example is the highly automated function Highway Pilot with the use case safe stop in the same lane. This example focuses on fail-degraded motion control in case of a system failure.

The third example is connected functionality for more highly automated vehicles. This example focuses on predictive Vehicle Motion Control based on data services (e.g. road condition map) from the cloud.

## Example 1: Evasive Steering Support – Vehicle Motion Control for assisted functions

In an internal car clinic at Bosch with 35 untrained drivers (see Fig.10), as well as in external studies, it was found that in case of an obstacle suddenly appearing on highways ~25% of the drivers initiate an evasion maneuver to avoid a collision.

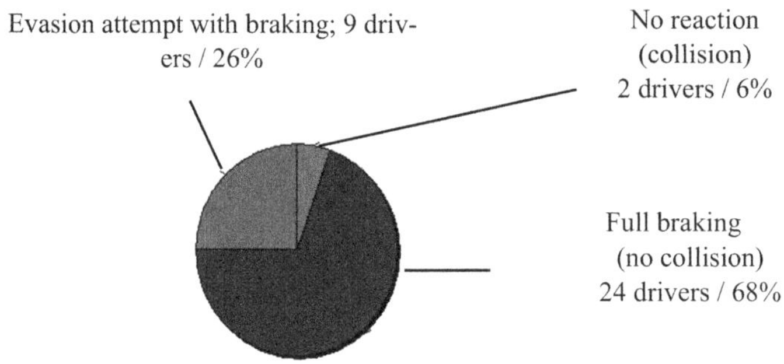


Figure 10: Bosch Study-reaction of untrained drivers on suddenly appearing obstacle

In the case of under reaction by the driver there will still be a collision with the obstacle during this evasion maneuver. In the case of overreaction by the driver, the vehicle is in danger of leaving the target lane or even the road. For this situation, "Evasive Steering Support" as an advanced driver assistance function, supports the driver in staying on a safe path in the case of an emergency situation. This function has been developed by Bosch and will be introduced into series production in 2015.

The function is realized by evaluating surround sensors accounting for data like time-to-collision, objects in the evasion target space and by monitoring activation of pedals and the steering wheel to discern the drivers' intention for an evasion maneuver.

Based on these inputs an optimal trajectory is determined and controlled via Vehicle Motion Control also generating additional steering torque and or additional (differential) brake torque after the driver starts the evasion maneuver (see Figure 11).

Via the intelligent actuator management of Vehicle Motion Control it is possible to use different actuators for the same use case. In the example of Evasive Steering Support, Vehicle Motion Control masters coordinated access to brake and/or steering systems for emergency evasion support.

Evasion Support via the steering system only, leads to a good (natural) driver feedback and no unintended deceleration.

Evasion support by brake only leads to a good haptic feedback in critical situations and reduced influence of the driver's steering behavior on optimal trajectory control.

Evasion support through a combination of steering and brake leads to increased control by removing dynamics from the maneuver through the use of braking in addition to steering.

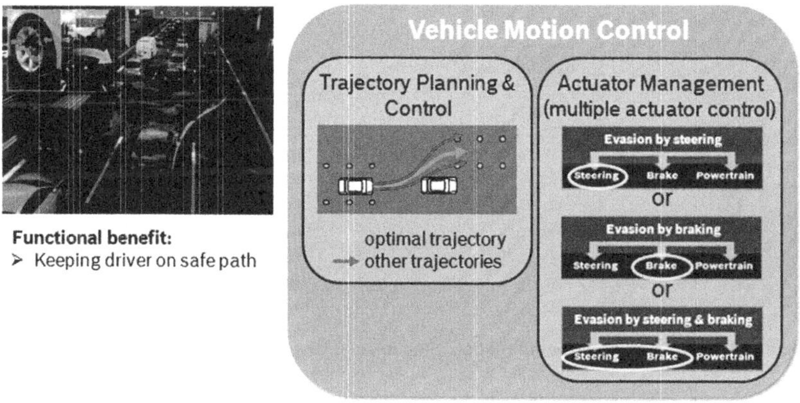


Figure 11: Evasive Steering Support using potential of different actuator systems

As shown in this example, Vehicle Motion Control enables an optimal combination of different actuator capabilities which influence vehicle dynamics.

## Example 2: Highway Pilot with Safe Stop in the same lane – Vehicle Motion Control for Automated Functions

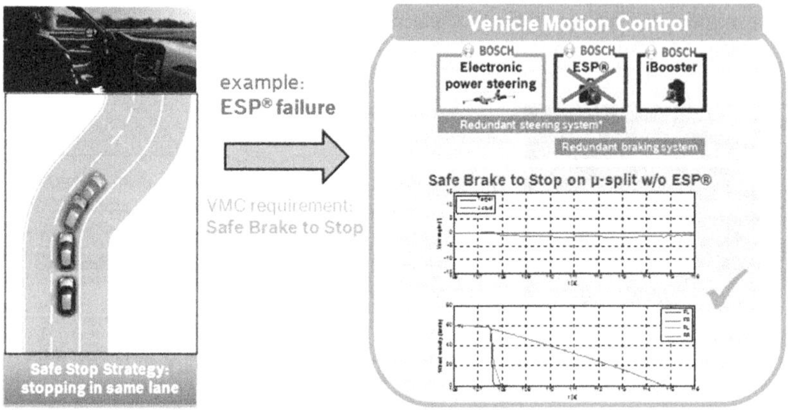


Figure 12: Vehicle Motion Control for Safe Stop in same lane in case of ESP® failure

For highly automated functions like Highway Pilot, guidance of vehicle on highways must be achieved by the system without permanent supervision by the driver.

Therefore, in case of a system failure, fail-degraded motion control including planning and executing safety trajectories is needed to bring vehicle to a safe state without driver intervention

In case of a failure of the Electronic power steering for example the lateral guidance can be realized within a certain operating range through ESP® via wheel individual brake intervention.

Figure 12 shows the Bosch redundancy concept for braking and steering and also the system reaction "safe stop in the same lane" in case of an ESP® failure.

To master this system reaction during an ESP® failure, deceleration is realized through the iBooster. The measured trace in Figure 12 shows that not only safe brake to stop on homogeneous roads, but also the challenging use case of modulated safe brake to stop on µ-split without ESP® can be realized with iBooster.

## Example 3: Predictive Motion Ctrl. based on Road Condition Map – Vehicle Motion Control for Connected Functions

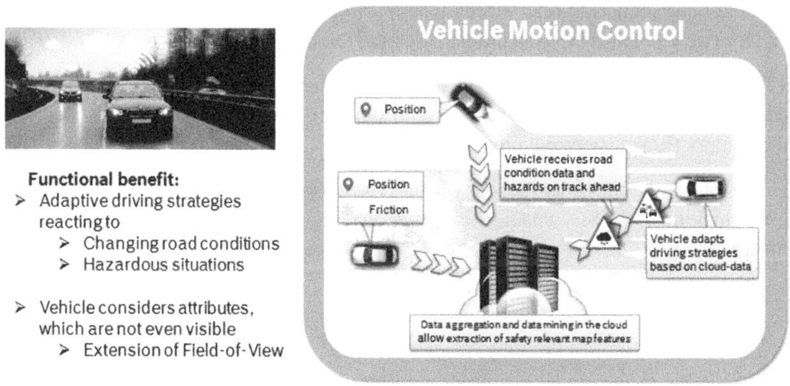


Figure13: Predictive Vehicle Motion Control using connectivity data

In the future, inclusion of cloud-based data services will quite literally extend the view of Bosch equipped automobiles (see Figure 13).

Interconnectivity will make it possible for a vehicle motion control system to develop predictive driving strategies which account for roadway characteristics and occurrences along a given route which may not be physically perceived by either the car or the driver. This type of information could for example bring about an autonomous decision from the VMC to move over into a safe lane to avoid an oncoming wrong-way driver on the other side of a blind curve. This type of information can also be used for example to slow an automobile down to a safe speed as it approaches a patch of black ice; information derived from the friction coefficient generated by the wheels of other automobiles which have already negotiated the dangerous area. This type of predictive strategy will be made possible, through the networking of automobiles, infrastructure and aggregation of data available to us from other sources of information, for example weather services, outside of our traditionally self-contained automobiles…virtually anywhere and in real time.,

At the latest, the introduction of highly automated driving functions will necessitate a reality which today may sound only visionary. We do not see a safe and comfortable future for highly automated driving without predictive functions and interconnectivity. However, we do see the possibilities opened by today's communication infrastructure and technologies making highly automated driving a reality sooner rather than later.

# 5 Summary

Trends like automated driving, connectivity as well as the evolution of actuators influencing vehicle motion (e.g. electrification of powertrain) lead to new requirements as well as new potential for planning and execution of driving tasks.

Increasing number of driving functions requires an abstraction layer "Vehicle Motion Control" which handles complexity by coordinating and controlling access to different actuators (brake, steering, powertrain systems).

Vehicle Motion Control as the system of systems enables cross-domain vehicle dynamics making it the key enabler for future driving functions with respect to agility, comfort & safety.

# Future consumer protection demands on vehicle safety

Andreas Rigling, ADAC e. V. Technik Zentrum

**This manuscript is not available according to publishing restriction.
Thank you for your understanding.**

# Model-based development methods – What can chassis and powertrain development learn from each other?

Bernhard Schick,
Global Business Unit Manager Calibration & Virtual Testing Solutions,
AVL List GmbH Graz

Michael Paulweber,
Director Global Research & Technology Management,
AVL List GmbH Graz

# Abstract

The biggest challenge for today's vehicle development is the increasing number of vehicle variants for different markets, which have to be developed in ever shorter cycles. In addition, customers and governments have high demands when it comes to comfort, driving pleasure, consumption, security and CO2 emissions. The powertrain and the chassis domain have developed their own methods to solve these difficulties. Although there are some important differences, the question arises how one domain can profit from the methods of the other domain. For example, the strict emission legislation in the powertrain sector lead to advanced model-based calibration and testing methods, which could be useful also for other domains.

This paper shows how model-based development methods, which have been established in the powertrain domain for years, can be used for chassis/vehicle dynamics development. The focus of this paper lies on "model-based testing", model-based validation" and "model-based calibration". These methods can be used successfully on system test benches, for MIL, SIL, HIL and in pure simulation environments and help to solve complex calibration and validation tasks. They deliver results that are ready to be used creating transparency to take the right decisions. Time- and cost-intensive iterations and unnecessary evaluations can be avoided by automated, model-based processes. The knowledge gained is condensed in easy understandable behavior models, which can be re-used at any time. This makes the complexity much easier.

# The chassis of the all new Audi Q7

C. Jablonowski, V. Underberg, M. Paefgen

# Introduction

High customer expectations in growing markets worldwide have set a challenging task for the chassis development of the Audi Q7. Not only is it supposed to offer the typical driving dynamics of an Audi, but also the goal was set to offer best in class comfort in order to provide a luxurious ride, especially for long distance trips. Besides these objectives, the chassis also has to contribute to achieving the $CO_2$-emission-limits implemented by governments around the world. Finally, the vehicle stability is to be ensured during critical manoeuvers, since rollover is an issue with Sports Utility Vehicles (SUV) in general.

In order to meet those requirements, Audi has defined three fields of action in chassis development that have been worked on with great emphasis:

- Weight reduction in all chassis components
- State of the arte driver assistance systems
- Intelligent Chassis

This paper will give insights in those three fields and illustrate some of the innovations that made it into series production. Prior to this, it will outline the technical targets of the chassis performance derived from the strategic objectives. In the end it will compare the resulting performance with the Q7 predecessor and other competitors.

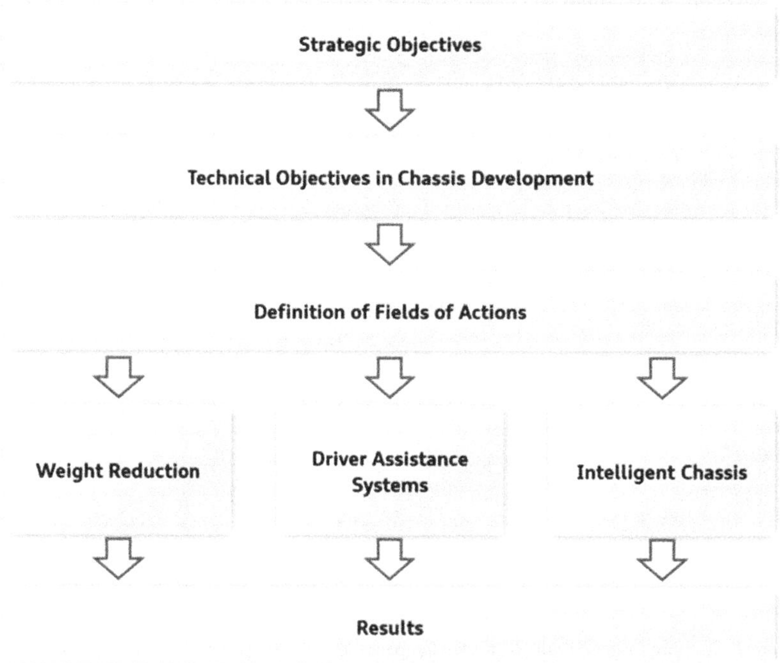


Figure 1: Structure of Paper

## Strategic Objectives & Technical Targets

Besides the regular evolutional improvements to the chassis, the strategic main goal of the all new Q7 was offering the best comfort in the SUV C-segment. Especially the long distance comfort, the interior noise level and a smooth ride were the main focus during the definition of the chassis properties.

Based on those strategic objectives, a thorough analysis of the key competitors delivered a range of technical targets in the form of characteristic figures and their optimal value. The following targets are for a car with air suspension. All competitors were benchmarked with a comparable specification.

For the comfort, the following targets were determined:

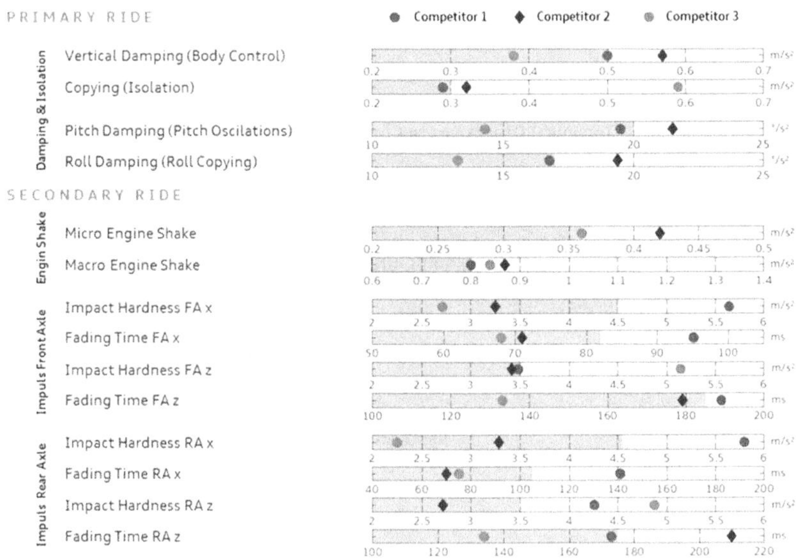


Figure 2: Comfort Targets for the New Q7 Compared to Key Competitors

The characteristics shown in Figure 2 represent the point of emphasis for the chassis development of the comfort. The grey, blue and green dots represent the results of measurements conducted with key competitors. The green area is the target range for each characteristic figure.

Especially noteworthy are the targets for the categories “damping & isolation” and “engine shake”. Within the former, the emphasis was on isolation (green area left of best competitor), accepting a good average in regards to body control- two opposed characteristics. In the latter category, a performance better than the competitors for both micro and macro engine shake was targeted. In general, a comfortable setup in comparison to the chosen competitors is aspired, represented by the green target areas tending to be left of the competition.

Similar to the comfort objectives, the dynamic performance was analyzed aswell, although it was compared to its predecessor:

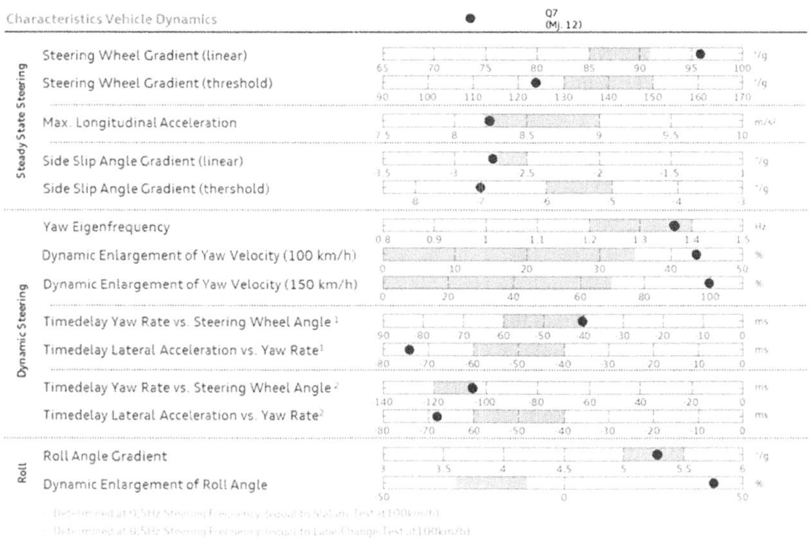


Figure 3: Dynamic Performance Targets of the New Q7 Comapred to its Predecessor

The red dot symbolizes the results for the former Q7. The characteristics have been optimized in the direction of a regular passenger car. The target was, to procure the feeling through an outstanding handling and agility for the class.

While those targets can partly be influenced by traditional chassis setup, further innovation was necessary to reach the ambitious goals. Therefore, a range of fields of action has been defined to fulfill the technical targets on the one hand and add to the strategic objectives in addition to that. Those fields will be discussed in the following chapter.

# Fields of Action

In order to improve the chassis of the new Q7 without giving up any of the qualities of its predecessor, a number of fields of action have been defined for chassis development that add to the overall performance in regards to comfort and driving dynamics. In this chapter, three main fields that contributed largely to the result will be explicitly highlighted.

## Weight Reduction

Decreasing the overall weight has an extensive impact on the driving characteristics, even more so, when it is in components that contribute to the unsprung masses. While the new Q7 has a total weight reduction of up to 325kg, the chassis alone has contributed approximately 73,5 kg to that.

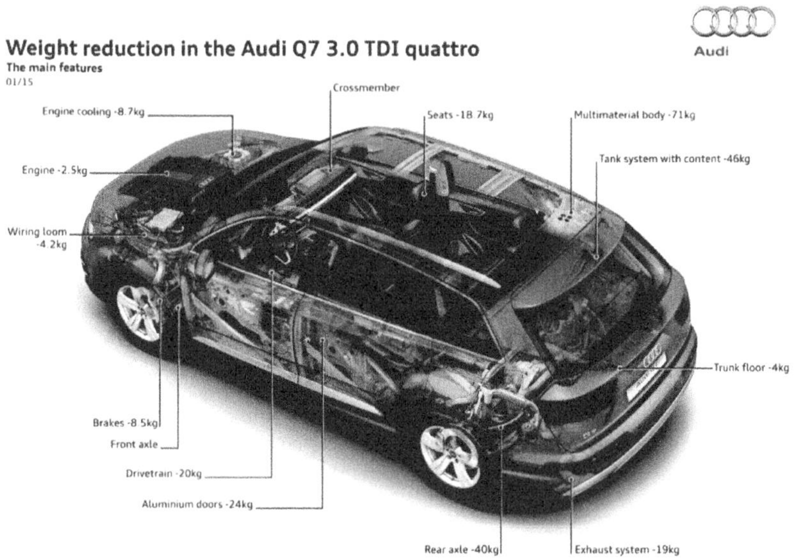


Figure 4: Weight Reduction in the new Audi Q7

The weight of the brakes could be reduced by 8,5kg using aluminum brake calipers and the reduction of the total vehicle weight, since a smaller brake could be used.

The front axle is 25kg lighter, extensively using aluminum for most rods and the hinge bearings, hollow hinge shafts and an anti-roll bar with variable wall thickness depending on the requirements. The axle carrier is using a material mix of aluminum and steel, depending on the application so that the lightweight construction can combine high performance with reduced weight.

The rear axle is a completely new design, saving a total of 40kg of weight. It has been changed from a double wishbone to a multi-rod layout, making it possible to separately absorb lateral and longitudinal forces and therefore increasing comfort by using optimal bearings for each force. It has also been designed to offer space for a rear-wheel steering system. Only through the extensive use of multi-material design, the weight reduction in combination with the increased requirements was possible.

The results of the lighter vehicle weight is an increase in both, ride comfort and dynamic performance. Also the stability has increased through a lower center of gravity, reducing the tendency of the vehicle to roll.

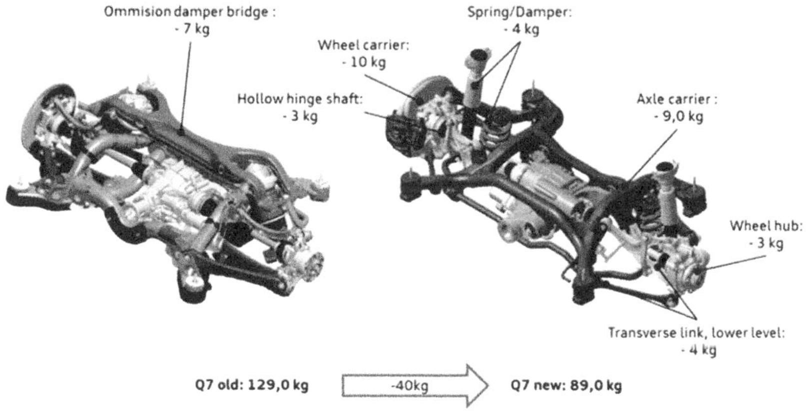


Figure 5: Weight Reduction of the Rear Axle

## Driver Assistance Systems

For an outstanding comfort the new Audi Q7 does not only have an improved ride, but also a considerable range of driver assistance systems, offering the widest choice in its class.

The different functions are clustered into three groups for facilitated customer access. Those groups are:

- **Tour** for mid-and long-distance journeys outside of the urban environment
- **City** for the mobility inside of the urban environment
- **Parking** especially for support during challenging parking tasks

Each group accesses a different set of sensors including ultra sound and radar sensors, cameras around the car and a night vision camera (using heat detection). By combining those data sources depending on the function, a wide range of possible application can be generated. While a lot of those functions greatly improve the driving safety, in this paper we will focus on those that enhance comfort.

One of the main features of the group "Tour" is the adaptive cruise control, whose main function is to automatically keep the distance to the car driving in front. In the new Audi Q7, the performance of this system has been widely increased. It now covers a speed range of 0-250 km/h, offering a stop-and-go function that enables the car to come to a complete stop if necessary and then accelerate again after a short confirmation by the driver. In the speed range of 0-65 km/h the system also can perform the steering task as long as the traffic is not running fluently. In order to do so, it detects not only the markings on the street but also the other vehicles around it. Based on this information, it then chooses "the right line", depending on the situation.

In the group "City, the trailer-assistant offers a great increase in transport. The Audi Q7 is the optimal vehicle for trailer use. Since reversing with a trailer is difficult to handle, especially in an urban environment, the trailer-assistant offers help when required. The hitch is equipped with a sensor that measures the angle between the car and the trailer and the car calculates the optimal steering wheel angle for the desired path (chosen over the Audi Multi Media Interface). All the driver has to do during reversing is to brake and accelerate while the rest is controlled by the assistance system.

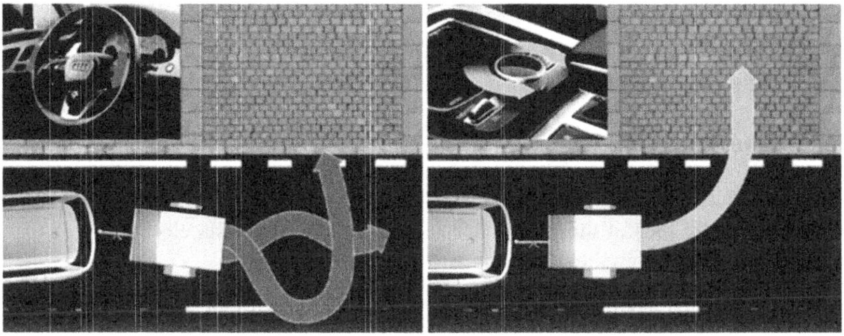

Figure 6: Audi trailer assistant for easier manoeuvring with attached trailer

The perfect addition to the group "City" is the group "Parking". The parking assistant uses eight ultra-sound sensors around the car to guide it during parallel and transverse parking manoeuvers. In addition to that, several cameras create a 180° view of the front and rear area as well as a bird-view perspective for better overview.

In summary, there is a whole range of driver assistance systems that the customer can choose from depending on the use case of her or his vehicle. By offering support in challenging or inconvenient situations the comfort of using the vehicle is increased while at the same time, the risk of costly damage or even harming passengers or people in the surrounding traffic is reduced.

## Intelligent Chassis

Within the chassis development there is often need for compromise, since different properties diverge in their requirements to the chassis configuration and setup. The best example is comfort and dynamic performance, but also vehicle stability is of importance when creating the conceptual chassis design, especially for SUVs.

Adaptive components offer the possibility to change their behavior according to the requirements of different states of driving and active components can implement forces or change parameters of the chassis that are usually not present or constant. As a next step, the behavior of each of those components can be controlled centrally instead of individually, so that the optimum overall vehicle behavior can be achieved. In this paper, we call this an "Intelligent Chassis", as it enables the vehicle to recognize the state it is in and then react accordingly. The control of all functions is localized in the Electronic Chassis Platform EFP. It controls all systems that influence the control of the vehicle`s body

In the following, this paper will highlight some of the adaptive and active components of the new Q7's chassis that contribute to improved comfort while at the same time increasing vehicle stability and dynamic performance.

## Rear Wheel Steering

The new Q7 is equipped with rear wheel steering, a technology unique in the SUV-segment. The tie rods of the back suspension are connected to an electric motor that makes it possible to turn the wheels by angles up to five degrees.

For better maneuverability at low speeds, a small angle opposite to the direction of the front wheels is applied. Consequently, the turning circle can be reduced noticeably. At higher speeds, the rear wheels follow the direction of the front wheels in order to increase steering response and improve the vehicle's stability during evasive maneuvers.

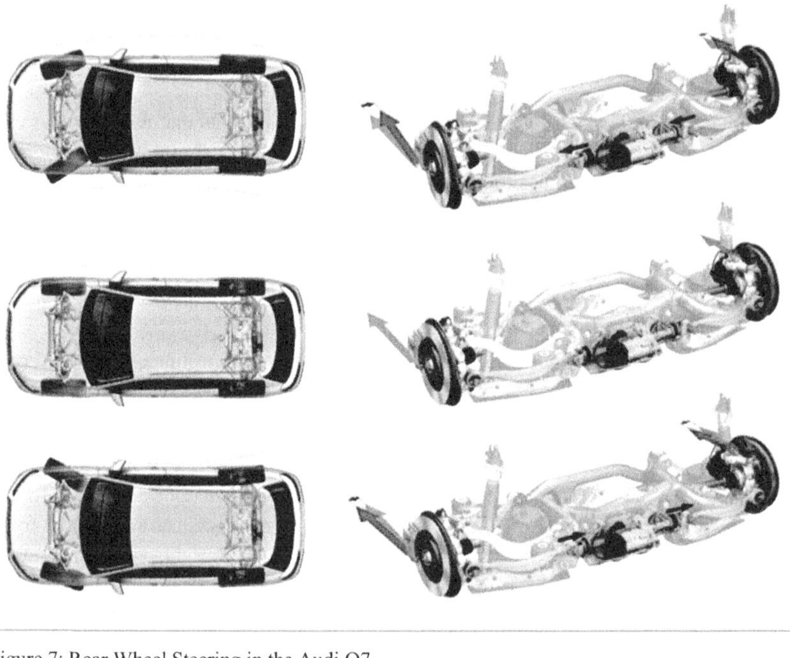

Figure 7: Rear Wheel Steering in the Audi Q7

## Air Suspension

The adaptive air suspension enables the Q7 to lift or lower its level according to load, speed and use case. When loaded heavily, it increases the pressure so that the spring stroke in both directions is comparable to the unloaded vehicle. This ensures that the damper makes sufficient displacement to work effectively, increasing both ride comfort and safety in critical situations.

Additionally, the level can be lower at higher speeds in order to reduce air resistance and stability, or increased when the vehicle is used offroad, so that an increased ground clearance can be provided.

Finally, the air pressure of the air suspension is included in a load calculation in the EFP. This way, other systems can use this information to adapt their setup, improving the overall behavior under load.

## Adaptive Damping

One of the highlights of the Q7's intelligent chassis is the adaptive damping control. It has three completely new features, that considerably improve the overall performance of the car.

Firstly, the adaptive damping reacts to the vehicle load calculation provided by the EFP. With passive damping, high loads lead to unparalleled oscillation of the vehicles body as well as an underdamped body in general. With the new adaptive damping, the setup is adjusted to meet the new requirements implemented by loading the vehicles trunk an offers optimal performance.

Secondly, the damper has a temperature detection algorithm that calculates the temperatures of the damper over the currents needed. It then compensates the different viscosity of the damper oil at low temperatures, offering the best comfort at all conditions.

Finally, the adaptive damping considers the dynamic behavior of the damper by calculating a damper model in real time. While regular dampers show a hysteresis in their behavior, triggered by the excitation earlier in time, the new adaptive damping includes this effect in its calculations and therefore improves the precision of the applied forces by up to 500 Newton. Especially on bad roads with high-frequency profiles the comfort can be increased noticeably.

Additionally to those functions, the adaptive damping is in communication with the Electronic Stability Control over the EFP. When the ESC detects a situation that might be critical for vehicle rollover, the damping is stiffened in milliseconds, so that body roll can be minimized and stability is improved. This is especially the case, when the active roof load detection senses a mounted roof box, because the ESC then uses a different set of parameters that trigger an earlier intervention. Vice versa, a softer damping can be chosen for regular use cases, improving the ride comfort further.

# Results

After the development was finished, another set of measurements was taken in order to compare the targets with the actual vehicle.

The following results are the summary of those measurements.

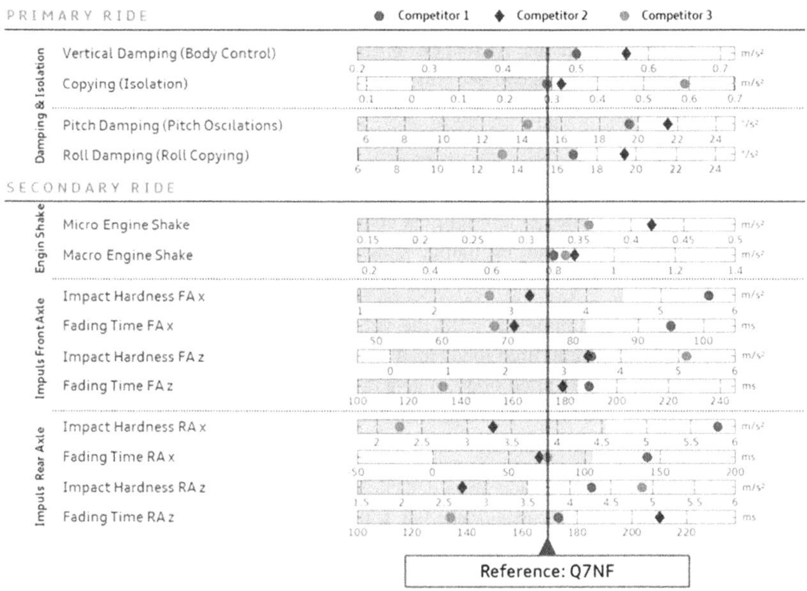


Figure 8: Comfort Characteristics of the New Audi Q7

The presentation has been changed for the comfort characteristics so that the red line represents the new Q7 and the target areas as well as the competitors are aligned accordingly. Most of the targets have been met, resulting in an outstanding comfort experience.

Similarly, the dynamic performance characteristics have been determined and are shown in Figure 9 in comparison to the Q7 predecessor. Again, most of the technical targets have been met. Accordingly, the handling and agility of the new model is outstanding. The few target that could not be fulfilled are part of the compromise necessary to achieve the strategic objective of developing a best-in-class comfort.

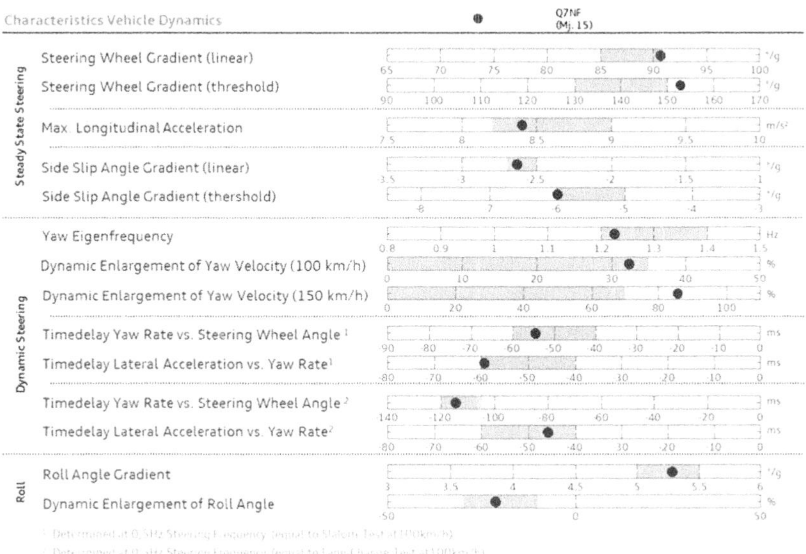


Figure 9: Dynamic Performance Charactersistics of the New Audi Q7

Only through intensive support by objective measurements, a considerable set of innovation as well as consequent use of light weight construction was is possible to make a big step in all aspects of the chassis characteristics. The result however is a luxurious SUV that offers a smooth ride and long distance comfort on the one hand and an extraordinary handling while approaching the agility of a regular passenger car.

# Network topology for chassis – potential of ethernet-based systems

Trenkel, Kristian; Wunner, Patrick

{Kristian.Trenkel, Patrick.Wunner}@isyst.de

iSyst Intelligente Systeme GmbH, Nordostpark 91, 90411 Nürnberg

# Introduction

The increasing number of functions in modern vehicles like automobiles induces an increasing number of information to be interchanged between the diverse electronic control units (ECU) in the vehicle. Modern systems, like advanced driver assistance systems or infotainment systems, demand broadband communication channels. Beside the traditional bus systems, like CAN, MOST or FlexRay, Ethernet becomes more and more important for the in-vehicle data interchange, too.

The amount of data produced by sensors in the car increase fast, too. The sensors itself become more and more complex. That is why sensors with analog or PWM interfaces are replaced by sensors with digital interface like SPI, SENT or PSI5. With such an interface the sensor is able to provide more data including diagnostic information.

Beside the pure bandwidth also a deterministic behavior and low latency of the transmission are required by different applications like active suspension systems and distributed closed loop control systems. For these applications CAN and FlexRay are widely used. But also in this domain Ethernet based networks can become an option.

# Retrospect

In the past a small number of electronic components were used in the chassis domain. The development starts with wheel speed sensors used for Anti-lock braking system. The first systems used direct connections to each other.

With the increasing number of ECUs the integration of bus systems started in the early 90's. The first widely used bus system was CAN. The ECUs were connected by one single CAN bus. With the rising number of ECUs and functions the limitation of the bandwidth of the CAN bus led to the integration of several CAN buses in one car. These buses were connected by gateways as you can see in the figure 1.

With the rising demand for bandwidth and more and more safety relevant functionality a new bus system was needed. With the focus on applications with higher requirements on availability and deterministic behavior like drive by wire the FlexRay standard was developed.

The first practical use of the FlexRay in a car was an active suspension in 2006. Later on, FlexRay was used by different OEMs in the chassis domain. X by wire applications are not realized with FlexRay up to now.

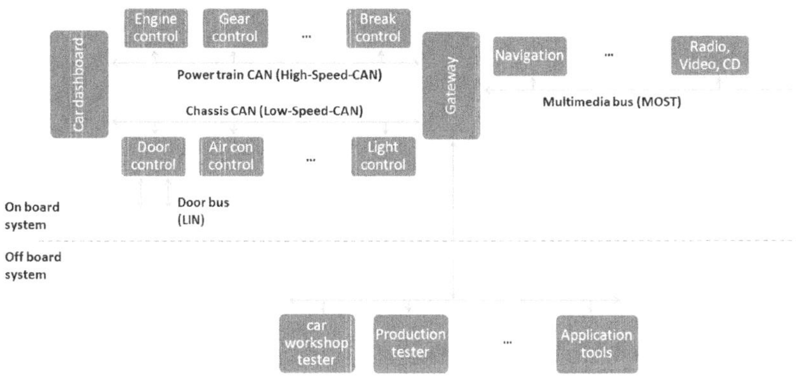


Figure 1. Typical bus system architecture (based on p.1 [1])

# Situation Today

In today's cars a number of different bus systems are combined. Usually several CAN buses (e.g. comfort), one FlexRay bus (e.g. chassis), one MOST (e.g. infotainment) and some LIN buses (e. g. central locking system) are used. The backbones of the cars are the several CAN buses. FlexRay is not able and will not be able to replace the CAN bus. The topology depends on the OEM of the car. But there is normally one central gateway like it is shown in figure 2. This gateway is connected to all the bus systems in the car and is responsible for the routing of the signals and messages between these different buses. In a typical grown E/E architecture there are additional ECUs with gateway functionality. That's why the E/E architecture of modern cars is very complex and the communication relation between the different systems is nearly not maintainable.

In addition to the mentioned bus systems there are some recent systems for the connection of sensors or sometimes smart actors, too. The two most popular systems are PSI5 and SENT. Both are primarily developed for the connection of sensors to the ECU. With these sensor interfaces it is possible to connect more complex sensors (e. g. different sensor signals) to the ECU. Furthermore it is possible to have a wide range of diagnostic functionality inside the sensors. These sensors behave like individual ECUs. And this leads to an additional increase of the complexity of the E/E architecture and to a new challenge for the development and the test of such systems.

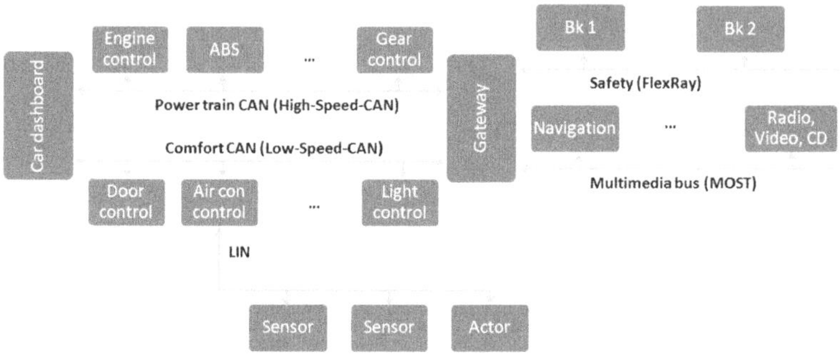


Figure 2. Typical bus system architecture today (based on [2])

For typical diagnostic access to the car and the import of software updates the CAN bus is used. With up to some GB of software in a modern car the time for software update process is very high. That's why some of the OEMs are using already standard Ethernet for the software update process in their cars. In this case there is a Ethernet link connected directly to the central gateway and the gateway distributes the diagnostic access and data on the different bus systems inside the vehicle.

# Ethernet based Systems – SOME/IP and BroadR-Reach

## SOME/IP

SOME/IP is a UDP/TCP-based network protocol that has been developed by BMW in the scope of a promoted project. It is the only solution which complies with all automotive requirements and is compatible to AUTOSAR [3]. The advantage in comparison to other protocols is that needed data are only sent from the host to the client when the client is subscribed to the service. For communication some standard components from the Ethernet stack are used. In Figure 3 the location of the SOME/IP (-SD) protocol in the OSI model is shown. IP and TCP/IP layers are used unmodified. The MAC layer is extended by a Virtual LAN (VLAN) tag. The physical layer in contrast has been substituted. SOME/IP (-SD) is not a real time protocol because it is just based on standard Ethernet layers. Actually the real-time requirements are met by a moderate covered Quality of Service (QoS).

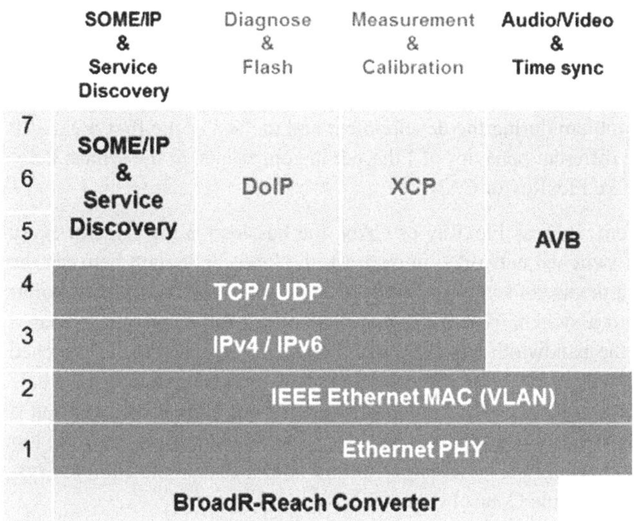


Figure 3. OSI model and classification of protocols

Currently the used cycle times for SOME/IP messages are around 100 ms. For hard real-time requirements the AVB protocol is intended.

### BroadR-Reach

BroadR-Reach [4] is a new physical layer for automotive Ethernet and is standardizes as IEEE 100BASE-T1. It was developed by the company Broadcom and the standard is provided by the OPEN Alliance SIG [5]. Instead of the shielded CAT cable with four or eight wires an unshielded drilled cable pair is used. The advantage of this solution is a bidirectional connection with the bandwidth of 100 MBit/s using a PAM3 modulation. The EMC characteristics of BroadR-Reach are developed according to automotive requirements, too.

## Current Development

### Practical Usage of SOME/IP

In the current year 2015 the first cars inducing SOME/IP based Advanced Driver Assistance Systems will come to the market. BroadR-Reach is used as physical layer.

Based on the AUTOSAR integration of the SOME/IP in Version 4.1 the seamless integration of SOME/IP in the software of ECUs is possible. Based on the experience of the first projects SOME/IP integration was optimizes in the AUTOSAR version 4.2.

One of the biggest problem during the development and the test of the first SOME/IP based ECUs was the different behavior of Ethernet in comparison to the typical automotive bus systems like FlexRay or CAN.

In a shared bus system, such as FlexRay or CAN, the bus load is the same at every physical location. In switched networks, in contrast, the load can variant between the single branches. With the use of switched Ethernet every node has a 100 MBit/s point-to-point connection to a switch. This is an enormous advantage compared to shared bus systems, where the bandwidth has to be split between the nodes. If the switched network is well designed, every node can send and receive data exploiting the whole 100 MBit/s bandwidth. Additionally a 1 GBit/s connection between a host and a switch on the same board can be realized to increase the efficiency of the network. For high load of the whole network it is necessary to find bottlenecks in the complex system of switches and Electronic Control Units (ECUs).

Currently there are no simulation tools available to simulate the whole behavior of the Ethernet based automotive network including the characteristics of SOME/IP. The support for the testing of such networks is limited in the available tools, too. The problem will be discussed later in this article.

## Ethernet as Backbone

For an efficient integration of Ethernet based networks in cars a new E/E architecture is needed. With the current usage of Ethernet in separate parts of the E/E architecture it is not possible to use all the advantages.

One possible solution for the new E/E architecture and a migration strategy for parts of the current bus systems is shown in figure 4. The idea is the usage of Ethernet as a backbone between headunits of the different subsystems and the central diagnostic access in the architecture. The headunits itself implement a gateway between traditional bus systems like FlexRay or CAN and the Ethernet backbone. The biggest advantages are:

1. Further use of subsystems
2. Increased performance for diagnostic access and flashing ECUs
3. Future-proofed architecture extendable to fully Ethernet based vehicles

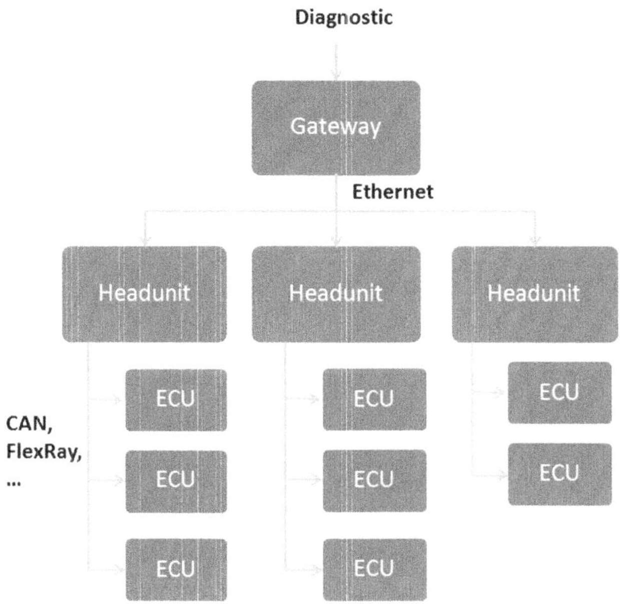


Figure 4. E/E architecture with an Ethernet based backbone

## Simulation and Test of SOME/IP based Networks

There are several Ethernet-Test solutions available on the market that can be used in a HiL environment. The dSPACE Ethernet Blockset [6] is mainly used for unit and function test. The Vector CANoe.Ethernet [7] is a standalone solution which can be controlled from a control desk to expand the HiL tests. For the test of existing nodes they work properly, but for developing whole systems or ECUs they are not as flexible as needed. There are no generic tools for the development and simulation of SOME/IP based networks, too.

To solve both problems – simulation and testing – a SOME/IP implementation using the network simulation framework OMNeT++ is developed. The implementation enables the development and the simulation of complete SOME/IP based networks including the timing and the busload behavior. In addition the connection of the network simulator to a real network or to a single ECU for testing purpose is possible. Figure 5 shows the graphical illustration of a network simulation with three simulated ECUs and a connection to one real world ECU provided by the router in the simulation.

Beside the manual configuration of the simulation also the usage of FIBEX bus description file to generate the simulation automatically is possible. With the developed extension for OMNeT++ the development, simulation and test of SOME/IP based networks is possible including the test of the network with a wide range of testcases. Further information can be found in [8].

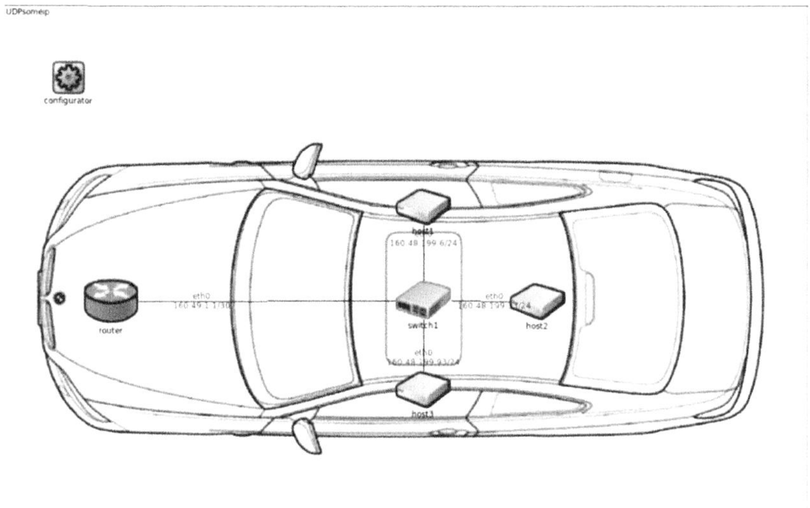


Figure 5. Network simulation with connection to a real world ECU

## Simulation and Test of Sensors

Similar to the simulation and test of Ethernet networks the simulation and test of sensors with digital interfaces like PSI5 and SENT is problematical, too. For the development and test it is necessary to simulate the sensor values and the behavior of non-conforming sensors. It is not possible to catch all behaviors and tolerance possible by real sensors. Compared to the simulation of analog or PWM signals the simulation of sensors with a digital interface is much more complex. The sensor provides, beside the sensor signal, more and more diagnostic and configuration information. The currently available sensor simulation methods do not provide the possibility of a real time simulation. Over the mostly used USB interface it is only possible to program fixed values or predefined signal sequences. The validation of the diagnostic functionality of the sensor in safety-related applications needs a real time simulation with fault injection possibilities. That is why the integration in Hardware in the Loop test systems is needed. There is only a small number of systems (e.g. dSPACE and iSyst) feasible for the described scenario.

Further discussions can be found in [9] and [10].

## SOME/IP and chassis

The current automotive Ethernet implementation based on SOME/IP provides only slow cycle times for the messages in the range of 100 ms. Because of the use of the UDP protocol the realization of cycle times like they are typical in the chassis domain (e. g. active suspension systems) with less the 5 ms, it seems to be not possible. With the experience from companies in the automation field, and their use of Ethernet based real time systems other solution are needed for systems with low cycle time and low latency in the range of 1 ms. For example Ethernet Powerlink (EPL)[11] is an automotive Ethernet system with cycle a time below 1 ms based on standard Ethernet hardware. The systems work typically with a proprietary low level protocol based on a Master-Slave concept.

For the use of Ethernet based networks in the chassis domain there are additional developments necessary. The application of AVB or Time-Sensitive Networking (TNS) as they are discussed in the next chapter can be possible solutions for Ethernet in the chassis domain.

# Future Development

## AVB

The term Audio Video Bridging (AVB) describes a number of standards that are developed by the Audio/Video Bridging Task Group (IEEE 802.1). The focus of the development targets at the distribution of synchronizing multiple streams in an Ethernet based network. The standards are:

1. 802.1AS – Timing and Synchronization for Time-Sensitive Applications in Bridged Local Area Networks
2. 802.1Qat – Stream Reservation Protocol (SRP)
3. 802.1Qav – Forwarding and Queuing Enhancements for Time-Sensitive Streams
4. 802.1BA – Definition of profiles for AVB systems

The practice of AVB is discussed for infotainment systems in the car. It is a candidate for the replacement of MOST.

Based on the 802.1Qav standard and the priority based forwarding and queuing mechanisms with Credit-based shaping, the realization of low latency network communication with guaranteed network latency of 2ms is possible. The extended OSI-Layers 1 and 2 can also be used for data communication with a cycle time below 1 ms beside the audio and video streams [12].

## TNS

The Time-Sensitive Networking (TNS) is a set of standards developed by the Time – Sensitive Networking Task Group (also IEEE 802.1). This group replaces the Audio/Video Bridging Task Group in 2012. There the extension of the AVB standards for real-time control streams which are used in automotive or industrial control facilities is in the focus. Therefore the following standards are developed:

1. 802.1ASbt – Timing and Synchronization
2. 802.1Qbv – Enhancements for Scheduled Traffic
3. 802.1Qbu – Frame Preemption
4. 802.1Qca – Path Control and Reservation
5. 802.1CB – Frame Replication and Elimination for Reliability
6. 802.1Qcc – Enhancements and improvements for stream reservation

For a low latency data communication the standard 802.1Qbv and it's time-aware shaper are very important. This standard and the time synchronization based on 802.1ASbt allow the time based distribution of Ethernet frames in the network topology and thereby a defined real time behavior of the Ethernet network. That is why an Ethernet based real time network for automotive and industrial control is possible [12].

# Conclusion

The currently used SOME/IP protocol does not allow applications with low latency and low cycle times are not possible. But the several actual activities in the development in the field of Ethernet networks in automotive applications lead to an increasing number of new and interesting applications. This opens up a new potential for innovations.

But this new bus system generates some new problems in development and test of these new applications. As mentioned, new test systems and tools are needed for an efficient development of the topology of such networks. For the testing new tools and standards are needed, too.

# Outlook

Based on the development of AVB and TSN new upper layer protocols and systems for low latency applications are possible. So totally Ethernet based E/E architectures for cars are feasible.

But in the near future E/E architectures with Ethernet backbones will be the first migration steps to Ethernet based cars.

The pivotal question is how reasonable is the usage of Ethernet based systems. This has to be answered during the next years by the development departments.

# References

[1] W. Zimmermann und C. Schmidgall, Bussysteme in der Fahrzeugtechnik: Protokolle, Standards und Softwarearchitektur, Wiesbaden: Vieweg + Teubner, 2011.

[2] elektroniknet.de, „Serielle Bussysteme im Automobil I," WEKA FACHMEDIEN GmbH, 23.04.2015. [Online]. Available: http://www.elektroniknet.de/automotive/sonstiges/artikel/273/2/. [Zugriff am 01 05 2015].

[3] AUTOSAR development cooperation, „AUTOSAR," AUTOSAR development cooperation, 2015. [Online]. Available: www.autosar.org. [Zugriff am 01 05 2015].

[4] broadcom, „BroadR-reach PHYs," broadcom, 2015. [Online]. Available: http://www.broadcom.com/products/Physical-Layer/BroadR-Reach-PHYs. [Zugriff am 01 05 2015].

[5] Open Alliance SIG, „Open Alliance," Open Alliance SIG, 2015. [Online]. Available: http://www.opensig.org. [Zugriff am 01 05 2015].

[6] dSPACE, „dSPACE – ethernet blocksets," dSPACE, 2015. [Online]. Available: http://www.dspace.com/de/gmb/home/products/sw/impsw/ethernet blocksets.cfm. [Zugriff am 01 05 2015].

[7] Vector Informatik, „CANoe.ethernet," Vector Informatik, [Online]. Available: http://vector.com/vi canoe ethernet de.html. [Zugriff am 01 05 2015].

[8] P. Wunner, K. Trenkel, S. Dengler und S. May, „Development and Testing of Automotive," in s *OMNeT++ Community Summit 2014*, Hamburg, 2014.

[9] K. Trenkel, „Echtzeitfähige Sensorsimulation für Entwicklung und Test," in s *AmE 2014 – Automotive meets Electronics – Beiträge der 5. GMM-Fachtagung*, Dortmund, VDE Verlag, 2014, pp. 147 – 150.

[10] K. Trenkel und F. Spiteller, „Sensorsimulation in Hardware-in-the-Loop-Anwendungen," in s Echtzeit 2014, Berlin Heidelberg, Springer, 2014, pp. 51-60.

[11] Ethernet POWERLINK Standardization Group, „Ethernet Powerlink," Ethernet POWERLINK Standardization Group, [Online]. Available: http://www.ethernet-powerlink.org. [Zugriff am 01 05 2015].

[12] W. Steiner, „Time-Sensitive Networking/TSN – Die nächste Generation von AVB," 27 05 2014. [Online]. Available: http://vector.com/portal/medien/cmc/events/commercial_events/Automotive_Ethernet_Symposium_AES14/AES14_09_Steiner_TTTech_Lecture.pdf. [Zugriff am 01 05 2015].

ISYST
Network topology for chassis – Potential of Ethernet based systems
Kristian Trenkel, Patrick Wunner
chassis.tech 2015 – 16.06.2015
TESTHAUS AUS ÜBERZEUGUNG

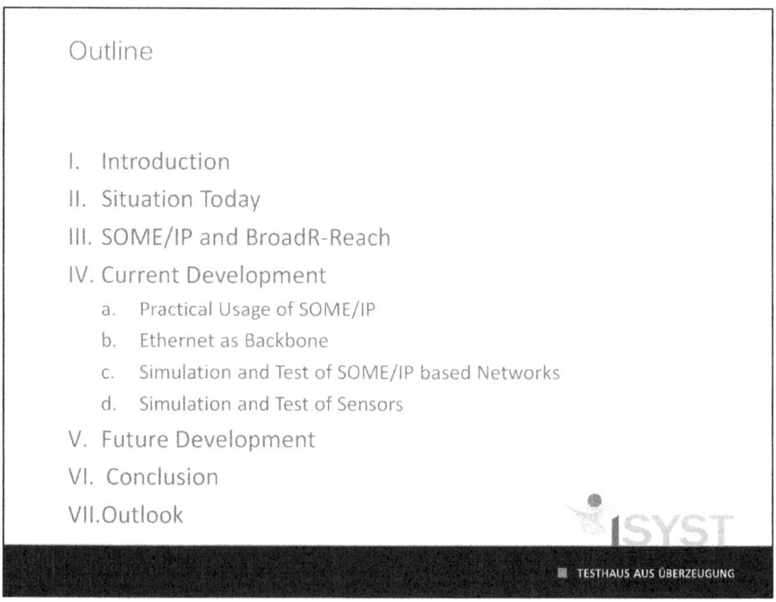
Outline
I. Introduction
II. Situation Today
III. SOME/IP and BroadR-Reach
IV. Current Development
a. Practical Usage of SOME/IP
b. Ethernet as Backbone
c. Simulation and Test of SOME/IP based Networks
d. Simulation and Test of Sensors
V. Future Development
VI. Conclusion
VII.Outlook
ISYST
TESTHAUS AUS ÜBERZEUGUNG

## Introduction

- Complex and grown E/E architectures
- Increasing number of functions
- Increasing requirements to the bus systems
  - Usage of Ethernet
  - New buses for sensors (e.g. SENT, PSI5)

➔ Further development
➔ Problems in simulation and testing

ISYST
TESTHAUS AUS ÜBERZEUGUNG

## Outline

II. Situation Today

ISYST
TESTHAUS AUS ÜBERZEUGUNG

## Situation Today

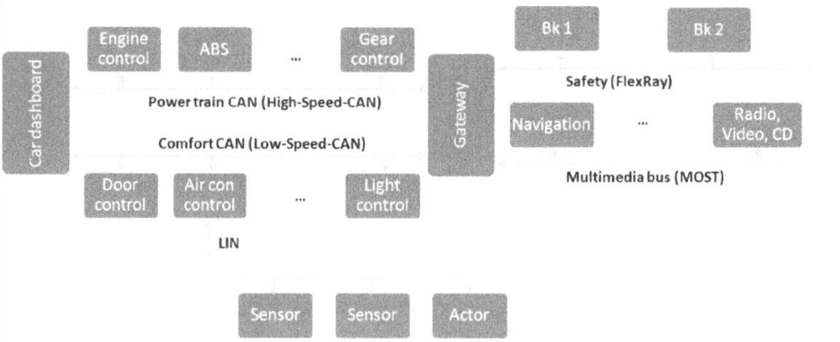


TESTHAUS AUS ÜBERZEUGUNG

## Outline

III. SOME/IP and BroadR-Reach

TESTHAUS AUS ÜBERZEUGUNG

## BroadR-Reach

- New physical layer for Automotive Ethernet
  - IEEE 100BASE-T1
- Developed by Broadcom
- OPEN Alliance SIG
- 100 MBit/s on unshielded drilled cable pair

ISYST
TESTHAUS AUS ÜBERZEUGUNG

## SOME/IP (-SD)

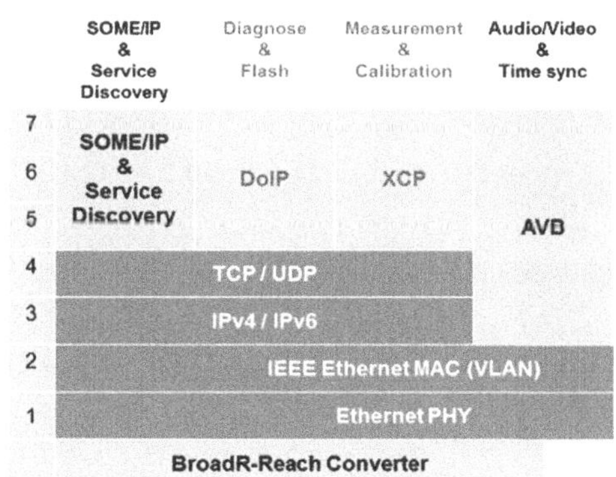


ISYST
TESTHAUS AUS ÜBERZEUGUNG

## Outline

IV. Current Development

a. Practical Usage of SOME/IP

ISYST

TESTHAUS AUS ÜBERZEUGUNG

## Current Development

Practical Usage of SOME/IP

- First car with SOME/IP integration and BroadR-Reach
- SOME/IP support in ATUSOAR with V4.1
- Optimized support with V4.2

→ Different Problems during the development and test

ISYST

TESTHAUS AUS ÜBERZEUGUNG

## Current Development

Practical Usage of SOME/IP

Problems:

- Ethernet – switched network, CAN and FlexRay shared buses
  - Different bus load
  - Different design strategies
  - ➔ Different architectures needed
- No final simulation tools and test tools
  - ➔ New solution based on OMNeT++

ISYST

■ TESTHAUS AUS ÜBERZEUGUNG

## Outline

IV. Current Development

b. Ethernet as Backbone

ISYST

■ TESTHAUS AUS ÜBERZEUGUNG

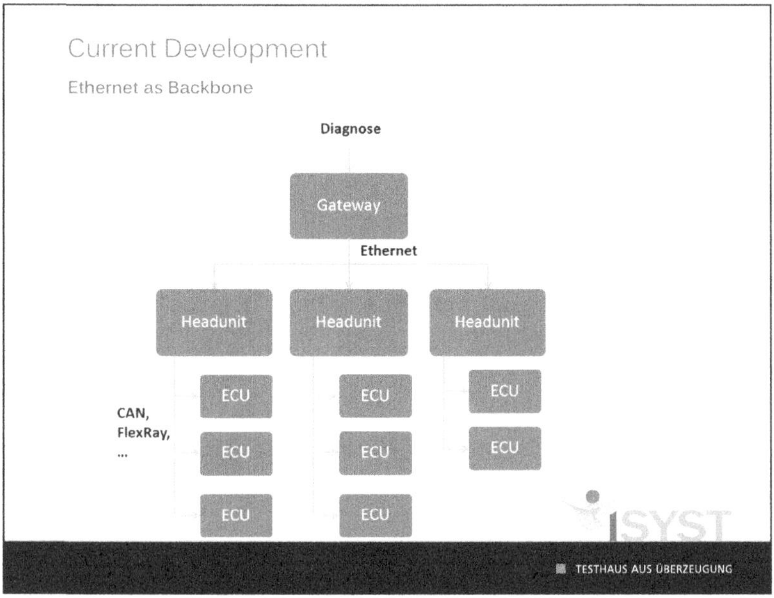
Current Development
Ethernet as Backbone
Diagnose
Gateway
Ethernet
Headunit
Headunit
Headunit
ECU
ECU
ECU
CAN,
FlexRay,
...
ECU
ECU
ECU
ECU
ECU
ECU
ECU
ISYST
TESTHAUS AUS ÜBERZEUGUNG

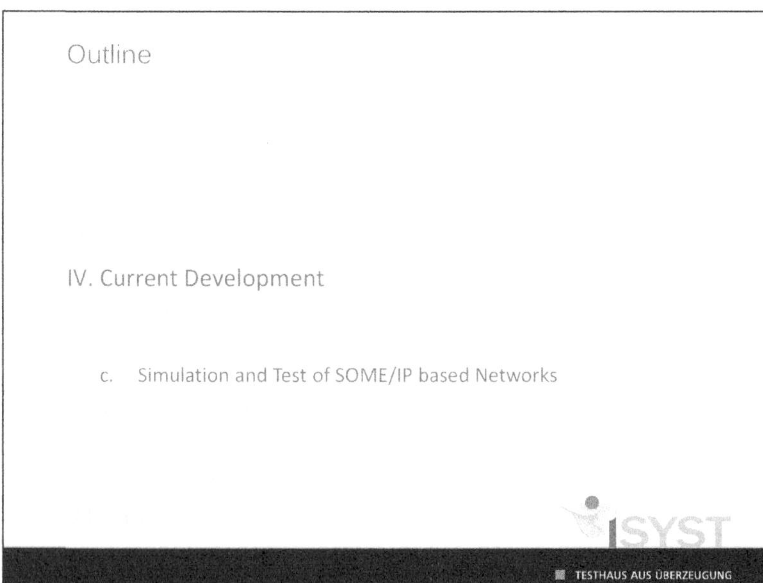
Outline
IV. Current Development
c. Simulation and Test of SOME/IP based Networks
ISYST
TESTHAUS AUS ÜBERZEUGUNG

## Current Development

Simulation and Test of SOME/IP based Networks

- Solution for simulation and testing
- Extension of network simulation framework OMNeT++
  - Implementation of SOME/IP
  - Pure simulation of networks
  - Connection to real world ECU(s)
- Creation of simulation by FIBEX

TESTHAUS AUS ÜBERZEUGUNG

## Current Development

Simulation and Test of SOME/IP based Networks

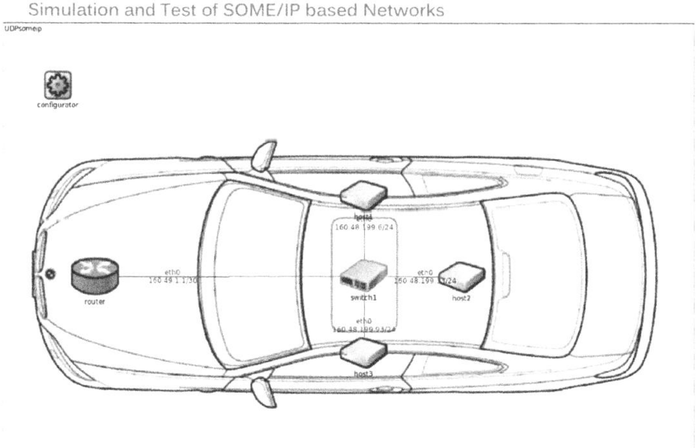


TESTHAUS AUS ÜBERZEUGUNG

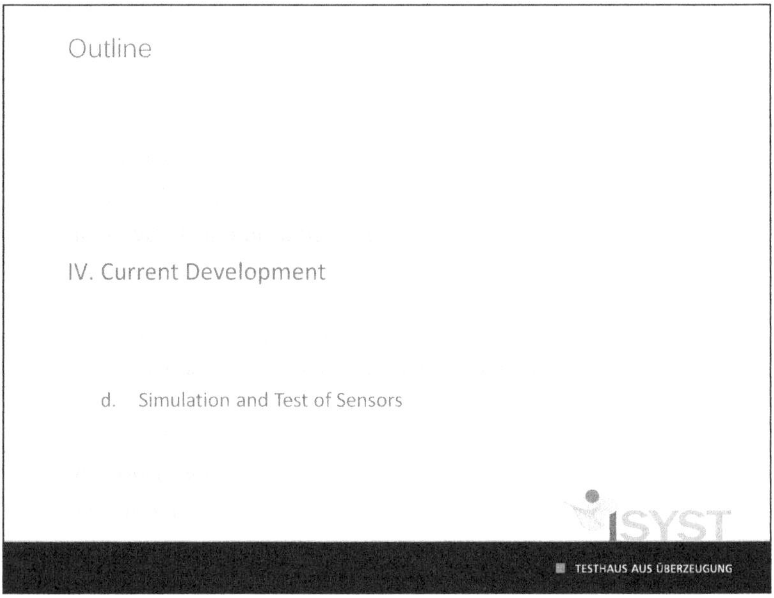
Outline
IV. Current Development
d. Simulation and Test of Sensors
ISYST
TESTHAUS AUS ÜBERZEUGUNG

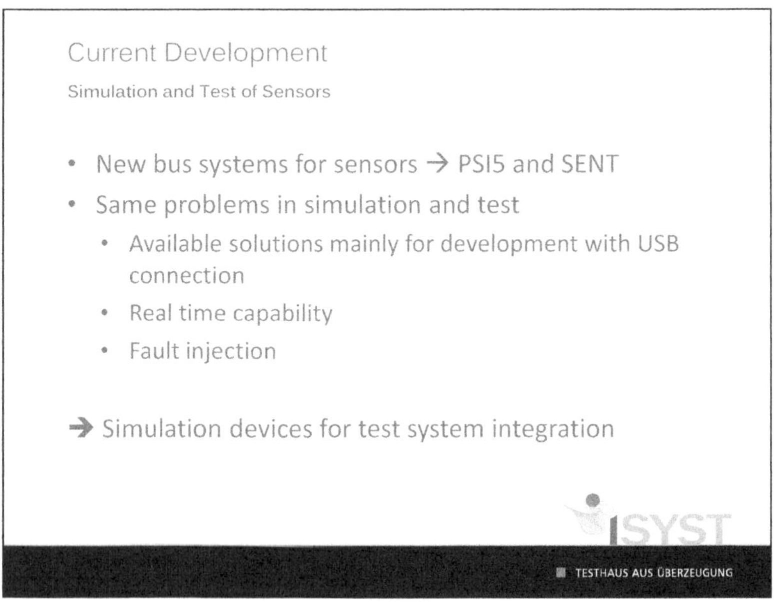
Current Development
Simulation and Test of Sensors
• New bus systems for sensors → PSI5 and SENT
• Same problems in simulation and test
• Available solutions mainly for development with USB connection
• Real time capability
• Fault injection
→ Simulation devices for test system integration
ISYST
TESTHAUS AUS ÜBERZEUGUNG

## Current Development

Simulation and Test of Sensors

- System integration by CAN
- Real time simulation of the bus system
- Fault injection on physical and protocol level

ISYST
■ TESTHAUS AUS ÜBERZEUGUNG

## Outline

V. Future Development

ISYST
■ TESTHAUS AUS ÜBERZEUGUNG

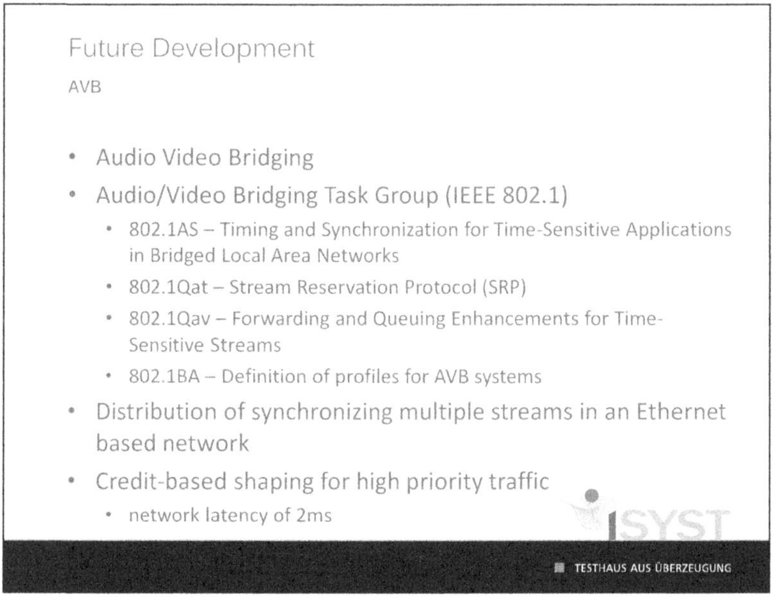
Future Development
AVB
• Audio Video Bridging
• Audio/Video Bridging Task Group (IEEE 802.1)
• 802.1AS – Timing and Synchronization for Time-Sensitive Applications in Bridged Local Area Networks
• 802.1Qat – Stream Reservation Protocol (SRP)
• 802.1Qav – Forwarding and Queuing Enhancements for Time-Sensitive Streams
• 802.1BA – Definition of profiles for AVB systems
• Distribution of synchronizing multiple streams in an Ethernet based network
• Credit-based shaping for high priority traffic
• network latency of 2ms
ISYST
TESTHAUS AUS ÜBERZEUGUNG

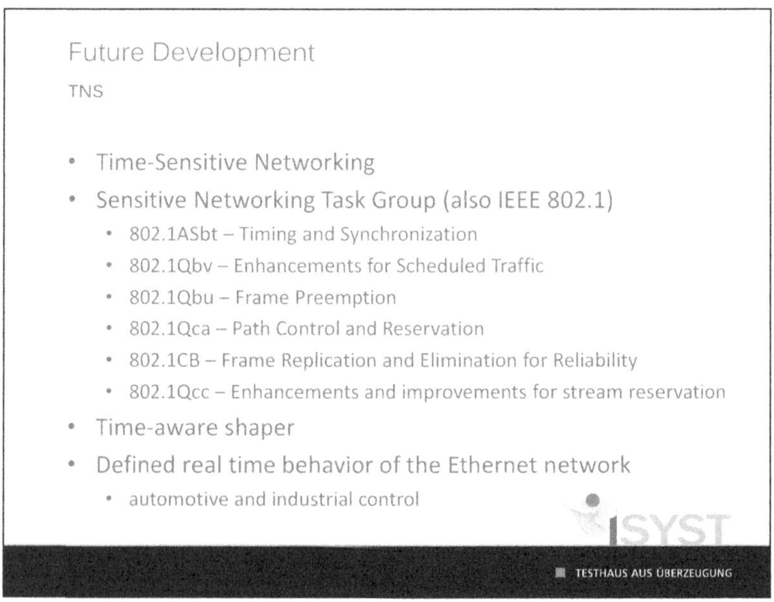
Future Development
TNS
• Time-Sensitive Networking
• Sensitive Networking Task Group (also IEEE 802.1)
• 802.1ASbt – Timing and Synchronization
• 802.1Qbv – Enhancements for Scheduled Traffic
• 802.1Qbu – Frame Preemption
• 802.1Qca – Path Control and Reservation
• 802.1CB – Frame Replication and Elimination for Reliability
• 802.1Qcc – Enhancements and improvements for stream reservation
• Time-aware shaper
• Defined real time behavior of the Ethernet network
• automotive and industrial control
ISYST
TESTHAUS AUS ÜBERZEUGUNG

## Outline

VI. Conclusion

ISYST
TESTHAUS AUS ÜBERZEUGUNG

## Conclusion

- SOME/IP(-SD) on UDP not usable for low latency and low cycle time applications
- Further development on protocols needed
  - Use experience from industrial domain (e.g. Ethernet Powerlink)
- Ethernet opens up a new potential for innovations
- New tools for development and testing needed

ISYST
TESTHAUS AUS ÜBERZEUGUNG

## Outline

VII.Outlook

ISYST

TESTHAUS AUS ÜBERZEUGUNG

## Outlook

- AVB and TNS provides real time capability for Ethernet based networks
- Totally Ethernet based E/E architectures for cars are feasible
- Pivotal question:
  - How reasonable is the usage of Ethernet based systems?

ISYST

TESTHAUS AUS ÜBERZEUGUNG

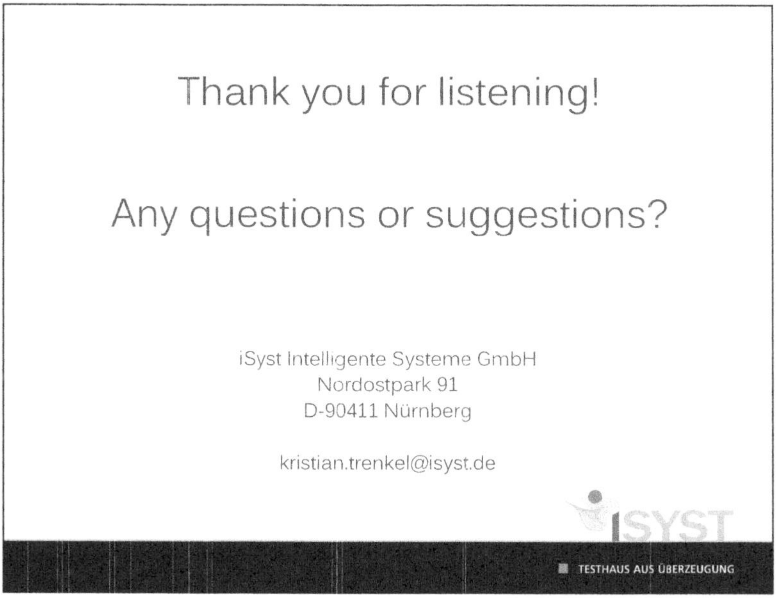
Thank you for listening!
Any questions or suggestions?
iSyst Intelligente Systeme GmbH
Nordostpark 91
D-90411 Nürnberg
kristian.trenkel@isyst.de
ISYST
TESTHAUS AUS ÜBERZEUGUNG

# Suspension design of the Visio.M electric research vehicle

Dipl.-Ing. Andreas Schultze
Thorsten Helfrich, M. Sc.
Prof. Dr.-Ing. Markus Lienkamp

Institute of Automotive Technology, Technische Universität München

# 1 Introduction

"Electric vehicles powered by electricity from renewable energy sources are an attractive option for mobility within the urban area and beyond. However, previous approaches lead to vehicles that either are too heavy and too expensive or do not meet mass-market safety requirements. Within the joint research project Visio.M scientists at the Technische Universität München (TUM) in cooperation with engineers from the automotive industry, have developed a concept to produce electric cars that are efficient, safe, and inexpensive. The project had a total volume of 10.8 million euros and was funded by the German Federal Ministry for Education and Research (BMBF)" [Vis15].

Part of the project was the development of a reliable chassis for the vehicle concept, which should offer good driving dynamics and ensure a behavior equivalent to normal cars. Also active systems like torque vectoring were included and developed. The process of the development and the resulting chassis systems are described in this paper.

# 2 Visio.M vehicle concept

The Visio.M is a vehicle concept focused on the second car market. It has a driving range of 166 kilometers in the NEDC and provides space for two people with luggage in the front and rear of the car. With a rated motor power of 15 kW a top speed of 120 km/h is reached, which is suitable for urban and suburban use. The traction battery with a capacity of 13.5 kWh is located behind the seats with a weight of 85 kg.

The car meets many requirements placed on normal cars, like infotainment or airconditioning. The car's high energy efficiency is facilitated by a low rolling resistance, aerodynamic drag and low weight. This was possible through a small frontal surface and use of carbon fiber as a structure material. The effective passenger safety had to be ensured. Therefore a concept for collisions with heavier vehicles was researched and developed. It includes active safety elements like structural airbags and an active lateral shift of the seats on side impacts. The key facts of the concept Visio.M are summarized in table 1.

Table 1: Key facts of the vehicle concept Visio.M

| **Vehicle feature** | **Characteristic** |
|---|---|
| Type | Electric lightweight vehicle |
| Passengers | 2 |
| Load capacity | 2 pieces of luggage |
| Range | > 100 km |
| Top speed | 120 km/h |
| Power at wheels | 15 kW |
| Empty weight | 450 kg (without battery) |
| Market | Central Europe |

# 3 Suspension Concept and Development

Following the definition of the vehicle concept, the suspension system had to be specified. In the first step the tire dimensions were determined. Due to the main requirements of the concept for light weight, low rolling resistance and a minimal drag, large and narrow tires are used for their advantages to fulfill the demands [Goe12]. Therefore tires with the dimension 115/70R16 are used, which were developed and produced by Michelin. Caused by the high height to width ratio and the tire's optimization on rolling resistance, cornering stiffness as well as the potential for lateral forces are limited.

Driving dynamics is one of the main costumer requirements in buying a car [Atz08]. Due to the limitations in longitudinal dynamics, lateral dynamics become more important. Therefore the suspension system has to be designed in a sporty way. The basis resulting from the vehicle concept is already good. The lightweight concept with its small size results in a low yaw inertia moment (approximately $J_{zz} = 540\ kgm^2$) and center of gravity (approximately $z_{sp} = 470$ mm). In contrast to the agile concept, the tires do not have such sporty characteristics. Therefore, this had to be compensated by the kinematics of the axles. A passive anti-roll bar is used only at the front axle to create an understeering behavior. This is caused by the slightly rear-biased load distribution of 45:55 in combination with identical wheel dimensions in front and rear. Due to the low center of gravity and subsequent small anti-roll torque, an anti-roll-bar was not necessary at the rear axle.

Further requirements are derived from the vehicle concept targets. Costs and weight have to be as low as possible. The small width of the car restricts the available space for the control arms. In the rear, for example, this is primarily caused by the required space for the traction battery. Trunk spaces are required as well in the front as in the rear of the car.

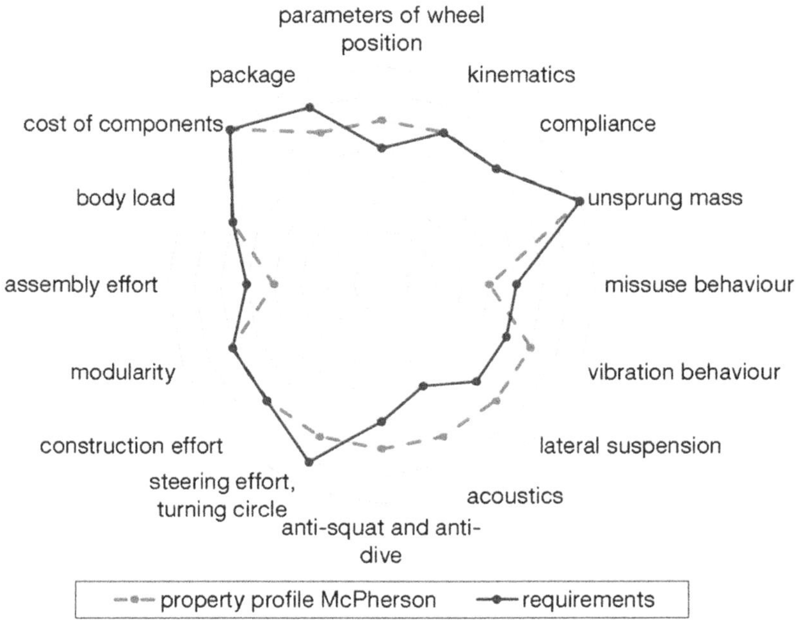


Figure 1: Comparison of McPherson axles to the requirements of the Visio.M

As shown in figure 1 the McPherson axle concept meets most of the requirements well. It is therefore used for the front and rear axle. The possible alternative twist beam axle in the rear could not be used for mainly two reasons: The first being the rear wheel drive concept, which could not be handled by the compliance of the twist beam axle. The other reason is the required space of the twist beam, which overlaps with the battery location in the Visio.M. The compartment for the energy storage needs to be as low as possible to lower the center of gravity.

Following the definition of the concept, the kinematics have been developed. The aims for toe and camber variations were adapted from a sports car with a similar weight distribution. These targets were scaled to the wheel travel of 120 mm. The toe variations were made as linear as possible over the whole wheel travel to prevent rising toe-in values during roll movements. Due to this the driving resistance while cornering is increased to a lesser extent. The static settings for toe are nearly zero at both axles to improve the efficiency of the car. Camber is set to zero in the front and -1.33° in the rear. These settings were chosen to improve response behavior and side slip stiffness of the rear axle in comparison to the front. After defining the targets, the kinematic points for all control arms were defined using iterative multibody simulations. The resulting characteristics for both axles is shown in figures 2 and 3.

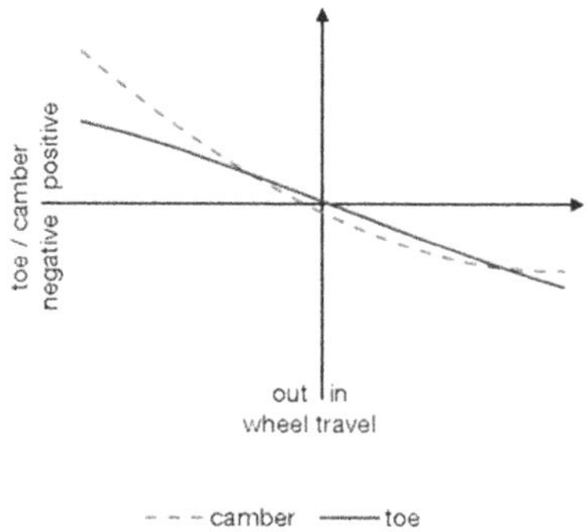


Figure 2: Characteristics of the front axle

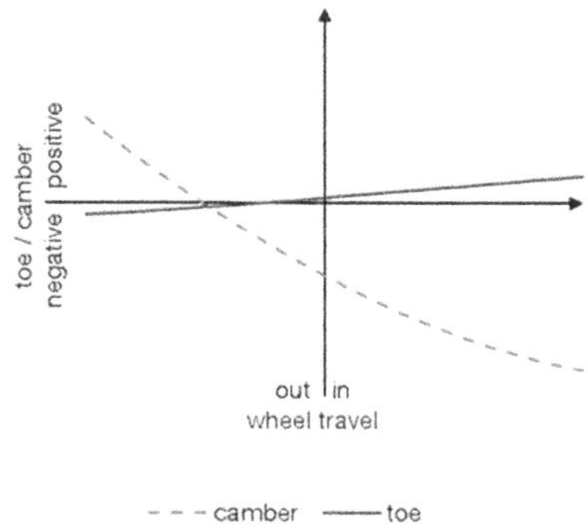


Figure 3: Characteristics of the rear axle

Caused by the package restrictions very short control arms are used. This leads to high sensitivities of the kinematics caused by manufacturing tolerances. The response of the kinematic behavior of the axles was analyzed via a multibody simulation model. In a first step the most sensitive direction of the tolerances for each point considered was evaluated. After that, in a design-of-experiment screening the most critical combination of the kinematic points was examined. The validity of the simplifications made were analyzed via a Plackett-Burman [Pla46] analysis. Besides the effects on the kinematics the impact on the driving dynamics was investigated. The result of the tolerance analysis was used to improve the manufacturing and assembling of the suspension system.

To achieve the objectives of weight and costs, the vertical dynamics are realized with the conventional components of a linear coil spring made of steel and a passive twin-tube damper. The vertical eigenfrequency was set at approx. $f = 1.4$ Hz with a difference of $\Delta f = 0.2$ Hz between front and rear to prevent pitching oscillations [Gra11]. For a constant vertical eigenfrequency over different loads the kinematic spring translation over wheel travel was designed progressive. An additional progression of the spring rate at higher wheel travels was realized by using an ancillary spring made of PU-foam. The damper characteristics were developed using multibody simulations. Different low- and high-speed characteristics were applied and rated using an assessment criterion, which regarded weighted body acceleration and wheel load differences [Gra11].

The suspension system contains several parts sourced from production vehicles alongside special developments, which are mainly all control arms and wheel carriers. In the beginning there were no experiences regarding the operational loads of such a light car. Because of this restriction, load assumptions were estimated in a first step. With these loads the first design of the parts was made and produced. In a first prototype, which was developed to ensure the functionality of the chassis and suspension system, the components were installed and equipped with strain gauges and accelerometers. The equipped car was driven on different test tracks to measure the operational loads aftwerwards, the measured data was evaluated with respect to maximum loads and force-time functions, which were accumulated in load spectrums. The durability of the control arms was analyzed with the new loads using finite element analysis. It was possible to improve structural strength and weight of the resulting constructions. In figure 4 the loads spectrum is shown in comparison to the Wöhler curve of the longitudinal control arm at the front axle. Due to the uncertainties in the only calculated Wöhler curve and the high maximum loads under unusual driving conditions, the distance between the curves is comparatively high for this control arm. The resulting construction of the right front axle is shown in figure 5.

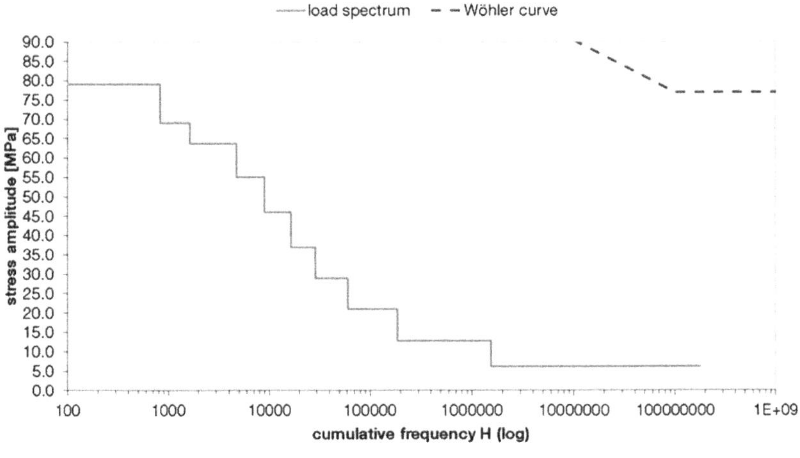


Figure 4: The load spectrum and the Wöhler curve of the front longitudinal control arm

Figure 5: Front axle

# 4 Steering System

The steering system of the Visio.M consists of a steering column and a steering gear provided by ThyssenKrupp Presta AG and tedrive Steering Systems GmbH. The low weight of the vehicle and the low positive scrub radius facilitates a purely mechanical steering system. No hydraulic or electric steering assistance is needed.

The steering column is a conventional system using two universal joints connected by a shaft, which may collapse in case of a frontal crash. It is adjustable longitudinally to accommodate the different body sizes of drivers. Due to the fixed eye-point concept, which fixes the longitudinal position of the driver in the vehicle, no vertical adjustability of the steering column is required.

The mechanical steering gear is optimized for weight using a split steel housing, executed as a welded construction, bellows instead of a solid casing and a segmented steering rack with a hollow connecting tube achieving a total weight of 3.6 kg, see figure 6 and [Ted15]. The durability for the steering system has been proven for a representative load spectrum of the Visio.M in endurance and miss-use tests.

Figure 6: lightweight steering gear (tedrive Steering Systems)

The steering ratio is i = 12.5 which, combined with the short wheelbase of 2.1 m, eliminates the need for changing the hand position on the steering wheel in city driving. A turning radius of 8.4 m was achieved. The handling is very agile for driving speeds of less than 60 km/h, which constitutes the main use-case for the Visio.M-vehicle, while maintain reasonable damping approaching the top speed of 120 km/h.

# 5 Braking System

The low gross weight of the vehicle provides several advantages for the design of the braking system. It uses lightweight motorcycle calipers with stainless steel discs at all four wheels. This is made possible due to the low maximum kinetic energy of the vehicle, which results from the low weight and top speed (figure 7). For comparison, a motorcycle using the same brake discs is shown. Furthermore, there is no need for a brake booster, the pedal force provided by the driver is sufficient for maximum deceleration. This contributes an additional 2 to 3 kg in weight savings.

Starting from the brake pedal with its ratio of 5:1 the structure of the system is chosen as follows: the brake pedal acts directly on the tandem brake master cylinder with its 2x19.5 mm diameter. The floating brake calipers are twin cylinder (32 and 30 mm diameters) on the front wheels and single cylinder (34 mm diameter) on the rear wheels. The discs have the same diameter of 296 mm at all four wheels and are drilled stainless steel with organic brake pads. The resulting brake force distribution is 62:38. The strong front bias was chosen due to the regenerative braking at the rear wheels, which in sum leads to a more balanced distribution.

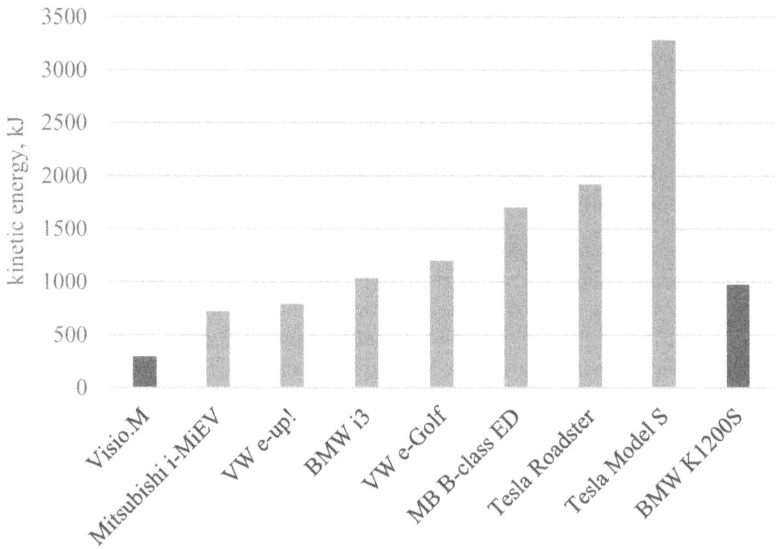


Figure 7: Comparison of kinetic energy (kerb weight) at top speed for several vehicles

# 6 Driving Dynamics Control

There are four systems which allow an active control of the Visio.M vehicle: anti-lock braking, electronic stability control, regenerative braking and torque vectoring. The anti-lock braking system was provided and adapted by the manufacturer, the other systems were developed at the Institute of Automotive Technology.

The ESC algorithm consists of several extended Kalman filters for driving state estimation, a non-linear single track model for calculation of driver demands, a sliding mode controller for yaw moment generation and a rule based arbitration to distribute the braking force according to the current driving situation. The ESC detects and acts in situations of severe under- and oversteer to stabilize the vehicle. It has been developed and tuned in multibody simulation and standard driving tests on high friction surfaces.

The regenerative braking system mostly acts through the lift-off of the throttle achieving a maximum deceleration of 0.21 g. Because different drivers prefer different decelerations on lift-off there are three alternative levels of regenerative braking of 0.05 g, 0.10 g and 0.16 g. The regenerative braking usually stays active on depression of the brake pedal to achieve a progressive feel of deceleration over pedal travel for the driver.

During cornering, the torque vectoring control works in concert with the regenerative braking system to generate a yaw moment that counteracts the destabilizing influence of the superimposed longitudinal and lateral forces on the rear tires. The regenerative braking therefore has to be disabled less often, resulting in a higher energy efficiency of the vehicle [Wie12]. Further increase of the lateral acceleration results in a reduction of the regenerative braking moment depending on the longitudinal slip of the rear wheels to avoid unstable situations. The system is switched off during ESC and/or anti-lock brake interventions.

Unfortunately, the passive brake system does not include a pedal force generator or an electrical vacuum pump. Therefore, using the ESC aggregate for brake blending results in a pretty strong reaction through the feel of the brake pedal and noise from the ESC pump. It is therefore only used in situations in which a fully charged battery or overheated electric motor prevent the regenerative braking. Here, the ESC aggregate is actuated to achieve the same deceleration on lift-off as during regenerative braking, as is expected by the driver.

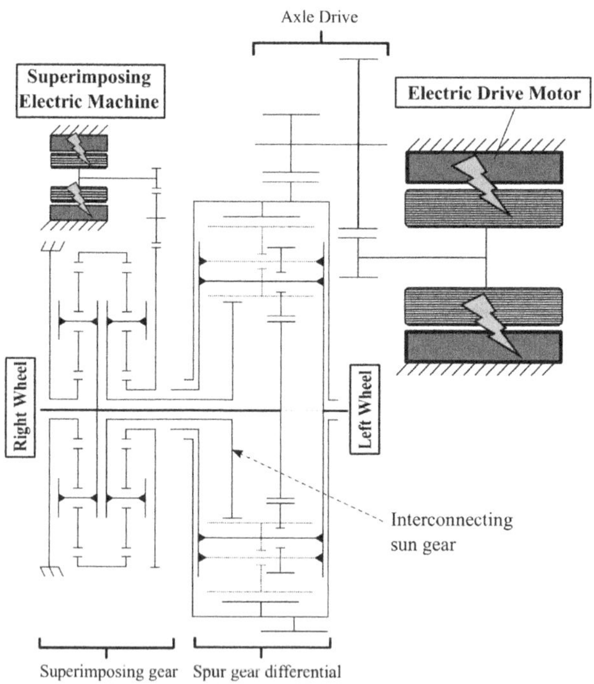


Figure 8: Structure scheme of the Visio.M powertrain [Gwi14]

The torque vectoring mechanism acts on the differential through an additional electric machine and a superimposing gear (figure 8). The high gear ratio allows a maximum torque difference of 576 Nm at the rear wheels [Gwi14]. It therefore facilitates a distribution of more than two thirds of the total available driving moment of 812 Nm and, correspondingly, a powerful influence on driving dynamics.

Additionally, to stabilization during regenerative braking the torque vectoring system is used for two applications: increasing agility e.g. during turn-in and stability during severe cornering. The algorithm for generation of the reference is based on dual non-linear single-track models for agility and stability, which allow for intuitive and quick tuning. The two models generate two references for the lateral vehicle behaviour. Depending on lateral acceleration, the algorithm switches and interpolates between these two [Gra14].

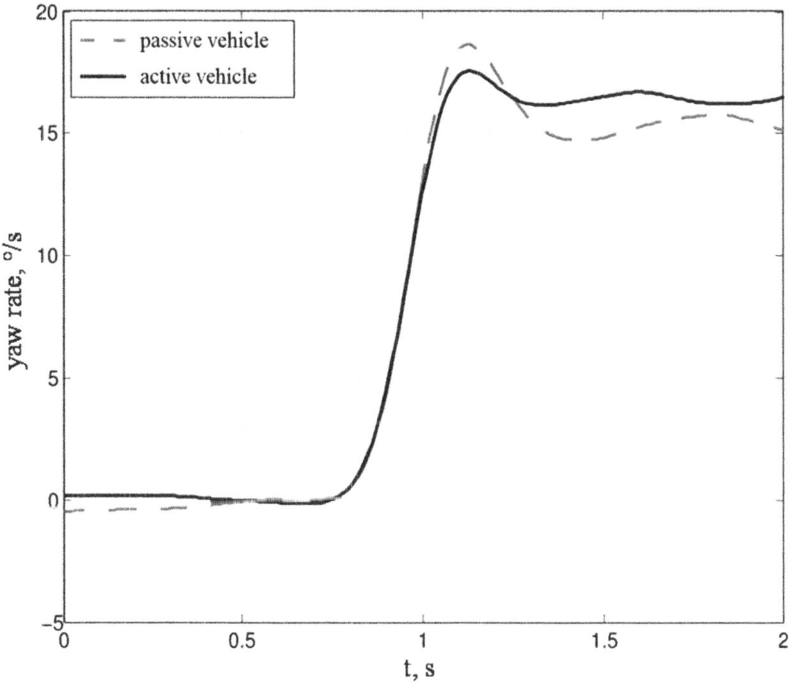


Figure 9: Yaw rate results of step-steer according to ISO 7401 [Gra14]

In constant radius cornering on a 55 m radius, the algorithm increases the maximum lateral acceleration by 0.36 m/s$^2$. In step-steer maneuvers the overshoot is greatly reduced in all domains (yaw rate, lateral acceleration, side slip angle) (figure 9). Also, the damping of the oscillations following the step excitation is increased.

# 7 Resulting Driving Dynamics and Testing

The vehicle behavior was developed using multibody simulations. An agile but save handling was intended. Due to the rear biased weight distribution in combination with the same tire dimension at front and rear axle, the car tends to oversteer in highly dynamic maneuvers. To optimize the controllability, the self-steering gradient is chosen to be comparatively understeering. The maximum stationary lateral acceleration is $a_y > 8$ m/s², similar to the state of the art for urban cars. The developed driving dynamics were analyzed in driving tests. The result for the stationary behavior is shown in figure 10.

The dynamic behavior of the Visio.M was also tested using the maneuver 'step steer'. It was driven with a velocity of $v_x = 80$ km/h and the steering wheel angle required for a stationary lateral acceleration of $a_y = 4$ m/s², as described in [Iso09]. The driving test results are shown in figure 11.

The resulting objective parameters confirm the targeted agile transient behavior of the driving dynamics concept. Compared with the literature, the Visio.M is located in the agile part of the normal area, regarding to [Zom91] (table 2).

Table 2: Parameters resulting from step-steer maneuvers

| Parameter | **Literature [Zom91]** | **Visio.M** |
|---|---|---|
| $T_{R,\dot{\psi}}$ | 0.12 – 0.30 | 0.12 |
| $T_{R,a_y}$ | 0.20 – 0.60 | 0.25 |
| $T_{\dot{\psi},max}$ | 0.20 – 0.50 | 0.29 |
| $T_{a_y,max}$ | 0.40 – 1.00 | 0.49 |
| $U_{\dot{\psi},bez}$ | 1.00 – 1.70 | 1.48 |
| $U_{a_y,bez}$ | 1.00 – 1.40 | 1.14 |

With the support of a professional test driver, the behavior in the 18 m slalom was tested as well. The car reached an average speed of more than $v_x = 65$ km/h.

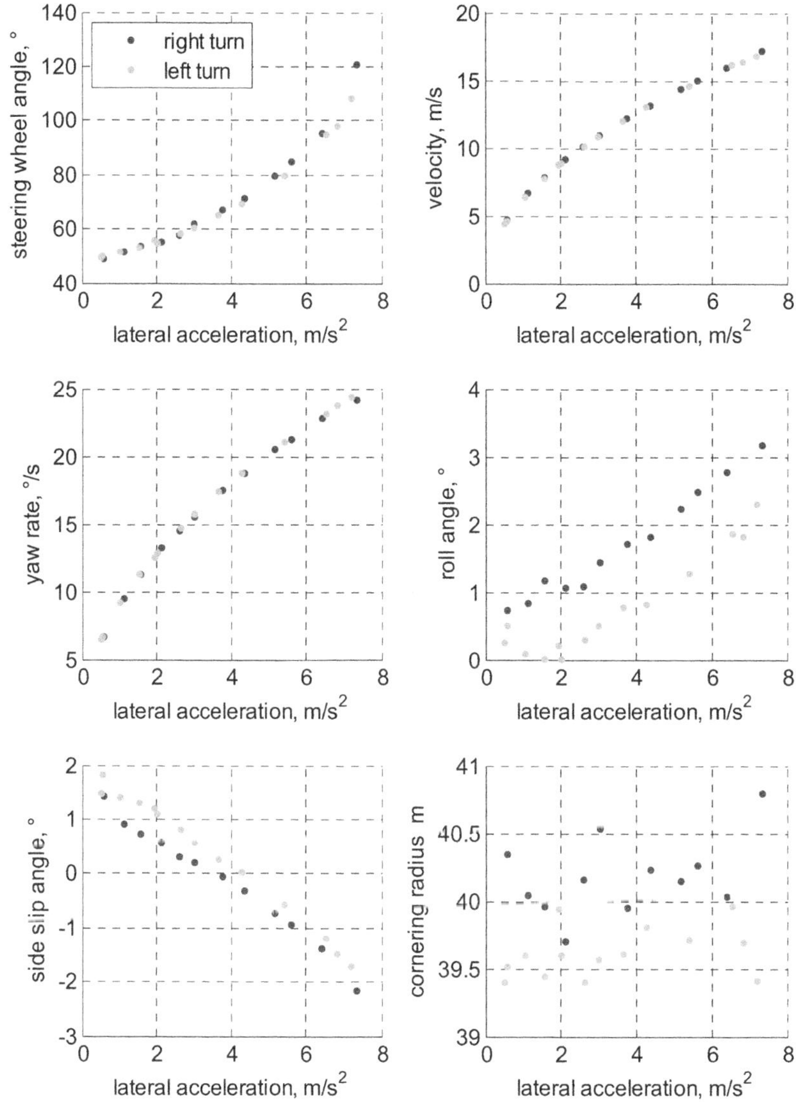


Figure 10: Results of the driving test 'constant radius cornering' with a radius of R = 40 m

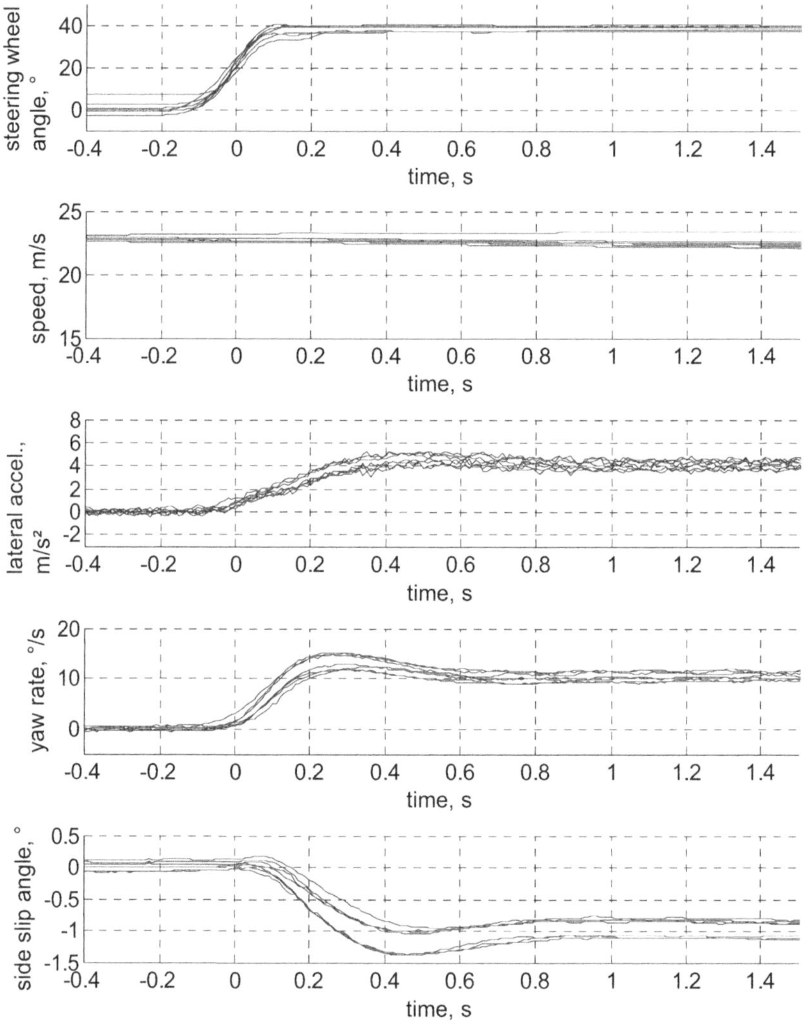


Figure 11: Results of the driving test 'step steer'

# 8 Summary and Outlook

In this article the authors present the development and design of the Visio.M suspension. It is shown how the basic axle concept was derived from the requirements of a small, electric city vehicle. The specific challenges for suspension development in this context were detailed, especially concerning weight and energy efficiency. The available systems for vehicle dynamics control and their respective strategies were described. Finally, the resulting driving dynamics were shown.

At the Institute of Automotive Technology further research is carried out to improve the energy efficiency of suspension systems. This includes the passive suspension, tires ([Fic15]) and integrated control to improve regenerative braking. Also, research of new control systems for interaction and integration of the available active systems is carried out.

The authors would like to thank all researchers and students that participated in the development of the Visio.M, the partners and sponsors of the project, and, especially, the German Federal Ministry of Education and Research for its generous funding.

# 9 References

[Atz08] Atzwanger, K.; Negele, H.-J.: Fahrdynamikstudie 2008 – Kundenorientierung in der Fahrwerkstechnik, in: Automobiltechnische Zeitschrift ATZ, issue 06/2008, volume 108

[Fic15] Ficht, A.; Lienkamp, M.: Rolling resistance modelling for electric vehicle energy consumption simulation, chassis.tech plus, München, 2015

[Gra11] Graf, M.; Wiesbeck, F.; Lienkamp, M.: Fahrdynamikauslegung des Elektrofahrzeugs MUTE, in: Automobiltechnische Zeitschrift ATZ, issue 06/2011, volume 113

[Gra14] Graf, M. Methode zur Erstellung und Absicherung einer modellbasierten Sollvorgabe für Fahrdynamikregelsysteme, Dissertation, München, 2014

[Goe12] Goerlich, M.; Volk, H.; Strzelczyk, M.; Wies, B.: A Tire Concept for low CO2-Emission and eCars, in: VDI Reports, No 2137, 2012

[Gwi14] Gwinner, P.; Otto, M.; Stahl, K.: Leightweight Torque-Vectoring Transmission for the Electric Vehicle Visio.M. Bayern Innovativ. Conference on Future Automotive Technology, Garching, 2014

[Iso09] International Standardization Organization: Road vehicles – Lateral transient response test methods – Open-loop test methods, 2009

[Pla46] R.L. Plackett and J.P. Burman, "The Design of Optimum Multifactorial Experiments", *Biometrika* 33 (4), pp. 305-25, June 1946

[Ted15] tedrive Steering Systeme GmbH Hompage, http://www.td-steering.com/lightweight.html

[Vis15] Visio.M Homepage: Press release ‚Project start Visio.M‘, http://www.visiom-automobile.de/en/aktuelles/news-single-view-en/?tx_ttnews%5Btt_news%5D=1&cHash=13327886f5ed63fca3522bdfa6c11c95, 2015

[Wie12] Wiesbeck, F.; Lienkamp, M.: Bewertung der fahrdynamischen und energetischen Rekuperationsgüte. Hybrid and Electric Vehicles Symposium. Braunschweig, 2012

[Zom91] Zomotor, A.: Fahrwerktechnik: Fahrverhalten, 2. Auflage, Vogel Verlag, Würzburg, 1991

# Development of an active motion system of tire contact point control

Hiroshi Shibuya
Hiroo Iida
Hiroyuki Kanayama
Daigo Fujii
Xabier Carrera Akutain
Kotaro Shima

TOYOTA MOTOR CORPORATION

## Abstract

This paper presents the possibilities of a direct wheel control system, which can optimize both cornering performance and turning radius. Tire contact patch position and toe angle can be controlled independently for the front wheels at both left and right hand sides using a robotic suspension system. Quasi-trivial modes such as different configurations of track width or wheelbase lead to a reduction on roll angle and weight transfer. A fully independent wheel control enables further a much more powerful cornering performance control. This paper describes the initial investigations on performance prediction.

## 1 Introduction

Short wheelbase and large wheel steer angle are preferable for parking manoeuvres. Reducing track width has the potential to decrease the aerodynamic drag. On the other hand, a wider track width contributes to decreasing roll angle and load transfer. These conflicting requirements are often difficult to combine satisfactorily. In this paper we introduce the concept and realization of a robotic suspension system, which allows a permanently variable geometry-suspension and thus maximize the vehicle dynamics performance.

# 2 Suspension Outline

## Suspension Structure

Three rotational actuators with pantograph arms are combined on each corner of the front suspension.

Besides the wheel steer angle, these actuators allow a continuous modification of the track width of the front axle, the wheelbase for both left and right hand sides, which means each of the front corners can behave fully independently.

The actuators in this initial phase of the project are driven by predefined mappings, which send the rotational command necessary for each actuator to shift the kingpin axis of the corner to the desired location in function of the selected mode and using the steering wheel angle as main input signal.

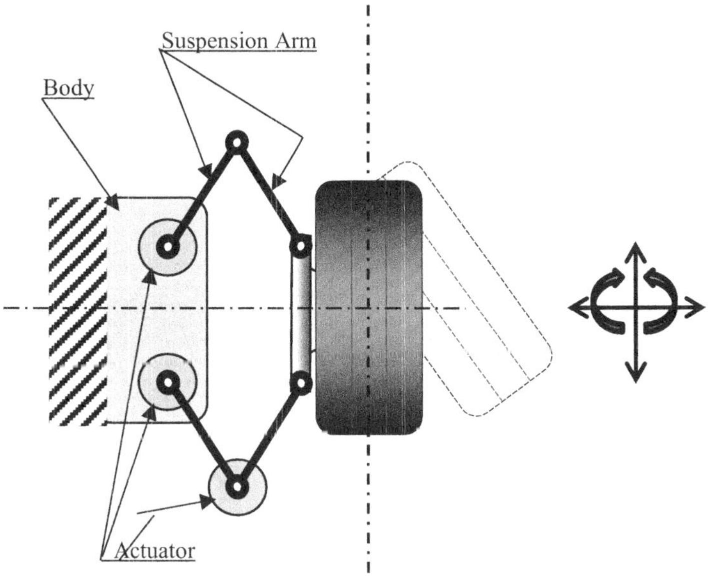


Fig.1 Suspension Mechanism

# 3 Suspension performance

## 3.1 Cornering performance improvement

A direct control of the kingpin axis location and orientation should intuitively be able to maximize the cornering performance potential of the vehicle. In order to quantify the merit of the system, a performance prediction study by simulation has been performed. Simulation software has been the used to perform several dynamics virtual tests.

### 3.1.1 Performance improvement with lateral position control

During the cornering phase, roll angle and lateral load transfer are generated due to the centrifugal force acting on the vehicle mass. Tire performance will decrease in proportion to the mentioned load transfer and it's very likely that the camber variations caused by the vehicle roll will further contribute negatively. These effects are essential to keep a safe cornering performance and can be critical for instance for small vehicles with a relatively high center of gravity.

A way to minimize this damage is by controlling the effective track width. A conspicuous move is to increase the width as in figure 2 (center). A less obvious solution is using an independent control (figure 2, right), so that the outside wheel is pushed out from its original location. We want to show how a better performance can be achieved without the need to maximize the track width.

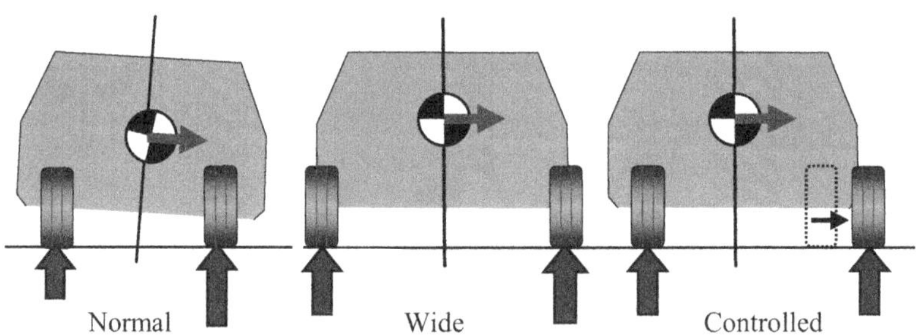

Fig.2 Conventional and developed system during cornering

### 3.1.2 Performance prediction

Roll Angle

The roll angle generated with a Slowly Increasing Steer test is shown in figure 3. The track width for outside wheel is proportionally extended with steering wheel angle in *Controlled* mode. In the *Wide* mode, the vehicle sees its track width stretched. The calculation result suggests that the roll angle in *Controlled* mode can be at a very close level to the minimum value achieved with *Wide* mode, and significantly smaller than in the *Normal* mode.

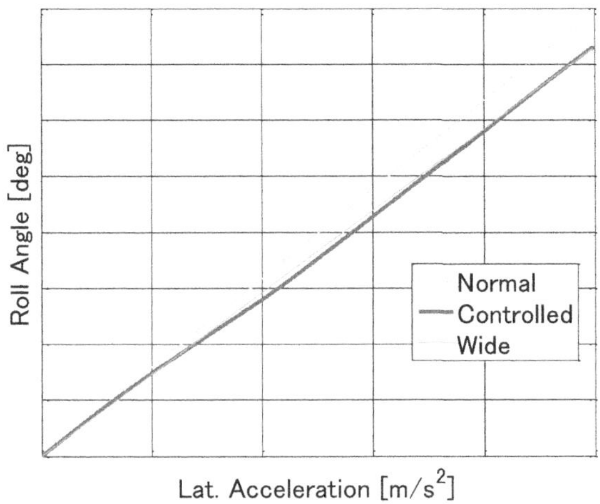


Fig.3 Roll Angle at Slowly Increasing Steer

Load Transfer

A more interesting result is the resulting load transfer, as the *Controlled* mode can minimize it best as shown in figure 4 for the same Slowly Increasing Steer test, thanks to a more equal load sharing of both front tyres.

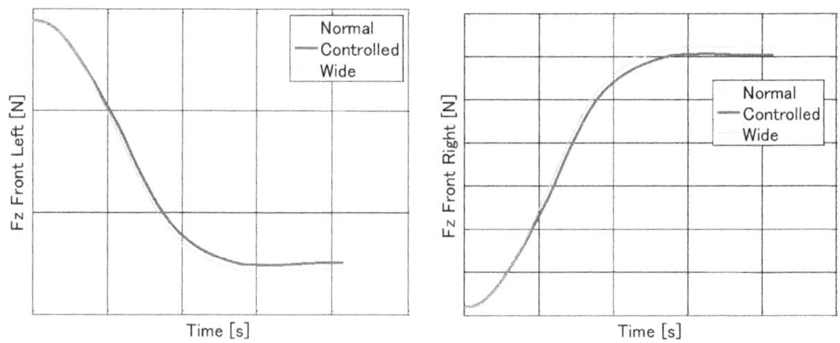


Fig .4 Load Transfer comparison for Slowly Increasing Steer

Maximum Lateral acceleration

This load reduction leads to a higher lateral acceleration of Controlled Mode, as shown in figure 5.

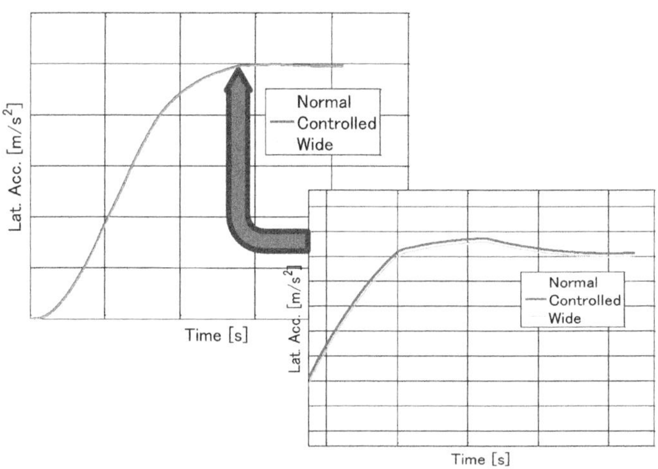


Fig.5 Lateral Acceleration at Slowly Increasing Steer

Step input response

Initial response improvement is also observed for a step angle input as shown in figure 6, where a significantly quicker yaw response is achieved by the *Controlled* mode.

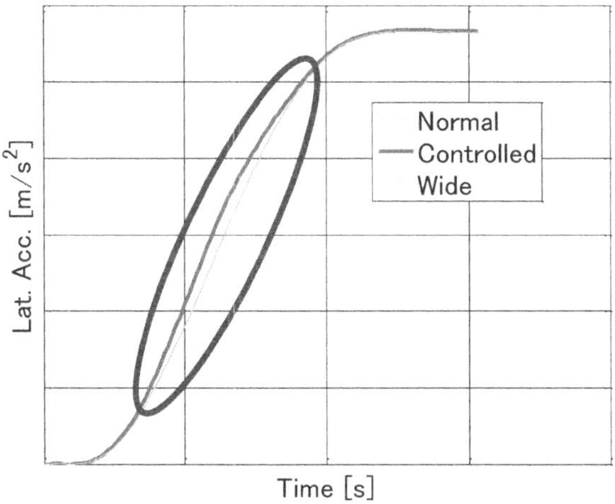


Fig.6 Response at Step Steering

# 4. Actual vehicle evaluation

Pic.1 Formula Test Car with New Suspension

Table.1 Spec of Formula Test Car with New Suspension

| | VALUE | UNIT |
|---|---|---|
| Weight | 780 | kg |
| Wheel Base | 2355<br>(Standard Condition) | mm |
| Tread (Fr/Rr) | 1488/1450<br>(Standard Condition) | mm |
| Drive System | Rear Wheel Drive | -- |

A racing Formula F20 was chosen to accommodate a prototype robotic suspension system. First test drives were devoted to ensure a safe operation of the system. Initial subjective evaluation of the test driver confirms the improvement of response time and additional grip in cornering for the *Controlled* mode.

# 5. Summary

1. Widening or shifting the front track width with a robotic suspension system is found effective to improve vehicle steady-state characteristics following lateral load transfer reduction.

2. Subjective evaluation with the experimental vehicle suggests a clear improvement of dynamic response.

3. Next steps include a detailed objective evaluation, creation of model based controls and optimization of transient dynamics.

# 6. References

(1) Asogawa Katsunori, Tomoki Hirabayashi, Atsushi Tamura, Yusuke Minagawa: Society of Automotive Engineers of Japan thesis book No. 27; variableness geometry vehicle of variable geometry system, report No.1

# The influence of the modeling depth of active chassis systems with respect to the development stage and their interaction with driving characteristics

Kilian Dettlaff, Prof. Dr.-Ing. Jochen Wiedemann
Institut für Verbrennungsmotoren und Kraftfahrwesen
der Universität Stuttgart, Stuttgart

Uli Schaaf, Ingo Scharfenbaum, Dr.-Ing. Andreas Wagner
AUDI AG, Ingolstadt

# Abstract

Active Chassis Systems (ACS) cause perceptible driving characteristics or help to maintain driving stability even in precarious situations. Their use enhances driving agility and comfort. The development of such systems needs to be well coordinated with the development process of the full vehicle. During different stages of the latter, knowledge in different extent needs to be achieved and transferred between manufacturers of the actuator system and the car. In terms of vehicle dynamics, top level system characteristics are of utmost importance even in early stages. A proper way of simulative representation by simple models is presented. With ongoing development, further insight into the actuator in terms of energy demand and occurring dynamic states (forces, torque) is necessary to develop connected subsystems such as power supply or axes of the final vehicle. Different degrees of abstraction for simulative representation are presented here as well. An exemplary study regarding the influence of parameter variation and modelling depths is given.


# Introduction

The introduction of the electronic Anti-lock Braking System (ABS) in 1978 and the Electronic Stability Control System (ESC) for regular production cars in the year 1995 can be seen as origins for mechatronic components in the chassis domain of a passenger car. In the following years comfort and agility enhancing systems have been developed. Nowadays, several different systems are available in various vehicle types. Their functionalities and technical progress are still ongoing and versatile variations as well as combinations of multiple systems enter the market. Their main benefit is improvement of customer relevant, thus perceptible, driving characteristics. This is realised by an early in-depth definition of the properties of the system during the full vehicle development process. As shown in the remainder of this study, knowledge of overall system behaviour as well as interaction with other systems (e.g. power supply of the vehicle) is of utmost importance to achieve the goals stated above.

With respect to engineering guidelines (cf. (VDI-Richtlinie 2206)) the conceptual layout of a mechatronic system is divided into software, electronic and mechanic branch of the final product. The paper at hand connects the development process of the individual actuator system with an exemplary full vehicle development process. In general it is considered to be true that a close linkage of the two development processes mentioned above enhances efficiency during development. The three categories of the actuator as a component of the full vehicle can be cascaded into the electronic and mechanic category of the latter, cf. Figure 1. At this point the assumption is made that the specific components of the actuator and their design are not within responsibility of the manufacturer of the vehicle. However, it is indeed the task of the manufacturer to define conceptual details, such as overall mechanic and electric behaviour. Furthermore it is the task of the OEM to develop high-level control algorithms to meet the designated attributes regarding vehicle dynamics with the effective use of the particular ACS. With in-depth definition of the properties of the actuator at an early development stage, these high-level control algorithms and other connected vehicle components can be designed and modified accordingly. Present article links both measurements of the full vehicle equipped with ACS and of the isolated actuator on a test rig to support the development process in its different stages regarding simulative investigations.

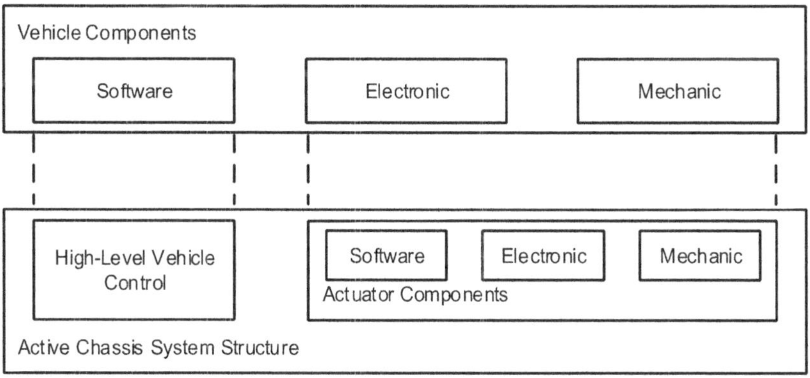


Figure 1: Relation of Components of the Full Vehicle as a Mechatronic System and of an Active Chassis System.

## Outline

The relation between actuator behaviour and vehicle driving attributes and necessary modelling depth for simulative analysis during development is the main focus of this research. As a general introduction, an exemplary development process in the automotive industry is briefly explained. It is followed by a generalized overview of available ACS for regular production cars and after that a suitable categorization is made. As an example of use of the present methodology, various active systems and their benefits regarding vehicle dynamics are presented. As metric and evaluation tool, the field of objective vehicle assessment in terms of chassis attributes is introduced. Consequently, two different model types with two levels of detail each are presented. With reference to the development stage introduced earlier, the actuator models are connected to different phases and their main benefits and output capabilities are displayed. Finally, the effects of the different modelling depths on full vehicle driving characteristics are shown and discussed. As an overall conclusion, this study gives recommendations regarding choice of model of ACS for simulative analysis during the development process.

# The Development Process in the Automotive Industry

The goal of any vehicle development venture is the Start of Production (SOP). Scheduling the project planning backwards from this point is best practice. From an initial project start decision onwards, the process is divided into several stages. Regarding the automotive industry, a top level division of this process can be made between concept and series development, divided accordingly into two branches of the "V-model", cf. Figure 2. An initial benchmark sets top-level goals at overall vehicle level, influenced by the predecessor and products of the competitor and is the starting point for the concept phase. In a top-down approach the requirements for the sub-systems and later on components are derived consequently. From this point onwards, the series development continues in a bottom-up methodology, i.e. components over sub-systems to the overall vehicle again.

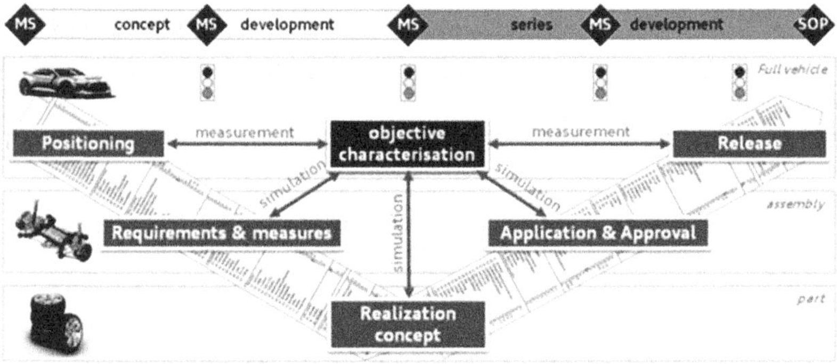


Figure 2: Schematic Overview of an Automotive Development Process
(Einsle & Fritzsche, 2013)

The present paper focuses on the left branch of this process model representation. At early stages of the process only a few and rather conceptual parameters (e.g. drivetrain concept) of the final vehicle are defined or derived from a predecessor and therefore available. Starting with less information, simulative analyses impel further development. The virtual representation of vehicles, their sub-systems and components in different degrees of abstraction (single-track model, two-track model or multi-body models) is part of state of the art for years now (Ammon, 1997; Rill,

1994). Regarding ACS, their dynamic behaviour needs to be represented in full vehicle simulation environments as well and different degrees of abstraction are presented in the following.

# State of the Art Regarding Active Chassis Systems

The market share of ACS has increased in the last few years. Starting with (semi-) active systems such as continuous damping control or adaptive air suspension, the ride comfort and driving agility has enhanced significantly. These systems are even available in segment C (medium)[1] cars nowadays.

For further ease of use, ACS are grouped as follows: First and foremost the group of safety relevant systems needs to be stated. The main representatives are stated above, ESC and ABS. The second group features enhancement of driving agility and comfort. Within this group, further categorization is made by their main directions of interaction with the vehicle. Using active steering systems as an example, they are classified within systems influencing the lateral dynamics of the car. Furthermore air suspension, continuous damping control, active roll stabilisation or active suspension are a part of the group of vertical systems.

## Lateral Active Chassis Systems

The Audi Dynamic Steering system is a representative example for lateral ACS. According to the driving situation, the toe angle of the front wheels, i.e. steering ratio, is influenced. The main states of the full vehicle, such as steering wheel angle and velocity, longitudinal velocity and other values obtained from the ESC, are taken into consideration for the high-level control algorithm. The advantages of this system are manifold. On the one hand, very convenient low speed manoeuvrability is achieved by realisation of a low steering ratio. This means that even small steering wheel angles of the driver are directly fed through to the front wheel and low steering wheel angles cause high turning behaviour of the car. On the other hand, high speed stability is realised. While driving at high speed, the steering ratio is increased, hence small steering wheel angles result in very low toe angles. The yaw amplification of the vehicle is reduced.

1 Classification with respect to Regulation (EEC) 4064/89

## Vertical Active Chassis Systems

Regular mass production cars are equipped with conventional stabilizer bars. They interconnect the wheel carriers of each axle. Their main functionality is the reduction of roll angle while cornering. They significantly influence the vertical forces between wheel carrier and chassis, thus contribute a lot to the vertical tire forces and therefore self-steering tendency of the vehicle. In order to make a car inherently safe (i.e. understeering towards the lateral limits of handling), the ratio of front and rear stiffness of the stabilizer bars is subject to certain limitations. With respect to the driving comfort domain, the stiffness of each stabilizer also needs to be within certain limitations to be able to cope with unwanted transmission of one-sided road excitation to the chassis (Heißing, 2011). The limitations, necessities and design conflicts above can be circumvented and the overall driving agility can even be expanded by the use of active stabilizer bars which are a prominent example for vertical active chassis systems. In contrast to conventional stabilizers that can be described as linear torsional stiffness, active stabilizers are additionally equipped with a force element producing externally controlled torque between left and right wheel carrier. With the use of control algorithms that mainly utilize full vehicle sensor measurements in real-time, the roll angle as well as roll rate can be influenced as a dominant feature. Additionally, with shifting the supporting external roll moment from one axle to the other (e.g. rear to front) the characteristics of the car in terms of agility and stability can be designed. By taking the driving situation as additional input utmost safety concerning the lateral limits of handling is ensured.

## Vehicle Chassis Systems Architecture and Interfaces

In contrast to passive vehicles (i.e. without ACS), many additional components are involved in a vehicle with additional actuators. Besides hardware changes of the physical interfaces in the chassis, supplementary electronic and software components are necessary. A schematic overview is given in Figure 3. The input of the driver is sensed by high-level control algorithms. In addition with measurements of the vehicle's motion, actuator target values are calculated. With realization of the target values in terms of forces or torques by the mechatronic actuator system, the motion of the vehicle is influenced. The resulting driving properties are perceptible by the driver.

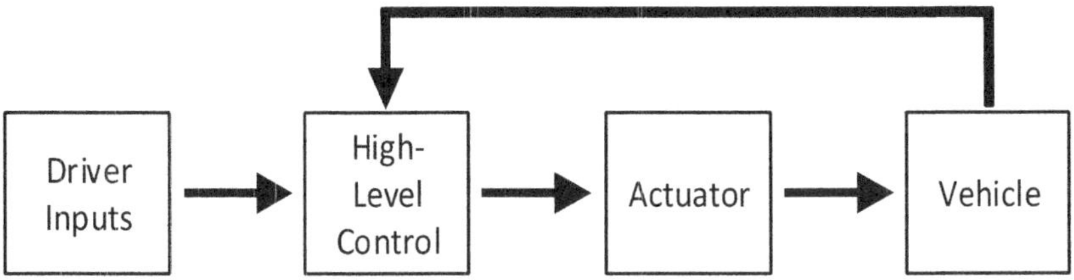


Figure 3: Generic Overview for Active Chassis Systems in Passenger Vehicles

The boundary between high-level control and actuator on the one hand and between actuator and vehicle on the other are of utmost interest in the following. The transport of signals or energy (electrical as well as mechanical) over this border gives fundamental input in terms of system identification and validation. Simulative representation of this system behaviour supports the development process in significant manner.

### Vehicle Dynamics Assessment and Driving Characteristics

The behavior and performance of a car regarding its vehicle dynamics can be assessed in different ways and over the years two main paths evolved. They are now the predominant measures in the development process. First and foremost the most customer relevant assessment method is subjective evaluation. It represents one individual's evaluation of a car sensed only by the human body of a trained driver. The second method, objective evaluation, uses individual driver independent measures: it is based on measurable vehicle reactions (i.e. motion) on defined and also measurable inputs to the vehicle. Objective methods for vehicle characterization are more and more utilized during the development process in early stages as they are not bound to a specific driver and their individual impression. Additionally, they feature utmost reproducibility and applicability to simulative results. Consequently, a well-chosen mixture of both methods needs to be utilized. As best practice, the share of subjective evaluation increases with the progress of vehicle development to gradually approximate to the perception of the customer. For the remainder only objective measures are applied.

## Overview of Active Chassis Systems Modelling Philosophies and Depths

The way of representation of a particular ACS in a simulation environment can be manifold. Depending on the desired accuracy and information regarding the model output, different approaches are possible. Moreover, necessary information such as physical parameters regarding the system itself is often not accessible. They are proprietary knowledge of the system manufacturer. In the following, the models are divided into two main groups which are again split into two subgroups. The reader is referred to (Isermann, 2007) for further information.

## Empirical Modelling of Active Chassis Systems

The empirical modelling approach describes the behaviour of the system e.g. with mathematical models. They are defined either by sufficient knowledge of the underlying process itself or by initial characteristic measurements of the system. The parameters of the system are identified based on measured inputs and outputs of the test object. For further reference methods for system identification are stated in e.g. (Ljung, 1987).

As a first modelling approach of the system, the linear time-invariant (LTI) representation is used. The main benefits are well-known system behaviour regarding the model and a small manageable amount of parameters. Additionally, the effect of parameter changes regarding static and dynamic system behaviour is predictable with ease due to linear system behaviour.

For further improvement regarding model accuracy, the linear parameter-varying (LPV) model philosophy is applied. The amount of main parameters stays the same, but their dependency of external signals (e.g. model input) turns them into a sort of look-up tables of arbitrary shape. The difference between the LTI and LPV models is shown in Figure 4.

Figure 4: Comparison of LTI and LPV Model Variant

## Linear Time Invariant Model Approach

By separating the model at its physical or other wisely chosen corresponding interfacing points, the behaviour of the system can be approximated by well-known transfer functions of low order. The order itself depends on the characteristics of the system, often identified on component test rigs per inspection.

This way of modelling is often used during the development of high-level vehicle control algorithms since the most important fact, the main behaviour of the system, is captured by this modelling philosophy with only a few parameters. For instance, a possible LTI representation of an active roll stabilisation system is a second order transfer function (Scharfenbaum, Fratini, & Prokop, 2013):

$$G_{PT2}(s) = \frac{K_P}{T^2 s^2 + 2\mathrm{DT}s + 1}$$

This rather simple representation targets one operating point from the system and loses validity[2] according to deviation from this operating point. To alleviate these inaccuracies the representation as follows is recommended.

### Linear-Parameter Varying Model Approach

The empirical modelling approach stated above can be expanded by varying parameters while maintaining the main structure of a transfer function. By taking a second order transfer function as example, the model parameters $K_p$, $D$ and $T$ are now dependent of the input signal $u(t)$. To be more precise in approximating the behaviour of the real system, the range of varying parameters can be limited. For example, a proportional gain parameter exceeding 1 by numbers does not cohere with the designated system behaviour. The parameters are identified with optimization routines. Due to the fact that the minimum of the defined cost function is given as optimal solution, highly discontinuous characteristics of the parameters can occur. For further improvement regarding ease of parameter interpretation, common form functions are used. For example, the shape of the time constant with respect to the operation point is constrained to a polynomial curve to a certain degree.

## Physical Modelling Philosophies

In contrast to the empirical modelling approaches above, the physical modelling philosophy gives more insight into the real behaviour of the actuator. They are based on physical models in different degrees of abstraction. Mechanical states such as torque, angular velocity and positions of the components get accessible. Additionally, energetic measures such as currents and voltages are attainable. This information is needed in a multi-disciplinary engineering team and serves well in a manifold development process. For example, the energy supply capacities can be stated. A stable and largely available high power output is needed to operate electro-mechanic chassis systems. Even at the early stages of the development process, rough energy demand estimations can alleviate possible problems in the late phases when the power supply of the car is under combined stress.

2 Under the assumption of non-linear system behavior with respect to the operation point

## Semi-physical Modelling Approach

As an extension regarding both the LTI and LPV approach, the semi-physical approach is explained in the following. For further reference, this method is used and validated in (Pellegrini, 2012; Unger, 2012) for variable damping control systems. The overall structure of the model is divided into two main segments: a physical and an experimental part. The latter is similar in methodology as either the LTI or LPV model above. The first is a classic physical model representation of a component of the ACS under investigation, cf. Figure 5. The serial arrangement of the two sub models is arbitrary since only the sub-system subject to further analysis needs to be modelled as physical model. For instance, the electro-mechanic behaviour of a vertical ACS (i.e. from the input signal to the motor shaft torque) can be modelled as either LTI or LPV and the following straight mechanic components can be modelled as second order dynamic system with conventional parameters such as inertia $J$, damping $d$ and stiffness $c$. For further enhancement, e.g. dissipative components of the system can be approximated by well-known friction elements such as a Stribeck Curve (Stribeck, 1903).

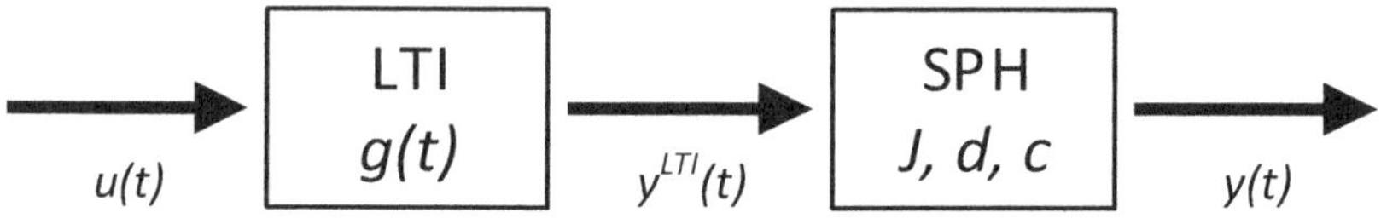


Figure 5: Model Structure of Semi-physical Modelling Approach

## Physical Modelling Approach

The most detailed philosophy covered in this paper is physical modelling of ACS. The actuator is divided into different subsystems to best knowledge, e.g. low-level actuator control, electric circuit and motor shaft dynamics. As a consequence, these subsystems can be modelled independently, cf. Figure 6. In dependence of component architecture knowledge, sometimes in reality more complex structures are approximated by simpler models. To be precise, an alternating current brushless motor can be modelled by a direct current motor without incorporating too much inaccuracy regarding electric and mechanic behaviour (Schröder, 2009). Despite existing requirements regarding modelling architecture of the actuator, data sheets and technical facts can be utilized to reduce the size of identification parameters. For example, inductivities and armature resistances of each strand can be used for further simplification of the model identification process.

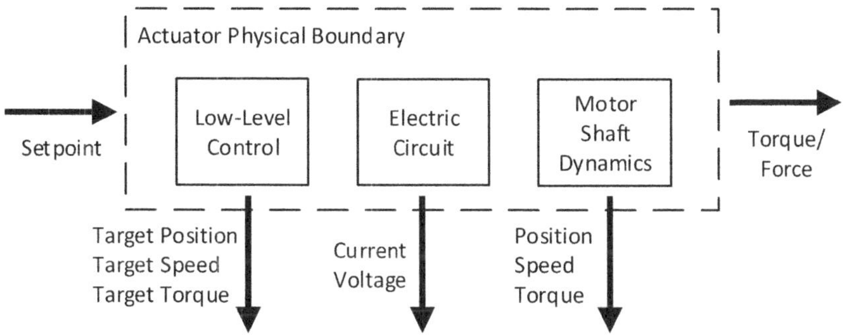


Figure 6 Physical Actuator Model Layout and Component Output

## Model Identification Methodology

All of the models above are identified based on the same source. A component test rig is used to isolate the behaviour of the system. In contrast to an identification methodology in the full vehicle, effects from the whole chassis are eliminated. As a drawback, influences e.g. from compliances in the joints of the wheel carrier are not captured. For the paper at hand these influences are also not modelled in the vehicle model and are subject to further research.

Different reference signals are applied to the actuator system on the test rig. They are grouped as follows (Ljung, 1987):

- Steady state profiles
- Transient profiles
- Stochastic profiles.

With the use of slow ramp profiles, effects like friction and stick-slip phenomena and general steady-state behaviour can be analysed. Regarding transient profiles, sinusoidal profiles with fixed frequency, sine-sweep signals or step profiles are used. The behaviour of the system with respect to different operating points is captured by the variation of amplitude for the transient category. While almost linear behaviour can be expected for system specific low amplitudes, most systems tend to show an increasingly non-linear behaviour with rising amplitude. The third category specifies either white or coloured noise signals or signals recorded by real road tests. The main advantage is that excitation of the system with a profile that is as close as possible to its technical use-case promises best matching attributes regarding validation of measurement and simulation. Additionally, nonlinear effects get visible with ease due to stochastic nature of the excitation.

The result of a real road profile excitation is given in Figure 7. The measured behaviour of the system is given as reference. For illustration purposes only the model results of one physical and experimental model are shown. Alignment is given for vanishing frequencies across all models, hence stationary accuracy is given. With increasing frequency, deviation between physical model and measurement occurs, while the experimental model still maintains validity. This is due to the fact that on the one hand, the physical model is identified mainly by the use of technical data sheets and a simplification of the electric circuit and that on the other hand friction phenomena are applied. The rather unexpected high validity for medium frequencies regarding the experimental model is explained as follows: only the model structure itself is fixed regarding the identification process. The parameters are then (within physical boundaries) subject to optimization. As a consequence, the model itself is fitted solely on the measurements and therefore high accuracy occurs. Towards high frequencies, the experimental model process tends to diverge from the underlying process of the real system represented in measurements. Again, this is due to the fact that the model structure and parameter set are fixed. The best solution of parameter sets is given for low to medium frequencies. Missing flexibility of the underlying process model causes these deviations. Further propagation to a LPV model allows improvement of fitting also in higher frequency regions. Regarding the physical model the benefits of a physical modelling philosophy strikes. With higher frequencies regarding actuator set point value, the angular velocity and torque increase likewise. Since physical limitations such as maximum torque or maximum speed of the electric rotor are incorporated in the model, the effects are visible in alignment of simulation and measurement towards higher frequencies.

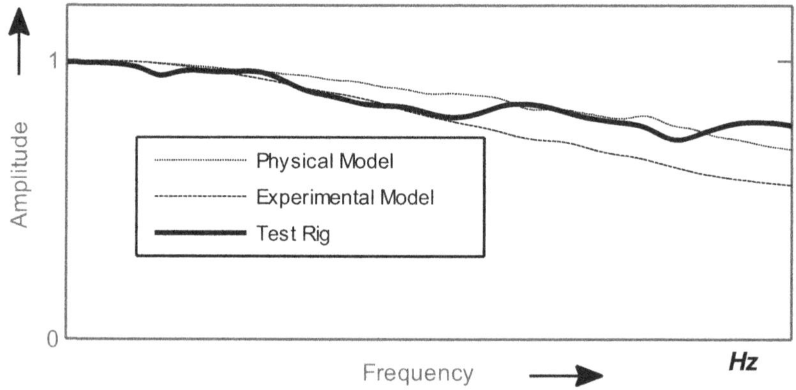


Figure 7: Transient behaviour of Different Models of an Active Chassis System in Comparison with Measurement

A step response for a medium step height is shown in Figure 8. For illustration purposes, only the highly transient part right after the beginning of the step is shown. Albeit not visible, all models as well as measurement reach stationary accuracy after time. The reference value differs from an ideal step due to actuator limitations, even though a very high gradient is chosen. Regarding response times of the models and measurement, the physical model tends to react nearly at the same time as the measurement, notwithstanding different initial gradients are visible. The measurement shows low initial reaction. The assumption of internal latencies regarding dynamic feedforward features and signal propagation time within the modules of the actuator will hold. Since these effects are explicitly not modelled in the physical model, the latter reacts faster to the input step. Regarding the experimental model, the results do not differ from the expected behaviour of a low order transfer function. Among all models, fastest response to the input signal is visible. This is caused by the fact that neither inertias nor friction are incorporated and instantaneous response is possible.

As an overall conclusion, differences between the models and measurement occur as expected. Regarding low frequencies, both main model groups show acceptable validity.

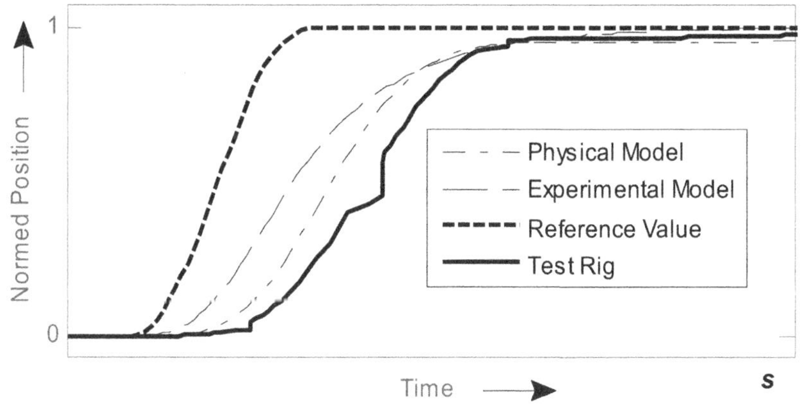


Figure 8: Step Response of Actuator Models and Measurement

# Achievable Accuracy and Matching with Development Stages

The different modelling philosophies stated earlier can be linked with different stages in the development process of a vehicle. Right after definition of the vehicle goals as a result from the benchmark, appropriate measures need to be found for fulfilment. One provision can be the use of ACS. Following ongoing development, a first estimation of actuator usage and performance needs to be made. At this early stage, experimental models can be used. Regarding model identification and parametrization, mainly two options are possible. On the on hand the system itself can be defined according to the specifications of a predecessor. The assumption of similar system behaviour regarding a new generation will hold. Additionally, slight modification of the parameters, thus system behaviour, is possible and adaption to evolved full vehicle goals is possible.

On the other hand, the attributes of the system can be defined from scratch or derived from vehicle goals. The analogy of the experimental model with known transfer functions alleviates definitions of system behaviour on a top-level while sufficient validity regarding chassis attributes is ensured. With propagation to a LPV model based on the behaviour of the predecessor, validity and accuracy regarding isolated system behaviour and driving characteristics is enhanced further. As an alternative, well-defined definition of attributes in dependence of the operation point by low order form functions can be used and an LPV model can be created without the necessity of a physical specimen. Then the desired model behaviour can be stated towards third party suppliers for creation of a first hardware prototype.

The semi-physical models come into play regarding detailed analysis of the ACS in combination with other parts of the chassis itself. As an example, the electrical part is represented as an experimental model while the mechanical part from the motor shaft onwards including an abstract friction model is implemented as physical model. A first analysis regarding the interaction of mechanic parameters such as stiffness or damping and acceptable friction is possible. Influences from the system towards other chassis components can be analysed.

With ongoing development additional information is necessary. As a prominent example, the electrical energy demand of a mechatronic system must be stated during development. The power supply architecture needs to be designed according to the characteristics of the load. At this stage, the physical model is utilized for an early approach regarding electrical load. With further model refinement the quality of information can be improved. Additionally, further insight into the internal states of the actuator is accessible, cf. Figure 6. This creates possibilities to further analyse occurring torques and velocities within the system. Conclusions can be drawn regarding mechanical behaviour and necessary energy demands. Strategies for energy efficient use of ACS can be developed.

# The Interaction of Modelling Depths and Vehicle Driving Characteristics

The different models stated above are analysed in a vehicle dynamics simulation environment. With reference to Figure 3, the inputs of the driver are represented by defined open-loop manoeuvres, thus eliminating the influence of the human driver in the following. Regarding the high-level control algorithm, original Model-in-the-Loop algorithms are used. The following actuator is then represented by the different models stated earlier. The real vehicle is substituted by a well-parametrized two track model. As a consequence, variation is only made on the actuator model depth level.

## Study Overview and Methodology

For this research a sports utility vehicle is used. Its characteristic high centre of gravity and high mass makes it suitable for the use of ACS. For example, the reduction of static roll angle and dynamic roll rate imparts agile driving behaviour to this vehicle class. Additionally, active steering systems can support further agility enhancement.

The simulation environment used in this study is a two-track model with full axle (elasto-) kinematic representation. They are parametrized by a kinematics and compliance test rig. The tire is represented by a semi-physical model approach and is also parametrized by a component test rig. Inertias and masses are derived from measurements of a pendulum test rig. Sufficient validity for the model has been ensured for both passive vehicles as well as those with ACS.

Steady-state and transient manoeuvres concerning the lateral dynamics of the car are chosen. The first manoeuvre category is used for the evaluation of steering tendency and static roll behaviour of the car. The ACS is subject to its low to even vanishing dynamic operating range. Regarding the second and therefore transient category, the vehicle is excited in the relevant frequencies for vehicle dynamics. The vehicle is stimulated up to steering wheel angles achievable by a human driver. Longitudinal influences are reduced as much as possible by maintaining a constant highway velocity. Both yaw and roll motion can be measured as a consequence. The results of real measurements and simulation are validated but omitted in the following. As a premise for this paper, the vehicle measurement methodology gives utmost reproducibility but further proof is not stated here in detail.

## Vehicle Simulation Results and Discussion

Vehicle characteristics are divided into different categories. The present paper uses separation as seen in e.g. (Wagner, 2014). First and foremost the category of stationary steer is stated albeit minor relevance regarding actuator properties is evident. Figure 7 makes it obvious that for vanishing frequencies a transfer characteristic near 1 for every model eliminates any influence from model variation for stationary driving manoeuvres. As a consequence, even an ideal actuator without any frequency dependent characteristics can be used. Referring to the second category, dynamic steer properties, significant influence is present regarding lateral ACS. Taking the damped Eigenfrequency or damping coefficient regarding yaw motion for instance, actuator model influence is significant. The influence of the varying time constant of a steering actuator is shown in Figure 9 as an example regarding parameter sensitivity of the LTI model. Significant deviations occur albeit minimal changes in the time constant of the actuator are made. This emphasises the necessity of thoroughly parametrized models from component test rig measurements. On the other hand, the additional degree of freedom in terms of design of driving characteristics by actuator property alteration is shown. Referring to a development process where no reference point of a predecessor is given, proper choice of only one model parameter causes large deviation on full vehicle level.

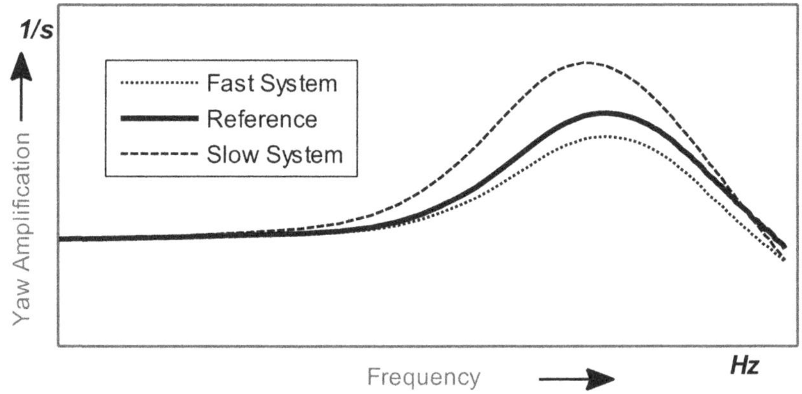


Figure 9: Influence of Actuator Model Parametrization on Driving Characteristics

As an exemplary result, the normalized roll angle amplification with respect to the steering wheel angle of a vehicle is shown in Figure 10. Three variants are presented. As an absolute reference the outputs of the control algorithm are directly fed into the vehicle model, thus no actuator is represented in the simulation. The roll angle is reduced almost over all frequencies. Towards higher frequencies the roll angle does increase. This is caused by full vehicle motion and chassis roll inertia. In contrast to the ideal actuator, differences between the experimental and physical model are present among all frequencies. In the lower dynamic range, the experimental model nearly behaves like the ideal actuator model. This is caused by an amplitude ratio near 1 and minor phase lag. The behaviour of the physical model shows differences in this frequency region. This is due to the fact that friction effects are modelled by a first approximation only and deviations are accepted. For low to vanishing frequencies and actuator movement from stand still these effects significantly dominate the movement of the active system and hence chassis motion. Towards higher frequencies, the physical model can uphold desired roll angle reduction longer compared to the experimental model in correspondence to the test rig measurements shown in Figure 7. The physical modelling philosophy shows ability to match highest frequencies.

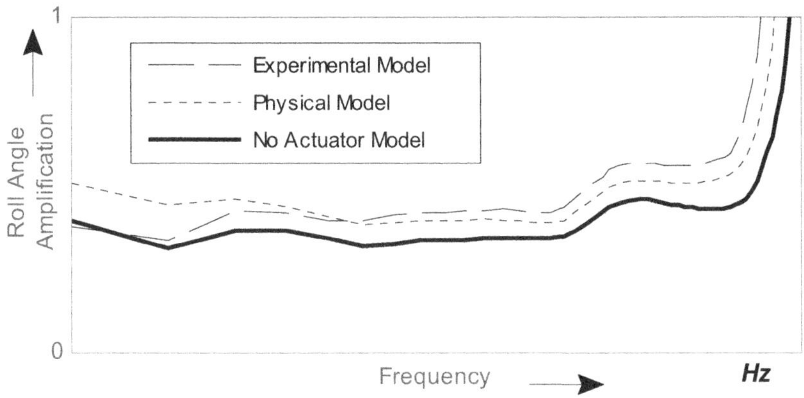


Figure 10: Differences Regarding Vehicle Characteristics from Different Actuator Models

# Conclusion and Benefit of this Research

As an overall conclusion the importance of appropriate modelling of active chassis systems has been shown. Depending on the current development stage suitable models need to be chosen. The first attempt, a LTI representation of the system, boosts achievable accuracy in terms of isolated test rig behaviour and driving characteristics, albeit only little insight into the actuator's behaviour is accessible. With further model improvement related vehicle components such as power supply can be coordinated for failure-free and operation with high performance during transient manoeuvres.

As stated earlier, simulative representation of vehicle motion has been part of the state of the art for years now. Appropriate modelling of mechatronic components has also been known for years. The main benefit of this study is the connection of both development processes of vehicle and system and their interaction in a simulative environment. The conflict of necessary information for further progress regarding the development can be eased with appropriate assumptions and ACS model use. The paper at hand gives recommendations for development stage specific model use as well as guidance regarding model creation at certain stages. With proper actual or desired knowledge of the system under investigation, high-level control algorithms can be developed efficiently, connected sub-systems of the vehicle can be designed accordingly and specifications towards suppliers can be stated.

## Acknowledgements

The author wants to thank AUDI AG for highly appreciated support and the helpful hands of students. Without them this study could not have been completed.

## References

Ammon, D. (1997). *Modellbildung und Systementwicklung in der Fahrzeugdynamik*. Stuttgart: B. G. Teubner.

Einsle, S., & Fritzsche, C. (2013). *Utilization of objective tire characteristics in the chassis development process*. ATZ. chassis.tech plus 2013 4. Internationales Müncher Fahrwerk-Symposium, München.

Heißing, B. (2011). *Fahrwerkhandbuch: Grundlagen, Fahrdynamik, Komponenten, Systeme, Mechatronik, Perspektiven* (3., überarbeitete und erweiterte Auflage). *Praxis ATZ/MTZ-Fachbuch*. Wiesbaden: Vieweg+Teubner Verlag / Springer Fachmedien Wiesbaden GmbH, Wiesbaden.

Isermann, R. (2007). *Mechatronische Systeme: Grundlagen* (2., vollständig neu bearb. Aufl). Berlin: Springer-Verlag Berlin and Heidelberg.

Ljung, L. (1987). *System identification: Theory for the user*. *Prentice-Hall information and system sciences series*. Englewood Cliffs, NJ: Prentice-Hall.

Pellegrini, E. (2012). *Model-Based Damper Control for Semi-Active Suspension Systems* (Dissertation). Technische Universität München, München.

Rill, G. (1994). *Simulation von Kraftfahrzeugen*. *Grundlagen und Fortschritte der Ingenieurswissenschaften*: Vieweg+Teubner Verlag / Springer Fachmedien Wiesbaden GmbH, Wiesbaden.

Scharfenbaum, I., Fratini, A., & Prokop, G. (2013). A Novel Method for the Development of an Idealised Active Roll Stabilisation System Model. doi:10.1109/SMC.2013.765

Schröder, D. (2009). *Elektrische Antriebe – Regelung von Antriebssystemen*. Berlin, Heidelberg: Springer Berlin Heidelberg.

Stribeck, R. (1903). *Die wesentlichen Eigenschaften der Gleit- und Rollenlager*. Mitteilungen über Forschungsarbeiten auf dem Gebiete des Ingenieurwesens, insbesondere aus den Laboratorien der technischen Hochschulen: Julius Springer.

Unger, A. (2012). *Serientaugliche quadratische optimale Regelung für semiaktive Pkw-Fahrwerke* (Dissertation). Technische Universität München, München.

VDI-Richtlinie 2206 (2004-06). Düsseldorf: VDI-Gesellschaft Produkt- und Prozessgestaltung.

Wagner, A. Potentials of virtual chassis development. In *14th International Stuttgart Symposium 2014* (pp. 535–547).

# Smart electromechanical system to improve vehicle handling and stability by toe and camber control on the rear wheel

Isabel Ramirez Ruiz – FERRARI

Marc Alirand, Nicolas Kieny – Siemens PLM

Federico Cheli – Politecnico di Milano

POLITECNICO DI MILANO

SIEMENS

# SMART ELECTRO MECHANICAL SYSTEM TO IMPROVE VEHICLE HANDLING AND STABILITY BY TOE AND CAMBER CONTROL ON THE REAR WHEEL

*Isabel Ramirez Ruiz – FERRARI*

*Marc Alirand, Nicolas Kieny – Siemens PLM*

*Federico Cheli – Politecnico di Milano*

1

POLITECNICO DI MILANO

## Active Kinematics Suspension

SIEMENS

The challenge to enhance the vehicle driving and handling with a state estimation and prediction system is presented by fusing a real time vehicle model capable of providing a good indication of vehicle stability and control, and a secondary model able to estimate the vehicle state from real sensors to correct the indications of the primary model.

The aim of the activity is to use predictions of unmeasured quantities (e.g. sideslip angle, camber angle, toe angle, etc.) provided by a vehicle model within a vehicle control strategy (e.g. active camber control). The model forecast behavior is closer to the real one and improve the comfort and linearity of the vehicle response.

One emerging active control system to improve vehicle handling is the novel Active Kinematics Suspension (AKS), which is realized by four electromechanical actuators (two per side) in a sophisticated multilink axle. So in linear range is possible to use actively vary the toe angle to reduce the delay lateral acceleration / yaw rate and to reduce slip on rear axle, but in the non-linear range camber angle control is definitely superior in performance of lateral force and also in traction.

The driver's benefits are an improved safety and higher driving pleasure.

## Building multi-disciplinary vehicle models

For design iterations in early phases of the design having access to modular models very efficient in CPU is essential.

Despite the inherent limitations of functional 15 dof vehicle models, Amesim was chosen for its modularity and its capability to manage multi domains.

A functional 15 dof vehicle model makes the separation between suspension kinematics and compliance effects coming from the bushings. Here it has been decided to pilot the elastic contributions to superimpose the kinematics contributions of toe and camber variations.

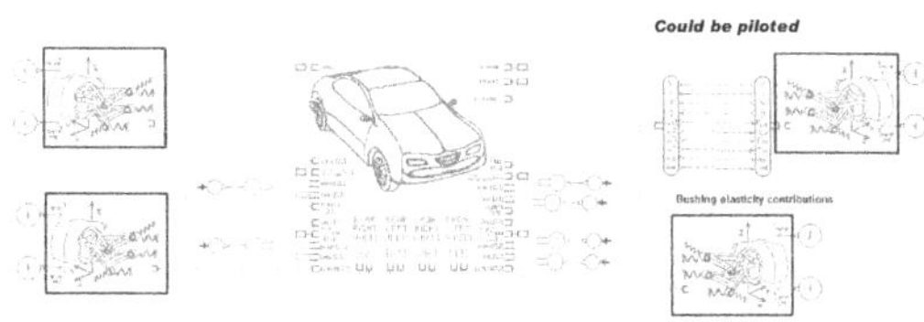


Limitations of such an approach are known but this is a preliminary analysis. Thanks to the process put in place, different model complexity levels will be achieved allowing checking the design and the control influences on driving dynamics.

## Pure Kinematics to vehicle

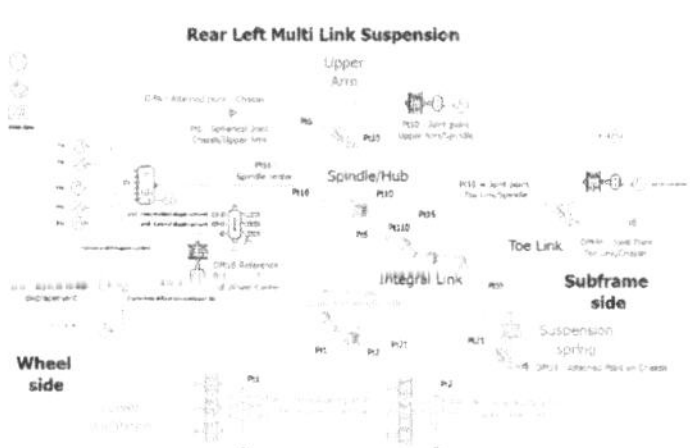


Thanks to a 3D mechanical approach (kinematics and bushings), all the parameters required for the functional 15 dof vehicle model have been generated.

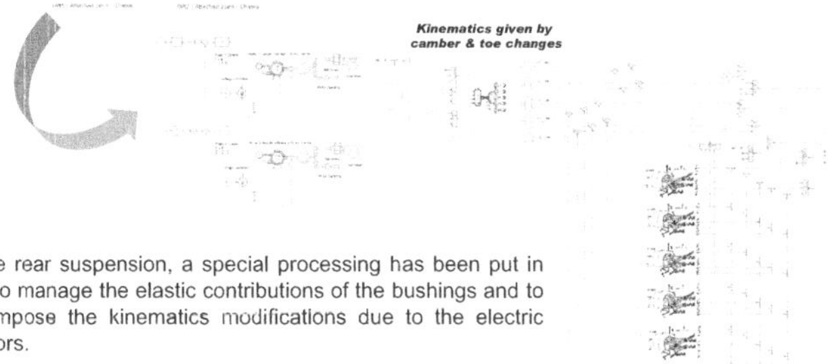


For the rear suspension, a special processing has been put in place to manage the elastic contributions of the bushings and to superimpose the kinematics modifications due to the electric actuators.

## Getting FEM result data to manage electric motor models

POLITECNICO DI MILANO

SIEMENS

The initial actuator model was a simple physical actuator with position control based on manufacturer data. JMAG RT is used for refining electromagnetic parts and thus nonlinearities such as material saturation and leakages.

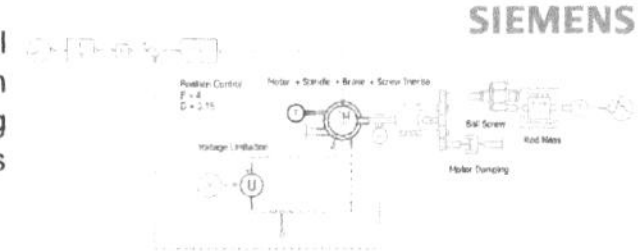

FEM analysis results has been exported in a .rtt file to be used in a JMAG-RT detailed PMSM model and used in cosimulation with Amesim

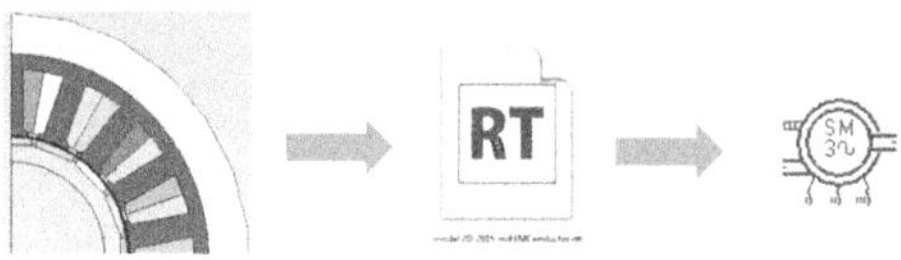


JMAG viewer enables getting control characteristics and parameters used in a flux oriented vector control.

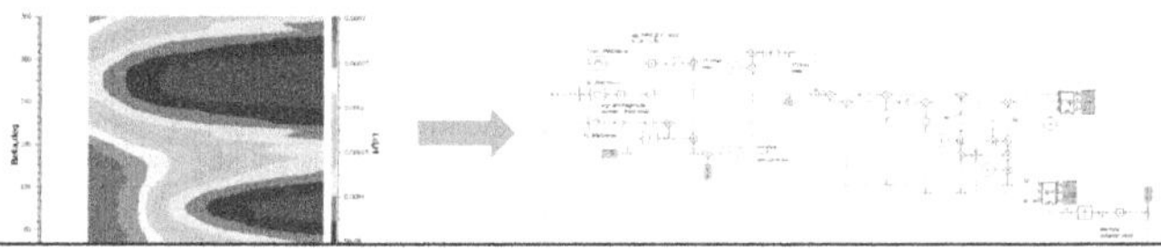

## The Active Kinematics Suspension (AKS) Controller

POLITECNICO DI MILANO

SIEMENS

The drive control logic is integrated into a single high level vehicle dynamics controller which also controls other active elements such as Active Kinematic Suspension.

A Sensor Signal Processing Model (SSPM) has been developed to estimate the vehicle states and calculating tire-road contact forces and vehicle sideslip angle. The methodological approach uses the equations of motion of the chassis applying the fundamental principles of classical physics: Newtonian method and Euler angles.

The AKS logic controls the camber and toe values based on a combination of vehicle model forecast and the SSPM predicted tire forces and orientations.

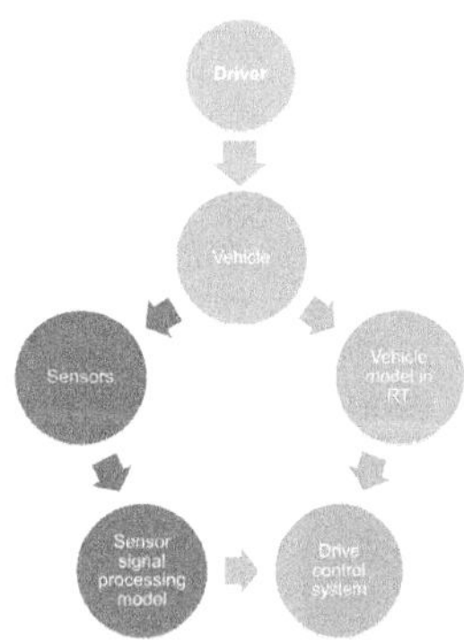

POLITECNICO DI MILANO

## The Active Kinematics Suspension (AKS) Controller

SIEMENS

Each control was applied to a neutral handling version of the normal vehicle (increasing front cornering stiffness).

The control logic is a function of the yaw rate.

The toe control steers the rear axle into the turn center, maintaining the side slip angle close to zero.

Camber control linearly increases negative camber until -6 deg (top wheel in) is reached on the outside wheel.

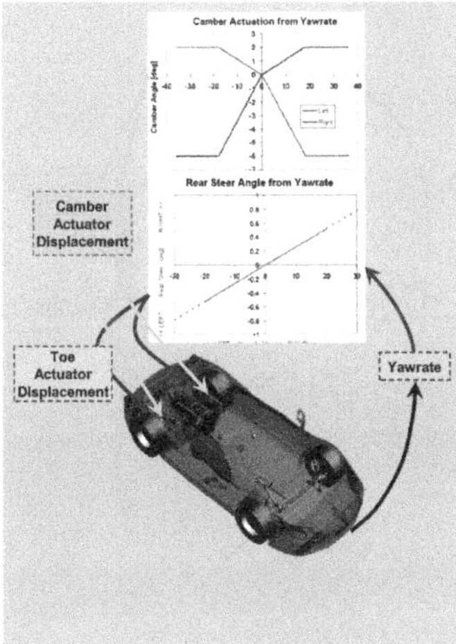

# Tire use and road safety – background to policy recommendations for new EU measures

Sven Jansen, Dr. Antoine Schmeitz
TNO Technical Sciences / Automotive, The Netherlands

Lars Akkermans
Transport & Mobility Leuven (TML), Belgium

# Abstract

TNO and TML performed a study commissioned by the European Commission regarding what measures on a European level can be taken in relation to the use of tyres to improve road safety. The study considers the use of Winter tyres, tread depth requirements, tyre inflation pressure maintenance and tyre ageing effects and damages. An assessment has been made of the tyre safety performance, current use of tyre and consumer awareness, tyre related accident statistics and existing regulations for the different tyre aspects (including TPMS). Results from analyses on accident records from Germany (GIDAS) and a NHTSA study provide estimation of safety benefits. Policy options are defined and estimations are made of the accident reduction on a European scale to assess the monetary benefits, which are compared to the associated cost of execution of the policies. The results indicate that extended enforcement of tread depth regulation and increase of tyre pressure maintenance are most cost effective. For Winter tyres it is cost effective to introduce a harmonised definition using the 3PMS performance criterion, and installing a dedicated tread depth requirement of 4 mm would be around break even. Promoting the use of Winter tyres, or extending the enforcement of the use of Winter tyres may not be cost effective as a policy option, however the estimated safety benefit is significant (i.e. 3 % reduction in the number of fatalities during winter conditions). Installing a harmonised tread depth for truck tyres, and organising tyre inspections between periodic vehicle inspections are expected to be beneficial.


# 1 Introduction

To assess various aspects of tyre use and quality related to road safety, a study [1] was commissioned by the EU Directorate-General for Mobility and Transport in 2014. The overall aim of the study is to propose policy options concerning the use of tyres for improvement of traffic safety. Centrally to this, is the idea that end users need to make the correct assessment in relation to the tyre condition in order to achieve a level of safety as high as possible. Within this context, the tyre condition is considered in relation to technical elements (tyre inflation pressure, tyre tread depth, tyre damage, tyre age, and meteorological influences) and one information element (driver awareness). The study addresses the topics of tyre usage and the impact on road safety (which are the potential safety improvements related to tyre usage from a technical perspective) and how can road users be supported to use tyres that have a better safety performance by policy options. The policy options are subject to a cost-benefit analysis.

Apart from an accident analysis, the study mainly reflects on existing studies from open data sources. By combining these insights and applying them on the study objectives, the relevant conclusions and recommendations are derived.

# 2 Tyre safety aspects

The main safety aspects that have been considered are the tyre grip level (maximum force that can be transmitted between tyre and road) and tyre failure in terms of sudden loss of inflation pressure (often referred to as blowout). The tyre grip level is mainly affected by meteorological influences in combination with the tyre condition (tread depth, ageing, inflation pressure), and tyre choice (Summer tyre or Winter tyre). A reduced grip level extends the stopping distance of vehicles and can result in vehicle instability (e.g. spin out). The risk of tyre blowout failure is related to inflation pressure being too low and tyre damage (including effects of ageing), which are strongly linked to tyre maintenance. When tyre blowout failure occurs it leaves debris on the road, the vehicle can become unstable, or may come to a halt at an unsafe location for changing tyres.

# 3 Tyre usage

Tyre usage concerns the condition of tyres as found on vehicles in Europe and their use in summer and winter conditions respectively. The tyre usage is assessed from several (statistical) data sources such as consumer awareness campaigns, periodic and roadside inspections and accident recordings that are either publically available, or have been provided by stakeholders. The tyre usage is assessed for passenger car tyres and for commercial vehicles (when available), and the following observations are made:

- Winter tyre use: The use of Winter tyres is affected by national legislation in EU member states and the occurrence of winter conditions. Member states with high occurrence of winter conditions typically mandate the use of Winter tyres for a fixed period. In these member states up to 90 % (some report 100 %) of the vehicles is equipped with Winter tyres during the mandated periods. For member states where the use of Winter tyres is recommended a large variation is found ranging from 30 % to 80 %.
- Tread depth: The share of vehicles that is equipped with tyres that have a tread depth close to or below the legal limits is estimated to between 10 % and 20 %. Tread depth of Winter tyres generally is more due to specific Winter tyre requirements in member states.
- Inflation pressure: Deviations in tyre inflation pressures are generally below the recommended settings, and sources indicate that up to 20 % of passenger cars have one or more tyres with a reduced inflation pressure that is considered dangerous. For heavy duty vehicles it is found that the inflation pressure of pulling vehicles is better maintained than trailers, probably due to the fact that the owner of the trailer is not always the user (i.e. rental trailers).

- Tyre damage and ageing: Reports from visual tyre inspections for damages indicate that about 30 % of vehicles with Summer tyres have at least one tyre that shows signs of ageing or other irregularities that increases the risk of tyre failure. Remarkably the results indicate that less than 10 % of the vehicle population with Winter tyres shows these types of damages, suggesting that Winter tyres are better maintained than Summer tyres, or that people using Winter tyres take better care of tyre maintenance in general.

# 4 Role of tyres in road safety

As mentioned above the main safety aspects of tyres are considered to be related to the grip performance of tyres and potential sudden loss of inflation pressure (e.g. tyre blowout failure). An overview of technical aspects of the tyre condition affecting these safety aspects is listed below:

- Winter tyre: Winter tyres are engineered for better grip in a lower temperature range than Summer tyres, and the tread design is optimized for traction on winter surface conditions. Generally the industry recommends the use of Winter tyres below 7 degrees Celsius however the precise temperature below which Winter tyres excel in grip performance depends on meteorological conditions and may be tyre specific. An increased tread depth of Winter tyres compared to the legal requirement of 1.6 mm improves grip on snow (however some Winter tyres also show good grip performance at 1.6 mm).
- Tread depth: Tyre grip performance on wet roads is better with increased tread depth as it increases the speed that can be driven without aquaplaning and shortens braking distance. A strong reduction in grip on wet roads is found for tread depths below the legal requirement.
- Inflation pressure: Both the grip performance and the risk of tyre blowout failure are affected by inflation pressure. A deviation in inflation pressure that is too large reduces the grip level and can lead to vehicle instability also on dry roads due to the reduced stiffness. Tyre blowout failure can occur due to heat generation from large tyre deformations that result from severe underinflation.
- Tyre damage and ageing: In general tyre damage poses a threat of tyre blowout failure. Secondly the mechanical properties of rubber change due to exposure to the environment (e.g. heat, UV, moisture), resulting in less strength, less flexibility, and in general less resistance to heat and mechanical damages. The major safety concern is the increase of the risk of tyre blowout failure, however ageing also reduces the tyre grip performance. Ageing is difficult to quantify as the extent of exposure is a mix of meteorological conditions, specific use of the vehicle (e.g. indoor parking) and tyre age.

# 5 Road accidents and tyre usage

The occurrence of road accidents is related to tyre usage from a causation study carried out using accident records from the GIDAS [2] database, and a study presented by NHTSA [3]. With the GIDAS database, only the causation for tyre grip accidents has been analysed due to the available population. The analyses have been done for different categories of temperature and road surface conditions that are relevant for tyre grip performance of summer and Winter tyres respectively. The population of specific tyre conditions in accidents is compared to a reference group that is assumed to be representative for vehicles on the road in general to assess the potential accident reduction. A similar approach has been used on the NHTSA study data; however the information presented contains fewer details about the accident conditions. That study relates the tyre tread depth, inflation pressure and existence of damages to accidents in general.

The results of the studies are used to assess the safety potential using improved tyres for specific accident conditions. An overview of the safety potential is provided in Table 1

Table 1: Summary of identified safety potential by improving tyres.

| Accident condition | Inadequate tyre | Improved tyre | Accident probability reduction / ARR*) | Source |
|---|---|---|---|---|
| Grip accidents on dry road, below zero | Summer tyre | Winter tyre | 45.8 % / 0.816 | GIDAS |
| Grip accidents on snow covered roads | Tread depth Winter tyre below 4 mm | Tread depth Winter tyre 4 mm or more | 56.1 % / 1.147 | GIDAS |
| Grip accidents on wet roads or snow covered roads (assumed) | Tread depth at or below 1.6 mm | Tread depth above 1.6 mm | 84.1 % / 3.722 | NHTSA |
| Tyre related accident (assumed mainly tyre blowout failure) | Incorrect inflation pressure | Correct inflation pressure | 35.1 % / 0.446 | NHTSA |
| Tyre related accident (assumed tyre blowout failure) | Damage | No damages | 85.9 % / 5.194 | NHTSA |

*) The ARR (Accident Reduction Rate) [1] indicates the potential reduction of accidents for 1 % increased share of improved tyres for the distribution used to calculate the accident probabilities.

From these results, it is concluded that the use of Winter tyres will reduce grip accidents in winter conditions, and maintaining a tread depth of more than 4 mm will reduce grip accidents in snow conditions. In general maintaining a tread depth above the legal requirement of 1.6 mm will reduce accidents, which are assumed mainly grip accidents on wet and snow covered roads. Regarding tyre blowout failure, it is found that avoiding tyre damage and incorrect inflation pressure will reduce accidents. Typically the combination of damage and incorrect inflation further increases the risk of an accident.

# 6 Harmonization use of snow tyres

The benefits of harmonised measures concerning the use of the most appropriate tyres for different climatic conditions are evaluated. In relation to the definition and legislation of Winter tyres, different important aspects need to be clarified.

Firstly, it needs to be noted that different definitions may apply to 'Winter tyres', corresponding to different technical properties: M+S marked tyres do not meet performance requirements in contrast to tyres marked with the Alpine symbol (3 Peak Mountain Symbol – 3PMS, see Figure 1), which indicates the tyre meets performance requirements on snow.

Figure 1: Alpine symbol (3PMS).

Therefor only the Alpine symbol ensures a specific performance on snow, as it follows UNECE Regulation 117, to be implemented in the EU through Regulation (EC) 661/2009, requiring for snow tyres a minimum level of performance threshold on snow (braking and traction). This distinction is also carried in the existing legislation in EU member states. Most EU member states only mention the use of M+S tyres. Specific mention of the "Alpine symbol" is only made in Germany and Sweden.

Secondly, existing legislation in EU member states comes in different varieties. Member state legislations may be formulated differently and define the following elements (a) the specific period of time (calendar dates), (b) meteorological conditions, (c) specific locations and (d) further technical requirements (tread depth requirements typically range between 1.6 and 4 mm for cars, and 5 mm for trucks).

Overall, the following distinction is made:

- No legislation: In member states without legislation the use of Winter tyres is generally recommended by road user groups when the occurrence of conditions of snow and ice is usual.
- Conditional: The legislation for conditional fitment is generally during occurrence of winter conditions (or mandated by road signs) within a fixed period. This type of legislation is typically used in member states with incidental winter conditions, or occurrence in specific regions of member states (e.g. mountain regions).
- Fixed period: In member states, typically with long duration winter conditions, the fitment requirement is for a fixed period regardless of the weather conditions.

Table 2 shows an overview of Winter tyre regulation in EU member states with an indication of the occurring winter conditions (n = none, i = incidental, s = severe, r = regional).

Table 2: Overview of current Winter tyre legislation in EU member states.

| No legislation | Conditional | Fixed period |
|---|---|---|
| • Belgium [i] | • Czech Republic (*) [s] | • Austria (**) [s] |
| • Bulgaria [s] | • France [r] | • Croatia [s] |
| • Cyprus [n] | • Germany [i] | • Estonia [s] |
| • Denmark [i] | • Luxembourg [i] | • Finland [s] |
| • Greece [r] | • Romania (*) [s] | • Italy [r] |
| • Hungary [s] | • Slovenia (*) [s] | • Latvia [s] |
| • Ireland [i] | • Switzerland [s] | • Lithuania [s] |
| • Malta [n] | | • Slovakia (**) [s] |
| • Netherlands [i] | (*) within a period | • Sweden [s] |
| • Poland [s] | | • Norway (**) [s] |
| • Portugal [n] | | |
| • Spain [r] | | (**) not all vehicle categories |
| • United Kingdom [i] | | |

This combination (different tyre requirements and different national legislations) requires a specific approach to assess the possible impacts of different policy options on safety. Within the current project, this analysis is built around the following elements:

- The identification of accidents under relevant conditions. Relevant meteorological conditions (based on the analysis of the European Food Agency Data (EFSA) and Deutsche Wetterdienst (DWD) databases) and current tyre usage statistics are used to enrich the CADaS (Common Accident Data Set) database and identify those accidents where the different policy options could have an influence.
- The assessment of the possible effects linked to different policy options is based on the analysis of information presented in the GIDAS database. As a result of the comparison of grip-related accidents between two vehicle categories (those with and those without the appropriate tyre use conditions), it is possible to identify risk reduction rates for changed tyre use behaviour. This is then extrapolated to the overall fatality statistics.

In particular, the following possible policy options are identified as a result of the analysis of current legislation in relation to the use of Winter tyres:

- Promoting the voluntary use of Winter tyres in all Member States (with and without existing legislation);
- An increase of the use of Winter tyres in those Member States with existing (mandatory or conditional) legislation as a result of improved enforcement activities;
- A shift of relevant legislation from M+S towards 3PMS performance requirements;
- A minimum thread depth for Winter tyres of 4 mm.

# 7 Tyre tread depth requirements

Passenger cars with tyres that have a tread depth below the legal required value of 1.6 mm are significantly more involved in accidents on wet roads. The NHSTA study indicates a reduction in accident probability of 86 % when using tyres with adequate tread depth. Increased tread depth generally will improve the tyre grip level (and safety) on wet roads and snow covered roads. As a result, organisations in different member states recommend changing tyres before the minimum tread depth is reached with a margin of about 1 mm or more. For passenger car Winter tyres a variety of requirements is adopted by member states up to 4 mm tread depth. A required tread depth for truck tyres exists in a large share of member states and ranges from 1.0 to 3.0 mm, and for truck Winter tyres at least one member state requires 5.0 mm tread depth. Also in some member states tread depth requirement for truck tyres do not exist, meaning trucks are allowed on the road without any tread left on the tyres.

Increasing the minimum tread depth requires changing tyres more often, leading to higher replacement cost and environmental burden due to waste increase. The average age of tyres will reduce, resulting in less ageing effects. Tyres with reduced tread depth have less rolling resistance, and thus earlier replacement may also increase fuel consumption of vehicles. If the change in minimum tread depth requires adapting the tyre wear markings it can lead to high cost for tyre manufacturers when tyre moulds need to be modified.

To ensure sufficient traction on snow, a minimum tread depth is required for truck tyres. This will reduce the risk of aquaplaning as well, which however is less likely to occur compared to passenger cars. Maintaining a minimum tread depth will also prevent too extensive wearing of the tyres that can result in tyre blowout failure.

# 8 Consumer awareness

Specific information on consumer awareness in relation to mobility, vehicle technology or tyre choices is very scarce. No studies were found that provided relevant information. Information about awareness of road users regarding tyres and safety was obtained from a stakeholder questionnaire. In general, it is indicated that end-users (passenger car) are still lacking awareness on the importance of tyre properties and tyre use in relation to safety. Professional drivers are more aware of this. In particular national administrations may also be lacking this know-how, although this differs between member states.

From other fields of practice where consumer awareness is linked to health and safety, it was found that long-term information campaigns may elicit a behavioural response away from undesired behaviour for between 10 and 20 % of the persons targeted. A good example on consumer awareness, in relation to the correct use of tyres, is present in Germany. In this example, consumers are reminded at different occasions (carwash, parking, etc.) on the importance of correct tyre use on safety, and are actively helped or engaged to check on the status of their own vehicles' tyres. In general consumers should be made aware of tyre safety behaviour when confronted with relevant situations such as purchasing, car maintenance, holiday planning, etc.

Such awareness campaigns are at least partially already conducted by the tyre industry, consumer groups and safety organisation to inform road users of the relevance of tyres for safety. However, these information campaigns are also mostly limited to the retail points. These campaigns typically address tyre inflation pressure, the use of Winter tyres and tread depth. Additionally information about tyres and road safety is provided using different media such as television and internet.

# 9 Tyre Pressure Monitoring Systems

Two categories of Tyre Pressure Monitoring Systems (TPMS) exist:

- Direct systems use devices that register the pressure inside the tyre. These systems typically are battery operated and require installation by specialists (and reset in case of mounting different specification tyres). The systems can be installed as an aftermarket solution, and generally need to be replaced when the battery life expires (5-10 year). Part of the systems is exposed to the environment and therefore can be prone to damage. For the use of Winter tyres a duplicate set of systems is required when these are mounted on separate rims. The end user can be provided with real-time inflation pressure readings in addition to the warning function, and has little interaction with the system. The cost for installation of direct systems is estimated at 50 Euros per wheel.
- Indirect systems rely on wheel speed signals from ABS to estimate underinflation of one or more tyres. These systems do not require additional hardware in the tyre wheel assembly, and are available only as ex-factory systems. The end user is required to reset the system when tyres are changed, or when inappropriate warnings could be generated (e.g. when adapting tyre inflation pressure for loading of the vehicle).

Safety related performance:

- Indirect TMPS require user actions to recalibrate the system in case inflation pressure settings are adjusted to loading of the vehicle. Scenarios exist where a sequence of incorrect execution of the calibration procedure can lead to unsafe inflation pressure. There are however no accidents known that can be related to these scenarios, and therefore it is not assessed as real threat.
- Detection of underinflation on all tyres simultaneously is challenging for indirect systems due to their working principles. This scenario however does not seem to be the most important safety concern. The primary threat is quick loss of inflation pressure of a single tyre due to puncture or other reasons that can lead to tyre blowout failure. The speed of detection for this scenario should be significantly less than 10 minutes as defined in the current regulation, however based on the technical working principles it is expected that both technologies can detect this event with adequate response time.

The user needs to take action on warnings of TMPS, and therefore there is a large factor of perception on usefulness of warnings. For systems providing inflation pressure real time, users can proactively take action to keep inflation pressures at the correct level, and then it is important that the pressure indication at the air filling station matches the TPMS indications. Based on the available information on consumer behaviour it cannot be assessed if an enhancement of the current performance requirements of TPMS will lead to a better maintenance of inflation pressure. However, user acceptance of TPMS is related to consumer awareness regarding the relevance of having correct tyre inflation pressures.

# 10 Stakeholder input

A list of about 130 stakeholders has been assembled including tyre suppliers, TPMS suppliers, vehicle OEM, road user organizations, safety institutes, national authorities, vehicle type approval organizations, etc. Almost 50 questionnaire replies have been received from organizations in more than 15 member states. The questionnaire respondents provide information mostly on the topics around "vehicle and tyre population", "tyre regulation and impact on use", and "tyre use and consumer awareness". The replies confirm that a large variation in driving conditions between member states exists. Generally the questionnaire answers indicate that the tyre usually is not the cause of an accident, but can have a significant influence on the outcome of the accident specifically for slippery road conditions. The introduction of tyre labelling is perceived as a positive step throughout Europe as it improves consumer awareness and challenges the tyre suppliers to make high quality tyres. Extensions of tyre labelling to indicate tyre performance for more operating conditions is suggested by many stakeholders.

A stakeholder meeting was held in Brussels with 41 attendants from 31 organizations, and the main conclusions are listed below:

- Winter tyre: A unified definition of a Winter tyre is put as a requirement for European harmonisation and stakeholders propose to use the 3PMS indication for the definition of a Winter tyre. The member states however should be able to define own legislation to mandate the use of Winter tyres. On a voluntary basis additional marking can be applied to indicate when the specific tyre is designed for ice conditions (i.e. 'Nordic' Winter tyre). Consumer awareness can be improved by providing more information on tyre performance under different conditions and consumers should be supported with information from member states when travelling across borders in the winter season.
- Tread depth: Based on the information provided, stakeholders conclude that insufficient evidence is available of a reduction of wet road accidents by an increase of tread depth to justify a revision of the current tread depth regulation for passenger car tyres. For passenger car Winter tyres an increased tread depth requirement would be supported when it is sufficiently based on evidence of improved safety and that it could lead to harmonisation across Europe. For truck tyres the introduction of an EU regulation is supported to achieve a level playing field, and also results in harmonisation across Europe.

- Inflation pressure: Stakeholders conclude that improvement of maintaining correct inflation pressure cannot be achieved by introducing TMPS alone, but also an increase in consumer awareness and some education in adjusting inflation pressures are also required. Air filling stations should be made widely available (free of charge) and calibrated to ensure that proper inflation settings can be maintained. A more active role of member states would be welcomed. Stakeholders indicate that new requirements for TPMS should be defined; however several stakeholders state that any revision should be based on evaluation of the current regulation. TPMS should be introduced in other vehicle categories.
- Tyre ageing: Tyre visual inspection is currently the only method to get indications about tyre ageing, and the need for of inspections is emphasized by stakeholders also regarding other tyre conditions aspects (e.g. damages, tread depth and inflation pressure). In practice, for most vehicles, tyres wear to the minimum tread depth before effects of ageing can be noticed.

# 11 Possible EU measures and their cost-benefit analyses

Measures that could be implemented at a European level were identified for which social cost-benefit analyses were executed. The identified measures focus on (1) the harmonised use of 'Winter tyres', (2) tyre tread depth requirements and (3) tyre pressure and monitoring systems. In addition, covering all three topics, the importance of improving consumer awareness was identified as a major concern warranting specific measures. As a result of preparatory work done in the current study, it was indicated that tyre age and ageing effects are of relatively little importance towards road safety and are difficult to objectify. In contrast, tyre damages play a more important role but precise quantification remains an issue. Therefore no specific measures on tyre ageing and tyre damages were analysed.

Eight scenarios were identified as input for the Social Cost-Benefit Analysis (SCBA) that was performed in the current study:

Harmonisation of the use of Winter tyres:

- Scenario 1: behavioural change through voluntary pick-up;
- Scenario 2: behavioural change through enforcement;
- Scenario 3: 3PMS in national legislations.

Tyre tread depth requirements:

- Scenario 1: behavioural change through voluntary pick-up (legal limit);
- Scenario 2: forced renewal of tyres below legal limit, depending on tyre tread depth, through enhanced enforcement (police enforcement or vehicle periodic checks);
- Scenario 3: 4 mm tread depth requirement for Winter tyres.

Tyre pressure maintenance:

- Scenario 1: change of performance requirements (UNECE regulation 64) on TPMS;
- Scenario 2: voluntary monitoring of tyre pressure and availability of air pressure equipment at petrol stations.

Throughout the study, the importance of consumer awareness was considered. Although little evidence could be found in scientific literature (within the working domain of road safety), findings in other domains (medicine, health and safety) and practical experiences were considered important enough to incorporate improvements in consumer awareness in scenarios that rely on voluntary changes or enforcement activities to change behaviour. The German example for improving consumer awareness on the use of tyres is considered a good practice. In particular, the following elements are of interest:

- Unified information, to be distributed by different organisations;
- Extending information exchange in terms of frequency and content.

The Social Cost-Benefit Analysis is based on information on road fatalities; however the calculation considers cost reductions from a fixed ratio of fatalities, injuries and material damages. The main conclusions for the different scenarios are presented in Table 3 (base year 2012).

Table 3: Summary of social cost benefit analyses for policy options.

| Policy | Included measures | Main cost elements | Benefit-cost ratio[1] | Safety benefits[2] |
|---|---|---|---|---|
| **Harmonisation of the use of Winter tyres** | | | | |
| Behavioural change through voluntary pick-up | Consumer awareness campaign | Consumer awareness campaign, purchase of tyres and rims | 0.40 | Between 7 and 14 fatalities avoided (out of 588) |
| Behavioural change through enforcement | Police enforcement<br>Application of regular periodic checks | Periodic inspection: time cost<br>Enforcement actions: police time cost<br>Purchase of tyres and rims | Between 0.33 and 0.44 | Between 8 and 16 fatalities avoided (out of 588) |
| 3PMS as standard in national legislations | Referral in national legislations to EC Regulation 117.2 and the so-called 3PMS standard | Tyre purchase costs (partial price increase)<br>Gradual replacement of production facilities<br>R&D costs | 1.89 | Between 14 and 28 fatalities avoided (out of 588) |

1 Values lower than 1 indicate that costs outweigh benefits. Values higher than 1 indicate benefits outweigh costs.

2 Values between brackets are estimated total of relevant fatalities (Values are for EU28 without Bulgaria for all measures and without Greece for Winter tyre related measures).

| **Tyre tread depth requirements** | | | | |
|---|---|---|---|---|
| Behavioural change through voluntary pick-up (valid for vehicles running Summer tyres below legal 1.6 mm limit) | Consumer awareness campaign | Campaigns (replacement cost of illegal tyres not considered) | Between 34.21 and 68.43 (assuming 10 to 20 % change) | Between 14 and 28 (out of 1584) |
| Behavioural change through enforcement (valid for vehicles running Summer tyres below legal 1.6 mm limit) | Police enforcement Application of regular periodic checks | Periodic inspection: time cost Enforcement actions: police time cost (replacement cost of illegal tyres not considered) | Between 2.63 and 5.25 (assuming 50 to 100 % change) | Between 71 and 142 avoided (out of 1584) |
| 4 mm tread depth requirement for Winter tyres assuming voluntary behavioural change | Consumer awareness campaign | Campaigns Tyre purchase cost (earlier renewal) | Between 0.88 and 0.92 | Between 2 and 3 (out of 330) |
| **Tyre pressure maintenance**[3] | | | | |
| Change performance requirements (UNECE regulation 64) on TPMS | Consumer awareness campaign Updating of TPMS | Campaigns TPMS purchase cost | Between 1.26 and 1.08 | Between 11 and 88 (out of 12815) |

3 SCBA for tyre pressure related measures include benefits as a result of fuel consumption improvement and CO2 emission reductions

# 12 General conclusions and proposed actions

Regarding actions for different tyre aspects the conclusions and proposed actions are listed below. Stakeholders are generally in favour of harmonisation of the tyre use legislation to ease transport across EU member states and to achieve a level playing field.

## 12.1 Conclusions

- Winter tyre: Following laboratory study results, it can be stated that Winter tyres have an improved performance relevant for road safety compared to Summer tyres for low temperatures and snow or ice road conditions. This is, to some extent, confirmed in accident data. The main interest is to define a standard for Winter tyres that is common for all member states. The unified Winter tyre definition should be based on the current 3PMS approval test. Also from a technological point of view, it can be stated that using the 3PMS test as a base is supported as it ensures a minimum level of performance for Winter tyres that exceeds the capability of Summer tyres. A more detailed assessment of tyres under specific conditions (e.g. wet grip low temperature, or ice testing) will allow consumers to choose tyres that match their use better, and increase awareness of the relevance of choosing the right tyres. Enforcement and consumer awareness campaigns can extend the use of Winter tyres under appropriate conditions.
- Tread Depth: Firstly, technical performance testing indicates an improved safety level on wet roads when tread depth is increased. The accident data used in the current study however indicates no benefit in terms of reducing the number of accidents by increasing the minimum tread depth. Also, stakeholders do not seem to provide sufficient support to revise the current minimum tread depth for passenger car tyres. Secondly, more accidents do occur when the tread depth is below the minimum tread depth (about 25 % of tyre related accidents), indicating that enforcement would bring safety improvements. For Winter tyres, a harmonized tread depth is preferred to a level that is to be chosen based on evidence of safety improvements. The results of the study suggest that 4 mm could be a suitable level based on accident analysis and existing national legislation in member states. For truck tyres a harmonized tread depth is preferred to ensure that cross border traffic is seamless. The results of the study suggest that 1.6 mm could be a suitable level based on existing national legislation in member states.
- Inflation pressure / TPMS: Poor maintenance of inflation pressure can lead to tyre underinflation which is unsafe. This can be prevented by TPMS, however the user is required to take action and in that respect consumer awareness is very relevant. Additionally the air filling stations need to provide correct pressure indication not to confuse the driver. The major safety concern of inflation pressure is quick defla-

tion that can lead to tyre blowout failure or vehicle instabilities. Tyre Pressure Monitoring Systems can detect such events, however the current regulation prescribes a detection time of 10 minutes while safety critical situation can take place over several seconds. Currently no accidents are known to result of faulty user interaction with the TPMS (e.g. resetting). Consumer awareness in general seems to be relevant to improving inflation pressure maintenance on vehicles.

- Tyre damage and ageing: The most important safety impact of tyre damage and ageing is that it can lead to tyre blowout failure. Only by visual inspection the condition of the tyre can be assessed for these aspects. A safety improvement can be achieved by more inspections between regular periodic vehicle inspections on the tyre condition, during which at the same time inflation pressure and tread depth can be checked. Introducing (dedicated) tyre inspections will increase consumer awareness.

## 12.2 Proposed actions for policies (in order of cost effectiveness):

A. Increase efforts for campaigning and enforcement of current tread depth legislation;
B. Increase efforts for campaigning of tyre inflation pressure maintenance;
C. Provide sufficient access to calibrated air filling stations;
D. Define a unified Winter tyre with the requirement to have the 3PMS marking;
E. Include a requirement for quick deflation detection in TPMS regulation with a short reaction time (e.g. several seconds rather than minutes);
F. Install a harmonized tread depth requirement for Winter tyres (e.g. 4 mm).
G. Increase efforts for enforcement of Winter tyre legislation;
H. Depending on the occurrence of wintery road conditions and member state-specific mobility demands increase awareness campaigns or install Winter tyre legislation.

Note: item G and H may not be cost effective as a policy option; however the estimated safety benefit corresponds to 3 % reduction of the number of fatalities under wintery conditions.

The following actions are not supported with a cost benefit analysis, but are expected to be beneficial:

I. Install a harmonized tread depth requirement for truck tyres (e.g. 1.6 mm);
J. Organise tyre inspections on a voluntary basis or regulated between periodic vehicle inspections.

## Acknowledgements

This study was commissioned by the EU Directorate-General for Mobility and Transport.

## References

[1] Sven Jansen, Antoine Schmeitz, Sander Maas, Carmen Rodarius, Lars Akkermans: Study on some safety-related aspects of tyre use. Helmond: TNO report TNO 2014 R11423, doi 10.2832/67191, 2014, http://ec.europa.eu/transport/road_safety/pdf/vehicles/study_tyres_2014.pdf2010.

[2] German In-Depth Accident Study (GIDAS): http://www.bast.de/EN/FB-F/Subjects/e-gidas/e-gidas.html.

[3] National Highway Traffic Safety Administration (NHTSA): Tyre-Related Factors in the Pre-Crash Phase. NHTSA Report No. DOT HS 811617, 2012.

# The influence of wheel and tire aerodynamics in WLTP

Felix Wittmeier, Timo Kuthada

# Abstract

The geometric shape of passenger car wheels can have a large influence on the aerodynamic drag of a vehicle as it has been shown already in different publications. In combination, tire and rim cause approximately 25% of the overall drag. Until now, the challenge for aerodynamicists is to find the rim / tire combination with the best aerodynamic characteristics, which then is used for the official $c_D$-value of the vehicle and for the determination of fuel consumption according to NEFZ regulations.

With the introduction of WLTP this procedure is going to change. To determine the fuel consumption of a vehicle every varying part of the equipment (standard or optional) has to be considered, so that every sold vehicle can have its own consumption value. Alternatively the worst case of all configurations can be used, but this would lead to a higher fleet consumption.

For this reason, the aerodynamic characteristics of wheels and tires gain more and more importance and especially the interaction between different rims and tires has to be considered. This paper will show some possibilities to influence the aerodynamics of passenger car wheels and their effect at WLTP.


# 1 Introduction

With increasing fuel costs and an increased environmental awareness, fuel consumption has become one of the most important development targets for passenger cars. Standardized driving cycles help to compare the fuel consumption of different vehicles on a common basis and form the foundation for determining the official fuel consumption. Of cause, every standardized cycle is only a modelling the real world driving and can never cover every single on-road situation.

With the currently used NEDC cycle, the gap between cycle consumption and the real world consumption of a vehicle has been increasing over the last years, as average speeds on the roads are increasing and manufacturers are using the entire margin to optimize the consumption especially for this driving cycle. For this reason, a new official cycle (WLTP) is about to be introduced, which has stricter boundary conditions and is adopted to today's average traffic situations. With this new cycle, the focus for different parts of driving resistance is shifted and topics like aerodynamics, that are strongly dependant from the vehicle speed, gain more importance.

# 2 Standardized Driving Cycles

Standardized driving cycles are the basis for judging fuel consumptions and emissions for different vehicles and are important when it comes to comparing different vehicles on a common basis. A driving cycle is basically a simplification of real world driving. It will never be able to determine the fuel consumption of every individual driver on the road and every traffic situation. However, when it comes to official consumption figures, its purpose is to build a common basis. With the ongoing vehicle development (e.g. in recent years the average power of a vehicle has been steadily increased), also the driving cycles have to be adopted.

## 2.1 NEFZ

The New European Driving Cycle (NEDC) was introduced in the year 1992 and is mandatory since 1997 for the determination of the fuel consumption of passenger cars in the EU. It consists of two different stages: The Urban Driving Cycle witch is repeated four times and the Extra Urban Driving Cycle (see **Figure 1**). The length of the cycle is around 20 min.

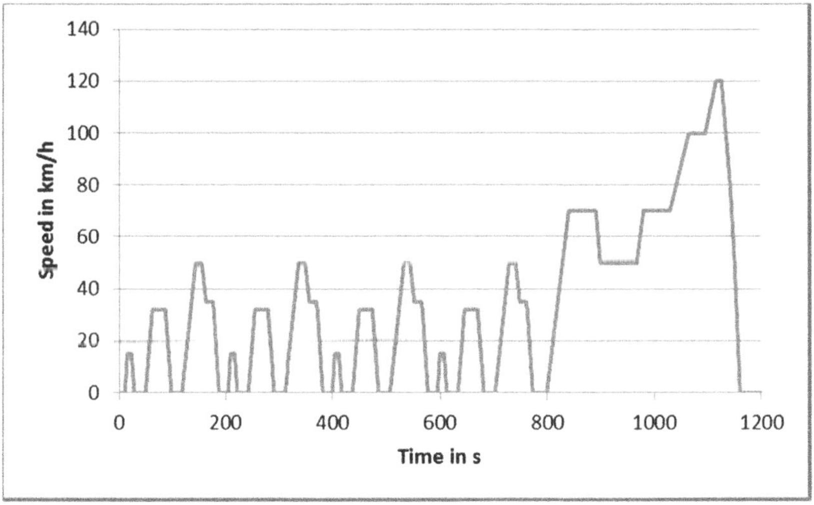


Figure 1: NEDC Driving Cycle [1]

The NEDC cycle nowadays is often reported to be no longer representative for the typical vehicle usage on European roads, which can for example be seen in **Figure 2**, where official NEDC values are compared with reported results from real world driving. One reason for this discrepancy is the low average speed during the cycle. The influence of aerodynamic drag for example, is proportional to the squared velocity and at an average speed of 33.3 km/h it has only minor effect. With roundabout 25% of idle-phases during NEDC, stop-start systems are highly effective in this cycle and are typically overrated compared to real world driving.

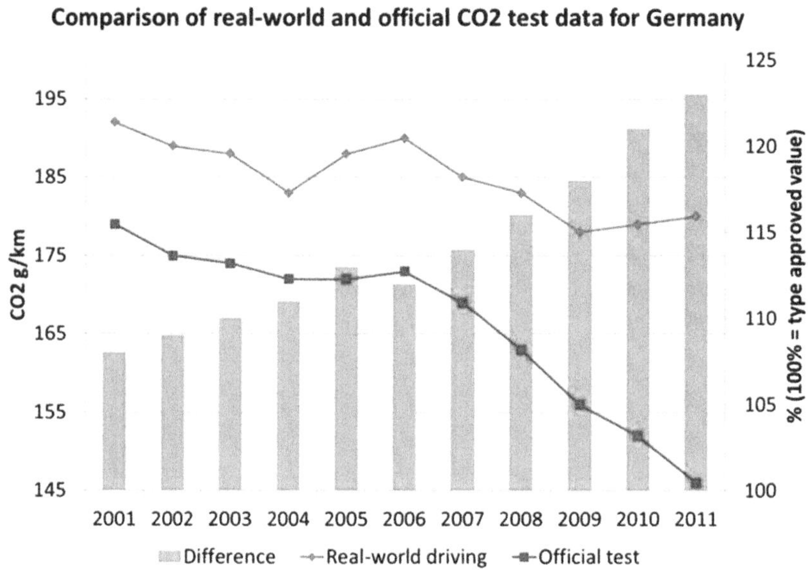


Figure 2: Comparison of NEDC results with real world driving [2]

Another reason for the discrepancy between real-world driving and the NEDC results can be found in the relatively tolerant boundary conditions of the cycle, allowing car manufactures to optimize their vehicle and testing conditions for minimal fuel consumption and minimal emissions. For example, most of the auxiliary loads can be switched off for the cycle and there is no need to recharge the battery during the cycle.

The determination of aerodynamic drag is carried out in a coastdown test, where for example all the gaps of the vehicle can be sealed to optimize the aerodynamics of the vehicle. Also, the aerodynamic drag is determined only for one vehicle configuration and optional equipment, such as larger tyres or roof rails is not being considered at all. Therefore, large aerodynamic effects are often neglected in the NEDC driving cycle.

## 2.2 WLTP

With the increasing gap between NEDC results and the real world driving emissions, the NEDC cycle is no longer able to represent the average passenger car usage in the EU. Therefore a working group by UNECE was founded to create a global standard for determining $CO_2$ emissions and fuel consumption. The result is the "Worldwide Harmonized Light Vehicles Test Procedure" WLTP, which is being developed at the moment. The test procedure covers a wide range of vehicles, put in different vehicle classes, including passenger cars (class 3b).

The cycle, defined for class 3b vehicles is plotted in **Figure 3** and consists of four parts:

- Cold start and low speed phase (589 sec)
- Medium speed phase (433 sec)
- High-speed phase (455 sec)
- Extra-high-speed phase (323 sec)

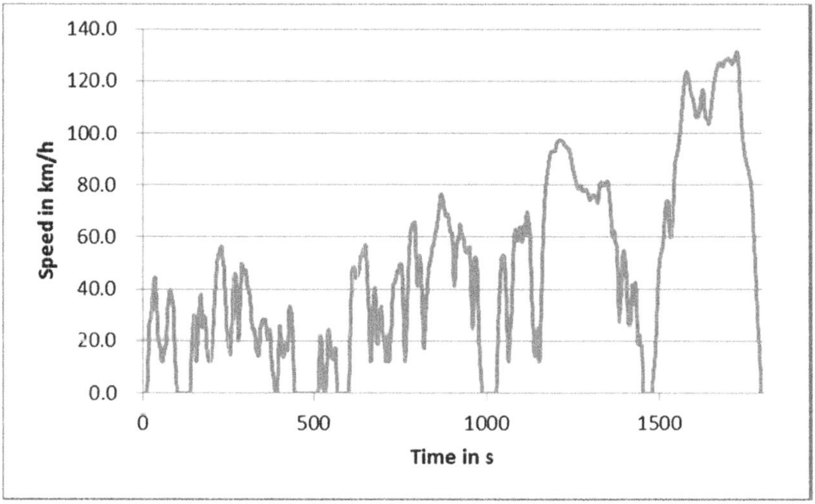


Figure 3: WLTP Class 3b Driving Cycle [3]

## 2.3 Comparison of NEFZ and WLTP

When comparing both cycles, it is obvious that the WLTP is much more dynamic with higher acceleration requirements and much shorter idle phases than the generic NEDC. Figure 4 shows the basic facts of both cycles compared to each other.

| | NEFZ | WLTP |
|---|---|---|
| Duration in s | 1181 | 1800 |
| Length in km | 10,92 | 23,27 |
| Average speed in km/h | 33,3 | 46,5 |
| Max. Speed in km/h | 120 | 131,3 |
| Idle in % | 24,9 | 13,1 |
| Max. Acceleration in m/s² | 1,04 | 1,67 |
| Max. Deceleration m/s² | -1,39 | -1,5 |

Figure 4: Comparison of WLTP and NEFZ [1, 3]

Not only is the maximum and average speed increased but also idle phases are reduced by 50%. These facts already result in an increased influence of aerodynamic drag in WLTP by roundabout 50% compared to NEDC [4].

Additionally the boundary conditions of WLTP are much more regulated, resulting in fewer possibilities to optimize the vehicle for the cycle run. It is therefore expected that WLTP will result in higher fuel consumptions and emissions than NEDC, especially when the vehicle uses techniques that are highly effective in the NEDC cycle.

When it comes to aerodynamics, the most important fact is that not only the basic vehicle is evaluated, but also the influence of optional equipment has to be considered. This can either be done by considering only the worst case, but this makes no sense since it would lead to a higher fleet consumption and could therefore result in penalty taxes. It is therefore important to consider the aerodynamic influence of each optional equipment available to the customer.

# 3 Influence of Tyres and Rims

Tyres and rims are responsible for around 25% of the aerodynamic drag of a vehicle [5]. Most of this influence is caused by the overall shape of the wheels, which can – for obvious reasons – not drastically be changed. But still, also the shape of both the rim and the tire that can be modified by the manufacturer results in quite a large influence to drag. For WLTP this influence is getting more and more into focus, because it will result in different $CO_2$ emissions and fuel consumptions for the specific vehicle.

## 3.1 Rim Influence

When buying a new car today, customers often have the possibility to select more than 10 different rim sizes and designs for their vehicle. In medium-class the range often varies from 15" wheels to 18" or 19". Additionally the width of the rims also changes. While the smaller 15" and 16" rims often have a width of only 6", the larger ones often are manufactured in widths of 7.5" or even 8".

Figure 5: Selection of different wheels (15"-19") for the VW Golf [6]

**Figure 5** shows part of the variety in rim sizes and designs for the VW Golf. It can be seen that not only the size varies, but also the design of the rims is quite different. While some wheels are nearly closed, which is typically good for aerodynamics, especially the larger ones have very thin spokes and they are often preferred by the customers.

With NEDC, this does not matter for the official fuel consumption of the car, because there is only one type approval necessary for each vehicle type. But with WLTP this is going to change. The fuel consumption has to be declared for each configuration chosen by the customer and this means, the aerodynamic influence has to be known for each set of wheels.

### 3.1.1 Aerodynamic Drag

When it comes to the aerodynamics of rims, it is typically beneficial to have a completely closed rim with a flat surface.

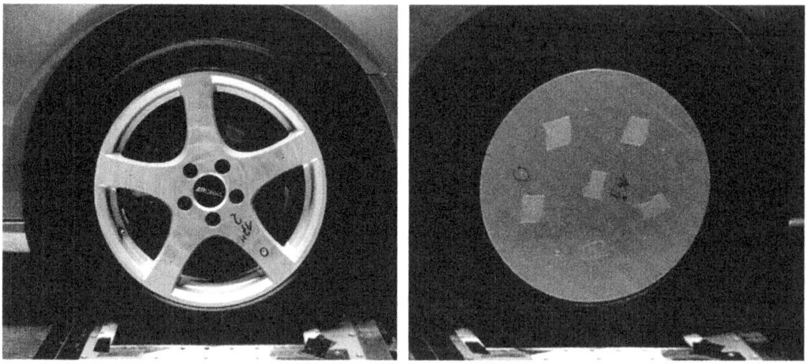

Figure 6: 17" open aluminium rim (left) and completely closed rim (right)

To investigate the magnitude of this influence, a 17" rim with an already smooth surface (see **Figure 6**) was compared to a completely covered rim in the IVK wind tunnel of University of Stuttgart. For this measurement, all 4 wheels of the vehicle were modified. The results show, that the aerodynamic drag of the vehicle is reduced by over 5% when covering the openings in the rim.

This influence gets even higher, when the aerodynamic characteristic of the baseline rim is worse than in the example above. Steel rims without wheel covers, for example, typically result in high aerodynamic drag. For a 16" steel rim, aerodynamic drag is typically reduced around 5% when a standard production wheel cover is attached. With a completely closed, flat wheel cover this influence can even be doubled (Figure 7)

Figure 7: 16" Steel wheel with different wheel covers [7]

This means, with an aerodynamically optimized wheel, the aerodynamic drag can be reduced or on the other hand, if the customer chooses a large wheel with worse aerodynamic characteristics, he will not only get higher fuel consumption on the road, but also higher fuel consumption at WLTP.

## 3.2 Tyre Influence

In difference to the rim, the possibilities in designing a tyre are much more restricted. Additionally, the design specifications for a tyre are mostly driven by topics as handling, rolling resistance, brake performance, noise etc. and the aerodynamic characteristics so far are more or less neglected. But still, the shape of a tyre has influence on the aerodynamic characteristics of a passenger car. This influence can be split in two areas. The first one is the design of the tyre, where already small geometric differences can result in effects to aerodynamic drag, the second one is the tyre width, which has a huge impact on aerodynamic drag.

When looking at the geometry of a tyre, there is especially one small part that will influence the aerodynamic characteristics of a tyre: The labelling on the side wall. Investigations have shown that only by modifying the small labelling on the sidewall of a tyre, the aerodynamic drag of a vehicle can be reduced by 1-2% [7]. In this case it is advantageous to have the labelling not raised from the sidewall, because this will typically result in a flow separation, but engraved into the sidewall. One example of a tyre with optimized aerodynamic characteristics is plotted in Figure 8 right. With this tyre, the aerodynamic drag of a wide variety of vehicles could be reduced compared to the standard production tyre in the left picture.

Figure 8: Standard production tyre (left) and tyre with optimized aerodynamic characteristics (right) [7]

The width of the tyre is the second important parameter influencing the aerodynamic characteristics of a vehicle. With a wider tyre, not only the frontal area is increased, but also the aerodynamic drag rises. This could be shown when comparing three different tyres on the same vehicle with sizes of 195/60 R16, 205/55 R16 and 225/50 R16. The results show an increase of 1% in aerodynamic drag between the 195 and the 205-tyre and an increase of 2% between the 205 and the 225-tyre.

Figure 9: Tyre shape for two tyres of the same type (205/55 R16) [7]

The same effect can occur even within the same tyre size, when the absolute dimensions of the tyres vary due to legal tolerances. In **Figure 9** two tyres are plotted, both with the size 205/55 R16. The measured dimensions of both tyres differ by 10 mm, resulting in an increased aerodynamic drag of around 2% with the wider tyre.

The effects of rims and tyre cannot be simply added, because both parts influence each other. If, for example, the shape of the tyre causes a large flow separation, this separation can shield the rim from the flow and reduce the aerodynamic impact of the rim design. On the other hand, the effect of the rim is typically much higher with an aerodynamically optimized tyre. So, in terms of drag reduction, both parts have to be optimized together to achieve the best overall result.

This shows that in combination rims and tyres can have quite some effect on the aerodynamic drag of a vehicle. Although the effects of rim and tyre cannot just be added together, the influence of the wheels can easily be more than 10% of the overall drag of a vehicle. With NEDC, this influence never showed up in the official data but with the new WLTP cycle, this will become an important part in the aerodynamic development, especially if all the available rim and tyre configurations have to be tested and evaluated.

# 4 Conclusions and Outlook

The aerodynamic influence of tyres and wheels can change the aerodynamic drag of a vehicle by more than 10%. So far, this influence was not considered in the official fuel consumption determination, because only a single type approval was necessary for each vehicle type and optional parts were not considered here.

With the transition from NEDC to WLTP not only the influence of aerodynamics within the cycle gains much more influence, but also optional equipment has to be considered. At this point the aerodynamic influence of the wheels becomes more and more important. Customers often choose to buy their car with larger wheels and wider tyres and both parts result in a higher drag value and therefore in a higher fuel consumption and more $CO_2$ emissions. This means, that in the vehicle development more work has to be spend not only to improve the drag of the baseline vehicle, but also to improve the aerodynamic characteristics of the optional equipment.

When it comes to rims and tyres, the large variety of options present the aerodynamicists with quite a challenge, because each combination of parts has to be evaluated concerning its aerodynamic properties on the vehicle.

# 5 Bibliography

[1] EC Regulation No 692/2008

[2] Jos Dings – Mind the Gap, Transport and Environment, 2013

[3] WLTP-09-04 GTR, Version 19.12.2014

[4] Lueginger, C.; Neuer Verbrauchszyklus WLTP, HDT Aerodynamik, 2014

[5] Wickern, G., Zwicker, K., & Pfadenhauer, M. Rotating Wheels – Their Impact on Wind Tunnel Test Techniques and on Vehicle Drag Results. *SAE Technical Paper 970133, 1997*

[6] VW vehicle configurator, www.volkswagen.de, 2015

[7] Wittmeier, F.: Ein Beitrag zur aerodynamischen Optimierung von PKW Reifen, Springer Verlag, ISBN: 9978-3-658-08806-4, 2015

# Towards a comprehensive approach for the sustainability assessment of a product: product social impact assessment

Marzia Traverso[1], Peter Tarne[1], Volkmar Wagner[1]

[1] BMW Group, Knorrstrasse 147, 80788 Munich, marzia.traverso@bmw.de

# Abstract

The importance of sustainability assessment along the product life cycle is getting more significant in companies and the scientific community. The BMW Group uses life cycle assessment (LCA) as tool for improving the environmental performance of its products. The definition of a sustainability assessment of a product life cycle says that three dimensions should be evaluated: environmental, economic and social.

The success of the BMW Group has always been built on long-term thinking and responsible action. The company has therefore established environmental and social sustainability throughout the value chain and along the entire product life cycle with a clear commitment to conserving resources, emissions reduction and, a comprehensive product responsibility.

Because there is still no standard available to assess social performance along the product life cycle, the BMW Group together with 11 other companies and Pré Sustainability started the Roundtable for Product Social Metrics in 2013. The main goal of the project is to develop a feasible and practicable methodology for assessing positive and negative social impacts of a product. A handbook to support LCA practitioners in assessing social performance quantitatively and qualitatively has already been published.

It is the first time that a group of companies has worked together towards a harmonized approach to assess social performance of a product life cycle using bottom-up approach.

Case studies for implementing this method were carried out by partners of the project to prove its feasibility and to identify its benefits and challenges. An example of a case study was the assessment of the social performance of a Run On Flat tire as mounted onto a BMW vehicle.

Benefits and challenges of the methodology as well as the case study were identified to move a step forward towards a standard for the social assessment of a product.

# 1 Introduction

The interest of governments and companies in sustainable consumption and production has been continuously increasing. It is shown by the numerous activities in several parts of the world, such as Rio+20, European Sustainable Production and Consumption Plan, Product Sustainability Roundtable, Sustainability Consortium, sustainability reporting and so on.

The main goal of most of these activities is to identify:

- Scientific and technical solutions to implement sustainable production;
- Methods and tools for **assessing sustainability performances** of companies and products to support sustainable consumption; and
- Tools and labels to support consumers towards a sustainable choice.

There are different methods to assess sustainability performance depending if the focus is at corporate/company, product, community, or consumer level.

At company level, we can find corporate social responsibility, sustainability reports as well as rating and ranking models such as the Dow Jones Sustainability Index (Windolph, S.E:, 2011).

A sustainability report can be developed according to several guidelines such as the Global Reporting Initiative (GRI) and the ISO norm 26000 (ISO 26000, 2010)

The Global Reporting initiative is a leading organization that promotes the use of sustainability reporting as main tool to measure and monitor the company activities and gear them towards more sustainable ones. This initiative developed guidelines to support companies in the development of sustainability reporting. These guidelines present a set of key indicators that measures the three pillars of sustainability: economical, social and environmental. G4 is the latest version of the GRI´s Sustainability Reporting Guidelines (GRI, 2000-2006).

Another approach to support sustainable production and consumption is a sustainability assessment at product level. When assessing the sustainability performance of products, it has to be ensured that the effects on all three dimensions of sustainability over the product's life cycle have to be captured and evaluated. A lot of literature has been written to assess the environmental performance of a product along its life cycle and the methodology, the life cycle assessment (LCA), was standardized with the ISO 14040 in 2006 (ISO 14040:2006, ISO 14044:2006). More efforts have to be made to assess the social performance as well the comprehensive assessment of sustainability along the product life cycle.

The BMW Group is considered a sustainability leader in the automotive industry. Its successful strategy is testified by the Dow Jones Sustainability Index which ranked it as the most sustainable company in the automotive sector again in 2014. As reported in the Sustainability Value Report 2013 (BMW Group, 2013), the BMW Group assesses its sustainability performance by considering the three pillars of sustainability. Its strategy involves the company itself with all plants as well as its products. Sustainability targets are identified and monitored in the BMW Group plants each year. At product level the BMW Group defines targets in the earliest phase of the product development and monitors them until the start of production by using LCA. The main objective is the reduction of the environmental impacts, measured as Carbon Footprint, of each new model compared to previous version. The LCA of a vehicle at the BMW Group is carried out according to the ISO 14040/44 and certified by TÜV SÜD as Third Party (BMW Group, 2014).

## 1.1 Product Sustainability Assessment

Life cycle sustainability assessment (LCSA) has been defined as a valid tool to identify, assess and measure environmental, economic and social impacts that a product can potentially have in both positive and negative terms along its life cycle. Even if not so many implementations (Traverso et al., 2012a, Traverso et al., 2012b) can be found and still several aspects have not found the agreement of the scientific community, this methodology is commonly accepted as the theoretical reference to assess sustainability performance of a product along its life cycle (UNEP/SETAC, 2011, Valdivia et al, 2012).

A first publication to guide an LCA practitioner in LCSA was published by the UNEP/SETAC Life Cycle Initiative (LCI) (UNEP/SETAC, 2011) at the beginning of 2012.

According to the definition a complete LCSA of a product is developed by contemporary implementation of LCA, life cycle costing (LCC) and social LCA (SLCA). But because they have a different level of standardization their parallel and contemporary carrying out assessing the same product still presents numerous challenges. The definition of LCSA is presented with the formal equation (Finkbeiner et al. 2010; Finkbeiner et al. 2008; Klöpffer 2008):

$$\text{LCSA} = \text{LCA} + \text{LCC} + \text{SLCA} \qquad [1]$$

Where LCA is life cycle assessment, standardized by ISO 14040-44 (ISO 14040 2006, ISO 14044 2006), LCC is life cycle costing which assesses the economic impacts, costs and revenues, along the product life cycle (Swarr et al. 2011, Hunkeler et al. 2008, Rebitzer et al. 2003), and social life cycle assessment (SLCA) which focuses on positive and negative social impacts on different stakeholder groups from the product (Benoit et al. 2011, UNEP/SETAC 2009).

## 1.2 Sustainable Supply Chain Management

The implementation of LCSA to a product life cycle is getting more difficult when the product assessed presents a complex and long supply chain and/or it consists of a large amount of components. This is the case with the assessment of a car, where a lot of components have to be considered and for most of them a long supply chain has to be included. That means the sustainability performance of each manufacturing step of each specific component has to be assessed. A lot of experience and have been collected in the last ten years for the assessment of the environmental impacts of a vehicle, but a lot of efforts have to be made for the implementation of the other two dimensions and the comprehensive LCSA method. However, the integration of supply chain management (SCM) and sustainability management is gaining in importance as pressure on companies to address these issues is also increasing. The pressure on companies is driven by different factors like consumer demand, stakeholder interest, and regulations (Paquette 2005; Srivastava 2007). To address these upcoming pressures, sustainable supply chain management emerged which integrates sustainability goals with supply chain management goals (Seuring & Müller 2008).

There exists a plethora of concepts to introduce sustainability to the supply chain, inter alia green design, reverse logistics, and cooperation (Srivastava 2007; Fortes 2009). A simple and accessible option of a company to influence its supply chain is via the business relations to its suppliers and customers. When faced with the decision to select a supplier, a company can resort to different strategies to increase the sustainability of the purchased goods. They can range from rather straightforward requirements regarding the product content to more complex measures such as trainings and audits of suppliers (Halldórsson et al. 2009; Hamner 2006). Furthermore, communication and generation of transparency along the supply chain in order to gain insight into the actual supply situation and to identify key improvement potentials are integral parts to sustainable SCM (Solér et al. 2009; Wittstruck & Teuteberg 2012; Wolf 2011).

## 1.3 Social Dimension of Product Sustainability

Currently, the social dimension of sustainability is mainly addressed at company level (Parent et al. 2012; Lehmann et al. 2013). The purpose of product related social assessment is the identification and assessment of positive and negative impacts of a product on stakeholders involved with the product's life cycle. When comparing it to environmental assessment, the practitioner of social product assessment is faced with additional challenges. For instance, the causal link between production processes and social impacts on stakeholders is often non-existent or not exhaustively investigated (Dreyer et al. 2006; Benoît & Mazijn 2009; Hutchins & Sutherland 2008). Instead of production processes, the determining factor of social impacts is the conduct of companies as they can more clearly enhance or impede the well-being of stakeholders (Dreyer et al. 2006; Benoît & Mazijn 2009). Figure 1 shows the difference between the modes of interaction for environmental and social impacts.

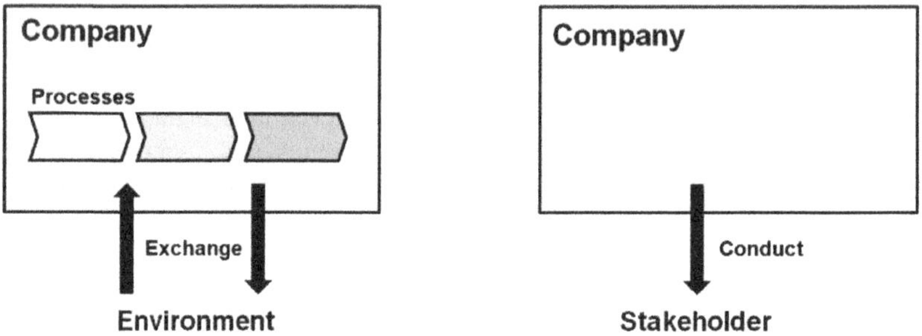


Figure 1: Environmental impacts (left) are determined by the physical input and output flows of processes. Social impacts (right) are determined by the conduct of a company towards a stakeholder, e.g. workers. (adapted from Dreyer et al. 2006)

This fact, however, makes it more difficult to determine the social impacts of a product as the effects of a company's conduct can only be determined at company level. There still exists no consensus on how to relate company specific impacts to the product itself (Dreyer et al. 2006). As the field of research on social assessment is still very young, there is little agreement on the methodology and on which indicators to use. Furthermore, there are not many case studies that subject the existing approaches to a practical test (Lehmann et al. 2013). However, The UNEP SETAC Life Cycle Initiative published a handbook on the assessment of social impacts of products, the "Guidelines for Social Life Cycle Assessment of Products", that is aiming at securing compatibility with the environmental LCA (Benoît & Mazijn 2009). The supporting Methodological Sheets were published in 2013 by UNEP/SETAC LCI to complete the guidelines by suggesting a certain number of indicators for each stakeholder group (Benoít Norris et al. 2013).

## 1.4 Goal of the Paper

The main goal of this paper is the presentation of a new initiative supported by the BMW Group for the development of a harmonized methodology for the assessment of the social impacts along a product's life cycle.

# 2 Roundtable for Product Social Metrics

The Roundtable for Product Social Metrics is a working group of experts from large companies that decided to join forces to tackle the practical dilemma that workable solutions for the social assessment at product level remains a complex issue and has not yet been addressed.

The working group started in 2013 and has structured its intended work load into three phases. This initiative represents the first time that a group of eleven leader companies sat together to develop a bottom-up approach for a feasible and practicable method to assess the social dimension of sustainability at product level.

The BMW Group joined this initiative as founder member in the beginning of 2013.

At the end of the first phase the first draft of the Handbook of the Product Social Impact Assessment was presented and a consultation process was realized among a group of experts from external organizations to revise the Handbook. This version of the Handbook assessed the social performance of a product relative to three main stakeholder groups Workers, Consumers, and Local Community. These stakeholder groups were selected from the main five stakeholder groups identified by the UNEP/SETAC LCI guidelines. Around 20 topics and 80 indicators were identified by using the main literature on social product assessment (UNEP/SETAC, 2013, GRI4, Benoît-Norris et al. 2012, UNEP/SETAC 2009).

In the second phase, the Handbook was finalized and published in September 2014. The comments of the consultation group were incorporated and 6 case studies have been implemented for testing and verifying the feasibility and validity of the methodology and its indicators. The case study of the Run On Flat tire as mounted onto a BMW vehicle was carried out by BASF, the BMW Group and Goodyear.

The third phase of the Roundtable is still running this year and further case studies are currently developed. The main target of the third phase is to face several challenges in the implementation that were identified at the end of phase 2 and could not be addressed due to the available time budget. Challenges are, for instance, the realization of a more supportive communication tool and a detailed description of the implementation steps of the methodology. These are going to be published on the website of the project to support the users of the Handbook step by step in their first implementation of a Product Social Impact Assessment.

As we presented in this paper, a huge amount of literature has been published in the last ten years on the sustainability assessment of product life cycles, and particular attention was paid to better define the SLCA methodology. No agreement has yet been found regarding the definition of a common and scientifically valid set of indicators to measure the social impact of a product. Even if suggestion of possible indicators are reported by Dreyer et al. (2006), Jørgensen et al. (2008), and by the Methodological Sheets (UNEP/SETAC, 2013), no final proposal of an indicator set has been produced. This was the main reason that brought together 11 companies led by PRé Sustainability consulting to develop a handbook for supporting user of SLCA. They did not want to reinvent the wheel which is why they started by consulting the literature on this topic and by comparing the vast array of indicators published in the scientific literature with the priorities of each company involved and with the current data availability. As result of the first two phases a set of 30 quantitative indicators and 41 qualitative was proposed for the assessment.

# 3 Product Social Impact Assessment to a Run On Flat

The Handbook developed by the Roundtable presents two approaches for the assessment of social impacts of products: a quantitative approach and a scales-based, i.e. semi-quantitative, approach. When quantitative data is at hand an allocation of company level data to product level data can be performed. If only qualitative data is at hand, no product specific information can be retrieved. In the following, the quantitative approach shall be outlined a bit further in order to exemplify the assessment procedure:

Figure 2: Quantitative and Qualitative Approach

The first step of both methodologies is the definition of the goal and scope of the assessment. We need to define the functional unit of the product and the system boundaries that determine what is going to be considered in the assessment. In the Run On Flat case study we identified a tire as functional unit and as system boundary we considered the manufacturing of the main four raw materials (data collected by BASF and Goodyear), the manufacturing of the tire (Goodyear), the assembly of car and tire at the BMW Group, and the use phase (BMW Group). We have not considered the entire life cycle of the product because no data was available for all steps, in particular for the end of life. For this pilot project we decided to focus on the quantitative assessment to verify its applicability and feasibility. The quantitative method presents more challenge because we need to collect quantitative data at product and/or site specific level.

For each of these steps we succeeded to collect primary data. We could not assess the entire life cycle because some steps were not under the direct control by members of the Roundtable, and as the methodology was applied for the first time it was not easy to involve external partners.

The indicators presented in the Handbook were classified according to the three main stakeholder groups: Workers, Consumers, and Local Community. A set of indicators has been proposed in the Handbook, and for each of them a referencing process in order to evaluate the distance-to-target to a pre-defined reference point.

As referencing points two main scenarios have been considered:

- An ideal "ethical" scenario, defined by an ideal performance of the indicators, e.g. 0 average rate of incidents during the reporting period, 100% workers who are paid a living wage; and
- A minimum scenario for all those indicators for which an ideal scenario was not definable, e.g. numbers of health and safety trainings per worker given during the reporting period.

For the life cycle inventory was necessary to collect data at product level. However, even if we have site specific data but more than one product is produced at the respective plant, we need to allocate the site specific value to the different products of that site in order to find out what impact the product of interest is responsible for. This is carried out in the allocation step by using the amount of working time needed to manufacture the specific product in relation to the entire volume of working hours at a company or site. Data are either quantitative (number or ratio) or qualitative. In our pilot, only quantitative data were collected and as it was related to the site, the allocation step was necessary. Once we collected all data and obtained the values per product, it was possible to build a sum in order to retrieve values for the entire life cycle comprised of the respective impacts at every life cycle step (or supplier). The next step was the referencing step. For each indicator the obtained value at product level was compared to the pre-fixed reference value in order to determine the respective distance to target. It allowed us to understand if the product had a good or bad performance compared to the reference scenario.

Because the pilots in the second phase were carried out among all members of the Roundtable, it was not necessary to verify the data. However, when conducting a general case study, qualitative and quantitative data should be verified by means of certificates or other documents that can prove their validity. The requirements for this certification or documentation necessary for the validation of the inventory data are going to be defined in the third phase. Thanks to the pilot project of the Run On Flat, it was possible to define the indicator *Experienced Well-being* for the consumers. It was defined with a semi-quantitative indicator which expresses the consumer's well-being, in this case the experience of the driver associated with the car but attributed to the tires. This specific indicator cannot be transferred to other products. For instance, in another case study evaluating the oil pan of a BMW vehicle, the specific product did not affect the driver and in general the use phase of the car and therefore the stakeholder category Consumer was not considered.

The final values obtained at indicator level and after the referencing phase were able to support a decision making process by identifying which the opportunities and the main impacts were, generated by the product, for which stakeholder group, and in which steps of the product life cycle they occur. This was considered the first step to a constructive and cooperative process in which a company and its suppliers can work together to improve the social conditions of the stakeholder groups that are mainly affected by their product.

# 4 Conclusion and Discussion

The methodology presented allows identifying the most relevant social impacts and in which phase of the product life cycle they occur. The transparent data interpretation throughout the referencing phase provides direction for decision-making and can support development processes as well as demonstrating future product improvements. Furthermore, it was found that already the data inventory process could facilitate discussions with supply chain partners and represents a tool for the engagement on important social topics within the company across functional groups. Due to consideration of both positive and negative impacts, the methodology is a valid tool to proposed action and constructive discussion throughout the supply chain. Because both qualitative and quantitative approach can be used, the obtained results can facilitate the communication on the social benefits of products to both expert and non expert of sustainability issues.

The pilot showed that the quantitative approach can generate information related to a product's contribution to society (e.g. how many safety training hours are associated with the production and assembly of a tire throughout its life cycle) (Fontes et al, 2014). These data points are not normally available, but support the understanding of the overall importance of a product and provide a full picture of improvement opportunities beyond environmental and economic aspects in order to generate more sustainable solutions (Fontes et al, 2014).

Data availability remains one of the biggest challenges in this topic, but we are confident that this experience will lead the main software developers and database providers for LCA to move a step forward and collect data on social impacts of a product along its life cycle. Who has been pioneer of LCA knows that the same challenge was faced for LCA twenty years ago.

# 5 References

Benoít Norris, C., Norris, G.A. & Aulisio Cavan, D., 2013. Social Hotspots Database – Supporting Documentation. , p.79.

Benoît-Norris C., Vickery-Niederman G., Valdivia S., Franze J., Traverso M., Ciroth A. and Mazijn B. (2011). Introducing the UNEP/SETAC methodological sheets for subcategories of social LCA. The International Journal of Life Cycle Assessment, 2011-05-31, Springer Berlin / Heidelberg, ISSN: 0948-3349, pag. 1- 9

Benoît, C. & Mazijn, B., 2009. *Guidelines for social life cycle assessment of products*, UNEP/Earthprint.

BMW Group, 2014. Environmental Certification BMW i3. http://www.bmwgroup.com/com/en/_common/_pdf/Enviromental_Certification_i3.pdf. Accessed 22 February 2015

Dreyer, L., Hauschild, M. & Schierbeck, J., 2006. A Framework for Social Life Cycle Impact Assessment (10 pp). *The International Journal of Life Cycle Assessment*, 11(2), pp.88–97. Available at: http://dx.doi.org/10.1065/lca2005.08.223 [Accessed October 30, 2013].

Environmental management — Life Cycle Assessment — Principles and framework (ISO 14040); ISO: Genova, Switzerland, 2006.

Environmental management — Life Cycle Assessment — Requirements and Guidelines (ISO 14044); ISO: Genova, Switzerland, 2006.

Finkbeiner, M. et al., 2010. Towards Life Cycle Sustainability Assessment. *Sustainability*, 2(10), pp.3309–3322. Available at: http://www.mdpi.com/2071-1050/2/10/3309 [Accessed November 10, 2013].

Finkbeiner, M., Reimann, K. & Ackermann, R., 2008. Life cycle sustainability assessment (LCSA) for products and processes.

Fontes J., Bolhuis Andrea Karin Bogaers, Peter Saling, Richard van Gelder, Marzia Traverso, Jacobine Das Gupta, Henk Bosch, Dave Morris, Dave Woodyard, Lynn Bell, René van der Merwe, Markus Laubscher, Marcel Jacobs, D.C., 2014. Handbook for Product Social Impact Assessment. Available at: http://product-social-impact-assessment.com/wp-content/uploads/2014/08/Handbook-for-Product-Social-Impact-Assessment.pdf.

Fortes, J., 2009. Green Supply Chain Management: A Literature Review. *Otago Management Graduate Review*, 7, pp.51–62.

Global Reporting Initiative (2000-2006). Sustainability Reporting Guidelines version 4.0.

Halldórsson, Á., Kotzab, H. & Skjøtt-Larsen, T., 2009. Supply chain management on the crossroad to sustainability: a blessing or a curse? *Logistics Research*, 1(2), pp.83–94. Available at: http://dx.doi.org/10.1007/s12159-009-0012-y [Accessed October 23, 2013].

Hamner, B., 2006. Effects of Green Purchasing Strategies on Supplier Behaviour. In J. Sarkis, ed. *Greening the Supply Chain SE – 2*. Springer London, pp. 25–37. Available at: http://dx.doi.org/10.1007/1-84628-299-3_2.

Hunkeler D, Lichtenvort K, Rebitzer G (Hg): Andreas Ciroth, David Hunkeler, GjaltHuppes, Kerstin Lichtenvort, Gerald Rebitzer, Ina Rüdenauer, BengtStehen (Lead authors) (2008), Environmental Life Cycle Costing. SETAC Publications, 2008.

Hutchins, M.J. & Sutherland, J.W., 2008. An exploration of measures of social sustainability and their application to supply chain decisions. *Journal of Cleaner Production*, 16(15), pp.1688–1698. Available at: http://www.sciencedirect.com/science/article/pii/S0959652608001455.

ISO 26000:2010 – Guidance on social responsibility, International Standard Organization

Jørgensen, A. et al., 2008. Methodologies for Social Life Cycle Assessment. *The International Journal of Life Cycle Assessment*, 13(2), pp.96–103.

Kloepffer W 2009. Life Cycle Sustainability Assessment of Products. Int. J. Life Cycle Assess. 13 (2), pp 89-95.

Klöpffer, W., 2008. Life cycle sustainability assessment of products. *The International Journal of Life Cycle Assessment*, 13(2), pp.89–95. Available at: http://dx.doi.org/10.1065/lca2008.02.376 [Accessed November 10, 2013].

Lehmann, A. et al., 2013. Social aspects for sustainability assessment of technologies—challenges for social life cycle assessment (SLCA). *The International Journal of Life Cycle Assessment*, 18(8), pp.1581–1592. Available at: http://dx.doi.org/10.1007/s11367-013-0594-0 [Accessed November 2, 2013].

Paquette, J., 2005. *The Supply Chain Response to Environmental Pressures*. Massachusetts Institute of Technology. Available at: http://hdl.handle.net/1721.1/34530.

Parent, J., Cucuzzella, C. & Revéret, J.-P., 2012. Revisiting the role of LCA and SLCA in the transition towards sustainable production and consumption. *The International Journal of Life Cycle Assessment*, 18(9), pp.1642–1652. Available at: http://dx.doi.org/10.1007/s11367-012-0485-9 [Accessed November 2, 2013].

Rebitzer G, Hunkeler D (2003) Life Cycle Costing in LCM: ambitions, opportunities, and limitations, discussing a framework. Int. J. Life Cycle Assess. 8 (5), 253-256.

Seuring, S. & Müller, M., 2008. Core Issues in Sustainable Supply Chain Management – a Delphi Study. *Business Strategy and the Environment*, 17, pp.455–466.

Solér, C., Bergström, K. & Shanahan, H., 2009. Green supply chains and the missing link between environmental information and practice. *Business Strategy and the Environment*, 19(1), pp.14–25. Available at: http://dx.doi.org/10.1002/bse.655 [Accessed October 29, 2013].

Srivastava, S.K., 2007. Green supply-chain management: A state-of-the-art literature review. *International Journal of Management Reviews*, 9(1), pp.53–80. Available at: http://dx.doi.org/10.1111/j.1468-2370.2007.00202.x [Accessed October 18, 2013].

Swarr TE, Hunkeler D, Kloepffer W, Pesonen HL, Ciroth A, Brent AC, Pagan R 2011 Environmental Life Cycle Costing: A Code of Practice. SETAC book 978-1-880611-87-6, 98 pp, February 1.

Traverso, M., Finkbeiner, M., Jørgensen, A. and Schneider, L. 2012a. Life Cycle Sustainability Dashboard. Journal of Industrial Ecology, 16: 680–688. doi: 10.1111/j.1530-.2012.00497.x

Traverso,M., Asdrubali,F., Francia, A., Finkbeiner M., 2012b. Towards life cycle sustainability assessment: an implementation to photovoltaic modules. International Journal of Life Cycle Assessment, DOI 10.1007/s11367-012-0433-8.

UNEP/SETAC, 2011. Towards a Life C ycle S ustainability A ssessment: Making informed choices on products J. H. (Chalmers U. Sonia Valdivia (UNEP), Cássia Maria Lie Ugaya (Technological Federal University of Parana and ACV Brasil), Guido Sonnemann (UNEP) & Authors, eds., ISBN 9789280731750

UNEP/SETAC 2009. Guidelines for Social Life Cycle Assessment of Products. United Nations Environment Programme ISBN: 978-92-807-3021-0.

Valdivia, S. et al., 2012. A UNEP/SETAC approach towards a life cycle sustainability assessment—our contribution to Rio+20. The International Journal of Life Cycle Assessment, pp.1–13. Available at: http://dx.doi.org/10.1007/s11367-012-0529-1.

Wittstruck, D. & Teuteberg, F., 2012. Understanding the Success Factors of Sustainable Supply Chain Management: Empirical Evidence from the Electrics and Electronics Industry. *Corporate Social Responsibility and Environmental Management*, 19(3), pp.141–158. Available at: http://dx.doi.org/10.1002/csr.261 [Accessed October 29, 2013].

Wolf, J., 2011. Sustainable Supply Chain Management Integration: A Qualitative Analysis of the German Manufacturing Industry. *Journal of Business Ethics*, 102(2), pp.221–235. Available at: http://dx.doi.org/10.1007/s10551-011-0806-0 [Accessed October 22, 2013].

Windolph, S.E:, 2011. Assessing Corporate Sustainability Through Ratings: Challenges and Their Causes in Journal of Environmental Sustainability. Volume 1 – 2011.

# A vehicle lateral control approach for collision avoidance by emergency steering maneuvers

Dipl.-Ing. Martin Keller, research assistant,
Institute of Control Theory and Systems Engineering, TU Dortmund,
Otto-Hahn-Straße 8, 44227 Dortmund,
Martin.Keller@TU-Dortmund.de

Dr.-Ing. Carsten Haß, TRW Automotive GmbH, Engineering Manager,
Integrated Active + Passive Safety Technologies,
Hansaallee 190, 40547 Düsseldorf, Carsten.Hass@TRW.com

Dr.-Ing. Alois Seewald, TRW Automotive GmbH,
Global Director Advanced Steering
and Integrated Active + Passive Safety Technologies,
Hansaallee 190, 40547 Düsseldorf, Alois.Seewald@TRW.com

Univ.-Prof. Dr.-Ing. Prof. h.c. Dr. h.c. Torsten Bertram,
Head of Institute, Institute of Control Theory
and Systems Engineering, TU Dortmund,
Otto-Hahn-Straße 8, 44227 Dortmund,
Torsten.Bertram@TU-Dortmund.de

# 1. Introduction

Many accidents lead back to the driver. It is assumed to be beneficial that driver assistance systems and/or automated driving functions ease tension off the driver or release him from the task of driving in order to reduce traffic fatalities. The development of driver assistance systems started with ABS (Anti-Lock Braking System) in 1978, TC (Traction Control) in 1987 and ESC (Electronic Stability Control) in 1995. These systems are known for their contribution on the reduction of traffic fatalities over the last years. But still far too many cars crash every day. The aforementioned systems do not take environmental information into account, which limits the collision avoidance potential. A system that senses its environment by radar and/or camera sensors is the AEB (Autonomous Emergency Braking) which is supposed to further reduce the amount of accidents in the future as the take rate increases. Although assisting the driver in emergency situations by braking is beneficial in some situations, it also has its drawbacks. The problem is that emergency braking can only be initiated if the obstacle is detected for sure. Nowadays sensors can only provide limited safe ranges. This means that at typical highway speeds the collision with a stationary obstacle can only be mitigated by braking. A swerving maneuver is a viable alternative to braking or the combination of both if the required space is not occupied by other obstacles. The emergency steering maneuver offers the advantage that the last-point-to-steer is even closer to the obstacle than the last-point-to-brake with growing vehicle speed, which yields a higher collision avoidance potential. The disadvantage is that the maneuver is far more difficult than full braking. A subject study [1] in a driving simulator has proven that the average skilled driver is not capable of steering the vehicle properly around obstacles in the majority of occasions. This is why an advanced driver assistance system (ADAS) is useful that supports the driver by steering torque overlay and, as an option, by braking interventions. The improvement on the collision avoidance behaviour of the driver has been proven by the authors in the aforementioned study [1].

Assistance systems for emergency situations have been subject to research and in early development projects before. In [2], [3] and [4] a sigmoidal function is used for the collision avoidance path and an active front steering system for lateral vehicle guidance. In [5] a trapezoidal acceleration profile is used to obtain an evasive path. In addition to steering torque overlay differential braking is used for collision avoidance in [6]. The previous mentioned publications describe rather simple path planning algorithms. The advantage of these algorithms is their low computational burden. As a result real time capability is not an issue. Nevertheless more complex and adaptable trajectory planning algorithms have been developed, see for example [7], [8] or [9]. More recently the two tasks trajectory planning and vehicle control have been united by model predictive approaches in [10], [11], [12] and [13]. The remainder of this paper is organized as follows. Section two describes the used development environments, sensors and actuators.

Section three gives an overview of the systems architecture. The various parts are then explained in detail in the following sections, which is path planning in section four, lateral vehicle control in section five and longitudinal control in section six. Finally the results obtained from test drives in a prototype test vehicle and a fixed base driving simulator are presented and discussed. The paper ends with the conclusion.

## 2. Development Environment

In order to increase efficiency during the development, a scheme of three steps is used. At first a multibody vehicle dynamics simulation is used to develop basic concepts and structures which are suitable for the task of collision avoidance. In the second step a fixed base driving simulator is utilized to further develop, tune and evaluate the control system. In figure 1 a picture of the driving simulator is shown. The simulator consists

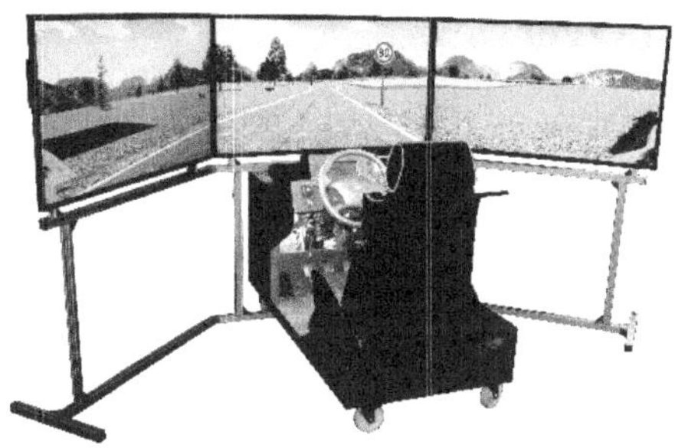

Figure 1: Picture of the fixed base driving simulator.

Figure 2: Picture of the prototype test vehicle during a swerving maneuver.

of a vehicle dynamics simulation, an environment and traffic simulation, a mockup (including the pedals and the actuated steering wheel) and a rapid prototyping system similar to the one used in the prototype test vehicle. The different parts communicate via CAN BUS as in the vehicle. Testing in the driving simulator does not constitute a risk to drivers or the vehicle, which is very important because the emergency maneuver is highly dynamic and involves some degree of risk even on closed test tracks. To sum up the driving simulator allows for pretesting using the same tools as in the test vehicle, which is shown in figure 2. This saves costs and time. Finally the pretested function is tested in the prototype test vehicle, which is an Opel Insignia re-equipped with TRW's EPS-BD (Electrically Powered Steering System - Belt Drive) and TRW's EBC450 slip control brake system. Also TRW's forward looking sensors are mounted to the vehicle. In this case a camera sensor and a radar are used. The developed functionality can be adapted to any other passenger car, which is equipped with similar sensors and actuators.

## 3. System Overview

The Emergency Steering Assist (ESA) consists basically of two parts. The first one is path planning, which generates a collision free and smooth path. The second part is the vehicle control system. This part shall guide the vehicle along the predefined path and also communicate the systems intervention to the driver. In figure 3 the block diagram of the emergency steering assist is shown. The controller output is the steering torque

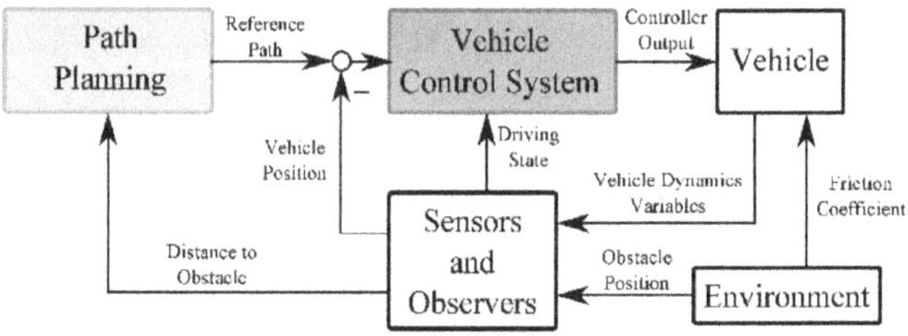


Figure 3: Block diagram of the emergency steering assist.

overlay and (optional) the reference deceleration. The steering torque overlay is applied by an EPS, which can be a belt drive or a column drive. The deceleration is provided by the brake system. As mentioned before, the used sensors need to detect obstacles in front of the car and provide information about the vehicle driving state. Apart from camera and radar typical vehicle dynamics sensors used for the electronic stability control (ESC) are required. Because the side slip angle is usually not available on the CAN BUS, an observer is utilized. The observer consists of a nonlinear single track model and an extended Kalman filter (EKF). Based on the model the vehicles position is estimated. In figure 4 the single track model is shown from the top view. The state space model consist of the form

$$\dot{\boldsymbol{x}} = \boldsymbol{f}(\boldsymbol{x}, u) \tag{1}$$

with $\boldsymbol{x} = (\beta \quad \dot{\Psi} \quad \Psi \quad X \quad Y)^T$, $u = \delta$ and $\delta = \delta_f i_s$. Herein $i_S$ is the steering ratio.

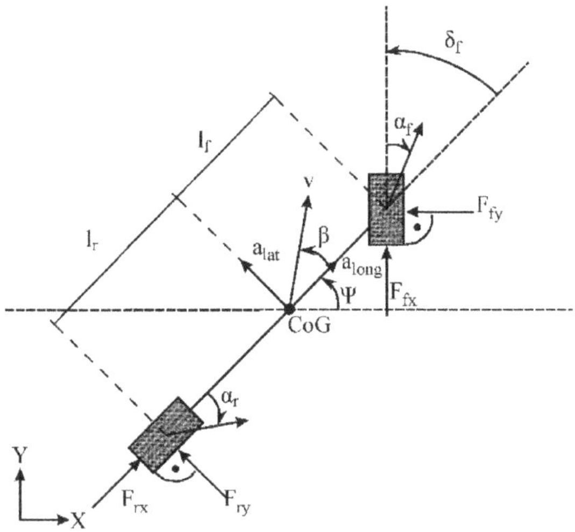


Figure 4: Top view of the single track model.

## 4. Path Planning

In order to avoid the obstacle, a fifth order polynomial is calculated by utilizing the boundary conditions given by the traffic situation which is depicted schematically in figure 5.

The path starts with a straight segment which corresponds to the look-ahead time $t_{la}$ of the control system. Although this seems disadvantageous at first sight, it actually enables the controller to accurately guide the vehicle along the path. After that, the polynomial, which is beneficially shaped, continues the path. The fifth order polynomial is given by

$$y_{pol}(x) = a_5 x^5 + a_4 x^4 + a_3 x^3 + a_2 x^2 + a_1 x + a_0. \tag{2}$$

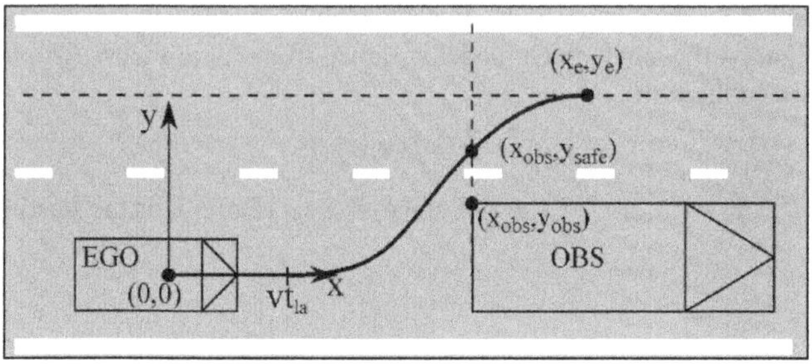


Figure 5: The considered traffic situation.

Initially the path is calculated to the point $(x_{obs}, y_e)$, followed by iteratively increasing the coordinate $x_e$. It is checked that the path proceeds to the next lane in a safe distance to the obstacle at the point $(x_{obs}, y_{safe})$. The advantage of this rather simple path planning approach is its real time capability. In addition the path is smooth and feasible, because it is two times continuously differentiable at the transitions between the straight segments and the polynomial.

# 5. Lateral Vehicle Control

In the following subsections the lateral vehicle control system is explained, starting with the general structure and detailing the different parts subsequently.

## 5.1 General Structure

In figure 6 the general structure of the lateral controller is outlined. It consists of five parts. The first controller is the steering wheel angle (SWA) controller. It calculates the

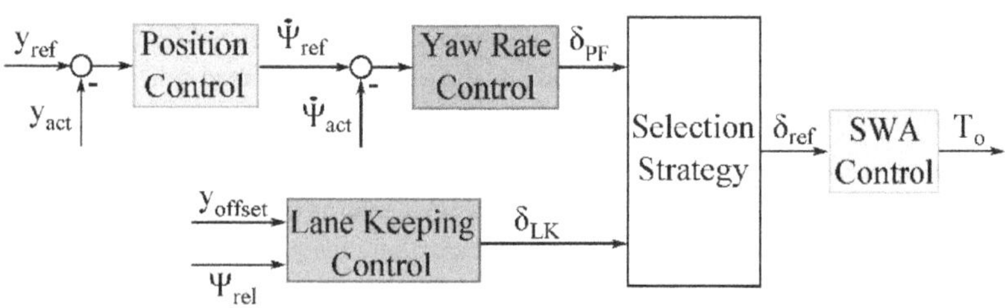


Figure 6: Block diagram of the general structure of the ESA controller.

steering torque overlay based on the reference SWA $\delta_{ref}$ and tries to follow that reference. At the same time the torque overlay communicates the "right way to steer" to the driver. The reference SWA is chosen by the selection strategy from the lane keeping controller or from the path following controller, which consists of the yaw rate controller and the position controller. The cascaded structure of the path following controller was chosen because of the rather high order and the structure of the plant model. The selection strategy that selects between path following and lane keeping is useful, because it corresponds to the two tasks of avoiding an obstacle in the first place and keeping the vehicle on the road afterwards. Note that the two parts of the maneuver require different characteristics which requires switching between two controller types.

## 5.2 Steering Wheel Angle Control

The steering wheel angle controller is the only part that differs between the function in the test vehicle and the simulator. This is mainly because the driving simulator is equipped with a standard BLDC motor as a steering actuator and not with a modern EPS.

The steering system in the driving simulator is identified as a linear third order system using least squares optimization. Because only the SWA is measured an observer is designed to estimate the unknown states. This allows to apply a state controller, which was designed according to Riccati.

In the vehicle application a reference dependent feedforward controller is supplemented by a PD feedback controller. The steering rate is fed back utilizing a dead band proportional controller. All controllers depend on the vehicle's velocity. Figure 7 shows the

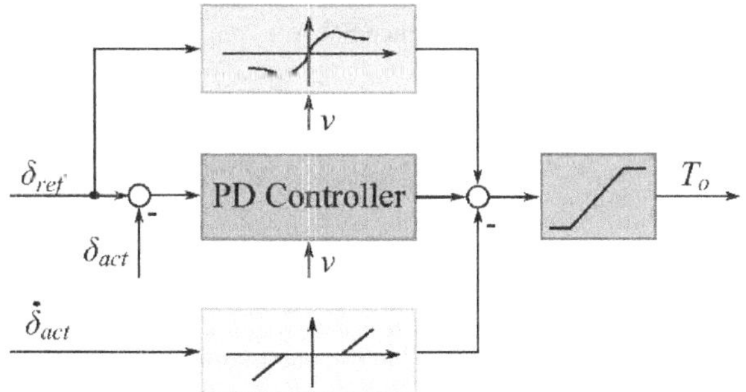


Figure 7: Block diagram of the SWA controller in the prototype test vehicle.

block diagram of the SWA controller. Its output the torque overlay is limited in order to achieve safety situations on the test track. The limitation allows the driver to overrule the system in the case of failure, but it is still enough to perform autonomous emergency maneuvers in order to show and evaluate the systems capabilities and behaviour. For production cars applications the limitations may need to be chosen more strictly in dependence on the OEM.

### 5.3 Yaw Rate Control

The yaw rate controller, shown in figure 8, utilizes a similar structure as the SWA con-

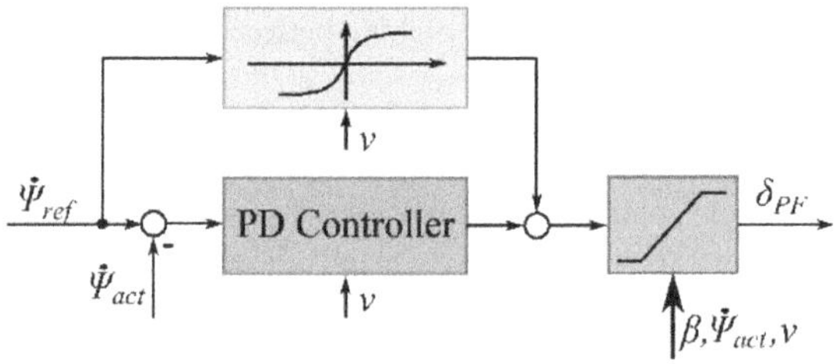


Figure 8: Block diagram of the yaw rate controller.

troller. But its limitations are far more complex and depend on the vehicles driving state. In general the vehicle dynamics system is nonlinear, where the main nonlinearities are called understeer and oversteer. Between the two driving states the dynamics can be regulated by linear controllers as suggested before. To account for the nonlinearities the authors suggest to utilize corresponding nonlinear controller output saturations. The first saturation shall prevent understeer. Because the side slip angle of the vehicle is estimated, it is allowed to use the single track model kinematics, which yields the maximum and minimum values for the reference SWA.

$$\delta_{max} = \alpha_{f,max} + \arctan\left(\frac{l_f\dot{\Psi} + v\sin(\beta)}{v\cos(\beta)}\right) \tag{3}$$

$$\delta_{min} = \alpha_{f,min} + \arctan\left(\frac{l_f\dot{\Psi} + v\sin(\beta)}{v\cos(\beta)}\right) \tag{4}$$

The values $\alpha_{f,min}$ and $\alpha_{f,max}$ correspond to the maximum tire force that can be generated in lateral direction.

In order to avoid oversteer the single track model is used to determine whether the current driving state is stable or unstable. This is exemplary shown in figure 9 for a velocity of $v = 100$ km/h and a SWA of $\delta = 90\ °$. The states which belong to the steady state circular drive, are calculated by solving the single track model's differential equations starting from the equilibrium. These states form a cloud of points, which is approximated by a polynomial

$$h = \sum_{i=1}^{5} c_i p^i \tag{5}$$

using regression technique. We define two conditions of which one has to be fulfilled to call the state stable. The first condition is fulfilled if the vehicle states do not change or (to account for numeric inaccuracies) their derivative is under a certain value.

$$(|\beta| \quad |\dot{\Psi}|)^T < (\beta_{\lim} \quad \dot{\Psi}_{lim})^T \tag{6}$$

This indicates the steady state circular drive $h$ is reached. In figure 9 these trajectories are drawn as solid lines. In addition we also define trajectories (drawn as dotted lines) stable that intersect with the states indicating the steady state circular drive.

$$(\beta \quad \dot{\Psi})^T = h \tag{7}$$

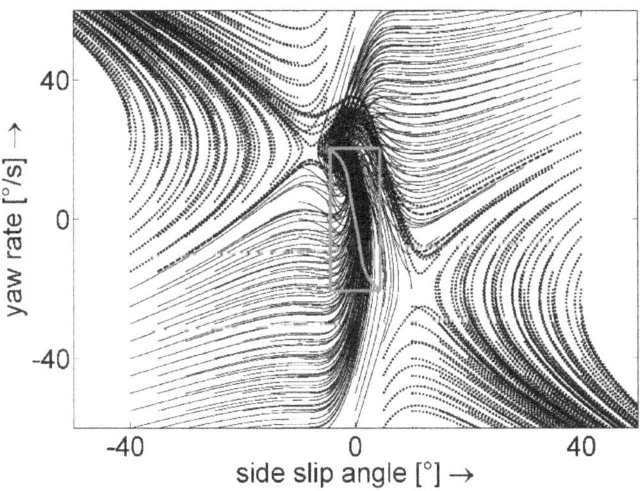


Figure 9: Stable (drawn as solid and dashed lines) and unstable (drawn as dotted lines) state trajectories. The polynomial approximating the steady state circular drive is printed as grey line. The maximum values of the circular drive arc represented by the grey frame.

This procedure yields a look-up table indicating whether a given driving state is stable or not. The maximum and minimum SWAs, that yield a stable driving state, are determined based on the data of the look-up table. This provides two additional look-up tables, one for the maximum and one for the minimum SWA limitation.

## 5.4 Position Control

The position controller (figure 10) also consists of a feedforward and a PD feedback

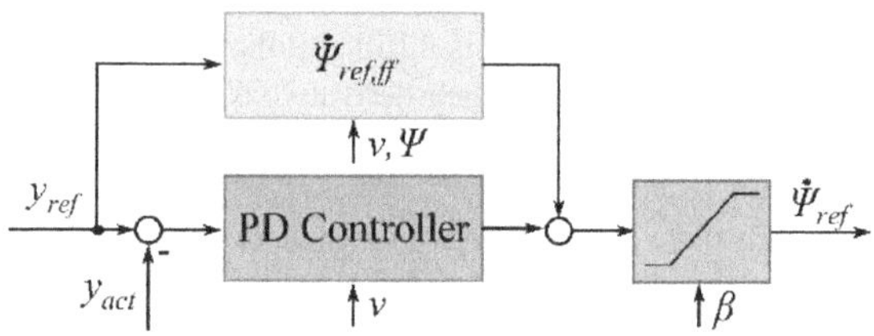


Figure 10: Block diagram of the position controller.

controller, which reduces the control deviation at the look ahead point in front of the vehicle. The feedforward is calculated from the reference path.

$$\dot{\Psi}_{\text{ref,ff}} = \frac{\mathrm{d}}{\mathrm{d}x}\left(\arctan\left(\frac{\mathrm{d}y_{ref}}{\mathrm{d}x}\right)\right) v\cos(\Psi) \tag{8}$$

Note that the double differentiation is not a problem because $y_{ref}$ is a planned quantity and not a noisy measurement.

The differentiation of the derivative part of the PD controller can be avoided, because the derivative of the reference can be calculated analytically and the derivative of the actual position is already available from the position estimation. Using the same method as described for the saturations of the yaw rate controller, additional two look-up tables are derived limiting the reference yaw rate in dependence on the side slip angle.

## 5.5 Lane Keeping Control

The lane keeping controller (figure 11) is activated after the lane is changed and the vehicle is in a safe driving state. Now the vehicle needs to be hold on the road or in its lane. Although this is a rather simple task, the authors assume that the driver still needs assistance shortly after an emergency situation because he might still be shocked. On the one hand the offset to the middle of the lane is controlled. The controller applies the commonly used bath tube profile consisting of a dead zone around small deviations and

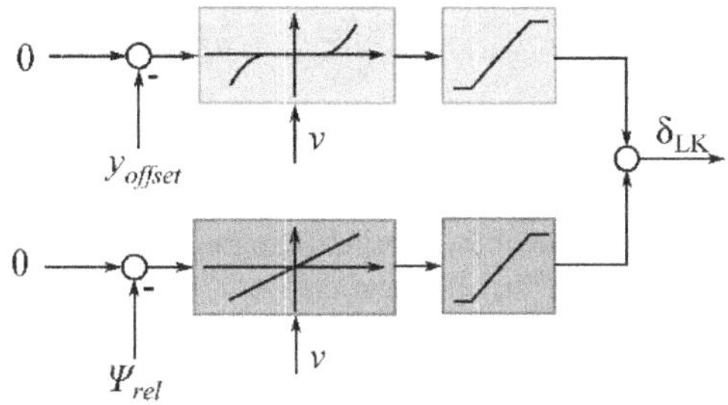


Figure 11: Block diagram of the lane keeping controller.

a sharply increasing controller output for larger deviations. On the other hand the yaw angle relative to the lane is controlled by a P controller yielding sufficient performance. Both controllers are limited in their output. Compared to the path following controller the lane keeping controller is tuned much more towards comfort. This could also be achieved by switching the parameters of the path following controller. But the control structure of the lane keeping controller is supposed to be more pleasant for the driver, which is mainly due to the dead zone. Human drivers do not always drive exactly in the middle of a lane, which corresponds to the proposed controller. Otherwise the driver may get annoyed, if he is constantly pushed in the centre of the lane.

### 5.6 Selection Strategy

The selection strategy is rather simple. At the beginning of the maneuver the path following controller is applied. Two conditions need to be fulfilled to switch to the lane keeping controller. The first one is, that the vehicle has to be stable and the second one is that the fifth order polynomial has been passed.

## 6. Longitudinal Vehicle Control

The paper at hand focuses on lateral vehicle control. Nevertheless the proposed lateral controller can be supplemented by a longitudinal controller in order to conduct combined emergency steering and braking maneuvers. In the following two strategies for longitudinal control are presented. One strategy for a combined maneuver is to define the lateral vehicle control as dominant and subordinate the longitudinal control. The total acceleration $a_{max}$ of a vehicle is supposed to be limited to a circle. Utilizing this circle and the current lateral acceleration $a_{lat}$ yields the reference longitudinal acceleration

$$a_{long,ref} = \sqrt{a_{max}^2 - a_{lat}^2}. \tag{9}$$

This means that the acceleration potential of the vehicle, which is not used for lateral control, is left for longitudinal control.

Sensing the environment is one of the most challenging tasks in the field of driver assistance systems and automated driving functions. As long as there is no full registration of the environment, one has to rely on the driver's perception. Braking is not beneficial in every emergency situation. For example, consider driving on the highway and a faster car is approaching from behind. If in this situation an emergency swerving maneuver is initiated, braking might lead to a collision with the overtaking vehicle. This is why the decision on braking and/or steering is left to the driver. But once the driver has made the decision he gets assistance in order to improve his steer and brake inputs. For the swerving maneuver this implies that the driver has to initiate the maneuver and the torque overlay must be limited in rate and amount. To transform these ideas to the longitudinal control there are no braking interventions as long as the driver does not brake by himself. If the driver braking input exceeds a certain value, the braking effort is increased in order to use the maximum deceleration as mentioned above.

# 7. Results

## 7.1 Fixed Base Driving Simulator

In the following subsections emergency maneuvers at different speeds and with different intervention strategies are presented.

### 7.1.1 Sole Steering Maneuvers at Different Speeds

The following figures show the evolution of important dynamic quantities during autonomously driven emergency steering maneuvers. In figures 12, 13, 14 and 15 emergency maneuvers with an initial velocity of about 55 km/h, 85 km/h, 115 km/h and 145 km/h respectively are shown. The results document that the reference path is tracked well, no collision occurs and the vehicle drives within the road limits. The acceleration diagrams show the criticality of the maneuver, while the stability phase plane diagram proves the car stays stable. The velocity shows only a small reduction. The SWA and yaw rate controller track their references sufficiently accurate. A higher performance is prevented by the limited actuator dynamics.

### 7.1.2 Combined Steering and Braking Emergency Maneuvers

In figure 16 an autonomously driven swerving maneuver is shown that utilizes the full deceleration potential of the vehicle. As a result the vehicle comes to a stop next to the obstacle at the end of the maneuver. As before the reference path is tracked very well and there is no collision with any obstacle. Despite the criticality of the situation the vehicles driving state remains stable. The acceleration potential of the vehicle is exploited. The initial velocity of the vehicle is 130 km/h and decreases while the maneuver is executed. In figure 17 an assisted maneuver is shown. Due to the influence of the driver the tracking of the reference path decreases a bit compared to the autonomous maneuvers. The total acceleration potential of the vehicle is exploited when the driver hits the brake pedal. Due to the driver's inappropriate control inputs the side slip angle exceeds 5°. But the vehicle does not spin, although no ESC is implemented here. In contrast to the autonomous maneuver the velocity does not necessarily decrease to zero. At the end of the maneuver the driver lifts his foot of the brake pedal and as a results there is no further (unnecessary) deceleration, because the obstacle has already been passed and a normal driving situation is reached.

## 7.2 Prototype Test Vehicle

In the following results obtained from the prototype test vehicle are presented. Figures 18 and 19 show important quantities of autonomously driven emergency steering maneuvers. Similar to the results from the driving simulator the reference path is tracked well. Because of the limitation of the steering torque overlay the performance of the SWA and yaw rate controller decreases a bit compared to the results obtained with the driving simulator. The lateral acceleration reaches 1 g, which indicates the criticality of the maneuver. Because no braking interventions are used here the velocity remains nearly constant. In every maneuver the vehicle behaves a little bit different. To show the spread of the accuracy of the path tracking figure 20 shows the estimated vehicle positions of multiple test drives. The spread is rather low at the beginning of the maneuver and increases over time. Note that this is intended because the lane keeping controller is designed to intervene only slightly.

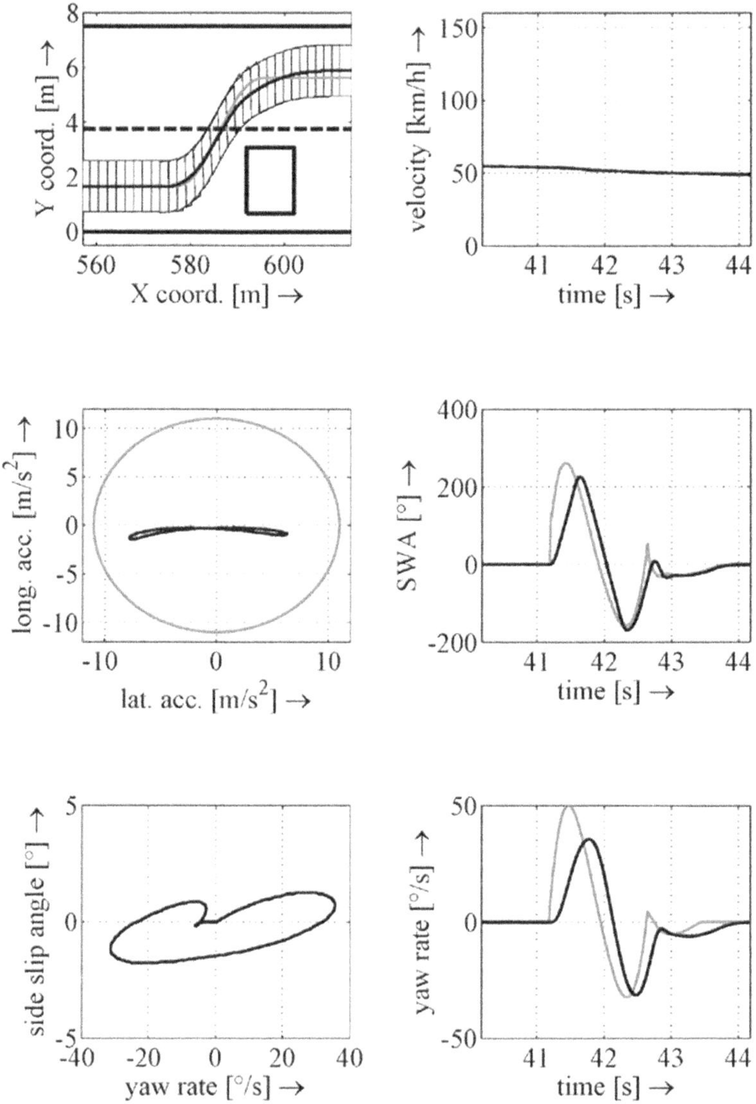


Figure 12: Emergency steering maneuver with an initial speed of 55 km/h. Reference values are printed in grey while actual values are printed black.

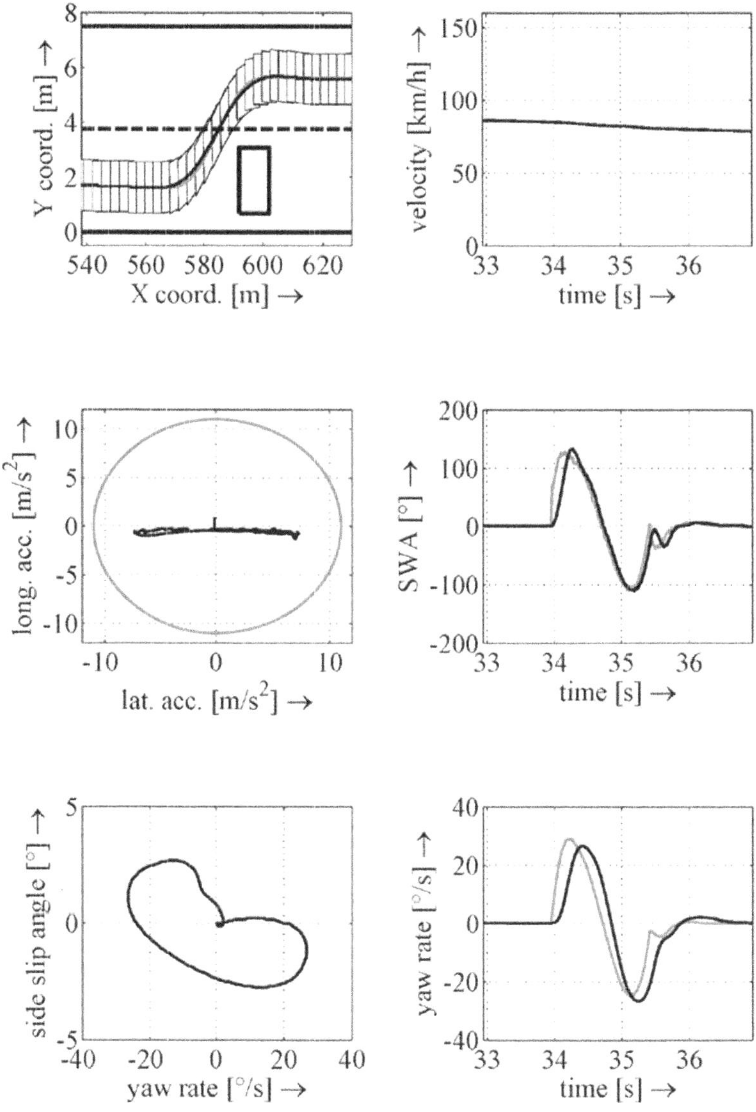


Figure 13: Emergency steering maneuver at 85 km/h initial speed. Reference values are printed in grey while actual values are printed black.

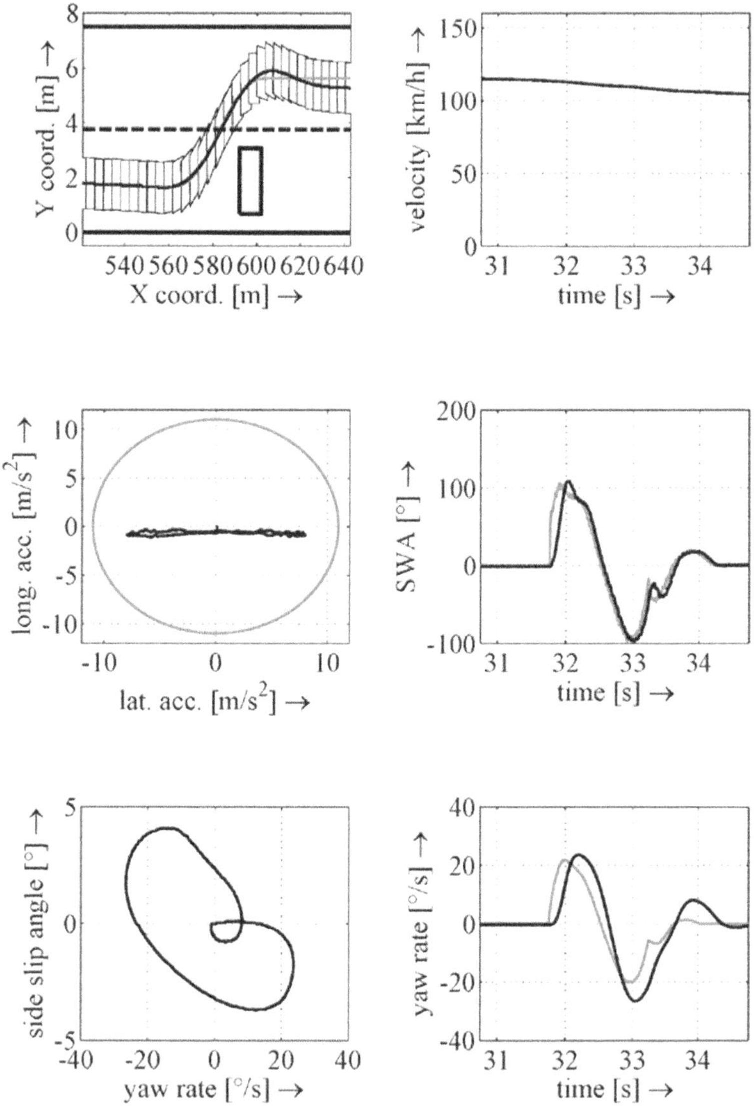


Figure 14: Emergency steering maneuver at 115 km/h initial speed. Reference values are printed in grey while actual values are printed black.

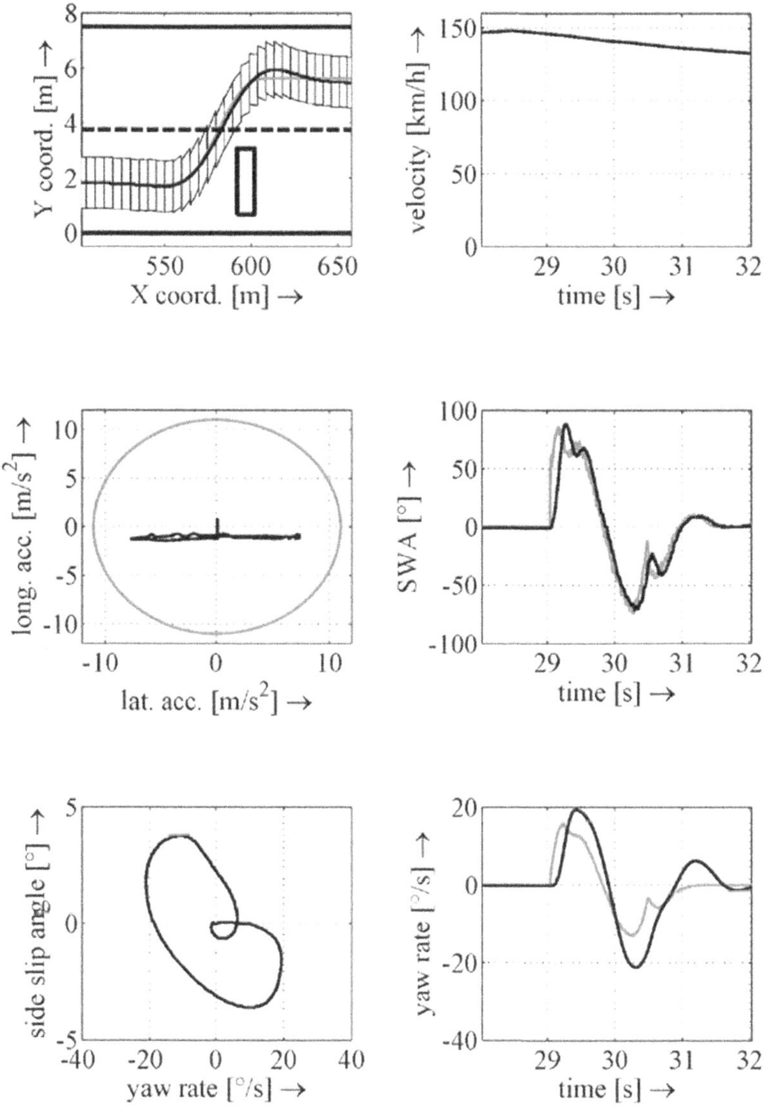


Figure 15: Emergency steering maneuver at 145 km/h initial speed. Reference values are printed in grey while actual values are printed black.

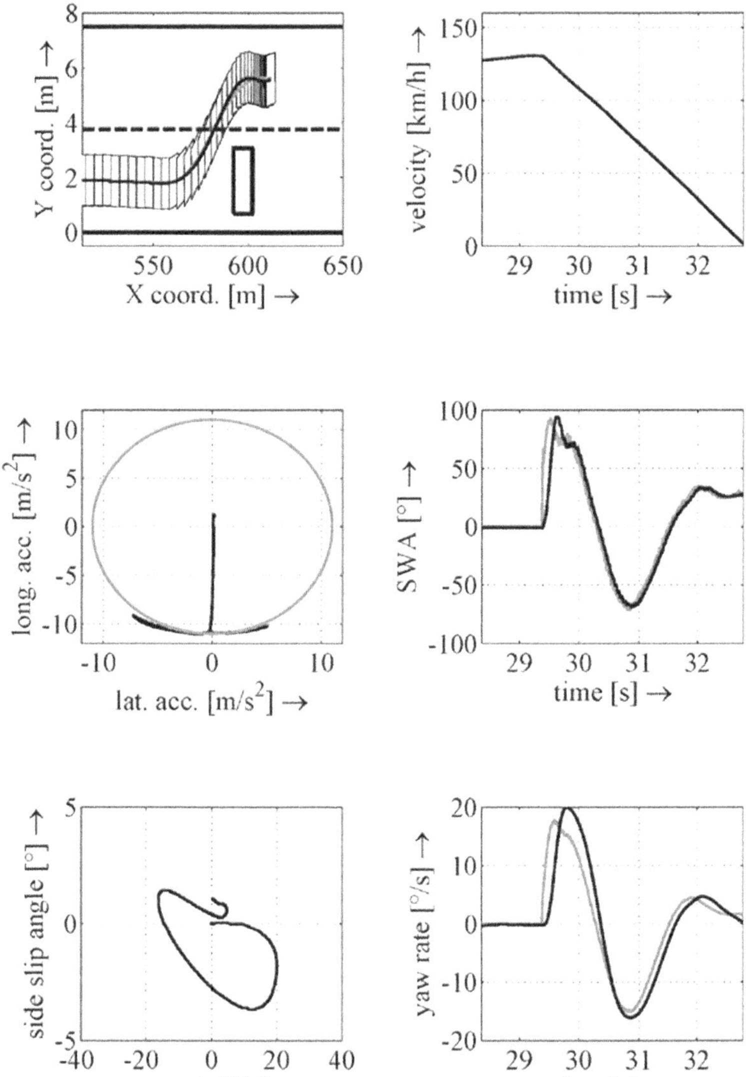


Figure 16: Combined steering and braking maneuver using maximum deceleration at an initial speed of 130 km/h. Reference values are printed in grey while actual values are printed black.

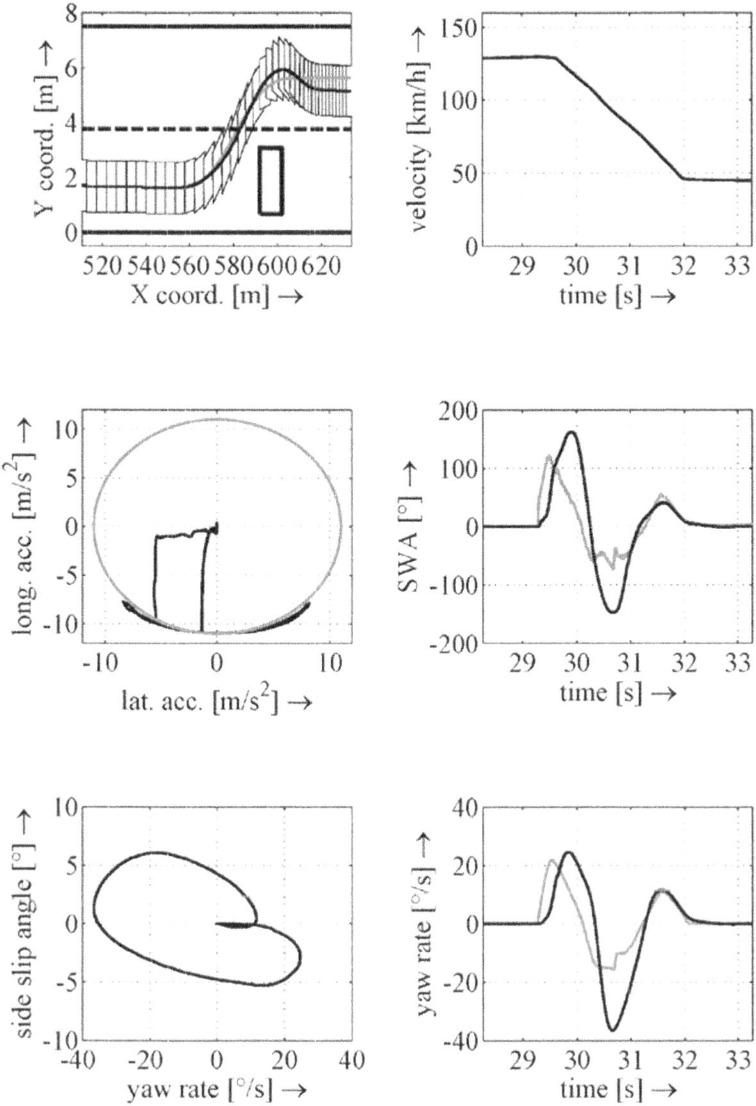


Figure 17: Assisted steering and braking maneuver using the driver controlled longitudinal control. Reference values are printed in grey while actual values are printed black.

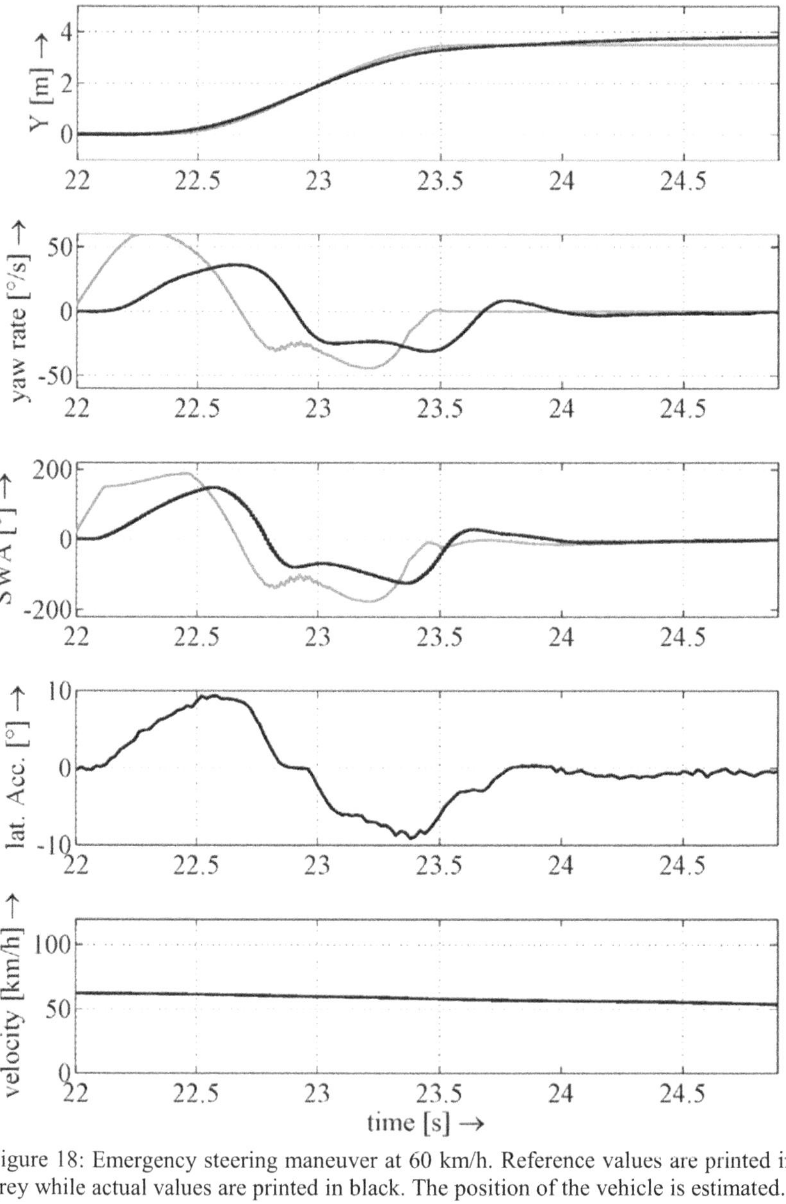


Figure 18: Emergency steering maneuver at 60 km/h. Reference values are printed in grey while actual values are printed in black. The position of the vehicle is estimated.

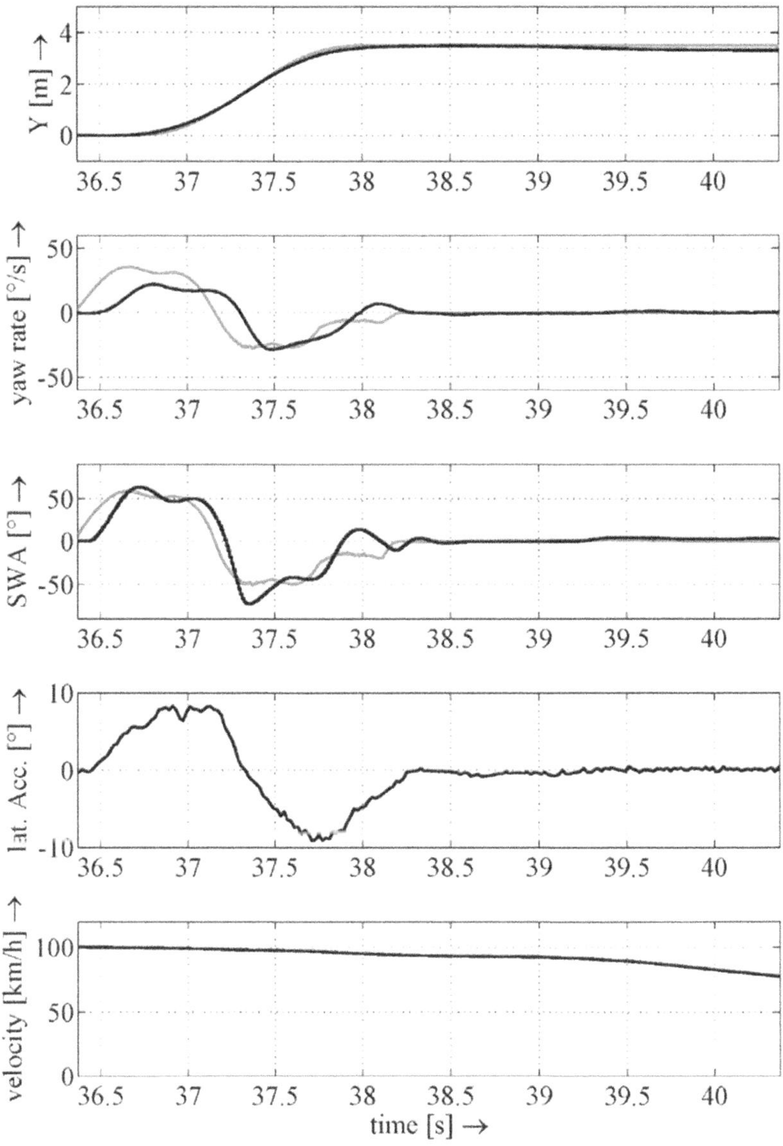


Figure 19: Emergency maneuver at speed of 100 km/h. Reference values are printed in grey while actual values are printed in black. The position of the vehicle is estimated.

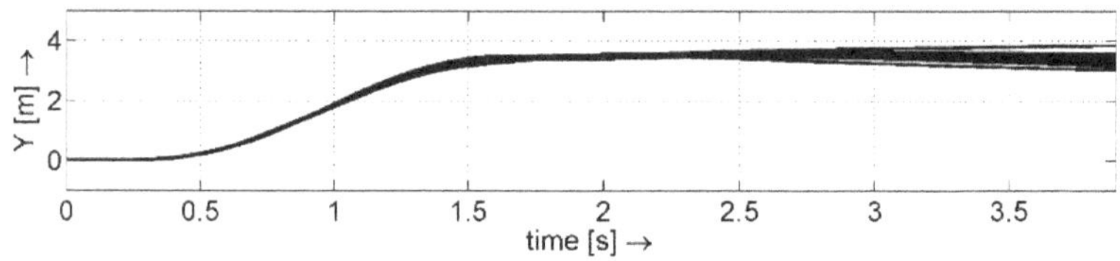


Figure 20: Spread of the path tracking.

# 8. Conclusion

The paper at hand proposes an efficient and real time capable path planning approach yielding a smooth and feasible evasive path, which is two times continuously differentiable at the transitions from the straight segments to the polynomial. The lateral vehicle controller consists of cascaded controllers which are supplemented by the nonlinear saturations to account for the nonlinear plant dynamics. The subordinated longitudinal control seems appropriate as a driver controlled amendment to the emergency steering maneuver. The results from the driving simulation and the test vehicle prove the capabilities of the proposed concept. Future research will deal with subject studies conducted in the driving simulator in order to compare different approaches towards emergency driving support.

## Literature

[1] M. Keller, C. Haß, A. Seewald and T. Bertram, "Driving Simulator Study on an Emergency Steering Assist", IEEE International Conference on Systems, Man and Cybernetics, 2014, pp.3039-3044

[2] M. Schorn, "Quer- und Längsregelung eines Personenkraftwagens für ein Fahrerassistenzsystem zur Unfallvermeidung", Dissertation, TU Darmstadt, Germany, 2007

[3] M. Schorn, U. Stählin, A. Khanafer and R. Isermann, "Nonlinear Trajectory Following Control for Automatic Steering of a Collision Avoiding Vehicle", American Control Conference, 2006, pp.5837-5842

[4] U. Stählin, "Eingriffsentscheidung für ein Fahrerassistenzsystem zur Unfallvermeidung", Dissertation, TU Darmstadt, Germany, 2008

[5] D. Soudbakhsh, A. Eskandarian and J. Moreau, "An emergency evasive maneuver algorithm for vehicles", IEEE Conference on Intelligent Transportation Systems, 2011, pp. 973-978

[6] J. Choi, K. Yi, J. Suh and B. Ko, "Coordinated Control of Motor Driven Power Steering Torque Overlay and Differential Braking for Emergency Driving Support", IEEE Transactions on Vehicular Technology, Vol. 63, NO. 2, pp. 566-579, February 2014

[7] C. Schmidt, F. Oechsle and W. Branz, "Research on trajectory planning in emergency situations with multiple objects", IEEE Intelligent Transportation Systems Conference, 2006, pp.988-992

[8] M. Keller, F. Hoffmann, C. Haß, A. Seewald and T. Bertram, "Planning of Optimal Collision Avoidance Trajectories with Timed Elastic Bands", IFAC World Congress, 2014, pp. 9822-9827

[9] J. Ziegler, P. Bender, T. Dang and C. Stiller, "Trajectory Planning for Bertha - a Local Continuous Method", IEEE Intelligent Vehicles Symposium, 2014, pp. 450-457

[10] E. Bauer, F. Lotz, M. Pfromm, M. Schreier, S. Cieler, A. Eckert, A. Hohm, S. Lüke, P. Rieth, B. Abendroth, V. Willert, J. Adamy, R. Bruder, U. Konigorski and H. Winner, "PRORETA 3: An Integrated Approach to Collision Avoidance and Vehicle Automation", at Automatisierungstechnik Vol. 60, pp. 755-765, December 2012

[11] J. V. Frasch, A. Gray, M. Zanon, H. J. Ferreau, S. Sager, F. Borrelli and M. Diehl, "An Auto-generated Nonlinear MPC Algorithm for Real-Time Obstacle Avoidance of Ground Vehicles", European Control Conference, 2013, pp. 4136-4141

[12] S. Schmidt und R. Kasper, "Ein hierarchischer Ansatz zur optimalen Bahnplanung und Bahnregelung für ein autonomes Fahrzeug", at – Automatisierungstechnik Vol. 60, pp. 743-753, December 2012

[13] M. Keller, C. Haß, A. Seewald und T. Bertram, "Ein modellprädiktives Planungs- und Fahrzeugquerregelungsverfahren zur Kollisionsvermeidung durch Notausweichmanöver", GMM VDE/VDI Automotive meets Electronics, 2015, pp. 8-13

# Collision avoidance with combined braking and steering

Dipl.-Ing. Carlo Ackermann, Jakob Bechtloff, M.Sc.,
Prof. Dr.-Ing. Dr. h.c. R. Isermann
Technische Universität Darmstadt
Institute of Automatic Control and Mechatronics

# Abstract

Among various driver assistance systems with environmental sensors the first emergency braking systems are available. To avoid collisions with a vehicle driving in front of the own vehicle also an evasive maneuver is possible to avoid the collision. Particularly at higher velocities this maneuver is characterized by a later intervention time.

To allow a still later intervention, a combination of steering and braking is possible. If the evasive maneuver and the braking maneuver are combined, it is possible to further reduce the evasive length.

In this contribution two control strategies will be presented for a combined intervention. Both methods have the goal to make use of the friction potential of the tires as much as possible. The first control strategy uses the accelerations which occur during the maneuver, the second control is a tire-individual slip-control. Thus it is possible to significantly reduce the evasive length in the speed range between 70 and 160 km/h compared to the classical evasive maneuver, which will be shown with simulation.

With such a system, collisions can be prevented to an even later stage.


# 1 Introduction and motivation

For series production vehicles the first collision avoidance systems are now introduced. In general braking systems which intervene at the last possible point can decelerate the vehicle up to 10 $m/s^2$, [1], [2]. With larger velocities the braking distance increases. At higher velocities swerving needs less distance to the object than braking [3]. Then it is possible to avoid the collision by an evasive maneuver at a later time.

Active steering systems [4] as used for Lane Keeping Support [5] allow inputs for lateral control. Emergency systems for automatic swerving are under development. Investigations are carried out in [6] and [7]. In [8], [9] and [10] a system for an emergency swerving was developed and tested in a real vehicle within the PRORETA 1 Project. This contribution extends former theoretical and experimental research for swerving at higher speeds and takes into account a combined braking intervention. Especially for velocities smaller than 140 km/h this can reduce the maneuver length significantly.

Hence it will be described how an additional braking intervention can reduce the evasive distance. The goal is to utilize the road-wheel traction as optimal as possible.

The contribution begins with a brief overview for generating the evasive trajectory and lateral control. Then longitudinal braking control is considered. Two different methods are presented. The different methods are then compared with each other at different velocities. The contribution closes with a conclusion and an outlook.

All results are simulated with Matlab and IPG CarMaker, a well known tool for driving dynamic simulations.

# 2 Lateral control

For the evasive maneuver a trajectory is needed. This trajectory is then driven by the steering controller. The control variable is the steering angle.

## 2.1 Trajectory planning

Trajectories for an evasive maneuver are nearly S-curved. A function describing such a curve is a sigmoidal function. In the following a sigmoidal function as follows is considered, which needs only three parameters, [3]:

$$y(x) = \frac{B}{1 + e^{-a(x-c)}}.$$

$B$ describes the maneuver width, $a$ the slope and $c$ the position of the turning point and so $2c$ becomes the total length of the maneuver. The minimum and the maximum of the sigmoidal function are in the infinity. Because of that an additional parameter $y_{\mathrm{tol}}$ is introduced. Hereby is $y(0) = y_{\mathrm{tol}}$ and $y(2c) = B - y_{\mathrm{tol}}$.

The goal is to find the parameters under the condition of minimum c under the condition that boundings of acceleration and jerk are not violated. In the following the parameters are estimated for two cases: First for maximum acceleration and secondly for maximum jerk. Then the appropriate parameters are selected. The following boundary conditions apply:

$$y(0) = y_{\mathrm{tol}},$$

$$y(2c) - B - y_{\mathrm{tol}},$$

$$|a_{\mathrm{Y}}(x)| \leq a_{\mathrm{Y,max}},$$

$$|b(x)| \leq b_{\mathrm{max}}.$$

It is now assumed that the vehicle has a constant speed $v$. Limiting the lateral acceleration the parameter $a$ can be calculated with the equations following from [3]:

$$a_{\mathrm{aY}} = -\frac{(p+1)^2\sqrt{-pB\left(a_{\mathrm{Y,max}}pB - v^2p^2 + v^2\right)a_{\mathrm{Y,max}}}}{pB(a_{\mathrm{Y,max}}pB - v^2p^2 + v^2)}$$

with

$$s_1 = 9v^4 + 3a_{\mathrm{Y,max}}Bv^2 + 2a_{\mathrm{Y,max}}^2B^2,$$

$$s_2 = \cos\left(\frac{1}{3}\tan^{-1}\left(\frac{3v^2\sqrt{81v^8 + 27v^4a_{\mathrm{Y,max}}^2B^2 + 3a_{\mathrm{Y,max}}^4B^4}}{27v^6 + 9v^2a_{\mathrm{Y,max}}^2B^2 + 4a_{\mathrm{Y,max}}^3B^3 + 27v^4a_{\mathrm{Y,max}}B}\right)\right),$$

$$p = \sqrt{2}\frac{s_2}{\sqrt{s_1}}\left(6v^2 + 2a_{\mathrm{Y,max}}B + \frac{4a_{\mathrm{Y,max}}^2B^2}{3v^2}\right) + \frac{2a_{\mathrm{Y,max}}B}{3v^2} + 1.$$

Considering the limitation of jerk, the parameter $a$ can be calculated with the following expression:

$$a_b = \frac{1}{6v^3}\left(\sqrt[3]{s_3} + \frac{B^2b_{\mathrm{max}}^2}{\sqrt[3]{s_3}} + Bb_{\mathrm{max}}\right)$$

with

$$s_3 = \frac{b_{\mathrm{max}}}{B}\left(86v^6 + B^4b_{\mathrm{max}}^2 + 24v^3\sqrt{1296v^6 + 3B^4b_{\mathrm{max}}^2}\right).$$

Because $a$ defines the gradient of the sigmoidal function and the lateral acceleration is smaller for smaller gradients, the smaller value of $a$ is used for the trajectory. The accompanying parameter $c$ can then be calculated in both cases as follows:

$$c = \frac{1}{a}\ln\frac{B - y_{\mathrm{tol}}}{y_{\mathrm{tol}}}$$

For smaller velocities the maximum jerk defines the value of $a$ and $c$. For higher speeds the lateral acceleration is the limiting value.

Because of the assumption of a constant speed $v$ a braking intervention is not considered. With combined braking, the velocity is decreased and the resulting lateral acceleration is smaller than the given maximum lateral acceleration. With this additional degree of freedom it is possible to plan a shorter trajectory and reduce the maneuver length.

## 2.2 Controller structure

For driving the vehicle along the planned trajectory a two-degree-of-freedom structure is used, consisting of a feedforward and a feedback controller.

### Feedforward control

Because of the precalculated trajectory it is possible to determine the desired yaw rate for each point of the trajectory. Using the one-track-model a desired steering angle depending on the yaw rate can be determined:

$$\frac{\delta}{\dot{\psi}} = i_S \frac{l + v^2 \mathrm{SG}}{v}.$$

SG is the so-called self-steering gradient and describes the steering behaviour depending on the velocity:

$$\mathrm{SG} = \frac{c_{\alpha,\mathrm{r}} l_\mathrm{r} + c_{\alpha,\mathrm{f}} l_\mathrm{f}}{c_{\alpha,\mathrm{f}} c_{\alpha,r} l}.$$

### Feedback control

In [11] different controllers were investigated. Finally it ended up in the relatively simple PD-Controller, which is also used here.

As the one track vehicle model shows, the lateral behaviour depends on the velocity. Therefore $M$ different controllers depending on the velocity of the vehicle are applied.

The controller outputs are weighted with membership or activation functions and summarized as shown in Fig. 1. This approach uses local linear models [12].

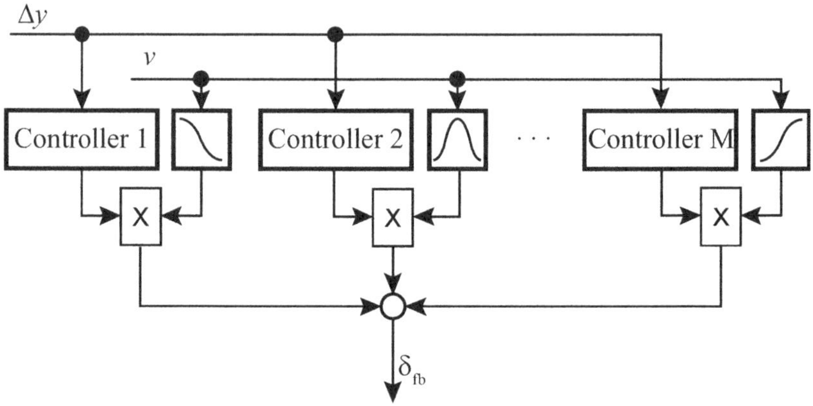


Figure 1: Local linear controller network

The output of the local linear controller network is obtained from the weighted controller outputs:

$$\delta_{\mathrm{fb}} = \sum_{i=1}^{M} \delta_{\mathrm{LLM},i} \Phi_i(v)$$

using

$$\Phi_i(v) = \frac{\mu_i(v)}{\sum_{j=1}^{M} \mu_j(v)} \quad \text{with} \quad \mu_i(v) = \exp\left(-\frac{1}{2}\frac{(v-c_i)^2}{\sigma_i^2}\right)$$

as activation function.

In order to compare the results an evasive maneuver without braking intervention is shown in Fig. 2. The maneuver width is chosen to $B = 3.6$ m and the maximum lateral acceleration is limited to $a_{\mathrm{Y,max}} = 7\ \mathrm{m/s^2}$.

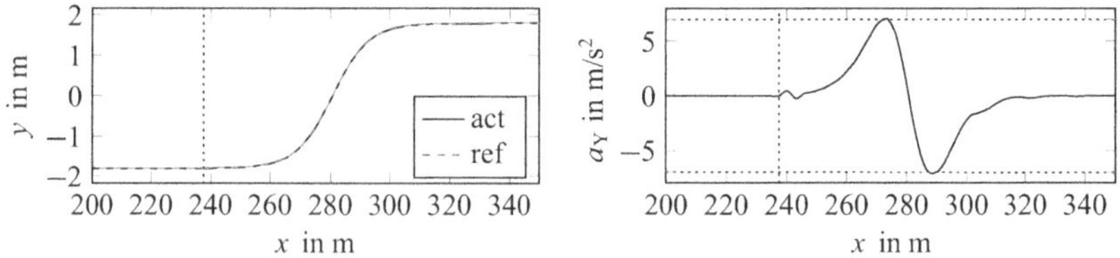


Figure 2: Reference trajectory

As a measure for the evasive distance the distance from the beginning of the evasive maneuver until crossing the middle line is chosen. In the case of the shown reference trajectory this distance is 40.5 m.

# 3 Longitudinal braking control

In addition to the lateral control a longitudinal control for the braking maneuver is needed. Therefore two different approaches will be considered. One is based on accelerations and extends a method presented in [11], the other one is based on slip.

## 3.1 Maximal acceleration control

The amount of longitudinal and lateral acceleration is limited by the maximum wheel-forces and depends on the road condition. This behavior is depicted by the following equation (Kamm's circle):

$$a_{\mathrm{X}}^2 + a_{\mathrm{Y}}^2 \leq (g\mu_{\mathrm{max}})^2.$$

Therefore the maximum possible longitudinal acceleration becomes for simultaneous braking and swerving:

$$a_{\mathrm{X,max}} = \sqrt{(g\mu_{\mathrm{max}})^2 - a_{\mathrm{Y}}^2}.$$

During the evasive maneuver the total acceleration of the vehicle is controlled to be $a_{\mathrm{max}}$. Therefore the lateral acceleration $a_{\mathrm{Y}}$ is measured and the reference value for longitudinal acceleration $a_{\mathrm{X,ref}}$ is calculated. $a_{\mathrm{max}}$ is chosen in a way that the physical limitations are not exceeded:

$$a_{\mathrm{X,max}} = \sqrt{a_{\mathrm{max}}^2 - a_{\mathrm{Y}}^2} \leq \sqrt{(g\mu_{\mathrm{max}})^2 - a_{\mathrm{Y}}^2}.$$

The vehicles longitudinal acceleration during the braking and swerving maneuver is then controlled to the reference value $a_{\mathrm{X,ref}} = a_{\mathrm{X,max}}$.

### Brake control

The interventions are carried out through the brake pedal position $\beta_{\mathrm{brake}}$.

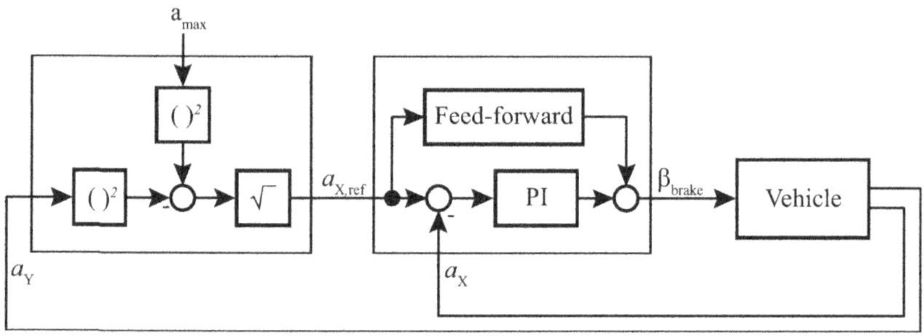


Figure 3: Brake control structure

Fig. 3 shows the controller structure. Because of the simple structure no complex modifications to the brake system or the sensors are needed. Only the sensors for measuring the accelerations must exist, which is fulfilled in most of the modern vehicles. As controller a two-degree-of-freedom structure can be used again. The feed-forward part calculates the brake pedal position in dependence from the given acceleration. Differences are corrected by the PI controller.

## Results

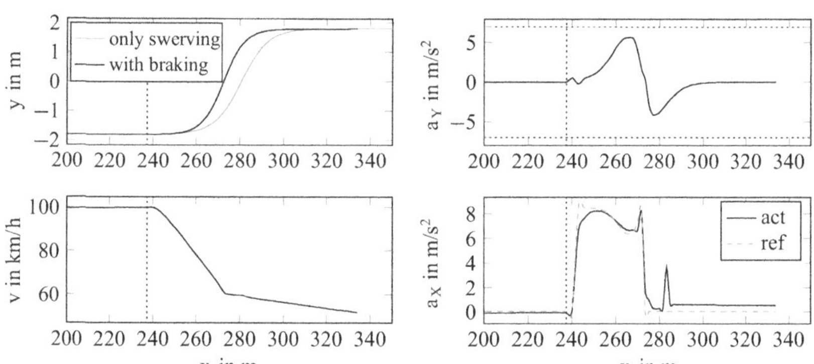


Figure 4: Simulation results for maximal acceleration control and simultaneous braking and swerving

Fig. 4 shows the results for a maneuver driven at 100 km/h. For comparison a reference trajectory for a maneuver without braking is depicted. Due to the braking maneuver the evasive distance is now 32.3 m. The lateral acceleration is significantly below 7 m/s$^2$. That means that even though the lateral acceleration is smaller, the length of the maneuver can be shortened. This is due to the high deceleration in the first phase of the evasion. With increasing lateral acceleration the braking intervention is reduced. After changing the lane, the braking intervention is deactivated automatically to have a high road-friction reserve for stabilization.

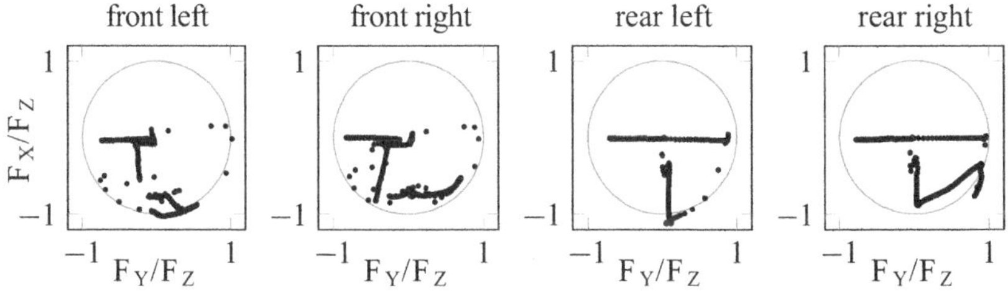


Figure 5: Kamm's circles for maximal acceleration control

Fig. 5 shows the utilization of the road-friction capability. While the rear wheels have a good utilization there is some potential at the front wheels. Because only one control variable and one manipulated variable are taken into account a wheel individual control is not possible in this case. Therefore in the next section another method is considered.

## 3.2 Maximal slip control

Because of the relation between wheel-force and wheel-slip now a method is presented which uses the values of lateral and longitudinal slip for the amount of road friction utilization. The advantage of this method is, that all wheels can be individually taken into account.

## Brake control

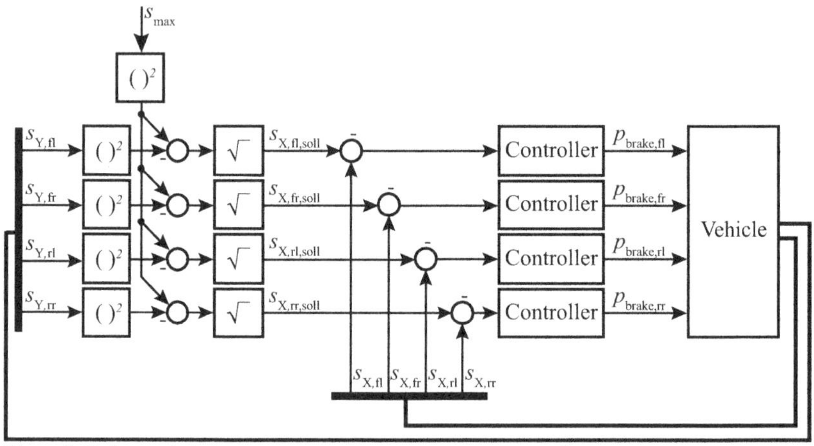


Figure 6: Brake control structure

On the basis of the Kamm's circle, which relates longitudinal and lateral force to the maximum wheel force, a similar method can also be applied for the slip. The total slip consists of longitudinal and lateral slip:

$$\lambda = \sqrt{\lambda_X^2 + \lambda_Y^2}$$

If the maximum possible slip is known and the lateral slip can be measured, it is possible to calculate the maximum of the longitudinal slip:

$$\lambda_{X,max} = \sqrt{\lambda_{max}^2 - \lambda_Y^2}$$

Because maximum deceleration is desired, the calculated longitudinal slip is the reference value for the slip control. So optimal road friction utilization is achieved. The structure is illustrated in Fig. 6.

Braking each wheel individually, it is not possible to use the brake pedal position. Individual braking pressures are adjusted by the valves of the ESC system.

For controlling the longitudinal slip again a two-degree-of-freedom structure is used. The manipulated value for each wheel is then determined from the comparison of reference and actual slip.

## Results

In Fig. 7 the results of the slip control are depicted. The evasive distance is 29.6 m. Also with this method the maneuver length can be reduced significantly. During high lateral acceleration the braking is reduced also to ensure that enough road-friction for lateral direction is provided.

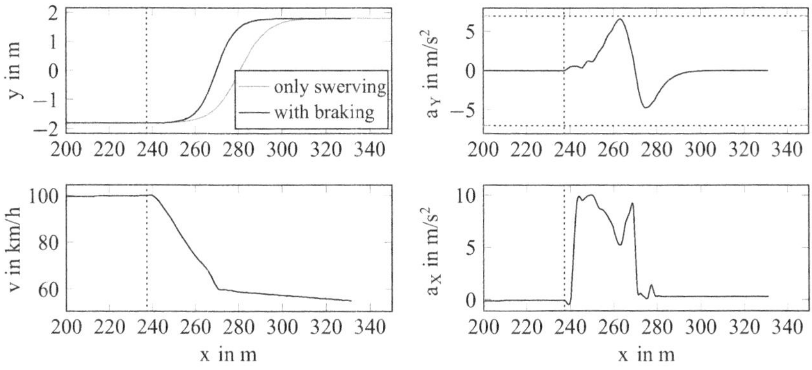


Figure 7: Simulation results for maximal slip control

The Kamm's circles in Fig. 8 show that the road friction capability is now utilized very well. Especially at the front wheels the force vector moves at the edge of the circle, so that these wheels have an optimal road friction utilization.

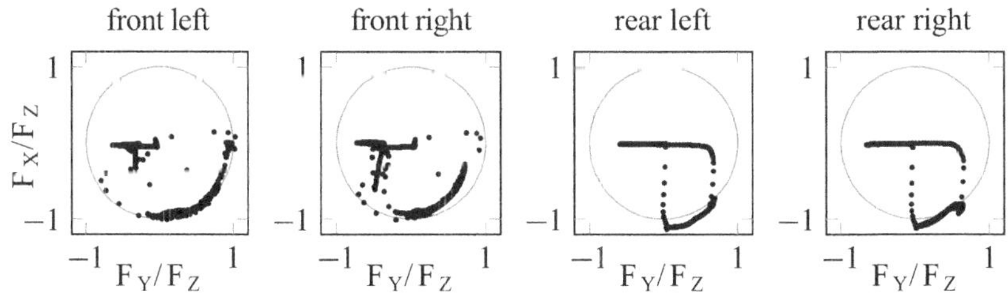


Figure 8: Kamm's circles for maximal slip control

# 4 Comparison

In the next two sections a comparison between the both presented methods and a comparison for different velocities with and without additional braking is shown.

## 4.1 Different methods

Fig. 9 shows the driven trajectories. While looking at the lateral acceleration it is noticeable that the evasive length is smaller with a combined braking intervention. The velocity of the vehicle is reduced significantly during braking. The method with slip control generates a higher longitudinal deceleration because of better force utilization. So the velocity is decreasing faster.

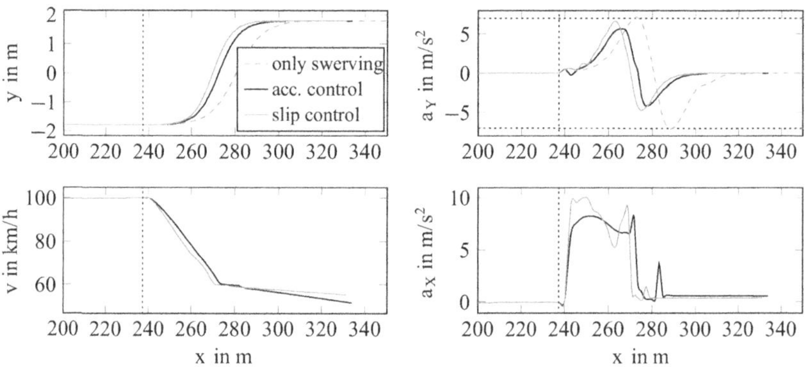


Figure 9: Comparison between the two methods

Both methods deliver a clearly shorter evasive length. Table 1 shows the distances. The deceleration control reduces the length by 20%, the slip control by even 27%. The difference between both methods seems at a first stance not very high. But the saving of 2.7 m could be crucial if an accident is prevented or not. The difference of slip control to speed control is approximately 8%.

Table 1: Comparison between the evasive distances

| **Method** | **Distance** | **Saving** |
|---|---:|---:|
| Without braking | 40.5 m | - |
| Acceleration-control | 32.3 m | 20 % |
| Slip-control | 29.6 m | 27 % |

However, the slip control is the much more complex method. For the acceleration control only sensors for lateral and longitudinal acceleration are required. But for the slip control detailed knowledge of the lateral slip of each tire is necessary. This must be estimated using appropriate methods. In highly dynamic situations, such as in the method considered here, however, it is not always ensured that the estimates are highly accurate.

When such a system will be prepared for road use, it must be considered if the extra effort for the slip control is justified with a relatively small improvement or whether it is sufficient to use the acceleration control. Even with this method it is possible to achieve significant improvements in the evasive distance.

## 4.2 Different velocities

Until now a velocity of 100 km/h was considered. In a next step it will be investigated which benefits result in dependence on the speed. Therefore the distance without braking and with combined braking for different velocities are compared. Table 2 shows the results.

Table 2: Comparison between the evasive distances for different velocities

| Velocity | Braking | Only swerving | Combination | Saving |
|---|---:|---:|---:|---:|
| 30 km/h | 4.1 m | 11.6 m | n.n. | n.n |
| 50 km/h | 11.3 m | 20.3 m | 11.5 m | 43 % |
| 70 km/h | 22.2 m | 28.2 m | 17.9 m | 37 % |
| 100 km/h | 45.4 m | 40.5 m | 29.6 m | 27 % |
| 130 km/h | 76.7 m | 52.8 m | 40.2 m | 24 % |
| 160 km/h | 116.2 m | 65.0 m | 54.3 m | 16 % |
| 200 km/h | 181.6 m | 81.3 m | 73.7 m | 9 % |

In the low speed range braking is still more suitable. At medium speeds the combination of braking and steering yields major benefits. At very high speeds, however, these benefits decrease. Thus at a speed of 200 km/h only a saving of 9% of the distance is possible whereas the saving at 100 km/h is about 27%.

This can be explained as follows: Considering the evasive trajectory in time domain at different velocities there are only slight differences. So the evasive maneuver takes regardless of the speed the same time. Also the longitudinal deceleration is at a similar value for different velocities. So the difference velocity between the initial and the final speed of the braking operation is independent of the vehicle speed and nearly constant. However this difference velocity is relatively to the vehicle velocity higher at low speed. Because the maneuver length is linearly related to the speed this results in a higher efficiency of the additional braking intervention at slower speeds.

# 5 Conclusion and outlook

In order to allow an even later intervention for collision avoidance, a method was presented, which combines braking and swerving. The evasive maneuver is carried out with priority and follows a predetermined trajectory. The remaining traction reserves can then be used to decelerate the vehicle by braking. Two different methods were presented. First the occurring lateral acceleration is measured and used to determine the usable deceleration. Secondly a wheel-individual slip control is designed. The lateral slip is determined for every wheel and the remaining slip reserve is calculated. Using a slip controller each wheel is then braked individually. Simulations have shown that an additional braking intervention especially at medium speeds can reduce the evasive distance significantly.

The method was developed by simulations in the CarMaker simulation environment. In a next step the algorithm must be implemented in a real-time hardware and tested in a real vehicle. Furthermore a determination of the road conditions should be included. The current implementation allows the adaption to the road by varying the maximum lateral acceleration and the maximum total acceleration respectively the total slip. However this should be adapted automatically by estimating the expected road-friction-coefficients.

# References

[1] H. Winner, „Frontalkollisionsschutzsysteme," in *Handbuch Fahrerassistenzsysteme*, Wiesbaden, Vieweg + Teubner, 2012, pp. 522-542.

[2] A. van Zanten und F. Kost, „Bremsenbasierte Assistenzfunktionen," in *Handbuch Fahrerassistenzsysteme*, Wiesbaden, Vieweg + Teubner, 2012, pp. 356-394.

[3] U. Stählin, Eingriffsentscheidung für ein Fahrerassistenzsystem zur Unfallvermeidung, Bd. 683, Düsseldorf: VDI-Verlag, 2008, pp. XII, 140 S.

[4] T. Raste, „Fahrdynamikregelung mit Brems- und Lenkeingriff," in *Handbuch Fahrerassistenzsysteme*, Wiesbaden, Vieweg + Teubner, 2012, pp. 395-403.

[5] J. E. Gayko, „Lane Keeping Support," in *Handbuch Fahrerassistenzsysteme*, Wiesbaden, Vieweg + Teubner, 2012, pp. 554-561.

[6] U. S. Lages, Untersuchungen zur aktiven Unfallvermeidung von Kraftfahrzeugen, Bd. 446, Dusseldorf: VDI Verlag, 2001, pp. v, 153.

[7] C. Ameling, Steigerung der aktiven Sicherheit von Kraftfahrzeugen durch ein Kollisionsvermeidungssystem, Als Ms. gedr. Hrsg., Bd. 510, Düsseldorf: VDI-Verl., 2002, pp. VII, 130 S.

[8] R. Isermann, R. Mannale und K. Schmitt, „Collision-avoidance systems PRORETA: situation analysis and intervention control," *Control Engineering Practice,* Bd. 20, Nr. 11, pp. 1236-1246, 2012.

[9] E. Bender, M. Darms, M. Schorn, U. Stählin, R. Isermann, H. Winner und K. Landau, „Anti Collision System Proreta – On the Way to the Collision Avoiding Vehicle," *ATZ Worldwide,* Bd. 109, Nr. 5, pp. 32-35, 2007.

[10] E. Bender, M. Darms, M. Schorn, U. Stählin, R. Isermann, H. Winner und K. Landau, „Anti Collision System Proreta – On the Way to the Collision Avoiding Vehicle," *ATZ Worldwide,* Bd. 109, Nr. 4, pp. 20-23, 2007.

[11] M. Schorn, Quer- und Längsregelung eines Personenkraftwagens für ein Fahrerassistenzsystem zur Unfallvermeidung, Bd. 651, Düsseldorf: VDI-Verlag, 2007, pp. x, 148 S..

[12] A. Fink, O. Nelles und M. Fischer, „Linearization Based and Local Model Based Controller Design," 1999.

# Driver assistance for trucks – from lane keeping assistance to smart truck maneuvering

Alexander Gaedke, Roland Greul, Stefan Kanngießer, Nicolas Boos

Robert Bosch Automotive Steering GmbH, Schwäbisch Gmünd

# Introduction

In June 2013 Bosch Automotive Steering introduced with the Servotwin the first electro-hydraulic steering system worldwide for heavy commercial vehicles. The Servotwin represents the innovative combination of a recirculating ball power steering gear Servocom with an electronic drive and control unit (see Figure 1). Now for the first time already proven EPS steering functionalities from passenger cars are available in the commercial vehicle sector, e.g. active return for smooth mid position alignment.

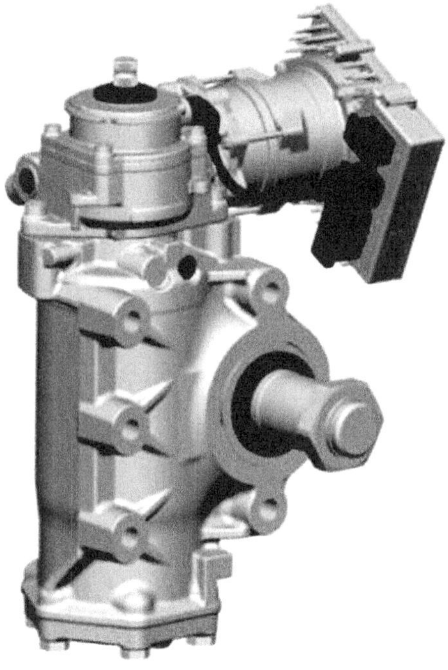

Figure 1: Servotwin

In addition to increased steering precision and improved steering comfort, the use of the second generation of the Servotwin enables the implementation of driver assistance functions on commercial vehicles.

This paper will give an overview of the second generation of Servotwin and the related driver assistance functions Lane Keeping Assistance and Smart Truck Maneuvering.

# Servotwin

## *Mechanical design of the Servotwin*

Figure 2 illustrates the structure of the Servotwin steering system focusing on the essential components needed for driver assistance functions. The Servotwin is basically made up of two components: the established recirculating ball power steering gear Servocom (hydraulic unit with sector shaft in Figure 2) and an electric motor with control unit, which is attached by a worm gear to the steering shaft. This electric drive portion can apply an additional steering torque to generate the basic steering functions as well as all kinds of driver assistance functions. Thus, the manual steering effort of the driver can be changed according to the driving situation. Nevertheless, the mechanical connection between the steering wheel and the steering linkage is maintained. If the electronic system fails, hydraulic steering effort assistance is always retained.

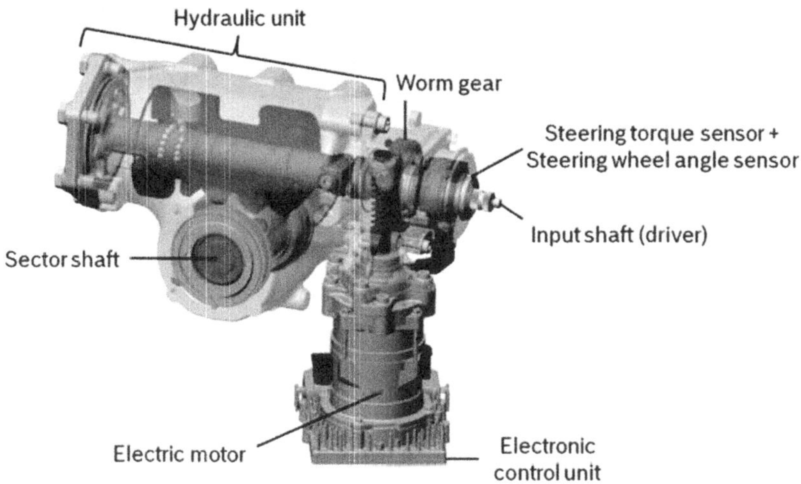


Figure 2: Structure of the Servotwin

## *Electronical operation of the system*

Figure 2: Structure of the ServotwinA core element of the Servotwin steering system is the electric control unit (ECU) with its electric motor, sensors and CAN-Communication bus. With these components the implementation of well known and proven in use passenger car EPS steering functions for commercial vehicles is possible.

The functional principle of the electronic components can be described as follows: The applied steering torque is detected by the steering torque sensor at the input shaft (see Figure 2) and is transmitted to the ECU of the Servotwin. Figure 3 depicts the ECU. Taking into account further vehicle signals, e.g. speed and load condition from CAN, the ECU determines within a few milliseconds the steering assistance required and controls the electric motor.

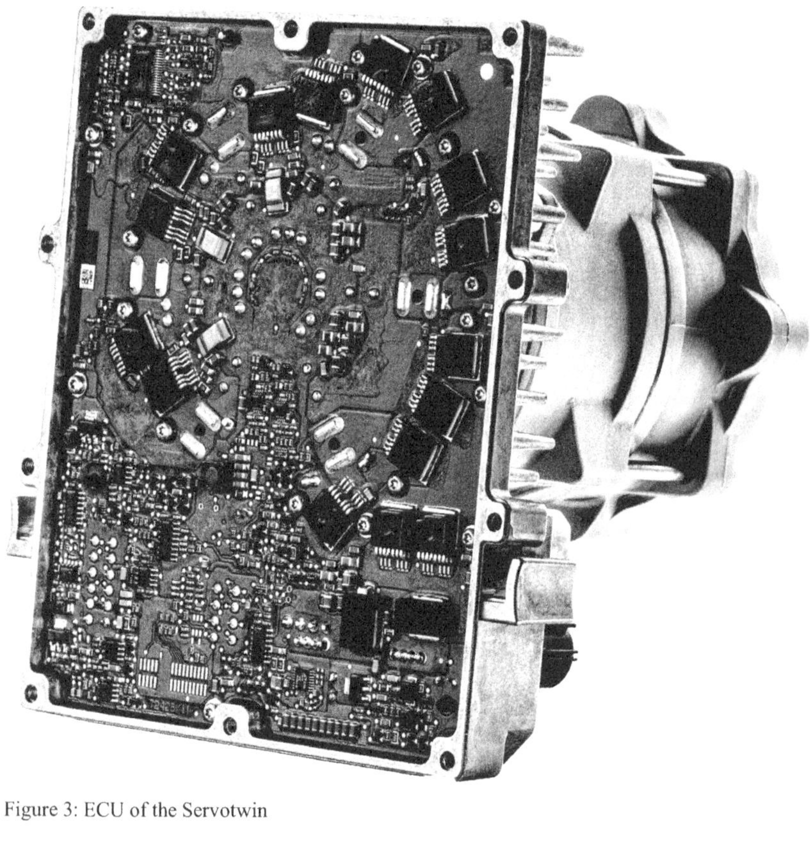

Figure 3: ECU of the Servotwin

The main electronic steering features of the Servotwin are listed in Table 1.

Table 1: Main electronic steering features of the Servotwin

| Functionality | **Technical background** |
|---|---|
| Speed-dependent steering effort | The Servotwin permits precise, fast steering activities at high speed and at the same time ease of handling of the vehicle at low speed. |
| Active return to straight ahead | After cornering, the wheels are smoothly returned to center position by the electric motor. |
| Cross wind compensation | In the event of cross wind the Servotwin gives a recommendation to the driver. |
| Realization of driver assistance functions | Necessary interfaces for additional steering torque, motor torque and angle control are available. |

Focus of this paper is related to the new possible driver assistance functions. Therefore the following three interfaces / control variables are introduced:

- Nominal steering wheel angle (respectively nominal road wheel angle)
- Additional steering torque
- Additional motor torque

Figure 4 presents these interfaces / control variables in the driver-vehicle-control loop.

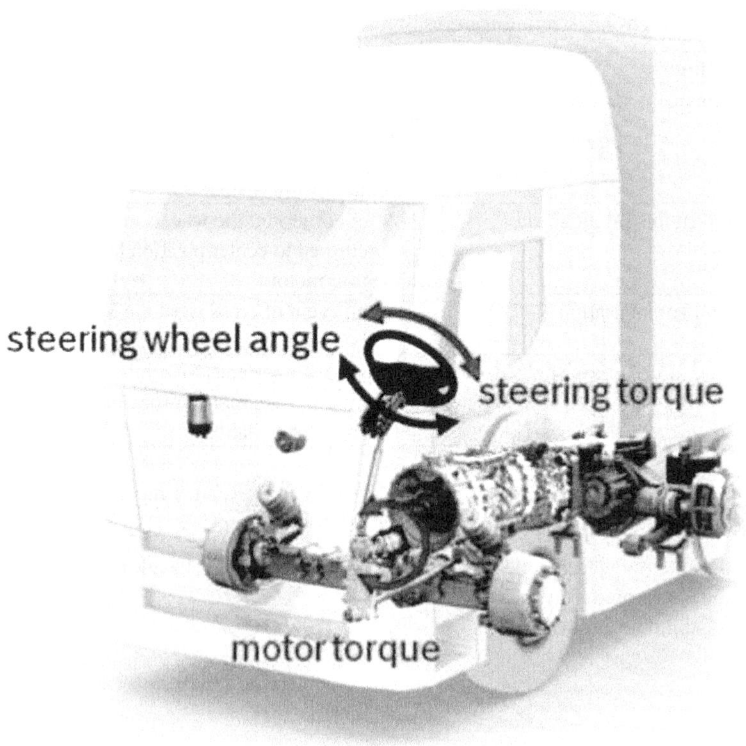


Figure 4: Interfaces / control variables for driver assistance functions

Driver assistance functions can be implemented on the Servotwin ECU, but it is also possible, that the control variables are calculated on separate ECUs and then transmitted via CAN to the Servotwin ECU.

### *Legal aspects for steering systems*

Steering systems have to fulfill several requirements to get the road release approval for the steering system. These requirements are defined in the regulation ECE R79. In the context of driver assistance systems the following requirements have to be considered especially, most important parts are highlighted:

- *Section 2.3.4.:* "Advanced Driver Assistance Steering System" means a system, additional to the main steering system, that provides assistance to the driver in steering the vehicle but in which the **driver remains at all times in primary control of the vehicle**. It comprises one or both of the following functions [1]:
- *Section 2.3.4.1.:* "Automatically commanded steering function" means the function within a complex electronic control system where actuation of the steering system can result from automatic evaluation of signals initiated on-board the vehicle, possibly in conjunction with passive infrastructure features, to generate continuous control action in order to assist the driver **in following a particular path, in low speed manoeuvring or parking operations** [1].
- *Section 2.3.4.2.:* "Corrective steering function" means the discontinuous control function within a complex electronic control system whereby, for a limited duration, changes to the steering angle of one or more wheels may result from the automatic evaluation of signals initiated on-board the vehicle, in order to **maintain the basic desired path** of the vehicle or to influence the vehicle's dynamic behaviour.
  Systems that do not themselves positively actuate the steering system but that, possibly in conjunction with passive infrastructure features, simply warn the driver of a deviation from the ideal path of the vehicle, or of an unseen hazard, by means of a tactile warning transmitted through the steering control, are also considered to be corrective steering [1].
- *Section 5.1.6.:* Advanced driver assistance steering systems shall only be approved in accordance with this Regulation where the function does not cause any deterioration in the performance of the basic steering system. In addition they shall be designed such that the **driver may, at any time and by deliberate action, override the function** [1].
- *Section 5.1.6.1.:* Whenever the Automatically Commanded Steering function becomes operational, this shall be indicated to the driver and **the control action shall be automatically disabled if the vehicle speed exceeds the set limit of 10 km/h by more than 20 per cent or the signals to be evaluated are no longer being received**. Any termination of control shall produce a short but distinctive driver warning by a visual signal and either an acoustic signal or by imposing a tactile warning signal on the steering control [1].

# Lane Keeping Assistance

Lane Keeping Assistance (LKA) is a driver assistance function, which provides an additional steering torque in a continuous and harmonic way. Basic idea is to interactively support the driver in his lane keeping task. The following chapter will give an overview about the motivation behind LKA and the function itself.

*Motivation*

The long term target of road traffic safety is to achieve a highway system with no fatalities or serious injuries in road traffic, the so called Vision Zero. In Figure 5 the statistic of road traffic accidents in Germany in 2013 is shown, categorized in five different types of accidents [2]. Lane departure represents with 39% the highest percentage of road accidents in Germany. Thus, lane keeping functions could have a significant contribution in reducing road traffic accidents. For this reason Bosch Automotive Steering started in March 2014 to develop the steering function Lane Keeping Assistance (LKA) for commercial vehicles. LKA not only addresses road traffic safety, but also aspects of driving comfort by reducing drivers' stress.

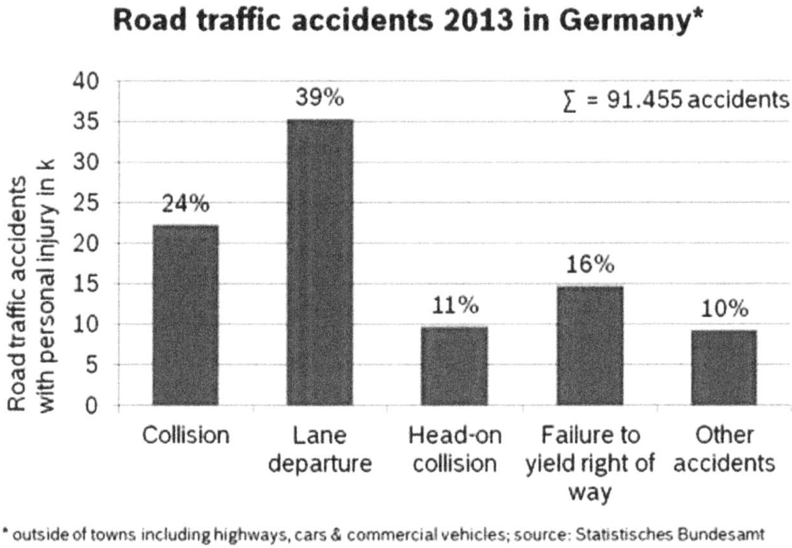


Figure 5: Road traffic accidents in Germany in 2013

*Functionality*

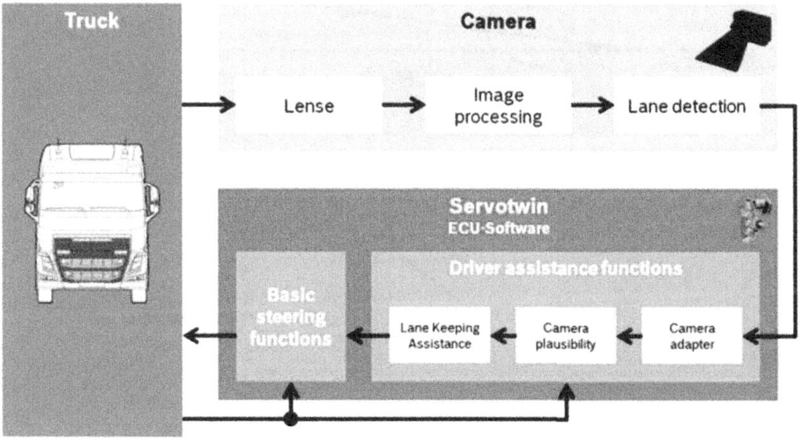


Figure 6: Control loop of LKA

Figure 6 describes the control loop of the LKA system. It consists of three components *Truck*, *Camera* and *Servotwin*.

The *Camera*, which is mounted in the *Truck*, determines from the video data and various vehicle data (e.g. vehicle speed and yaw rate) the course of the lane markings, as well as the vehicle position and orientation referred to. The required environment signals for LKA are:

- Description of the lane
    - Curvature $\chi$
- Description of the vehicle position
    - Lateral deviation to lane marking $\Delta y$
    - Yaw angle to lane marking $\Psi rel$

In Figure 7 the required environment signals from the *Camera* are exemplary shown on left lane marking. This data is transmitted to *Servotwin* by CAN bus.

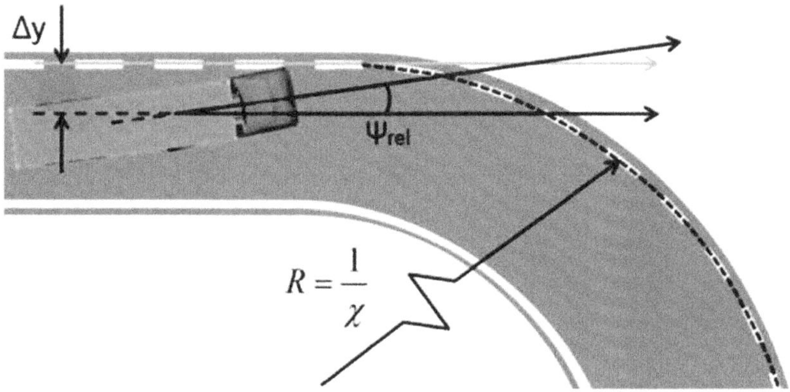


Figure 7: Required environment signals for LKA

Within the *Servotwin* the camera data is first converted to the data format required for LKA (*Camera adapter*) and then checked for plausibility (*Camera plausibility*), see Figure 6 .

Finally in the *Lane Keeping Assistance* module, the additional steering torque of LKA is calculated. The detailed structure of the *Lane Keeping Assistance* module is shown in Figure 8. Based on the adapted and checked camera and vehicle data the additional steering torque of LKA is calculated in several steps.

First of all various nominal signals are calculated in *Calculation nominal signals*. Depending on the course of the road and the vehicle position within the lane two nominal trajectories are determined. The *Feedforward* trajectory defines the nominal position of the vehicle within the lane, e.g. lane center or with adequate offset. In case of having a deviation between the nominal and the actual position a *Feedback* trajectory will be calculated, which describes the vehicle trajectory back to the *Feedforward* trajectory/position. In the next step each trajectory is evaluated on an adaptive forecast point to get nominal values for road wheel angle and yaw rate. After the calculation of nominal signals the shaping of the additional steering torque is realized. Therefore in *Calculation additional steering torque* a raw additional steering torque of LKA is calculated. In *State machine + fault handling* the raw additional steering torque is finally adapted and provided to the driver assistance interface. The controlled additional steering torque of LKA is a haptic steering recommendation for the driver at the steering wheel. If the driver follows that steering recommendation / additional steering torque, the vehicle will automatically remain on the *Feedforward* trajectory. The driver can at any time easily override the LKA function and remains at all times in primary control of the vehicle.

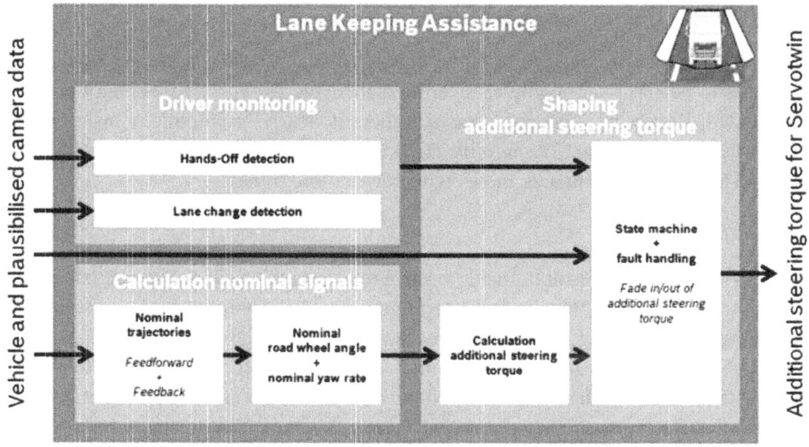


Figure 8: Structure and algorithm of *Lane Keeping Assistance* module

The final shaping depends on the driving situation and the driver activity (*Driver Monitoring*). For example the raw additional steering torque is faded out in the following situations:

- Camera fault / implausible data
- Exceed / fallen below allowed vehicle speed
- Detection of lane change by driver intention, indicater, etc.
- permanent, constant lateral deviation (temporal threshold)
- Detection Hands-Off
- high lateral acceleration
- high steering angle velocity
- high steering torque
- Intervention of vehicle dynamic control systems

# Smart Truck Maneuvering

Smart Truck Maneuvering (STM) is a tablet based maneuvering assistant system for trucks with one or more trailers. With STM the driver is able to remote control the truck from outside with a tablet. The following chapters will give you an overview about the motivation behind STM and the function itself.

*Motivation*

Reversing a truck with one trailer is the biggest challenge while driving a truck. The driver has to perform very complex steering maneuvers and his view behind the truck is restricted. Reversing a truck with two trailers is nearly impossible. Crashes often occur while maneuvering. To simplify the process of maneuvering Bosch introduced STM. STM is able to maneuver trucks with one or more trailers.

*Control Loop*

With STM the driver does not directly control the steering wheel and the pedals of the truck. The driver inputs the nominal lateral movement of the truck by introducing the trailer angle with an app on a tablet. The trailer angle is the angle between the last trailer and the rest of the truck. This nominal trailer angle is than transmitted to a trailer angle controller which is implemented on the Servotwin. The trailer angle controller computes the required steering angle to reach this trailer angle and the Servotwin controls this steering angle. The steering angle results in lateral movement of the truck and trailer. The driver monitors this movement and corrects the nominal trailer angle if necessary. Figure 9 displays the control loop.

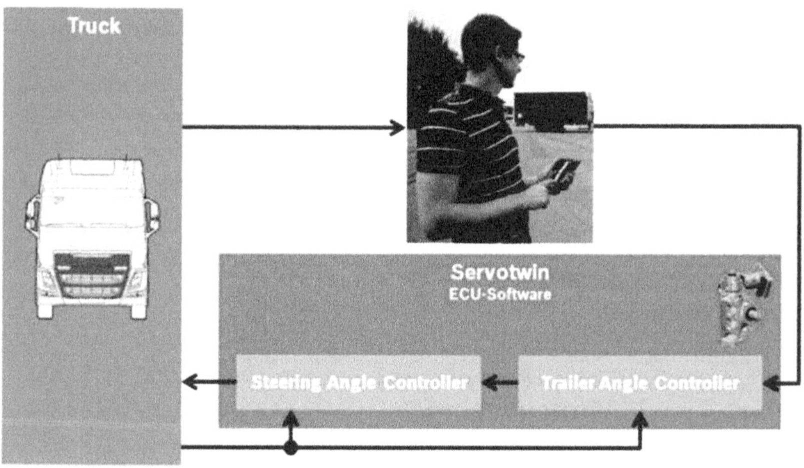


Figure 9: Control Loop of Smart Truck Maneuvering

### *Trailer Angle Controller*

The trailer angle controller calculates the required steering angle to achieve the nominal trailer angle inputted by the driver. The trailer angle controller is scalable and works with one or more trailers. Figure 10 displays the one-track model of a semi-truck with semi-trailer and tandem-trailer.

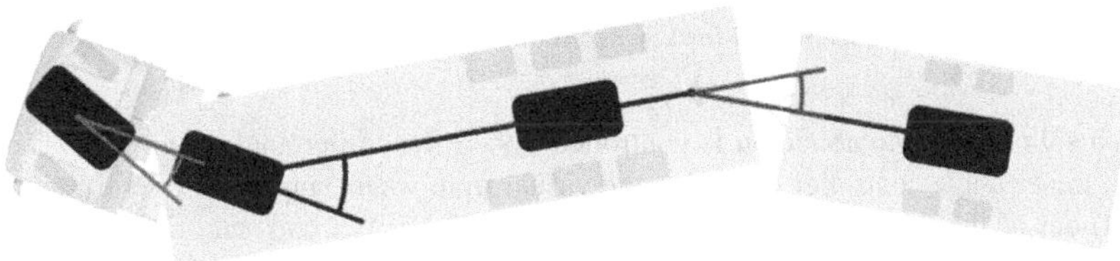

Figure 10: One-track model of semi-truck with semi-trailer and tandem-trailer

A trailer angle controller that calculates the front wheel angle (green) directly from the nominal trailer angle between semi-trailer and tandem-trailer (blue) is very complicated because of the complex model of the vehicle combination and is not easily scalable. Because of that a different approach is used. The vehicle combination is split into two smaller vehicle combinations that each consist of only one trailer and one tractor. The first split vehicle combination consists of the tandem-trailer as trailer and the semi-trailer as tractor. The semi-truck plays the role of a virtual front wheel of the semi-trailer. Figure 11 displays the one-track model of the split vehicle combination of tandem-trailer and semi-trailer.

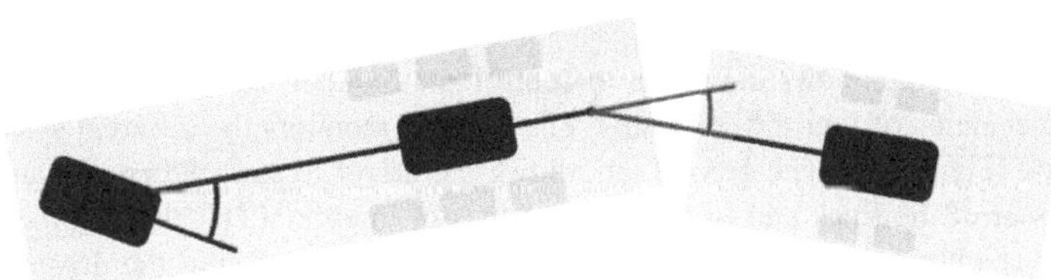

Figure 11: One-track model of semi-trailer with tandem-trailer

The second split vehicle combination consists of the semi-trailer as trailer and the semi-truck as tractor. Figure 12 displays the one-track model of the split vehicle combination of semi-trailer and semi-truck.

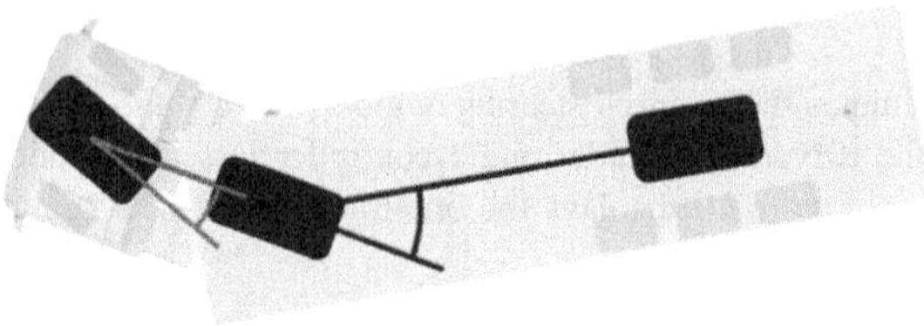

Figure 12: One-track model of semi-truck with semi-trailer

Each split vehicle combination is controlled by an own trailer angle controller. The first trailer angle controller computes the virtual front wheel angle of the semi-trailer (red) out of the nominal trailer angle between tandem-trailer and semi-trailer (blue). The second trailer angle controller computes the front wheel angle of the truck (green) out of the trailer angle between semi-trailer and semi-truck (red). The trailer angle between semi-trailer and semi-truck is equal to the virtual front wheel angle of the semi-trailer. Figure 13 displays the cascade of the two trailer angle controllers.

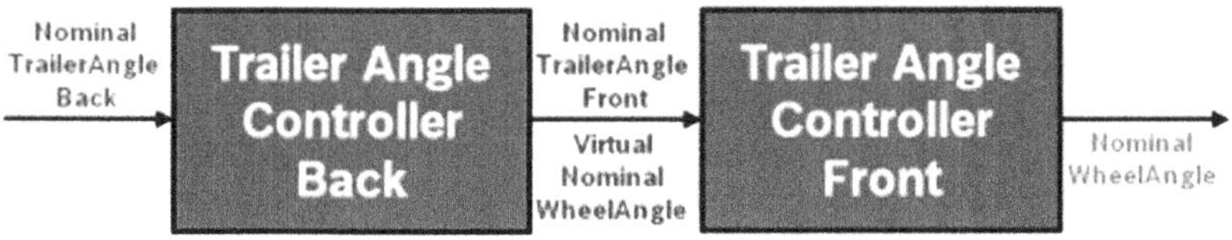


Figure 13: Cascade of the trailer angle controllers

### *App*

The STM-App allows the driver to control and monitor the vehicle combination. The app contains of four different tabs. The first tab monitors the different systems of the truck (Figure 15). The second tab allows the driver to start the engine and air compressor of the truck and to check the values of important signals as voltages and angles (Figure 14). The third tab displays the current position of the driver around the truck and the cameras that are selected because of the position of the driver (Figure 17). The fourth tab displays the live video feeds of the selected cameras and allows the driver to remote control the truck (Figure 16). The driver is able to preselect the driving direction and the driving speed. With touching the active part of the vehicle combination (semi-truck for forward and tandem-trailer for reverse) the vehicle combination starts to move with the preselect speed. By laterally moving the active part the driver is able to input the nominal trailer angle. Removing the finger from the tablet stops the vehicle combination.

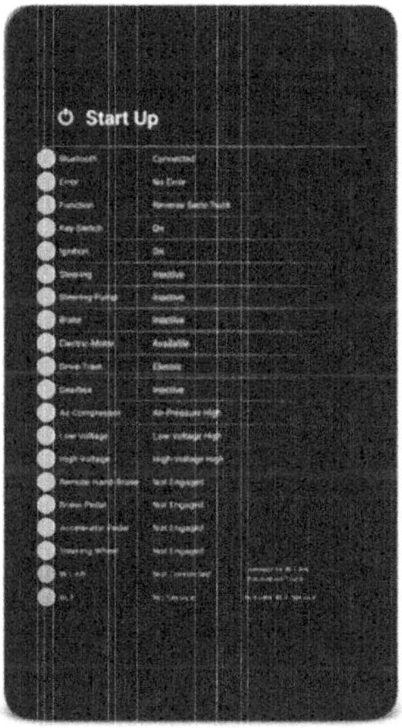


Figure 15: App – StartUp

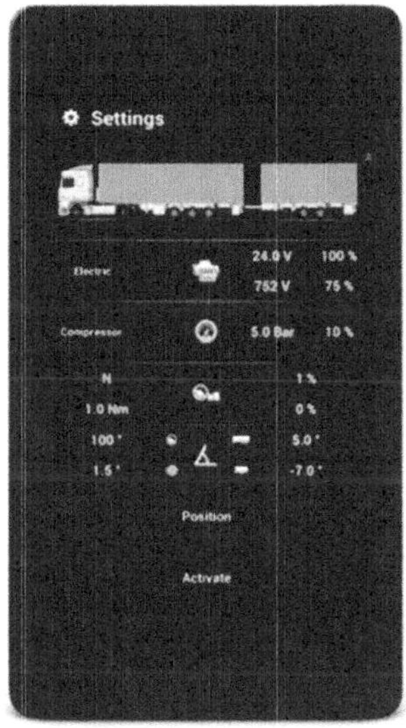


Figure 14: App – Settings

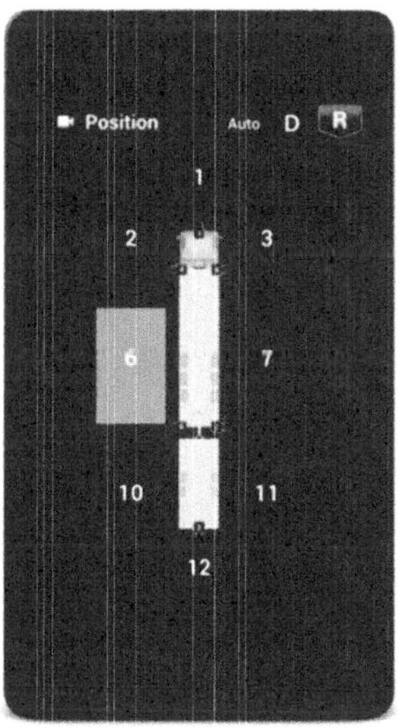


Figure 17: App – Position

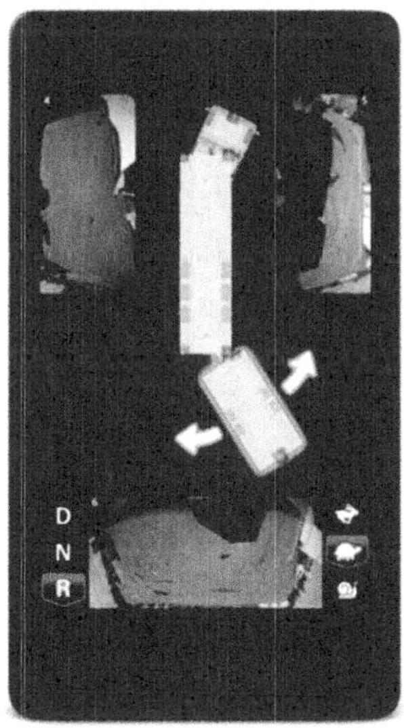


Figure 16: App – Control

### *Development*

App and trailer angle controller were developed and tested within 6 months. The main development was done with the simulation environment IPG TruckMaker. Figure 19 displays the development cycle and Figure 18 displays the simulator. At first the simulator was used to drive the truck manually to identify the behaviour of the truck while reversing with multiple trailers. Out of these insights the trailer angle controller was developed. Later the trailer angle controller and app were tested with the simulator. Because of testing the whole system with the simulator the first integration of the function into the real truck happened within two days. Without the simulator it would have not been possible to develop the function and app within 6 months.

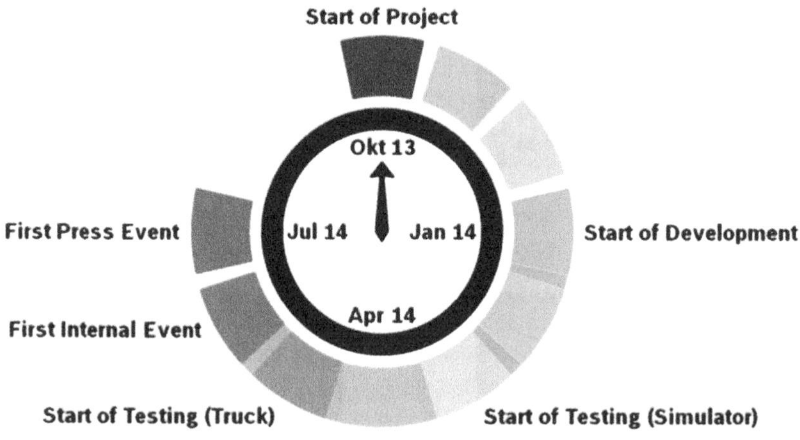


Figure 18: Development Cycle

Figure 19: Simulator with IPG TruckMaker

# Conclusion

The Servotwin enables for the first time the implementation of steering functions in commercial vehicles to improve comfort and safety, which are already state of the art in passenger cars. Especially the second generation of Servotwin offers all necessary interfaces to realize driver assistance functions.

With Lane Keeping Assistance (LKA) a driving assistance function was introduced, which interactively supports the driver in his lane keeping task by providing an additional steering torque. This steering functionality allows the driver a comfortable lateral guidance and has the potential to reduce road traffic accidents caused by unintentional lane departure. The development of LKA for trucks means a demanding challenge, although comparable functions exist already in passenger cars. The main differences are:

- less room within the lane for control
- high vehicle weight and inertia (vehicle reaction)
- electro-hydraulic steering system
- stiffness of connection from steering wheel to the road wheels (no rack and pinion steering system)

To understand the requirements and master the challenges a test vehicle was build up and proband studies were done. More user experiences are planned to optimize the LKA function.

With Smart Truck Maneuvering (STM) a tablet based maneuvering assistance systems for trucks was introduced. It helps to reduce drivers' working hours and crashes while maneuvering. It enables everyone to reverse trucks with one or more trailers and simplifies the process of coupling trailers.

# Literature

[1] United Nations (2005): ECE Regulation No. 79, Revision 2. P.6-7 and p.13-14

[2] Statistisches Bundesamt (2014): Verkehr Verkehrsunfälle – Fachserie 8 Reihe 7 – 2013. p.72

# Development of a driving dynamics-oriented suspension design during the early concept phase

Karthik Vemireddy, M.Sc
Dipl.-Ing. Torben Dittmar
Univ.-Prof. Dr.-Ing. Lutz Eckstein

Institut für Kraftfahrzeuge, RWTH Aachen University (ika)

Dipl.-Ing. Lars Hesse
Dipl.-Ing. Peter Rettweiler

Forschungsgesellschaft Kraftfahrwesen mbH Aachen (fka)

## Abstract

During the initial phase of concept development, the goals for suspension design are generally limited to achieving certain kinematic and elasto-kinematic characteristic curves. The evaluation of the resultant vehicle dynamics characteristics generally takes place at a later stage. The tools as well as tool chains, which are typically used for evaluating the influence of suspension parameters on driving dynamics, are expensive in terms of time, required effort and computing power. Also nonconformity with desired goals at a later stage may result in additional investment of time and effort in large magnitudes for the corrective measures leading to an unnecessary prolongation in the development process. Thus, being able to directly evaluate the influences of suspension kinematics on vehicle dynamics during concept phase, and optimise either the suspension characteristics or suspension parameters based on the vehicle dynamics requirements would result in a shorter and more cost effective development process.

This work focuses on the methodology developed at the Institute for Automotive Engineering, RWTH Aachen University that will help overcome the above mentioned shortcomings and accelerate the development process. The methodology is realized in the form of a comprehensive tool chain that allows direct evaluation of the influences of suspension parameters (e.g. hardpoint location of kinematic elements, bushing stiffness, etc.) as well as suspension characteristics (e.g. kinematic characteristics) on the vehicle dynamics behaviour. Later, an application example is presented to demonstrate the scope of the methodology and its results will be evaluated. Finally, the scope for further development of the methodology is also presented.


## 1 Introduction

During the early concept stages of passive suspension development, today the focus is mostly on achieving appropriate kinematic and elasto-kinematic characteristics. The resulting vehicle dynamic characteristics need to be reviewed at a later point in time. If the objectives for the entire vehicle have not been met, then corrective measures, which may take a large amount of time and effort to complete, are required.

A method is presented in this paper that will be able to speed up this process, which allows the suspension developer to carry out an initial evaluation of the expected full vehicle driving dynamic characteristics at an early stage in the design process and in doing so, achieve a higher degree of maturity of the suspension concept early on. A suspension analysis tool has been developed for this at the Institute for Automotive Engineering, RWTH Aachen University for investigating the kinematic and elasto-kinematic characteristics of suspensions, which is coupled to an efficient full vehicle driving dynamics model. An application example at the end of the paper will provide insight into the various advantages and possibilities of this method.

While in the past, different approaches regarding this matter have been published such as [1] demonstrating the possibility to optimise suspension hardpoints to achieve required kinematics and [2] directly optimizing vehicle dynamics in a multi body simulation setup, so far an integrated approach to combine suspension characteristic optimisation and vehicle dynamics response is unique for the approach illustrated in this paper.

## 1.1 Suspension Parameters

The analyses presented here concentrate on the influence of suspension parameters on the driving dynamics. Consequently, the focus is placed on accurately depicting the suspension characteristics in the vehicle dynamics model. This includes, amongst other things, observing the wheel position characteristics

- toe,
- camber,
- spring rates – main springs and stabilizer bars,
- roll and pitch pole position,
- track width and wheelbase changes,

the steering characteristics

- toe and
- camber,

the elasto-kinematic wheel position changes depending on the lateral and longitudinal forces as well as the lateral and longitudinal stiffness. The definition of these characteristics can be concluded from the relevant literature, e.g. [3].

## 1.2 Vehicle Dynamics Models

In literature, a multitude of different modelling approaches for analysing vehicle driving dynamic phenomena can be found. This wide spectrum includes everything from the simplest linear models with only a few degrees of freedom (for example, the linear single track model) to the most complex multi-body systems with flexible structures and complete control systems. Fig.1 shows an overview of the modelling approaches and sorts them according to increasing model complexity and effort required for parameterisation.

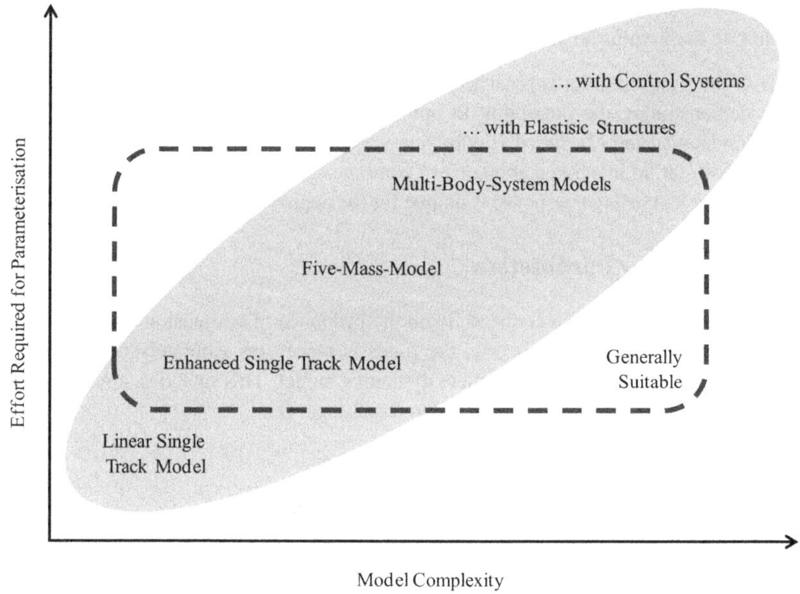


Fig. 1: Categorisation of the conventional modelling approach according to [4]

Generally models from the middle of the graph are best suited for analysing passive suspension concepts. These models are suited to predict vehicle dynamics with sufficient accuracy. Although the approach for the MBS-models allows the most accurate predictions, the effort for the calculations and parameterisation increases dramatically. Since many parameters, such as component masses and inertia, are not fixed during the concept phase, the achievable increase in accuracy is limited, while the increased computational effort is disadvantageous with the large number of calculations necessary for the sensitivity analyses and parameter optimisations. In particular, the experiences from [2] show that MBS-models may not be particularly suitable for such tasks due to the large computational effort. Therefore, they are not examined in any further detail.

Analyses on enhanced single track models and five mass models have shown that in order to achieve an acceptable degree of predictive quality, the amount of enhancements necessary is so extensive, that only a minimal advantage in terms of the computational complexity remains. Since using a five mass model allows for a much simpler way to draw conclusions about the impact of the suspension characteristics, the five mass model will be used in the following.

## 1.3 Driving Manoeuvres

To evaluate the effect of suspension kinematics on the vehicle dynamics behaviour, it is important to choose significant manoeuvres that provide the information useful to evaluate the impact of the kinematic influences on the vehicle dynamics behaviour. For the evaluations in this paper two basic driving manoeuvres are used:

- Constant Radius Cornering (ISO 4138),
- Step Steer Input (ISO 7401).

Open-loop manoeuvres are preferred, since they eliminate any influences that a driver model has and avoids complex modelling and tuning. Even though the lateral control has to be regulated for the constant radius cornering, only the vehicle is evaluated and the driver has no influence on the results. This approach is required in order to determine the characteristic velocity. The manoeuvre evaluation proceeds according to the process described in the standard. Additional information about the driving manoeuvres can be found in the corresponding standards and the related literature, such as [5].

# 2 Research Approach

The research approach presented here will outline the method and will describe the required features of the tools.

## 2.1 Methodology

The process presented here can be broadly classified into suspension level and full vehicle level analyses (see Fig. 2). In the kinematic suspension level analysis, using the geometric hardpoints and vehicle parameters, the suspension properties and characteristics are determined using appropriate load cases. The influence of these suspension characteristics on the vehicle dynamics are then analysed using appropriate manoeuvres in the dynamic full vehicle level analysis. This process constitutes the backbone of the approach.

Previous works have performed the optimisation at suspension and at vehicle levels. But the suspension geometry has not yet been considered in the vehicle level optimisation. The approach presented here will perform the optimisation of the suspension hardpoints with achieving the desired vehicle dynamics as the objective. The approach will be described in detail in Chapter 3.

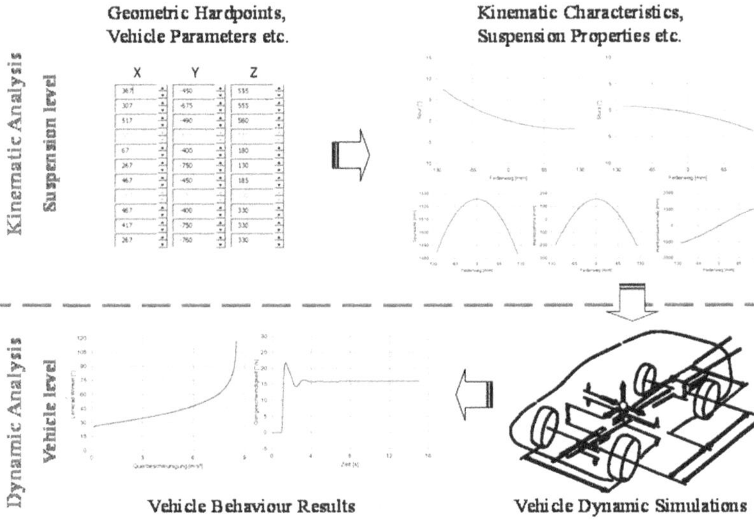


Fig. 2: Methodology of the tool chain

## 2.2 Requirements of the tool

The task of the tool chain here is to help analyse the influences of the vehicle parameters and the suspension properties on the vehicle dynamics behaviour. A tool chain that is capable of achieving these goals has to meet certain requirements and at the same time has certain demands, which have to be fulfilled to ensure its proper functioning.

The tool should be able to perform the kinematic analysis in such a way that it depicts all relevant effects of suspension parameter variations on vehicle dynamics with a sufficient level of accuracy. The vehicle model should be sufficiently detailed to consider all the effects of the suspension kinematics while determining the vehicle dynamics.

The tool chain has to be able to not just accurately determine the vehicle behaviour, but also be able to optimise the suspension parameters to achieve a desired driving behaviour. In order to perform the optimisations, an interface has to ensure that the data from one module is processed and transferred in a format that is acceptable to the other module. For performing optimisation using evolutionary methods, which involve numerous iterations, but yield better results especially for non-smooth problems like this application [6], the model should, apart from being accurate enough, permit fast computation.

# 3 Development of the tool chain

This chapter will elaborately describe the tool chain and the tools involved. The limitations and the challenges faced by predecessors, who have worked with optimisation of suspension characteristics and vehicle dynamics are taken into consideration while choosing the options for the realisation of this methodology.

## 3.1 Vehicle Model

As mentioned in chapter 1.2, the chosen full vehicle model works on the principles of a five mass model. The vehicle model allows considering the influences of all common design targets from suspension kinematics. The tyres are modelled according to the Magic Formula [7]. The vehicle model also has a driver for longitudinal and, in the case of constant radius cornering, also for lateral control.

The five mass vehicle model incorporates the suspension properties in form of characteristic curves. These curves provide the information about the position and the orientation of the wheel carrier for the corresponding wheel travel and tierod travel. These characteristic curves are calculated from the geometric suspension properties by using specially developed kinematics simulation software, see chapter 3.2.

An interface between the kinematic calculation and the vehicle model transforms the required suspension characteristics information into a format acceptable to the vehicle model. Routines automatically modify the vehicle model's parameters, based on the factors obtained from the optimisation algorithm, and consecutively perform the simulations of the various manoeuvres (refer to 1.3). In addition, automated evaluation routines provide the results for the relevant target values to the optimisation algorithm.

## 3.2 Suspension Kinematics Software

For the analysis at the suspension level, the in-house developed, axle kinematics simulation software 'aksis' is used. Using the geometric hardpoints of the suspension, the software calculates the required wheel kinematics using an approach by Matschinsky [3]. It is capable of providing generic wheel kinematic curves, motion ratios for spring and damper elements and also the wheel carrier and tire contact patch displacements for commonly considered load cases such as parallel wheel travel, opposite wheel travel, single wheel travel and steering. The characteristics are calculated using the equations of motion, by calculating the velocity of the wheel carrier based on vertical wheel travel and steering. The effective and quick calculation facilitates an immediate depiction and evaluation of parameter variations, making it not just very intuitive to use its graphical user interface, but also well-suited to be used for applications where calculations of a multitude of variants is necessary, such as sensitivity analyses or optimisation problems.

The tool provides the suspension related information required by the vehicle model in the form of wheel centre displacement maps in all six degrees of freedom and relative displacements of spring-damper elements as functions of wheel travel and tierod travel.

## 3.3 Optimisation

This chapter describes the optimisation strategies that can be used for the approach, the algorithm, criteria used for evaluating the dynamic behaviour and the fitness value, which quantifies the evaluation for the optimisation algorithm.

### 3.3.1 Optimisation Strategy

The method adopted for performing the vehicle dynamic and suspension analyses provides the possibility of choosing among various optimisation strategies. One such method is to perform the whole optimisation in a single step as shown in Fig. 3.

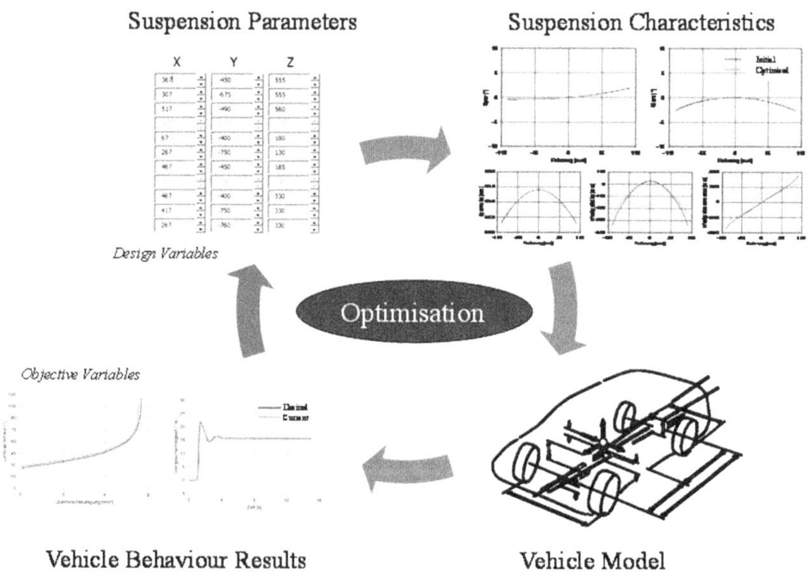


Fig. 3: Global, single step optimisation approach

In this method, the influence of the suspension parameters on the vehicle dynamics is directly evaluated. When attempting to optimise the dynamic driving behaviour of a vehicle configuration, it, however, quickly becomes apparent that such a global approach, which effectively necessitates enabling of all parameters at once while simultaneously optimizing all of the targets, is not effective. Even powerful optimisation methods such as evolutionary algorithms will have difficulties to converge on an optimal solution. Depending on the type of problem, various simplification approaches can lead to a more successful optimisation.

One of the possible simplified approaches would be a sequential optimisation, where individual parameters can directly be assigned to specific targets as main influencing factors, so that the targets can be optimised by varying the relevant parameters only. This approach assumes that a clear differentiation is possible and cannot be used in all cases.

Alternatively, the optimisation can be done in two stages, i.e. one optimisation at suspension level and another optimisation at vehicle level. The set point curves of the kinematic characteristics are derived from an initial optimisation of the driving dynamic targets (see Fig. 4).

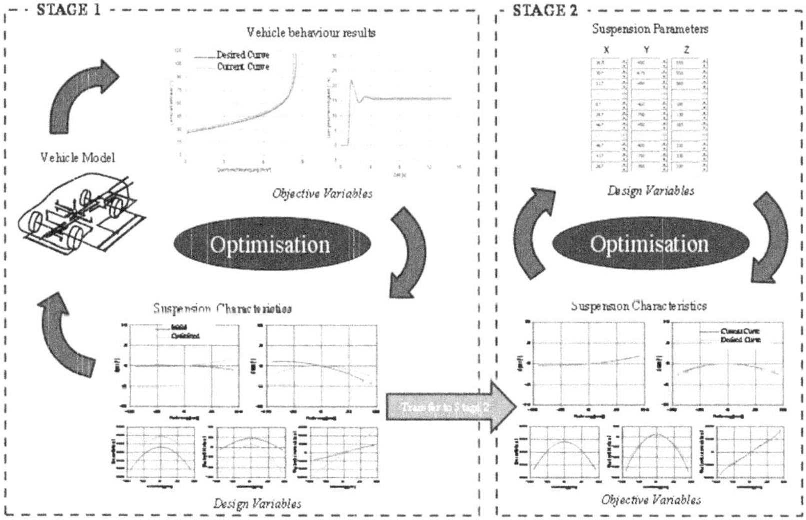


Fig. 4. Two stage optimisation approach

In the next stage, the hardpoints and suspension parameters are then optimised with these design targets obtained from the first optimisation. In a final step, again full vehicle simulations with the resultant curves from the suspension optimisation are performed to verify the ultimately achieved vehicle behaviour. To keep the number of design variables low, with this approach, it is recommended to represent the suspension characteristics at vehicle level through polynomials.

Due to the use of multiple stages, this optimisation method provides the target characteristic curves, providing also insights about the demands on the suspension characteristics for a desired driving behaviour. For the reasons mentioned in this sub-chapter, the multistage optimisation approach shall be used in this paper. Thus a schematic structure of the tool chain is chosen as illustrated in Fig. 5.

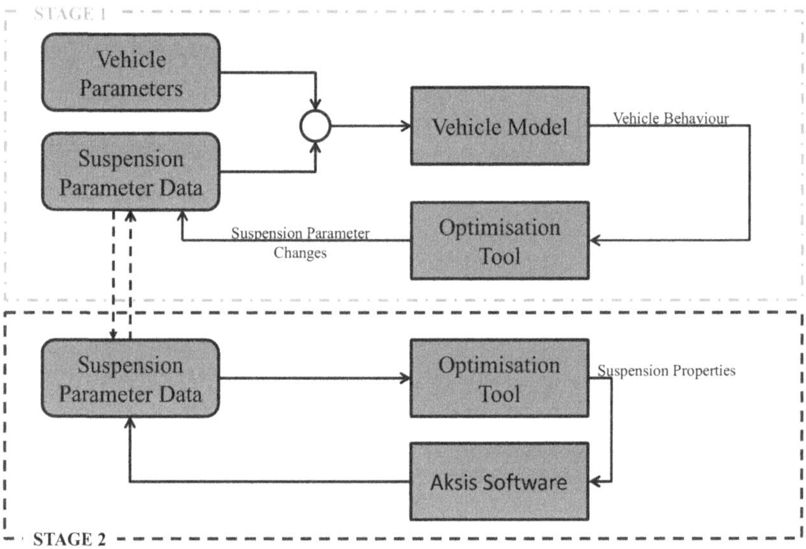


Fig. 5: Schematic view of the tool chain

### 3.3.2 Optimisation Algorithm

The optimisations at both the vehicle and suspension levels are performed using a Genetic Algorithm, which performs a random, yet directed search to locate a globally optimal solution. Hence for non-linear or non-smooth problems like the ones in this paper, genetic algorithms are more suitable than gradient based techniques as the search is not biased towards the locally optimal solutions [8]. At both suspension and vehicle levels, there are multiple objectives that have to be optimised. These objectives are reduced into a single objective using an *A Priori* weighted average.

Special routines automatically trigger the vehicle model to consecutively perform the simulations, import and evaluate the results with the help of the criteria mentioned in chapter 3.3.3. At the suspension level, the optimisation subroutines vary the individual points arbitrarily within the provided boundary limits to optimise the output curves.

### 3.3.3 Evaluation Criteria and Fitness Values

To evaluate the dynamics behaviour at the vehicle level, Constant Radius Cornering, as well as Step Steer Input are considered. The constant radius cornering is performed according to ISO 4138, on a circular track with a radius of 100 m. The transient manoeuvre is performed and evaluated according to ISO 7401. Additionally a ramp steer manoeuvre with a constant velocity of 100 km/h is performed to determine the steering amplitude required for a steady state lateral acceleration of 4 m/s². The step steer input manoeuvre is evaluated using peak response time and overshoot values for lateral acceleration and yaw rates. The characteristic velocity determined from the constant radius cornering is used as an evaluation criterion.

A scalar result value is determined for each of these evaluation criteria. For those criteria, which aren't inherently available as a single value, the mean value for the considered range is used, e.g., for self-steering gradient and roll gradient, the mean values for the lateral acceleration range of 0 to 4 m/s² are considered.

Absolute percentage deviation is considered as the individual fitness value for each criterion. The mean of fitness values for each manoeuvre is considered as fitness value for the manoeuvre. Weights are assigned to the mean fitness values, and their weighted sum is taken as the overall fitness value (1), which is minimized by the optimisation algorithm.

$$Fitness\ Value = \sum_i \left[ \omega_i \cdot \sum_k \left( \frac{f_{desired_{i,k}} - f_{current_{i,k}}}{f_{desired_{i,k}}} \right) \right] \tag{1}$$

The number of design variables is intelligently reduced to improve the performance and efficiency of the optimisation algorithm. The spring, damper, and stabilizer properties are taken with respect to wheel travel, so that their motion ratios remain constant and need not be optimised. For the considered manoeuvres and evaluations at vehicle level, some of these suspension parameters have little influence on the vehicle dynamics behaviour. Hence only those parameters, which have shown an influence in a sensitivity analysis, are considered for the optimisation. In this case, the displacement of the wheel carrier along the y axis and the rotation of the wheel carrier about x and z axes with respect to wheel travel; and the rotation of the wheel carrier about x and z axes with respect to steering are considered for optimisation as design variables.

The movements with respect to steering are considered for the front suspension only. The coordinate system here is defined according to DIN 70000 /ISO 8855.

A regression analysis is performed on the suspension curves to approximate them using polynomial curves in order to use the obtained coefficients as design variables in the optimisation. The coefficient of determination or $R^2$ (R-squared) is used as evaluation parameter for estimating the Goodness of Fit. A value of 0.99 is taken as criterion for evaluating the Goodness of Fit. As the curves for wheel travel and steering of a particular characteristic must pass through the same point in the design state, the constant terms in their equations must be the same, hence a single design variable should be used to represent the constant term in both the equations. The lower and upper bounds for the coefficients are created based on a predetermined solution space for the kinematic curves.

The vehicle model requires the suspension properties in form of a three dimensional characteristic map, with wheel travel and tierod travel as input parameters. As using 3D equations, even for the selected parameters mentioned above in this chapter, would increase the number of design variables drastically and also make the evaluation of the deviation between the desired and current maps very complicated during the second optimisation stage (see Fig. 4), the map is generated by using two separate sets of equations for steering and wheel travel at the design states and superimposing their values for the other wheel positions. Here, the relative movements with respect to the steering input are superposed onto the absolute values of the characteristics with respect to wheel travel.

In the second optimisation stage, the reference points are varied within set boundary limits so that the initially determined set point curves are as close to the actual value as possible. To evaluate the proximity and similarity between the desired and current curves, various evaluation criteria are considered. Known methods such as ratio of the sum of squares due to error (SSE) and total sum of squares (SST), correlation distances etc., were evaluated. As each of them had certain drawbacks and were prone to errors in certain situations, here, a method is applied, which is based on the idea presented in [1] and is further developed based on root mean square error (RMSE) with weightings and normalisation. The quality criterion using RMSE of the deviations of several reference points between the set point curve and actual value curve (see Fig. 5).

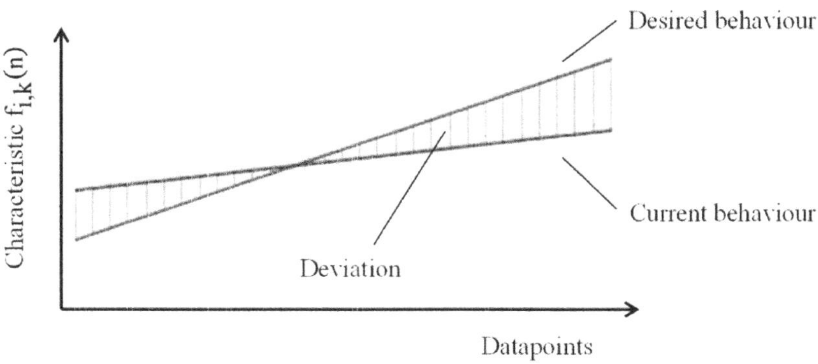


Fig. 6: Deviation of individual characteristic curve

For each objective curve (characteristic curve), the fitness value is calculated as the RMSE between the desired and current behaviour, as shown in Fig. 6. Each curve is divided into 201 data points and the deviation is calculated at all these points. Their weighted sum provides the $RMSE_{individual}$ for each curve, where the weight 'g(n)' emphasizes the influence of the selected wheel travel or tierod travel range on the overall value. This provides an opportunity to lay special emphasis on the more relevant wheel or tierod travel range 'm' about the design state in comparison to the extreme cases. The formula can be seen in (2), where the indices 'i' and 'k' represent the characteristic value and loadcase respectively.

$$RMSE_{Individual_{i,k}} = \sum_{n}^{m} g(n) \times \left[ \sqrt{\left( f_{desired_{i,k}}(n) - f_{current_{i,k}}(n) \right)^2} \right] \tag{2}$$

The overall fitness value is then determined from these individual deviations. Each individual deviation is normalized with a factor $\lambda_{i,k}$, as the characteristic curves have quite different magnitudes. This is important to avoid inaccuracies in the fitness value especially due to an undesired dominance of a less sensitive parameter with large magnitude over more sensitive small magnitude parameters. Here, following the conventional practice, the average value is not considered to avoid the cases of having division by zero or a very small value due to symmetry about the zero crossover point resulting in an unwarranted dominance. Hence the value is obtained by considering the absolute maximum value at half of the spring/tierod travel, which can be considered as a more sensible factor from a vehicle dynamics point of view. Equation (3) shows the normalizing factor, where 's' represents tierod travel for the steering and vertical wheel travel for the bounce travel.

$$\lambda_{i,k} = max\left\{\left|f_{desired_{i,k}}\right| - {}^{s}/_{2}, \left|f_{desired_{i,k}}\right| + {}^{s}/_{2}\right\} \tag{3}$$

The overall fitness value also considers the weighting of individual RMSE values to lay more emphasis on the preferred values. The total RMSE value is shown in (4)

$$RMSE_{total} = \sum_{i} \left[\frac{\sum_{k}\left(\omega_{i,k} \times RMS_{Individual_{i,k}}\right)}{\lambda_{i,k}}\right] \tag{4}$$

# 4 Application Example

To demonstrate the usefulness of the tool chain, a typical challenge due to varying customer preferences and market requirements is considered. A derivative of a D-segment reference vehicle with an extended wheelbase is created for emerging markets. The goal is to be as agile as the base variant in spite of being longer and heavier i.e., to keep the influence of this variation on the vehicle dynamics as low as possible. The suspension of the vehicle is optimised to achieve this goal, using the tool and criteria explained in chapter 3.

The reference vehicle has multilink suspensions at front and rear axles. The wheelbase of the new vehicle variant is extended by 150 mm and the curb weight is increased by 270 kg. These changes shift the centre of gravity by 33 mm in longitudinal direction towards the front axle and 111 mm in vertical direction. The inertias about the three axes about x, y and z axes are changed accordingly. The other parameters of the vehicle remain unchanged. The evaluation criteria for both the vehicles are shown in Tab. 1.

To illustrate the scope of this methodology, the tuning of the suspension parameters is done in two variants. In the first variant, only the suspension properties such as spring, damper and stabilizer properties are considered as design variables. In the other variant, also the suspension geometry hardpoints are optimised according to the two stage optimisation method (Fig. 4) as described in chapter 3.

Tab. 1: Evaluation Criteria for the reference vehicle and its derivative

| Criterion | Reference Vehicle | Longer Version |
|---|---|---|
| Characteristic Velocity [km/h] | 87 | 78 |
| Overshoot Value Lateral acceleration | 0.027 | 0.0004 |
| Overshoot Value Yaw Rate | 0.057 | 0.11 |
| Peak Response Lateral acceleration [s] | 0.71 | 0.55 |
| Peak Response Yaw Rate [s] | 0.37 | 0.37 |

The values of the reference vehicle shall be considered as the goals for the optimisation process and the longer version values before the optimisation process, would be considered as the initial values and hence will be referred as the same hereafter.

## 4.1 Optimisation of the Suspension Component Properties

In the first variant, the design variables are chosen in such a way that additional criteria that have a critical influence on the vehicle are also taken into account. The eigenfrequencies of the front and rear axle, the front and rear damper coefficients for rebound stroke, front stabilizer bar stiffness and stabilizer bar stiffness ratio are considered as design variables. The springs and stabilizer properties are considered to have linear characteristics. For the dampers, piecewise linear characteristics with two segments for each direction of travel are used. The threshold velocity at which the slope change takes place is set to 0.2 m/s. The maximum compression and rebound forces, the ratio of the compression and rebound coefficients are kept constant. All the characteristics are considered with respect to wheel travel. The lower boundaries for the design variables are set at 50% of the base vehicle values and the upper boundaries are twice that of the base values.

As mentioned in chapter 3.3.3, three manoeuvres i.e., ramp steer, step steer input and constant radius cornering are performed. The first manoeuvre is performed to determine the required wheel steering angle for the step steer input for each vehicle setup, i.e. in each optimisation loop. The other manoeuvres are used for creating the evaluation criteria for the optimisation. All the evaluation criteria are weighted equally.

This optimisation variant results in a characteristic velocity of 82 km/h. The overshoot and the peak response time for the lateral acceleration are 0.036 and 0.69 s respectively. For the yaw rate, the overshoot is 0.057 and the peak response time is 0.34 s. Comparison of the results with the desired and initial values shows that the values achieved are quite closer to the desired values.

Fig. 7 compares the yaw rate during the step steer manoeuvre for the desired, current and initial configurations of the vehicle. The overshoot value is achieved quite accurately, and the deviation of the peak response time is also significantly reduced. The gradient varies in the latter part of the step. It has to be taken into consideration that the optimisation has been performed with characteristic values and not the shape of the characteristic.

To achieve this, the vehicle eigenfrequencies at the front and the rear axles are reduced by 3 % and 35 % respectively. The damper coefficients for rebound stroke at the front and the rear axles are reduced by 20 % and 30% respectively. The front stabilizer bar stiffness is reduced by 38 % and the ratio of stabilizer bar stiffness is changed to 0.72.

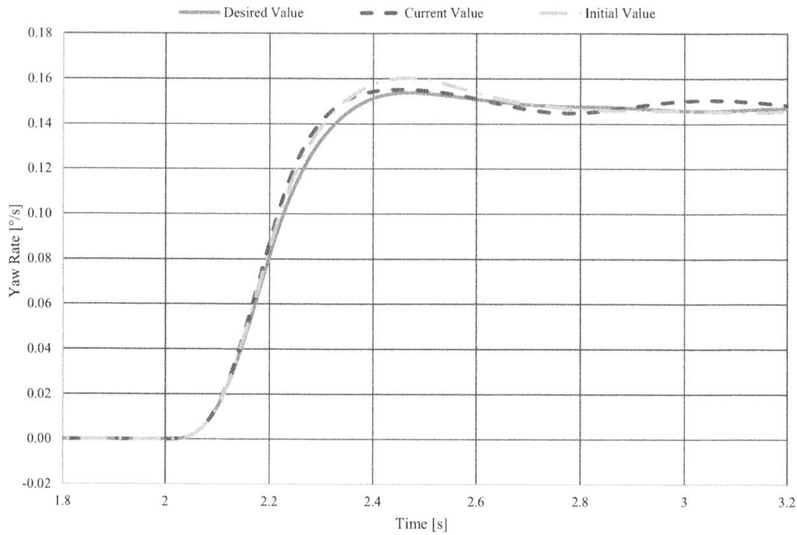


Fig. 7: Comparison of yaw rate for the first optimisation variant

## 4.2 Optimisation of the Suspension Geometric Properties

In the second variant, the optimisation is performed in two stages as described in chapter 3.3.1. Here, along with the spring, damper and stabilizer properties, the suspension characteristic curves are subsequently optimised to achieve the desired goals. The goals as well as the fitness functions are the same ones used in the first variant. In the first stage, as illustrated in Fig. 4, the optimised characteristic curves at the front axle are the displacement of the wheel carrier along the y axis for wheel travel and the rotation about the x and the z axes for both steering and wheel travel. For the rear axle, only the rotation about the x and the z axes are considered.

The polynomial equations for the chosen characteristics above are determined through regression analysis as explained in chapter 3.3.3. For the front axle, third degree polynomial equations represent the displacement along the y axis and the rotation about the z axis for wheel travel, and for the rotation about the x and the z axes for steering inputs. A second degree polynomial is used to define the rotation of the wheel carrier about the x axis for wheel travel. At the rear axle, for the vertical wheel travel, the displacement along the y axis and the rotation about the x axis can be accurately represented using a second degree polynomial. A third degree polynomial is required for the rotation about the z axis.

These equations would result in 29 design variables for the optimisation problem. As the location and orientation of the wheel must be the same at the design state for both steering and wheel travel, the constant values in the equations representing the same characteristics (such as the rotation about the x axis during steering and during wheel travel) will be equal. Hence, one design variable can be used to represent the constant term in both equations. To maintain the same static track, the constant terms for the displacement along the y axis are not optimised. Thus the total number of design variables is reduced to 25.

At the end of the optimisation at the vehicle level, the vehicle has a characteristic velocity of 87 km/h. During the step steer manoeuvre, the lateral acceleration exhibits an overshoot of 0.030 with a peak response time of 0.81 s, whereas the yaw rate had an overshoot of 0.05 and its peak response time was 0.37 s, which is exactly the same as that of the reference vehicle as it can be seen in Fig. 8.

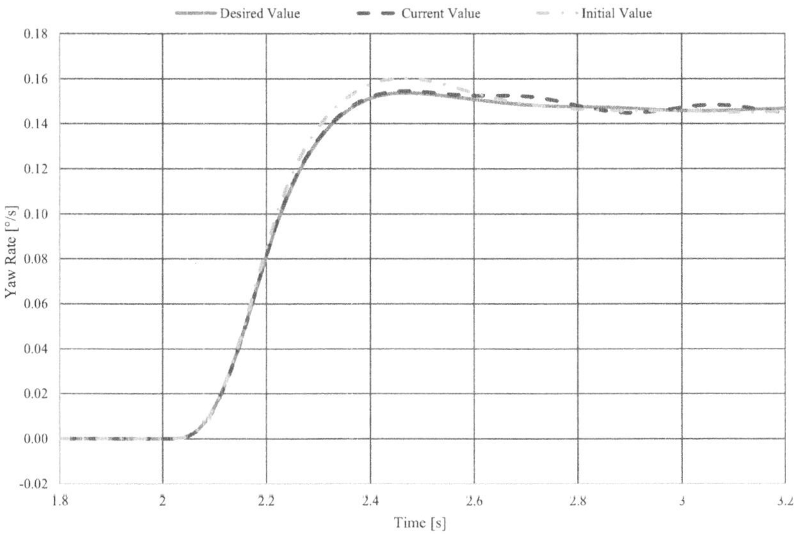


Fig. 8: Yaw rate after vehicle level optimisation in the two stage optimisation

In stage 2 (see Fig. 4), at the suspension level, the suspension geometric properties are optimised to achieve the suspension characteristics, which provide the above results. In the aksis software, only the characteristic curves, which are optimised at the vehicle level, i.e. in Stage 1, are considered as the objective values. The curves are optimised for a symmetric wheel travel of 100 mm and a tierod travel of 80 mm. The entire travel range for both the load cases is weighted equally (see Eq. (2)). The objective variables are also weighted equally (see Eq. (4)). The design variables, i.e. the hardpoints of the suspension links, are varied in a 50 mm cubical space with the start point at its centre. The wheel carrier hardpoint is not varied.

Some of the results from the optimisation at the suspension level are shown in the following figures. Fig. 9 presents the optimisation results for the lateral displacement of the wheel carrier, i.e. track change at the front suspension with respect to vertical wheel travel.

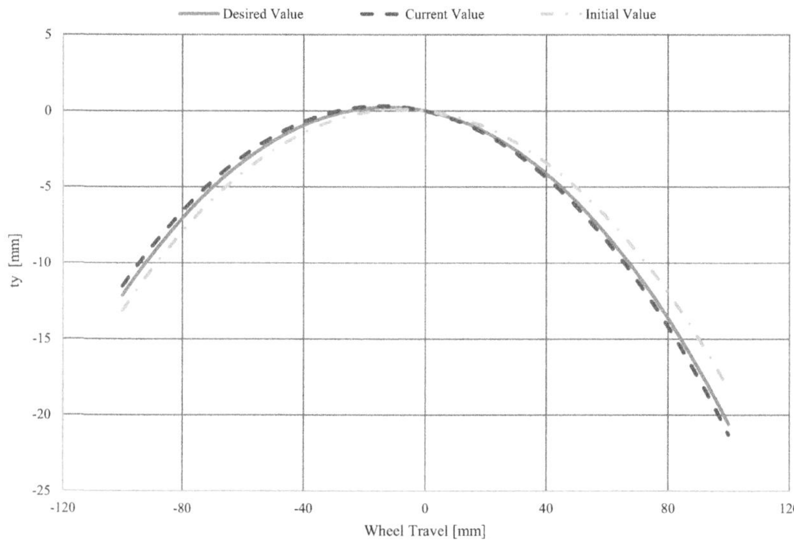


Fig. 9: Lateral displacement of the wheel for front suspension

Fig. 10 shows the rotation of the wheel carrier about the z axis, i.e. toe change with respect to the vertical wheel travel.

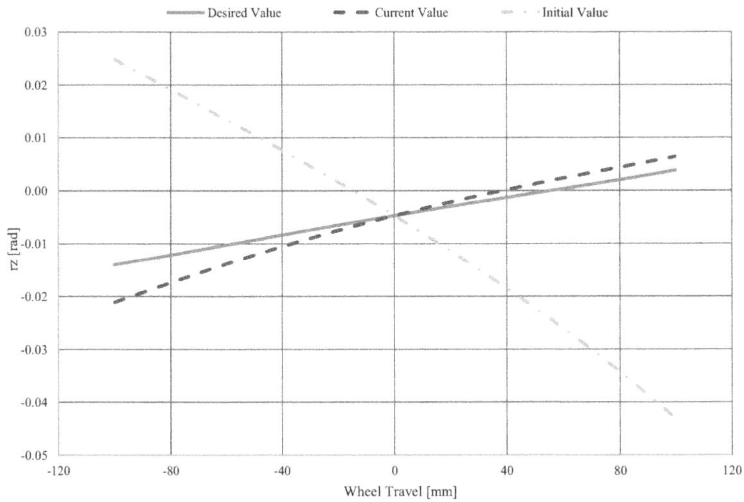


Fig. 10 Rotation of the wheel carrier about z axis for the front suspension

Fig. 11 shows the rotation of the wheel carrier about the x axis, i.e. camber change with respect to tierod travel.

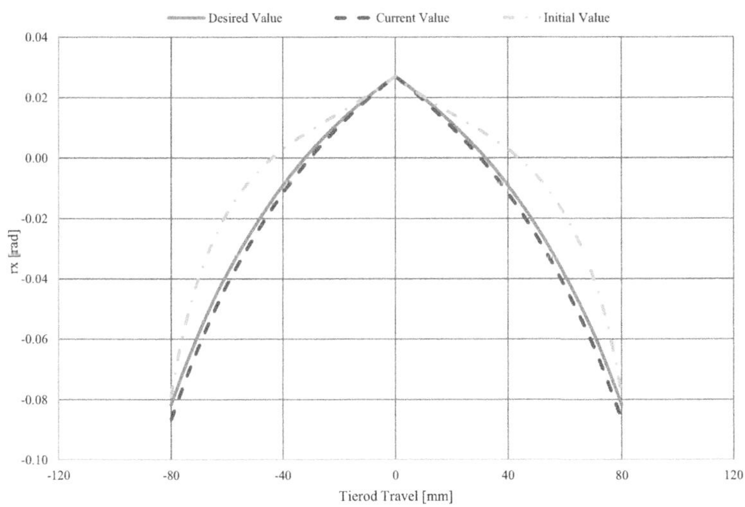


Fig. 11: Rotation of wheel carrier about the x axis for front suspension under tierod travel

The following figures show the results from the rear axle optimisation.

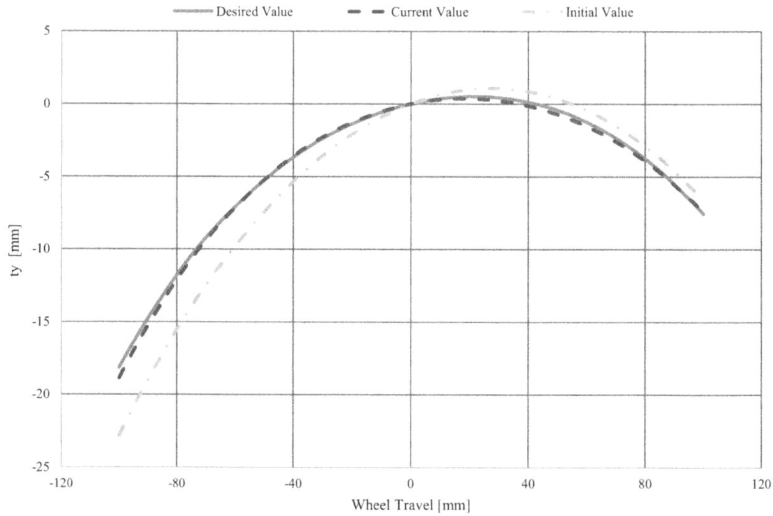


Fig. 12: Displacement of wheel carrier along the y axis for the rear suspension

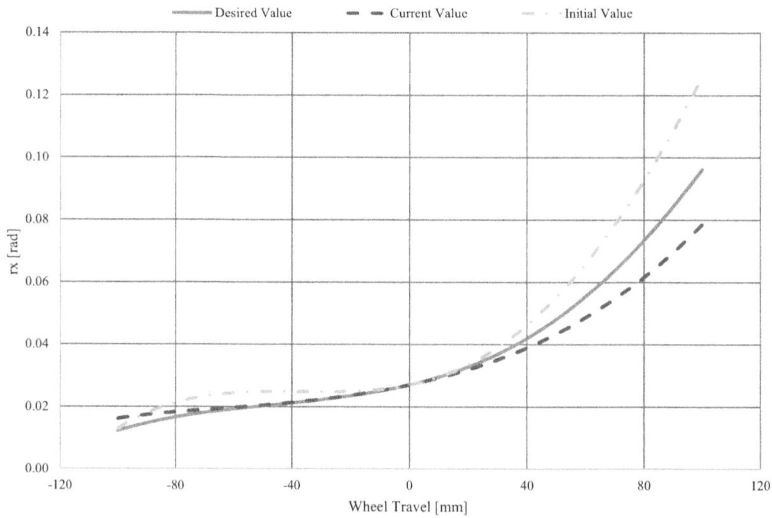


Fig. 13: Wheel carrier rotation about the x axis for the rear suspension

With the achieved kinematic curves, the final vehicle dynamic simulations are run to obtain the ultimately achieved vehicle dynamic behaviour. The outcome is a vehicle with a characteristic velocity of 87 km/h, with overshoot values of 0.029 and 0.057 and peak response times of 0.8 s and 0.37 s for lateral acceleration and yaw rate respectively. The desired yaw rate and the yaw rate after each optimisation stage is illustrated in Fig. 14. To achieve this, the suspension hardpoints at the front suspension were moved within a range of about 50 mm and within a range of 5 mm for the rear suspension.

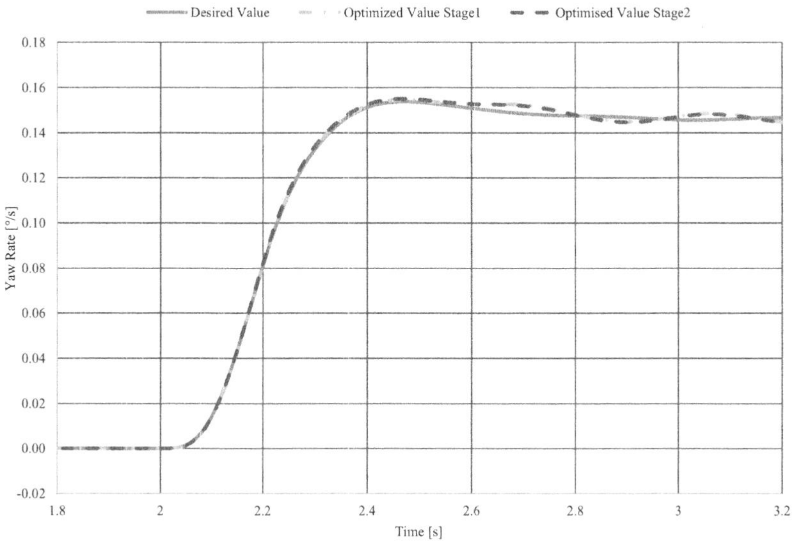


Fig. 14: Yaw rate of the vehicle after two stage optimisation

## 4.3 Evaluation of the Results

The optimisation of the suspension components has achieved the set goals with sufficient precision. There is a variation in the slope of the step. It should be noted that the optimisation function considers only characteristic values and not the curve trend. The optimisation including the suspension geometry has provided even better results, where the shapes as well as the characteristic values of the curves match quite well in the optimisation at the vehicle level. The optimisation at the suspension level uses a more complex objective function, as it has to evaluate the curves instead of individual characteristic values. The fitness function works quite well and the goals are achieved in almost all the cases. The ultimately achieved vehicle behaviour shows only small deviations from the results achieved in the intermediate step (stage 1) (see Fig. 14).

# 5 Summary

In an effort to overcome the shortcomings of the contemporary approaches of passive suspension development and to accelerate the development process, a new methodical approach has been presented. This method allow the direct evaluation of the influences of suspension kinematics on vehicle dynamics in the early concept phase, thus reducing the development time and resulting in a more cost effective development process.

To realize this approach, a tool chain has been created at the Institute of Automotive Engineering, RWTH Aachen University with the in-house developed axle kinematics simulation software and an efficient five mass full vehicle model. A two stage optimisation method is chosen, where the suspension characteristics are optimised to achieve the desired vehicle behaviour, which is evaluated using predefined criteria that are calculated by performing selected open loop manoeuvres. At the suspension level, the suspension geometric properties are then optimised to achieve the predetermined suspension characteristics from the vehicle level optimisation. With the achieved suspension characteristics, the final driving simulations are performed to obtain the ultimately achieved driving dynamics behaviour. This paper also presented an application example to illustrate the scope of the methodology, where the optimisation is performed in two variants. In the first variant, only the properties of the suspension elements are optimised and in the other variant, the suspension geometry hardpoints are also optimised. The results were then analysed.

The next steps in the development of the approach will be to increase the evaluation criteria and to incorporate elasto-kinematic analysis at the suspension level to broaden the scope of the tool. The methodology will later be implemented in one single simulation environment, so that the tool has a seamless interface between various modules and can provide a better user interface.

# References


[1] ALBERS, I.; Auslegungs- und Optimierungswerkzeuge für die effiziente Fahrwerkentwicklung; Dissertation RWTH Aachen; fka, Aachen, 2009

[2] NIERSMANN, A.; Modellbasierte Fahrwerkauslegung und –optimierung; Dissertation TU Braunschweig; Shaker, Aachen, 2012

[3] MATSCHINSKY, W.; Radführungen der Straßenfahrzeuge; Kinematik, Elasto-Kinematik und Konstruktion; Springer, Berlin Heidelberg New York, 3. Auflage, 2007

[4] MELJNIKOV, D.; Entwicklung von Modellen zur Bewertung des Fahrerhaltens von Kraftfahrzeugen; Dissertation Universität Stuttgart; Inst. A für Mechanik, Stuttgart, 2003

[5] MITSCHKE, M.; WALLENTOWITZ, H.; Dynamik der Kraftfahrzeuge; Springer, Berlin Heidelberg,4. Auflage, 2004

[6] N.N., Genetic Algorithm Optimisation Toolbox Documentation; THE MATHWORKS Inc.; Natick; 2015

[7] PACEJKA, H.-B., BAKKER, E.; The Magic Formula Type Model; Vehicle System Dynamics, Volume 21, 1993

[8] PAL, Shankar; Wamg, Paul; Genetic Algorithms for Pattern Recognition; CRC Press LLC, Boca Raton, 1996

# Development of a chassis model including elastic behavior for real-time applications

Vortragender:

Frédéric Etienne Kracht, M.Sc., wissenschaftlicher Mitarbeiter / research assistant, Universität Duisburg-Essen, Lehrstuhl für Mechatronik, 47057 Duisburg, Deutschland, 0203/379-3951, kracht@mechatronik.uni-duisburg.de;

Co-Autoren:

Yihai Zhao B.Sc., Masterand / master's degree candidate, Universität Duisburg-Essen, 47057 Duisburg, Deutschland, yi-hai.zhao@stud.uni-due.de;

Prof. Dr.-Ing. Dieter Schramm, Universitätsprofessor / university professor, Universität Duisburg-Essen, Lehrstuhl für Mechatronik, 47057 Duisburg, Deutschland, 0203/379-3275, schramm@mechatronik.uni-duisburg.de;

Dr.-Ing. Benjamin Hesse, akademischer Rat / academic councillor, Universität Duisburg-Essen, Lehrstuhl für Mechatronik, 47057 Duisburg, Deutschland, 0203/379-1604, hesse@mechatronik.uni-duisburg.de;

Dr.-Ing. Michael Unterreiner, Entwicklung Fahrwerk Vorentwicklung (FEV) / advanced chassis development, Dr. ing. h. c. F. Porsche AG, Porschestraße 11, 71287 Weissach (Böblingen), Deutschland, 0711/911-87087, michael.unterreiner@porsche.de.

# 1 Introduction

For the prediction of the dynamics of vehicles, various mathematical models have been used successfully for many years. The complexity of these models ranges from very sophisticated to rather straightforward approaches. The right to exist for simple models often results from the need for short computation times due to many iterations associated with optimization tasks or even from real-time applications. If the models are to be used as part of driving dynamics control systems these computations have to be performed on ECUs affordable for series applications.

As increased modeling accuracy generally gives results closer to the real behavior of the system it is always at the expense of extended computation time [1]. The ideal model to be strived for is therefore a tradeoff between runtime and simulation accuracy determined by the actual application. For these reasons classic active control systems in vehicles are basically oriented on basic mathematical models like the single track model of Rieckert and Schunck [1], these however proven to be sufficient for control systems as versatile as the Electronic-Stability-Program (ESP) and its variants.

Due to the trend to sophisticated active control systems and combinations of these systems as part of present-day and future vehicles, these straightforward approaches are way off what is needed for the future. In combination with the still existing limitations by ECUs this is a major challenge for the engineer in charge for the modeling.

During the development phases of a new vehicle normally high-speed computers are at hand to set up even very detailed simulation tasks. However, even then the need to reduce development cost which is associated with time, kicks in. Also the requirement for optimization tasks with more parameters and quick system adjustments in appropriate time asks for efficient simulation models [2].

For the aforementioned reasons most suspension components are modeled as rigid bodies for simulations and the bearings linking together the bodies as ideal joints without elasticity. However, for dynamic driving maneuvers, due to the bearing forces acting, the suspension components as well as the joints are subject to elastic deformations. This results in an elasto-kinematics behavior of the suspension thus changing the vehicle dynamics depended on the reaction forces acting in the bearings between the suspension components [3]. In many cases these elasticities are utilized by the suspension designer to add self-steering features to the driving behavior of the car.

This paper aims at suspension models considering elasto-kinematic behavior of the suspension. Therefore it is necessary to determine the impact of the effects due to component elasticity along with elastic joints (bushings) and to derive from that optimized vehicle dynamics models [4]. This guarantees that only relevant elasticitics are modeled. The second step is to define an approach for fast (real-time) simulations.

This is feasible if the elasto- and the rigid body kinematics are considered separately, and simple physical approaches are used.

In this paper the elastic properties of the bushings are modified in a way to reflect the structural elasticity of metallic components. In addition a physical object-oriented approach is demonstrated. These approaches enable models suitable for advanced control tasks including active suspension elements. In advanced chassis development they allow rapid adjustments and optimizations, saving time and costs. Moreover, such models allow for a fast adaptation of suspension parameters in driving simulators.

# 2 Elasticities in suspension system

Due to the reaction forces in the suspension, the suspension components are deformed elastically. This results in an elasto-kinematics behavior of the suspension and changing the dynamic behavior of the vehicle.

## 2.1 Elastic components

The car chassis consists of a variety of components including typically rim and tire, the suspension linkages, bearings, parts of the steering linkage, brakes, stabilizers, drive shafts and wheel hubs. This paper focusses primarily on the suspension linkage. In this case, the elasticity can be basically divided into two types:

- the structural elasticity of supporting components normally made from metal
- and the (partly) elastic bushings.

The structural elasticity results from the normally elastic deformation of the metal components, such as e.g. the links, the wheel carrier and the tie rod. The elasticity of the bushings in turn results from their partly rubber-like properties. This creates, in addition to the rigid body degrees of freedom (DoF), up to three additional rotary and translational DoF, respectively, combined with spring and damping forces.

## 2.2 Types of models

### 2.2.1 Linear approach with lateral tire stiffness

The simplest approach to allow for elasticities in the wheel suspension is to calculate the lateral wheel force $F_y$ via the generalized cornering stiffness $c_\alpha$ and the slip gle $\alpha$ of the correspondent wheel. The cornering stiffness is often known from measured data as a function of the normal force $F_z$. Thus the lateral force is given by:

$$F_y = c_\alpha(F_z)\,\alpha. \tag{1}$$

The major advantage of this basic approach is its real-time capability. Elasticity is covered by one additional elastic DoF per wheel. This combines all structural components with significant elastic properties (in particular the tires). The *dynamic* contribution of the elasticity of the components is not taken into account. Normally this way of modeling is used in single-track and basic two-track models [1], [2], [5]. For high dynamic driving maneuvers these simplified approaches are no longer adequate. Applications, which demand for the simulation of the system along with more detailed investigations of some parts, need a more sophisticated modeling approach.

### 2.2.2 Finite Element Analysis

The simulation of systems with elastic bodies is covered by the Finite Element Method (FEM). The FEM is based on the discretization of a structural element by a typically large number of elements, describing the part's (material) behavior. By defining the element interactions, a model is created that allows the analysis of deflection, deformation and other material based items. Within FEM analysis, partial differential equations are to be solved for each structural suspension component. The finite elements are mathematically characterized by the location of nodes, shape functions and material properties along with boundary conditions which restrict the motion of the nodes. Forces and torques are introduced as external and internal loads on the nodes. The FEM enables simulations that show the reaction to forces applied to a body. Depending on the choice of the number and structure of elements there are sometimes over 500 DoF per element [1]. Due to the large number of elements, the simulation needs long computation times compared with the Multi Body System (MBS) approach.

The FEM enables an accurate prediction of the static and dynamic effects of elasticity. Due to the numerous DoF, however, real-time capability is generally not guaranteed.

### 2.2.3 Flexible Multibody System Analysis

Investigations on vehicle dynamics are predominantly based on the method of Multibody Systems [1]. Here, the system is represented by a number of rigid bodies. These are linked together by bearings modeled as ideal joints. The massless force elements (e.g. springs, dampers, servomotors) are represented by applied forces. The main advantage of MBS is the possibility to efficiently simulate even large systems composed of many bodies with a comparatively low expenditure of computation time. This method is widely used despite the restrictions implied by the rigid body assumptions.

In MBS analysis, the equations of motion are described using Newton's and Euler's equations and incorporating additional constraints resulting from joints applying the principle of d'Alembert or Jourdain [1]. The dynamics of the MBS is thereby fully described via the kinematic differential equations and the equations of motion:

$$\dot{\boldsymbol{y}} = \boldsymbol{K}(\boldsymbol{y})\boldsymbol{z}, \tag{2}$$

$$\boldsymbol{M}(\boldsymbol{y})\dot{\boldsymbol{z}} = \boldsymbol{q}(\boldsymbol{y}, \boldsymbol{z}), \tag{3}$$

with the generalized coordinates $\boldsymbol{y}$ and the generalized velocity $\boldsymbol{z}$. The mass matrix $\boldsymbol{M}(\boldsymbol{y})$ represents the mass and inertial properties of the system whereas the vector of the generalized forces and torques $\boldsymbol{q}(\boldsymbol{y}, \boldsymbol{z})$ accounts for the applied forces [1].

The extension of MBS with deformable bodies leads to flexible multibody systems. The additional DoF due to elasticity can be represented e.g. by deformable beams or as FE models. The single linear modes are determined by Craig-Bampton reduction in modal analysis and then integrated as an elastic component [6]. Thus the flexible MBS analysis is a suitable modeling method for the deformations and resulting strains and stresses. Furthermore, the individual linear system modes are integrated. By reducing the modes, the DoF are reduced; however, still a few hundred DoF remain.

Based on the scope of the investigation task a simulation model has to be designed as a tradeoff between complexity and performance. It is to be observed, that especially investigations on the set of standard MBS simulations and FEM is often needed. The overall goal is to simulate the system behavior. The deformation of single parts within the system however contributes to that behavior to a great extent. Therefore both, deformation and dynamics have to be considered. The modeling method is therefore in general only suitable as a high accuracy reference model.

### 2.2.4 Characteristic Map-based elasticity

Modeling elasticity by characteristic maps is a mathematical modeling method, which works according to the I/O principle. Based on a (e.g. by experiments) validated MBS model various characteristic maps are created to describe the elasticity [7]. Then these characteristic maps are integrated into a simplified twin-track model with complex suspension or other simplified vehicle models. It is thus possible to simulate drive maneuvers in a faster way with an almost equivalent accuracy as the MBS model.

The disadvantage of this modeling technique is the expense to create the characteristic maps as well as the black box character. A change of one parameter of a suspension causes an adjustment of the MBS resulting in a need for new characteristic maps. Furthermore, such a model is not suitable for optimization of elastic parameters.

## 2.3 Influence of elasticity on vehicle dynamics

Simulations were carried out with a flexible suspension model in ADAMS/View, in which bushings have linear stiffness, flexible metal components were built based on modal theory, tire is considered as rigid body (see Figure 1). In order to reduce the computational effort of elasticity calculation, it is necessary to consider the components with significant influence only. The method proposed shall be exemplified at a double wishbone suspension system. The bushings are divided into lower and upper pair for the upper and the lower wishbone connections to the chassis and investigated separately. In addition to the bushings with rubber-like behavior, the structural elasticity of the components lower wishbone, upper wishbone, tie rod and wheel carrier are incorporated. The influence of each elastic component on the suspension parameters is studied by building the models with various assumptions, Table 1. The comparisons are carried out for longitudinal, lateral and vertical load, respectively.

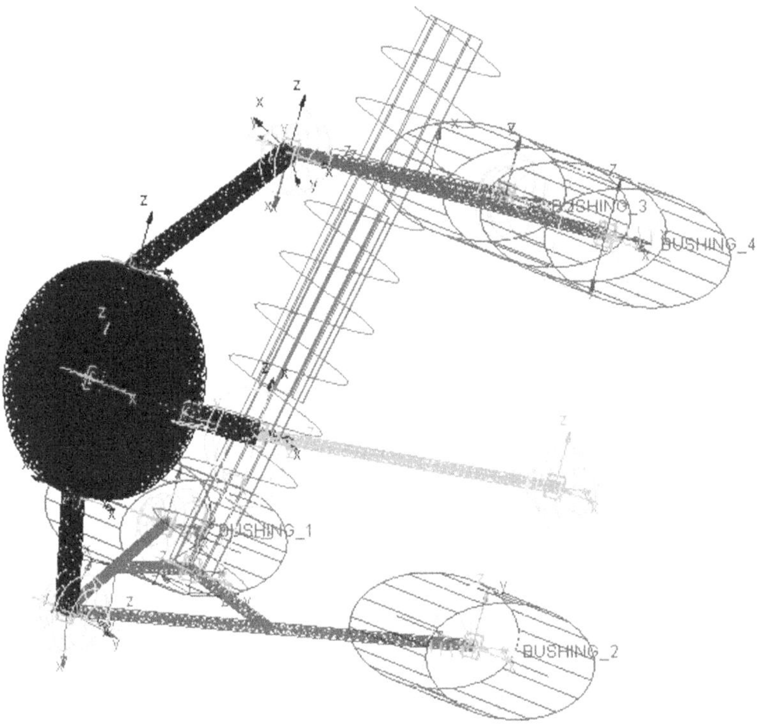


Figure 1 ADAMS Model

The parameter change caused by each elastic component is calculated by subtracting the reference value from the result of the corresponding rigid MBS model. The rigid model (*Model_rigid*) and the model equipped with lower and upper bushings (*Mod-*

*el_lub*) are used as reference for analyzing the effect of bushing and metal components elasticity, respectively. Without large motion the effect of elastic components depends linearly on the external load. Therefore only the slopes (deg/kN) of the parameters variations are used to evaluate the suspension compliance in that case.

Table 1 Model descriptions

| Models | Descriptions |
|---|---|
| *Model_rigid* | Suspension model with rigid joints and rigid metal components |
| *Model_lb* | *Model_rigid* with a pair of bushings on lower wishbone |
| *Model_ub* | *Model_rigid* with a pair of bushings on upper wishbone |
| *Model_lub* | *Model_rigid* with bushings both on lower and upper wishbone |
| *Model_lwb* | *Model_lub* with elastic lower wishbone |
| *Model_uwb* | *Model_lub* with elastic upper wishbone |
| *Model_wc* | *Model_lub* with elastic wheel carrier |
| *Model_tr* | *Model_lub* with elastic tie rod |
| *Model_ce** | *Model_lub* with elastic lower wishbone and wheel carrier |
| *Model_ce* | Complete elastic model |
| *Model_su* | Superposition of *Model_lb, Model_ub, Model_lwb, Model_uwb, Model_wc* and *Model_tr* |

Figure 2 shows the effect of the elastic components under longitudinal load varying from -5 kN to 5 kN. A positive slope indicates a tendency to increase under traction, while a negative value means upward tendency of the parameters under braking. It can be seen that the main influence comes from the bushings and the lower wishbone. Table 2 illustrates the comparison between the parameter alternations measured in the complete elastic model and the effect of that caused by each elastic component. It is obvious that the elastic effects basically satisfy the superposition principle with an error of less than 7%. This result verifies that an analysis of the model with one additional elastic component can predict its effect in a complete elastic system with high accuracy.

All elastic components increase the camber angle under positive lateral force which appears during right cornering on left wheel. This phenomena, as a consequence of a tilting torque generated on tire contact point, is true for all double wishbone suspensions. Superposition principle is also satisfied in this case with small error (Table 3).

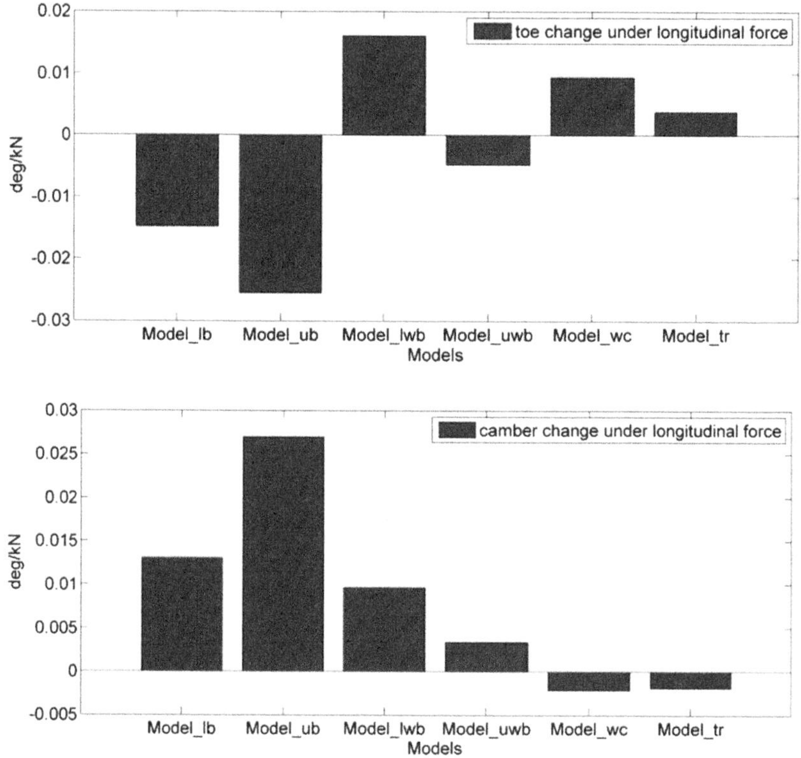


Figure 2 Influence of elastic components under longitudinal load

Table 2 Superposition property of elastic components under longitudinal load

| Parameters | *Model_ce* | *Model_su* | Difference |
|---|---|---|---|
| | [deg/kN] | [deg/kN] | % |
| Toe change | 0.015 | 0.016 | 6.7 |
| Camber change | -0.052 | -0.049 | 5.8 |

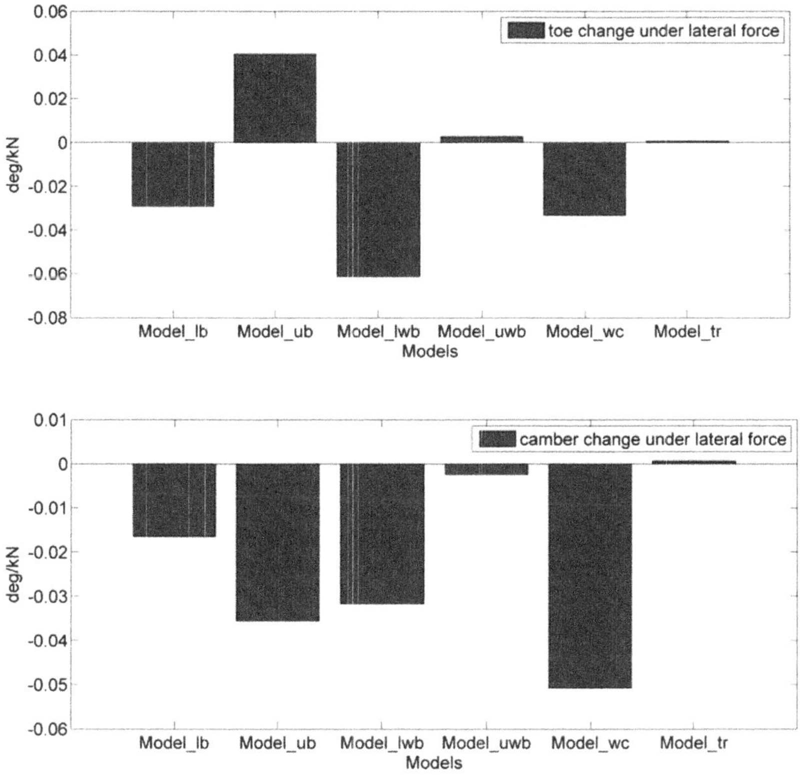


Figure 3 Influence of elastic components under lateral load

Table 3 Superposition property of elastic components under lateral load

| Parameters | *Model_ce* | *Model_su* | Difference |
|---|---|---|---|
| | [deg/kN] | [deg/kN] | % |
| Toe change | -0.079 | -0.079 | ~0 |
| Camber change | -0.135 | -0.136 | 0.7 |

In comparison with the change of parameters under longitudinal force, they are more sensitive to the compliance of elastic components when lateral force is applied (Figure 3). All elastic components increase the camber angle under positive lateral force appearing during right cornering on the left wheel. This phenomena, as a consequence of a tilting torque generated on tire contact point, is true for all double wishbone suspensions. Superposition principle is also satisfied in this case with small errors (Table 3).

The influence of wheel compression is studied by depicting the suspension parameters as functions of wheel travel. To some extent, the value of wheel travel reflects the vertical force applied at the tire contact point. However, if the lower wishbone is considered flexible, the relation wheel travel vs. vertical force changes significantly due to bending deformation. Here the influences of elastic components under wheel compression are investigated using vertical force as reference. The flexibility of the lower wishbone has a dominant effect on the suspension when a vertical force is applied, Figure 4. This influences the vehicle performance significantly since it is closely related to the vehicle's static toe and camber. Comparison of *Model_ce* vs. *Model_su* is illustrated in Figure 5. The mean difference between the models is less than 7% (Table 4).

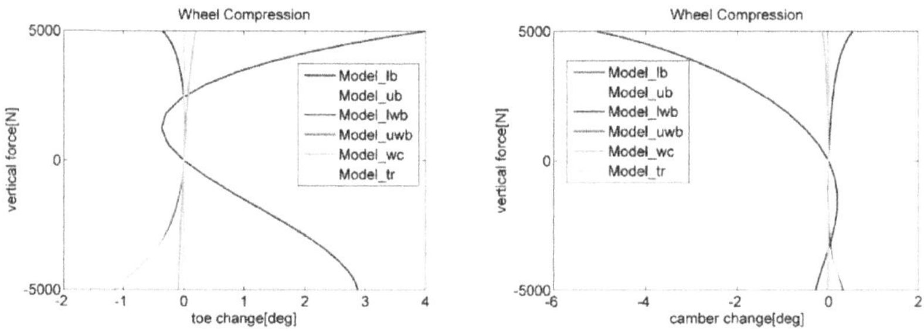


Figure 4 Influence of elastic components during wheel compression.

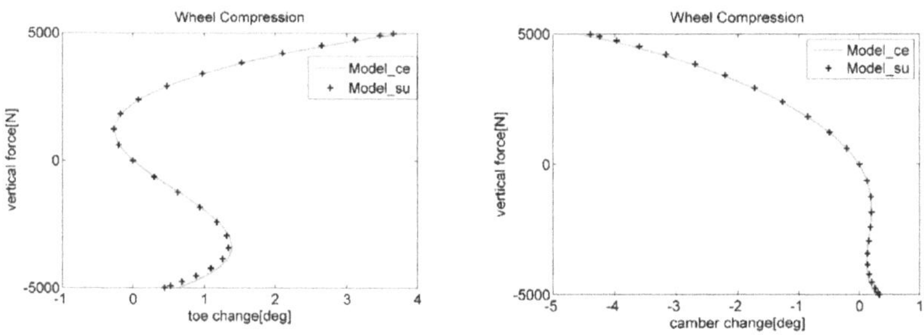


Figure 5 Comparison of the influence of elastic components during wheel compression

Table 4 Superposition property of elastic components under vertical load

| Parameters | Mean difference |
|---|---|
| Toe change | 6.3% |
| Camber change | 7.0% |

The results show that the elasticities of tie rod and upper wishbone have very limited influence. It is reasonable to omit them to increase model efficiency. To further simplify the model, the wheel carrier can be also considered as rigid. However, significant error will be generated when a lateral load is applied. The differences between the simplified models and complete elastic model are shown in Table 5 to Table 7.

Table 5 Simplification under longitudinal load

| Models | Toe change [deg/kN] | Difference % | Chamber change [deg/kN] | Difference % |
|---|---|---|---|---|
| *Model_ce* | -0.015 | | 0.052 | |
| *Model_ce** | -0.013 | 13.3 | 0.048 | 7.7 |
| *Model_lwb* | -0.014 | 6.7 | 0.054 | 3.9 |

Table 6 Simplification under lateral load

| Models | Toe change [deg/kN] | Difference % | Chamber change [deg/kN] | Difference % |
|---|---|---|---|---|
| *Model_ce* | -0.079 | | -0.135 | |
| *Model_ce** | 0.082 | 3.8 | -0.132 | 2.2 |
| *Model_lwb* | 0.014 | 111.7 | -0.054 | 60 |

Table 7 Simplification under vertical load

| Models | Toe difference | Camber difference |
|---|---|---|
| *Model_ce** | 0.5% | 1.2% |
| *Model_lwb* | 12.6% | 4.8% |

The described suspension model is assembled into a full vehicle model as front suspension in ADAMS/Car to explore the influence of elastic components on the vehicle's behavior. A constant radius cornering (ISO 4138) is performed to observe the vehicle's handling characteristics. Once the acceleration reaches 0.5g, the elasticities of bushings cause an understeer tendency, while the models with elastic metal components have an opposite effect. This is explained by their effects on static camber. Results of simplified models *Model_ce** and *Model_lwb* are quite similar as that of complete elastic model.

# 3 Elasto-kinematics

*Elasticity* is the reversible deformation of solid bodies under stress. The deformation depends on geometry and material of the solid bodies of the suspension under the impact of external forces, which occur during driving maneuvers [8]. In turn this deformation results also in an elasto-kinematic movement of the bodies.

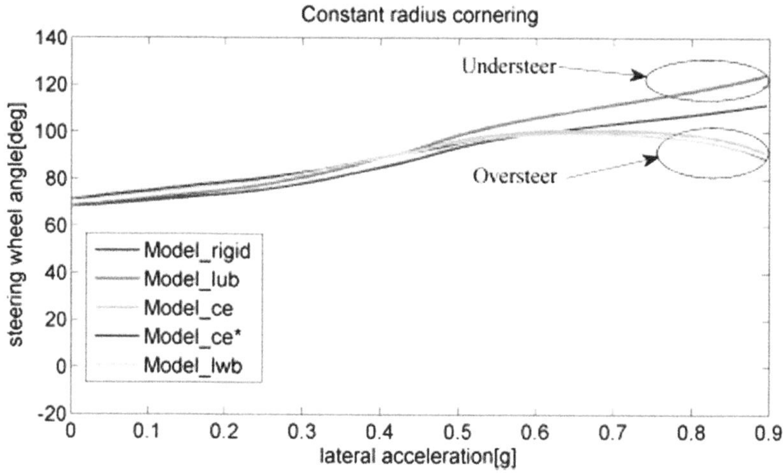


Figure 6 Constant radius cornering

As shown, it makes sense to simplify the system by concentrating on the significant parts, here upper wishbone and bushings. First the elasticity of the bushings is be modeled. Then, it will be shown that the structure of elasticity of the metallic components can be integrated in the bushings. This is done by adapting the spring stiffness. A second method is an object-oriented physical modeling of the structural elasticity.

## 3.1 Modelling of the complete elasticity in bushings

In this section the elastic properties of the bushings which connect the wishbones and vehicle body are considered in the model [9]. As illustrated in Figure 7, displacements between vehicle body and wishbone endpoints caused by bearing forces are allowed. The connecting bearings on the vehicle body $L_{20}$, $L_{30}$, $U_{20}$ and $U_{30}$ have fixed coordinates. The bushing model is composed of three translational and rotational springs. Linear stiffness is utilized for the bushings. Other assumptions are:

- the wheel is rigid and fixed to wheel carrier,
- elasticity of the metal components are not considered,
- all the components are taken as massless,
- bearing has a linear stiffness and is installed with zero preload,
- linear stiffness is assigned to the suspension spring.
- The DOF of the mechanism are calculated as follows:

$$\begin{aligned} \mathrm{DOF} &= \underbrace{5\cdot(6)}_{5\cdot\text{bodies}} - \underbrace{3\cdot(6-3)}_{3\cdot\text{spherical joints}} - \underbrace{1\cdot(6-2)}_{1\cdot\text{universal joint}} - \underbrace{1\cdot(6-1)}_{1\cdot\text{translational joint}} \\ &= \underbrace{1}_{\text{wheel vertical motion}} + \underbrace{1}_{\text{steering motion}} + 10 \end{aligned} \tag{4}$$

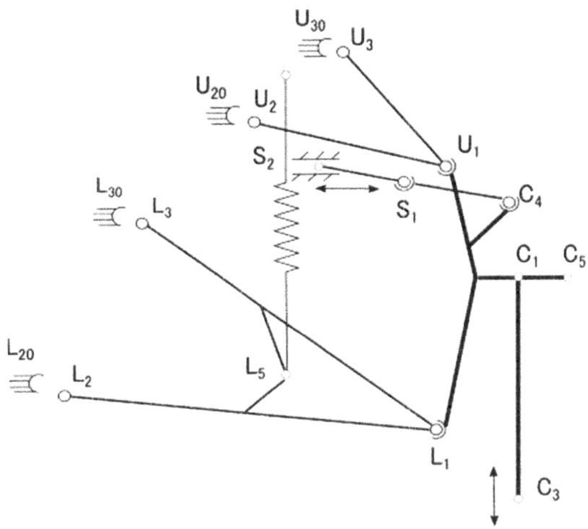


| Joint | Index |
|---|---|
| Spherical | $L_1$ $U_1$ $S_1$ |
| Universal | $C_4$ |
| Translational | $S_2$ |
| Bushing | $L_2$ $L_3$ $U_2$ $U_3$ |

Figure 7 Elasto-kinematics model of double wishbone suspension

### 3.1.1 Modeling and Simulation using Analytical Method

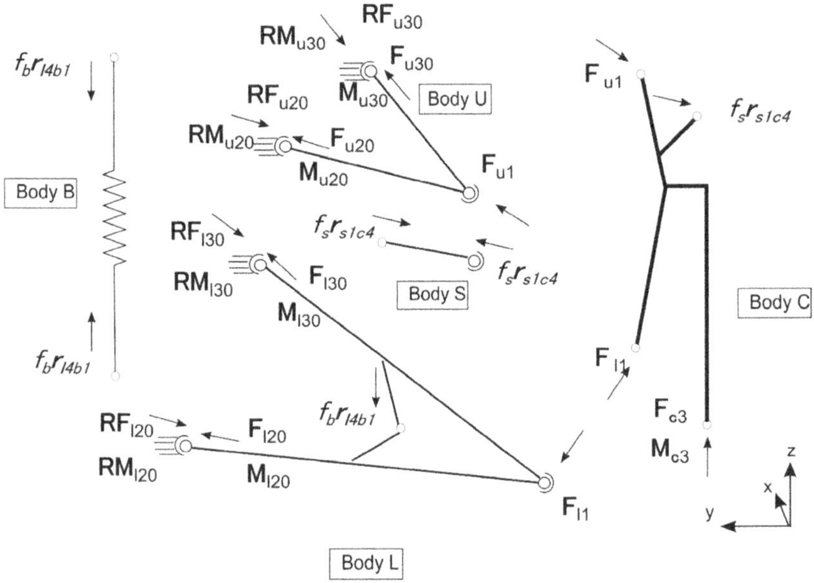


Figure 8 Reaction forces and torques

Unlike kinematic analysis, the motion of an elasto-kinematics model is not only constrained by joints but also influenced by forces. The nonlinear equation system should include the bearing forces analysis as well. The bushing is represented here by 3 translational and 3 rotational springs. In Figure 8 the directions of bearing forces and movements are illustrated. Since the vertical motion of the wheel is generated from $C_3$, the displacement matrix for the wheel carrier can be expressed as [10]:

$$\boldsymbol{D}^c = \begin{bmatrix} \boldsymbol{R}^C & \boldsymbol{r}_{C3} - \boldsymbol{R}^c \boldsymbol{r}_{C30} \\ \boldsymbol{0} & \boldsymbol{1} \end{bmatrix}, \tag{5}$$

where $\boldsymbol{0}$ is the 1x3 zero matrix, $\boldsymbol{1}$ is the 1x3 unit matrix, $\boldsymbol{R}^c$ is the rotation matrix of wheel carrier. The relation between initial coordinates and instantaneous coordinates of the points on wheel carrier can be obtained:

$$\begin{bmatrix} \boldsymbol{r}_{L1} & \boldsymbol{r}_{U1} & \boldsymbol{r}_{C4} \\ 1 & 1 & 1 \end{bmatrix} = \boldsymbol{D}^c \begin{bmatrix} \boldsymbol{r}_{L10} & \boldsymbol{r}_{U30} & \boldsymbol{r}_{C40} \\ 1 & 1 & 1 \end{bmatrix}. \tag{6}$$

For body L and body U the constraint equations can be written as:

$$\begin{bmatrix} \boldsymbol{r}_{L2} & \boldsymbol{r}_{L3} & \boldsymbol{r}_{L4} \\ 1 & 1 & 1 \end{bmatrix} = \boldsymbol{D}^l \begin{bmatrix} \boldsymbol{r}_{L20} & \boldsymbol{r}_{L30} & \boldsymbol{r}_{L40} \\ 1 & 1 & 1 \end{bmatrix}, \tag{7}$$

where the displacement matrix for body L is:

$$\boldsymbol{D}^l = \begin{bmatrix} \boldsymbol{R}^l & \boldsymbol{r}_{L1} - \boldsymbol{R}^l \boldsymbol{r}_{L10} \\ \boldsymbol{0} & \boldsymbol{1} \end{bmatrix}. \tag{8}$$

Likewise, three Euler angles constitute the rotation matrix $\boldsymbol{R}^l$. The constraint equations for body $U$ are established in a same way and given by:

$$\begin{bmatrix} \boldsymbol{r}_{U2} & \boldsymbol{r}_{U3} \\ 1 & 1 \end{bmatrix} = \boldsymbol{D}^u \begin{bmatrix} \boldsymbol{r}_{U20} & \boldsymbol{r}_{U30} \\ 1 & 1 \end{bmatrix}, \tag{9}$$

where the displacement matrix for upper wishbone is:

$$\boldsymbol{D}^u = \begin{bmatrix} \boldsymbol{R}^u & \boldsymbol{r}_{U1} - \boldsymbol{R}^u \boldsymbol{r}_{U10} \\ \boldsymbol{0} & \boldsymbol{1} \end{bmatrix}. \tag{10}$$

For body $S$:

$$\boldsymbol{r}_{S1C4}^T \boldsymbol{r}_{S1C4} = \boldsymbol{r}_{S10C40}^T \boldsymbol{r}_{S10C40}. \tag{11}$$

The relation between the bearing force and displacement at the points $L_2$, $L_3$, $U_2$ and $U_3$ are determined by the stiffness matrix of the bushings:

$$\begin{bmatrix} \boldsymbol{F}_{i0} \\ \boldsymbol{M}_{i0} \end{bmatrix} = \begin{bmatrix} \boldsymbol{k}_{ti} & \boldsymbol{0} \\ \boldsymbol{0} & \boldsymbol{k}_{ri} \end{bmatrix} \begin{bmatrix} \boldsymbol{d}_{ti} \\ \boldsymbol{d}_{ri} \end{bmatrix}, \tag{12}$$

where $i$ denotes the index of bushings ($L_2$, $L_3$, $U_2$, $U_3$); $\boldsymbol{k}_{ti}$ and $\boldsymbol{k}_{ri}$ are the translational and rotational stiffness matrix of the bushings in global coordinate system, respectively; $\boldsymbol{d}_{ti}$ and $\boldsymbol{d}_{ri}$ are translational and rotational displacement between the vehicle body and wishbones:

$$\boldsymbol{d}_{ti} = \boldsymbol{r}_{i_0 i}. \tag{13}$$

The rotational displacement about $x$, $y$ and $z$ axis can be derived by the elements of rotation matrix [11]:

$$\boldsymbol{d}_r^i = \begin{bmatrix} \arctan\left(-a_{23}^j / a_{33}^j\right) \\ \arctan\left(a_{13}^j / a_{33}^j\right) \\ \arctan\left(a_{21}^j / a_{11}^j\right) \end{bmatrix}, \quad j = l, u. \tag{14}$$

The reaction forces and torques at the connecting points on vehicle body read:

$$\boldsymbol{RF}_{i0} = -\boldsymbol{F}_{i0}, \tag{15}$$

$$\boldsymbol{RM}_{i0} = -\boldsymbol{M}_{i0}. \tag{16}$$

The static equilibrium equations of the suspension system are expressed as:

$$\boldsymbol{RF}_{u30} + \boldsymbol{RF}_{u20} + \boldsymbol{F}_{u1} = \boldsymbol{0}, \tag{17}$$

$$\boldsymbol{RM}_{u30} + \boldsymbol{RM}_{u20} + r_{u30u1} \times \boldsymbol{RF}_{u30} + \boldsymbol{r}_{u20u1} \times \boldsymbol{RF}_{u20} = \boldsymbol{0}. \tag{18}$$

For body $L$

$$\boldsymbol{RF}_{l30} + \boldsymbol{RF}_{l20} + \boldsymbol{F}_{l1} + f_b \boldsymbol{r}_{l4b1} = \boldsymbol{0}, \tag{19}$$

$$\boldsymbol{RM}_{l30} + \boldsymbol{RM}_{l20} + \boldsymbol{r}_{l30l1} \times \boldsymbol{RF}_{l30} + \boldsymbol{r}_{l20l1} \times \boldsymbol{RF}_{l20} + \boldsymbol{r}_{l4l1} \times f_b \boldsymbol{r}_{l4b1} = \boldsymbol{0}, \tag{20}$$

where $f_b$ is a scale factor determined by the spring stiffness and spring length:

$$f_b = -k(|\boldsymbol{r}_{l4b1}| - |\boldsymbol{r}_{l40b1}|)/\boldsymbol{r}_{l4b1}. \tag{21}$$

For body C

$$(-\boldsymbol{F}_{u1}) + (-\boldsymbol{F}_{l1}) + (-f_s \boldsymbol{r}_{s1c4}) + \boldsymbol{F}_{c3} = \boldsymbol{0}, \tag{22}$$

$$\boldsymbol{r}_{c3l1} \times \boldsymbol{F}_{c3} + (-\boldsymbol{r}_{u1l1} \times \boldsymbol{F}_{u1}) + (-\boldsymbol{r}_{c4l1} \times f_s \boldsymbol{r}_{s1c4}) + \boldsymbol{M}_{c3} = \boldsymbol{0}. \tag{23}$$

A nonlinear equation system constituted by 67 equations is given above. Newton-Raphson is employed here to solve the non-linear equation system due to its rapid convergence speed. After solving the equation system, the characteristic parameters of suspension can be obtained according to their definitions.

$$\text{Toe} = \arctan\left(\frac{a_{12}^c}{a_{22}^c}\right), \ \text{Camber} = \arctan\left(\frac{a_{32}^c}{a_{22}^c}\right) \qquad (24)\ (25)$$

### 3.1.2 Simulation Results

*Wheel compression*

During vertical motion of the wheel, the bushings are deformed.. This results in a different behavior of the suspension vs. the rigid model. Validation is carried out by comparing the results with that of an elasto-kinematic model in ADAMS/Car with the same bushings. The results are shown in Figure 9 and Table 8.

*Steering*

Since pure steering motion generates only slight forces, very limited differences are observed in comparison with the kinematic model (Figure 10 and Table 9). The MATLAB model results for steering differ from that in ADAMS/Car, notably the camber angle. The reason is that in MATLAB model tire is assumed to be fixed to the wheel carrier, in ADAMS/Car, however, relative motion between the two parts is allowed.

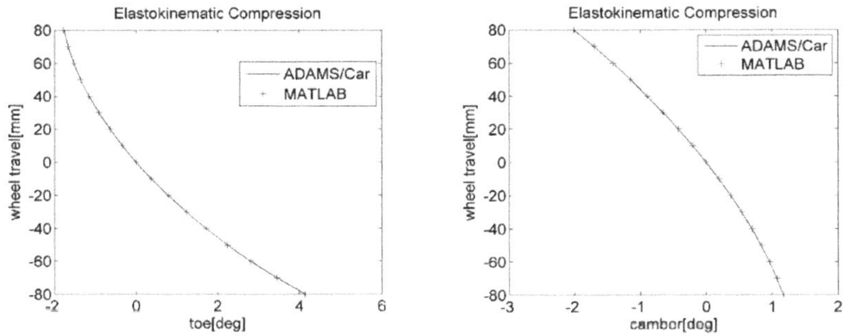


Figure 9 Elasto-kinematics wheel compression

Table 8 Elasto-kinematics wheel compression

| | MATLAB | | ADAMS/Car | |
|---|---|---|---|---|
| Wheel Travel [mm] | Toe [deg] | Camber [deg] | Toe [deg] | Camber [deg] |
| 80 | -1.78 | -2.01 | -1.77 | -2.01 |
| 40 | -1.15 | -0.89 | -1.14 | -0.89 |
| 0 | 0 | 0 | 0 | 0 |
| -40 | 1.72 | 0.70 | 1.72 | 0.69 |
| -80 | 4.15 | 1.18 | 4.16 | 1.16 |

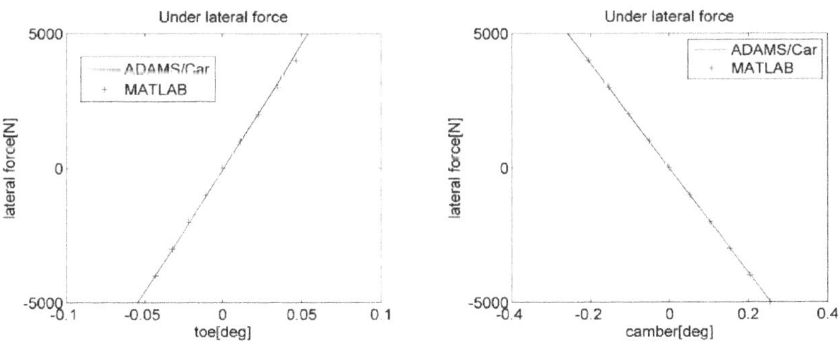


Figure 10 Elasto-kinematics steering

Table 9 Elasto-kinematics steering

| | MATLAB | | ADAMS/Car | |
|---|---|---|---|---|
| Rack Travel [mm] | Toe [deg] | Camber [deg] | Toe [deg] | Camber [deg] |
| 70 | 31.30 | -1.51 | 31.20 | -1.33 |
| 30 | 12.89 | -0.95 | 12.89 | -0.93 |
| 0 | 0 | 0 | 0 | 0 |
| -30 | -12.58 | 1.43 | -12.59 | 1.39 |
| -70 | -29.84 | 4.60 | -29.88 | 3.95 |

### *Lateral load*

The lateral compliance property of the suspension system was obtained by applying a varying lateral force on the tire contact point. The result of MATLAB model is identical with that of ADAMS/Car model (see Figure 11).

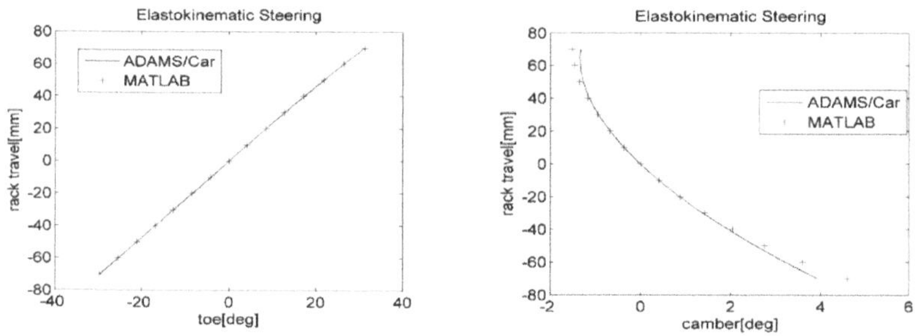


Figure 11 Lateral Load

### *Longitudinal load*

Longitudinal load is caused by breaking or traction. The behavior of the wheel under varying longitudinal force within the range of -5 kN and 5 kN is shown in Figure 12.

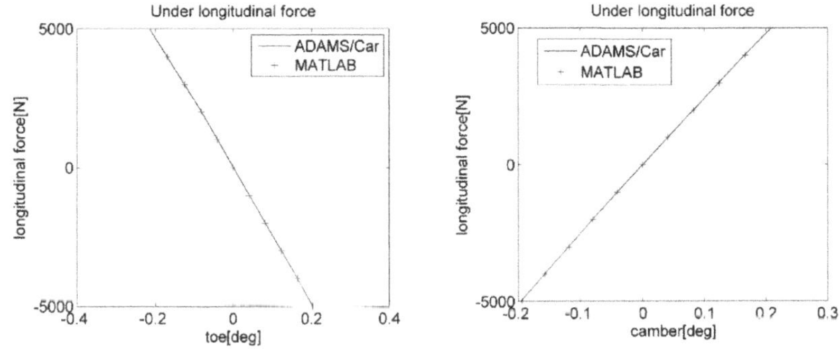


Figure 12 Longitudinal Load

## *Computation time comparison*

Efficiency comparison between MATLAB model and ADAMS/Car model is shown in Table 10. It can be seen that simulation of an elasto-kinematics model in ADAMS/Car did not need more time compared with that of a kinematic model (model without bushings). The size of the equation system is not significantly increased. For the analytical program in MATLAB, however, the size of the nonlinear equation system is dramatically increased when the suspension motion is also influence by forces (17 kinematic equations vs. 67 elasto-kinematics equations). Even so, ADAMS/Car model is relative inefficient compared with the simplified model programmed in MATLAB.

Simulations were performed on a computer with Intel Core2 2.10GHz processor. Owing to reasonable simplifications, the presented suspension model programmed in MATLAB has an outstanding performance in terms of efficiency. By contrast, a detailed suspension model in ADAMS/Car contains a vast number of components, thereby resulting in a large Jacobian matrix. More time is required for each iteration of the Newton algorithm. Moreover, during suspension analysis in ADAMS/Car many parameters are evaluated no matter if they are requested.

Table 10 Elapsed time of kinematic model

| Number of steps | MATLAB<br>[s] | ADAMS/Car<br>[s] |
|---|---|---|
| 50 | 0,73 (25,3%) | 2,89 |
| 100 | 1,39 (27,5%) | 5,06 |

### 3.1.3 Bushing optimization

The modeling of a flexible body requires FEM technique which is complex and costly since it involves a vast number of DoF to the suspension system, while deformation of bushing can be precisely estimated by few springs. A more efficient and economical model can be derived if additional bushing stiffness is used to simulate the influences caused by structural elasticity. The parameter changes under lateral and longitudinal force are listed in Table 11. The global optimization algorithm Pattern Search is adopted to minimize the differences between the results of the complete elastic model and that of the model equipped with bushings only. The objective function reads:

$$f(\boldsymbol{K}) = \omega_1\big(p1_{ce} - p1_{lub}(\boldsymbol{K})\big) + \omega_2\big(p2_{ce} - p2_{lub}(\boldsymbol{K})\big) + \cdots$$
$$\cdots + \omega_3\big(p3_{ce} - p3_{lub}(\boldsymbol{K})\big) + \omega_4\big(p4_{ce} - p4_{lub}(\boldsymbol{K})\big) \qquad (26)$$

where $\boldsymbol{K}$ is a 24x1 stiffness matrix of 4 bushings, $p1, p2, p3$ and $p4$ are toe and camber change under lateral and longitudinal force respectively. $\omega_1$, $\omega_2$, $\omega_3$, $\omega_4$ are coefficients to convert multi-objective into single-objective. Results of the optimization are shown in Table 11, where *Model_lub** is the model with optimized bushings. It appears that *Model_lub** has similar compliance properties in longitudinal and lateral directions as *Model_ce*. However, if the compliance property of the suspension in vertical direction is included in the objective function, the bushings fail to represent the influences of flexible metal components. One reason is that for large motions, the stiffness matrix of metal components in the global system changes with their orientations whereas the stiffness matrix of bushings in global system remains unchanged no matter how the suspension moves.

Table 11 Optimization results

| Models | *Toe/la* [deg/kN] | *Camber/la* [deg/kN] | *Toe/lo* [deg/kN] | *Camber/lo* [deg/kN] |
|---|---|---|---|---|
| *Model_ce* | -0.079 | -0.135 | -0.015 | 0.053 |
| *Model_lub* | 0.011 | -0.052 | -0.041 | 0.041 |
| *Model_lub** | -0.079 | -0.106 | -0.15 | 0.053 |
| Improvement | 100% | 34.9% | 100% | 100% |

## 3.2 Physical-oriented modeling of elasticities

With the tools of continuum mechanics the motion of deformable bodies due to external forces and torques will be described. Here, the relationship between stress and strain is defined by material laws. However, in order to represent the time dependence of the state variables, an accounting is needed. Therefore the storage-elasto-mechanics structure is used. Here, the state-changes are calculated in the elasto-mechanics-elements. In the storage the deformation energy is balanced. This results from the bearing forces and the existing strain that occur. The method uses the structure of the different properties of the states, which are now defined.

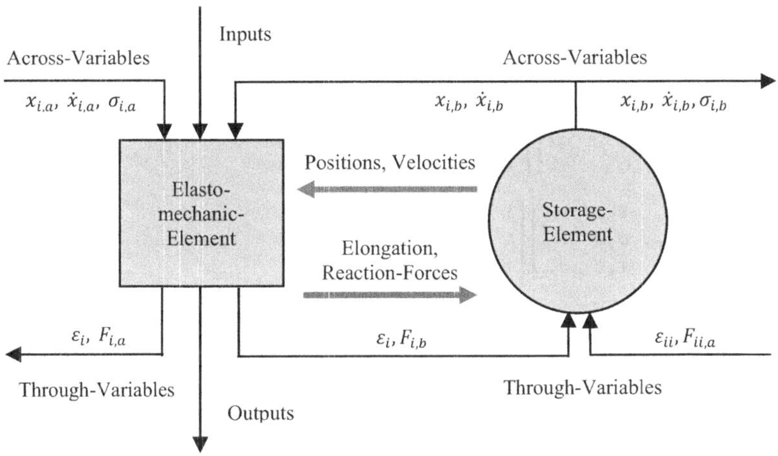


Figure 13 Model structure of elasto-mechanic- and storage-elements

The state variables are classified into two types, which are the through- and the across-variables. The through-variables are the forces $F$ and the strains $\varepsilon$, which can be measured directly at one point. The across-variables need a potential between to two measuring points. This is the position $x$ or velocity $\dot{x}$ and the stress $\sigma$ which occur at the beginning and at the end of a component.

The schematic structure used is shown in Figure 13. It consists of two basic elements, the elasto-mechanic-element and the storage. In the following the elasto-mechanic-elements is symbolized as squares and the storages as circles. The index $i$ stands for the component that is represented by this element. The index $ii$ refers to the subsequent component which is not shown. Also, the previous storage is not shown graphically.

By the elasto-mechanic-elements the bearing forces are determined at the beginning and at the end of the component and the strains. This is done using Hooke's law of isotropic and homogeneous body of a triaxle stress state. The deformation behavior of an isotropic body is described by the modulus of elasticity $E$ and the Poisson's ratio $v$, whereby the shear modulus is also calculated [12]:

$$\boldsymbol{\varepsilon}_i = \begin{bmatrix} \varepsilon_{xx} & \frac{1}{2}\gamma_{xy} & \frac{1}{2}\gamma_{xz} \\ \frac{1}{2}\gamma_{xy} & \varepsilon_{yy} & \frac{1}{2}\gamma_{yz} \\ \frac{1}{2}\gamma_{xz} & \frac{1}{2}\gamma_{yz} & \varepsilon_{zz} \end{bmatrix} = \frac{1+v}{\mathrm{E}} \begin{bmatrix} \sigma_{xx} & \tau_{xy} & \tau_{xz} \\ \tau_{xy} & \sigma_{yy} & \tau_{yz} \\ \tau_{xz} & \tau_{yz} & \sigma_{zz} \end{bmatrix} - \frac{v}{\mathrm{E}}\left(\sigma_{xx} + \sigma_{yy} + \sigma_{zz}\right) \begin{bmatrix} 1 & 0 & 0 \\ 0 & 1 & 0 \\ 0 & 0 & 1 \end{bmatrix}. \quad (27)$$

The applied forces $\boldsymbol{F}_\sigma$ and torques $\boldsymbol{M}_\sigma$ caused by the tension are calculated over the cross section $A$, which is assumed averaged over the complete length, and the moment of inertia of area $I$ divided by the distance of outer fiber $a$:

$$\boldsymbol{F}_\sigma = \begin{bmatrix} \sigma_{xx} & \tau_{xy} & \tau_{xz} \\ \tau_{xy} & \sigma_{yy} & \tau_{yz} \\ \tau_{xz} & \tau_{yz} & \sigma_{zz} \end{bmatrix} \begin{bmatrix} A_x \\ A_y \\ A_z \end{bmatrix}, \quad (28)$$

$$\boldsymbol{M}_\sigma = \begin{bmatrix} \sigma_{xx} & \tau_{xy} & \tau_{xz} \\ \tau_{xy} & \sigma_{yy} & \tau_{yz} \\ \tau_{xz} & \tau_{yz} & \sigma_{zz} \end{bmatrix} \begin{bmatrix} I_x/a_x \\ I_y/a_y \\ I_z/a_z \end{bmatrix}. \quad (29)$$

With this applied forces it is possible to calculate the reaction forces $\boldsymbol{F}_i$ and torques $\boldsymbol{M}_i$ at the beginning and the end of the component by using of the Newton-Euler equation and the general coordinates with the mass $m$ and inertia tensor of the element $\boldsymbol{\theta}$. Further the translational and rotational Jacobian matrices ($\boldsymbol{J}_T$, $\boldsymbol{J}_R$) and acceleration components ($\overline{\boldsymbol{a}}$, $\overline{\boldsymbol{\alpha}}$) are used

$$\boldsymbol{F}_{i,a} = m\boldsymbol{J}_T\ddot{\boldsymbol{q}}_{i,a} + m\overline{\boldsymbol{a}}_{i,a} - \boldsymbol{F}_\sigma, \quad \boldsymbol{F}_{i,b} = m\boldsymbol{J}_T\ddot{\boldsymbol{q}}_{i,b} + m\overline{\boldsymbol{a}}_{i,b} - \boldsymbol{F}_\sigma, \quad (30)$$

$$\boldsymbol{M}_{i,a} = \boldsymbol{\theta}\boldsymbol{J}_R\ddot{\boldsymbol{q}}_{i,a} + \boldsymbol{\theta}\overline{\boldsymbol{\alpha}}_{i,a} - \boldsymbol{M}_\sigma, \quad \boldsymbol{M}_{i,b} = \boldsymbol{\theta}\boldsymbol{J}_R\ddot{\boldsymbol{q}}_{i,b} + \boldsymbol{\theta}\overline{\boldsymbol{\alpha}}_{i,b} - \boldsymbol{M}_\sigma. \quad (31)$$

In the storage element the balance of the energy has to be calculated with the Integral of the specific linear strain energy $\overline{\boldsymbol{u}}$. For simplification it is assumed an average of the length $l_0$

$$\boldsymbol{u} = \int_0^t \overline{\boldsymbol{u}}_{i,b}(\tau) - \overline{\boldsymbol{u}}_{ii,a}(\tau)d\tau, \quad (32)$$

$$\overline{\boldsymbol{u}} = \boldsymbol{F}_\sigma\boldsymbol{\varepsilon}_i \begin{bmatrix} l_{0x} \\ l_{0y} \\ l_{0z} \end{bmatrix}. \quad (33)$$

Out of the linear strain energy $\boldsymbol{u}$ it is possible to determine the strain at the endpoint of the component:

$$\boldsymbol{u} = \frac{1}{2}\boldsymbol{\sigma}_{i,b}\begin{bmatrix}\varepsilon_{xx}\\ \varepsilon_{yy}\\ \varepsilon_{zz}\end{bmatrix}. \quad (34)$$

On the other hand the time integral is used to calculate the momentums and with this the velocities of the endpoint

$$m\dot{\boldsymbol{x}}_{i,b} = \int_0^t \boldsymbol{F}_{i,b}(\tau) - \boldsymbol{F}_{ii,a}(\tau)d\tau, \quad (35)$$

$$\boldsymbol{\theta}\dot{\boldsymbol{\varphi}}_{i,b} = \int_0^t \boldsymbol{M}_{i,b}(\tau) - \boldsymbol{M}_{ii,a}(\tau)d\tau. \quad (36)$$

## 3.3 Potential for improvement

Driven by the rising complexity of mechatronic systems in modern vehicles, simulations are not limited only to the development process any more. Various systems, especially out of the sector of advanced driver assistance systems, use simulations to enrich the sensor information or to detect or predict different vehicle states. Unlike during simulations during the development process (that are often used for desk research) the used models have to simulated in real-time. While todays simulation hardware (e.g. in vehicle simulators) has high computation performance, which allows for large systems and detailed models to be simulated in real-time, the computational power installed in a modern vehicle is comparatively small. Running a FEM based simulation on the system or even parts of the systems seems to be impossible for that reason. To be able to still take deformation of different parts into account during a dynamics simulation, so called look-up tables are used that describe the deformation under different conditions.

With these reduced model approaches the simulation time can be optimized with constant accuracy of the interested variables such as reaction forces, stresses and deformations. Also the model variant hedge would take less time. Among them is to understand if a coupe, a four-door vehicle or a convertible would be expected on vehicle dynamics, vehicle stability and chassis rigidity. The same applies to the impact with and without sunroof or with 4 or 8-cylinder engines (weight!). Also a sensitivity analysis for comfort and vehicle dynamic is possible. For example, the influence of different steering rigidity could be studied on the driving dynamics.

# 4 Summary

Modeling and simulation of vehicle dynamics plays a major role during the development process of modern vehicles. The possibilities of simulations used range from rough approximations to really sophisticated models being able to predict the behavior of the real system in great detail and accuracy. Based on the area of interest a simulation is intended for different simulation and modeling techniques are used.

Within the illustrated influence analysis on the example of the double wishbone suspension was found that in addition to the bushings the elasticity of the wheel carrier and the lower wishbone are essential for the total deformation. These results here with good accuracy by additive superposition of the individual deformations. The deformations are strongly manoeuvres-dependent.

Through two simplified approaches for elasticity description, it is possible to reach fast simulation times. On the one hand the structure elasticity of the metallic components was placed in the bushings by optimizing the spring stiffness. Through this approach, good accuracy can be achieved. On the other hand, a physical modelling method has been developed, which is based on the object-oriented modeling. This approach is currently still in the testing phase.

The approach described is capable to create models that can be used for more advanced control tasks (especially for active elements in chassis). In the advanced chassis development it allows rapid adjustments and optimizations, which save time and reduces costs. Moreover, the use of such a model would allow a fast adaptation of suspension parameters in racing simulators.

# 5 Literature

[1] D. SCHRAMM, M. HILLER, R. BARDINI: *Vehicle Dynamics: Modeling and Simulation.* 405. Springer Verlag: Berlin, Heidelberg, 2014. – ISBN 978-3-540-36045-2

[2] M. UNTERREINER: *Modellbildung und Simulation von Fahrzeugmodellen unterschiedlicher Komplexität.* Duisburg-Essen, Lehrstuhl Mechatronik, Dissertation. 2013

[3] B. HEIßING, M. ERSOY: *Chassis Handbook.* 591. Vieweg+Teubner Verlag: Wiesbaden, 2011. – ISBN 978-3-8348-0994-0

[4] Y.Y. TONG: Vehicle dynamic simulations based on flexible and rigid multibody models. In: *SAE Technical Paper* (2000).

[5] M. UNTERREINER, D. SCHRAMM: Modelling of a Twin-track Vehicle Model with Modular Wheel Suspensions. In: *In: Applied Mechanics and Materials (AMM), Vol. 165, Trans Tech Publications, Schweiz, 2012,ISBN: 978-3-03785-413-6, 978-3—03813-740-5, S. 214-218* 165 (2012), S. 214-218.

[6] R. SCHWERTASSEK, O. WALLRAPP: *Dynamik flexibler Mehrkörpersysteme: Methoden der Mechanik zum rechnergestützten Entwurf und zur Analyse mechatronischer Systeme* 476. Vieweg+Teubner Verlag, 2014. – ISBN 978-3322939760

[7] H.-P. LANG: *Kinematik-Kennfelder in der objektorientierten Mehrkörpermodellierung von Fahrzeugen mit Gelenkelastizitäten.* Universität Duisburg-Essen, Fortschrittberichte VDI Reihe 12 Nr. 323, Dissertation. 1997

[8] W. SCHIEHLEN, P. EBERHARD: *Technische Dynamik- Modelle für Regelung und Simulation ; mit 4 Tabellen und 44 Beispielen.* XI, 251 S. Teubner: Stuttgart u.a., 2004. – ISBN 978-3-519-12365-1

[9] M. BLUNDELL, D. HARTY: *The multibody systems approach to vehicle dynamics.* 1-518. Elsevier Butterworth-Heinemann: Oxford, 2004. – ISBN 978-0-7506-5112-7

[10] C.H. SUH, C.W. RADCLIFFE: *Kinematics and Mechanisms Design.* University of Michigan: Wiley, 1978.

[11] J. MEISSONNIER, J. FAUROUX, G. GOGU, C. MONTEZIN, Geometric identification of an elastokinematic model in a car suspension, Proceedings of the Institution of Mechanical Engineers, Part D: Journal of Automobile Engineering, 2006, pp. 1209-1220.

[12] H. ESCHENAUER, W. SCHNELL: *Elastizitätstheorie – Grundlagen, Flächentragwerke, Strukturoptimierung.* Wissenschaftsverlag: Mannheim, 1993. - ISBN 3-411-16263-5

# Lightweight design in subassemblies with changing design spaces to find an overall weight optimum for series-produced cars

Dipl.-Ing. (FH) Gerhard Steber M.Sc., BMW Group Munich

Prof. Dr.-Ing. Roland Lachmayer, Leibniz University of Hanover

# Introduction

A car needs to fulfill an enormous number of requirements. A production car cannot be developed by a single team therefore it is necessary to split the car into different assemblies such as body, axle carrier or engine. These assemblies are processed by different teams. The requirements of the complete car need to be translated into different specification sheets for the assemblies. In this way the teams are able to work efficiently and independently between the synchronization points. Each team creates a concept model and analyzes the properties with the goal to fulfill the requirements. At the synchronization points the properties are analyzed for the complete car. All deviations to the requirements are recorded and set to a list of points respectively a lightweight strategy. The responsible assembly for these points is subjectively identified by experienced engineers. After some iterations all requirements are fulfilled.

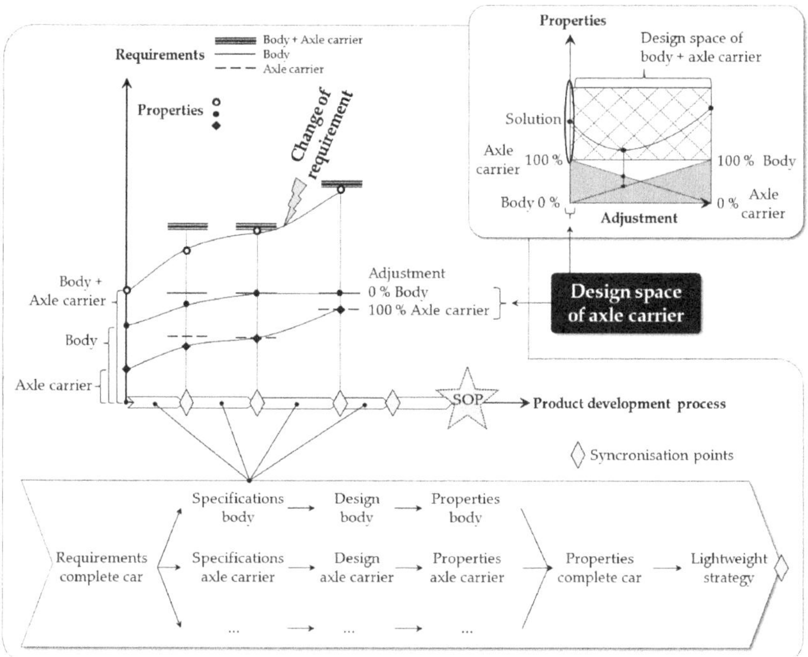


Fig. 1: Development process of a production car

In reality this process is not straight forward. In the concept phase the knowledge of the complete car behavior is little and is reflected in a coarse description of the requirements [1]. This implies external changes [2] of the requirements in order to find the best solution. Internal changes [2] for example a new manufacturing technology affects the assemblies as well. The changes and the interactions between all assemblies make it very difficult for the engineers to identify the ideal lightweight strategies for the assembly.

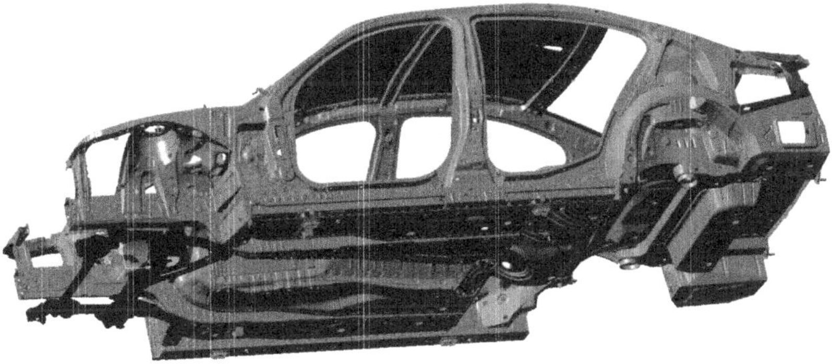

Fig. 2: Body (grey) and axle carrier (black) of a production car

A very strong interaction exists between the body and the axle carrier (Fig. 2). Changes for this system affect directly the mechanical properties such as stiffness, strength or acoustics. With commercial computation technology it is possible to analyze these properties separately but the interaction between the assemblies and the different disciplines cannot be objectified. Without a deeper understanding of both assemblies combined to a particular system it is only possible to adjust one of both assemblies. It is not obvious how the interaction of both together would lead to a partly adjustment of both assemblies regarding an ideal lightweight strategy. This development process leads to a lot of iterations until the best possible solution for the body axle carrier system and the complete car is found.

This paper focuses on the possibilities that the axle carrier team has to accelerate the product development process of their own assembly and in the interaction with the body assembly. To identify the lightweight strategy at the synchronization points it is necessary to make the assemblies more transparent. This can be done with significant data which describe the behavior of the assembly and the history of the changed requirements. This data enables the engineers to derive a more efficient lightweight strategy and reduce the development time in order to find the best possible solution for the system.

The contributions of this paper are:

1. A method to accelerate the development process to find an overall lightweight package for a complex system between an axle carrier and a body.
2. A method to systematically find the best possible solution for an axle carrier by identifying the behavior of the assembly in a system.
3. A method to react to changing requirements (e.g. package, technology, material, stiffness, etc.) during the product development process.
4. A method to support the different teams with the most significant data for more transparency.

For a clear understanding of the method the current development process of an axle carrier will be described. This is followed by basic explanations of the method and its different evaluation blocks. Two different requirement changes will be applied to a simplified example and described in detail. Afterwards a real axle carrier with changing stiffness requirements will be exercised. The paper will conclude with a discussion of the new process and the next steps to go even further.

# State of the art for the development of an axle carrier

The axle carrier assembly and the body assembly are produced in separate plants and mounted together in the assembly line. This separation is reflected in the divisions of the car companies. One division is responsible for the plant of the axle carrier, the other one for the plant of the body. The body is one of the most dominating parts of the car due to the heavy weight, the costs and the expensive plant structure. Changing the body is connected with big efforts and high costs. The axle carrier in contrast is just a "mounting part" of the body and thereby less significant. In the concept phase the dominating teams for the body, engine and chassis geometry start with first concept models. The axle carrier team has to deal with the remaining installation space. The most challenging task is to fulfill all requirements (e.g. stiffness, strength) in that installation space.

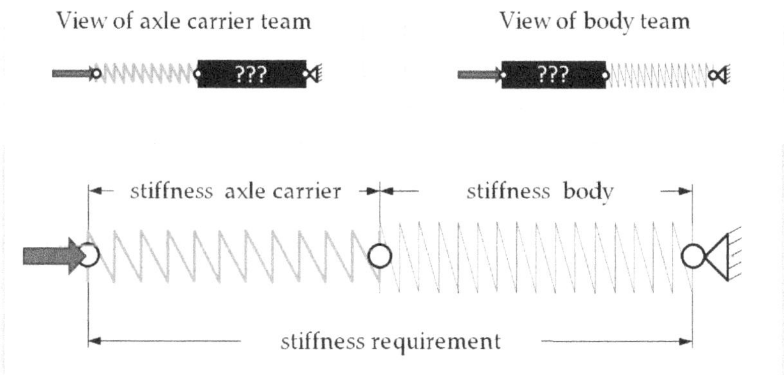


Fig. 3: Abstraction of the stiffness functionality between a body and axle carrier

The geometry of an axle carrier is mainly driven by the installation space and the stiffness requirements. A combination of the axle carrier and the body defines the stiffness and is comparable to two connected springs (see Fig. 3). The stiffness is analyzed separately by each team. It is computed with large finite element (FE) models including the axle carrier and the body. At a synchronization point the performance of both joined assemblies is evaluated. Missed requirements of the axle carrier are reflected to the axle carrier team. In a local feedback loop the axle carrier team tries to fulfill the requirements until the next synchronization point. The first step of this loop is to adapt the design of the axle carrier. Afterwards the new design gets analyzed with the FE model of the axle carrier and the body, which remains in the state of the previous synchronization point. The analyzed results focus on the unfulfilled requirements and weight. The experienced engineers will derive a strategy for the next local feedback loop and also react to internal changes. In this time the body team operates a local feedback loop for the body and starts working on internal changes as well. At the next synchronization point both adjusted assemblies are rejoined, evaluated and lightweight strategies are derived – the global feedback loop between two synchronization points is closed.

# Method

The method is derived for the designers and simulation experts of an axle carrier. The axle carrier has a fixed connection to the body. The method will be a first step for an automatic geometry optimization of an axle carrier, considering the change of external and internal conditions, the available manufacturing technologies and the interaction with a body assembly. In the product development process of VDI 2221 [3] the general idea is applicable from the middle of phase two[1], for the principle solution until the end of phase three[2], for the definitive layout.

As mentioned before the stiffness requirements and the installation space have the main influence on the geometry. These two will be in focus of this paper. In other requirements such as crash, acoustics, durability, costs etc. will be investigated in future work to extent the method step by step.

The introduced process assumes an optimized axle carrier assembly as initial design, which only needs to be adjusted as one of the conditions changes. The first step after a change is to find the best possible axle carrier with different lightweight strategies. In a second step the pervious change is interpreted against the new properties of the axle carrier. This interpretation helps to make a meaningful statement whether the material is optimal used in the axle carrier or better used in the body.

In the following sections the single steps of the process in Fig. 4 will be explained in Detail.

1 Concept development phase
2 Series development phase

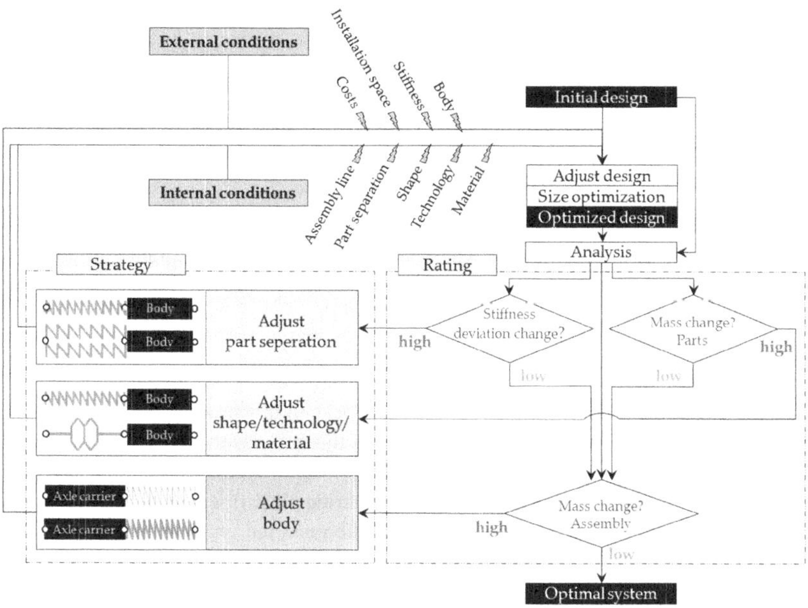


Fig. 4: Flowchart of local feedback loop

Table 1: Definition of variables and indices

| Variables | |
|---|---|
| m | mass |
| t | thickness |
| c | stiffness |
| f | rel. stiffness deviation |
| X | rating |

| Indices | |
|---|---|
| m | assembly number |
| n | part number |
| i | name of stiffness requirement |
| j | iteration number |
| init | initial |
| opt | optimized |

| | |
|---|---|
| **Example 1**: $m_{m,n,init,j}$ | → *initial* mass of part *n* in assembly *m* for iteration *j* |
| **Example 2**: $f_{i,opt,j}$ | → *optimized* rel. stiffness deviation of the requirement *i* for the iteration *j* |

*Starting point*

The method starts with a change of the internal or external conditions. Internal conditions can be a new part separation, shape, material, technology or assembly line and can be defined by the axle carrier team. External conditions are stiffness, installation space, costs or other requirements which are not defined by the axle carrier team.

*Adjust design*

The design only has to be adjusted when the installation space is changing or due to a pervious applied lightweight strategy on the assembly. Only the shape has to be adapted, the thickness will be automatically found in the next step.

*Size optimization*

The size optimization [4] is one of the crucial blocks in this procedure. The goal is to synthesize the geometrically design variables. In this process the design variables are the thicknesses of the single parts $t_{m.1}$, $t_{m.2}$, $t_{m.3}$ ... $t_{m.n}$. Thereby the lowest possible weight $m_{m,opt}$ is found for the axle carrier, considering all stiffness requirements $c_1$, $c_2$, $c_3$ ... $c_i$ and the manufacturing constraints for the maximal and minimal thicknesses $t_{min}$, $t_{max}$.

This new development process has the advantage to give a precise statement about the weight of an axle carrier which fulfills all stiffness requirements. The traditional analysis just delivers results for the stiffness under a fully geometrical predefined axle carrier. Instead of searching of an adequate thickness to fulfill all requirements, the size optimization automatically synthesizes the ideal thickness to fulfill all requirements. Therefore a lot of iterations can be saved and the engineers gain time to focus on a systematic weight reduction.

*Analysis*

The analysis evaluates the significant model data of the size optimized design. The data is compared to the initial design, in order to see how the design reacts on the previous change. Statements of the change can be done by the mass distribution, the thickness change and stiffness deviation. With the help of intermediate points a more precise analysis can be done for the nonlinear behavior of the axle carrier. For a changed shape the computation of the intermediate points could be time consuming and the engineers need to decide whether this is appropriate or not.

The significant data are:

- mass change of the assembly m with $m \in \mathbb{N}$

  $$\Delta m_{m,opt} = m_{m,init} - m_{m,opt} \tag{1}$$

- mass change of assembly parts n with $n \in \mathbb{N}$

  $$\Delta m_{m.n,opt} = m_{m.n,init} - m_{m.n,opt} \tag{2}$$

- thickness change of the parts n with $n \in \mathbb{N}$

  $$\Delta t_{m.n,opt} = t_{m.n,init} - t_{m.n,opt} \tag{3}$$

- relative stiffness deviation change of the requirements i with $i \in (A, B, C, \ldots Z)$

  $$f_{i,opt} = (c_{i,opt} / c_{i,init}) - 1 \tag{4}$$

With intermediate points the data can be illustrated in a graph where the abscissa shows the normalized change from 0 % to 100 % and the ordinate represents the different analyzed data.

### *Rating*

The ratings interpret the analyzed data and decide the next steps for the lightweight strategy. The ratings are on two levels. The first level forces the axle carrier team to find the best possible solution for the assembly. The second level rates the pervious change and permits the axle carrier team to get in interaction with the body team. It is possible to skip the first level when the change should be clearly handled by the body team.

The different ratings are:

- rating of the assembly mass change:

  $$m_{m,opt} / m_{m,init} < X_{m,Assmebly} \quad \text{with} \quad X_{m,Assmebly} > 1 \tag{5}$$

  This number states if the previous change has improved the weight of the axle carrier or not. A smaller rating than zero is a weight improvement, a larger rating a degradation. The factor $X_{m,Assmebly}$ decides if the change belongs mainly to the axle carrier or to the body.

- rating of the part mass change:

  $$m_{m.n,opt} / m_{m.n,init} < X_{m.n,Part} \quad \text{with} \quad X_{m.n,Part} \geq 1 \tag{6}$$

  These ratings compare the initial weights of the parts to the optimized weights. A high weight increase in a part might show that the part is not appropriate designed to the new stress state. The factor $X_{m.n,Part}$ decides if the part should be redesigned.

- rating of the stiffness deviation change:

  $$|(c_{i,opt} / c_i) - 1| < X_{i,Stiffness} \quad \text{with} \quad X_{i,Stiffness} > 0 \tag{7}$$

  The coefficient $c_{j,opt} / c_i$ shows whether the design is feasible or infeasible. If at least one of the thicknesses reach the maximal valid thickness the design is infeasible. If it is feasible, it shows how much the other stiffness requirements are violated. As soon as all coefficients are close to zero, the optimum is reached. The factor $X_{i,Stiffness}$ decides if at least one deviation is too high and the structure has a poor part separation.

The ratings X are dependent on many factors such as the quantity of the change, the costs, the schedule for the project and sometimes on personal experience of the engineers. For example a large change will have a big influence on the weight. Expensive manufacturing technologies will influence the costs. A short time span reduces the possibilities to react. The exact evaluation of the rating factors will be part of future work.

*Strategy*

The strategy decides the treatment to save weight. Dependent on the rating the corresponding strategy is chosen. The different strategies are combined with additionally analysis or optimization methods [5] on specific parts. The lightweight strategies are the adjustments of the part separation, the shape, the technology, the material and the body.

The separation of the parts is used as soon as the rating for the relative stiffness deviation is too high for at least one requirement. This means that the size optimizer is dominated from one requirement and chooses an appropriate thickness for the parts. If other kinematic points with lower requirements use the same parts the stiffness is too high and the structure is overdesigned. The approach to lower the stiffness at these points is to search for a new part separation. This is done with the help of a free size optimizer. The results of the optimizer show areas with similar thicknesses, which can be with respect to the possibilities of the manufacturing technology applied to the assembly.

The adjustment of the shape, manufacturing technology or the material is used as soon as one part reacts very sensitive in the mass or thickness change compared to the other parts. For a better understanding the stresses are analyzed. For example a modification of the shape could change the stress state and higher bending stresses appear in the size optimized part. A useful treatment is to adjust the profile by changing the diameter or by choosing a more appropriate manufacturing technology. Another material should be used, if the manufacturing limits are reached and the installation space is fully used. These treatments lower the stresses and use the material more efficient.

The adjustment of the body is only applied if all accomplished adjustments on the axle carrier show an unacceptable effect for the previous change. The body needs to evaluate the optimized weight, but in the first attempt with the unchanged axle carrier. The optimized body weight with the initial axle carrier is compared to the optimized axle carrier with the initial body. This comparison shows clearly whether the material is better used in the body, in the axle carrier or in a combination of both.

*Iteration process*

The process runs in several loops until all ratings are passed. The main purpose of the first loop is to analyze the effect of the change between the initial design and the design from the size optimizer. The results show directly the introduced strategy which is to apply at first and also which further steps arel follow.

In the next chapter this iteration process will be applied to a principle example.

# Principle example

In order to make the previous claims understandable a principle model will be introduced. Two different cases will be put into practice: one changes the installation space, while the other one changes the stiffness requirements. For simplification both changes have similar effects on the model.

The principle model as shown in Fig. 5 represents the behavior between the axle carrier assembly 1 and the body assembly 2. Two load cases are defined and both assemblies are fixed. Assembly 1 can change the technology from a simple circular profile to a casting profile and is able to divide the single part into three parts with individual thicknesses. The change of the installations space is represented by the changed hatched circle.

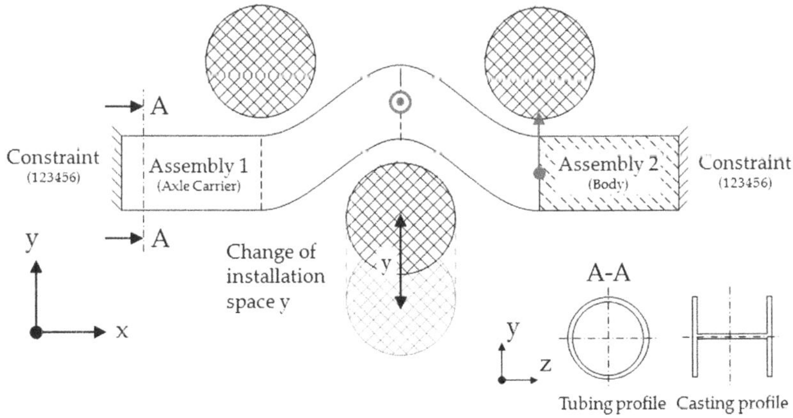


Fig. 5: Principle model demonstrating an axle carrier body system

All ratings and strategies are applied step by step in four loops. In the first loop the basic change is rated. The second one investigates the evaluation of the part splitting. The third one analyzes the different technologies and the last loop compares the results from the first three loops to the optimization of the body.

### *Basic rating of a change*

The initial design consists of a long tube (Fig. 6). One section of the tube belongs to the assembly 1 (axle carrier) and the other belongs to the assembly 2 (body). The hatched part symbolizes the black box behavior of the assembly 2. The responsible team for the assembly 1 is not allowed to change assembly 2. The initial design fulfills all requirements $c_{Z,init}$, $c_{Y,init}$ and the relative stiffness deviation $f_{Z,init}$, $f_{Y,init}$ is zero for both load cases.

The general process starts with an external or internal requirement change. The two case studies which will be exercised are: on the left side, the change of the installation space $y_{change}$ and, on the right side, the change of the stiffness requirement $c_{Z,Change}$ for one load case. The purpose of the first loop is to get an impression of the change and the influence it has on the assembly 1.

The structure has a nonlinear behavior. In order to see these nonlinear effects intermediate points are calculated. Case 1 and case 2 are normalized by the "change". The initial design has a 0% change and the optimized design has a 100% change ($y_{change}$ in case 1; $c_{Y,Change}$ in case 2).

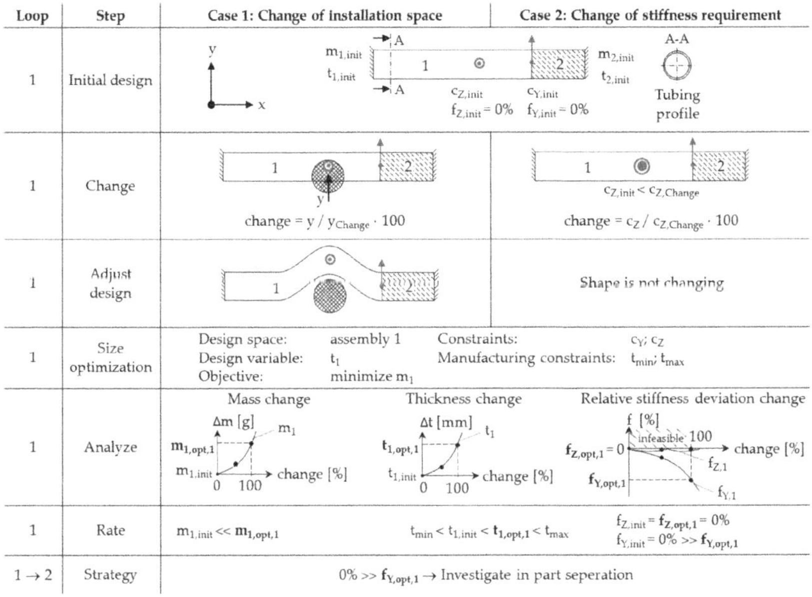


Fig. 6: Procedure for the rating of different requirement changes on the principle model

The first step for the designer is to change the shape of the initial design in order to fit in the new installation space, as in case 1. In case 2 there is no shape change. The design is the input model for the size optimization. The job of the optimizer is to minimize the mass $m_1$ of assembly 1 by finding the lowest thickness $t_{1,opt}$, under all required constraints, such as stiffness $c_Y$, $c_Z$ and manufacturing constraints $t_{min}$, $t_{max}$. For intermediate points the optimizer has to run at least twice.

The analyses display the significant results, which are the graphs for the:

- mass change of assembly 1, $m_1$
- mass change of the parts built in assembly 1 (not shown in loop 1)
- change of the thickness, $t_1$
- change of the relative stiffness deviation $f_Y$, $f_Z$ for the requirements, $c_Y$ and $c_Z$

The ratings of these graphs are crucial for the next steps. The first analyzed graph is the mass change $m_1$ of the complete assembly. The high increase of the mass $m_1$ is a first hint that assembly 1 might not be the right choice to react to the changes. The second graph to examine is the relative stiffness deviation change for $f_Y$ and $f_Z$. The relative stiffness deviation change $f_{Z,opt,1}$ is zero and fulfills the requirement $c_Z$, while the other one $f_{Y,opt,1}$ is clearly lower than zero. This behavior of the new design belongs to the integral design of the assembly. The stiffness requirement $c_Z$ drives the thickness of the parts and is thereby overdesigned for the other stiffness requirement $c_Y$, which means automatically the existence of potential to safe weight. The consequent next step is to investigate in a different part separation for the assembly 1.

### *Investigating in part separations*

In loop 2 (Fig. 7) the single part of assembly 1 is separated in three equal parts. The new design variables of the size optimizations are now the three thicknesses $t_{1.1}$, $t_{1.2}$, $t_{1.3}$ for the single parts 1.1, 1.2, 1.3. The analyses in loop 2 show an additional graph for the changing part masses $m_{1.1}$, $m_{1.2}$, $m_{1.3c}$. The improvement can be evaluated in the assembly mass $m_{1,opt,2} < m_{1,opt,1}$, showing a lower weight in comparison to the weight of the first loop. This is validated in the graphs concerning the relative stiffness deviation change showing that $f_{Y,opt,2}$ is now closer to zero than in loop 1. Noticeable are the results of part 1.3. The mass $m_{1.3,opt,2}$ and the thickness $t_{1.3,opt,2}$ of part 1.3 increases more significantly as in part 1.1 and 1.2. Due to the new requirements also the stress state in the parts changes. This could be seen by comparing the stress results of the initial and the new design. In this example the stress state of part 1.3 has changed and a profile with a more suitable section modulus (e.g. casting profile) is needed. Considering the restrictions from the installation space and the impossibility of manufacturing a bigger diameter of the pipe, only a different technology (e.g. casting) can show optimization potentials in this example.

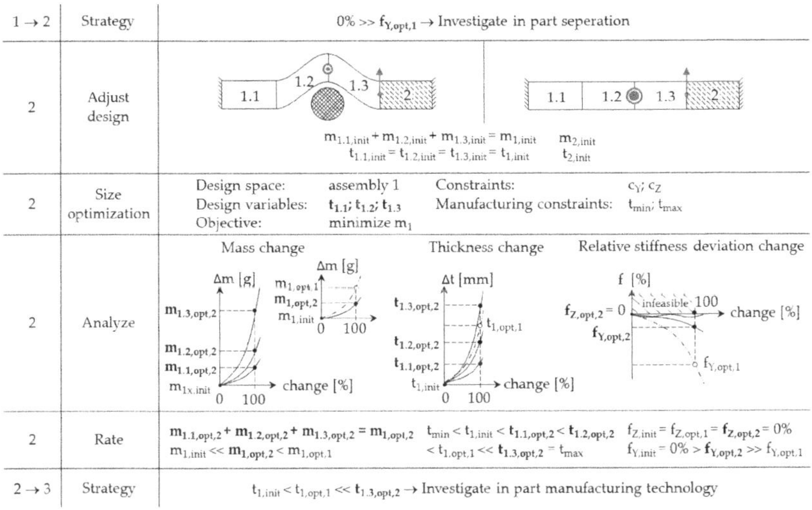


Fig. 7: Procedure for investigating in part separation in the principle model

## *Investigating in different manufacturing technologies*

In loop 3 (Fig. 8) part 1.3 is exchanged from a tubing profile to a casting profile, fitting the stress state in this section in a more appropriate way. The design variables $t_{1.1}$, $t_{1.2}$, $t_{1.3}$, constraints $c_Y$, $c_Z$ and objective $m_1$ are the same as in loop 2. The results in the analysis show that part 1.3 is clearly lighter, part 1.1 and 1.2 are just slightly heavier than in loop 2. This is also reflected in the thickness changes, although part 1.3 is difficult to compare. In sum, the new manufacturing technology adds another weight improvement to assembly 1 $m_{1,opt,3} < m_{1,opt,2} < m_{1,opt,1}$. As for the relative stiffness deviation change the results are inversed. This inversion is compare-able to find the root with an iterative bisection method. The inversion in combination with the improvements proof that the optimum of assembly 1 is nearly reached.

Finally, taking the coefficient for the mass of the optimized assembly 1 and the initial mass the number is still high compared to the initial change $m_{1,opt,3}/m_{1,init} >> 1$. The hint in the first loop was correct. In order to fulfill the stiffness requirement $c_Y$ and $c_Z$, assembly 2 should be optimized.

| | | |
|---|---|---|
| 2 → 3 | Strategy | $t_{1,init} < t_{1,opt,1} << \mathbf{t_{1.3,opt,2}}$ → Investigate in part manufacturing technology |
| 3 | Adjust design | 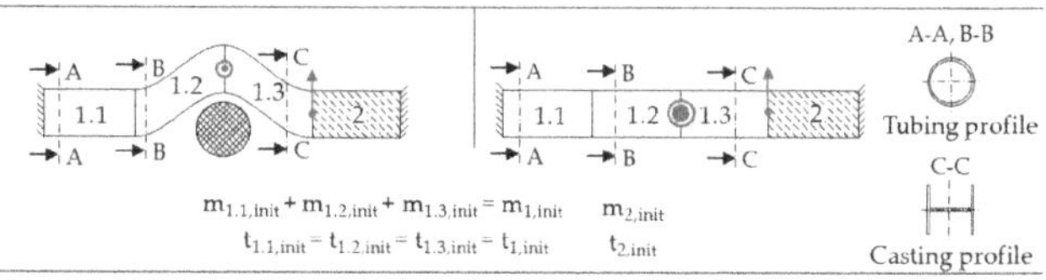 |
| 3 | Size optimization | Design space: assembly 1 Constraints: $c_Y$; $c_Z$<br>Design variables: $t_{1.1}$; $t_{1.2}$; $t_{1.3}$ Manufacturing constraints: $t_{min}$; $t_{max}$<br>Objective: minimize $m_1$ |
| 3 | Analyze | 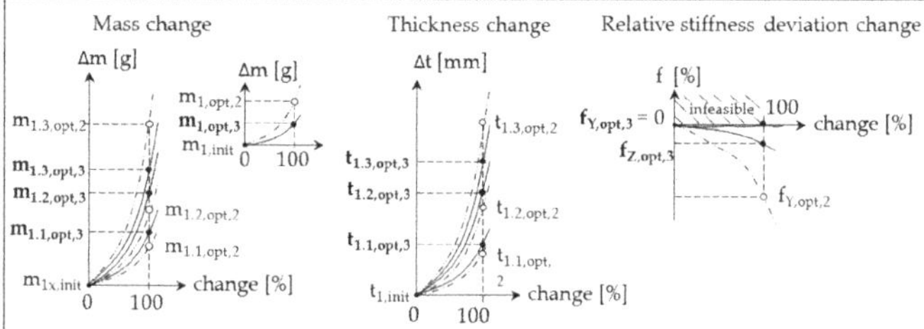 |
| 3 | Rate | $m_{1.1,opt,2} < \mathbf{m_{1.1,opt,3}}$<br>$m_{1.2,opt,2} < \mathbf{m_{1.2,opt,3}}$<br>$m_{1.3,opt,2} >> \mathbf{m_{1.3,opt,3}}$<br>$m_{1,init} << \mathbf{m_{1,opt,3}} < m_{1,opt,2} < m_{1,opt,1}$<br>$t_{1.1,opt,2} < \mathbf{t_{1.1,opt,3}}$<br>$t_{1.2,opt,2} < \mathbf{t_{1.2,opt,3}}$<br>$t_{1.3,opt,2} >> \mathbf{t_{1.3,opt,3}}$<br>$f_{Z,init} = f_{Z,opt,1} = f_{Z,opt,2} = 0\% \approx \mathbf{f_{Z,opt,3}}$<br>$\mathbf{f_{Y,opt,3}} = f_{Y,init} = 0\% > f_{Y,opt,2} >> f_{Y,opt,1}$ |
| 3 → 4 | Strategy | $\mathbf{f_{Y,opt,3}} = 0\%$; $\mathbf{f_{Z,opt,3}} \approx 0\%$; $\mathbf{m_{1,opt,3}}/m_{1,init} >> 1$ → Investigate in the connecting assembly |

Fig. 8: Procedure for investigating in manufacturing technologies in the principle model

## *Investigating in the connecting assembly*

In loop 4 (Fig. 9) assembly 2 is optimized from the responsible team. In case 1 with the changing installation space, the team of assembly 1 delivers the changed shape with the initial thickness $t_{1,init}$ of loop 1. In case 2 the initial design is directly used. Assembly 1 is the non-design space (black box) for the optimization. The design space, the design variable $t_2$ and the objective $m_2$ belong to assembly 2. The graphs in the analysis show the better use of mass $\Delta m_{2,opt,4} << \Delta m_{1,opt,3} < \Delta m_{1,opt,2} < \Delta m_{1,opt,1}$ in the assembly 2 in order to fulfill the stiffness requirements $c_Z$ and $c_Y$. Additionally the graph of the relative stiffness deviation shows more potential to save weight. With the information on the behavior of assembly 1 the system optimum can now be found.

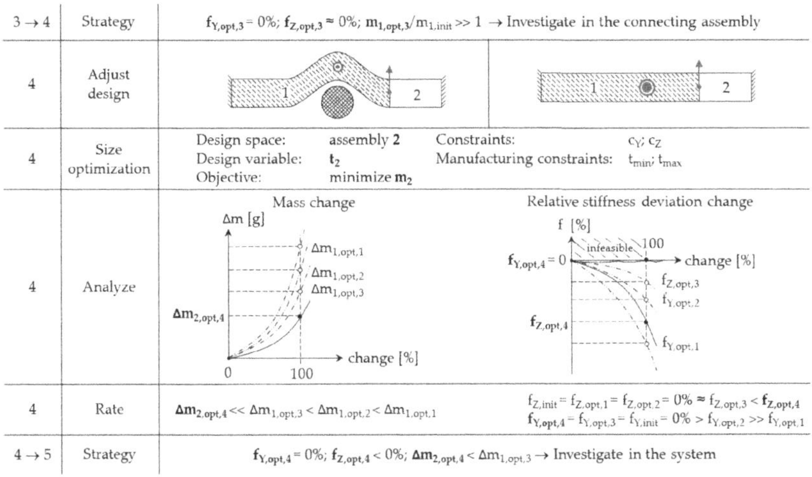


Fig. 9: Procedure for investigating in the connecting assembly in the principle model

# Application on an axle carrier

The method analyzed above is applied to a steel axle carrier of a mid size sedan car. The axle carrier is directly mounted to the body. For a static determined system the constraints are applied to the body. The forces are directly applied on the axle carrier, see Fig. 10. The different load cases for the stiffness requirements are measured directly at the kinematic points of the axle carrier. Each load case exists for the left side and the right side, for a better overview Fig. 10 shows just one side.

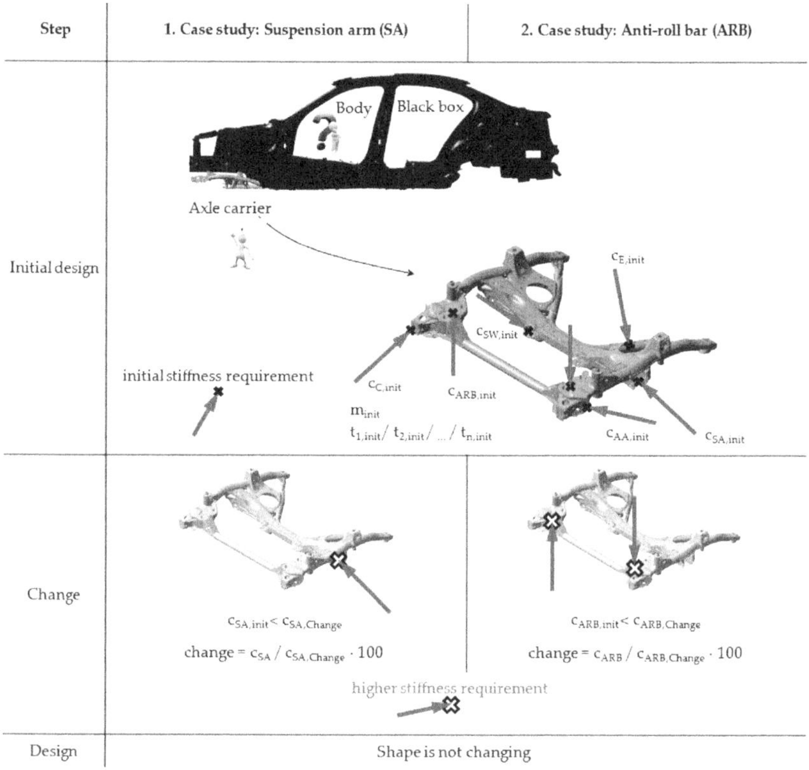


Fig. 10: Stiffness requirements on a mid size axle carrier body system

Two separate case studies are demonstrated, both changing the stiffness requirement. In each case only one stiffness requirement is raised while all other remain unchanged. In this example the significant data from one loop is evaluated and the corresponding treatments are established. One case leads to an adjustment in the axle carrier and the other in the body. The two case studies are illustrated in Fig. 9. The first case study raises the stiffness requirement of the suspension arm $c_{SA}$, the second of the anti-roll bar $c_{ARB}$. The anti-roll bar is responsible for the torsional stiffness of the body axle carrier system.

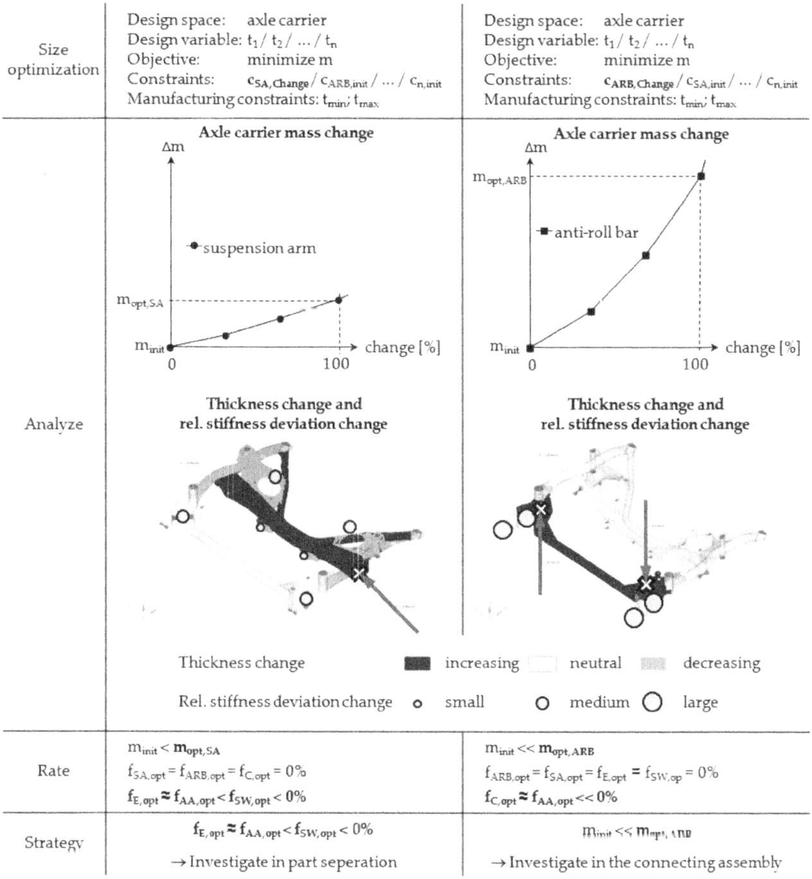


Fig. 11: Mass change, thickness change and relative stiffness deviation change of the axle carrier by changing different stiffness requirements

As described in the previous chapter, the initial design of the axle carrier can be used directly for the optimization. The shape does not have to be adjusted, after there is no change in the installation space. The design variables are thicknesses $t_{1.1}$, $t_{1.2}$, $t_{1.3}$, ... $t_{1.n}$ of the different parts. The constraints for the size optimizer is the changed stiffness requirement $c_{i,Change}$ and the old initial requirements $c_{i,init}$. Two intermediate points for the new requirement are calculated in order to see the nonlinear effects (Fig. 11). In the first case study the mass $m_{1,opt,SA}$ increases slightly. In comparison, the mass

$m_{1,opt,ARB}$ of the second case study increases much more with comparable stiffness changes. This noticeable difference in the second case study leads to an adjustment in the body, which is understandable since the body is mainly responsible for the torsional stiffness of the system.

The thickness change and the relative stiffness deviation change will be illustrated directly on the axle carrier (Fig. 11). The colors mark the significant increase or decrease of the single part masses. The sizes of the dots mark the relative stiffness deviation change of the requirements. In the first case study all connecting parts between the kinematic point of the suspension arm and the mounting points to the body are thicker due to the stiffness increase. The stronger parts are capable of taking more energy and compensate the task of other parts. In this way it is possible to reduce the mass in some parts. Similar to the principle example, the new stiffness requirement dominates the structure and influences other stiffness requirements. In the first case study three kinematic points have a small to medium deviation. Consequently the right strategy to decrease the weight is a slightly different part separation. The second case study has a big influence on the front parts of the axle carrier, where many requirements have to be fulfilled. Changing the stiffness of the anti-roll bar influences directly the other kinematic points. Almost all parts in the front are affected in the same way, which means that only small geometrical improvements are possible, but as already detected in the evaluation of the axle carrier mass the body needs to take care of the torsional stiffness.

# Conclusions

The introduced method needs a new thinking for the design engineers and the simulation experts. Instead of analyzing a model and set the strategy to fulfill the requirements, the size optimizer synthesizes the ideal thickness distributions as shown in the examples.

The analysis of the optimization results showed the most significant data which helped to get a better understanding of the axle carrier and also the system behavior. With this new knowledge simple and efficient lightweight strategies could be found to design the optimal system existing of two direct connected assemblies. These strategies accelerated the process as soon as internal or external conditions changed.

This paper focused on the requirements of the stiffness, considering however that in the automotive sector there are many other factors such as crash, durability or costs, which need to be taken into account to evaluate a rating factor for a specific decision. This factor rates the ratio between the requirement change and the assembly change, supporting the engineers with the decision which assembly and which part has to change and how much. This factor will be in focus of future work.

History forced automotive engineers to optimize assemblies separately, after computation power has been the bottleneck and it has not been possible to optimize a whole car for all requirements at once [6]. The future computation technology will deliver more efficient hardware and software in order to bring all parts back together for a more precise evaluation of all requirements on a complete car. The progress on this topic will go hand in hand with upcoming computation technology.

# References


[1] K. **Ehrlenspiel:** *Integrierte Produktentwicklung*, Carl Hanser Verlag, München/Wien, 2003

[2] **C. Köhler:** *Technische Produktänderungen – Analyse und Beurteilung von Lösungsmöglichkeiten auf Basis einer Erweiterung des CPM/PDD-Ansatzes*, D. Bähre und H. Bley, Saarbrücken, 2009

[3] **VDI 2221:** *Methodik zum Entwickeln und Konstruieren technischer Systeme und Produkte*, VDI-Handbuch Konstruktion, Beuth Verlag GmbH, Berlin, 1993

[4] **A. Schumacher:** *Optimierung mechanischer Strukturen – Grundlagen und industrielle Anwendungen*, Springer-Verlag, Berlin Heidelberg, 2005

[5] **S. Vajna, Chr. Weber, H. Bley, K. Zeman:** *CAx für Ingenieure – Eine praxisbezogene Einführung*, 2. Auflage, Springer-Verlag, Berlin Heidelberg, 2009

[6] **H.-H. Braess. U. Seiffert:** *Vieweg Handbuch Kraftfahrzeugtechnik*, 7. Auflage, Springer Fachmedien, Wiesbaden, 2013

# Objektive Ratingmethode für Handling- und Komfortkriterien für den Einsatz im Fahrversuch und in der Simulation

# (Objective method for rating ride and handling criteria in simulation and vehicle testing)

Joachim Ecker, Dr. P. Schöggl, E. Bogner, M. Oswald,
AVL List GmbH, Austria

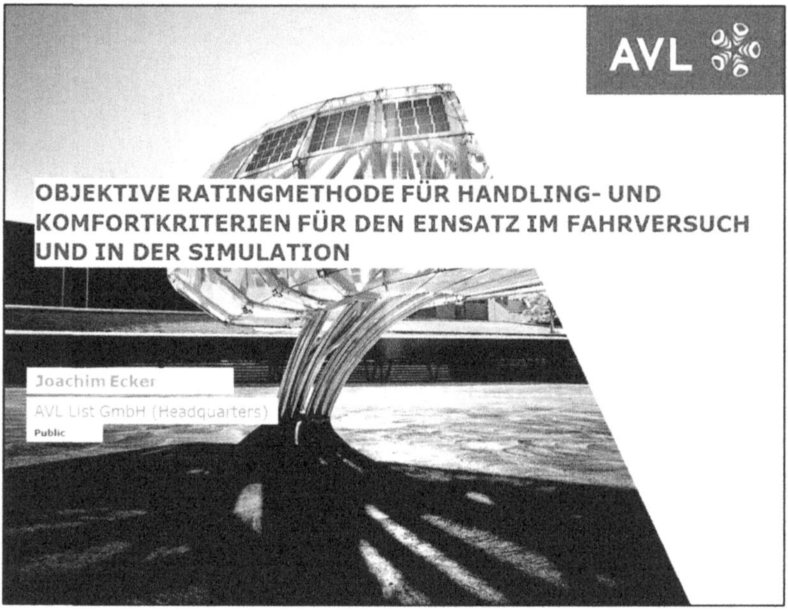


Die AVL List GmbH beschäftigt sich in Kundenprojekt im Rennsport und im Serienfahrzeugbau seit vielen Jahren mit der Optimierung von Fahrwerkssystemen hinsichtlich Performance, Handling und Fahrkomfort.
Im Rahmen dieser Tätigkeiten galt und gilt es, Methoden anzuwenden die die Qualität der Funktion eines Fahrwerks wiedergeben. Ist diese im Rennsport meist einfach über die Rundenzeitdifferenz darstellbar, stellt sich die Anforderungen im PKW Bereich komplexer dar, da hier in der Abstimmung ein Kompromiss aus agilem Handling, Fahrstabilität und Fahrkomfort gefunden und bewertet werden muss.

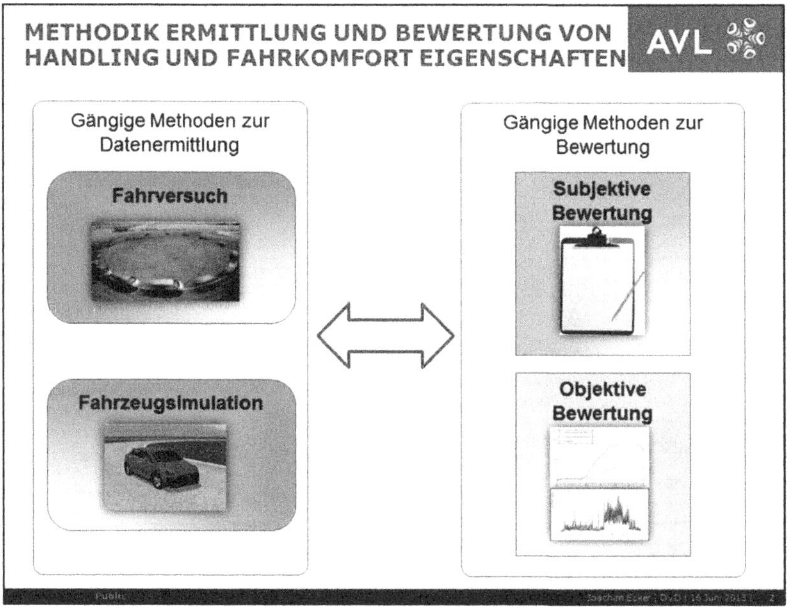


Im Rahmen der Untersuchungen hinsichtlich Fahrwerksoptimierung gilt die Fahrzeugsimulation zur Prüfung des Einflusses von Fahrwerksparametern mittlerweile als etablierte Methode die parallel zu den Versuchsfahrten im realen Fahrzeug im Entwicklungsprozess durchgängig zur Anwendung kommt.
Im realen Fahrversuch ist die subjektive Beurteilung durch erfahrene Testfahrer nach wie vor als höchste Instanz anerkannt. Die Ergebnisse werden in den meisten Fällen mit Noten zwischen 1 und 10 beschrieben. Das Resultat gibt schnell und plakativ Stärken und Schwächen wieder, ist aber zur Detailanalyse eher weniger geeignet. Neben der subjektiven Methode existieren auch objektive Methoden auf Basis von gemessenen oder simulierten Signalen die die Fahrwerksfunktion beschreiben sollen. Die Fahrmanöver zur objektiven Methode sind streng definiert und bieten eine hohen Wiederholgenauigkeit und liefern Ergebnisse im Form von Diagrammen und großen Anzahl von Kennwerten, geben aber oft nur einen Ausschnitt des Gesamtbildes der Fahrwerksfunktion wieder.

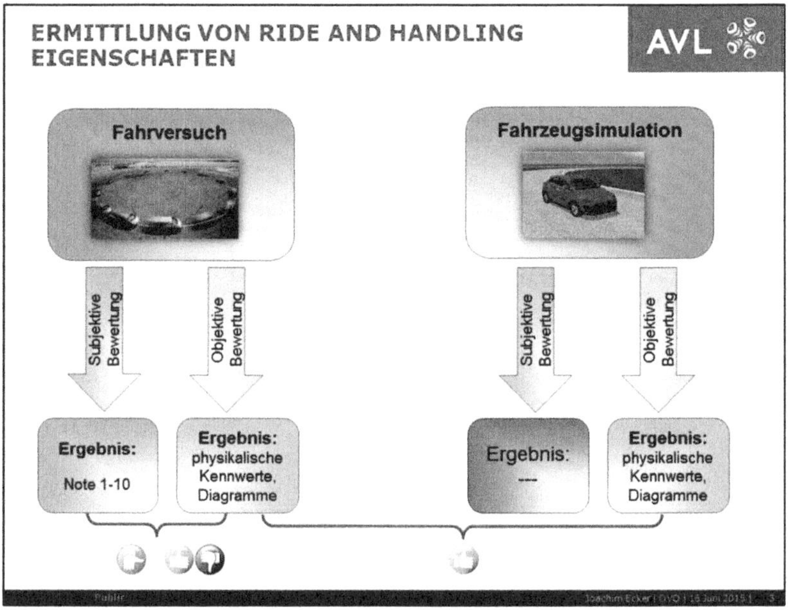


Betrachtet man den Überblick über die existierenden Bewertungen angewendet auf die gängigen Methoden zur Erheben von Handling und Fahrkomfortdaten ergibt sich das Bild, dass die objektiven Methoden zwar in Versuch und Simulation angewendet werden können und dann auch gute Korrelationen zeigen, jedoch aber die Vergleichbarkeit und damit auch die Korrelation zur subjektiven Methode nicht gegeben ist.
Die subjektive Methode wiederum kann nicht auf Simulationsergebnisse angewendet werden.

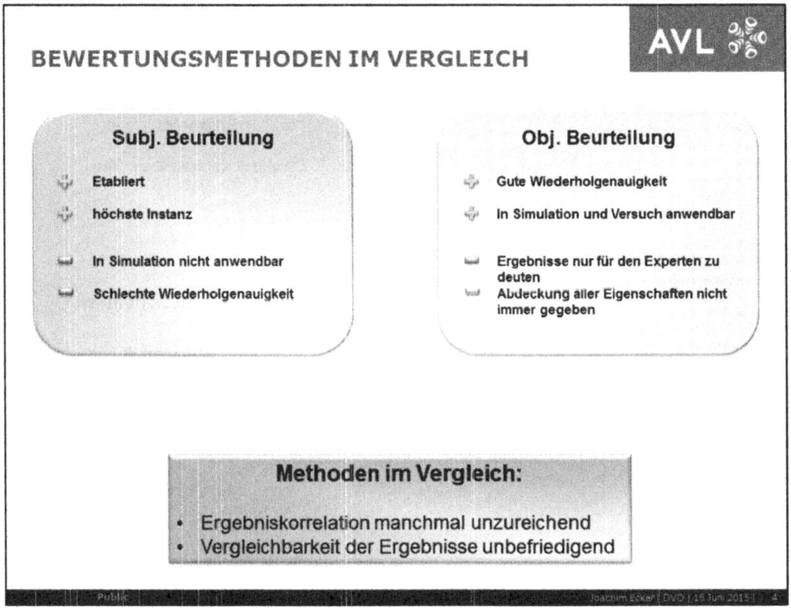


Vergleicht man beide etablierten Methoden über Vor- und Nachteile, erkennt man schnell, dass hier eine technologische Lücke existiert.

Im Rahmen der Projektarbeit mit diversen Kunden zeigte sich auch dass in der Fahrwerksentwicklung Schwächen subjektiv wahrnehmbar waren, die sich über die gängigen objektiven Methoden nicht darstellen ließen, da die strenge Manöverdefinition Fahrzustände in denen diese Phänomene zu beobachten sind, ausschließen.

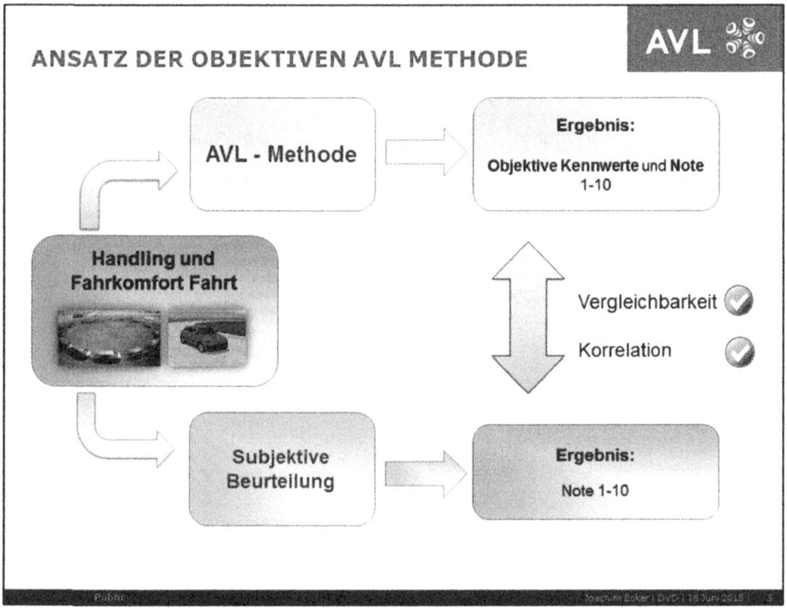


Inspiriert von dieser technologischen Lücke, begann man bei der AVL über eine neue objektive Bewertungsmethodik nachzudenken.

Jedenfalls musste die Anwendbarkeit auf Daten der Fahrzeugsimulation gegeben sein, des weiteren musste die Vergleichbarkeit mit subjektiven Beurteilungen möglich sein.

Die neue Methodik musste daher auf Basis von gemessenen oder simulierten Fahrzeugbewegungsgrößen, Fahrzustände erkennen, bewerten können und sich dabei an der Freiheit der Gestaltung der Fahrmanöver der subjektiven Bewertungsfahrt orientieren.

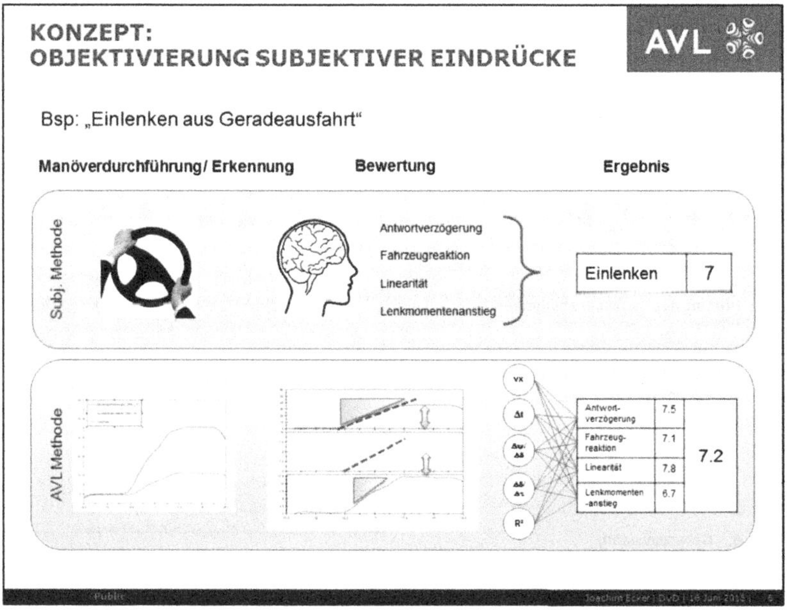


Als Basis für das Entwicklungskonzept der neuen Methodik diente der klassische Bewertungsbogen für die subjektive Beurteilung.

In der subjektiven Beurteilung, führt der Versuchsfahrer beispielsweise ein Einlenkmanöver aus Geradeausfahrt durch. Aus seiner Erfahrung heraus bildet er aus der Summe der Eindrücke von Ansprechzeiten, Verstärkungen und Rückmeldungen des Fahrzeugs seinen subjektiven Eindruck und bewertet diese Eigenschaft mit einer Note.

Die AVL Methode setzt es sich zum Ziel, genau diesen Vorgang auf Basis von Verarbeitung von Messsignalen abzubilden. Die Aufgabe bestand zuerst im selbstständigen Erkennen von bewertbaren Fahrzuständen und dann in der Bewertung für sich. Beim Bilden der Bewertungsfunktion galt es aus Aussagen erfahrener Versuchsfahrer jenen Kriterien die den subjektiven Eindruck beeinflussen zuzuordnen, je nach Ausprägung richtig zu bewerten und die Gewichtungen für die Gesamtbewertung aufzustellen.

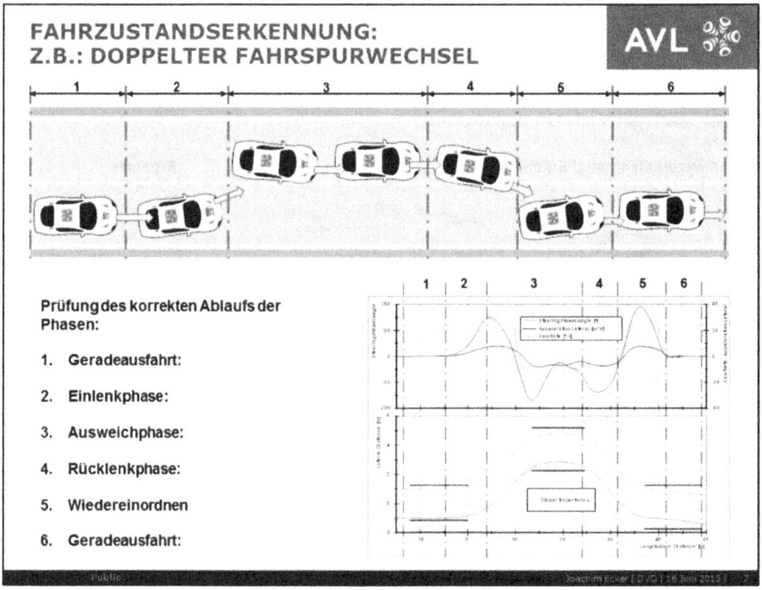


Erkennung der Fahrzustände im Fahrzeug Handling:

Als Beispiel für die Funktion der Eventerkennung soll der doppelte Fahrspurwechsel dienen, da dieser sehr oft im subjektiven Bereich zur Anwendung kommt, da er sehr viele Handlingeigenschaften eines Fahrzeugs plakativ wiedergibt.

Das Fahrmanöver beginnt mit einer Geradeausfahrt, mit einem Lenkimpuls wird das Fahrzeug auf die Ausweichspur gelenkt und muss dort mit einem Gegenlenkimpuls stabilisiert werden. Dies führt zu einem je nach Fahrzeug ausgeprägten Überschwingen des Aufbaus. Um von einem doppelten Fahrspurwechsel zu sprechen wird sofort wieder auf die Ausgangsspur gelenkt, da man im Falle dass die Ausweichphase zu lange gewählt wird und das Fahrzeug sich dort stabilisiert hat von zwei einzelnen Spurwechseln sprechen muss.
Im Falle des doppelten Spurwechsels wird jedoch ein Teil des Impulses in die nächsten Phasen mitgenommen und kann beim Wiedereinordnen zu heftigem Überschwingen des Fahrzeugs führen. Es folgt wieder eine Phase des Stabilisierens und wieder eine Geradeausfahrt. Prüft man Fahrzeugbewegungsgrößen per definiertem Programmcode auf Einhalten dieser Phasen, erkennt man automatisiert einen doppelten Spurwechsel.

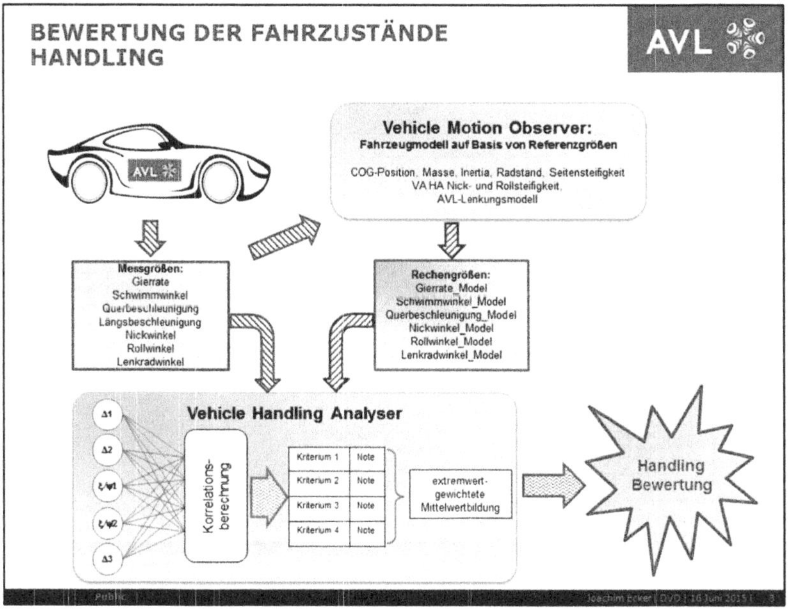


Bewertung der Handling Fahrzustände:

Die Bewertung der erkannten Handling Fahrzustände erfolgt auf Basis von einer Ist/Sollwert Differenz gegenüber Referenzdaten. Diese Referenzdaten werden vom „Vehicle Motion Oberserver“ gebildet, der auf Basis eines vereinfachten physikalischen Fahrzeugmodells Sollgrößen bildet. Die Parametrierung des Modells erfolgt mit physikalischen Parametern die technologisch erreichbar scheinen, jedoch in Summe aus Kostengründen und aus Gründen des trade offs zum Fahrkomfort von gängigen Fahrzeugen nicht erreicht werden. Die Abweichungen zu diesen Referenzgrößen bilden Parameter die die Basis für die Bewertung von Fahrzustandskriterien bilden. Als Bespiel sei wieder der doppelte Fahrspurwechsel erwähnt. Mit Ausnahme der Geradeausfahrt zu Beginn und am Ende werden alle Phasen getrennt bewertet. So lassen sich beispielsweise aus dem erwähnten Überschwingen in der Ausweichphase eine Note für die Steuertendenz und für das Rollüberschwingen des Aufbaus als Kriterienbewertung gewinnen.

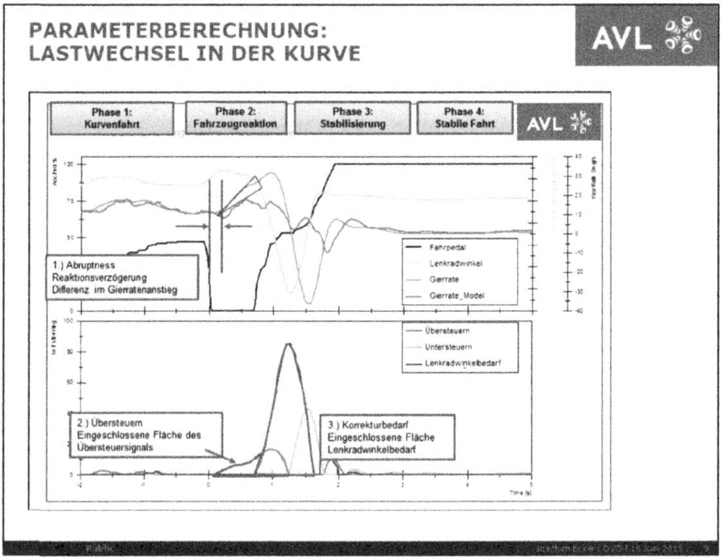


Parameterberechnung:
Als Beispiel soll hier ein Lastwechselmanöver in der Kurvenfahrt dienen, da hier die Vorgänge und zugehörigen Parameter besonders plakativ dargestellt werden können. Im gegeben Fall wird in der Kurvenfahrt das Fahrpedal (schwarz) schlagartig ausgelassen und zusätzlich noch etwas zugelenkt (rosa) um ein kritische Fahrsituation wie zB. eine Autobahnausfahrt, in die zu schnell eingefahren wurde, nachzubilden. Man erkennt dass es nach einer etwa einer halben Sekunde zu einer starken Zunahme der Istgierrate (orange) kommt. Die Referenzgierrate (pink) zeigt diese Tendenz nicht.
Der Zeitverzug des Anstieges sowie die Differenz der Gradienten aus Istgierrate und Referenzgierrate bildet ein Maß für die Heftigkeit der Fahrzeugreaktion.
Die Istgierrate ist also größer als die Referenzgierrate die das Referenzmodell aus Fahrerlenkradwinkel und Fahrgeschwindigkeit bildet, was bedeutet das das Fahrzeug aufgrund des Lastwechsels übersteuert. Das Integral über die Differenzkurve der beiden Gierraten ist ein Parameter und ein Maß für die Übersteuertendenz.
Anhand des Lenkradwinkelsignals (rosa) erkennt man dass der Fahrer zusätzlich zum Wiederaktivieren des Fahrpedals auch ein Gegenlenkmanöver zur Fahrzeugstabilisierung einleitet. Aus Fahrgeschwindigkeit und Lenkradeinschlag errechnet man einen Gierratenwunsch des Fahrers der als Differenzfläche zur Istgierrate ein Maß für den anliegenden Korrekturaufwand darstellt.

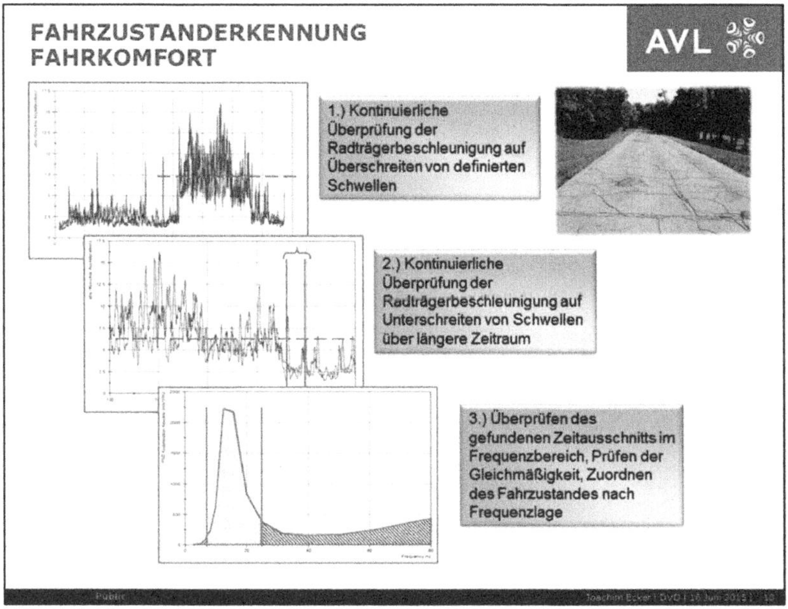


Erkennung der Fahrzustände im Fahrzeug Fahrkomfort:

Zur Demonstration für eine Fahrzustandserkennung im Fahrkomfort soll„ Smooth Ride" als Manöver dienen. Die Fahrzustand stellt sich für gewöhnlich beim Überfahren alter Betonstrecken ein. Auf solchen Strecken trifft man neben der Grundrauigkeit auf Plattenstöße und Aufbrüche durch Frost und durch Überlast. Je nach Fahrgeschwindigkeit stellt sich ein Anregungsband zwischen 10- 25 Hz bei mittlerer Amplitudenlage ein.
Diese Fahrbahn erzeugt am Radträger kontinuierliche Beschleunigungen. Überwacht man diese Signale auf Schwellenüberschreitung und nach erfolgter Überschreitung auf Unterschreitung über einen definierten Zeitraum gewinnt man ein Zeitintervall innerhalb dessen eine Beschleunigungsniveau an den Rädern angelegen ist. Analysiert man diese Signale im Frequenzbereich, kann man je nach Verhältnis der Frequenzinhalte und nach Prüfung auf eine gewisse Gleichmäßigkeit das Zeitintervall einem Komfortereignis zuordnen oder auch im Fall dass die definierten Anforderungen nicht anliegen den erkannten Fahrzustand verwerfen und nicht weiter bewerten.

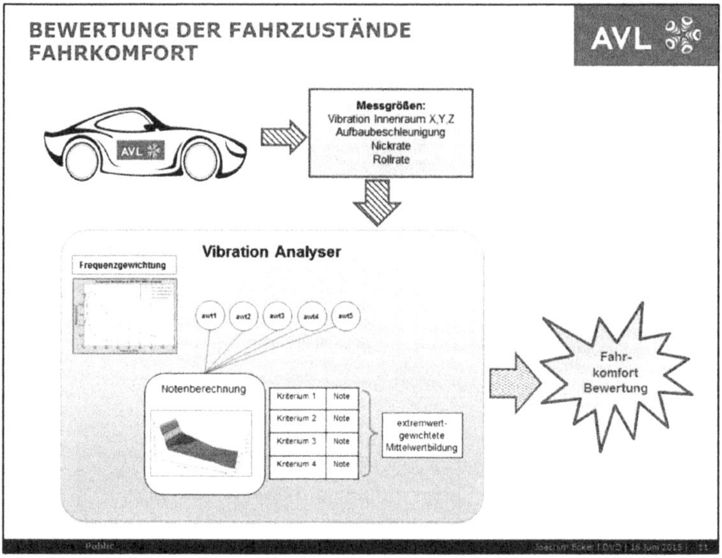


Bewertung der Fahrkomfort Fahrzustände:

Im Fahrkomfort wird kann der Komforteindruck bei mittleren und hohen Anregungsfrequenzen auf Basis von Beschleunigungssignalen diverser Sensoren in Innenraum beschrieben werden. Da die Sensibilität des Menschen für das Empfinden von Vibrationen stark von der Frequenz abhängt, müssen diese Signale vor der Verrechnung im Frequenzbereich frequenzgewichtet werden um dann in den Zeitbereich rücktransformiert, als Parameter in Form von frequenzgewichteten RMS oder VDV Werten zur Verfügung zu stehen.
Da im Fahrkomfort die exakte Form der Fahrbahn unbekannt bleibt, da der Reifen zwischen Fahrbahn und Beschleunigungssensor eine Unbekannte darstellt, kann man im Fahrkomfort nicht auf ein ideales Fahrzeugmodell zurückgreifen. Die Notenbildung gelingt daher über „Correlation Maps" die aus umfassenden Korrelationsuntersuchungen auf Basis subjektiver Eindrücke erfahrener Testfahrer gebildet wurden.
Die berechneten Vibrationsparameter werden fahrzustandsabhängig kombiniert. Die Kombinationsvorschrift ergab sich aus den angesprochenen Korrelationsuntersuchungen. Die Bewertung gelingt über die Maps aus der Parameterkombination und der Fahrgeschwindigkeit.
Der „Vibration Analyzer" Funktionsblock ergibt sich aus Funktion für die Frequenzgewichtung und den Correlation Maps für alle Fahrzustände.

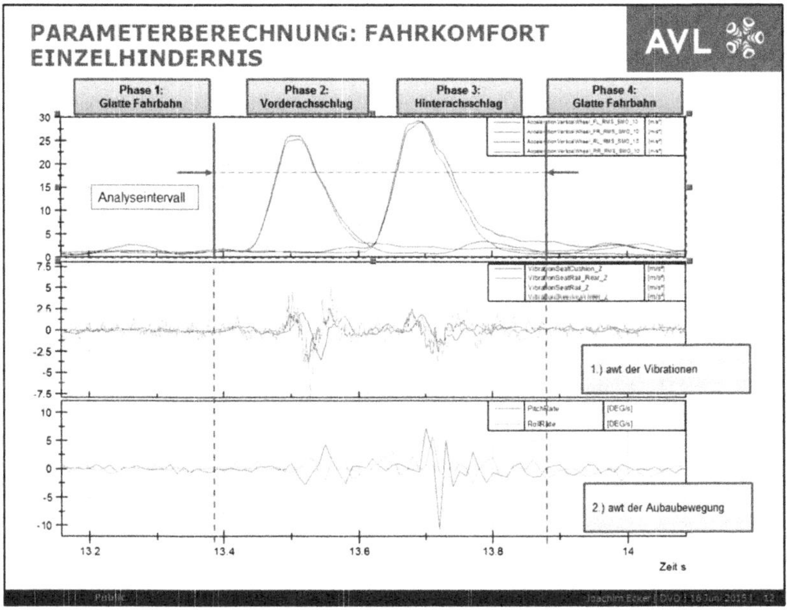


Beispiel für die Berechnung der Parameter:

Als Beispiel soll hier der Fahrzustand „Einzelhindernis oder „Impact Harshness" dienen. Hier kann man je nach Fahrgeschwindigkeit und Radstand des Versuchsfahrzeuges zwei Einzelschläge an den Radbeschleunigungssensoren der Achsen detektieren. Das Zeitintervall kurz vor dem Vorderachsschlag bis kurz nach dem Hinterachsschlag gibt den Analyseausschnitt vor. Der vorgegebene Zeitbereich wird frequenzgewichtet und zu RMS oder VDV Werten verrechnet.

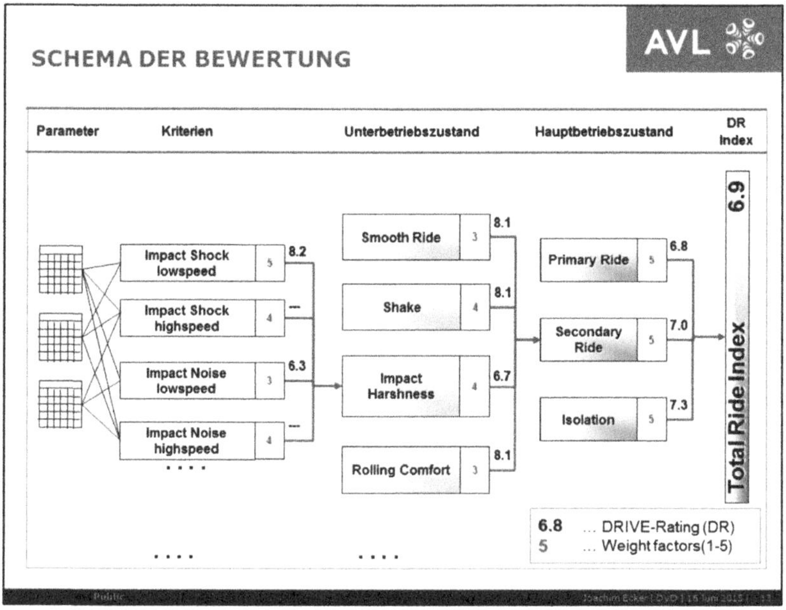


Das Bewertungsschema: Von den Parametern über Kriterien, Unterbetriebszustände und Hauptbetriebszustände zur Gesamtnote.

Von den wie zuvor beschrieben berechneten Parametern werden Kriteriennoten errechnet. Im Fall von „Impact Harshness" sind das ein Kriterium für den mechanischen Schlag und ein Kriterium für das Schlaggeräusch. Aus beiden Kriteriennoten wird die Bewertung für den Betriebszustand „Impact Harshness" gebildet. Da sich nicht alle Kriterien gleich stark auf die Summe der Eindrücke niederschlagen wird die Note mit einen fixen Gewicht versehen.
Nach dem gleichen Schema wird aus den Noten der anderen Unterbetriebszustände die Note für den Hauptbetriebszustand „Secondary Ride Comfort" gebildet. Die Gesamtbewertung ergibt sich analog dazu aus den Noten für „Primary Ride", „Secondary Ride" und „Isolation".

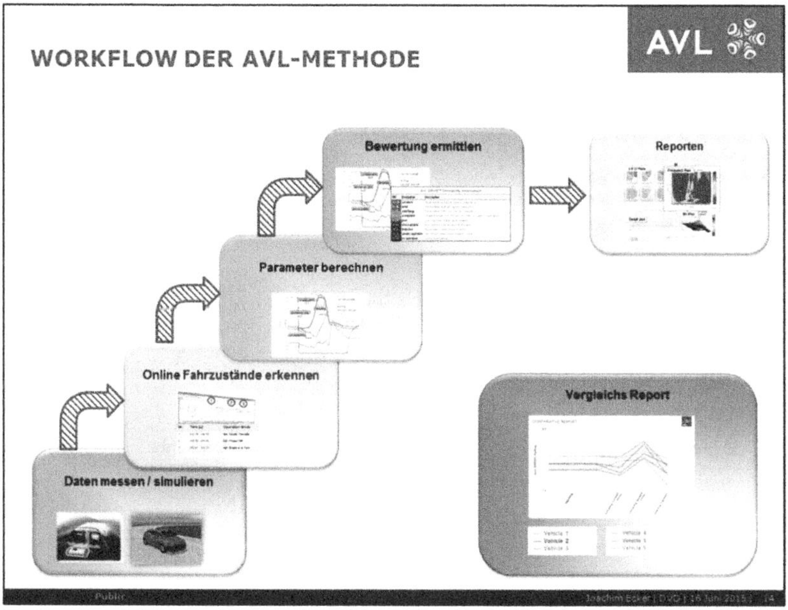


Übersicht über den Workflow der AVL Ratingmethode:

Die AVL Ratingmethode für Bewertung für Handling und Fahrkomfort lässt sich anhand des folgenden Workflows nochmals einfach und plakativ darstellen.

Um objektiv bewerten zu können, ist es notwendig gemessene oder simulierte Fahrzeugbewegungsgrößen vorliegen zu haben.
Auf Basis der Daten werden vom Softwarecode Fahrzustände automatisch erkannt, Parameter berechnet und Noten gebildet.
Hält man sich an einen Fahrzyklus ist garantiert, dass man alle geforderten Fahrzustände in der notwendigen Datendichte erhält. Alle erkannten und bewerteten Fahrzustände sind in einer Datei hinterlegt die mit einem Reporting Tool schnell und plakativ als Ergebnis in allen Details dargestellt werden können.
Nutzt man die Methode für Benchmarking oder begleitet ein Fahrzeug während der Entwicklung lassen sich auf diese Weise sehr einfach Berichte erstellen die die Unterschiede übersichtlich darstellen.

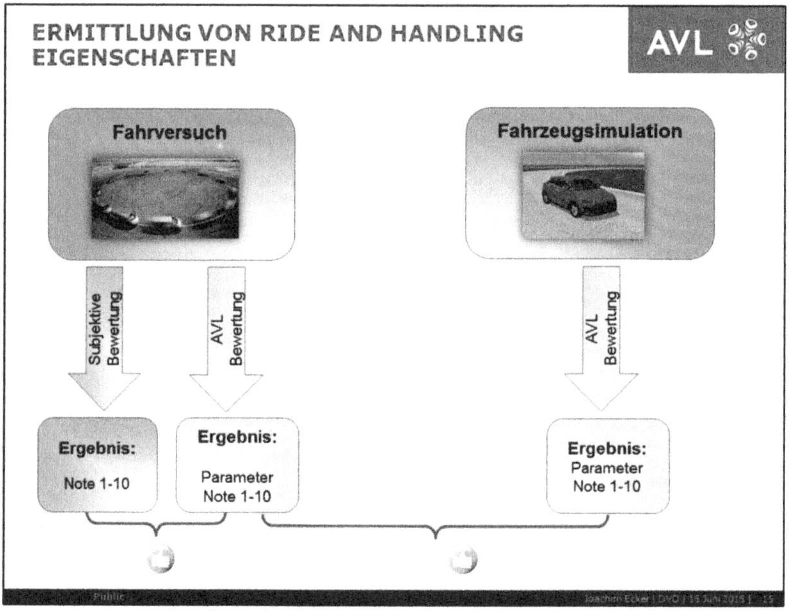


Die AVL-Methode schließt die technologische Lücke in der Bewertung von Handling und Fahrkomforteigenschaften.

Es ist möglich auf Basis von Messdaten oder simulierten Fahrzeugbewegungsgrößen Handling und Fahrkomforteigenschaften mittels physikalischen Parametern und Noten in der SAE Skala von 1-10 objektiv zu bewerten und ermöglicht damit den direkten Vergleich zu subjektiven Fahreindrücken.

In frühen Phasen der Fahrzeugentwicklung ist somit schon eine Bewertung im Maßstab 1-10 möglich. Entwicklungsziele können in diesem sehr plakativen Maßstab im Lastenheft definiert werden und mit einer durchgehenden Methode im gesamten Entwicklungszeitraum geprüft werden.

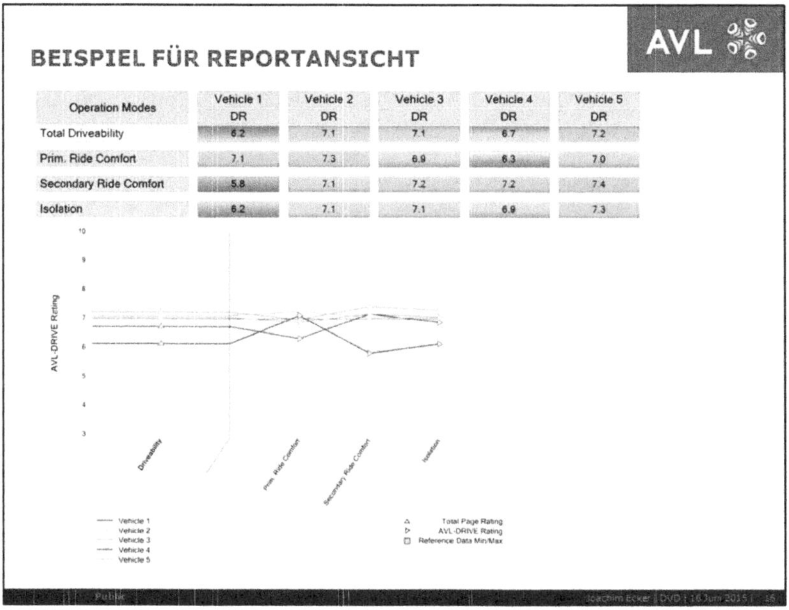


Bespiel für einen vergleichenden Bericht:

Zum Vergleich stehen 5 Fahrzeug der Kompaktklasse zur Verfügung. Die Ansicht aus dem Bericht zeigt zu Beginn übersichtlich die Ergebnisse der Hauptbetriebszustände „Primary Ride", „Secondary Ride" und „Isolation".

Man erkennt sofort die Auffälligkeit von Fahrzeug 1 (blau) das im tieffrequenten Komfort „Primary Ride" sehr gut bewertet wurde, jedoch bei höheren Anregungsfrequenzen deutlich schlechter als die anderen Fahrzeuge abschneidet.

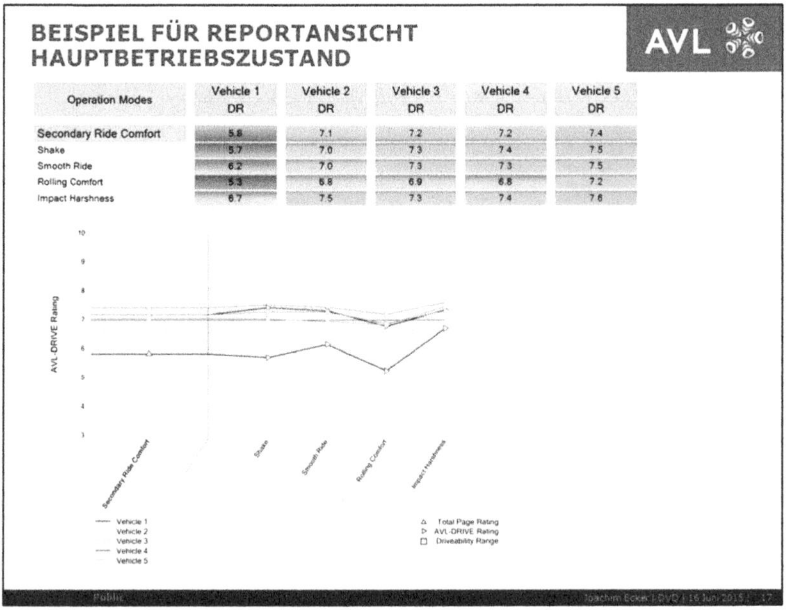


Betrachtet man die Detailansicht zum Hauptbetriebszustand „Secondary Ride Comfort“ ist gut ersichtlich dass Fahrzeug 1in allen Unterbetriebszuständen schlechter bewertet wird als die anderen Fahrzeuge, besonders aber im „Rolling Comfort“.

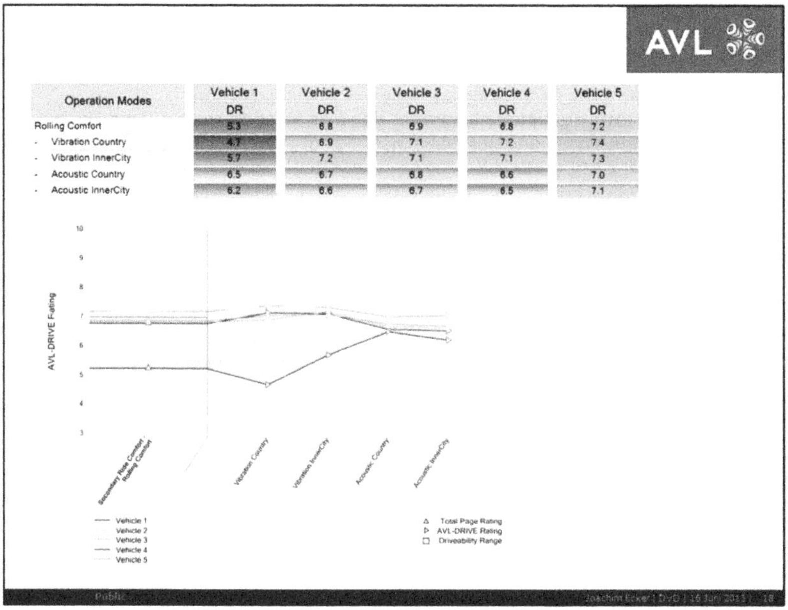

| Operation Modes | Vehicle 1 DR | Vehicle 2 DR | Vehicle 3 DR | Vehicle 4 DR | Vehicle 5 DR |
|---|---|---|---|---|---|
| Rolling Comfort | 5.3 | 6.8 | 6.9 | 6.8 | 7.2 |
| - Vibration Country | 4.7 | 6.9 | 7.1 | 7.2 | 7.4 |
| - Vibration InnerCity | 5.7 | 7.2 | 7.1 | 7.1 | 7.3 |
| - Acoustic Country | 6.5 | 6.7 | 6.8 | 6.6 | 7.0 |
| - Acoustic InnerCity | 6.2 | 6.6 | 6.7 | 6.5 | 7.1 |



Betrachtet man den Detailanalyse des Abrollkomforts die sich aus Bewertungen zur mechanischen Vibration und der Geräuschentwicklung zusammensetzt, dass das Geräusch mit Noten über 7 im Rahmen bleibt, jedoch die mechanischen Vibrationen als sehr störend bewertet werden.

Die Aufteilung auf zwei Geschwindigkeitsfenster Inner City – 50 kph und Country 80-kph ergibt, dass die Tendenz bei 80kph gegenüber 50 kph nochmals etwa 0.5 Punkte schlechter ist.

## ZUSAMMENFASSUNG:

AVL

- Die aktuell existierenden Methoden zur Handling und Fahrkomfort Beurteilung können Simulationen und Messdaten nur bedingt gleichwertig und vergleichbar zu subjektiven Ergebnissen bewerten

- Der AVL Bewertungsansatz nutzt objektive Daten aus Simulation um Fahrzustände online angelehnt an subjektive Methoden zu bewerten und schließt die technologische Lücke

- Benefit: Die Methoden werden vom SOD bis zum SOP angewendet werden. Es gibt eine objektive Bewertungsinstanz im gesamten Entwicklungsprozess für Simulationen oder Fahrzeugmessungen. Entwicklungsziele können auf Basis von SAE Bewertungen 1-10 definiert werden und objektiv geprüft werden.

# Virtual chassis tuning with emphasis on the damper characteristics – a method for optimal integrative damper adjustment by means of vertical and lateral dynamics simulation and evaluation criteria

Dipl.-Ing. Florian Klinger
Priv.-Doz. Dr.techn. Johannes Edelmann
Ao.Univ.Prof. Dr.techn. Manfred Plöchl
Technische Universität Wien, Institut für Mechanik und Mechatronik

Dipl.-Ing. (FH) Stefan Jeindl
Dipl.-Ing. Bernhard Angrosch
MAGNA Steyr Engineering, Department of NVH & Driving Comfort

# 1 Introduction

Vehicle suspension design is primarily characterised by conflicting targets concerning ride comfort, demanded handling performance qualities and save driving. Recent developments in the field of active and semi-active vehicle suspensions – aiming to dissolve these conflicts as far as possible – derive both from technological innovation as well as from corresponding theoretical studies. Although many active systems have been introduced to premium cars in recent years, the importance of well-designed passive systems is likely to remain for the foreseeable future, in particular for economy-class vehicles.

Multibody system (MBS) simulation models and numerical simulation techniques have proven to be appropriate tools in the early development process of new vehicle models or of derivatives from an existing baseline vehicle. By this means, chassis components can be determined to a certain level of maturity, resulting in a virtual prototype vehicle. Although the design of the characteristics of a passive suspension damper has been considered an ultimate goal in virtual prototyping, making expensive field tests redundant, realization granted rather limited success, and most often, damper tuning is still laboriously carried out on test tracks with prototype dampers and cars to a large extent. Above all, the complexity involved must be considered as a main reason, subjective–objective evaluation, involved model simplifications, conflicting expectations on damping effects, which not only influence the vehicle behaviour in all principal directions, but also the perception and comfort of the driver (and passenger). Damper tuning involves long-time experience of test engineers, and the idea of deriving "optimal damper" characteristics by simulation might be considered unrealistic to some extent. Choosing "representative" manoeuvres, "characteristic" excitations, setting "mathematical criteria" for performance evaluation and optimization, already implies an intrinsic compromise of conflicting demands, clearly revealing the comprehensive and difficult task related to virtual damper design.

Despite these challenges, the general need for a virtual prototype (MBS model) of a full vehicle compels an alternative. At an early stage of vehicle development simple, but still realistic damper characteristics are required to be included in a MBS model of the front and rear axle. The interaction of passive springs and dampers, bump-stops, and anti-roll bars influences demanded ride comfort, handling and directional stability properties, and their characteristics need to be accounted for. The present paper indicates a practicable way to derive passive suspension damper characteristics based on a nonlinear damper force–stroke velocity relationship. It is not intended to propose a straightforward damper design process for an optimized damper, nor is it pretended that a way has been found to overcome the above mentioned difficulties. However, a practicable way is presented that aims to reflect the design process of a test engineer in a simulation environment. It will be discussed exemplarily, how distinguished sections of the nonlinear force–stroke velocity spline of the damper are affected by the choice of excitation, either from the road or from the driver, or by the considered objective evaluation criteria. Conversely, varied sectional characteristics will give indication on the expected damper performance. Parameter variation is based on the design of experiments (DoE) method, which mimics to some extent the design process and freedom of design of the test engineer. Although DoE analysis hardly allows for a comprehensive understanding, as basic models can provide, the method can be of valuable help when the level of complexity becomes involved as in the case of suspension damper design. Based on the considered type of vehicle and demanded vehicle performance specific damper characteristics might be found preferable compared to others, and last but not least, reasonable damper characteristics result for a virtual prototype of the vehicle.

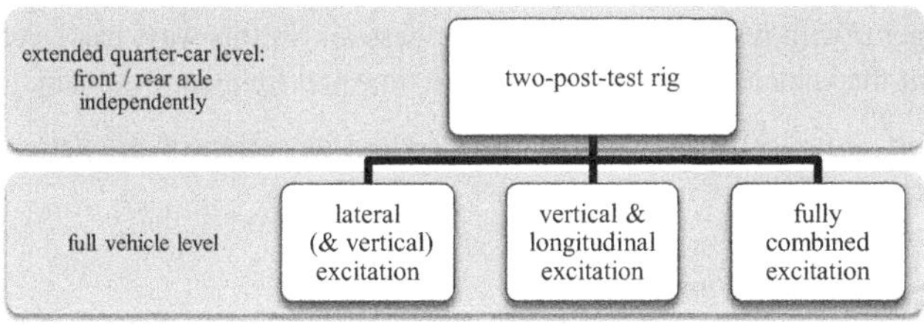


Figure 1: Overview of a simulation-based damper characteristics design procedure

Figure 1 shows a general overview of the presented simulation-based damper characteristics design procedure corresponding to the sections of the paper. Above all, a virtual two-post-test rig is used to basically analyse damper characteristics suitable for the front and rear axle of a specific vehicle, recapturing also some main results from scientific literature. An MBS model of the respective axle, accounting for the demand of a full MBS prototype vehicle model, loaded with proportional weight of the vehicle, is excited based on given random road profiles or single disturbances like bumps. Next, a full vehicle model is considered to test the front and rear damper design in combination w.r.t. the above excitations from the road. Finally, lateral excitations from the steering input of the driver are addressed. Only a horizontal flat road without excitation from the road profile, and no other external excitations, as from side wind gusts, are addressed here. The interrelation between damper characteristics and respective evaluation criteria on ride comfort and handling performance referring to literature is illustrated by some examples. Fully combined driver–road–environment excitation is not covered. The paper is closed with a short summary.

# 2 Linear quarter-car model

Fundamental ride comfort and handling properties of vehicles may be assessed by applying simplified models of the vehicle. Considering tyre–road and wheel–suspension–vehicle body interactions only, basic understanding may be gained by studying the dynamics of the linear quarter-car model in Figure 2. Besides neglecting kinematic properties of the wheel suspensions, a symmetric excitation from the road is assumed as well as particular mass and inertia properties of the vehicle body must apply [1, 2] to extract the basic quarter-car from a full vehicle model. As a consequence, the wheelbase filtering effect, requiring at least a half-car model, is not considered [3]. To account for a simplified (linear) tyre–road contact model, stochastic road profiles may be filtered considering the tyre contact patch dimensions [4]. The excitation from the road is usually described by Gaussian white noise ground velocities, or equivalently a step in road displacement or an impulse in road velocity [1]. The linear quarter-car model includes the sprung mass $m_2$, which is supported by a linear spring $c$ and linear damper $d$ with respect to the unsprung mass $m_1$. In this way, basic vertical ride properties of the vehicle can be studied in both time and frequency domain.

Ride comfort is typically assessed by the standard deviation of the vertical vehicle body acceleration, [4],

$$\sigma_{\ddot{z}} = \sqrt{\frac{1}{T}\int_T \ddot{z}^2(t)\,\mathrm{d}t} \rightarrow \min. \tag{1}$$

Although this objective – due to the lack of information on the frequency content in the vertical body acceleration – may be considered as a rough quantification only, it is appropriate to characterise basic ride comfort properties of a vehicle [1].

Road holding, and related handling and safety properties, [5], are typically assessed by the standard deviation of the normalised dynamic wheel load,

$$\sigma_{F_z} = \sqrt{\frac{1}{T}\int_T \frac{(F_z(t) - F_{z0})^2}{F_{z0}^2}\,\mathrm{d}t} \rightarrow \min. \tag{2}$$

The longitudinal and lateral tyre forces are essentially affected by the vertical tyre load (changes). The loss of cornering performance (decrease of average lateral tyre force at fixed sideslip angle) on uneven roads is due to the curvature of the cornering stiffness vs. the wheel load characteristic and hence the change of tyre relaxation length with load (changes). For the transfer of a longitudinal force between the tyre and an uneven road the arising variation of the effective rolling radius and respective slip velocity plays an important role, [6].

Several other performance criteria have been proposed to assess ride comfort and handling properties of vehicles, for instance to (also) account for rattle space demands or vertical jerk of the vehicle body [1].

Varying stiffness and damping coefficients within (here unrealistically large) given limits the resulting objectives in Eqs. (1) and (2) can be illustrated by Figure 3. The non-dominated solutions are collected in a black line, the Pareto-optimal set. For all other (dominated) solutions the value of one objective can be reduced without increasing the value of the other objective. A proper relation between damping coefficient and stiffness is necessary in order to achieve a non-dominated solution in the conflict diagram. Basically, rather large damping coefficients combined with rather large stiffness coefficients are required for good road holding (upper, left end of Pareto-optimal set), whereas small values for both damping and stiffness coefficients are essential for improved ride comfort (low, right end of Pareto-optimal set).

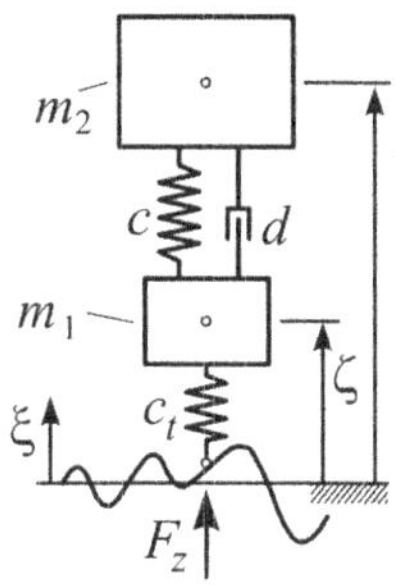


Figure 2: Quarter-car model

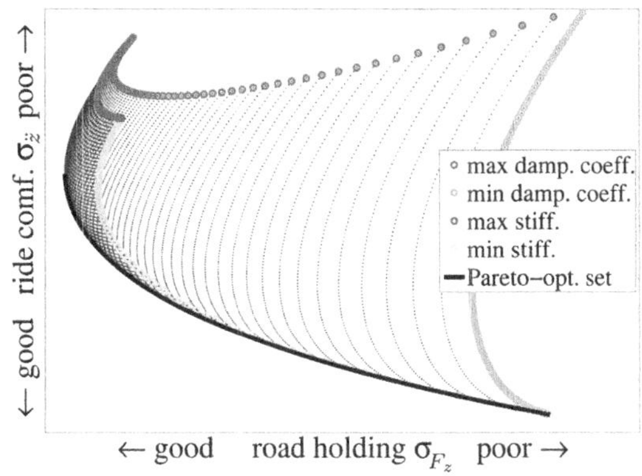


Figure 3: Conflict diagram: ride comfort vs. road holding

# 3 Virtual two-post-test rig

To obtain appropriate load spectra for component design and stress analysis as well as vehicle ride and handling properties at an early stage of vehicle development, MBS simulation models of vehicles are applied. A number of vehicle and suspension parameters needed to set up a proper MBS vehicle model including major nonlinearities are already (roughly) known at this early stage. However, corresponding nonlinear damper characteristics are not yet defined at this stage of vehicle development, but have considerable impact on the required simulation results. Thus, an MBS nonlinear extended "quarter-car" model is set up in order to obtain reasonable nonlinear damper characteristics by means of DoE and numerical optimization, if favoured. A full nonlinear model of the considered axle on a virtual two-post-test rig, loaded with proportional weight of the vehicle, is employed in the simulation environment, Figure 4, considering symmetric excitation on both the left and right wheels. For example, the front axle comprises MacPherson struts with elastic bump-stops and bushings. The vertical characteristics of a Magic Formula (MF) type tyre model are considered here, neglecting stiffness effects of a rolling tyre at high speeds. The full MF tyre model is used later for lateral manoeuvres on a horizontal flat road. Instead of a detailed rheological damper model, e.g. [7], a nonlinear, but sectionally linear, (a)symmetric damper force–stroke velocity relationship is assumed, see Figure 5.

Figure 4: MBS model of two-post-test rig with front axle and fixed engine

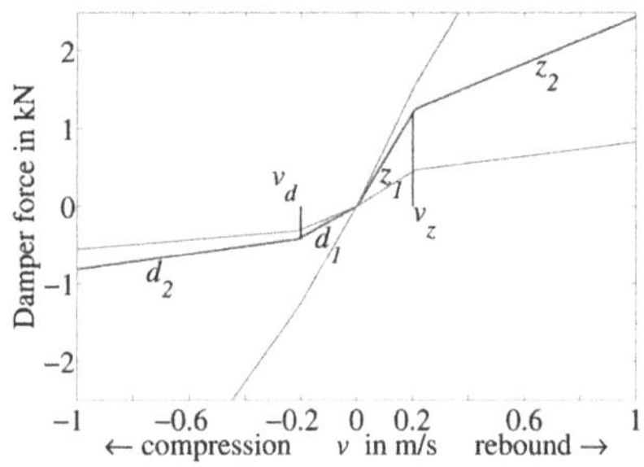


Figure 5: Damper force–stroke velocity relationship and span of 81 damper variants corr. to Figure 7

This presumed generic basic shape of the damper spline can be parameterised by six parameters: the slopes of the sectionally linear characteristics at lower and higher velocities of the damper stroke w.r.t. both compression and rebound ($d_1$, $d_2$, $z_1$, and $z_2$) as well as the corresponding transition velocities $v_d$ and $v_z$.

Alternatively, the four individual sections of the damper spline are also defined in a relative manner to allow for straightforward scaling of the damper characteristics. In the range of lower velocities of the damper stroke ($v_d < v < v_z$), an “average” damping coefficient $D = (d_1+z_1)/2$ is introduced; the larger $D$, the more energy will be dissipated. The ratio of the gradients of the damper forces in the compression and rebound range is defined by the “asymmetry” parameter $A = d_1/z_1$, typically between 1/2 and 1/3. The slopes of the damper spline at higher stroke velocities are scaled w.r.t. the corresponding damping coefficient at lower stroke velocities by $s_d$ and $s_z$, with $d_2 = s_d\ d_1$ and $z_2 = s_z\ z_1$. Damper splines are centrally symmetric for $A = 1$; $s_d = s_z$; $v_d = v_z$.

## 3.1 Stochastic road excitation

A stochastic vertical excitation from a road of class C (medium roughness), [2], is applied to the respective loaded nonlinear axle model on the virtual two-post-test rig, to simulate a 1000 m test drive with constant forward speed of 80 km/h. In Figure 6 resulting power spectral densities (PSD) of the vertical body acceleration are depicted (a) as well as corresponding PSDs of the dynamic wheel load (b) for varied average damping coefficients $D$ of five centrally symmetric damper splines. Small damping corresponds to the red line, large damping to the line in magenta. Resonance frequencies related to "body-heave" and "wheel-hop" can be identified at about 1.2 Hz and 12 Hz, respectively.

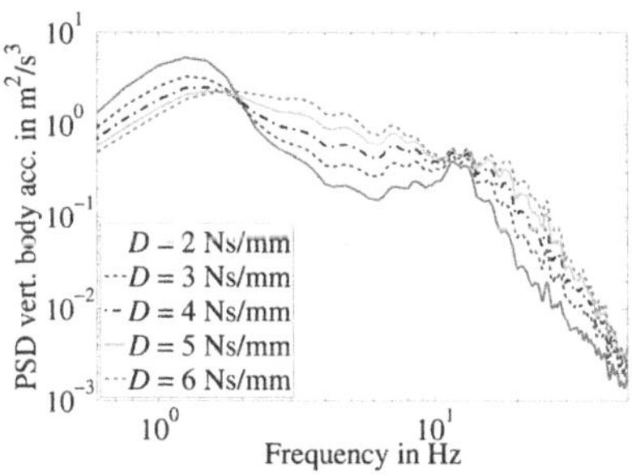


Figure 6 (a): PSD of vertical body acceleration for five centrally symmetric damper splines; $A = 1$; $s_d = s_z = 0.25$; $v_d = v_z = 0.2$ m/s

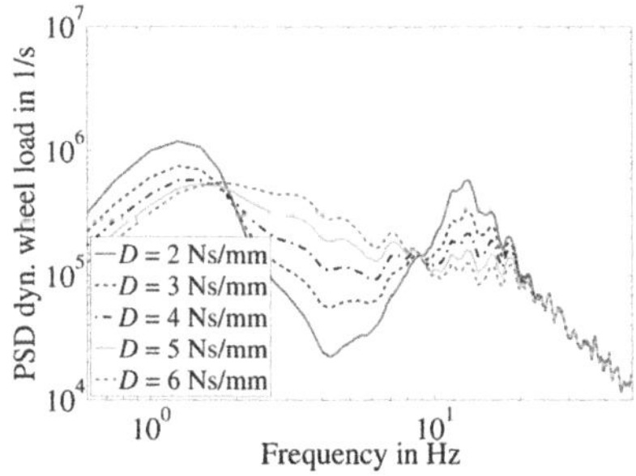


Figure 6 (b): PSD of dynamic wheel load for five centrally symmetric damper splines; $A = 1$; $s_d = s_z = 0.25$; $v_d = v_z = 0.2$ m/s

The PSD of vertical body acceleration may be evaluated to assess basic ride comfort properties w.r.t. stochastic road excitation, where in general a corresponding equivalent damping ratio of about $\sqrt{2}/2$ in the range of both body-heave and wheel-hop resonance is desirable, [4]. The damping ratio of a linear system may be estimated from the width and height of the resonance peak [8].

In agreement to findings from investigations on the transfer behaviour of the linear quarter-car model, no damper characteristics in Figure 6 (a) and (b) shows small amplitudes over the full range of frequencies. This conflict becomes even more serious, if one needs to find a compromise between ride comfort, Figure 6 (a), and safety, Figure 6 (b).

In the linear case, "invariant points" can be found by inspecting the transfer functions of the linear quarter-car model from (harmonic) vertical road excitation displacement to vertical body acceleration at three excitation frequencies, [9], that cannot be altered by only modifying the damper characteristics. Similar, but not invariant points appear in the corresponding PSD depicted in Figure 6 (a) at excitation frequencies in the range of the body-heave resonance frequency at about 2 Hz for the nonlinear axle model, as well as in the range of the wheel-hop resonance frequency at about 12 Hz and 14 Hz. However, from inspecting the linear transfer function from vertical road excitation velocity to vertical body acceleration, it is found that adding even a small amount of tyre damping may alter the invariance around the wheel-hop frequency and thus lead to a significant change in the vertical body acceleration in this frequency range, [11]. This fact may be of particular interest as tyre damping can vary from a few percent at low forward speeds to near zero at high forward speeds, [12]. Regarding the linear transfer function from vertical road excitation acceleration to vertical body acceleration, [13], the inherent isolation properties of the quarter-car model above body-heave resonance w.r.t road excitations become even more obvious.

Considering the transfer behaviour of the linear quarter-car model from vertical road excitation displacement to dynamic wheel load, two further invariant points can be identified, e.g. [14]. Again, similar points can be observed in Figure 6 (b) at approximately 2 Hz and 9 Hz.

Increased damping, see Figure 6 (a) and (b), decreases both the vertical body acceleration and dynamic wheel load amplitudes in the range of body-heave and wheel-hop resonance frequency. However, corresponding amplitudes in the excitation frequency range from 4 Hz to 8 Hz, which are found to be most relevant for human sensitivity to vibrations, [15], will increase. Loosely speaking, this behaviour may be called a waterbed effect.

By calculating both the standard deviation of the vertical body acceleration $\sigma_{\ddot{z}}$, Eq. (1), as well as the standard deviation of the dynamic wheel load $\sigma_{Fz}$, Eq. (2), the above five centrally symmetric damper splines can be mapped in a conflict diagram, similar to Figure 3, or, by means of DoE, with more general damper configurations in Figure 7.

Three different values for each parameter of a damper spline, average damping coefficient $D$, asymmetry parameter $A$, as well as scaling factors for the two slopes of the damper characteristics at higher stroke velocities, $s_d$ and $s_z$, have been selected, leading to 81 different damper variants in Figure 5. Transition velocities $v_d$ and $v_z$ remain unchanged. In order to identify the influence of respective parameter changes on the performance objectives for both ride comfort and road holding in a systematic manner, a new illustrative approach is introduced here.

In Figure 7, three different damper variants are interconnected by a line. Within each triple the parameter $s_z$ increases in the direction of the arrows, while all other parameters remain unchanged. Thus, the influence of changing parameter $s_z$ on the performance objectives for ride comfort and road holding can be observed.
Sets of varied parameter $D$ – where again all other parameters remain unchanged – can be identified by different line colours, where blue, green and red lines correspond to the lowest, medium and highest value of $D$, respectively.
The size of the plotted circles corresponds to the three variants of parameter $s_d$, where the smallest circles belong to the lowest value and the largest circles to the highest value of $s_d$, respectively.

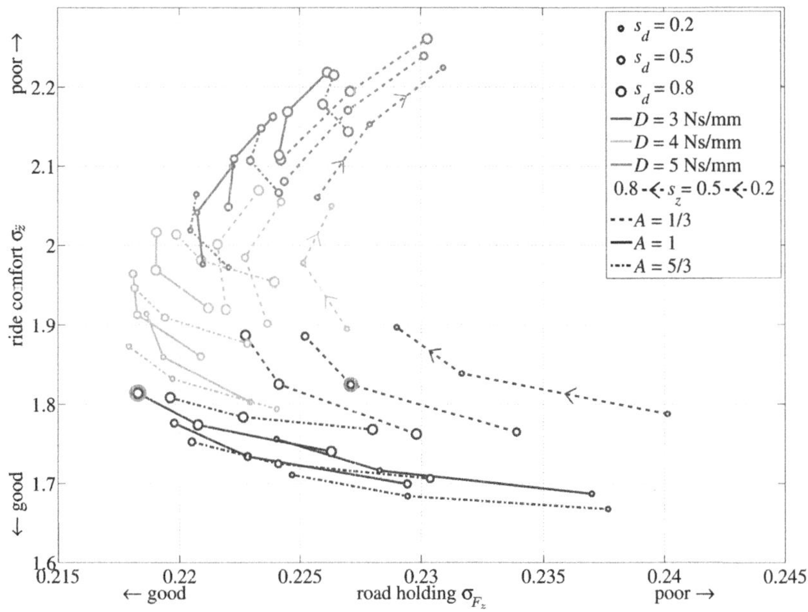


Figure 7: Conflict diagram with 81 different damper variants for random road profile input

Finally, three different line styles represent the three different values of parameter $A$, where dashed, solid and dash-dotted lines represent the minimum, medium and maximum value of $A$, respectively.

In this way, expected effects on the performance objectives of both ride comfort and road holding by changing individual damper parameters can be easily identified in a straightforward and systematic manner. Assuming, for instance, an initial damper setting, pointed out by the magenta circle, road holding can be improved, while ride comfort remains unchanged, pointed out by the orange circle, by changing the parameter $A$ from the lowest to the medium value, the parameter $s_d$ from the medium to the largest value, and parameter $s_z$ from the medium to the largest value. Parameter $D$ remains unchanged. A DoE map of only 81 damper variants already gives a good idea of the vertical dynamic behaviour of the considered nonlinear front (or rear) axle.

To work out more details, the conflict diagram for the considered suspension on the two-post-test rig is studied by means of 25 variants of the average damping coefficient $D$, for five different asymmetry parameters $A$ each, while all other parameters of the damper spline remain unchanged ($s_d = s_z = 0.25$; $v_d = v_z = 0.2$ m/s). To reveal the influence of the asymmetry parameter $A$, five specific variants with the same average damping coefficient $D$ have been marked in Figure 8 (a) by individual symbols. In particular, the damper spline representing the "typical" asymmetry ratio of $A = 1/3$ between compression and rebound, e.g. [16], shows poor performance w.r.t. both standard deviations of vertical body acceleration $\sigma_{\ddot{z}}$ and dynamic wheel load $\sigma_{Fz}$ compared to variants with larger values of $A$. For all values of $D$ considered in this DoE map, centrally symmetric damper splines, $A = 1$, or even splines with $A = 4/3$, show best performance and represent a Pareto-optimal set with respect to the considered parameter variations.

Considering stochastic road excitation input, this result becomes clearer, when inspecting the time histories of suspension travel, Figure 8 (b). Different relative vertical offsets of the suspension travel can be observed due to different asymmetry parameters of the damper splines. Since the mean of the damper force is not zero for asymmetric damper characteristics, this leads to a lowering of the car body for $A < 1$, and to a lift for $A > 1$, see e.g. [16]. As a consequence "the effect of hitting the stops is to some extent equivalent to stiffening the suspension spring […], and as the system becomes effectively stiffer, its damping factor decreases […] and the equivalent linear system is too stiff and too lightly damped to perform well in comparison with a purpose designed system.“ [17].

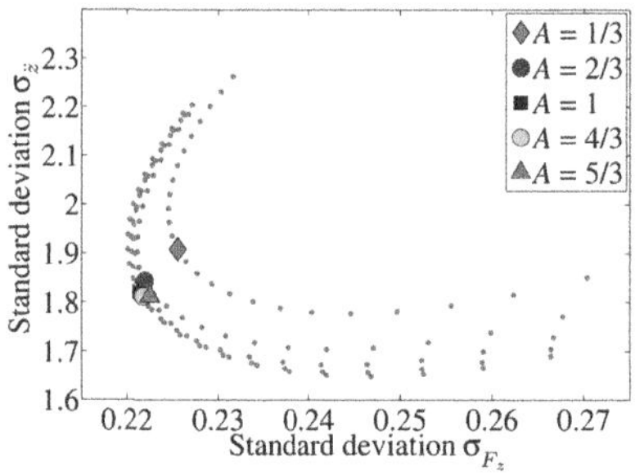


Figure 8 (a): Conflict diagram for 125 different damper variants for random road profile input

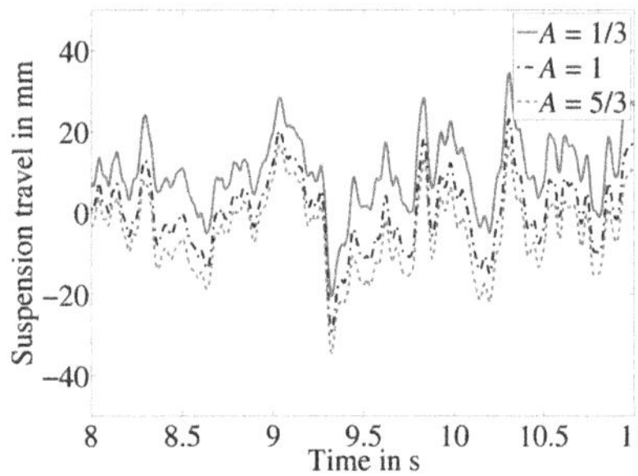


Figure 8 (b): Suspension travel (positive for compression) for one centrally symmetric ($A = 1$) und two asymmetric ($A = 1/3$; $A = 5/3$) dampers

With respect to the design of the damper characteristics at higher stroke velocities, nine damper splines with fixed $D = 4$ Ns/mm and $A = 1$ and varied parameters $s_d$ and $s_z$ between 0.2 and 0.8 are shown in Figure 9 (a). Evaluations are plotted as points in the conflict diagram Figure 9 (b). In addition, solutions from a full factorial design are depicted, where each of the parameters $D$, $A$, $s_d$, and $s_z$ has been varied in five increments, resulting in 625 variants of the damper characteristics (grey dots).

Regarding the damper spline with steepest slope in both compression and rebound at higher stroke velocities ($s_d = s_z = 0.8$, red line and diamond, respectively), the corresponding assessment in the conflict diagram yields non-Pareto-optimal properties, as overall energy dissipation over one circle is too high. Decreasing the slope in rebound to $s_z = 0.2$ (blue line and square, respectively), results in significantly reduced $\sigma_{\ddot{z}}$, and thus improved ride comfort, while road holding properties decrease. Additional decrease of the compression slope at higher stroke velocities to $s_d = 0.2$ (magenta line and triangle, respectively) leads to a significant improvement in ride comfort, while road holding properties decrease only little.

Easy inspection of the green, full lines ($A = 1$, $D = 4$ Ns/mm) of the basic DoE map in Figure 7 confirms the above results by varying $s_z$ on lines or $s_d$ with circle sizes. There, it becomes also obvious, that one may find a solution on the Pareto-optimal set, only by variation of high-speed damping parameters $s_d$ and $s_z$, green line and star in Figure 9.

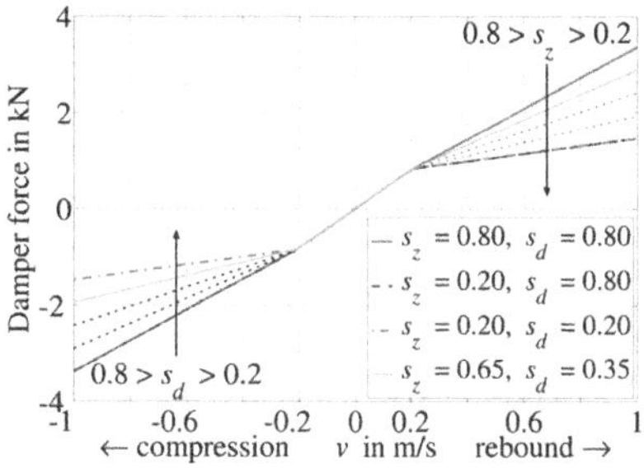


Figure 9 (a): Damper splines with $D = 4$ Ns/mm, $A = 1$, $v_d = v_z = 0.2$ m/s and variable $s_d$ and $s_z$

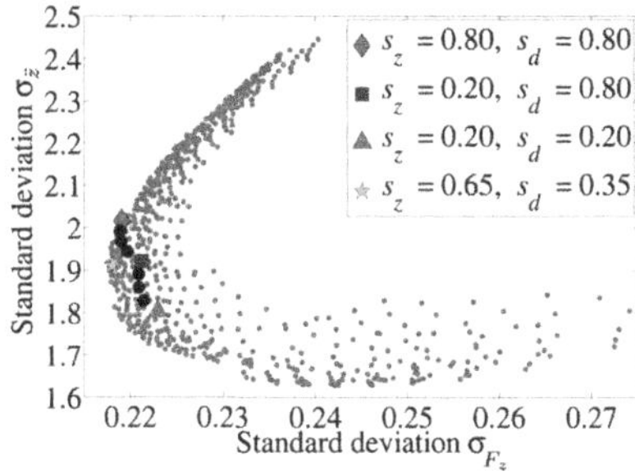


Figure 9 (b): Conflict diagram with $D = 4$ Ns/mm, $A = 1$, $v_d = v_z = 0.2$ m/s and variable $s_d$ and $s_z$

The variation and effects of changes in transition velocities $v_d$ and $v_z$ with fixed other parameters on the above performance criteria can be analysed in analogy to Figure 9. For the considered axle, Pareto-optimal solutions can be found for slightly larger values of $v_z$ compared to $v_d$.

## 3.2 Single road disturbance

Regarding ride comfort, single obstacle excitations are of particular interest. On the one hand, the driver recognises approaching obstacles and is thus particularly aware of obstacle-induced vibrations, on the other hand, he/she is sensitive to unexpected, and thus unpleasant singular shocks. Although road holding properties are of interest too w.r.t. safety issues, ride comfort properties are predominant.

Single obstacle excitations are characterised by large excitation amplitudes compared to respective stochastic excitations, starting at excitation frequencies below body resonance frequencies up to frequencies well above wheel-hop resonance. In general, arbitrary contours may be regarded. For subsequent basic investigation, smoothly shaped obstacle contours, represented by a scaled section of a cosine-function, are employed, [2]. Specifying the forward speed to 40 km/h, obstacle length and height have been selected to excite the vertical vehicle body motion at determined frequencies according to Figure 6:

- "Long Bump": body-heave resonance (1.2 Hz); length:10m; height: 0.125m
- "Medium Bump": isolation range (here 5 Hz); length: 2.2m; height: 0.05m
- "Short Bump": wheel-hop resonance (12 Hz): length: 0.9m; height: 0.025m

Vehicle response to single road disturbances is generally assessed in time domain, [2, 16], evaluating time histories of body and wheel vertical motions (rattle space), as well as tyre contact forces. However, also statistical measures are analysed. In subsequent DoE studies, several evaluation criteria are regarded.

Considering vertical body acceleration, in addition to the standard deviation of the vertical body acceleration, Eq. (1), the positive maximum value within the selected time range is evaluated, Eq. (3) [18], as well as the root mean quad, Eq. (4) [19].
Vertical vehicle body displacement w.r.t the nominal position is assessed by its respective maximum value, Eq. (5), as well as the corresponding standard deviation, Eq. (6). In addition, for obstacles exciting the body-heave resonance (e.g. Long Bump), the 3rd positive peak after striking the obstacle, Eq. (7), is evaluated, see Figure 10 (a). This measure should represent the test driver's demand that "one post-pulse oscillation is OK, but then low frequency oscillations should have already decayed" [18].
Dynamic wheel loads are assessed by the respective standard deviation, Eq. (2), as well as by the minimum value within the selected time range of the considered manoeuvre, Eq. (8). In total, eight criteria are evaluated in order to assess respective damper properties, where appropriate dampers should minimize above criteria.

$$\ddot{z}_{\max} = \max\{\ddot{z}(t)\} \quad (3) \qquad \mathrm{RMQ} = \sqrt[4]{\frac{1}{T}\int_T \ddot{z}^4(t)\,\mathrm{d}t} \quad (4) \qquad z_{\max} = \max\{z(t)\} \quad (5)$$

$$\sigma_z = \sqrt{\frac{1}{T}\int_T z(t)^2\,\mathrm{d}t} \quad (6) \qquad z_{\mathrm{nod}} \text{ of 3}^{\mathrm{rd}} \text{ positive peak} \quad (7) \qquad \Delta F_{z,\min} = \frac{F_{z0} - F_{z,\min}}{F_{z0}} \quad (8)$$

In Figure 10 (a) time histories of the vertical body displacement for three centrally symmetric damper splines with varied average damping coefficient $D$ are depicted when travelling over the Long Bump including the contour of the bump (corr. to

40 km/h). The damper with medium value $D = 4$ Ns/m corresponds to the damper marked by a black square in Figure 8 (a) and represents a compromise configuration between ride comfort and road holding in a respective conflict diagram for stochastic road excitations. This damper spline also satisfies the $z_{nod}$-criterion, Eq. (7), that is of particular interest in this manoeuvre. Due to considerable deflections of the suspension for considered damper variants while climbing the obstacle, the vertical body displacement overshoots w.r.t. the shape of the obstacle (rebound [20]). Although considerable vertical movement of the vehicle body can be observed, only low velocities of the damper stroke arise. To sum up, parameter $D$ has major impact on the vehicle response at the Long Bump, asymmetry parameter $A$ as well as transition velocities $v_d$ and $v_z$ to some extent, while the high-speed damping parameters are of minor importance.

Crossing the Medium Bump, damper stroke velocities up to 1 m/s occur. Thus, a wide range of the damper spline is covered. Higher damper stroke velocities pronounce the impact of coefficients $s_d$ and $s_z$ on the above objective criteria, while the importance of the location of the transition velocities $v_d$ and $v_z$ decreases.

Exciting the wheel-hop resonance by the Short Bump, asymmetry parameter $A$ shows major influence on the vertical motion of the vehicle, see e.g. [2]. This becomes evident from Figure 10 (b), where $A$ is varied, while $D$ and all other parameters remain constant. Dampers with $A = 1/3$ result in both considerably reduced vertical body-heave as well as reduced vertical body acceleration (not depicted here). Considering potholes, respective asymmetric dampers lower the vehicle body compared to centrally symmetric dampers, however advantages remain when leaving the bump.

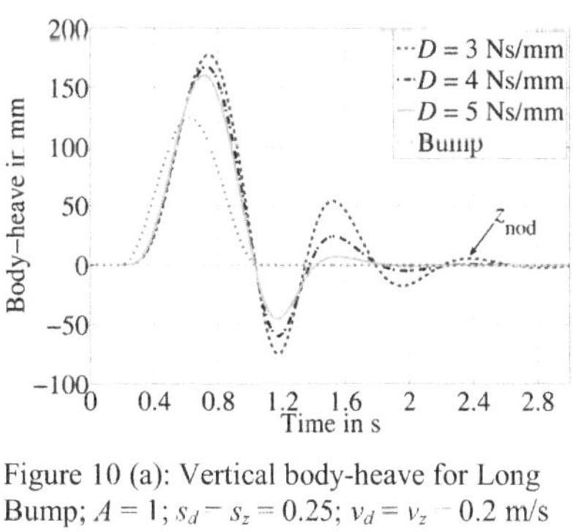


Figure 10 (a): Vertical body-heave for Long Bump; $A = 1$; $s_d = s_z = 0.25$; $v_d = v_z = 0.2$ m/s

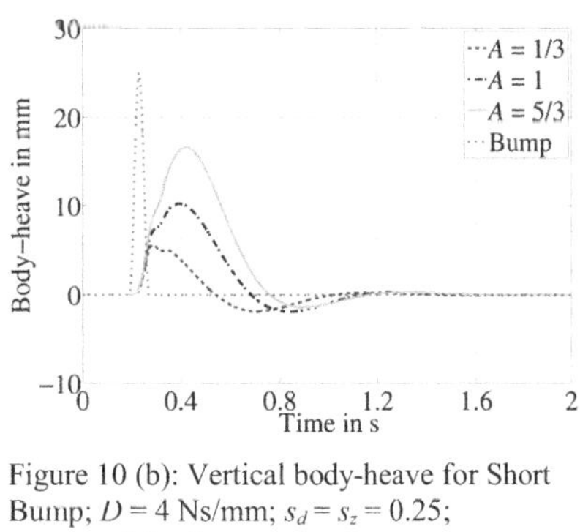


Figure 10 (b): Vertical body-heave for Short Bump; $D = 4$ Ns/mm; $s_d = s_z = 0.25$; $v_d = v_z = 0.2$ m/s

For selection of optimal damper characteristics considering single road disturbances, several objectives Eqs. (1) – (8) have been introduced with regard to ride comfort and road holding. A systematic approach is required to conclude on the performance of individual dampers as well as on the influence of changes in the individual damper design parameters. By means of DoE methods, linear Pearson correlation coefficients can be derived, that relate single parameters of the damper spline to individual performance objective values. Positive correlations indicate, that increasing a damper parameter value causes an increase of the respective performance objective value – and thus degradation of the vehicle performance –, while decreasing of the respective damper parameter value decreases the corresponding objective value – and improves vehicle performance. Correspondingly, negative correlations cause opposite relations.

For the Long Bump, in Figure 11 (a) correlation coefficients between damper parameters $D$, $A$, $v_d$ and $v_z$ and performance objective values in Eqs. (1) – (8) are presented, derived from a full factorial design of in total 625 variations. Considering damping parameter $D$, a strong, negative correlation can be observed for all performance objectives. Thus, increasing the average damping coefficient $D$ has major impact on the considered performance objectives and improves the overall performance of the vehicle w.r.t. the considered manoeuvre. Similar conclusions can be drawn on increasing $v_d$ and $v_z$. However, as values of respective correlation coefficients are much smaller, the impact on the objectives is less pronounced compared to changes in $D$. Asymmetry parameter A improves some of the criteria and degrades others; in general, it is of minor importance.

Running over the Medium Bump, opposite correlations are observed (not depicted). Increased parameter $D$ degrades both ride comfort and road holding properties, similar to findings from stochastic road excitations, see e.g. Figure 6.

In Figure 11 (b) correlation coefficients for damper parameters $D$, $A$, $s_d$ and $s_z$ are plotted for the Short Bump manoeuvre. Changes in sign for different performance objectives indicate the need for trade-offs between different demands made on the damper characteristics. Desired large values of $D$ for small dynamic wheel loads ($\Delta F_{z,\min}$, $\sigma_{Fz}$) contradict demanded low damping $D$ for improved ride comfort. Remarkably, damping parameter $D$ has no significant impact on the vertical displacement of the car body ($z_{\max}$, $\sigma_z$). Instead, a strong, positive correlation with asymmetry parameter $A$ can be observed, thus small values of $A$ improve respective objectives. However, negative correlation between $A$ and $\Delta F_{z,\min}$ indicates unloading of the wheel, for too small values of $A$, when leaving the bump.

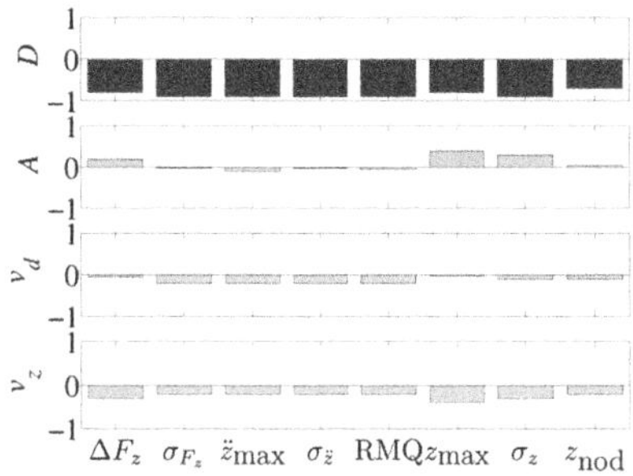


Figure 11 (a): Correlation coefficients for Long Bump

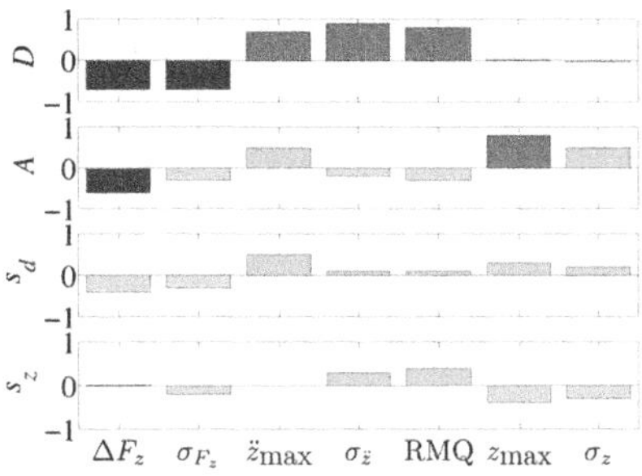


Figure 11 (b): Correlation coefficients for Short Bump

Based on the above gained insight into correlations between different performance objectives and damper parameters, a combined performance objective to assess ride comfort can be proposed. This performance objective is composed of the standard deviation $\sigma_{\ddot{z}}$, Eq. (1), and the positive maximum value of vertical body acceleration $\ddot{z}$, Eq. (3), normalized by the standard deviation and positive peak value of the road excitation, respectively,

$$\mathrm{RC} = w_1 \frac{\sigma_{\ddot{z}}}{\sigma_{\ddot{\xi}}} + w_2 \frac{\ddot{z}_{\max}}{\ddot{\xi}_{\max}}, \tag{9}$$

where both contributions may be scaled in order to meet a desired – subjective – balance. Weights $w_1 = 1/4$ and $w_2 = 3/4$ are suggested.

In Figure 12 (b), this combined ride comfort objective RC is plotted vs. the standard deviation of dynamic wheel loads $\sigma_{Fz}$ into a conflict diagram, derived from above calculated 625 variants. Damper characteristics of five Pareto-optimal solutions (coloured symbols) with priorities ranging from ride comfort to road holding are depicted in Figure 12 (a). The black damper spline with focus on ride comfort, rather than road holding, represents a "typical" asymmetry parameter $A$ of about 1/3 between compression and rebound, with very small average damping parameter $D$. The green damper spline shows similar properties at a larger value of $D$, while for the magenta spline an asymmetry parameter $A$ of about 1/2 can be observed with further increased $D$. Damper splines with focus on road holding, rather than ride comfort, tend to be centrally symmetric, blue line, or $A$ is even above 1, red line.

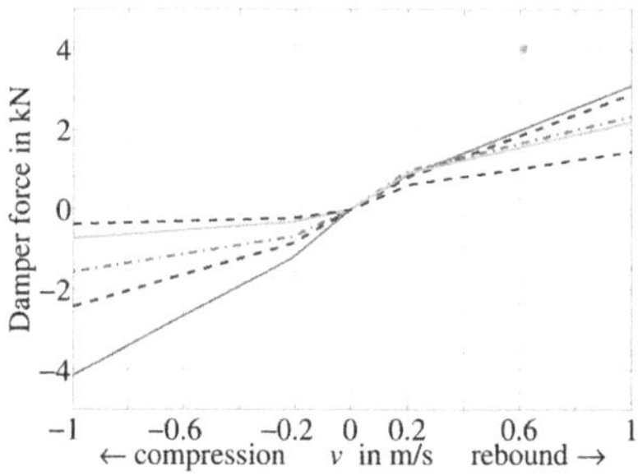


Figure 12 (a): Damper splines of five specific dampers corr. to Figure 12 (b)

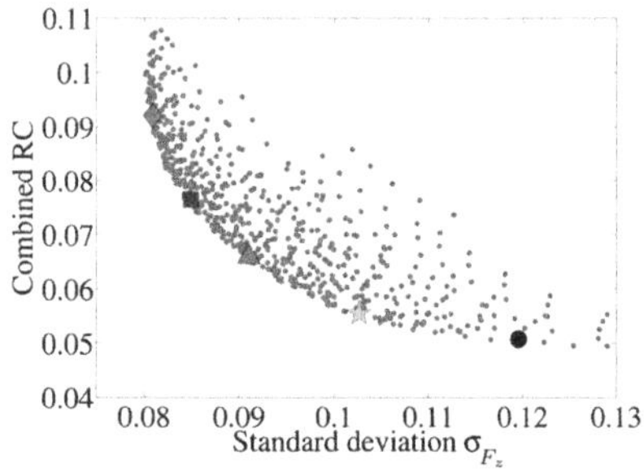


Figure 12 (b): Conflict diagram based on Short Bump with five dampers

# 4 Full vehicle simulation on straight road

Above, based on respective MBS axles on a virtual two-post-test rig, separate front and rear damper designs can be found by applying stochastic and single road disturbance inputs and mentioned objectives. It is assumed that front and rear damper design can be determined independently from each other, [1]. Despite rather detailed models of the axles and a wide spectrum of road induced excitations considered, additional aspects need to be addressed. With full vehicle simulation the position of driver between the axles and the pitching motion, adding to longitudinal and vertical acceleration, can be taken into account. A detailed complex physical tyre model allows for more realistic vertical, and also longitudinal tyre forces. If the spatial wavelength of the road excitation is smaller than 2–3 times the contact patch length, the envelopment properties of the tyre and the variation in effective rolling radius, that occurs when the tyre rolls over a short obstacle, must be taken care of in the applied tyre model [6]. In this way, full vehicle simulation, with possible coupling effects from front and rear (damper) suspension characteristics, and analysis by DoE methods, will help to check and fine-tune found damper designs.

The manoeuvres from the previous sections are repeated. The stochastic vertical road excitation of the full vehicle may be superimposed by excitations from braking actions of the driver, resulting in more pronounced bounce and pitch motions. Besides ride comfort, the ability of longitudinal (braking) force transfer considering variation of wheel load and effective rolling radius is evaluated. W.r.t. running over the bumps, a "harmonic" setup of front and rear dampers can be assessed by evaluating bounce and pitch motion.

As an example, Figure 13 (a) and (b) shows the PSD of the vertical acceleration at the driver's seat rail and the PSD of the vehicle pitch angle acceleration for stochastic road input. Three different front/rear damper setups are depicted. These setups include individually reasonable damper designs with comparable levels of combined damping, but opposed distributions between front and rear. By inspection, one may asses the characteristics at resonance frequencies (corresponding equivalent damping ratio), the amplitudes in the excitation frequency range from 4 Hz to 8 Hz, [15], and the area under the graphs. The pitch angle accelerations are important, as they are the primary source of longitudinal vibrations [13]. The wheelbase effect of varying degree can be observed in both plots.

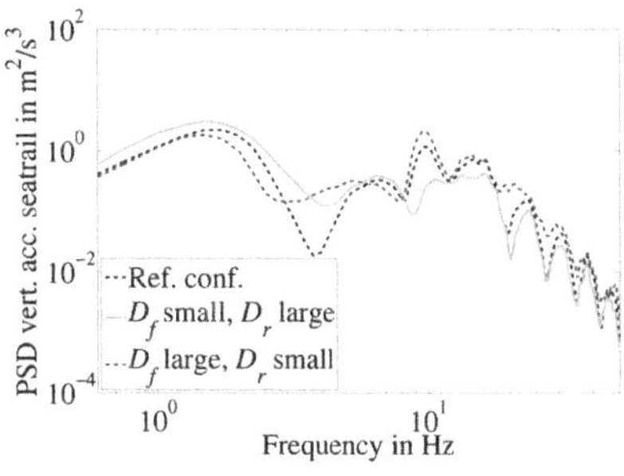


Figure 13 (a): PSD of vertical acceleration at the driver's seat rail for stochastic road input with three different front/rear damper setups

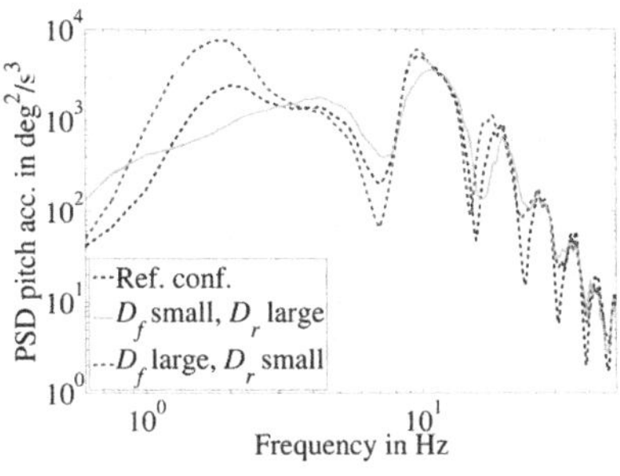


Figure 13 (b): PSD of vehicle pitch angle acceleration for stochastic road input with three different front/rear damper setups

# 5 Lateral excitation from the driver

In contrast to the excitation of the vehicle from road input, in particular in vertical and longitudinal direction, this section emphasizes the lateral excitation of the vehicle at constant speed, which may derive from steering actions of the driver, or lateral disturbances, such as side wind gusts. Involved changes of vehicle body attitude (pitch and roll) and heave motion depend on numerous vehicle parameters, e.g. height of body centre of gravity, elasto-kinematic properties of the suspensions, design of anti-roll bar, spring and damper characteristics, tyre characteristics etc., but also on the considered manoeuvre. Related vertical forces control both the vehicle body motions, which are preferably small for the ease of the driver's vehicle control, as well as the horizontal tyre forces and the resulting vehicle motion. Consequently, the front and rear dampers affect the vehicle body motion and the transient behaviour of the vehicle as a reaction to steering inputs or after a disturbance event, and appropriate damper tuning is a relevant part in suspension design for demanded lateral dynamics.

Subjective handling evaluation will not be addressed in particular here, and manoeuvres with a full MBS vehicle model on a horizontal flat road are considered only, although superimposed excitations from road roughness can be relevant in particular in the limit region [10]. The design of the damper spline influences the lateral and yaw dynamics of the vehicle, however, effects on the vehicle body motion are more distinctive. Ideally, the driver's observations of the path ahead will be from a stable platform to minimize the complexity of the view, implying the need for minimum pitch response to road roughness input [21] and minimum roll from steering input. A large overshoot of roll angles may cause unrequired steering actions by the driver [2], and large body motions may result in corresponding accelerations which are another source of discomfort for the driver.

Even for severe steering actions and resulting large vehicle body motions damper forces will mainly work in the lower stroke velocities range. Consequently, damper splines may be considered bi-linear with $s_d = s_z = 1$. For variation of $D$ and $A$ at the front and at the rear axles, respectively, full factorial design includes again 625 damper variants. Considered manoeuvres include standard manoeuvres, such as step-steer (omitted here), sinusoidal-steering and sine-with-dwell, to essentially account for vehicle reaction, vehicle responsiveness and limit behaviour, respectively. In particular comfort-related aspects of vehicle handling will be addressed in the following.

The roll angle and its derivatives are in particular important, as these states are sensed by the driver either as useful feedback from the vehicle or uncomfortable disturbance. For transient manoeuvres the damper is supposed to avoid overshooting roll angles and to allow for "harmonic" roll motions. The design of the dampers determines the (damped) course of the roll angle to its maximum after an initiated steering. Consequently, not the maximum roll angle but the roll rate is decisive.

First, sinusoidal steering manoeuvres with 0.5 Hz steering angle oscillations are simulated and evaluated. Forward velocity is about 100 km/h and resulting lateral accelerations of 5 $m/s^2$ are large enough to assess damper influences on handling performance, but not too large to get lost in the nonlinear suspension and handling characteristics.

The first objective is based on the first half wave of the sinusoidal steering angle and resulting roll motion. Figure 15 (a) shows the peak value of the roll rate w.r.t. the peak value of the steering angle, as an objective for the roll reaction, and the cross correlation between roll and steering, as an objective for the time lag due to damping. Figure 15 (a) confirms that the criteria are diametrical: a larger average damping coefficient (asymmetric or not) and therefore reduced roll rates, yields a delayed roll reaction. The distribution of the requested overall damping to the front and rear axle is less important in this respect. Changing the anti-roll bar stiffness by ±20% w.r.t. a reference configuration on both axles reveals more effect on the roll rate reaction than on the time lag.

The second objective is based on the first full wave of the sinusoidal steering angle. The roll motion is evaluated in Figure 14 (b) by the roll index proposed in [22], which is composed of both peak values of roll angle, roll rate and roll acceleration related to the lateral acceleration. The yaw velocity gain is used to characterize the agility of the vehicle. A corridor marks the range for appropriate combinations of agility and roll reaction performance for various damper designs. It becomes obvious that the roll index can be manipulated by the average damping coefficient $D$, or by the stiffness of the anti-roll bar, in a similar way.

The final objective considers the steady-state oscillation and evaluates the linearity and time lag between roll and yaw motion w.r.t. the steering angle. Fourier-transformation of the respective time signals and correlation analysis may be employed.

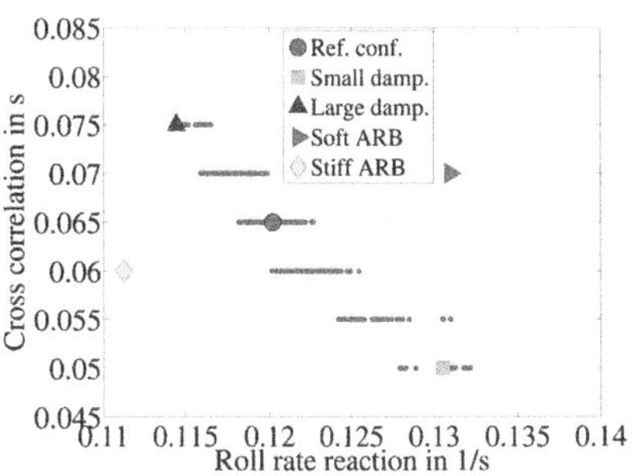


Figure 14 (a): Roll rate reaction vs. time lag (cross correlation)

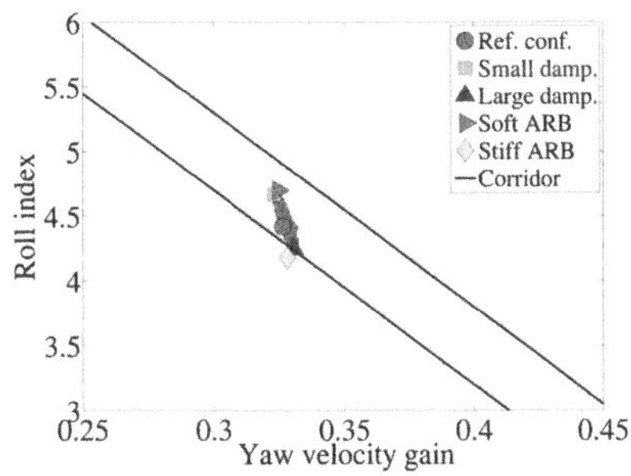


Figure 14 (b): Yaw velocity gain vs. roll index

As a second manoeuvre, the NHTSA sine-with-dwell manoeuvre is considered. Due to its characteristic steering excitation, this manoeuvre turned out to be appropriate to evaluate the combined tuning of front and rear damper w.r.t lateral vehicle dynamics. A steering angle amplitude, four times the steering angle required for 3 m/s$^2$ lateral acceleration at 80 km/h at steady-state cornering, is selected.

Because no loss of yaw stability of the vehicle has to be expected for "reasonable" suspension designs, tyre sideslip angles, tyre friction potential utilisation as well as vehicle sideslip angles are regarded instead. As an example, in Figure 15 (a) peak-to-peak values of the vehicle sideslip angles are plotted for 625 damper variants w.r.t. damper parameters $D_{front}$, $A_{front}$ and $D_{rear}$, $A_{rear}$ of the front and rear damper splines, respectively. For convenience, two surfaces spread out all maxima and all minima of vehicle sideslip peak-to-peak values w.r.t. $D_{front}$ and $D_{rear}$. Small peak-to-peak values may be desired, thus the bottom surface represents a Pareto-optimal surface, while the top surface represents adverse damper variants. Obviously, large values of $D_{rear}$ result in generally larger vehicle sideslip peak-to-peak values.

Kinematic properties of the vehicle suspensions as well as nonlinear spring and damper characteristics cause pitch oscillations at the considered manoeuvre. In Figure 15 (b), respective peak-to-peak values of the pitch angles are shown for varied asymmetry parameters $A_{\text{front}}$ and $A_{\text{rear}}$, in analogy to Figure 15 (a); extreme peak-to-peak values are again collected in two surfaces. Considering both surfaces, similar asymmetry parameters $A_{\text{front}}$ and $A_{\text{rear}}$ of the front and rear damper spline are – at least for the considered vehicle – favourable.

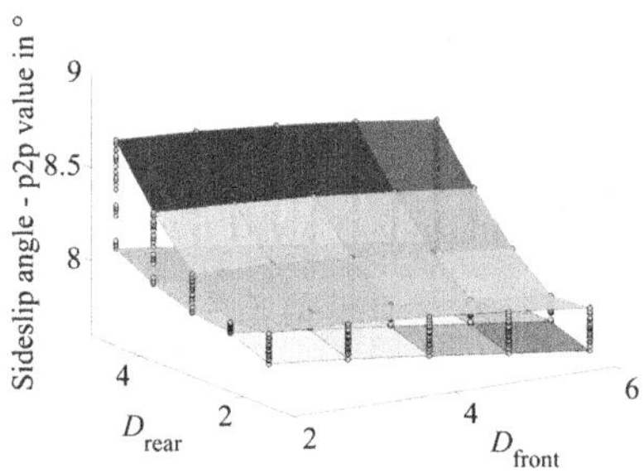


Figure 15 (a): Peak-to-peak values of sideslip angles

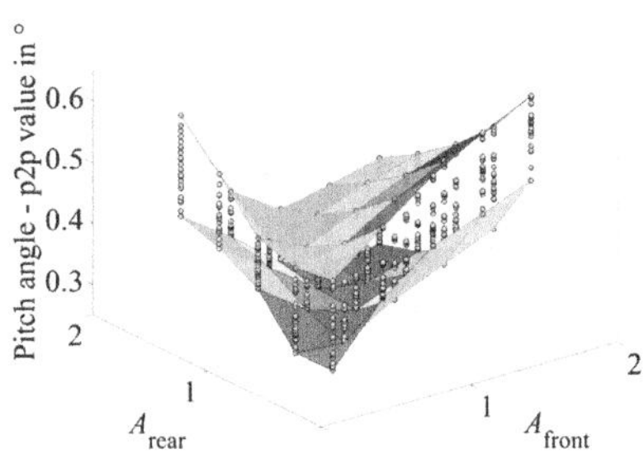


Figure 15 (b): Peak-to-peak values of pitch angles

# 6 Conclusions

To conclude, some considerations in particular on (a)symmetry of damper splines based on the above findings will be given.

For stochastic road excitations, centrally symmetric damper characteristics – at least for lower stroke velocities – will be of advantage for both ride comfort and road holding. A small gradient is basically preferred, as an increased level of damping will result in increased vertical body accelerations, which may be reduced by flattening out the spline at higher stroke velocities without degrading road holding too much. However, more or less pronounced asymmetric damper characteristics are typically found in most of today's vehicles, which have evolved through many iterations primarily based on intuition and decades of experience [1]. Already in 1969, Thompson wrote w.r.t. symmetry properties of dampers: "The analogue computer studies show that unsymmetrical damping may be employed with significant advantage in the reduction of shocks due to isolated obstacles and that an optimum damping ratio $r$ may be selected whilst preserving the same average damping rate." [23]

For long bumps (long road undulations) exciting vehicle body motion, large damping in the lower stroke velocity range are preferable, either from the compression or rebound characteristics of the damper spline (asymmetry is not essential). In contrast, for bumps exciting frequencies between body-heave and wheel-hop resonance, small damping is essential, both for ride comfort and road holding. This conflict can be relaxed by degressive damper characteristics in the higher stroke velocity range for higher excitation frequencies. For not degrading ride comfort, enough damping can be realized by asymmetry, with larger gradient for rebound than for compression. Short bumps exciting frequencies at the wheel-hop resonance or above reveal the conflict of opposed ride comfort and road holding performance for this frequency domain. Again, for not degrading ride comfort, asymmetric damper characteristics are appropriate.

The above conclusions have been drawn from considerations applying a quarter car or MBS axle model on a virtual two-post-test rig. Finally, some short remarks based on full vehicle simulation are given. After independent damper design of the front and rear axle on the two-post-test rig, full vehicle tests are required for final tuning and balancing both axle suspension dampers. For standard steering manoeuvres, regarding steering excitations from the driver, mostly the lower stroke velocity range is utilized, transition to higher velocities might be tuned w.r.t. the damper stroke–wheel travel ratio. Basically, large damping is beneficial for smooth body motions, constraint by a too "sluggish" vehicle response (roll or yaw motion) w.r.t. steering actions. In this respect asymmetry is of minor importance, however, asymmetric properties can be useful when balancing adverse effects of supporting the body on its axles in transient manoeuvres. Balancing energy dissipation and damping forces in series with forces from the suspension spring requires a careful compromise.

Current activities concentrate on a further in-depth analysis of effects of combined vertical-longitudinal(-lateral) full vehicle excitation on damper design.

## Acknowledgment

The authors acknowledge the support of this research by The Mathworks GmbH, Germany, National Instruments Ges.m.b.H., Austria, and Altair Engineering GmbH, Germany, for providing software licenses and support.

## References

[1] D. Hrovat: Active and Semi-Active Suspension Control. In: G. Mastinu, M. Plöchl (eds.): *Road and off-road vehicle system dynamics handbook*, CRC Press, 2014.

[2] M. Mitschke, H. Wallentowitz: *Dynamik der Kraftfahrzeuge*, Springer, 2014.

[3] R.S. Sharp and D.A. Crolla: Road Vehicle Suspension System Design – a review. *Vehicle System Dynamics* **16**, 1987, 167–192.

[4] D. Ammon: *Modellbildung und Systementwicklung in der Fahrzeugdynamik*, B.G. Teubner, 1997.

[5] B. Mashadi and D.A. Crolla: Influence of ride motions on the handling behaviour of a passenger vehicle. *Proc. IMechE Part D: J. Autom. Eng.* **219**, 2005, 047–1058.

[6] H.B. Pacejka: *Tire and Vehicle Dynamics,* Butterworth-Heinemann, 2012.

[7] A. Simms and D. Crolla: The Influence of Damper Properties on Vehicle Dynamic Behaviour. *SAE-paper 2002-01-0319*, 2002.

[8] H. Dresig, F. Holzweißig: *Maschinendynamik*, Springer, 2011.

[9] P.C. Müller, W.O. Schiehlen: *Linear vibrations*, Martinus Nijhoff, 1985.

[10] B. Mashadi and D.A. Crolla: Influence of ride motions on the handling behaviour of a passenger vehicle. *Proc. IMechE Part D: J. Automobile Engineering* **219**, 2005.

[11] J.A. Levitt and N.G. Zorka: The Influence of Tire Damping in Quarter Car Active Suspension Models. *J Dyn Syst–T ASME* **113**, 1991, 134–137.

[12] D. Hrovat: Survey of Advanced Suspension Developments and Related Optimal Control Applications. *Automatica* **33**, 1997, 1781–1817.

[13] T.D. Gillespie: *Fundamentals of Vehicle Dynamics,* SAE, 1992.

[14] S.M. Savaresi et. al.: *Semi-Active Suspension Control Design for Vehicles,* Butterworth-Heinemann, 2010.

[15] ISO 2631-1: Mechanical vibrations and shocks – Evaluation of human exposure to whole-body vibration, Part 1, ISO, 1997.

[16] B. Richter: *Schwerpunkte der Fahrzeugdynamik,* TÜV Rheinland, 1990.

[17] R.S. Sharp and S.A. Hassan: An Evaluation of Passive Automotive Suspension Systems with Variable Stiffness and Damping Parameters. *Vehicle System Dynamics* **15**, 1986, 335–350.

[18] M. Rosenmaier, personal communication with F.K., J.E., M.P., S.J., Feb. 2015.

[19] M.J. Griffin: Discomfort from feeling vehicle vibration. *Vehicle System Dynamics* **45**, 2007, 679–698.

[20] B. Heißing, H.J. Brandl: *Subjektive Beurteilung des Fahrverhaltens,* Vogel, 2002.

[21] R.S. Sharp: Vehicle dynamics and the judgement of quality. In: J.P. Pauwelussen (ed.): *Vehicle Performance*, Swets & Zeitlinger, 1999.

[22] S. Botev: *Digitale Gesamtfahrzeugabstimmung für Ride und Handling,* VDI, 2008.

[23] A.G. Thompson: Optimum Damping in a Randomly Excited Non-Linear Suspension. *Proc. IMechE Part D: J. Automobile Engineering* **184**, 1969, 169–184.

# Importance of body rigidity in the transient stage of the maneuver

C. Gagliano[1]
T. Geluk[2]

[1] Honda R&D Americas. Inc.
[2] Siemens Industry Software NV. – LMS Engineering Services

# 1 Introduction

Several studies have been performed underlining the importance of body rigidity for full vehicle performance, using a numerical/simulation approach (see [1] and [2]), a test-based approach ([3], [4] and [5]) or a combination of those ([6] and 7]).

In the studies where a body evaluation is performed to identify possible body weak-points or the improvement potential – this is mostly done based on the body deflection in the steady state part of handling maneuvers.

In this paper, the focus is on the influence of body rigidity on the vehicle performance in the transient stage of the maneuver. A comparison of the results in the transient stage with those in the steady state part of the maneuver shows that different body stiffness characteristics are important in the transient stage compared to the steady state.

The impact of changing body stiffness characteristics on the transient vehicle behavior is evaluated in detail by identification of the transient time-domain body load build-up with and without modifications. These changed objective results enable the establishment of links with the drivers' subjective evaluations.

Based on the expert driver subjective feedback, the transient body load build-up and body deflection, it is shown that the body flexibility is a limiting factor for optimal suspension usage in vehicle dynamic maneuvers. Enhanced body stiffness characteristics enable better usage of the suspension setup, which is confirmed by the ratings of the expert driver.

# 2 Background and Method Description

To enable improved body design for optimal full vehicle performance, it is needed to understand the relationship between body stiffness metrics and objective handling performance. At a higher level there is a challenge to connect objective handling performance results to the subjective perceptions of the driver.

Using established testing technologies it proves to be difficult to get an objective quantification of the body influence on the vehicle performance. Also, in the past, full vehicle multi-body-simulation is done with a flexible representation of the car body. However, it has shown to be difficult to quantify the body influence on the vehicle performance using these simulation results.

Ideally, one would be able to analytically define body stiffness targets for good transient handling performance. However, until now, no body target setting approaches are available for the transient stage.

In this investigation, both the relation of the body stiffness towards the transient vehicle performance as well as the connection to the subjective feedback is identified. This is done using the BodyFlexbility methodology, which allows quantifying and understanding the influence of body rigidity on the vehicle's performance both in the transient and steady state stage. This approach is based on two main components, time-domain load identification and time-domain body deformation estimation.

Time-domain load identification for vehicle dynamics is performed by instrumentation of highly sensitive strain-gauges around all suspension to body connection points. Load calibration measurements are performed at each of those connection points, to identify the strain to force calibration matrix. For each body configuration that is evaluated the strain to force calibration matrix is identified. Using this calibration matrix in combination with operational strain-data acquired on the track allows an inverse identification of the time-domain body-loads in a vehicle dynamic maneuver (see [3] and [8]). These time-domain body-loads show clear changes when different body stiffness characteristics are evaluated and therefore provide detailed insight in changing vehicle dynamic performance.

The time-domain body deformation is acquired by application of the body loads to a modal model of the car body. This is done in the modal domain using

$$m_i\ddot{q}_i + c_i\dot{q}_i + k_i q = \{\psi_i\}^T \{F(t)\}$$

with $m_i$, $c_i$ and $k_i$ the modal mass, damping and stiffness for mode $i$, $\{\psi_i\}^T$ the transposed modal vector, $\{F(t)\}$ the vector of the time-domain loads and $q_i$ the modal responses. In this study, a CAE model of the car body is used to create a reduced modal model with output at all suspension to body interface points and further visualization points enabling animation of the full body deflection during the handling maneuvers.

# 3 Vehicle Evaluation in Base Body configuration

The vehicle under evaluation is a minivan type vehicle with a McPherson Front Suspension and a 4-link rear suspension. Both the front and the rear subframe are floating, i.e. they are connected to the car body through rubber mounts.

A subjective evaluation is done with the vehicle under evaluation. One of the main feedbacks of the expert drivers is 'the delayed rear axle response'. This subjective evaluation will be the main focus in this study.

The next step is the objective evaluation, which is done by performing vehicle dynamic maneuvers on a test-track. A steering-robot is used to maximize measurement repeatability, especially in the transient part of the maneuvers. A step-steer maneuver, performed at 100kph, and 0.3g lateral acceleration is used to acquire the operational data for the time-domain body load identification. Examples of the identified loads are shown in Figures 1 and 2, with a set of lateral loads and a set of vertical loads acting on the Front body during the step steer maneuver.

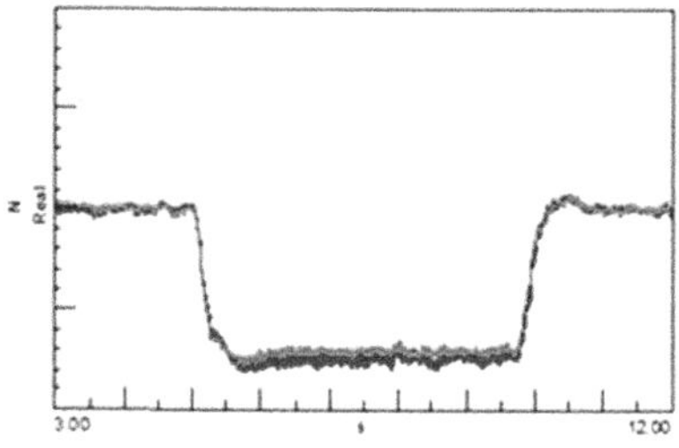

Figure 1: Front Subframe Lateral Loads

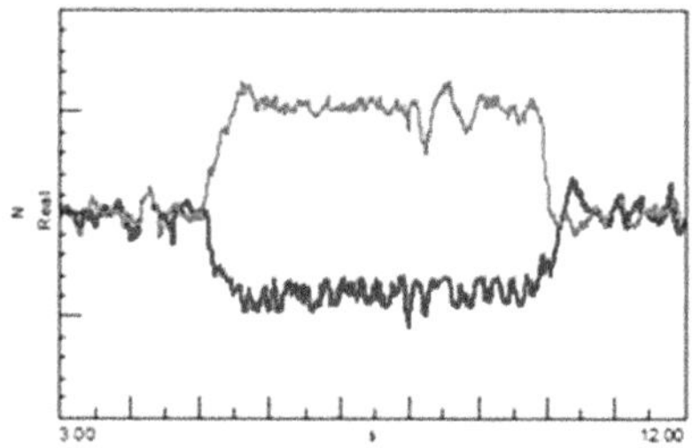

Figure 2: Front Strut Vertical Loads

By normalizing both the body loads and the steering angle input signal, a clear indication for the transient load build-up is obtained, see Figure 3.

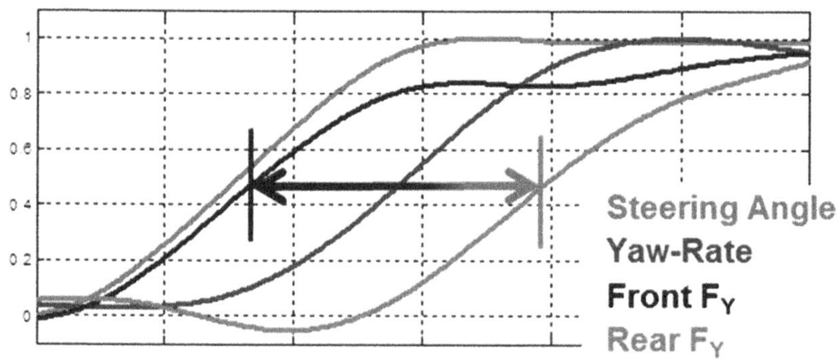


Figure 3: Front vs Rear axle transient load build-up

After the applied steering angle input, the front lateral loads build-up quickly. As it takes time before the rear tires start building up lateral load, the rear body loads show a clear delay with respect to the front body lateral load build-up. In section 4 the impact of changing body stiffness characteristics on this transient load build-up and delays with respect to the steering angle input will be discussed.

By application of the identified body loads to the modal model of the body, the time-domain body deformations are calculated. Examples of the deformation results are presented in Figure 4 and 5, with in Figure 4 a top-view of the vehicle body, in deformed state, during the step-steer maneuver. As a dominantly lateral load-case is applied to the body, the resulting deformation shape of the body shows a relevant amount of lateral bending.

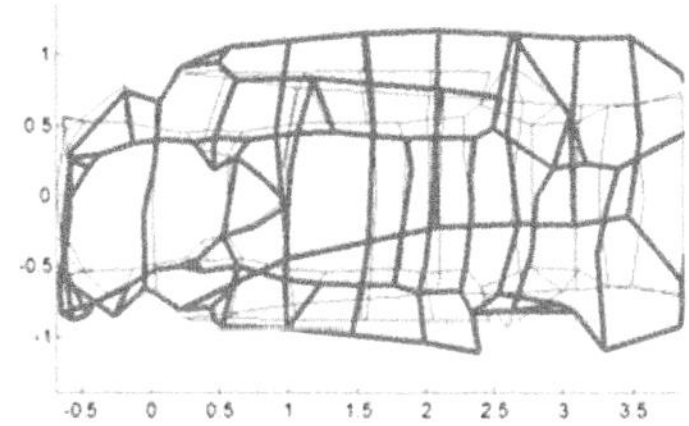


Figure 4: Total Body Deformation during the Step-Steer maneuver

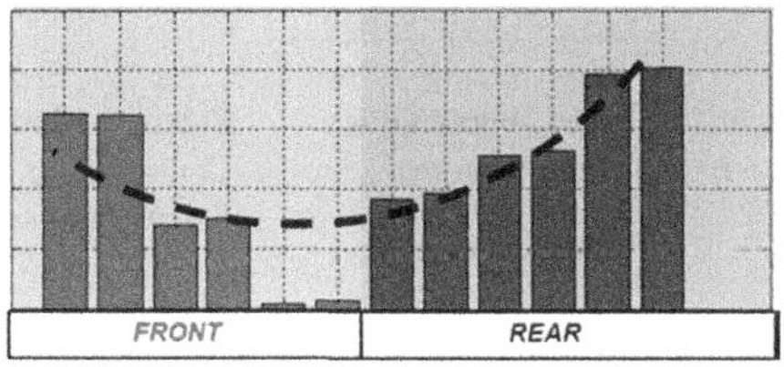


Figure 5: Interface point lateral deformation in Step Steer maneuver

This is further highlighted in figure 5, where the lateral deformation of the hard-points at the Front and Rear suspension are shown.

As the deformation calculation is done using a modal model, the total body deflection can be decomposed into individual mode contributions to the deformation. With that, the importance of individual modes (as lateral bending, torsion, or local flexibility at the interface points) is evaluated.

The time-domain mode contributions – an indication how strong a certain body mode is excited by the body loads – in Figure 6 show that 2 modes are strongly excited, which is clear from the amplitude level of the red and green contribution curve. More interesting however, is the large time-difference between these 2 dominant mode contributions. This indicates that a part of the body stiffness characteristics (mode A, Figure 7) is excited – and therefore relevant – from the early transient onwards, while another part of the body stiffness properties (mode C, Figure 8) is excited only after a large time delay, at the end of the transient stage of the maneuver. Resulting from this, the body deformation in the transient stage will be strongly different from the deformation during the steady state part of the maneuver.

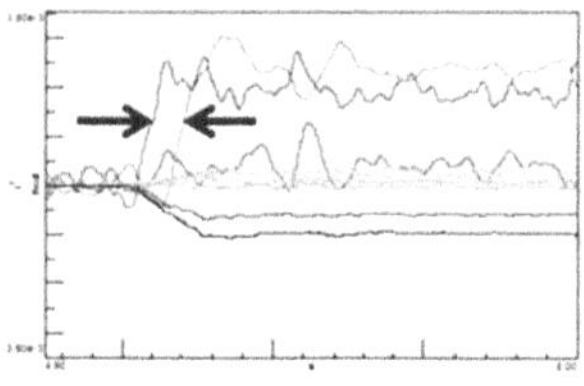

Figure 6: Mode Contributions A – E in the Step Steer maneuver

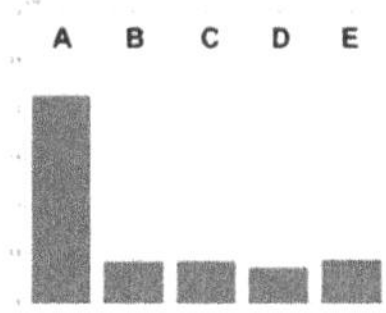


Figure 7: Mode Contributions A – E in **transient** stage

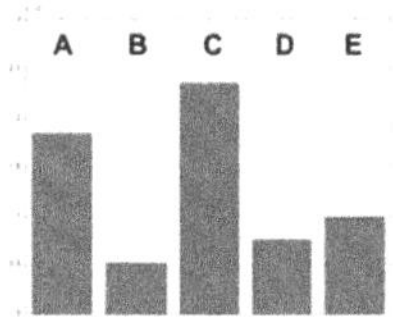


Figure 8: Mode Contributions A – E in **steady state**

Our aim is improving the transient vehicle performance and identification of the influence of body stiffness characteristics on that transient performance. This means that the focus should be on the body stiffness characteristics of mode A, not on the characteristics of mode C, even when this is stronger excited in the steady state part of the maneuver.

# 4 Vehicle Evaluation in Modified Body configuration

The base vehicle evaluation in Section 3 indicated a relatively high rear body lateral flexibility, which is shown in both figure 4 and 5. To reduce the rear body deflection during handling maneuvers a modification package, see figures 9 and 10, is applied that increases the rear body rigidity. Both the subjective and objective evaluations are also performed with this Modified Body condition.

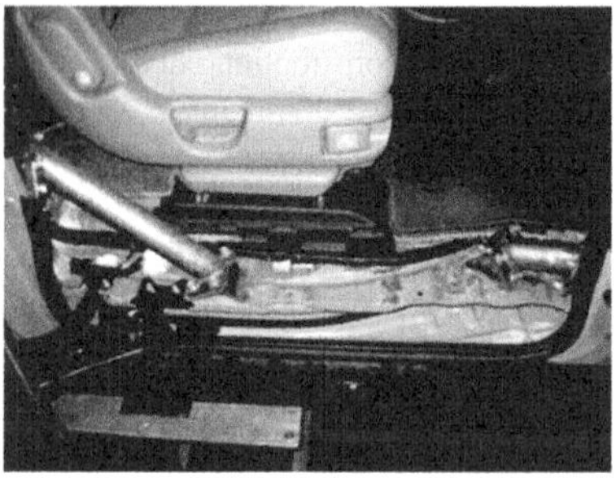

Figure 9: Rear Body Modifications

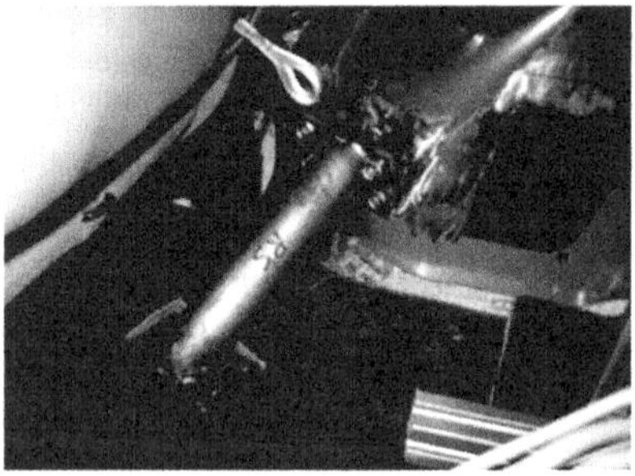

Figure 10: Rear Body Modification

To quantify the impact of the modification package on the body stiffness characteristics, the static hard-point stiffness values are compared in Base and Modified condition. The stiffness change resulting from the modifications, see Figure 11, proves that the main impact of the modifications is on the rear body hard point stiffness.

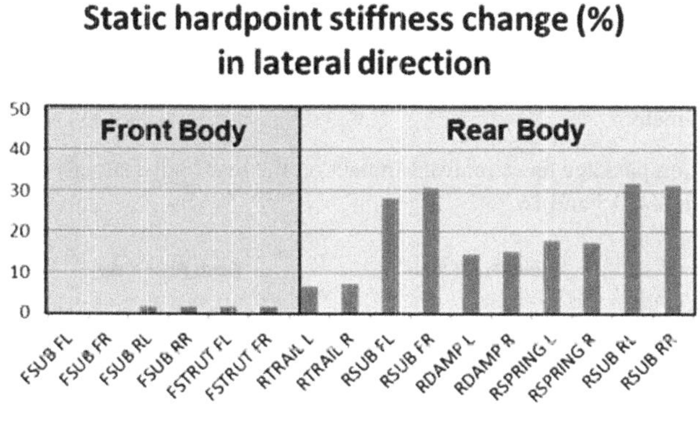


Figure 11: Impact of body modification package on lateral hard-point stiffness at the Front and Rear body.

Identical vehicle dynamic maneuvers are performed with the Modified body condition. Comparison of the lateral acceleration and yaw-rate responses Base and Modified body condition shows minimal differences as an effect of the modification package (Figures 12 and 13), only the yaw-rate shows an indication for a higher level at the peak of the transient.

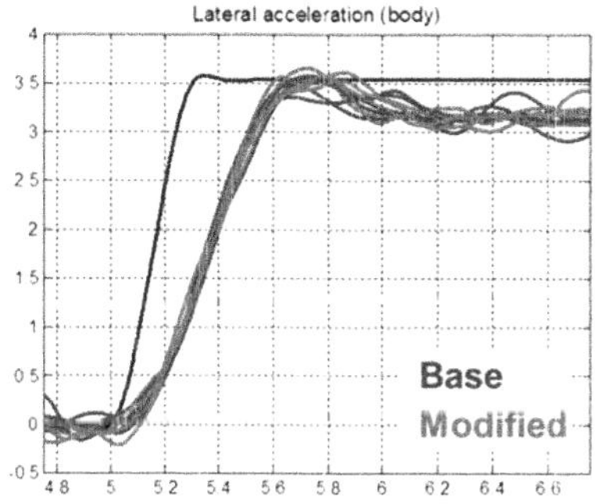


Figure 12: Lateral Acceleration, Step Steer maneuver for Base and Modified body condition

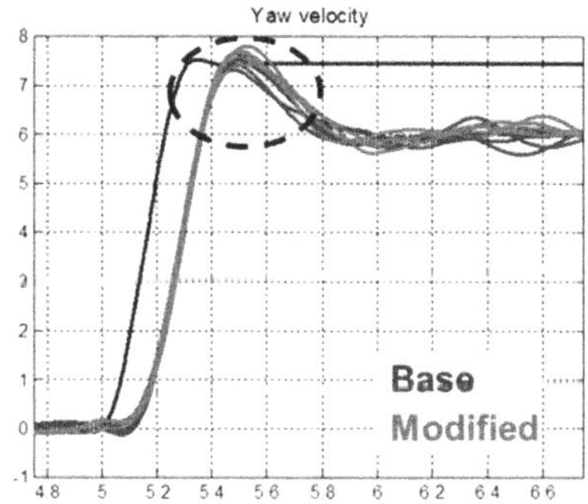


Figure 13: Yaw-rate, Step Steer maneuver for Base and Modified body condition

That the impact of body modifications on the global vehicle parameters such as the lateral acceleration and yaw-rate is small is known. Those parameters result from all body forces that are acting. When a subset of the body forces changes in amplitude or timing, the combined impact of all forces on the lateral acceleration or yaw-rate can still be small compared to the changes on individual force level. Therefore the individual body loads can provide much more insight in changing vehicle performance as the global vehicle parameters.

The applied modification package has a relevant impact on the rear body (lateral) load distribution; see Figures 14, 15 and 16.

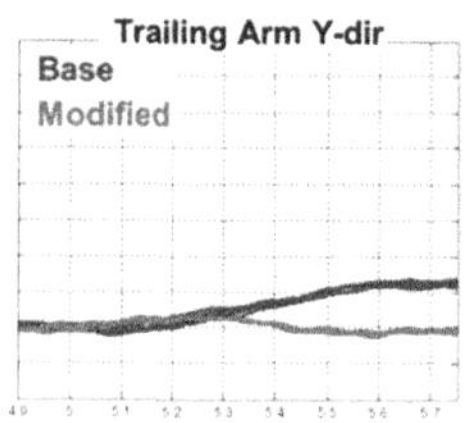


Figure 14: Transient load build-up Trailing Arm $F_Y$

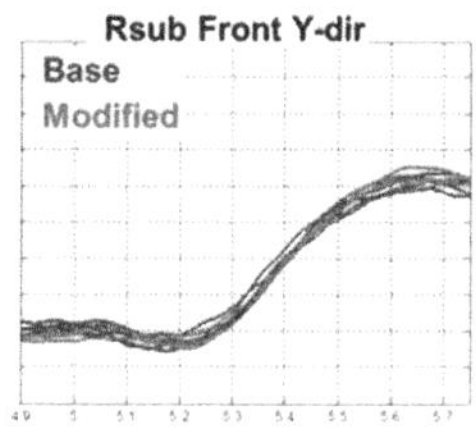


Figure 15: Transient load build-up Rear Subframe Front $F_Y$

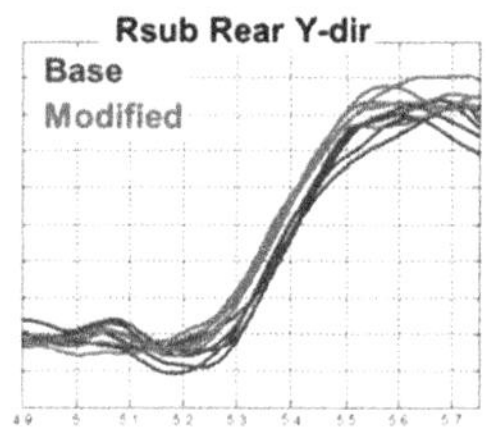


Figure 16: Transient load build-up Rear Subframe Rear $F_Y$

The most dominant rear body lateral force – at the rear subframe rear connection – is building up much faster (~20ms) in modified body condition. With a nearly similar Front axle load build-up and the faster Rear axle reaction, the 'gap' between the Front and Rear axle response is reduced, see Figure 17. The result from the subjective evaluation of the vehicle with Modified body condition indicates a 'quicker rear body reaction', which is directly in line with the objective results from the transient body load build-up.

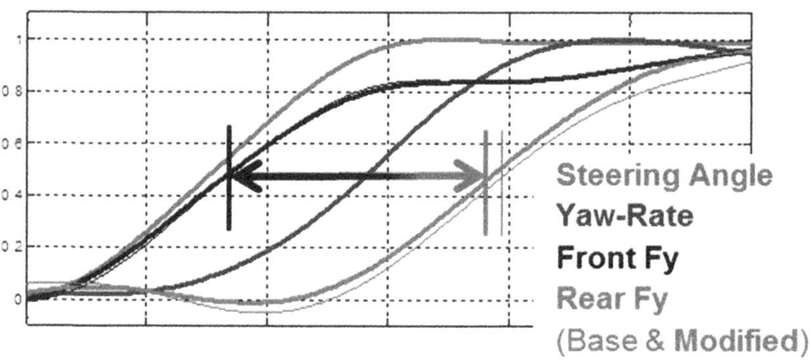


Figure 17: Faster Rear Axle load build-up for Modified body condition

In addition to this, the load comparisons indicate that in Base condition there is a lateral force component from the Trailing arm, which is not desired. In Modified condition, that lateral force becomes nearly zero, while the lateral force from the Rear Subframe – to which the lateral suspension links are attached – is increasing, also in Steady State condition.

Due to the increased body rigidity the transfer of load from tire towards the vehicle body is now done far more efficiently, resulting in faster and higher rear subframe lateral loads. As the rear suframe is mounted (floating), the changed load distribution – with the main change at the rear connection – also influences the rotational motion of the rear subframe with respect to the vehicle body. This could influence the yaw-gain, which is supported by the slight increase of the yaw-rate amplitude as observed in Figure 13.

Next to a clear influence on the body load distribution in the lateral direction, there is also a significant impact on the vertical load distribution, see Figures 18 – 20. When evaluating the steady-state part – although influenced by road undulations – the impact of the modification package on the steady state load level is minimal, see Figure 18. The transient stage however, has a clearly faster and more linear load build-up for the Rear Right body side (Figure 19), and a slower and also more linear load-build-up for the Rear Left body side (Figure 20). As this concerns a Step-Steer maneuver to the Left, this implies that the suspension side at the outside of the turn (the loaded side) can build-up load faster and more linear. These results further stress the importance of the body rigidity for the transient stage vehicle dynamic performance.

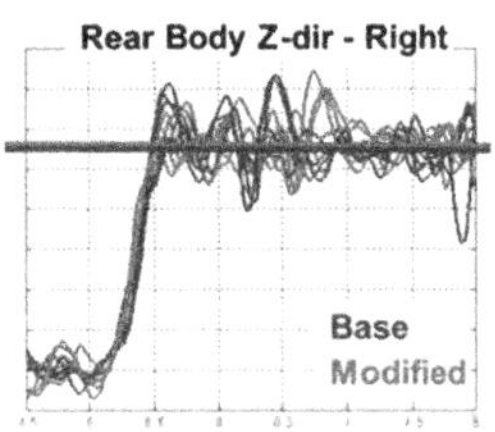


Figure 18: Rear Body $F_Z$, minimal changes in Steady State

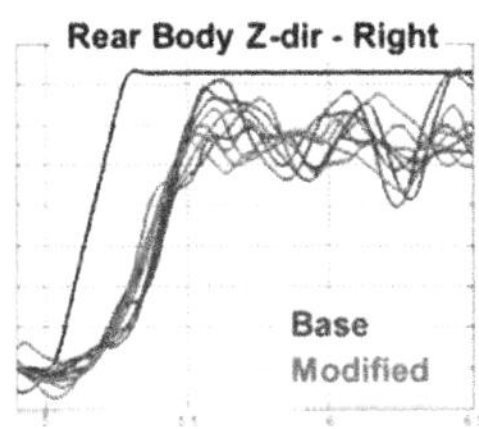


Figure 19: Rear Body $F_Z$, faster transient build-up right side

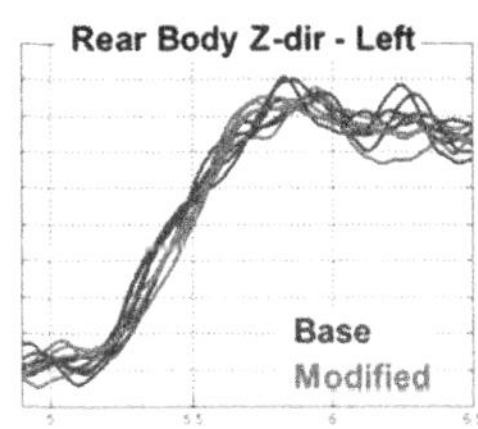


Figure 20: Rear Body $F_Z$, slower transient build-up left side

Both the findings for the lateral and the vertical transient body load-distribution indicate that the body flexibility is placing a limit on the suspension performance. Due to the Base body deflection, the rear axle is not able to respond quickly. Next to that, the lateral load distribution to the body is not optimal as a part of the rear axle lateral load is distributed through the Trailing Arm. After improvement of the body rigidity – based on the body deformation analysis in the transient stage, both of these limitations are solved, resulting in improved subjective rating of the vehicle performance.

# 5 Conclusions

An evaluation of the car body deflection in in a Step Steer maneuver has shown that different body stiffness characteristics play a role in the transient stage compared to the steady state stage.

Results from both the objective evaluation using body loads and the subjective evaluation indicate that the transient stage deflection of the body limits the suspension performance. The rear axle is not able to respond quickly, while both the lateral and vertical transient load distributions from tire to the car body are not optimal.

Using the applied methodology, the improvement potential for the body – relevant for the transient performance – is identified, enhancing the body target setting process. Comparison of the transient body load build-up for the vehicle in Base and Modified body condition enables an objective identification of the changed vehicle transient performance. Through enhanced body rigidity, the rear axle is able to respond faster, which is confirmed by subjective evaluations. In addition, the lateral and vertical load distributions to the body are significantly improved.

By focussing on both the transient body load build-up and the transient body deflection, a much better understanding of the mechanisms between the changing body rigidity and the vehicle performance is acquired compared to evaluations of the steady state part of the maneuver.

# 6 References

[1] E. Sampo, A. Sorniotti, A. Crocombe, 'Chassis torsional stiffness: analysis of the influence on vehicle dynamics", SAE Technical Papers, 2010

[2] E. Sampo, Modelling chassis flexibility in vehicle dynamics simulation", Phd thesis, University of Surrey, UK, 2011

[3] E. Plank, A. Guellec, T. Geluk, P. Mas, 'Body load identification and weak spot analysis to evaluate different body concepts for better balancing of vehicle dynamics and NVH', Chassis.Tech 2009

[4] J.H. Park, J.S. Jo, T. Geluk, G. Conti, J. Van Herbruggen, 'Improving the vehicle dynamic performance by optimizing the body characteristics using body deformation analysis', Chassis.Tech 2010

[5] H. Kyogoku, J. Nakajima, M. Okabe, T. Geluk, F. Daenen, 'How subjective evaluation by drivers is affected by car body stiffening – proposal of the hypothetical mechanism', Chassis.Tech 2014

[6] M. Nagahisa, K. Kusaka, M. Minakawa, T. Nirei, The influence of body flexibility on the handling characteristics", Honda R&D Co., Ltd. Japan, 1999

[7] K. Kusaka, M. Nagahisa, T. Nishimura, An influence of body lateral bending on handling characteristics", Honda R&D Co., Ltd. Japan, 2001

[8] S. Dom, T. Geluk, K. Janssens, H. Van der Auweraer ''Transfer Path Analysis: accurate load prediction beyond the tradi-tional mount stiffness and matrix inversion methods", SAE Paper 2014-36-0799, SAE Brazil International Noise and Vibration Colloquium 2014, November 4-5, Brazil

# Contribution to the objective evaluation of roll dynamics

Dipl.-Ing. A. Apfelbeck, Dipl.-Ing. M. Schwarz, Dipl.-Ing. S. Wegner,
Dr.-Ing. R. Henze, Prof. Dr. F. Küçükay

# 1 Introduction

In the early phase of the automotive development, the basic design of a vehicle defines the desired driving behaviour of the future series-production vehicle. The process relies mostly on simulation tools and experience from predecessor models, mainly because no physical vehicle models exist at this stage of the development.

To be able to assess the driving dynamics of the future vehicle based on simulation results, relationships between specific characteristic values (CV) of the vehicle dynamics and subjective evaluations are required. These are usually obtained by correlating the subjective ratings of driver and passengers with characteristic vehicle data that was measured in different manoeuvres.

The roll dynamics of a vehicle have a significant influence on the driver's perception of the vehicle's driving characteristics. Despite that fact, prior research has mainly been focused on classical driving dynamics aspects such as the steering properties or the yaw dynamics of vehicles, resulting in a lot of successful studies in the field of vehicle dynamics objectification. Compared to that, the roll dynamics of vehicles have been comparatively neglected. While more recent studies have examined some roll dynamics aspects (cf. [3], [4], [5]), there are still certain characteristics that have not been analysed yet.

In this paper, three different yet complementing approaches to find new conclusions and CVs for the objective description of the roll dynamics will be presented. One approach is based on the modification of the mass properties of a test vehicle, varying only roll inertia while the other vehicle parameters remain unchanged. This way, the influence of roll inertia on driving dynamics can be analysed and a quantitative relationship between the parameter change and subjective ratings can be established. The second and third approach combine the BMW Group dynamic driving simulator and a test vehicle with tuneable control systems to carry out the examinations. The variants for the studies are prepared in simulation using a nonlinear double-track model, which reproduces the driving dynamics of the test vehicle. Thus, the driving dynamics of the test vehicle can be rated in both the simulator and the actual vehicle allowing the comparison and verification of study results.

The three approaches will be explained in the following paper.

# 2 Examination of Isolated Roll Inertia Changes

Target values and areas for basic properties (e.g. mass distribution, aerodynamic coefficients) of a future series-production vehicle are fairly common in the automotive development process. These parameters have the most immediate effect on the handling characteristics of a vehicle and form the basis for any further tuning of the vehicle. If no target values existed, it would not be possible to monitor and control these vehicle parameters in the early, simulation based phase of the development.

The importance of some properties and the direction of their gradient towards an optimized driving performance is fairly evident, e.g. the positive impacts of a reduced height of the centre of gravity. However, a methodical approach would require to compare vehicle variants where only the one parameter that should be studied has been varied. Creating these variants is usually nontrivial and requires special effort but is necessary for the criticality assessment of parameter changes.

While target values for the yaw inertia of vehicles exist in the automotive industry, so far there has not been any published study investigating the quantitative effect of isolated roll inertia changes on handling characteristics and driving performance. To investigate this relationship, an approach has been chosen which only modifies the roll inertia of the test vehicle while leaving its other parameters constant.

The most fundamental vehicle properties can be divided into geometrical properties (front/rear track widths and wheelbase), mass properties (vehicle mass, vehicle inertia about the three axes $J_{xx}, J_{yy}, J_{zz}$ and position of the centre of gravity $l_f$, $l_r$, $h_{CoG}$) and aerodynamic properties (represented by the aerodynamic coefficients). Because only the mass properties are relevant for the roll inertia examinations, geometric and aerodynamic properties will be neglected henceforth.

Due to their interdependencies, an individual mass parameter cannot be trivially modified. If, however, information (specifically mass, inertia and position) about all the individual components of a vehicle is available, then the mass properties listed above can be calculated for the fully assembled vehicle using the parallel-axis theorem for the moments of inertia

$$J_\mathrm{n} = J_\mathrm{o} + md^2 \tag{1}$$

where $J_n$ is the moment of inertia relative to the new axis, $J_o$ is the moment of inertia relative to the old axis and $d$ is the perpendicular distance between the two axes, and the centre of gravity R can be calculated using

$$\boldsymbol{R} = \frac{1}{M}\sum_{i=1}^{n} m_i \boldsymbol{r}_i \tag{2}$$

where $m_i$ are the individual masses with their positions $\boldsymbol{r}_i$ and $M$ is the total mass of the vehicle.

These relations were used to implement the roll inertia variants: First, the vehicle mass was reduced by as much as possible to then insert several mass packages at dedicated positions to reattain the initial total vehicle mass. For this procedure, the amount and the mounting positions of the mass packages were determined using numerical optimization and equations (1) and (2) to ensure that only the roll inertia of the vehicle was altered.

## 2.1 Vehicle Modification

The vehicle used for the study is a BMW X5 (model year 2013) with a mass of about 2300 kg. To prepare it for the modification, as many components as possible were removed from its inside. Among these parts were the rear seat bench, the passenger seat, the window lift motors etc. The only vehicle parts which had not been tampered with were those necessary for future driving activities. Eventually, the vehicle weight could be reduced by approximately 200 kg. Then, a numerical solver was used to determine the best combination of position and amount of the mass packages to restore the original vehicle mass. The constraints used in the optimization were:

- Reasonable mounting positions that would not impair the roadworthiness of the car or disturb the driver were selected manually.
- All vehicle parameters should remain unchanged except for roll inertia and pitch inertia. Pitch inertia had not been constrained in the optimization, because its influence on manoeuvres in the lateral plane with stationary longitudinal dynamics is insignificant and can hence be neglected.
- Because switching between the final variants would be done manually, the individual masses had to have a size that could still be handled under ergonomic aspects. The individual size of the masses was thus set to either 5 kg or 10 kg.

The results of the optimization have been implemented in the BMW X5 that has been mentioned above. An example of the vehicle interior after the removal of the interior parts, with the mounts already installed, and some mass packages is shown in figure 1.

Figure 1: Vehicle interior after modification.

The roll inertia range of the variants, which were identified in the optimization and implemented in the test vehicle, is shown in figure 3 and will be presented in the following.

Overall, a roll inertia spread of 15% has been achieved, an 8% increase and a 7% decrease compared to the reference variant. Furthermore, deviations of the other mass properties from the reference are within a tolerance of 2%, as shown in figure 2, and the absolute change of the lateral position of the centre of gravity is below 5 mm. These values were determined in the BMW oscillatory test facility.

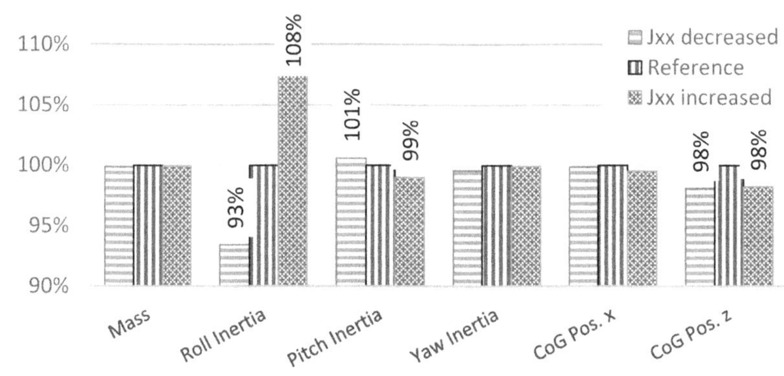


Figure 2: Measured mass properties of the implemented vehicle variants.

Figure 3 shows the roll inertia spread of several present-day SUVs and some of their predecessors. Each data point represents a vehicle from either BMW or a competitor that may also differ in engine and extras and has been measured at BMW Group. As can be seen in figure 3, the realized roll inertia spread of the three variants covers the reference group of SUVs, thus is relevant for the study, despite being on a slightly higher level, which resulted from the chassis modification.

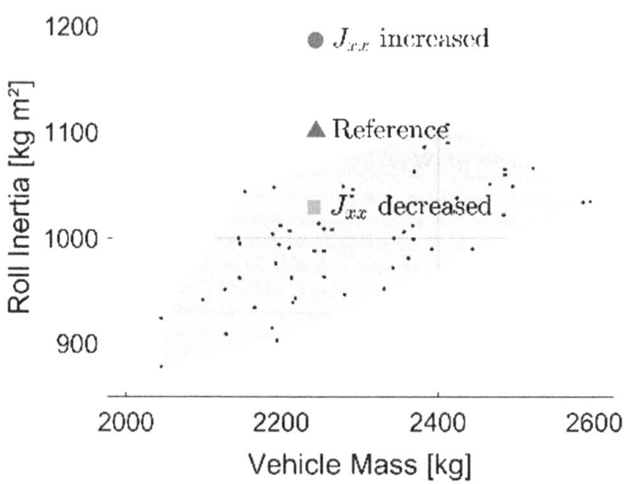


Figure 3: Roll inertia spread of different SUVs and the three roll inertia variants.

## 2.2 Modelling and Effect Prediction

The roll dynamics of a conventional suspension vehicle with springs, dampers, and anti-roll bars can be modelled as

$$J_{xx}\frac{d^2\varphi}{dt^2} + k_1\frac{d\varphi}{dt} + k_2\varphi = m(h_{CoG} - h_{RC})a_y \tag{3}$$

where $\varphi$ is the roll angle, $J_{xx}$ the roll inertia, $h_{CoG}$ the height of the center of gravity, $h_{RC}$ the height of the roll center, $m$ the vehicle mass, $a_y$ the lateral acceleration, $k_1$ a combination of damper parameters, and $k_2$ a combination of spring and anti-roll bar parameters. Comparing (3) to the differential equation of a second order linear system with eigenfrequency $\omega$, damping $\zeta$, gain $K$, input $u$ and output $y$

$$\frac{d^2y}{dt^2} + 2\zeta\omega\frac{dy}{dt} + \omega^2 y = K\omega^2 u \tag{4}$$

provides relationships between the vehicle parameters and the properties of the generic second order system (4):

$$K \propto m(h_{CoG} - h_{RC}) \tag{5}$$

$$\omega \propto \frac{1}{\sqrt{J_{xx}}} \tag{6}$$

$$\zeta \propto \frac{1}{\sqrt{J_{xx}}} \tag{7}$$

According to equation (5), the roll gain $K$ is not influenced by roll inertia $J_{xx}$ (cf. equation (5)). Roll eigenfrequency $\omega_0$ and roll damping $\zeta$ decrease when roll inertia $J_{xx}$ is increased. Hence, using relations (6) and (7), the effect of the increased roll inertia $J_{xx}$ can be estimated to be a decrease of about 3.8% of both roll eigenfrequency $\omega$ and roll damping $\zeta$. The effect of the decreased inertia $J_{xx}$ can be estimated to be an increase of about 3.7% of both $\omega$ and $\zeta$. Because system (3) behaves like a damped harmonic oscillator, the phase angle between roll angle and lateral acceleration should increase if roll damping is decreased.

## 2.3 Frequency Domain Analysis

To objectively assess the vehicle performance, the Bode plot and the average velocity of the vehicle in the ISO double lane-change are used. The Bode plot was determined from the continuous sine steering (CSST) manoeuvre at a velocity of $v_x = 100\ km/h$ and a lateral acceleration of $a_y = 4\ m/s^2$. Magnitude and phase plots of the transfer function

$$G(s) = \frac{\varphi(s)}{a_y(s)} \tag{8}$$

from measured data of the implemented vehicle variants are shown in figure 4.

As expected from the theoretical relations (5) to (7), roll gain $K$ remains unchanged when roll inertia $J_{xx}$ is modified and roll damping $\zeta$ behaves inversely proportional to roll inertia $J_{xx}$. Differences of eigenfrequency $\omega$ cannot be discerned in the measurement data, because measurement inaccuracies are greater than the actual change. So, overall the vehicle dynamics expected from theory and simulation match its measured dynamics.

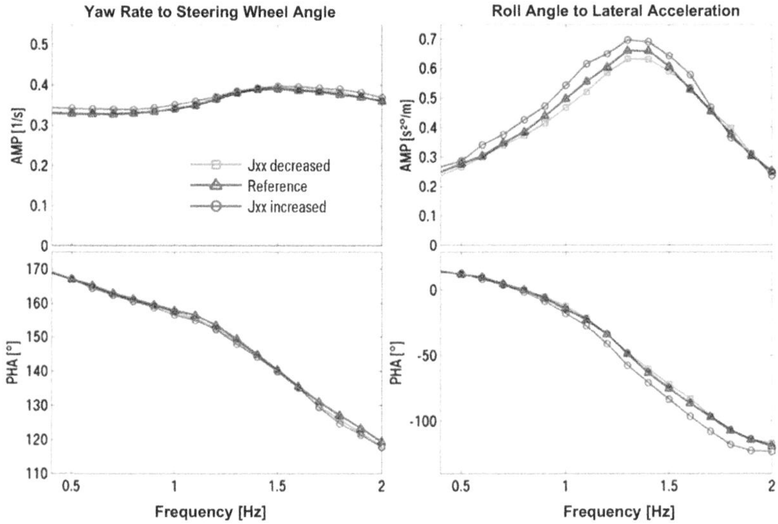


Figure 4: Bode plot of measured roll inertia variants.

## 2.4 ISO Double Lane-Change Results

Next, the quantitative effects of an isolated roll inertia change on driving were examined. The double lane-change manoeuvre according to ISO 3888-1:1999 was selected for this purpose, because it is a standardized test frequently used in the automotive industry and the automotive press, thus highly relevant and comparable. Resulting CV of the manoeuvre is the average vehicle speed required to complete the double lane-change, as defined in the above ISO norm.

The measurements were executed under identical environmental conditions by two BMW test drivers. For each variant several measurements were completed and the test results averaged. During the manoeuvre, the tire conditions were carefully monitored as their influence on the test results must not be neglected.

The final average speeds of the three variants determined in the ISO double-lane change are shown in figure 5. As can be seen from this graph, the driving performance of the vehicle does decrease when roll inertia is increased. The speed differences are at about the same level as the difference between a basic passive suspension setup and a suspension setup with active control systems (semi-active dampers, active anti-roll bars, but no rear wheel steering) or the difference between ordinary and performance tires.

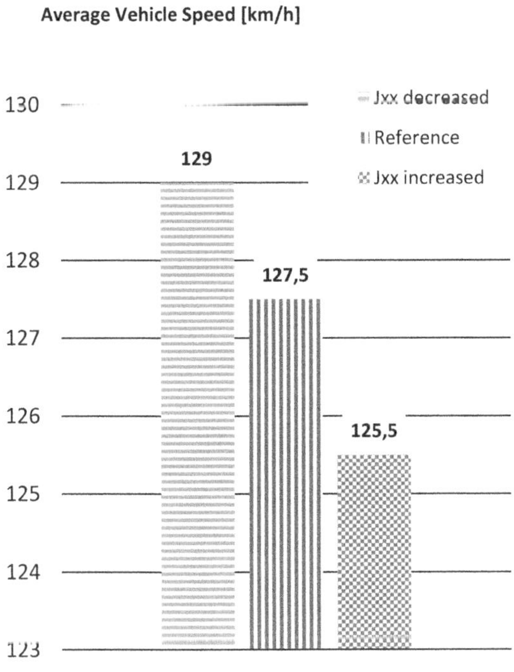


Figure 5: ISO double lane-change results of the three vehicle variants.

In terms of subjective driving impression, there is a notable increase in phase lag between steering input and roll dynamics and a reduction of steering accuracy with increased roll inertia. Additionally, the maximum roll inertia variant shows a clearly discernible two-phase character, that additionally reduces subjective rating of the variant.

Considering the clear effects of an increased roll inertia on handling characteristics, it is highly recommendable to at least monitor the roll inertia of a vehicle during its development process.

As mentioned above, the resulting impact on driving performance can be very close to the effect of the other two major levers in terms of driving dynamics: control systems and tire properties. Their benefits can easily be nullified if optimization potential regarding roll inertia is carelessly ignored or unused. This is especially important against the background of the competitive situation in the automotive industry.

# 3 Analysis of Roll Dynamics Using a Driving Simulator

Another tool used for driving behaviour assessments is the BMW Group dynamic driving simulator. It is a vehicle chassis mounted on a platform with six linear actors and can be moved in any of its six degrees of freedom, cf. figure 6. Aside from the general benefits of a driving simulator, it is employed for two purposes in the context of roll dynamics objectification:

- Preparation and preliminary assessment of vehicle variants for the examinations in the control system vehicle.
- Study of vehicle behaviour which cannot or only with significantly higher effort be implemented in the real vehicle.

The simulator maps the previously calculated driving dynamics to the driver's position in the simulator.

Figure 6: BMW Group dynamic driving simulator.

The vehicle dynamics are calculated using a simulation model of the control system vehicle presented in the final part of this paper. This way, study results obtained through different tools can be compared. The general approach used in this context will be presented on the example of the simulator study which has already been conducted: An open loop approach was chosen for the manoeuvre presentation, i.e. the simulator automatically follows predetermined vehicle trajectory. This guarantees the following extra advantages:

- High mapping quality of previously calculated driving dynamics.
- Less limitations of modelling depth of vehicle because no requirements to real-time capability.
- Full manoeuvre reproducibility regardless of the test subject.
- Test subject can focus on evaluation of vehicle dynamics instead on driving task.

Three aspects of roll dynamics were investigated:

- Influence of roll gain on vehicle assessment. Five variants were used to rate vehicle variants with conventional and inverted (i.e. vehicle body leans towards the inside of a turn) roll dynamics and a variant with levelled vehicle body.
- Phase angle between lateral acceleration and roll angle (deliberately ignoring the influence of steering on roll dynamics; five variants).
- Settling of the roll angle (three variants).

Implementing the inverted roll behaviour, the phase angle range, and the roll settling variants in a test vehicle would not have been possible due to hardware and software limitations, which is why they were assessed in the driving simulator. The test programme consisted of three assessment rounds, during which the roll dynamics aspects were varied independently. Despite that, each test subject was asked to rate the magnitude of all three criteria as well as their overall liking of the vehicle variant, but did not know that only one criterion had been modified. This way, interactions between the three aspects could be analysed. The magnitude of each criterion was rated to check if a test subject could differentiate the variants and if they did indeed evaluate the correct criterion. Furthermore, each variant was assessed at least twice to check the assessment quality of the test subject. Reference variants were used at the beginning, in the middle and at the end of each run to identify trends and habituation effects of the test subject. Additionally, the variants were permuted for different test subjects to avoid series effects.

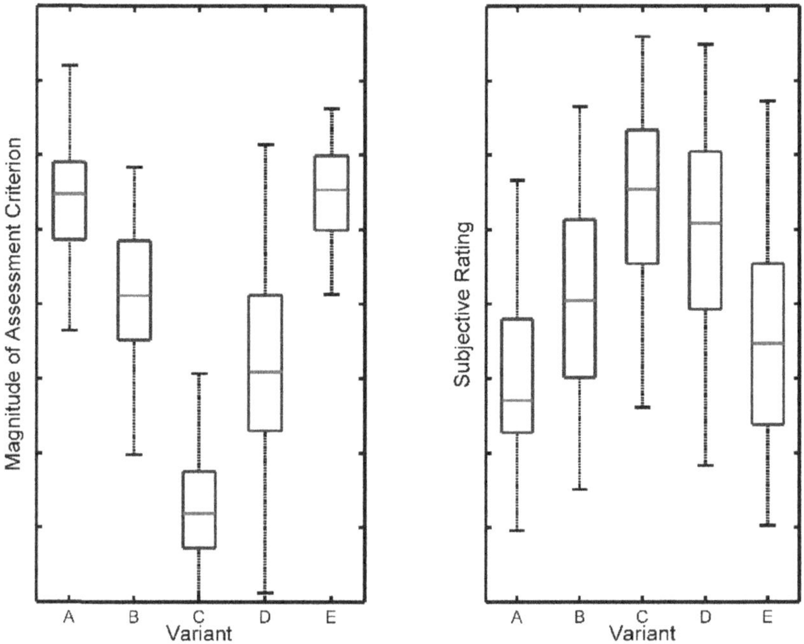


Figure 7: Results of the driving simulator study about roll gain level.

The variants were successively arranged for the assessment run. After each variant, the simulation was paused for a short time for an immediate rating by the test subject. Three consecutive lane changes with a frequency of approximately 0.3 Hz and a lateral acceleration of about $a_y = 2\,m/s^2$ to simulate a customer-relevant maneuver were presented for each variant. 41 test subjects, both normal and expert drivers, participated in the study.

Exemplarily, the results of the roll gain variation are depicted in figure 7. As the left side shows, the test subjects were able to significantly distinguish the five variants. The level of roll movement of all variants was identified correctly, although the inverted behaviour of the middle variant created a stronger perception of roll dynamics. The right side in figure 7 presents the subjective ratings of the variants. The identical roll angle amplitudes of both conventional and inverted behaviour are not significantly distinguished by the test subjects.

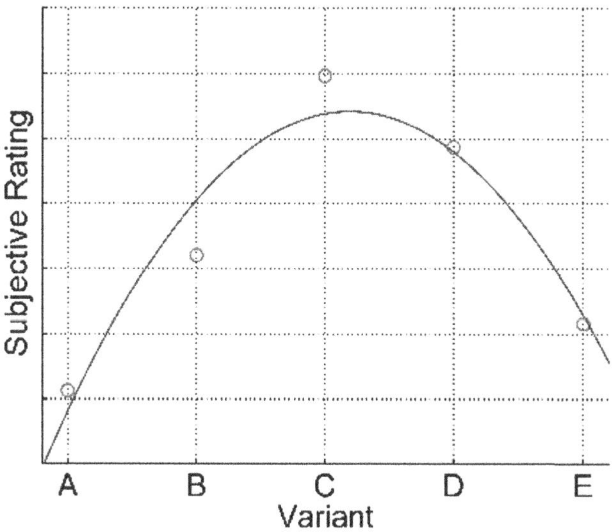


Figure 8: Correlation of maximum roll angle with subjective rating

However, only comparing the mean values of the assessments, the conventional vehicle behaviour is rated better than the inverted roll dynamics. The variant with the levelled vehicle body received the overall best rating although the difference to the conventional variant with little roll movement is small. Figure 8 finally shows the relationship between maximum roll angle amplitude and subjective rating. A quadratic correlation curve was used to include both the conventional and the inverted roll dynamics behaviour in the plot.

The following can be concluded from these results:

- Lately, there has been research toward the implementation of an inverted roll dynamics into series-production vehicles. At least for transient driving manoeuvres like a lane change drivers do not seem to like that behaviour. This might not be the case for quasi-steady state manoeuvres which have not been investigated in this context. Considering the high development efforts and low amount of elongated turns driven at high speeds, thus causing high roll angles and relevant in this context, the implementation of such a vehicle behaviour seems questionable.
- Drivers seem to best like small roll angles close to none. Rating improvements are small once the roll angle falls below a certain threshold. This is true for the inverted roll dynamics as well.

# 4 Analysis of Roll Dynamics Using Control Systems

In addition and complementary to the driving simulator examinations, the roll dynamics of vehicles are examined in a real vehicle. Compared to the driving simulator, a wider frequency range and higher vehicle dynamics can be studied and the driver experiences the closed-loop behaviour of the vehicle. Keeping its mass properties constant, its roll dynamics can be changed through a modification of its suspension, either by replacing hardware components in case of a conventional suspension or, if control systems are available, by modifying their parameters.

Main benefits of using control systems to create the vehicle variants compared to a purely mechanical suspension are:

- The effort to switch between variants (no removal and installation of hardware necessary) is significantly reduced.
- Test subjects can compare different variants without delay because switching between variants can be done instantaneously.

In addition, the results of a study using control system variations can be easily applied to other vehicles with the same systems because no hardware changes are required. By now, controlled anti-roll bars and dampers are fairly common, at least in luxury class cars. These systems can be used to manipulate the roll angle (anti-roll bar) and the roll rate (dampers) of the vehicle. Also, they are sufficient for the study of the following roll dynamic aspects:

- Magnitude and phase angle between lateral acceleration and roll angle,
- initial roll movement,
- roll gradient (which is equivalent to the steady-state roll angle) and
- roll damping.

To efficiently prepare and conduct a study about the above-listed aspects, a four-step approach is recommendable and is currently employed for the preparation of another study:

- Multivariate simulation using the control system parameters with the biggest influence on the roll dynamics.
- Script-based identification of variants which only vary in the aspect of investigation.
- Objective measurements of the selected variants to ensure that they can be sufficiently distinguished both in the measurement and by the test subject, and to verify the simulation results.
- Performing the study to rate the selected variants and correlate the results.

The first two steps are essential, because the resulting vehicle behaviour cannot always be predicted in advance due to nonlinear effects of control system parameter changes. Iteratively adapting parameters and measuring the corresponding variants would require too much effort in most cases.

Furthermore, the control system studies are used to verify the results of the roll inertia variation (which is equivalent to a roll damping variation) as well as the results of phase angle and initial roll movement variations examined in the driving simulator. This way, various tools are used to verify the study results and ideally, if results correspond, future tasks can be performed with the tool that generates the least effort.

Currently, the variants for the study are being prepared. An example of variants identified in simulation and verified through objective measurements is shown in figure 9 where two variants for the assessment of roll gain with almost identical phase angles, at least in the customer-relevant frequency range below 1 Hz, have been plotted.

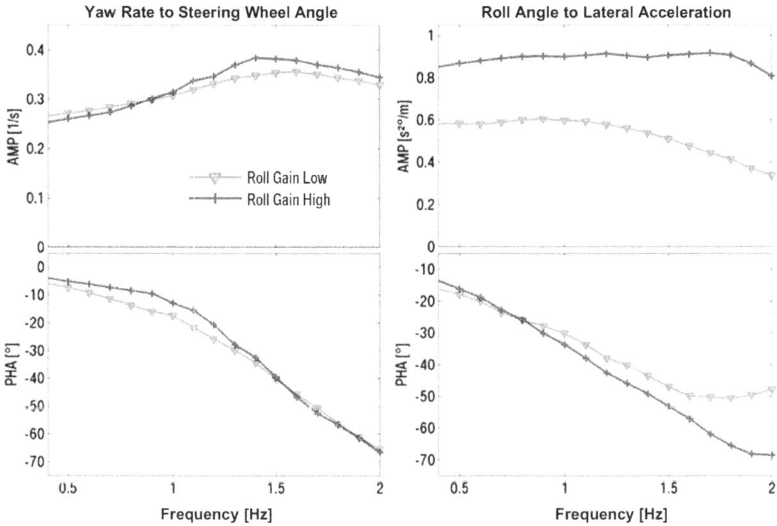


Figure 9: Bode plot of measured roll gain variants.

Figure 10 shows three measured variants for the assessment of roll damping with a clear distinction between the individual roll gain peaks. Once the vehicle assessments from this variation are available, they can be compared to the roll inertia study presented above, as the effect on the vehicle dynamics is the same in both cases.

The same overall approach in regards of control system parameter variation will be applied to the remaining roll dynamic aspects to generate a set of at least three variants for each aspect of examination to allow for conclusions about the drivers' preferences.

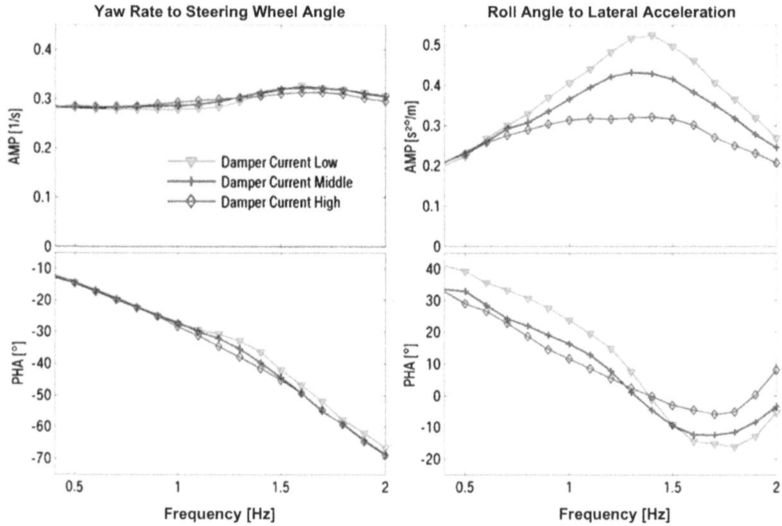


Figure 10: Bode plot of measured roll damping variants.

# 5 Summary

This paper presented three different approaches to create vehicle variants to determine driver ratings for the objective evaluation of a vehicle's roll dynamics.

The first section explained how to modify a vehicle so that only its roll inertia changes with the other vehicle properties remaining constant, which is equivalent to a roll damping variation. Then, the implementation results of three numerically identified vehicle variants were illustrated. Objective measurements verified the theoretically expected behaviour of these variants. Finally, the influence of roll inertia changes on the vehicle performance in the ISO double-lane change and on the subjective driving impression was presented.

The BMW Group dynamic driving simulator and a test vehicle with control systems are used for the examination of further roll dynamics aspects. The relevant vehicle variants are identified from a single simulation model whose vehicle dynamics largely coincides with the test vehicle dynamics. Depending on the assessment criterion the better suited tool can be chosen or, if necessary, both can be used to verify study results. The general approach was shown on the example of a simulator study which has already been conducted. Finally, the preparations of the coming examinations in the control system vehicle were briefly explained.

# References

[1] Riedel, Andreas; Arbinger, Roland: *Subjektive und objektive Beurteilung des Fahrverhaltens von Pkw.* FAT-Schriftenreihe Nr. 139, Bundesanstalt für Straßenwesen, Bergisch Gladbach, Forschungsvereinigung Automobiltechnik e.V.,1997.

[2] Henze, Roman: *Beurteilung von Fahrzeugen mit Hilfe eines Fahrermodells.* Schriftenreihe des Instituts für Fahrzeugtechnik, TU Braunschweig, Bd. 7. Aachen: Shaker, 2004 (also dissertation Technische Universität Braunschweig 2004).

[3] Botev, Stefan: *Digitale Gesamtfahrzeugabstimmung für Ride und Handling.* Fortschritt-Berichte VDI, Reihe 12, Nr. 684. VDI Verlag GmbH, 2008 (also dissertation Technische Universität Berlin, 2008).

[4] Kraft, Christian: *Gezielte Variation und Analyse des Fahrverhaltens von Kraftfahrzeugen mittels elektrischer Linearaktuatoren.* Karlsruher Schriftenreihe Fahrzeugtechnik, Bd. 5. Karlsruhe: KIT Scientific Publishing, 2011 (also dissertation Karlsruher Institut für Technologie 2010).

[5] Stock, Gregor: *Handlingpotentialbewertung aktiver Fahrwerkregelsysteme.* Schriftenreihe des Instituts für Fahrzeugtechnik, TU Braunschweig, Bd. 25. Aachen: Shaker, 2011 (also dissertation Technische Universität Braunschweig 2010).

[6] Hofstetter, Michael: *Untersuchung der Auswirkung simulativer und physikalischer Variationen der Masseeigenschaften eines Personenkraftwagens auf seine Wankdynamik.* Salzburg, Paris-Lodron-Universität Salzburg, bachelor thesis, 2014.

# Improved prediction of ride comfort characteristics by considering suspension friction in the automotive development process

Dipl.-Ing. Christian Angrick, Prof. Dr.-Ing. Günther Prokop

Institut für Automobiltechnik Dresden, Lehrstuhl für Kraftfahrzeugtechnik, TU Dresden

Dr.-Ing. Peter Knauer, Dr.-Ing. Andreas Wagner

Audi AG

# Abstract

Ride comfort of passenger vehicles is affected by several subsystems and their properties. For reaching specific objective ride comfort targets the suspension is able to influence the response of the body in a wide range. However, the characteristics of the subsystem have not been fully investigated. In the following research the static and dynamic characterization of a suspension is conducted. For this purpose, especially friction properties are in the focus, not being established in conventional multi body models for simulation of ride comfort. A methodology for determining the dependencies of friction of multiple parameters is depicted and corresponding results are shown. Additionally the dynamic response under different conditions is analysed. The results are evaluated and procedures for considering and obtaining required properties in the development process are given. On the basis of this study the definition and parameterization of simulation models in development of vehicle properties can be improved, providing new potentials for reaching specific ride comfort targets.


# Introduction

In future development, the importance of a vehicles ride comfort will increase in comparison to driving dynamics. On the one hand, this is due to a lower level of knowledge, on the other hand due to the desire of having a complete systematic design process. However, for reaching the aim of a valid prediction of the successor vehicles properties, the simulation of the vehicles vibrational behaviour needs to be accurate enough in each phase of the design process. In conventional applied conceptual or component models, the properties of the suspension are mostly represented by springs, dampers and levers, with latter defining the kinematic properties. In this context, friction or static hysteresis has most commonly be investigated on component level, by analysing dampers (Dixon 2007; Duym 2007; Lizarraga et al. 2008), bushings (Berg 1998; Sedlaczek et al. 2011) and ball bearings (Dahl 1968; Dahl 1977). Investigations on subsystem level have been carried out for specific tasks (Nakahara et al. 2001; Yabuta et al. 1981). Still, the properties and effects due to suspension friction have not been fully understood, yet.

In the following work, the full process starting with a method for determining friction properties of suspensions, over characterizing the static and dynamic responses, up to modelling and estimating the effects of friction on full vehicle level are presented. By doing so, the connections between subsystem and full vehicle as well as component level are established, which forms the basis for considering friction properties within the development process.

# Full vehicle targets and chassis development process

The automotive development process is commonly represented by a v-model, distinguishing between full vehicle, subsystem and component level (cf. Figure 1). The process can be divided into the regions of target cascading (left branch of the model) and verification (right branch of the model). In former case, the vehicle properties are derived on the different levels of detail, starting at full vehicle level and resulting in component parameters. Afterwards the results are verified in the series development, in which the components are assembled on subsystem and full vehicle level. The results are compared with the specifications determined during the cascading process. In the following, the derivation of parameters will be in focus of the investigation.

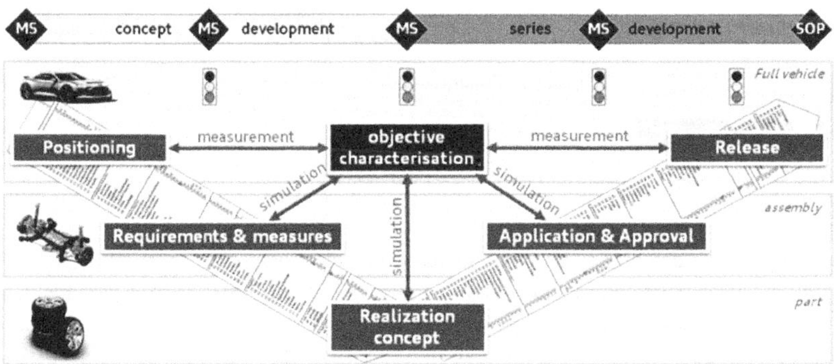


Figure 1: Development of vehicle attributes based on the v-model (Einsle and Fritzsche 2013)

Beginning on full vehicle level, objective ride comfort targets need to be defined, which serve as a reference for deriving subsystem parameters. In the current work, the vehicles ride comfort shall specifically not be summarized in one value, but instead be individually defined by characteristic values for different vibrational ranges. For instance, these are determined by analysing the frequency dependent transfer behaviour of the body, the step response, when a cleat is passed or the response on stochastic roads. By doing so, different characteristic values, like the natural frequency or the peak height of the transfer function of the vehicle body are obtained. By conducting a benchmark and identifying these objective values for the own vehicle and competitors, development targets are defined for the successor (cf. Figure 2). In turn, these serve as a reference for deriving vehicle properties.

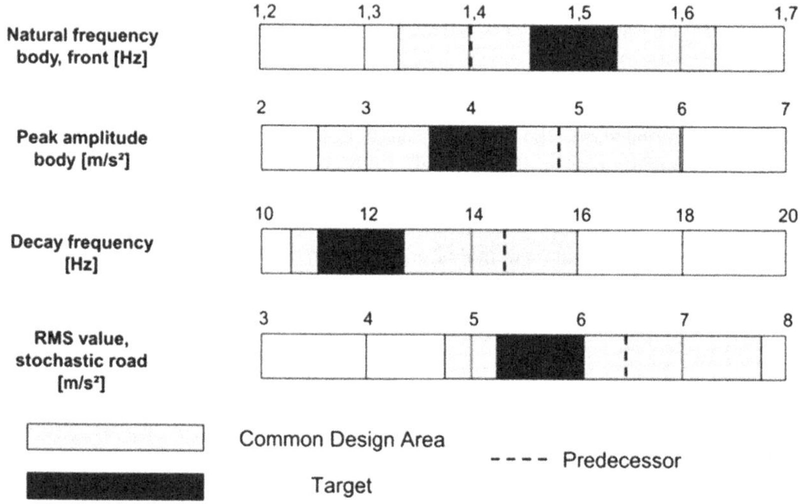


Figure 2: Definition of objective targets in ride comfort

By applying the depicted process, a change in the characteristic values allows for the definition of changes in associated subsystems. In turn, the change of a parameter in the subsystem results in changes in corresponding components. Although this process can be defined for different subsystems, like tire, suspension or engine, especially the suspension will be in focus of the following investigation.

Prospectively, for meeting new upcoming requirements, like a higher model variety, higher functional demands or reduced development times (Rauh 2003) as well as corresponding strategies, like the usage of standardized modules, an efficient realisation of the design process becomes necessary. Therefore frontloading (Hab and Wagner 2013) is desired, which allows for analyses in early phases of the development process. For enabling this, the application of simulation models increases, when prototypes are still not available.

As the availability of vehicle parameters changes with development time, also simulations models with different complexity are necessary. In early phases, when only full vehicle targets as well as singular concept parameters, like wheelbase or trackwidth, are known, simulation models are necessary, which allow for a simulation of the full vehicle based on subsystem parameters. This becomes necessary, as for instance the axle concept is not determined, so that component parameters are still unknown. In driving dynamics this is realized by applying parametric single- or dual track models (Heißing et al. 2011 and Schimmel 2010) based on characteristic curves. Thereby changes in subsystem parameters can be derived on basis of full vehicle targets. An example would be lateral or cornering stiffness of the suspension, without knowing the stiffness of individual bushings, as the concept is still not defined. Based on this, in subsequent phases of the process, component parameters can be derived from subsystem parameters by using conventional multi-body approaches (Heißing 2011). In this context, the effectiveness of this whole process depends on the quality of the applied simulation models.

In ride comfort however, the derivation process is still not fully defined, as there exists no coherent model approach for deriving subsystem parameters from objective full vehicle targets. For enabling this, the different subsystems also have to be defined independently of the given axle concept, which will be addressed for the suspension throughout the remaining research. However, in this case also the quality of the simulation models is of main importance, when an accurate prognosis of a successor has to be reached. Therefore an understanding of the static and dynamic behaviour of the suspension becomes necessary.

The conventional method for representing these properties is either to use generalized stiffness and damping elements, expressing the dynamic behaviour of the axle in vertical direction or to model each component with individual parameters and considering kinematic hard points of the given levers. In this context, friction or static hysteresis of the suspension as well as corresponding components has not yet been fully investigated. Therefore, especially the characteristics of the static suspension hysteresis as well as its dependencies lies in the focus of the following analysis. By doing so, the significance of this property on full vehicle level, the definition of corresponding subsystem parameters as well as its composition from different components is enabled, allowing for an integration within the chassis development process.

In the next chapter the method for identifying corresponding characteristic values on subsystem level with an appropriate axle test rig is depicted.

# Static and dynamic characterization of a multi-link suspension

For obtaining static and dynamic properties of the given suspension, the suspension test rig of the INSTITUTE OF AUTOMOTIVE TECHNOLOGY at TU DRESDEN is used. General properties of the test rig will be described on the basis of Figure 3.

Figure 3: Axle test rig for static and dynamic characterisation of suspensions, adapted from (Bindauf et al. 2014)

The first interface of the test rig is given between wheel hub and wheel, while on the opposite side the interface is given between the body-sided mountings and body. In former case, the wheel is replaced by an equivalent system, which possesses the same mass and inertia properties and connects to coupling rods, which apply an excitation to the wheel hub. On body side, the respective mounts are connected to a stationary testing frame, representing a rigid vehicle body.

The wheel replacement system can be excited in longitudinal and vertical direction at the wheel centre. Thereby arbitrary displacements are specified independently for the left and right side, which serve as input signals throughout the current investigation.

Additionally, displacements can be measured between the upper and lower connections of the damper. Longitudinal and vertical forces are obtained at the wheel side with appropriate load cells within the coupling rods, but are also recorded on the body side with load cells at the different mountings. By specifying the transfer behaviour between displacement and force at the wheel, it becomes possible to characterise the suspension on subsystem level, independent of the currently given concept. On the other hand, when considering the body sided load cells, a transfer path analysis is enabled, allowing for a breakdown of the overall force on component level. The orientation of the described measured values are consistent with the vehicle coordinate system, defined in DIN 70000. A deflection of zero millimetres corresponds to the curb load of the vehicle. (Bindauf et al. 2014)

In case of the current research, a multi-link suspension is mounted on the test rig for determining characteristic properties. At the lower side of the suspension, the connection between wheel hub and body is realized by two levers with ball bearings at the wheel side and bushings on the body side. At the front lever also a spring strut is connected between lever and body. On the upper side of the wheel hub, two additional levers, with an equivalent arrangement of ball bearings and bushings, are establishing a connection to the body. Furthermore a bushing is located between anti-roll bar and body.

## Static characterisation of the suspension

Characteristic static properties are obtained by conducting ramps of displacement at the wheel side. Thereby a constant velocity of one millimetre per second is maintained in a branch of the measurement, so that effects due to viscous damping or inertia forces of the wheel replacement system can be neglected. By preconditioning the axle in both directions, a maximum reproducibility of the results is reached (Mullins 1948; Mullins 1950).

For dynamic measurements a sinus sweep excitation with continuously increasing frequency is applied, for efficiently obtaining frequency-dependent characteristics of the suspension.

Both excitation types are conducted with different amplitudes and variants. The variants are given as follows:

- Standard Version
- Removal of the anti-roll bar
- Removal of the bump stop
- Removal of the spring strut – due to the axle concept this results also in a removal of the bumpstop as well as the anti-roll bar
- Replacement of the damper by a version without damping fluid

By doing so, the influences of different operating conditions, as well as singular components on the overall hysteresis can be identified.

In Figure 4 the wheel-based vertical force against deflection is depicted for two amplitudes. When examining the characteristic curves, the individual branches can be divided into two regions. First of all, when a reversal point of the displacement is passed, a transition area becomes obvious, in which the opposing force due to static hysteresis is generated. The corresponding force will be called friction force. Afterwards in the second region, this counterforce remains relatively constant until the next turning point is reached.

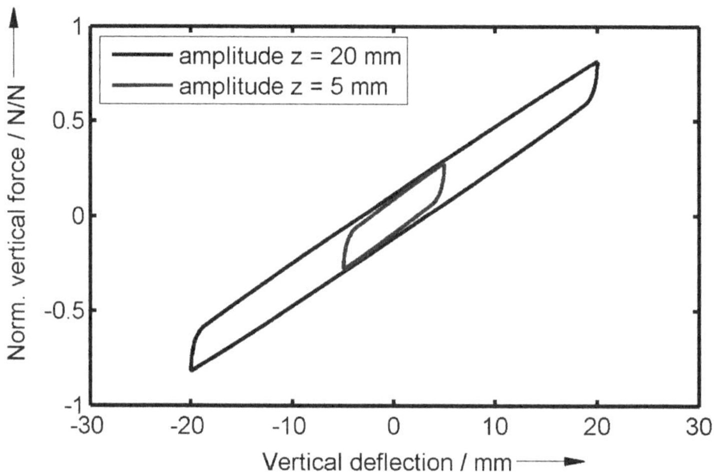


Figure 4: Characteristic static force-displacement-hysteresis at amplitudes of five and twenty millimetre

By averaging the force values of compression and rebound, the stiffness respectively elastic properties of the suspension can be determined, allowing a separate examination of stiffness and static hysteresis. It should be noted, that this is only valid when the deflection has passed the transition area. In case of the stiffness, characteristic values on subsystem level are found in the suspension travel, the constant stiffness in the linear region, the point at which the bump progression begins and the characteristics of the progression. For the static hysteresis, characteristic values on subsystem level are found in the characteristics of the transition area and the width of the static hysteresis at zero deflection.

In Figure 4 an increase of the width of the static hysteresis with higher deflection becomes obvious. Therefore the amplitude dependence of the static hysteresis is investigated. The tendency of the absolute amount as well as the relative amount (related to the overall force) against amplitude is depicted in Figure 5. The amount of friction force corresponds to the half of the hysteresis width at zero deflection.

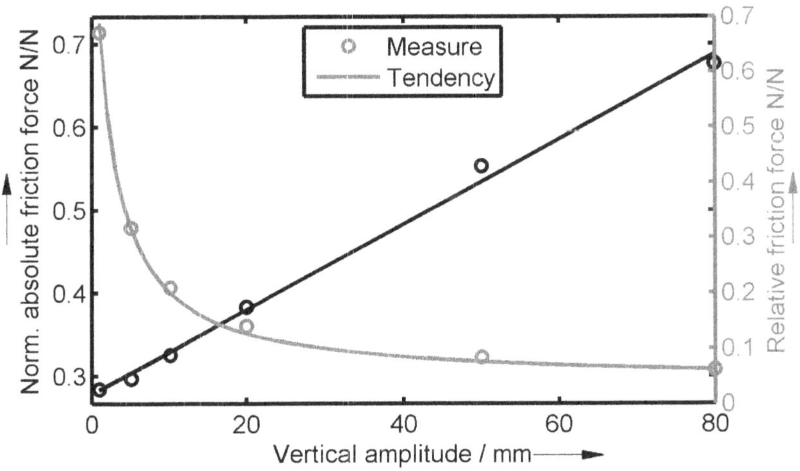


Figure 5: Absolute and relative amount of static hysteresis with respect to vertical wheel deflection

In Figure 5 the amount of the constant friction force has been identified for amplitudes, which are higher than the length of the transition area. Latter has been found to be nearly constant over excitation amplitude with a value of about one millimetre. Additionally to the discrete values per amplitude, a linear relationship has been identified between absolute static hysteresis and amplitude, which is valid under the given conditions. An appropriate relationship is given in equation 1, where $m_h$ and $n_h$ are the increase and offset of a linear fitting of the absolute friction force $F_h$ against amplitude.

$$F_h = m_h * z + n_h \tag{1}$$

When examining the relative friction force against amplitude, an inversely proportional relationship becomes obvious, as depicted in Figure 5. Therefore also the significance of static suspension hysteresis decreases with increasing amplitude. The depicted tendency is based on the linear relationship of the absolute hysteretic force presented in equation 1, with respect to the overall force, which additionally considers spring forces. The inversely proportional behaviour is summarised in equation 2, where $F_h/F_{full}$ corresponds to the relative friction force, $z$ to the current amplitude, while $c$ is equivalent to a linearized stiffness of the axle.

$$\frac{F_h}{F_{full}} = \frac{m_h * z + n_h}{(m_h + c) * z + n_h} \tag{2}$$

In this case, a characteristic value for the amplitude dependency can be found, when the upper relationship is examined under the condition, that the amplitude z tends towards infinity. In this case the quotient becomes $m_h/(m_h + c)$, which corresponds to a constant value describing the tendency of the suspension to change its friction force over amplitude with respect to the elastic properties of the axle.

After characterising static hysteresis on subsystem level, it is of fundamental interest, how the different components contribute to the subsystem properties of the wheel-based static hysteresis. This forms the basis for the derivation of subsystem to component parameters.

For this purpose, the change in static hysteresis due to the modification of different parts has been analysed. The resulting static hysteresis is depicted in Figure 6.

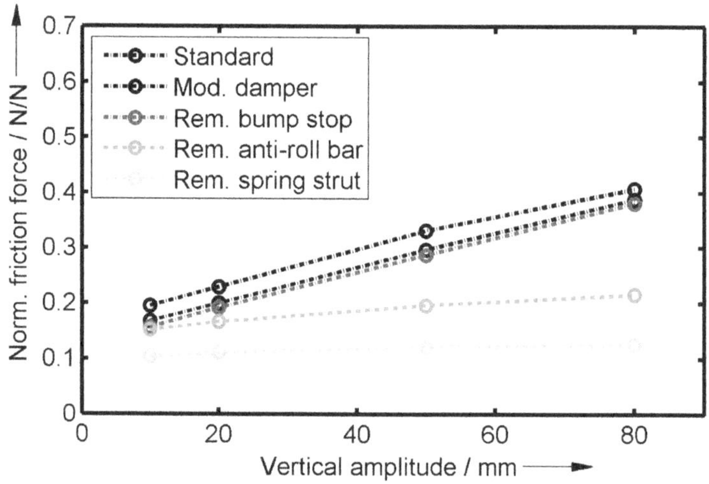


Figure 6: Contribution of different modifications on the wheel-based static hysteresis

As can be seen in the Figure, the use of a damper without damping fluid or the removal of the bump stop slightly influence the static hysteresis. However, when the static hysteresis is analysed at higher deflection, at which the bump stop is deformed, a higher static hysteresis is observable, which result of the individual properties of the component. When analysing the removal of the anti-roll bar, the hysteresis width at ten millimetre is slightly affected. Still, the differences increase with higher deflection. As the resulting amount of static hysteresis is nearly constant over amplitude, it can be stated, that the amplitude dependency of the suspension probably results of the anti-roll bar bushings. It has been found, that the respective rotational static hysteresis of the bushing, transferred to a vertical force on the wheel centre is equivalent to the upper difference, which confirms this assumption.

Finally, when the spring strut removed (which results in an additional removal of bump stop and anti-roll bar), a nearly constant decrease over amplitude is observable. It has to be taken into the account, that in this case also the static preload of the axle is decreased. Still, a constant amount of friction force remains, which results of the contribution of the individual ball bearings and rubber bushings, connecting the levers to wheel hub and body. Therefore it can be stated, that the damper is not the dominant element in the current multi-link suspension.

Based on the upper analysis, the contribution of the different transfer paths on the overall static hysteresis can be analysed. The contribution over amplitude is depicted in Figure 7.

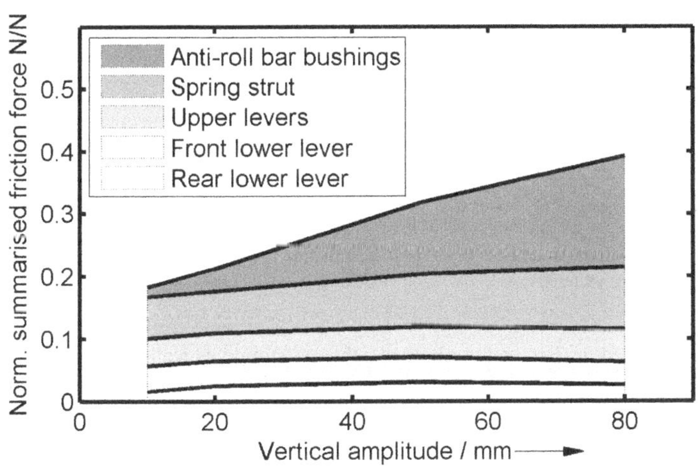


Figure 7: Contribution of individual components on the wheel based static hysteresis

In Figure 7 it is observable, that the three transfer paths, determined by the different levers with ball bearings and bushings, as well as the spring strut are contributing a nearly constant amount of friction force to the overall hysteresis. The amplitude dependent static hysteresis, which results of the anti-roll bar bushings is also observable in this Figure as well.

## Dynamic characterisation of the suspension

After analysing the response due to static excitation, the dynamic properties of the suspension are examined. In this context especially the frequency and amplitude dependency of the suspension is of main importance. For this purpose, the wheel centre is excited in vertical direction by a sinus sweep with continuously rising frequency. The rate of frequency increase is maintained low enough to reach steady state conditions in the resulting forces. Concerning the obtained measurement data, the magnitude and loss angle of the transfer function between output force and input deflection is analysed. These are determined by applying a short time fourier transform (Welch 1967). In this case, the nonlinear properties of damper and topmount concerning compression and rebound are linearized.

In Figure 8 the frequency response is depicted for amplitudes of five and ten millimetre.

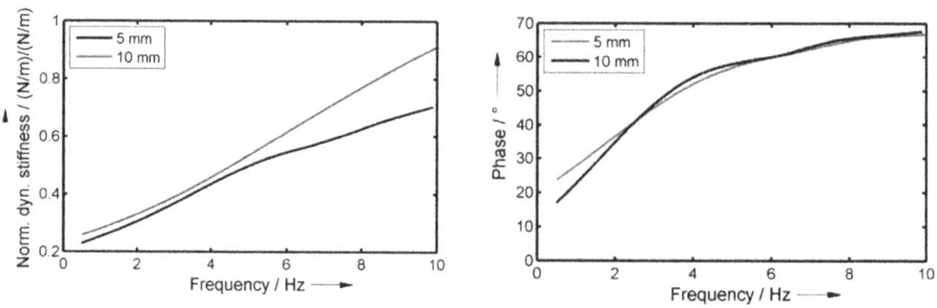


Figure 8: Dynamic stiffness and phase of the transfer function in vertical direction of the suspension (a = 5 mm; 10 mm)

As already described in (Bindauf et al. 2014), the magnitude as well as the loss angle are rising with increasing frequency. At higher frequencies and ten millimetre amplitude, a lower increase as well as a saturation of the loss angle are observable, which results of the deformation of the topmount and nonlinearities of the damper. (Bindauf et al. 2014)

When analysing the magnitude at the lowest frequency, the relations determined for static excitation can also be found in the dynamic response. The magnitude of the stiffness increases with declining amplitude, which corresponds to a higher relative contribution of the static hysteresis with declining deflection, cf. Figure 5. On the left side of Figure 8 it is observable, that this effect is still existent with rising frequency.

Depending on the amplitude, also a small difference in the loss angle is observable at the lowest frequency. This can also be attributed to the static hysteresis, which already results in an offset of the loss angle at static deflection (Sedlaczek 2011). With rising amplitude, the relative amount of friction is decreasing, which corresponds to a decline in the offset of the loss angle.

For estimating the influence of the determined characteristic subsystem parameters on the full vehicle behaviour, the suspension friction has to be integrated in an appropriate modelling approach. This will be described in the following chapter.

# Modelling of suspension friction on subsystem level and implementation in a full vehicle model

When modelling the static hysteresis, the relation between force and displacement at wheel centre is used as reference. Thereby the independency of the currently given multi-link concept is maintained. A modelling of the individual static hysteresis of the individual components, according to Figure 7, is also possible on component level, but will not be in focus of the following approach.

For modelling the overall static hysteresis, different modelling approaches can be applied. Parallel to the chosen friction element, an amplitude-dependent nonlinear spring is added for reproducing the elastic properties of the suspension, determined with respect to Figure 4. The quality criteria for evaluating measurement against simulation are found in the shape of the transition area, the reproduction of the subsequent constant friction force and the amplitude dependency.

In context of the given research, appropriate black-box models are given by Berg, Nakahara or Dahl (Berg 1998; Nakahara et al. 2001; Dahl 1977). In former case, the hysteresis shape is modelled with a mathematical shape function, applying an algebraic relationship between force and displacement, which relies on the state of the system in the last turning point. The model by Nakahara is based on multiple parallel Prandtl-elements, which in combination reproduce a smooth transition between force and displacement. The Dahl model however, which is originally intended for modelling the solid friction of ball bearings, consists of a differential equation, without utilizing switching functions.

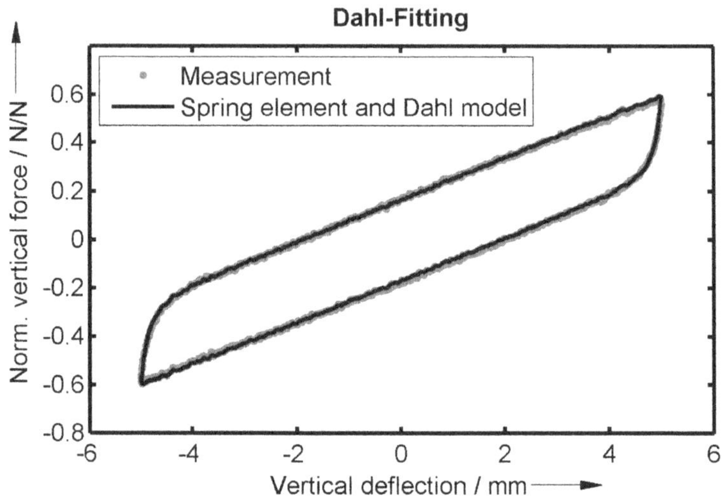


Figure 9: Fitting of the Dahl model on the static force-displacement hysteresis at an amplitude of five millimetre

In the present case, it has been found, that all of the given modelling approaches are able to accurately reproduce the force-displacement-hysteresis at the wheel centre. In Figure 9, this has been depicted on the example of the Dahl model at an amplitude of five millimetre.

Due to its continuous formulation and therefore highest numerical stability as well as simulation speed, the Dahl friction is applied in the following investigation. Additionally Coulomb friction, according to equation 3, is used to model the force-displacement-hysteresis without transition area. This allows to distinguish between effects due to different friction properties.

$$F_{fric} = F_h * sign(v) \tag{3}$$

## Consideration of the amplitude dependency

The described modelling approaches are able to reproduce an amplitude dependent behaviour, as long as the deflection is smaller than the width of the transition area. However, as already described with respect to Figure 4, the hysteresis shape consists of a transition area of approximately one millimetre, so that with higher deflection longer regions with a constant width are generated. The absolute width of the hysteresis area still increases with rising deflection, cf. Figure 5. For being able to reproduce this behaviour, the parameters of the friction models need to change with amplitude as well. In case of the Dahl friction, this becomes necessary for three parameters.

The described dependency is realised by adjusting the parameters at occurring turning points, based on the current deflection. For preventing a sharp transition between different states of the parameters, the transition between the current value $y$ and the target value $y_t$ is conducted using a PT1-element, according to equation 4.

$$T_1\dot{y} + y = y_t \tag{4}$$

Overall it can be stated, that by adding the proposed friction models to a conventional modelling approach, the simulation of the static axle behaviour can be improved.

## Modelling of the dynamic transfer behaviour

After analysing the influence of static hysteresis under static conditions, the effects on the dynamic transfer behaviour of the suspension is analysed. For this purpose, the proposed model, consisting of a nonlinear spring and a friction element is extended by a nonlinear damping element. Latter is based on the characteristic curve of the damper with consideration of the damper ratio between damper and wheel. A comparison between measurement and simulation for an amplitude of ten millimetre is given in Figure 10.

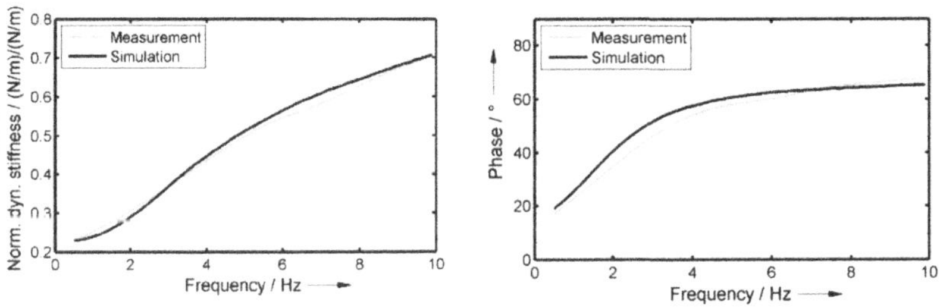


Figure 10: Comparison of the measured and simulated dynamic response of the suspension (a = 10 mm)

As can be seen in the Figure, the simulation with implementation of the proposed friction model is able to accurately reproduce the transfer function determined in dynamic measurement.

Based on the model, furthermore the contribution of different elements on the overall wheel based force can be analysed. This relationship is exemplified in Figure 11 for an amplitude of five millimetre.

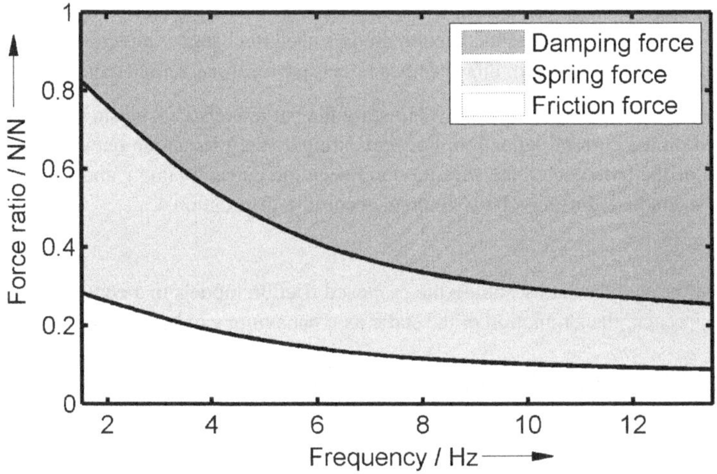


Figure 11: Contribution of individual model elements on overall force (a = 5mm)

It is observable, that at low frequencies the spring force has the highest share on the overall force. The ratio due to damping force and static hysteresis are in the same order of magnitude. At zero frequency the damping force would be zero. With rising frequency, the damping force increases and dominates the force generation while the ratio of spring force and static hysteresis are both decreasing. With decreasing amplitude, the share of friction is rising, compared to both other elements.

The presented verification of the simulation models on subsystem level provides the basis for analysing the influences of suspension friction on full vehicle level in a corresponding model. By doing so, the relationship between full vehicle targets and friction parameters on subsystem level can be identified. This will be addressed in the following chapter.

# Impact of friction parameters on objective ride comfort targets

When analysing the influence of friction properties on full vehicle level, the different results from simulations without friction, Coulomb friction and Dahl friction will be compared. Thereby, it can be analysed how different levels of detail in the parameter affect the full vehicle behaviour, which allows the choice of an adequate approach for a specific task.

As already depicted at the end of the last chapter, the implementation of suspension friction results in a generally higher simulation quality under static and dynamic excitation on subsystem level. Therefore, also the prognosis of full vehicle targets is generally improved.

## Validation on a four-post test rig

A consequence of this relationship can be analysed on the example of a four-post measurement. In this case, the powertrain has been fixed on the body and the vertical excitation is directly applied to the centre of the wheel hub. Therefore, the front axle of the system can be examined without considering the remaining subsystems. For simulating the full vehicle behaviour, a subsystem concept model is used, which considers generalised properties of the suspension, like stiffness or friction, as described in the previous chapters.

In Figure 12 the transfer function between the front dome and wheel centre is depicted for measurement and simulation, when the wheel centre is excited with an amplitude of one millimetre. This amplitude has been chosen, as the effects due to friction are more distinct than at higher amplitudes.

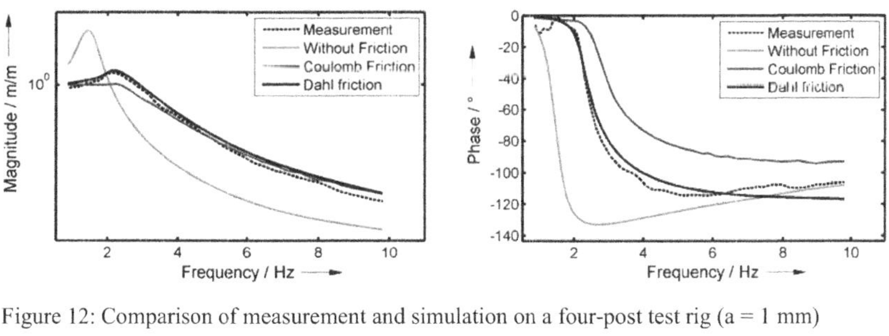


Figure 12: Comparison of measurement and simulation on a four-post test rig (a = 1 mm)

In the depicted example, there are relatively high differences observable, when comparing measurement and the simulation without friction. In this case, the natural frequency of the body is occurring at lower frequencies, while in case of the measurement, the maximum shifts towards the "copying" area and is lower compared to the simulation.

It has to be noted, that the result obtained by the simulation without friction is nearly similar to an equivalent complex component model, based on multi body simulation. Thereby, it can be derived, that the same difference is occurring when neglecting friction in more complex models.

Examining the simulation result with coulomb friction, clearly a more accurate reproduction of the measured tendency is achieved. When analysing Dahl friction, the simulation quality is even higher. This is valid for both amplitude as well as phase between both signals. Therefore also the parameters, which have been identified for reproducing the transition area, are of importance when a maximum accuracy of the simulation has to be reached.

Due to the nonlinear behaviour of the friction, which results in a force response with multiple frequencies, instead of the individual excitation frequency (Khalil 2002), not all relationships can be presented in the frequency response plot, depicted in Figure 12. Therefore a part of the output acceleration in time domain is depicted in Figure 13, whereby the time signals, containing all occurring frequencies, can be compared.

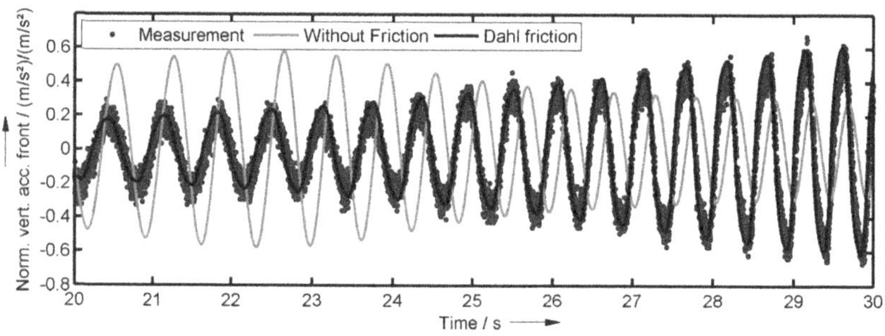


Figure 13: Comparison between measurement and simulation in time domain

As can be seen in the Figure, also in time domain the simulation considering friction elements more resembles the measurement. While the response without friction already shows a natural frequency at lower frequencies, the friction maintains a small relative movement between the masses. With rising frequency, both measurement and simulation with friction are showing a similar increase.

## Analysis of influences on objective ride comfort targets

As already stated above, the conducted investigation allows for a higher simulation quality on full vehicle level, which is associated with an improved prediction of ride comfort targets. For being able to derive these characteristics to subsystem level with additional consideration of friction, the relationship between full vehicle behaviour and friction parameters have to be defined. On the basis of the full vehicle targets, defined in the second chapter, the influence of friction on full vehicle properties will be exemplified for the frequency dependent behaviour of the body as well as a cleat response.

In former case, the transfer function between body acceleration and road displacement, as depicted in Figure 14, is analysed. For simulation, in this case also a full vehicle subsystem model is used, consisting of wheel and body masses as well as generalised suspension properties, as already described in the last section. Wheel and road are connected with an individual tire spring, neglecting the tire damping. A sinus sweep with a constant velocity amplitude of 30 millimetre per second is used as excitation. Thereby, the main influences shall be illustrated.

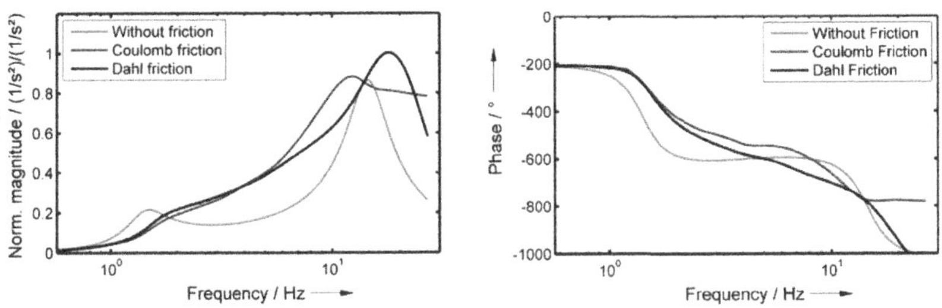


Figure 14: Influence of friction parameters on frequency dependent body vibrational behaviour (v = 30mm/s)

In Figure 14 similar influences, as already determined with respect to Figure 12, can be identified. The natural frequency rises, when friction is considered. This can mainly be attributed to two effects. As the friction mechanism can be considered as a nonlinear damping constant (Nakahara et al. 2001), a damping force with decreasing damping coefficient over frequency, is added to the system. Thereby lower frequencies are more affected than higher frequencies, shifting the natural frequency towards higher frequencies as well. Secondary the additional element between body and wheel causes a higher coupling, so that these bodies tend to vibrate more on the tire spring as a rigid combination (Yabuta et al. 1981). The thereby resulting new natural frequency is determined by the sum of body and wheel mass in combination with the vertical tire stiffness, which is higher than the uncoupled body natural frequency and lower than the uncoupled wheel natural frequency.

The magnitude in the copying area is increased with higher damping, so that the peak occurring due to the natural frequency of the body cannot be clearly separated from this area in the current transfer function.

Examining the wheel natural frequency, two different effects are observed, when Coulomb and Dahl friction are examined. As the Coulomb friction is predominantly acting as the nonlinear damping described above, the wheel natural frequency is shifted towards the already described natural frequency determined by the car mass and tire vertical stiffness. In case of the Dahl friction however, the imaginary damping value is partially converted to a real stiffness value, due to the transition area. Thereby this element acts as an additional spring-damper-system, which increases the wheel natural frequency. The described tendencies are also observable when examining the phase.

After analysing the frequency dependent behaviour of the body due to harmonic excitation, subsequently the response due to a cleat excitation is examined. Concerning ride comfort targets, especially peak-2-peak value, vibration frequency and decaying behaviour are of main interest. The system response for a step height of 3 mm is depicted in Figure 14.

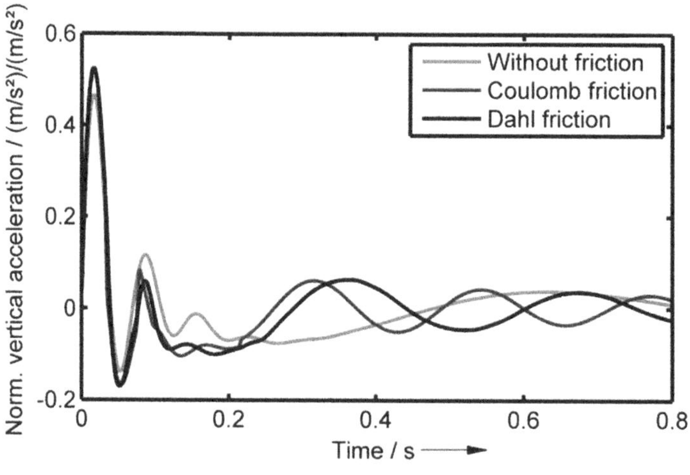


Figure 15: Influence of friction parameters on step response of the body

When looking at Figure 15, it is observable, that consideration of friction results in a higher Peak-2-Peak value in the first cycle of the vibration, associated with the wheel natural frequency. Due to the combination of several occurring frequencies, a distinct vibration frequency can only be identified in the subsequent area. In this case, predominantly the body vibrates in its natural frequency, showing a behaviour associated with the previously identified relationships, according to Figure 13. Thereby with higher friction, the vibration frequency is increased, compared to a system without friction. Also it has to be noted, that with increasing time and decreasing vibration amplitude, the relative amount of friction increases, so that the coupling between body and wheel as well as the vibration frequency increase. Therefore due to the friction, the body vibration frequency is increased during the decay process.

# Implementation within the development process

In the previous chapters, the characteristics and dependencies of suspension friction on subsystem level as well as the impact on full vehicle ride comfort have been determined. Based on this, in the following chapter the consideration of the determined friction properties within the development process is illustrated.

Basically, two main aspects are of importance in this case. First of all, it is necessary to define suspension friction on different levels of detail of the v-model, but also the acquisition of corresponding parameters needs to be specified. In this context, especially the subsystem level will be in focus.

## Derivation of properties onto subsystem level

As already depicted in the second chapter, initially objective ride comfort characteristics are determined on full vehicle level and compared for current vehicles within one class of different brands. By doing so, new targets are defined for the own vehicle, to achieve optimum results for the successor.

In this context, simulation models on subsystem as well as component level are able to assist in determining parameters, which have to be customised for reaching the specified targets. By using appropriate full vehicle test rigs, like a kinematics and compliance test rig (cf. Figure 16), simulation models for driving dynamics as well as ride comfort can already be parameterised, without necessarily measuring individual components. Thereby, general concept independent suspension properties are derived and different modifications can be simulated. This is also valid for vehicles from competitors.

Figure 16: Kinematics and compliance test rig

As such methods are already established within the development process, a kinematic and compliance test rig can also be used for determining the friction parameters depicted in the previous chapters, without increasing expenses. Instead of exclusively determining the stiffness parameters in a given direction, also a friction model can be fit on the determined force-displacement-hysteresis (cf. Figure 9).

The additional application of dynamic test rigs allows for the determination of dynamic parameters (cf. Figure 8), in equivalence to the parameter acquisition with a static kinematics and compliance test rig. The correct determination of dynamic properties is thereby only enabled, if friction is considered properly. For instance, the dynamic behaviour in vertical direction is predominantly defined by the properties and mounting of the damper. By properly considering friction properties, the respective parameters can be determined on the given dynamic axle test rig, without individually dismounting and measuring the damper (cf. Figure 17).

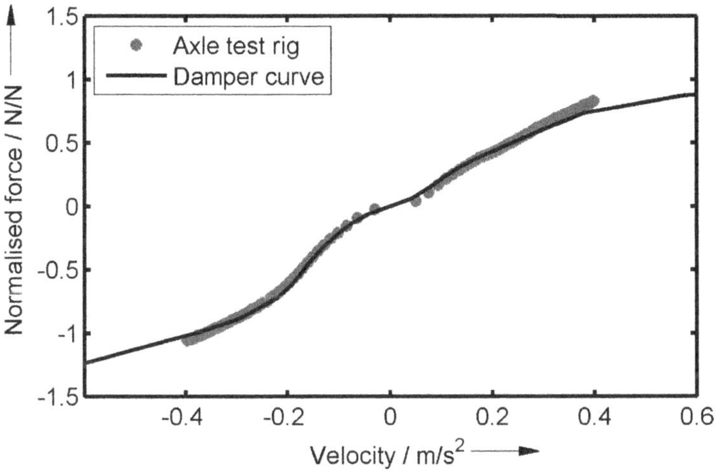


Figure 17: Determination of damper properties from a dynamic suspension measurement

On subsystem level, corresponding characteristic values for the friction have to be defined, allowing for a derivation of respective properties. In this context, the parameters addressed in the upper chapters, like hysteresis width, characteristics of the transition area or amplitude dependency are initially suggested. The shape of the transition area can thereby be defined using the parameters of the Dahl model.

Alongside with the remaining subsystem parameters, a derivation of full vehicle characteristics on these newly defined properties is conducted. Due to this approach, on the one hand the simulation results are improved, so that the estimation of influences of all parameters is enhanced as well. On the other hand, the interdependencies between friction and remaining parameters is reproduced more realistically as well.

An example would be damper tuning at low amplitudes. When friction is not considered, viscous damping possesses a dominant influence on the peak of the transfer function magnitude. However, as friction has a higher influence on this characteristic value under these conditions, the smaller impact of the viscous damping is not correctly reproduced, until static hysteresis is considered.

Finally, the newly defined parameters increase the potential for systematically designing friction properties, instead of merely considering the effect as unknown influence within the target cascading process.

### Derivation of properties on component level

After subsystem parameters have been defined by using the methods described above, the cascading process is continued onto component level. By previously defining and determining subsystem properties, the choice of an appropriate axle concept is enabled, based on the magnitude and characteristics of the specified characteristic values. Without the subsystem level, the choice of an adequate concept, based on full vehicle targets, is impeded. By characterizing different axle concept concerning their fundamental friction properties, the specified parameters can be incorporated in the selection process. Due to knowledge of the contribution of individual components on the wheel-based static hysteresis, according to Figure 7, the static hysteresis of every component can be derived. Although the empiric Dahl model has appeared as appropriate on subsystem level, at which the suspension is considered as black box, on component level a more physical approach can be appropriate. As the simulation quality is also improved on this level, when friction is considered, the prognosis of full vehicle targets is enhanced as well, similarly to the advantages described for the subsystem level.

## Summary and Outlook

In the present work, suspension friction was investigated concerning its characteristics, dependencies as well as influences on full vehicle behaviour and the automotive development process. For this purpose, initially a method for static and dynamic measurement and characterization of a suspension on an appropriate test rig was presented. In course of the realization, the characteristics and dependencies of the force-displacement-hysteresis were analysed.

The shape of the hysteresis can be characterized by a transition area and a range, in which the friction force is nearly constant. While the length of the transition area is regarded as nearly constant over amplitude, the width of the hysteresis loop increases. However, the ratio of the hysteretic force against the overall force, which additionally considers elastic spring forces, is declining, so that the influence of friction is decreasing as well. Potential characteristic values on subsystem level have been suggested. By modifying different components, the respective influences on the overall wheel-based force are analysed. It can be seen, that the overall friction is a composition of the properties of individual bushings and the damper, while the latter is not identified as dominant influence in the current multi-link suspension. Additionally, the amplitude dependency is predominantly attributed to the anti-roll bar bushings.

On basis of the static analysis, the dynamic response of the suspension is identified. Magnitude and phase of the transfer function increase, due to the viscous behaviour of the damper. Still, the effects of the friction can be examined in the offset at zero frequency of both characteristic values at different amplitudes.

After characterising the measurement data, different concept-independent modelling approaches on subsystem level are presented, which are able to accurately model the static behaviour of the suspension. In this context, the Dahl model of friction is chosen for the static hysteresis. When extending the approach for reproducing dynamic suspension behaviour, also a high correlation between measurement and simulation becomes obvious.

Based on the higher modelling quality on subsystem level, due to the consideration of friction, an improved full vehicle prediction is enabled. This is confirmed by comparing measured and simulated body accelerations of a full vehicle on a four-post test rig. The differences in the quality of results due to different levels of detail are presented.

Afterwards, the impact on full vehicle targets is examined. The natural frequency of the body is shifted towards higher frequencies, as the friction operates nonlinearly within the frequency range. The peak of the magnitude partially moves to the copying area, so that it becomes less distinct when examining accelerations. Additionally the compound of body and wheel tend to vibrate more intensively on the wheel vertical stiffness, as the coupling of both bodies within the suspension is increased. Still, the wheel natural frequency increases due to the higher real part of the transfer function. This also affects the step response. The peak-2-peak value of the vibration increases, while also the natural frequency of the decaying process is increased.

Finally, the implementation of the findings within the development is illustrated. Due to the higher modelling quality, the prediction of full vehicle targets as well as the influences of different parameters on full vehicle level are improved. By using static kinematics and compliance test rigs, which already have been established in course of the target cascading, the necessary parameters can be obtained without higher expenses. The newly defined set of parameters enables further potential for a specific design of suspension properties as well as an adequate choice of axle concepts.

In a future prospect, the presented friction characteristics have to be analysed for further axle concepts. Thereby the generality of the findings is evaluated, while also a catalogue with general suspension properties for different axle concepts can be developed. By doing so, the selection of axle concepts within the development process is even more improved. Additionally, it has to be investigated, which properties are actually necessary on the different levels.

## References


Berg, M. (1998). A NONLINEAR RUBBER SPRING MODEL FOR VEHICLE DYNAMICS ANALYSIS. *Vehicle System Dynamics: International Journal of Vehicle Mechanics and Mobility, 29*(S1), 723-728.

Bindauf, A., Angrick, C., & Prokop, G. (2014). Fahrwerkscharakterisierung an einem hochdynamischen Achsprüfstand. *Automobiltechnische Zeitschrift, 12/2014*, 76-81.

Dahl, P. (1968). A Solid Friction Model. *Space and Missile Systems Organization Air Force Systems Command.*

Dahl, P. R. (1977). Measurement of Solid Friction Parameters of Ball Bearings. *Space and Missile Systems Organization Air Force Systems Command.*

Dixon, J. C. (2007). *The Shock Absorber Handbook* (Bd. 2). John Wiley & Sons.

Duym, S. W. (2000). Simulation Tools, Modelling and Identification, for an Automotive Shock Absorber in the Context of Vehicle Dynamics. *Vehicle System Dynamics, 33*(4), 261-285.

Einsle, S., & Fritzsche, C. (2013). Utilization of objective tire characteristics in the chassis development process. *4th International Munich Chassis Symposium.*

Hab, G., & Wagner, R. (2013). *Projektmanagement in der Automobilindustrie* (Bd. 4). Springer Gabler.

Heißing, B., Ersoy, M., & Gies, S. (2011). *Fahrwerkhandbuch* (Bd. 3). Vieweg+Teubner Verlag.

Khalil, H. K. (2002). *Nonlinear Systems* (Bd. 3). Prentice Hall.

Lizarraga, J., Sala, A. J., & Biera, J. (2008). Modelling of friction phenomena in sliding conditions in suspension shock absorbers. *Vehicle System Dynamics, 46*(S1), 751-764.

Mullins, L. (1948). Effect of stretching on the properties of rubber. *Rubber Chemistry and Technology, 21*(2), 281-300.

Mullins, L. (1950). Thixotropic Behavior Of Carbon Black in Rubber. *The Journal of Physical and Colloid Chemistry, 54*(2), 239-251.

Nakahara, J., Minakawa, M., Gipser, M., & Wimmer, J. (2001). A Modelling Approach to Suspension Friction. *AutoTechnology, 1*(3), 54-56.

Rauh, J. (2003). Virtual Development of Ride and Handling Characteristics for Advanced Passenger Cars. *Vehicle System Dynamics: International Journal of Vehicle Mechanics and Mobility, 40*(1-3), 135-155.

Schimmel, C. (2010). *Entwicklung eines fahrerbasierten Werkzeugs zur Objektivierung subjektiver Fahreindrücke.* Dissertation, Technische Universität München.

Sedlaczek, K., Dronka, S., & Rauh, J. (2011). Advanced modular modelling of rubber bushings for vehicle simulations. *Vehicle System Dynamics: International Journal of Vehicle Mechanics and Mobility, 49*(5), 741-759.

Welch, P. D. (1967). The Use of Fast Fourier Transform for the Estimation of Power Spectra: A Method Based on Time Averaging Over Short, Modified Periodograms. *IEEE Transactions on Audio and Electroacoustics, AU-15*(2), 70-73.

Yabuta, K., Hidaka, K., & Fukushima, N. (1981). Effects of Suspension friction on Vehicle Riding Comfort. *Vehicle System Dynamics: International Journal of Vehicle Mechanics and Mobility, 10*(2-3), 85-91.

# Endurance tests of electronic suspension for motorcycles – a system approach

Prof. Dr.-Ing. Thomas Kuttner[1]; Dipl.-Ing. Frederik Harnischmacher[2]

[1] University of Federal Armed Forces, Department of Mechanical Engineering

[2] KTM AG

# Abstract

A test method for electronic suspensions for motorcycles has been developed and introduced in the industrial practice. This method bases on 2 poster rig tests using vehicle independent laser scans of road sequences. The motorcycle was fixed onto the 2 poster with several restraints. A failure detection check was developed to test the function of suspension and sensor. Suspension strokes and body accelerations measured at the 2 poster agree very well with independently measured road test data. Applications of this method are pointing towards operational strength of the semi-active suspensions, comparison between different suspension systems and testing of sensor malfunction.

# 1 Introduction

In vehicle development, testing of components and vehicles is an important matter. Test stands can produce fast and cost-efficient results with a high repeatability in an early stage of development.

Vehicle testing on test rigs like a 4 poster is a standard procedure in development of cars, trucks and their components. In the past, several attempts have been made for motorcycles. Only little work has been done in the upcoming field of electronic suspensions yet [1-5].

In order to test functions and durability of electronic suspensions, a test program consisting of component tests, vehicle tests on test rigs and field tests has been carried out. For a full vehicle test, several state-of-the-art techniques were adopted in a 2 poster test:

1. Fixation of the motorcycle to the test rig, conservation of the relevant degrees of freedom
2. Driver dummy in order to model the vibration behavior of the rider
3. Quality of input data (measured accelerations vs. displacements from road scans)
4. Condition monitoring during test and damage detection

With this developed test procedure the electronic suspension is tested in its vehicle environment, including sensors, control units and actuators. Magnitude and frequency of the loads correspond to the measured data, no scaling of the input data has been carried out. This ensures a high quality of results generated in the test. Furthermore, this process is independent from the vehicle itself.

The developed set up and procedure is a tool for functional and durability testing of electric suspensions of motorcycles. It can answer the questions:

- Does the electronic suspension fulfil the requirements regarding function and durability?
- How can we compare different hardware layouts and control strategies?
- Is the system fail-safe?

It is possible to make decisions in an early stage of product development based on these results. The last point allows carrying out highly dangerous tests on a test rig. All these results help to start the field testing with better tested products.

Although full vehicle testing on 4 poster test stands is a standard test procedure in the automotive industry, only little work has been done to transform these results to the motorcycle industry. Endurance testing of electronic suspension is usually carried out as component tests with stationary control signals. The combined use of vehicle environment, driver dummy, digitalized road and condition monitoring leads to a new quality of results and is a new technical approach.

# 2 The Semi-active Suspension in a Motorcycle

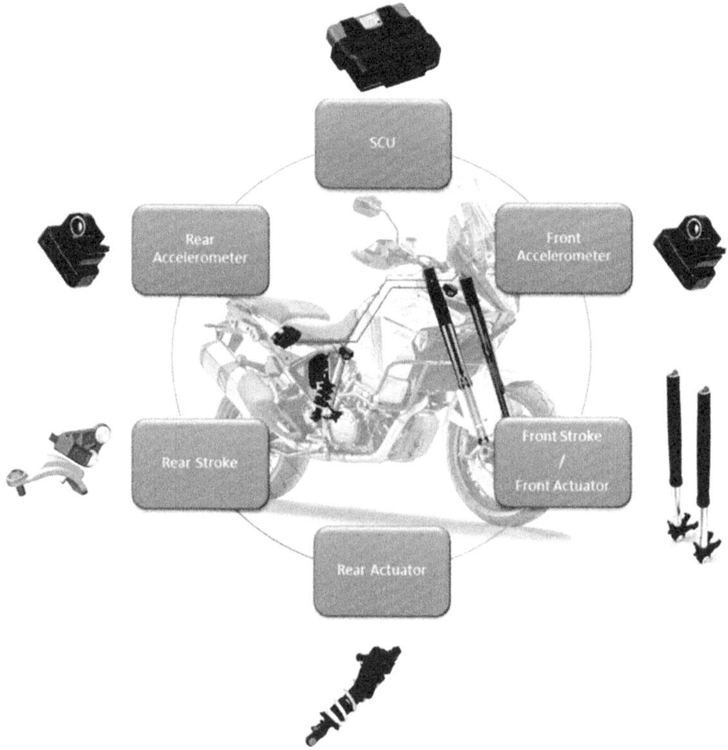


Figure 1: System, semi-active suspension

The 1290 Super Adventure is the first KTM motorcycle with a semi-active damping system. The system consists of the "Suspension Control Unit" (SCU), acceleration sensors, stroke sensors and two actuators (Figure 1).

The basic algorithm of the semi-active suspension calculates in real time damping requests for the front and rear actuator out of the primary sensor signals. Semi-active means, that the energy can only be dissipated.

The primary sensor signals are:

- Front body acceleration
- Rear body acceleration
- Front stroke
- Rear stroke

There was no experience about the durability of the components of this system, because this is a new technology for KTM. The target is to test all components on one test rig. In this case the structural durability is not relevant, only the so-called "functional durability" is important, because the focus is on the semi-active system.

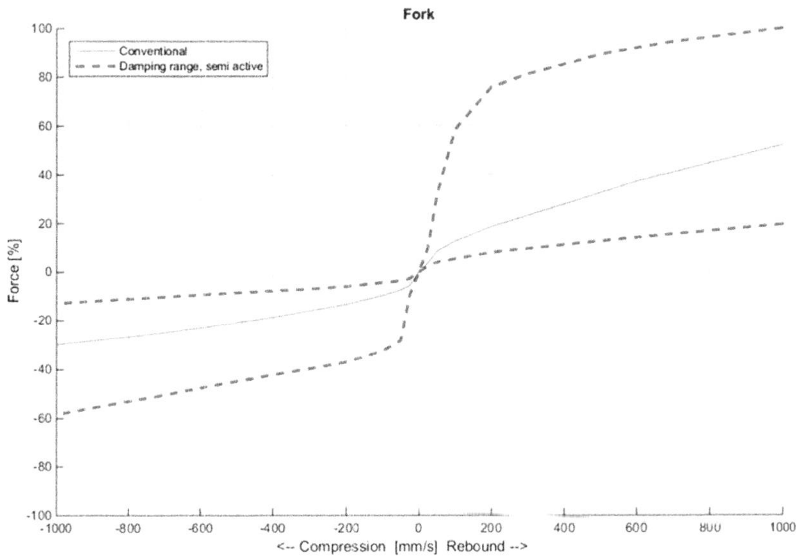


Figure 2: Comparison, conventional vs. semi-active fork

In Figure 2, there is a comparison, between a conventional motorcycle fork and a semi-active fork. A conventional suspension has only one damping curve, whereas a semi-active system has a wide damping range. So there are 2 possibilities to test a semi-active suspension [6]:

1. Constant current (medium damping curve similar to a conventional suspension)
2. Dynamic current

In order to test the functional durability of the valves it is essential to test the electronic suspension with a dynamic current so that there is a valve actuation and movement. This is the first and most important requirement for the test rig.

# 3 Test Design

Figure 3: Test rig, degrees of freedom

The requirements for the test rig are:

1. Realistic control of the semi-active suspension
2. Simulation of rider and payload
3. Use of road scan data

As mentioned above, the first and most important requirement is that the electronic suspension is tested with a realistic control strategy. In order to fulfil this requirement, all primary sensor signals must be available on the test rig. The relevant degrees of freedom of the test setup can be derived from the primary sensor signals:

- Front Stroke → displacement front wheel
- Rear Stroke → displacement rear wheel
- Front / Rear body acceleration → vertical displacement body / Pitch angle

The irrelevant degrees of freedom are:

- Lateral displacement
- Roll angle
- Yaw angle
- Steering angle

These degrees of freedom are restrained to hold the motorcycle on the test rig.

All other degrees of freedom (Horizontal displacement, front wheel rotation, rear wheel rotation) are not limited in order to keep the motion of the motorcycle as realistic as possible.

For fixing the motorcycle on the test rig, the horizontal movement is restrained by soft springs. The natural frequency of the system is smaller than 1Hz. The targets for the layout of the springs are [5]:

- Stabilization of the horizontal position
- No influence of the relevant frequency range for the test, 1Hz – 80Hz
- Minimal pulldown effect

The second requirement for the test rig setup is the rider and payload simulation. Especially for motorcycles, the mass of the rider and the mass of the payload have a significant portion on the overall mass of the system. An example is shown below.

- Motorcycle: 230kg
- Rider: 82kg
- Garment: 10kg
- Luggage: 2x15kg + 1x5kg

➔ Overall mass: 357kg

With these values, the portion of the rider and the luggage is approx. 36% of the overall mass. Therefore it is obvious that it is not valid to neglect these additional masses. The payload is only a mass whereas the rider is simulated by a mass coupled with spring and damper [7]. This system is designed to meet the values of the DIN 45676, which describes the input impedance and the frequency response of the human body. The concept and the dummy are shown below.

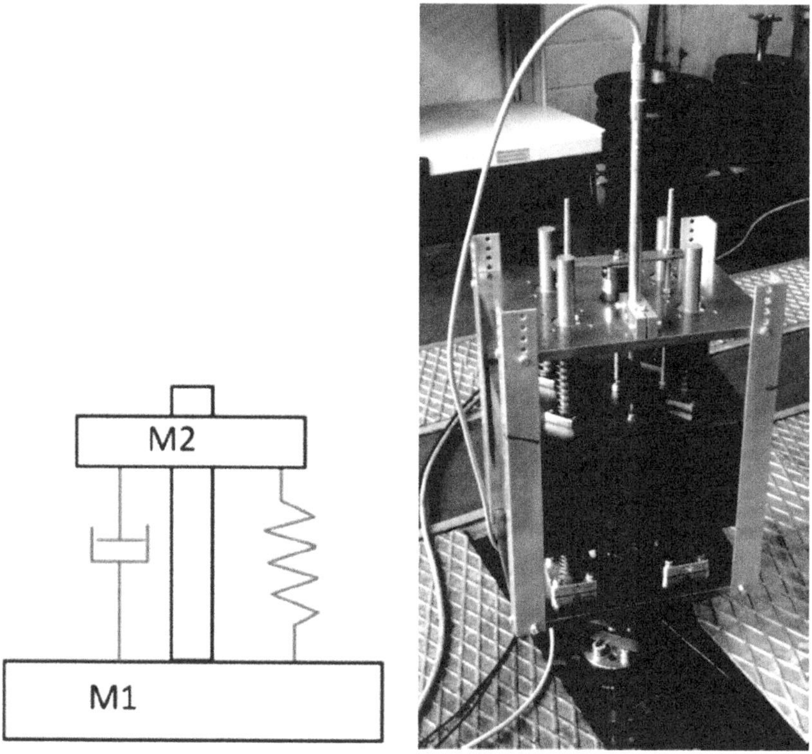


Figure 4: Rider dummy [7]

The last test requirement concerns the excitation signals. The aim is to use laser scans from real roads as excitation signals. The advantage is that these signals are independent from the vehicle. There is no need for road test data with this vehicle. So the whole testing procedure can be realized in a very early stage of vehicle development. The processing of the laser scans contains a high pass filtering and the calculation of the time signal with the corresponding vehicle speed and wheelbase. The high pass filter is necessary to restrict the maximum height difference of the scanned data to the stroke limits of the 2 poster. The test rig itself has a cut-off frequency of 80Hz.

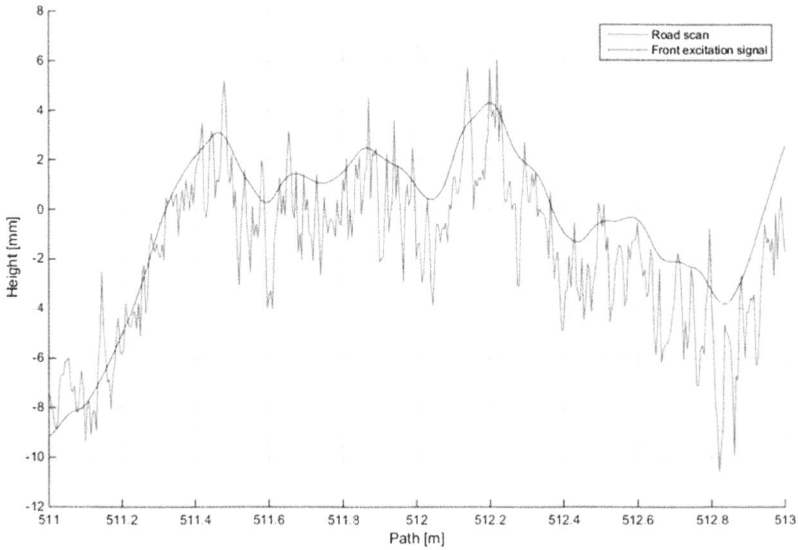


Figure 5: Comparison Road scan vs front excitation signal

In Figure 5, there is a comparison between the scanned data and the front excitation signal. At first, the frequency response function of the test rig was identified. With this function, the excitation signal converged to the target signal in less than 10 steps. For this proceeding standard iteration software was used. The excitation (target signal) shows a good accordance with the road data, even at lower frequencies. There is no need to introduce another low pass filtering stage in order to simulate the unwinding of the wheel, because the test rig already brings in a low pass filtering.

# 4 Test Rig and Testing Procedure

Figure 6 shows the 2 poster test rig with the motorcycle. The motorcycle is connected with a folding joint to the ground. The dead load of the rider dummy is connected rigidly to the frame and the handlebar. The payload mass is also coupled rigidly to the rear frame.

Figure 6: Test rig

The test cycle consists of a durability sequence which includes several road profiles with different speeds and a failure detection check in which malfunctions of the semi-active system should be detected. In order to avoid the attendance of an operator, the motorcycle is connected with a rip wire. If there is a mechanical damage, the wire is teared off and an emergency shut down will be activated. Furthermore there are safety light barriers which can also activate an automatic shutdown. The test chamber is equipped with a webcam, so the test can be surveilled from an extern computer. Also the test rig control can be remote controlled. The test cycle will be repeated until a malfunction occurs or the maximum number of cycles is reached.

Durability sequence:

This sequence should represent the typical riding profile of a big travel enduro, Figure 7. The single road profiles have variable speeds in order to cover the full frequency range.

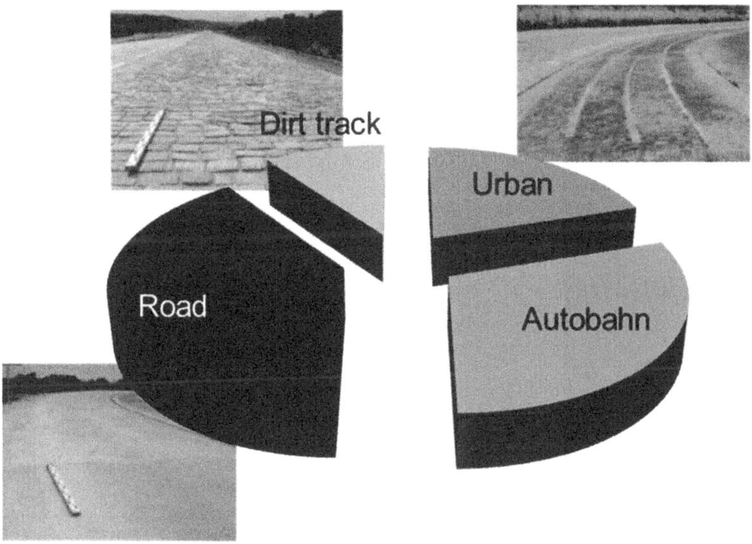


Figure 7: Percentage of the durability sequence

As the cooling effect of the airstream is missing, the profiles were arranged in a way, that the front and rear damper have enough time to cool down. So heavy profiles, like "dirt track" are followed by profiles, which are easy on the suspension like "autobahn". To speed up the cool down phases and to keep the dampers in a reasonable temperature range a temperature-controlled fan is used.

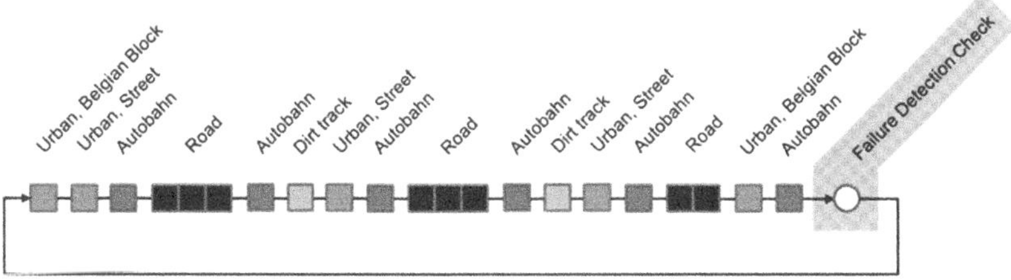


Figure 8: Test cycle

Failure detection check:

This test sequence should help to detect a malfunction of the semi-active system. Out-of-range errors, like the loss of control current or the complete breakdown of a sensor are detected throughout the whole testing procedure. Hidden errors like a reduced damping of the suspensions can hardly be detected during the durability sequence. The failure detection sequence is carried out at the end of each cycle (Figure 8) and contains three characteristic excitations:

- Step up / down, front wheel
- Step up / down, rear wheel
- Response to a sinus excitation

Figure 9 shows the rear stroke signal during the test sequence. If the damper is working correctly the signal should be very similar to the blue reference line. The second damper has a reduced damping and therefore the signal shows a clear deviation.

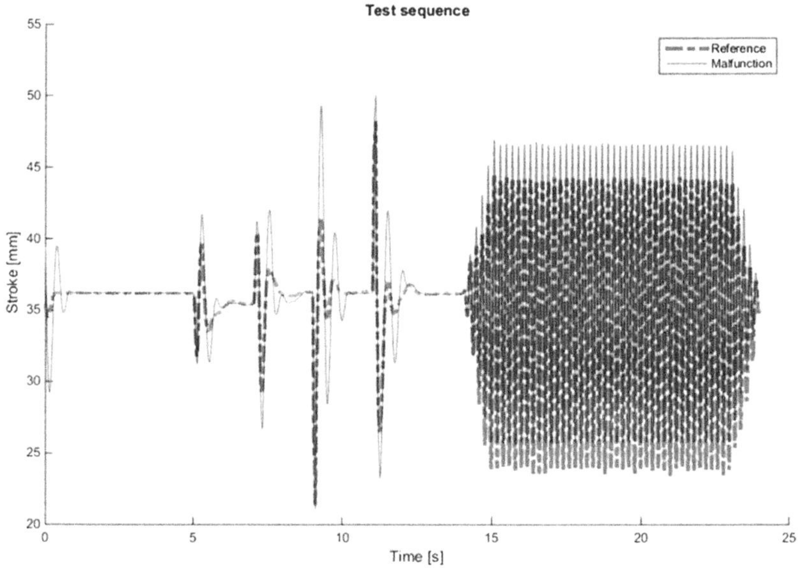


Figure 9: Test sequence, rear stroke

# 5 Summary and Outlook

One of the most important results is the verification of the excitation signals. For this verification, real test rides were compared to the same tracks on the 2 poster. It should be noted that the path of the vehicle on the test track does not match perfectly the line of the scanned data, because these two measurements were performed independently. The signals denoted "2 Poster" (orange) were measured at the test rig operating with a laser-scanned street profile. On the same track, a test ride with a motorcycle was performed (blue, "track"). So, this comparison is not between target and response signal of the test rig. In fact, it is a comparison between the test rig and an independent test ride (see Figure 10).

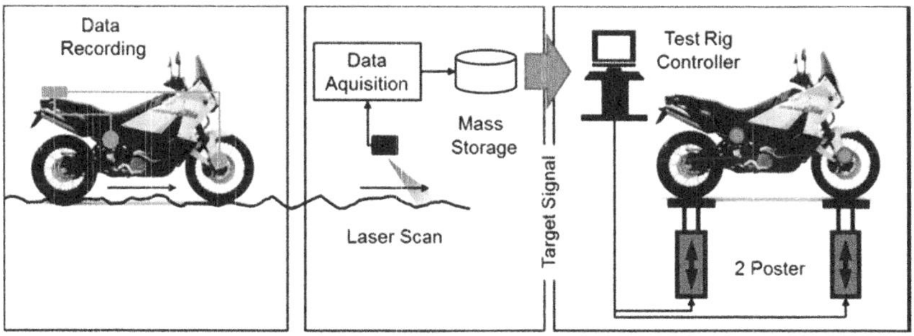


Figure 10: Verification excitation signal

In Figure 11, the spectrum of the front and rear stroke speed for one road profile is shown. The front spectrum as well as the rear spectrum of the 2 poster test matches the spectrums of the test rides very well. This global comparison already shows a good accordance. A closer look on the time domain verifies this.

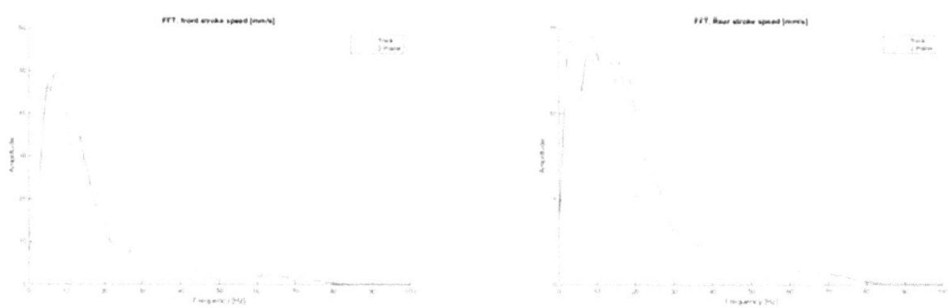

Figure 11: Comparison, test track vs 2 poster

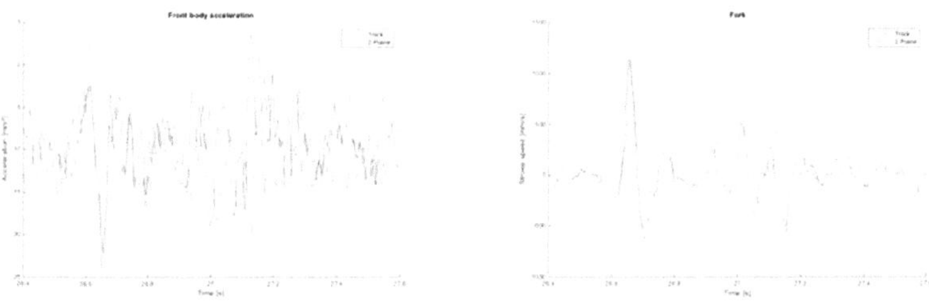

Figure 12: Comparison time domain, test track vs 2 poster, front

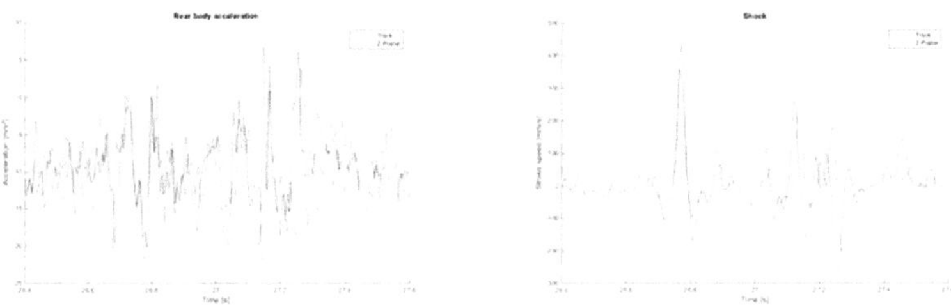

Figure 13: Comparison time domain, test track vs 2 poster, rear

In Figure 12 and Figure 13 is a comparison of the front / rear acceleration and stroke speed. The section shows a characteristic edge on the track. Both the acceleration and the stroke speeds from the rig match the signals from the test track pretty well.

Due to the fact, that the signals have a good accordance in the time and frequency domain, it is possible to use the road scan data directly instead of iterating on signals from road test data for this application.

With the help of this test rig, it is possible to test the following points:

- Operational strength of the semi-active suspensions
- Comparison between different suspension systems
- Testing of sensor malfunction
- Comparison of control strategies

For testing the interaction of components in a large system like a motorcycle, it is necessary to test the whole system either under road testing conditions or in a test rig. The use of this test rig leads to a significant reduction in testing time. A road testing under real conditions will take several months. In contrast to this, the described rig test gives the results after 400 hours of testing. However, road tests and rig tests are complementary to each other. Furthermore, testing all the components with individual tests would result in a fleet of testing benches working parallel.

Well defined test conditions make sure that the tests will be carried out with a high accuracy. Testing of different design layouts in an identical test environment is also possible and easy to perform.

Especially for mechatronic systems, the cross correlation between software and hardware is very important. This test rig provides an opportunity to highlight such things. On the one hand, a change in the hardware does not require a new control and excitation signal in the test rig. On the other hand, the influence of design changes (hardware, software) on the overall system behavior can be easily shown.

At the moment, only electronic suspensions are tested on the test rig. In order to test also structural parts the rider dummy has to be refined.

The rigid connection of the dummy to the motorcycle leads to a wrong application of the supporting forces. For example, damage calculation has revealed that the steering head has a higher relative Miner sum compared with road test data. So the application of the supporting forces has to be reviewed.

Another field is the testing of non-structural parts, like accessory, windscreens and cover parts. The test rig offers the opportunity to test a lot of parts simultaneously directly on the motorcycle.

# Literature

[1] Racco, D and Panetta, F: Restrain & Dummy Rider Setup for Actual Road Reproduction in Durability Test on MTS 2 Post. MTS seminar "Evaluating Bikes Performance & Durability", 2012

[2] Lin, K-Y et al.: Durability assessment and riding comfort evaluation of a new type scooter by road simulation technique. No. 2006-01-0730. SAE Technical Paper, 2006.

[3] Kharul, R V, et al.: Virtual testing and correlation for a motorcycle design. No. 2010-01-0925. SAE Technical Paper, 2010.

[4] Tuluie, R, and Gary S: Motorcycle suspension development using ride comfort analysis with a laboratory test system. No. 1999-01-3276. SAE Technical Paper, 1999.

[5] Schuh, M: Optimierung von Regelstrategien für den Straßensimulator. Masterarbeit, Universität der Bundeswehr München, 2013

[6] Bertemes, B: Prüfstandserweiterung für aktive Fahrwerkskomponenten, DVM Workshop Prüfmethodik für Betriebsfestigkeitsversuche in der Fahrzeugindustrie Plauen, 2012

[7] Streb, U: Entwicklung eines Sitz-Dummys für Motorräder. Bachelorarbeit, Universität der Bundeswehr München, 2013

# Availability and fail-safety approaches for electric power steering systems – trends and requirements

Eberhard Kübler, Thomas Pötzl, Dr. Thomas Frenz, Jürgen Sauler,
Robert Bosch Automotive Steering GmbH

# Abstract

The market for passenger car steering systems has more or less completely changed from Hydraulic to Electric Power Steering (EPS) systems in the recent years. This change in technology also changed the safety requirements for steering systems significantly.

Up to now the ultimate safety requirement to an EPS system is to avoid events like autosteer or blocked steering or other uncontrolled conditions in a potential failure mode of the electric system. This means that the shutdown of the steering support in an error mode is considered as the safe state of the EPS.

With the background of the investigation of highly or fully automated driving the requirements concerning the error mode will change significantly. This means the requirements for loss of assist avoidance will also increase and the shutdown of the steering support in the error state will not be feasible anymore.

The presentation shall explain the background of this development. Approaches how to prepare for the upcoming requirements and also potential solutions based on redundant architectures for the electronic system and the software will be shown and discussed.

# Innovative software functions to operate electric power steering systems in sports cars – Unterstützungskraftregelung (UKR)

Anton Uselmann, K. M. Krüger, Dr. C. Bittner, G. Rivera

Dr. Ing. h.c. F. Porsche AG, Weissach

# Introduction

When using electric power steering systems in sports cars there are additional requirements concerning steering feel and road feedback. Here off-the-shelf steering mechanics quickly maxes out. Also the software of many manufacturers of steering systems does not entirely cover these requirements. To fulfil these requirements, regardless of the steering system supplier, Porsche has developed an own control approach to increase road feedback as well as additional software modules for optimized tuning ability of steering feel.

Below first of all the importance of feedback for sports cars is explained and examined by simulation of electric power steering systems. Then several concepts for the control of electric power steering systems will be presented and examined on software functional level. The focus will be on the Porsche Unterstützungskraftregler (UKR, assist force feedback controller). Finally, objective measurements for the various concepts will be presented and evaluated.

# Requirements for a sports car steering system

The basic requirements for steering systems can be divided in three categories: comfort, sportiness and safety [2], [3].

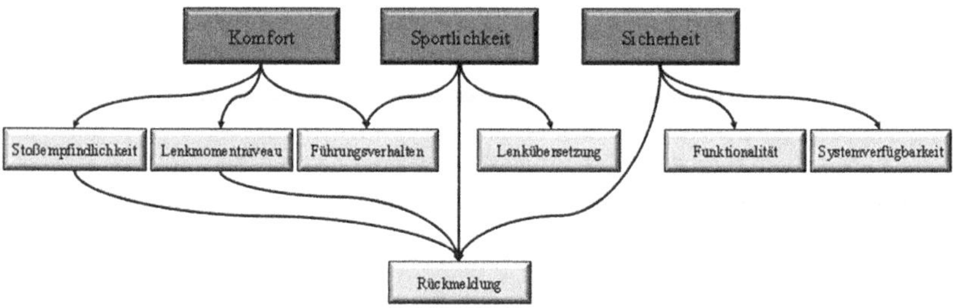


Figure 1: Basic requirements for steering systems

It quickly becomes clear that steering feedback is very important, particularly for sports cars [1].

In general steering feedback describes the stimulation of the road surface transferred through the axle and the steering gear up to the steering wheel to the driver.

In terms of comfort, the feedback must be tuned so that no unpleasant vibration is felt.

For sportiness, the feedback ensures direct contact of the driver to the road. Especially at the vehicle dynamics limits this is important. A transparent steering feedback allows the driver to approach the physical limits. Understeering for instance is revealed in time by decreasing steering wheel torque.

Also from security aspects, the feedback helps the driver. If the car oversteers, a self-stabilization of the vehicle is provided by the physically correct steering return, in accordance with the forces acting on the wheel.

Because of the great importance of steering feedback for sports cars, we will now take a closer look at it, regarding electric power steering systems.

# Steering feedback in electric power steering systems

Compared to hydraulic systems, electric power steering systems (EPS) offer high potential regarding adjustable functionality and implementation of characteristic perception. However there also exist concept-related limitations, in particular as to steering feedback. The high inertial mass from the power assist unit act as low-pass filter in the transmission of external road forces to the steering wheel torque.

To illustrate this effect, a simplified model of the steering system is configured according to Figure 2. The model describes a passive EPS reference system, which has been reduced to the rack movement.

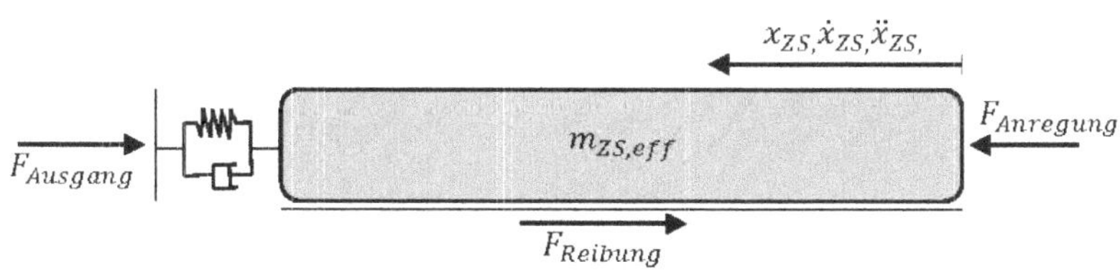


Figure 2: Frequency response of a passive electric power steering system in feedback path

The model is stimulated with forces of feedback relevant frequencies and amplitude ranges. The transmitted output force is evaluated.

The parameter $m_{ZS,eff}$ describes the whole mass reduced on the rack which needs to be overcome by rack stimulation. The high inertia of the power assist unit results in a mass of approximately 1200 kg in the example. In addition to the inertia forces the system friction $F_{Reibung}$ and the elasticity and damping of the torsion bar are taken into account in the simulation. Thus the differential equation of the system is as follows:

$$F_{Ausgang} = F_{Anregung} - F_{Reibung} - m_{zs,eff}\ddot{x}_{zs} - k_{DS}\dot{x}_{zs} - \tau_{DS}x_{zs} \quad (1)$$

For the calculation of the output force, the system according to Equation 1 is moved against a fixed steering wheel. The parameters are transferred from a reference steering system.

The model is stimulated using force sine sweeps in following variations:

- Excitation amplitude 0,2 to 10 kN
- Excitation frequency 1 to 30 Hertz.

The transfer behavior of the model is shown in Figure 3.

The analysis of transfer behavior illustrates the high degree of filtering effect by the inertial mass of an electric power steering system. At stimulation frequencies greater than six Hertz no or only very low forces are transmitted to the steering wheel, independent of the amplitude of stimulation. The excitation amplitudes below the system static friction also cannot be noticed by the driver.

These inherent limitations cannot readily be solved by mechanical measures. Because of modular systems and common part strategies, changes of system mechanics are often not desired.

To provide the driver with feedback information anyway, innovative software funcions have to be used.

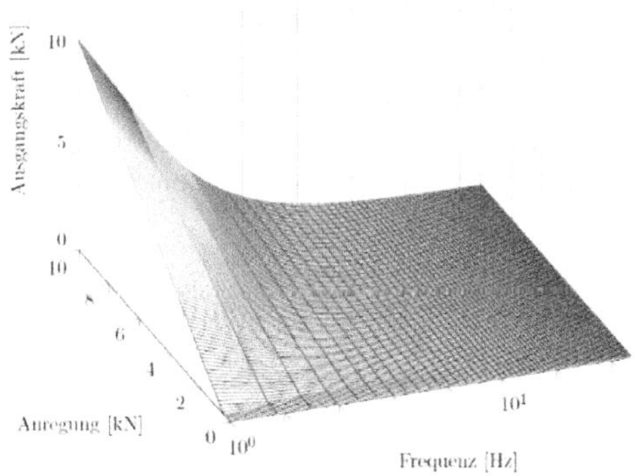


Figure 3: Frequency response of a passive electric power steering system

# Conceptual presentation of different control approaches in the EPS

Below several control approaches for electric power steering systems are briefly described, analyzed and evaluated with regard to their impact on steering feedback:

- Feedforward control of the power assist unit.
- Steering wheel torque control (HMR).
- Unterstützungskraftregelung (UKR).

The focus will be on high level software control. The underlying motor control loop of the low level software is not subject of the examination.

The functionality of the concepts is considered in two different directions: In the forward path and in the feedback path.

The function in the forward path describes the reaction of the steering system on stimulation by the driver, thus providing the assist forces of the steering system and change of direction of the vehicle and thereby the main task of the power steering.

Considered is a change of steering angle by the driver out of straight-ahead driving on a smooth road.

The feedback path examines the reaction of the steering system to stimulation from the road. This mainly takes into account the transfer of the road information to the driver. The system behavior in the feedback path is considered at higher frequency external stimulations. Specifically, crossing a curb while driving straight-ahead is under investigation (sudden stimulation from the road without any intervention by the driver).

## Feedforward control of the power assist unit

If the high-level function software has been established as feedforward control of the power assist unit, a set point for motor torque is calculated on the top level of the software and passed to the motor control loop of the low level software. The requested target torque is calculated depending on steering and vehicle measurement signals. So for example the assist force can be calibrated depending on the torsion bar torque $M_{DS}$ (which approximately equals the steering wheel torque $M_{Hand}$) and vehicle speed (boost-curve).

Initially, this concept is evaluated by the functionality in the forward path. The sequence of action in case of a stimulation of the steering system by the driver is shown in Figure 4.

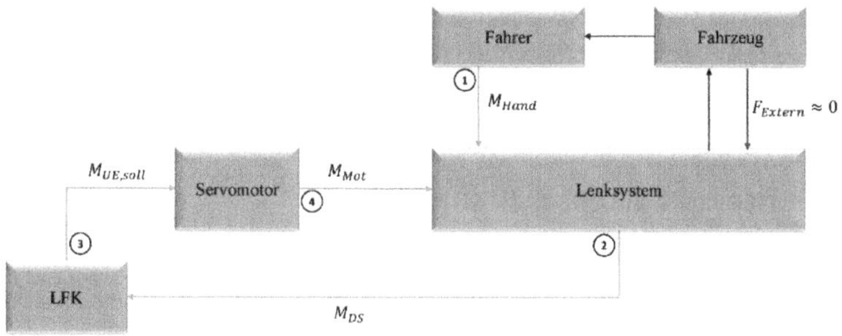


Figure 4: Functionality of a feedforward servo control in forward path

In step 1 there is a steering motion of the driver which results in a change of the measured torsion bar torque in the steering system software. Based on the changed torsion bar torque, the internal steering functions (LFK) calculate the calibration dependent motor torque set point $M_{UE,soll}$, which is passed to the servo motor control. The servo unit adjusts the assist torque $M_{Mot}$ and the desired steering response of the vehicle can be achieved.

With this functional principle, there are no disadvantages in the provision of the assist force. Proven, motor torque forming steering functions can continue to be used and thereby the desired steering feel can be created.

Following we have a look at the behavior of the concept at external stimulations. The functional flow while crossing a curb is shown in Figure 5.

At first the system is stimulated by an excitation impulse from the road (step 1). The amplitude and the frequency of this external force $F_{extern}$ are assumed so that a passive system would show only negligibly small changes in the EPS internal states when exposed to this stimulation, according to Figure 2.

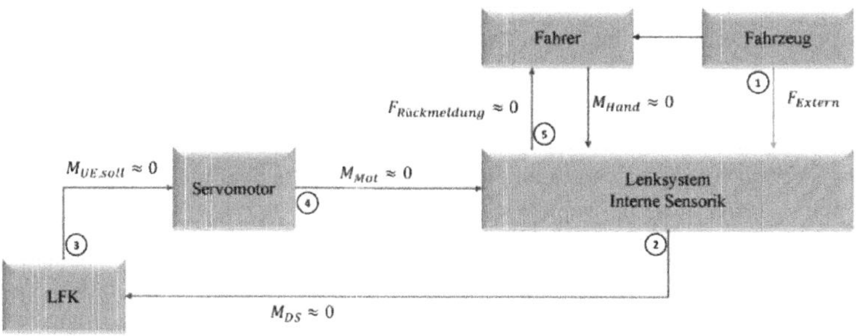


Figure 5: Functionality of a feedforward servo control in feedback path

The steering system is at rest at the time of stimulation: no steering wheel movement by the driver, no measurable torsion bar torque. Because the internal system states change only minimally by external forces, these measurement signals also remain almost unchanged by stimulation (step 2). No relevant motor torque set point is calculated in this situation because the steering functions directly depend on the internal measurement signals. The steering wheel torque $M_{Hand}$ remains unchanged during and after the external impulse since there is no interference by the actuator of the steering system.

Through this evaluation can be determined, that EPS systems with a feedforward servo control have the same limitations as a passive EPS steering system with regard to the frequency response in the feedback path. The feedback is directly related to the mechanical characteristics of the system and is as a rule inadequate.

EPS systems with a pure feedforward control of the servo unit are thus less suitable for vehicles with a high demand for naturalness and transparency in the steering feedback.

The advantages of such systems are low development cost and generic portability thanks to the simple concept. The function software remains – regarding system dynamics and stability – largely independent of changing steering mechanics and robust towards different vehicle setup.

Shall the characteristic impression of the overall vehicle, regarding sportiness and naturalness, be emphasized by the steering system, software solutions for increased road feedback must be integrated in the functional software layer. Two approaches are presented in the following sections.

## Steering wheel torque control

An approach to controlling the actuating forces on the high-level software level at an EPS system is the steering wheel torque control (Handmomentenregelung, HMR), as described in [1]. By calibration of functional modules a set point for steering wheel torque $M_{DS,soll}$ is defined and the torsion bar torque is adjusted to it. During calibration of the steering feel, the steering wheel torque set point is determined dependent on the external rack force and the vehicle speed.

The concept of steering wheel torque control requires an algorithm for calculating the external rack force in the software. Since the calculated force of the rack is used to represent the steering wheel torque set points $M_{DS,soll}$ in the basic assist and a large amount of steering feeling depends on it, the deviation of the computed rack forces from reality should be very low. The estimation of the external forces uses an observer approach.

In Figure 6, the functionality of such a software structure is shown in the forward path.

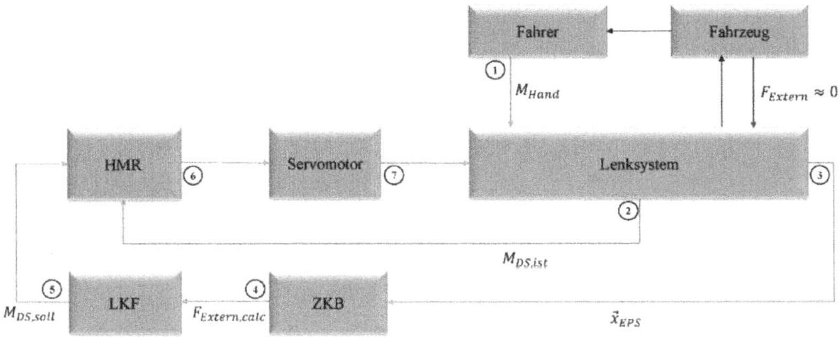


Figure 6: Functionality of steering wheel torque control in forward path

Again, the evaluation considers a situation with a sudden stimulation of the steering system by the steering wheel (step 1). At the time of the steering wheel stimulation the external force is assumed to be zero (step 3). The deflection results in a non-zero measured torsion bar torque $M_{DS,ist}$ and a steering wheel torque set point $M_{DS,soll}$ equal to zero, because no external forces apply at this moment and the steering wheel torque set point is calculated in the steering functions dependent on the rack forces(step 5). In step 6 the controller of HMR will regulate the steering wheel torque set point $M_{DS,soll}$ a set value and passes a steering wheel torque reducing motor torque set point $M_{soll,HMR}$ at the servo motor control. During the steering wheel deflection the HMR even leads to a more dynamic system reaction, the function of the servo assist in the forward path is definitely ensured by this concept.

To examine also the steering wheel torque control with regard to the feedback behavior, the concept is also evaluated according to the principle of feedback path. The corresponding functional flow is shown in Figure 7.

As previously described, in the first step the system is abruptly exerted by an external force. At this time the steering wheel torque equals zero Nm. Accordingly, the measured torsion bar torque is also approximately zero Nm. The effective external force $F_{Extern}$ can be calculated by the rack force estimator (resulting in $F_{Extern,calc}$) using the changing steering system and vehicle conditions. The steering functions are provided with the calculated external forces. In the next step, the function module generates a set point for steering wheel torque $M_{DS\ soll}$ and passes it as set point to the controller. Because the control variable $M_{DS,ist}$ equals zero Nm at this time, the steering wheel torque is changed by requesting the actuating variable $M_{soll,HMR}$ from the servo engine, depending on the calibration of the functional modules. Thereby the external excitation is recognized by the driver as road feedback.

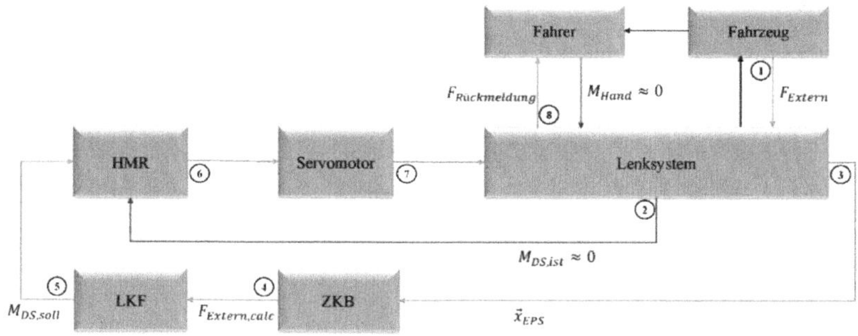


Figure 7: Functionality of steering wheel torque control in feedback path

With this concept, a transmission of the road stimulation to the driver is possible. The frequency content and the immediacy of the feedback information can be taken into account with such architecture in the development of the observer and controller synthesis. With regard to these criteria, feedback can be tuned separately from the calibration of the other functional modules. It must be noted that the subjectively perceived naturalness of this information is also strongly related to the amplitude progression of the steering wheel torque depending on road stimulation and accordingly stands in strong relation to the calibration of the steering wheel torque set point depending of the external forces. It is therefore not possible to entirely decouple the steering feel calibration of the representation of the feedback with this concept.

The transfer of this concept to a steering system with modified mechanics and to other vehicles is rather complex. The design of the rack force estimator and especially the tuning of the controller represent a significant development effort in the first implementation of the software concept on a new steering system or on an adaptation to another vehicle. But a brand, vehicle - and derivatives specific steering feedback for the customer and a unique selling point concerning noticeable vehicle dynamics can be achieved with this development.

## Presentation of UKR concept

With the Porsche's new control concept UKR the support force is adjusted at the rack dependent on the frequency. As a result, the by concept greatly reduced road feedback of EPS systems is brought back to the level of the traditional hydraulic steering systems [5]. At the same time, the numerous benefits of EPS are preserved.

On the basis of the concept of feedforward control of the servo unit, here not the motor torque set point $M_{Mot,soll}$, but the actually applied assist force FUE,soll at the interface servo transmission to rack is adjusted. The motor torque set point and the assist force are shown in Figure 8 as an example on an axis-parallel-powered steering system (APA-EPS).

Accordingly, the main difference from the conventional approach is to consider the dynamic forces of the assist unit in the assist force set point. These influences consist of the passive mass of the drive and transmission, as well as drive friction and are calculated via a rack force observer (ZKB). Because exactly these mass and friction represent the largest part of the mitigation of road feedback, maximum transparency of external force dynamics in the perception of the driver becomes possible by their calculation and consideration in the set point. For optimal representation of the transparency in the controller synthesis, the frequency range of the external stimulation is divided in useful and distracting information [4]. Accordingly the stimulation in the transfer path is amplified or suppressed. Through such a systematic controller tuning, the feedback ability taken of a hydraulic steering system is combined with the advantages of a flexible controllable EPS systems.

The structure of the UKR concept is illustrated in Figure 9. It becomes clear that only minimal intervention in the architecture of a feedforward controlled steering system is necessary. The controller adds only a small portion to the motor torque set point, compensating the inertia of the drive unit in the feedback path.

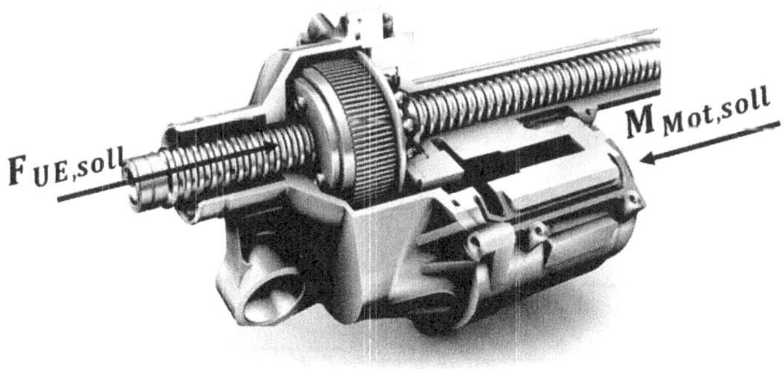


Figure 8: Assist force in the cross-section drawing (source: http://www.schaeffler.com)

In addition to UKR further Porsche software function modules come to use, as represented in Figure 9. They aim at a further optimization, brand-specific imprint and improved calibration ability of the steering feel, in terms of steering return, dampening and friction compensation. Notably the uncoupling of individual aspects of the steering feel is important. Relevant software function modules are described briefly below.

The intention of the friction compensation is the reduction of the friction of the steering system specifically for small steering angle corrections. The calibration depends on steering angle speed, torsion bar torque, vehicle speed and steering system temperature. No friction compensation shall be made for special situations such as, for example, vehicle yaw motion oscillations.

The damping depends on the same system states as the friction compensation, but forms a motor torque set point contrary to the rack direction of movement. It creates a "rich" steering feel, but it may be not exaggerated in normal driving situations. On occurrence of vehicle yaw motion oscillations higher damping forces are generated.

The steering return should be naturally and in accordance with the forces physically exposed to the wheel. Especially in the low vehicle speed range also a steering return straight ahead must be ensured. This is done through a special active return software module, which acts depending on steering wheel angle, torsion bar torque, vehicle speed and steering rack temperature and is calibrated according to the available axis self-aligning forces.

To shape the center steering feel and to adjust the general level of friction, the "dry friction" function is used. This basically damping function can also operate compensating for low temperatures.

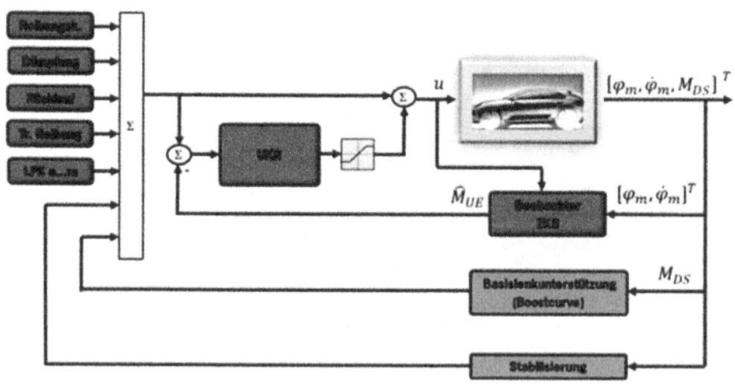


Figure 9: Structure of the UKR

As already in the previous EPS control concepts, the functionality of the assist force feedback control concept will be described in the distinction of forward and feedback paths.

Figure 10 presents the functionality of the UKR concept in the forward path. There is a movement of the steering wheel by the driver, without relevant forces from the road acting on the steering system at this moment (step 1). Because of the change of the internal steering states, the steering functions become active and $M_{UE,soll}$ requests a specific target assist torque set point in step 3. Parallel, the rack force observers estimate an assist force (step 4), which slightly lags behind the desired torque in the dynamics, due to the inertia considered in the model. Because to this small control deviation at the beginning of the steering wheel movement, a small dynamic control action in the form of the torque control set point $M_{UKR,soll}$ is given to the servo drive. The motor torque $M_{UE,soll}$ in step 6 still mainly consists of the feedforward control share and assists the driver in the steering action.

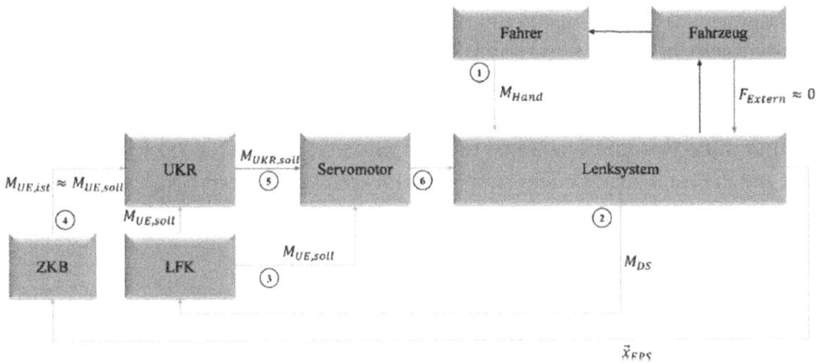


Figure 10: Functionality of the UKR in forward path

It is evident that the UKR concept in the forward path has a similar functionality as the principle of feedforward servo control, which is proven concept with regard to the functional architecture and calibration in the forward path. An optimization by controller is to speed up the servo engine in the initial displacement of the system.

Figure 11 represents the functional flow of architecture in the feedback path, when sudden rack stimulation from the road occurs.

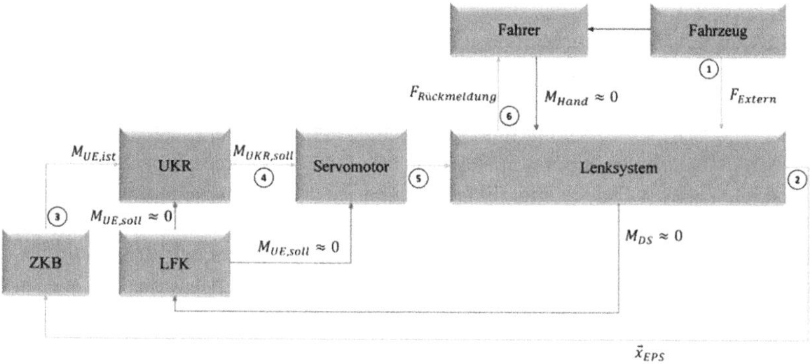


Figure 11: Functionality of the UKR in feedback path

As previously assumed, a force impulse takes place on the rack at a stationary steering wheel. Because there is no movement of the steering wheel, the measurable torsion bar torque $M_{DS}$ at this time is equals to zero Nm and the steering functions remain without output motor torque. The assist motor torque set point $M_{UE,soll}$, which is used as set value in UKR and shall be adjusted at the rack, is also at zero Nm.

In the ZKB, the actual assist torque $M_{UE,ist}$ is calculated as the control variable in step 3. This torque is calculated out of the infinitesimally small changes in the measurable servo conditions $\vec{x}_{EPS}$, which come along with the external stimulus (step 2) and comply with the blocking force consisting of the mass inertia and friction in the servo unit.

Now the control deviation thus resembles the (negative) actual assist torque. Incorporating the control gain, the UKR torque $M_{UKR,soll}$ is passed to the servo control as a set point, a nominal torque of zero Nm at the interface of gear to rack is adjusted and the external road stimulation can be felt by the driver (steps 4-6). The assist unit thereby gives way to the external stimulation, allowing this information to be passed to the driver.

As soon as the stimulation of the system overcomes the sticking friction in the rack and a minimal movement of the servo unit is caused, the feedback information can be transferred to the driver with adjustable characteristics. The frequency content, the fidelity of phase and the amplification of the forwarded information depends very much on the quality of the actual assist force. By a suitable controller structure and design of the frequency-dependent amplification, the robustness and stability of the system can be influenced.

A great advantage of this concept, concerning the system stability and robustness, is the integration of the feedforward force component, so that the controller just acts to correct and not to adjust the overall force level, as it is done in HMR. The height of the control deviation depends on the height of the resulting blocking forces from the servo unit and not from the calibrated level of support of the overall system. Therefore, the steering feedback can be calibrated largely independent of the other steering feel relevant function modules. Because the UKR concept is based on the concept of feedforward control, those benefits remain here, too. The disadvantages, however, can largely be compensated.

## Objective evaluation of the control approaches with regard to feedback

To objectively compare the presented control concepts, the systems are investigated on a steering system test bench to determine the transfer behavior depending on frequency.

For the experiment, the steering system is connected with a linear actuator on one side. The steering column with steering wheel remains freely oscillating. The linear actuator imposes a sine sweep with a constant amplitude of force of 600 N and frequencies from one up to 30 Hertz. The transfer characteristic between the impressed forces and the measured torsion bar torque is then calculated.

The evaluated amplitude response of the reference steering system with conventional motor feedforward control is shown in Figure 12 and Figure 13 in blue. It can be seen in the illustration that the amplitude gain already greatly decreases at a frequency of about two Hertz up to nine Hertz. At the frequency of 13 Hertz, a strong increase in amplitude can be seen. At this point, the active steering system has the resonance frequency.

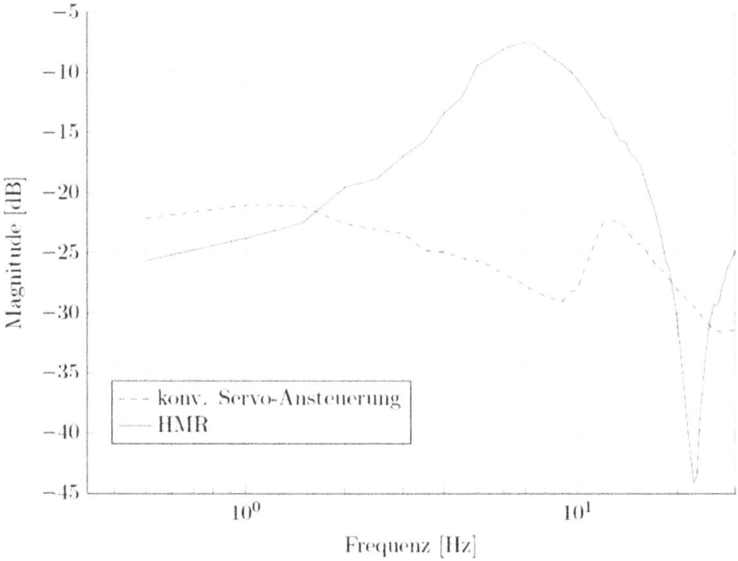


Figure 12: Transfer behavior HMR versus conventionally controlled system

The amplitude response of the HMR concept is pictured in red in Figure 12. It should be noted that the compared steering software concepts largely differ in the mechanics and the calibration environment, since they were developed by two different steering system suppliers. However, a qualitative comparison can be done anyway, because the HMR system, despite lower friction, has an approximately 30 percent higher inertial mass on the rack from the servo drive. It becomes clear, that the HMR system leads to a substantial increase of the amplitude gain up to a frequency of about 15 Hertz, thus covering the transmission of useful feedback information for the driver. As from 15 Hertz mainly interference information affect the steering system [4], the transmission in this range is clearly being attenuated by the frequency-related controller design.

The height and progression of the amplitude gain in the frequency range of useful information are adapted during the calibration process using different vehicle maneuvers and by changing the controller parametrization based on the needs of steering feel.

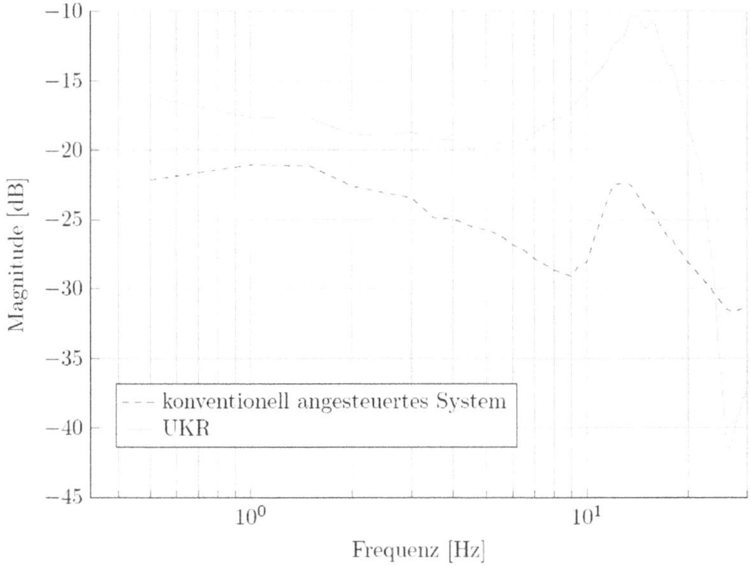


Figure 13: Transfer behavior UKR versus conventionally controlled system

Figure 13 shows the transfer behavior of the UKR steering system compared with conventional control. The two concepts were examined on the same reference steering system.

In the frequency range from zero to about seven Hertz, the amplitude gain is significantly increased through the use of the UKR. It is also clear that the progression in this range becomes nearly constant. This reduces the influence of the controller on the calibration of the steering feeling in the normal steering range.

In the spectrum from seven to 15 Hertz, the useful feedback information for the driver is located [4]. To forward these stimulations to the driver through the steering system, the amplitude gain in this range is increased significantly by the UKR. The height and the exact progression of this increase are determined in vehicle tests and taken into account in the design. The limits are set by the desired steering comfort and the stability of the system.

Above the frequency of 15 Hertz, the feedback is reduced with high gradient to filter out the unnecessary information in the steering wheel torque. This first of all improves steering comfort, but also the reference action and steering precision.

A different progression becomes evident in a comparison of the amplitude response of the HMR from Figure 12 with the UKR in Figure 13. This dissimilarity is based on the different design criteria for the different vehicles in which the controls are used and significantly different mechanics concepts of the steering system. However, both concepts show their ability to increase the feedback in the relevant frequency ranges considerably, compared to the feedforward control.

# Summary and Outlook

To establish a sports car steering feel with optimal road feedback independent of steering system suppliers, Porsche has developed its own control concept on function software level.

The crucial importance of the road feedback information in sports cars was shown.

Then, currently deployed control approaches including the new Porsche UKR were presented and evaluated on the basis of the behavior in the feedback path.

Finally, an objectification method for feedback evaluation was presented and the previously discussed control approaches were examined with respect to the transferability of external stimulation to the driver.

The new Porsche control concept UKR comes to use for the first time in the Porsche Macan GTS and is going to be deployed in future Porsche vehicle projects.

# References

[1]Hsu, H.; Harrer, M.; Gaedke, A.; Grüner, W.: *Die neue EPSapa im Porsche 911 – Anforderungen und Auslegung einer Lenkung für den Sportwagen.* In: chassis.tech plus, 3rd International Munich Chassis Symposium. Munich, 2012

[2]Pfeffer, P.; Harrer, M.: Lenkungshandbuch: *Lenksysteme, Lenkgefühl, Fahrdynamik von Kraftfahrzeugen.* 2., überarb. u. erg. Aufl. 2013, Springer Fachmedien Wiesbaden

[3]Reimpell, J.: Fahrwerktechnik: Lenkung. *Anforderungen, Mangel, Lenkarbeit, Lenkelastizität, Einzelteile der Lenkanlage, Lenkübersetzung, Lenkkinematik bei Starrachsen und Einzelradaufhängungen.* Würzburg: Vogel, 1984 (Vogel-Fachbuch Technik)

[4]Brunn, P.; Harrer, M.: *Objektivierung der Lenkungsrückmeldung.* In: TPA Wien, 2004

[5]Lunkeit, D.: *Regelungstechnische Potenziale zum Rückmeldungs- und Rückstellverhalten elektromechanischer Lenksysteme.* In: chassis.tech plus, 1st International Munich Chassis Symposium. Munich, 2010

# Steering System Fingerprint – a tool for steering system performance check and overview

Frank Esser, Supervisor Vehicle Dynamics

Methods & Tools, T. Vercammen,

Ford-Werke GmbH

# 1 Introduction

Since more than 15 years Vehicle Dynamics is a key attribute defining the character of Ford vehicle, starting in Europe with the first generation Ford Focus, Ford Puma and Ka in the late $90^{th}$. Key driver to get this character consistent into the vehicles was in the first step a common understanding about the way the character should look like and how trade-offs between the sub-attributes, especially between steering / handling and ride needs to be set.

This has been realized with the set up of a detailed verbal description by the experts, alignment with the management, and finally translation into easy understandable terms to communicate to customers and journalists.

On the other hand it was obvious that to realize the character consistently over time and cross carline, objective criteria would be very helpful. This wish leads to strong efforts transforming the verbal descriptions in objective criteria, metrics out of standardized open loop tests.

A 'steering fingerprint' as a summary of metrics out of this tests directly linked to subjective steering attribute characteristics has been established, knowing that these criteria would not be able to describe the steering character completely, but giving a consistent baseline for the final refinement, which still needs to be done supported by the knowledge and feedback of the application specialists.

The steering fingerprint has been refined over the years and adopted to the new steering technologies, especially EPS, getting implemented into the marked. It needs to be said that with implementation of EPS the amount of tests to be done to get the complete picture of the steering character has significant increased (e.g. much more vehicle velocities needs to be checked) and can only be covered for all derivatives by the support of CAE.

From the beginning on the idea has been there to break down the expected full vehicle characteristic into steering system and component requirement. Different approaches have been taken, but a clear brake down is still difficult. The need of refinement of every single component of the steering system was well understood, but a standard to check the sum up of the single component performances and their impact to the system performance was missing.

This gap should be closed by the development of the Steering System Finger Print (SSFP) shown in the report.

In summary the SSFP should be capable to:

- Sum up of performance component to system
- Comparable between carlines and systems
- Assessable before vehicle is available
- Should cover hardware and software
- Focus on performance criteria
  (error states covered in separate assessment)

# 2 Application in the development process

One of the major goals in the development of the SSFP was to make it applicable over the complete development of the vehicle and steering system. This effects the setup for the testing significantly.

For the first application will be fore *benchmarking*. Setting up the complete electric architecture is not possible for competition systems. Therefore a subset of the tests to be able to run power of has been developed. The power off setup can also be applied for new systems without a reasonable steering tune.

More information is needed for a proper *system selection* decision. To support this, power on tests needs to be added to the SSFP.

Same is true for the *performance break down* in the system, where both test set ups, power on and power off needs to be reviewed to draw the right conclusions.

Parallel to the implementation of the steering systems in prototype vehicles the systems will be installed on the SSTM and with the summary of the results in the SSFP and easy first *check of the development parts and systems* is possible.

Checks with the *final development systems* and tunes before releasing for production can be supported with the SSFP as well as a check of the *first mass production systems* and parts

These typical applications are summarized in the development 'V' shown in figure 1

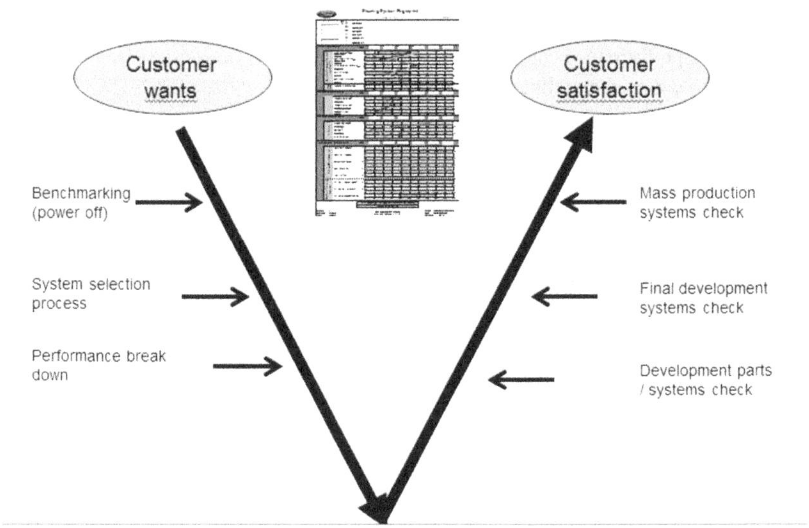


Figure 1: application of the SSFP in the development process

# 3 Test rig: Steering System Test Machine

The development of the common set of tests has been driven parallel on two test rigs, one in the Research and Innovation Center (RIC) at Ford in Dearborn (US) and one located in Lommel Proving Ground (Belgium), Fords main European Vehicle Dynamics test facility. The rig in Europe has been developed together with Anthony Best Dynamics (ABD) and is in use since 2006 for research work as well as for development support (figure 2).

Figure 2: Steering System Test Machine (SSTM)

The rig can be split in two big parts. On one hand the rack table to position the steering gear. Below the table two large electric actuators are placed capable to apply loads up to 9 kN and 30 Hz input frequency. One the other hand the column table enabling the position of the column exactly in vehicle position and being the baseline to apply input via the steering robot.

More details on the rig are published in the ATZ 2007.

One goal during the selection and definition of the tests was to make them not specific to one rig and keep them as generic as possible. Therefore back to back comparisons were made with the rig in the US, build by a different supplier. The generic set up of the tests should also help to cascade the requirements and findings to the suppliers to ease the common development and comparison of data.

# 4 Steering System Finger Print (SSFP) setup

## 4.1 Characteristics and details

A lot of different tests on complete steering systems and components have been developed and applied over the years. The goal for the SSFP was to standardize the test parameters and select the most important tests and metrics to describe the characteristics:

- Friction
- Compliance
- On Center behaviour
- Transmissibility

in a way that make comparison between systems and the link to the steering system performance as assessed by the steering experts easy. In addition the use of the well-known layout out of the full vehicle attribute assessment eases the communication to non-experts. The structure of the fingerprint can be seen in figure 3.

Two versions of the fingerprint are set up. One version is summarizing non-powered EPS (power off) and one the full operating system, getting all needed vehicle information provided to the system (e.g. velocity).

The approach for both is the same:

- On top the legend with a menu to select the systems to be compared
- The tests are grouped according the 4 characteristics mentioned above
- Generic or system specific targets can be defined
- Index numbers indicate how good the targets are met
- Summary stays on one sheet; link to the x-plots in the file

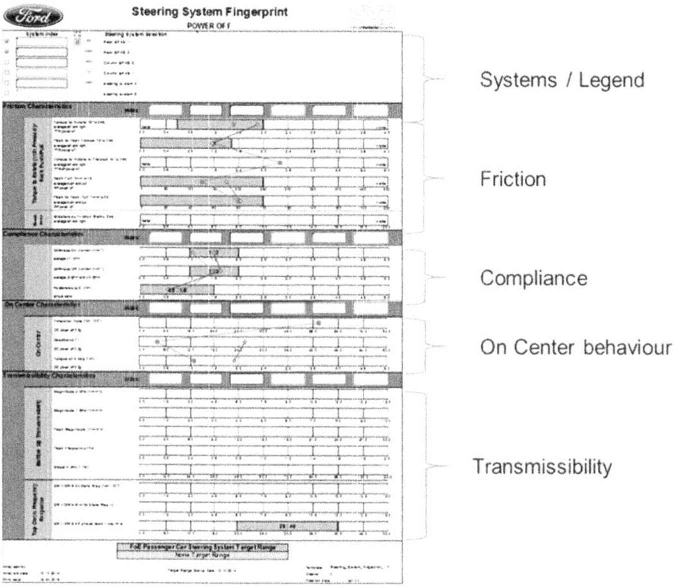


Figure 3: SSFP characteristics

## 4.2 Test details

As mentioned before a lot of different tests and test configurations have been used in the past and goal of the development of the fingerprint was to rely where possible on test set ups used in the past, already used during other tests (e.g. steering system test during full vehicle K&C steering tests) or already defined in specifications communicated to our suppliers. But for some characteristics new tests or metrics need to be developed and on some tests still different test set ups needs to be performed (e.g. different steering wheel angle input speeds or tie rod loads) to get the complete picture. In this case the setup being most representative is selected to be shown in the fingerprint.

The four characteristic and the relevant test scenarios are described in the next points.

### 4.2.1 Friction tests

To realize a clear feedback and a clean center feel, friction and hysteresis in the steering system has always been critical reviewed and specified with different component tests. With the setup of the complete system these tests have been extended using the same basic specifications.

A) Torque to rotate

- Power on / off; +/- 360 deg SWA
- Input via Steering wheel (5 – 30 deg /s)
- Metric: Avg torque; torque peap to peak
- Modification: with preload (1000N on, 200N off)

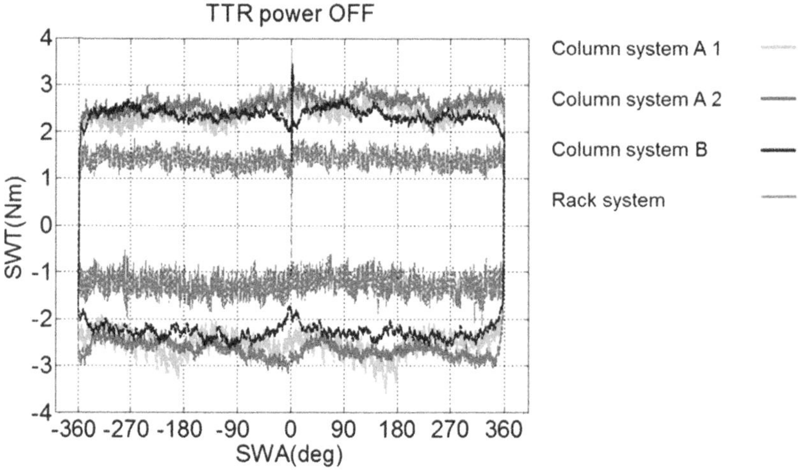


Figure 4: Torque to rotate test

B) Rack Pull test:

- Power off, 95% lock to lock
- Input via tie rod 1.0 mm/s – 4 mm/s
- Metric: Avg force, peak to peak force

C) Stick-Slip test

- Power on / off; +/- 5 deg SWA
- Input via Steering wheel 1.0 deg /s
- Up to 90 cycles (900 sec)
- Metric: torque peak to peak max; increase avg. torque

### 4.2.2 Compliance tests

A) System compliance

- Power off; +/- 8 Nm SW torque
- Rack mechanically blocked
- Apply rate 0,25 Nm/sec
- Metric: stiffness

    i. OnCenter (+/- 1Nm)
    ii. OffCenter (3-5Nm)
    iii. Hysteresis @ 0 deg SWA

- Modification: with preload (1000N on, 200N off)

### 4.2.3 On Center behaviour tests

- Power on / off
- Based on On Censter / Sine Steer road test

    i. 120 km/h – 0.2g
    ii. 50/75 km/h – 0.4g

- Vehicle specific and generic inputs (SWA, tie rod loads)

    i. Power off only generic

- Metrics (analog to road tests)

    i. OC torsional rate (Nm/ deg)
    ii. Hysteresis @ 0 Nm (deg)
    iii. Hysteresis @ 0 SWA (Nm)

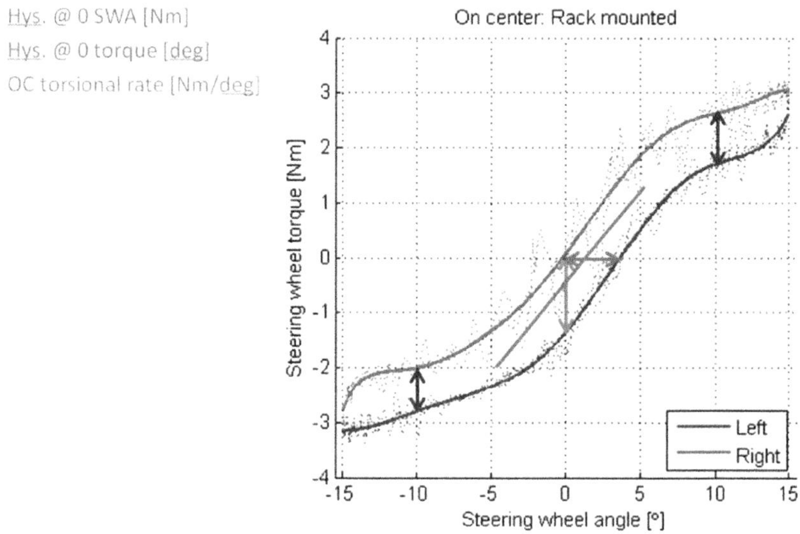


Figure 5: On Center test - rack EPS versus column EPS

### 4.2.4 Transmissibility tests

A) Bottom up transmissibility

- Power on / off
- Input via tie rod; sinusoidal sweep (2000N/0.1Hz → 200N/15Hz)
- Steering wheel locked
- Metrics

    i. Phase change / consistency
    ii. Magnitude @0.3 Hz and 1.5Hz

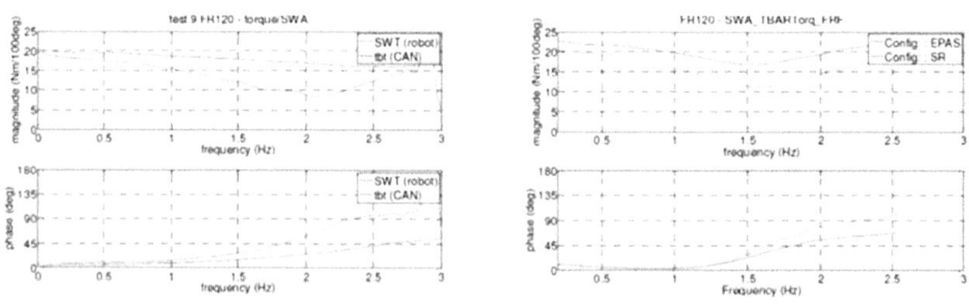


Figure 6: Transmissibility bottom up test - rack EPS versus Column EPS

B) Top down transmissibility

- Power on / off
- Input via steering wheel
- Input linked to FR road test input (carline specific tie rod force model)
- Metrics linked to FR vehicle test

# 5 SSFP examples

Only in the presentation.

# 6 Summary and Outlook

With the Steering System Finger Print (SSFP) a tool has been established to help comparing complete steering systems in an efficient way, analogue to what has been established on vehicle level already.

Baseline was a standardization of the important tests for friction, compliance, On center feel and transmissibility, together with a selection of the most relevant metrics.

Further refinement of the tests and metric might be needed with the application of the tool for various vehicle programs. In addition the target ranges needs to be verified and adopted where needed.

Communication to the steering suppliers, where not already happen, will help to get the needed data to use the SSFP as standard tool to review system performance in a quick summary.

# Model-based steering ECU calibration on a steering-in-the-loop test bench

H.-M. Koegeler, B. Schick, AVL List GmbH Graz

P. E. Pfeffer, A. Contini, Munich University of Applied Science

M. Lugert, T. Schöning, Hyundai Motor Europe Technical Center

# Introduction

Modern electric power steering (EPS) systems allow a flexible adaptation of the same steering hardware to different vehicle types by calibration of the corresponding control unit (ECU). As the steering behavior is one of the major factors influencing the end customers` buying decision and driving experience, the steering ECU calibration has to fit very well to the customer requirements. A growing number of functions in the steering ECU offer increasing possibilities to influence comfort, safety and steering feel and must be validated in different driving maneuvers (figure 1). However, several trade-offs of these characteristics have to be solved such as the driver's steering effort vs. the steering feedback while cornering.

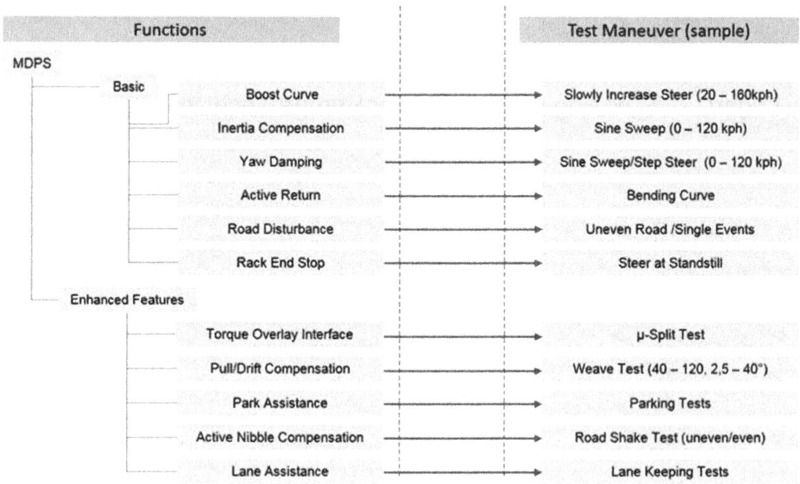


Figure 1: Functional hierarchy EPS (sample) and test maneuvers for tuning and validation

The resulting steering feel as important part of the driver's perception of a vehicle's handling characteristics is increasingly gaining importance as a customer-relevant criterion and as an element of the brand's DNA. A high-quality steering system will make the driver feel safe and comfortable; the aspect of driving pleasure is addressed by matching the driver input and the steering feedback with the vehicle's reaction. State-of-the-art simulation with CarMaker and hardware-in-the-loop (HiL) test rigs offer the possibilities needed for a reproducible and efficient tuning process in laboratory conditions. For example, the steering boost curve can be adjusted depending on the steering wheel torque and the vehicle velocity (figure 2). Hyundai Motor Europe

Technical Center (HMETC) has proven the feasibility of virtually tuning the steering logic of an electrically assisted power steering (EPS) system. Automation will be a major advantage of this fully objective tuning process [1], [2].

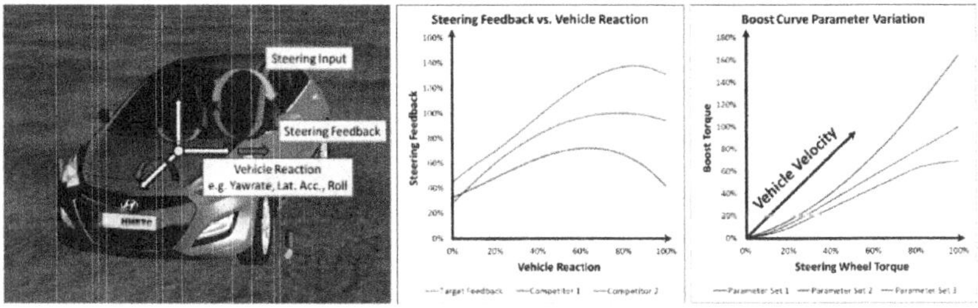


Figure 2: Steering input and feedback with vehicle reaction and the resulting task of boost curve calibration

# Model Based Tuning

Figure 3 shows a generalized development environment, which appears in an analog way for any Unit Under Test (UUT), like an engine, a gearbox or a power steering.

As shown in part a) of figure 3, in case of manual tuning, the engineer operates the UUT with a certain setting of control parameters in certain maneuvers. The engineer observes the behavior of the UUT and performs a judgment according to his experience. He or she defines the next setting with the intention to better approach the desired behavior.

This process might become complex in case of many relevant tuning parameters. The quality of the optimization depends on whether the engineer considers all the parameters that are relevant for the desired behavior and on the experience of the engineer. Further subjective decisions are not traceable in general and a reuse is not possible for future projects, e.g. if the parameters have to be tuned for different driving modes. Thus, in case of the manual tuning process, each variant has to be calibrated separately.

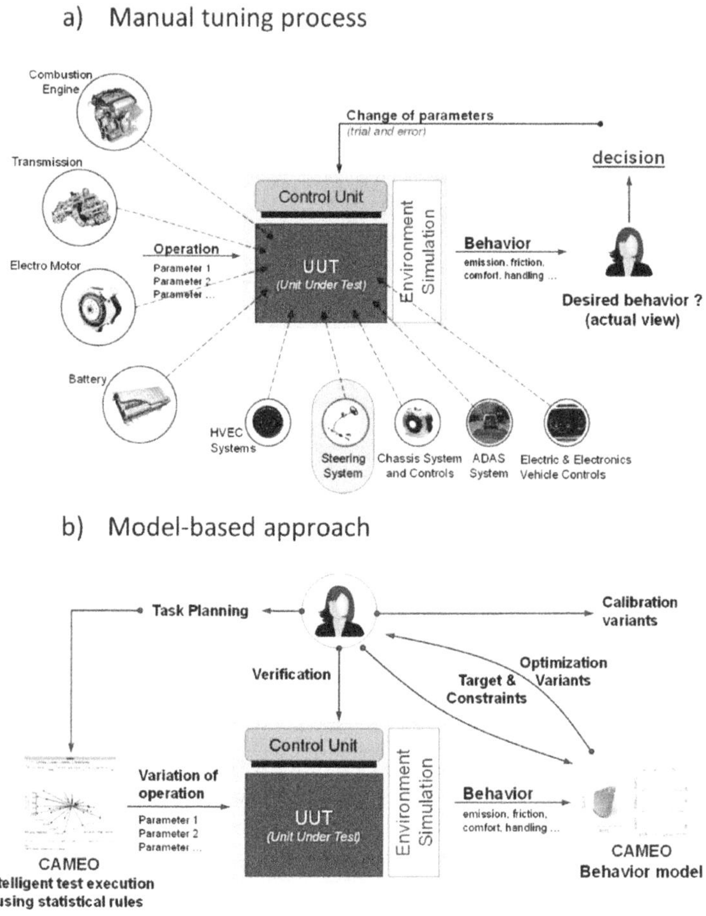


Figure 3: Calibration and optimization methodology in a generalized development environment

In part b) of figure 3, the model-based approach is shown: Here, the user starts with a task planning for a measurement series. The user aims to guide the experiments in a way allowing him/her to obtain behavior models for the evaluation criteria, which describe the UUT behavior for a wide range of possible calibration settings. These tests can be run fully automatically over several hours, like proven in the area of engine calibration e.g. [3, 4, 5].

The optimization and calibration process uses these models. Thus, the engineer gains several calibrations with different target(s). Multi-objective optimization and trade-off analyses offer additional mathematical support to the calibration engineer. In this paper, the model-based steering ECU calibration on a Steering-in-the-Loop test bench will be discussed by the example of the boost curve calibration.

Figure 4 shows the Steering-in-the-Loop test bench architecture. The system setup consists of two rack force actuators and a steering wheel actuator. In order to make sure that the mechanical behavior is comparable, the test subject is built on a stiff holder with the same fixing points and geometries as in a car. The electrical behavior is also the same as inside the car. The rest bus simulation is needed for the communication between the electrical control unit (ECU) of the steering system and the steering system test bench (STB). The communication between the ECU and the STB runs via a CAN or FlexRay interface on the real-time environment. The dSPACE real-time system serves to control the actuators and to run the virtual car model [6, 7].

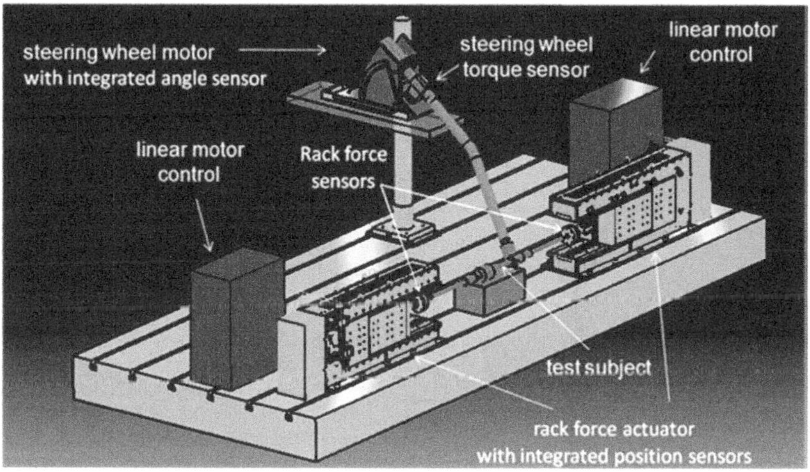


Figure 4: System setup of the steering system test bench at Munich University of Applied Sciences (MUAS)

The integration of a virtual car model into the real-time simulation environment enables the hardware-in-the-loop simulation. IPG CarMaker calculates the rack force for specific steering wheel angle inputs. The output quantities of the test subject are the resulting steering wheel torque and the rack position (Figure 5). The closed-loop simulation principle for the steer-by-torque control mode works complementary to the described steer-by-angle simulation.

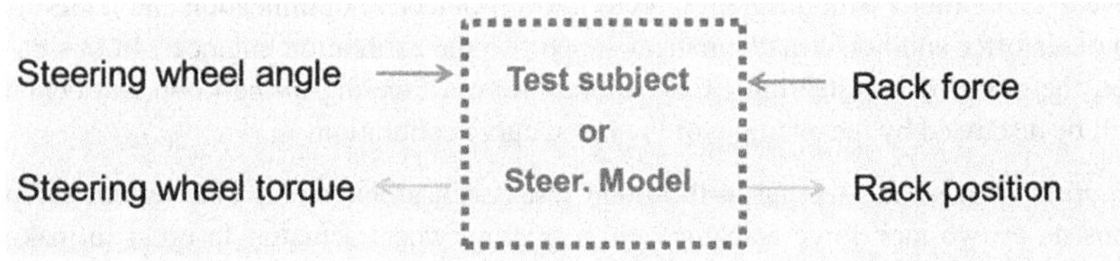


Figure 5: Steering system interface for steer-by-angle testing

The software to operate this Steering-in-the-Loop test bench runs on a workstation PC equipped with a real-time operating system.

In order to perform automated test runs according to figure 3b AVL-CAMEO is connected to both: The IPG CarMaker real-time vehicle and environment simulation as well as to the ECU application tool according to the ASAP 3 standard. The tuning parameters in the steering ECU can be overwritten for a certain maneuver by this application tool (figure 6).

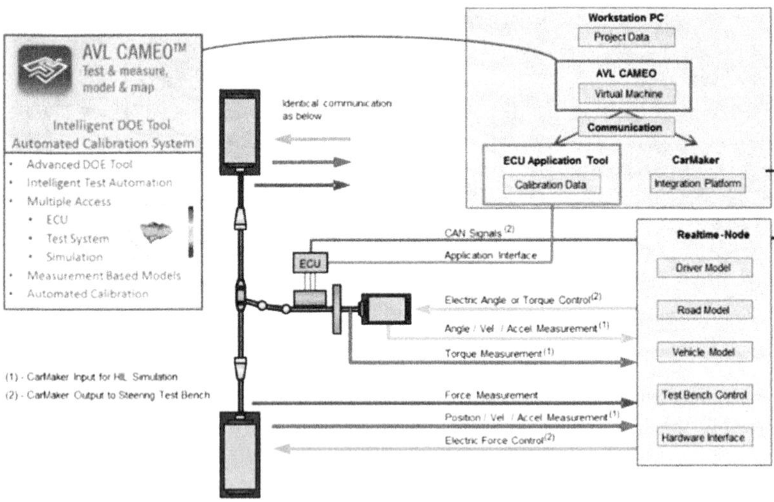


Figure 6: Hardware-in-the-Loop integration with AVL CAMEO

# Task Planning, Test Preparation and Test Execution

According to figure 1 and 2, the shape of the boost curve, which can be described by 5 to 9 parameters for each vehicle velocity, should be calibrated in a way that the driver feels a suitable steering feedback while he is driving a "Slowly increased Steering test" (SiS test).

The calibration engineer starts the Parameterization with the upload of parameters into the CAMEO database: on the one hand, the variation parameters, which can be set on the two systems "CarMaker" and the "ASAP 3 –System", and on the other hand the resulting key values, which characterize the driver feedback. Such key values will be transferred back to CAMEO after the maneuver has been finished. Using the same naming for the parameters, the calibration engineer parameterizes a DoE based test run in CAMEO, which triggers several SiS tests with different calibration settings. For the example shown, the 5 most relevant ECU parameters (depending on Var0 to Var4) were selected and the velocity of the vehicle in the SiS test was varied in a range from 90 to 160 km/h. The result of using the Design of Experiment techniques (DoE), which supposed high nonlinearity of the measured responses, was a list of 326 SiS-tests. It took less than 6 hours to run these tests (figure 7).

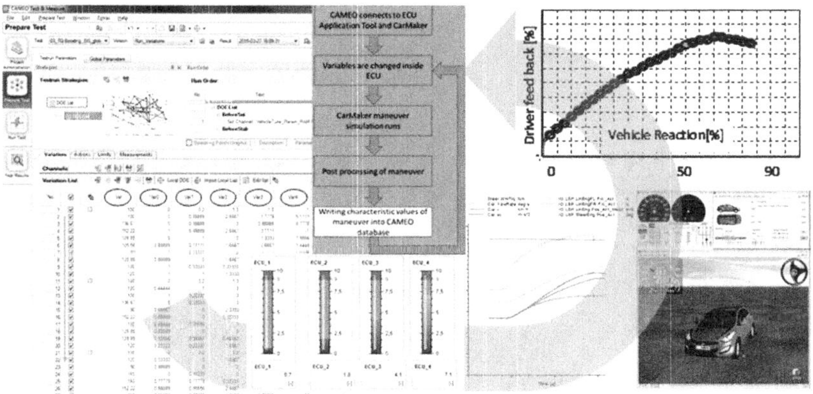


Figure 7: Test automation of the MUAS (compared to Figure 4) by AVL – CAMEO

# Raw Data Analysis

28 to 30 single steering feedback values at defined vehicle reaction situations are calculated out of the resulting low pass filtered curves during the SiS test. Figure 8 shows one of these values (Driver_feed_back_at_50%_vehicle_reaction) for the 326 SiS tests as an example. It can be seen that the reproduction of the settings (repetition points colored in green) causes only very small noise in comparison to the effect that is caused by different tunings. The reproducibility provides important information about the quality of the experiments. With very high probability, the reproducibility on the Steering-in-the-loop test bench is better than the one in the real car on proving grounds. The reasons for this are additional disturbances such as temperature changes, wind or the unevenness of the road which occur on the proving ground but not on the bench.

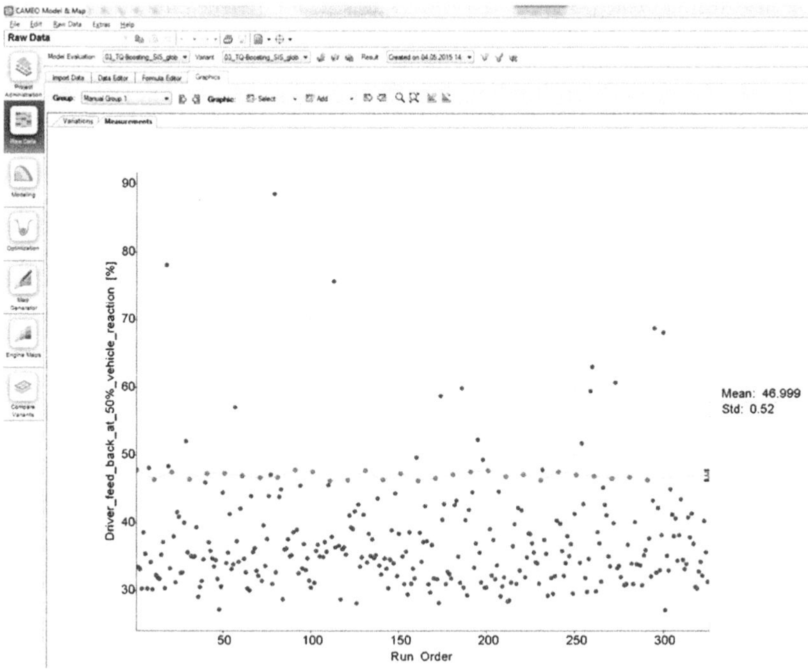


Figure 8: Raw data view of one driver feedback result with repetition point statistic

# Modelling

The target of the measurement series is to get behavior models. Thus, the next evaluation step is to build behavior models that will be the base for the following steps to optimize the tuning for the desired behavior. Figure 9 shows the result of that modeling process. The engineer gets support by an automatic suggestion of the model type taking into account several model quality criteria like "Goodness of fit" or "Significance compared to Repetition point noise" [8]. Independent from the model building process, the model visualization is important. As it can be seen on the left side of figure 9, a conventional 3D visualization can only show the dependency of the result as a function of two input variables (Var0 and Var1) in this example.

The intersection graph on the right side of figure 9 shows all six model dependencies (from velocity, var0, var1, var2, var3 and var4) around a certain cursor position. This graph gives the user an immediate impression which variable affects the desired behavior and how strong the influence is.

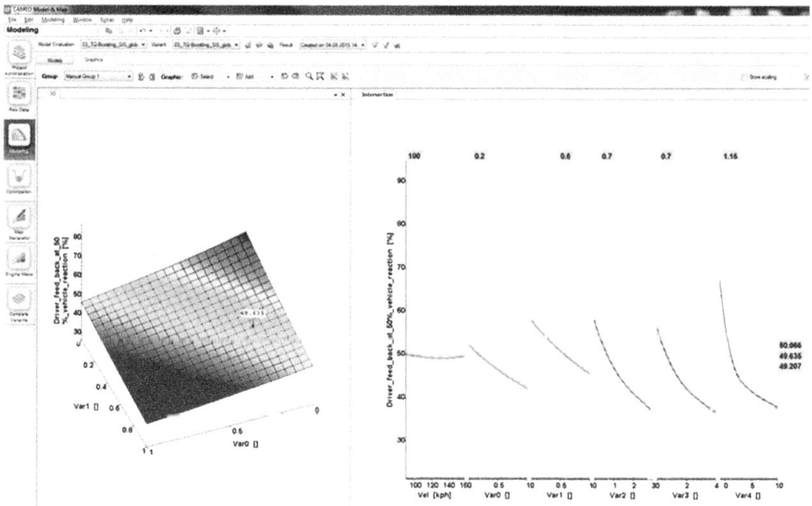


Figure 9: Behavior model of one driver feedback result in 3D mode and intersection graph with 95% confidence intervals for the model.

Only by using such behavior models in a manual way, the tuning process, which remains manual, can be run much faster and traceable than performing and evaluating new SiS tests for each setting. In addition, the calibration engineer can get support by algorithms that optimize several behavior targets simultaneously.

## Optimization

In our example we would like to get 4 driver feedback values (at 10%, 30% , 50% and 70% vehicle reaction in the SiS-test) under the constraint of certain target ranges fitting a desired company steering DNA. Moreover, the function of all 28 to 30 driver feedback values should show a smooth behavior as function of the vehicle reaction. This smoothness is ensured by minimizing the smoothness indicator as target function for the optimization. Figure 10 shows the corresponding suggestions delivered by CAMEO for the target speed 100 km/h.

The left upper part shows the model section of the driver feedback as function of the vehicle reaction. This is gained by using the 28 to 30 driver feedback values as one model with the vehicle reaction as additional input parameter. This view gives the calibration engineer the perfect feeling for the immediate result of his tuning.

The right side shows the model of the smoothness index as well as the four driver feedback models at defined vehicle reaction values together with its ranges as constraint values (D_f_b_xx%). The green bars on the x-axes show the design space, which is the area of interpolation.

Since it turned out, that it is not easy to meet all ranges with a smooth curve – especially the D_f_b_50% model tended to be too low – a multi-objective optimization was used to find the right decision. The picture on the bottom left shows the corresponding trade-off behavior: Out of 25000 randomly distributed settings within the design space, the gray ones do not fulfill all the constraints at the same time and the yellow ones do. The borderline of that definition is the so-called Pareto front, shown in blue. The calibration engineer can now select between highest "D_f_b_50%" and best "smoothness_index" on this basis. He can be sure that his choice will be the best obtainable compromise between both.

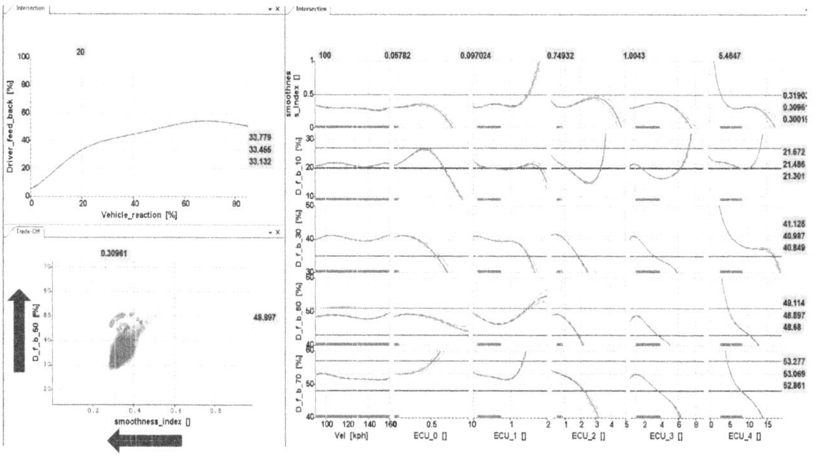


Figure 10: Optimization of the calibration setting at the vehicle velocity of 100 km/h

Similarly, the calibration engineer can derive traceable boost curve calibrations for different vehicle speeds out of this one measurement series.

# Conclusion

Virtual tuning of steering feel on a HiL test rig together with an optimization tool can be a powerful combination to provide an efficient process for the development of a brand's steering DNA. The classic approach of subjective steering tuning by expert drivers on proving grounds and public roads can be supported and accelerated by delivering parameter sets matching the main targets. The application of DoE studies with CAMEO increases the number of tunings significantly compared to the manual generation on the test rig. In addition, the number of targets and tests to match these targets can be increased. This allows the engineer to cover a wider range of steering feel in more driving situations. Achieving the desired steering characteristic by the use of the presented approach will underline the unique brand DNA for both, Hyundai and Kia cars.

# References

[1] T. Schöning (HMETC GmbH): Improvement of Steering Feel – Virtual Approach with HiL, IPG apply & innovate 2014

[2] A. Honisch, M. Lugert, T. Schöning (HMETC GmbH), S. Hakuli (IPG Automotive GmbH): Improvement of Steering Feel – Virtual Approach with HiL, ATZ 06, 2015, Volume 117

[3] Koegeler, H. – M.; Fürhapter A.; Mayer M.; Gschweitl K.; "DGI-Engine Calibration, Using New Methodology with CAMEO". In SAE NA, Capri – Italy, 23 – 27. September2001

[4] Castagna, E.; Biondo, M.; Cottrell, J,; Altenstrasser, H.; Beidl, Ch.; Koegeler, H.-M.; Schuch, N.: Multiple Tier 3 Engine Applications based on global modelling, MTZ 6/2007

[5] T. Fortuna, T.; Koegeler, H.-M.; Kordon, M.; Vitale, G.: DoE und darüber hinaus: Die Evolution des modellbasierten Entwicklungsansatzes – wie gesetzliche Vorgaben die Methodik verändern, ATZ 2/2015

[6] Schimpf R.; Pfeffer, P.: Entwicklung und Test von Lenksystemen auf einem Lenksystemprüfstand, VDI Wissensforum Reifen – Fahrwerk – Fahrbahn 2013

[7] Specka, F.; Contini, A.: Bewertung und Optimierung des Lenkverhaltens im Gesamtfahrzeug am „Steering-in-the-Loop" Prüfstand, VDI Wissensforum Reifen – Fahrwerk – Fahrbahn 2013

[8] Keuth N.; Pflügl H.: „Intelligent Neuronal Networks (INN) – a New Modeling Approach for the Global Model Based Optimization of Combustion Engines and Powertrain." Conference: DoE in Engine Development, Berlin 5/2007

# Test infrastructure for EPS steering systems – balancing between requirement-based, experience-based and free testing

Thomas Maur,

EPS Core System / System Integration & Testing,
TRW Automotive GmbH

# Introduction

Based on the increasing introduction of electric and micro processor technology during the last decades – not only in automotive industry – systematic analytical testing of software (SW) is developing as an independent & autonomous acting engineering discipline. These activities are embedded in the product development life cycle and mainly support the product release process.

This product development life cycle is split in different process activities. Referencing on the standardized V-Model® with ten engineering related process types, testing, verification & validation of SW functionalities covers five of ten of these process types (ENG.6 -10).

In order to identify the capability of all relevant SW development process types different types of rating references models are currently in use: CMMI® / A-SPICE® (see e.g.: HIS scope) / ISO 26262.

The process related work products (e.g. test content & schedule plan, work instructions, result & metric reports) have to follow different quality criterions.

Also the existence of a test method library and the usage of the right method during testing (e.g.: identification of right test case design) is one of the most important items to create cost & time efficient test activities.

As an example, this paper has a focus on the three top level test methods:

- Free testing
- Experience based testing
- Requirement based testing

In order to follow the A-SPICE® / ISO 26262 approach, requirement based testing is the preferred method, which a test organization has to support.

In this paper we report under which kind of circumstances “free & experience” based testing is recommended to be used, in order to develop at least requirement based testing.

At first, it would be helpful to know something about the EPS test infrastructure of our test organization. With this infrastructure we support all three test methods in a process conform test life cycle development, which is synchronized to the function & product life cycle.

The conclusion of this paper is a proposal under which kind of circumstances we give a higher priority to „free testing“, “experience based test” than “requirement based testing”

## Goals of Testing

Following short summarizing identify the main goals of testing:

- Provide evidence, that the product (delivery):
    - covers all related safety targets,
    - will functionally support all related use cases
    - support the next integration step without unexpected behavior,
    - fulfill all measurable product related performance,
    - has the reliability as expected.
- All relevant deviations from expected behavior are identified, documented and rated as product hand over of a delivery, regarding:
    - Relevant for safety
    - Reliability and
    - Performance.
- Confirmation of „workarounds“ of identified deviations
- Risk assessment based on:
    - the development life cycle state
    - planned systemic use cases against archived test deepness.

## List of the most important test rules

- Entire testing is not possible
- SW is per definition not fault free
- SW faults are systematic faults (!)
- Testing identifies the occurrence of faults
- In order to avoid costs, SW faults have to be detected as soon as possible in the SW development life cycle.
- Test repetition did not have any value
- Risk based Testing: balanced test budget (cost & capability) with archived test deepness
- ...

# Clarification

- Testing did not provide evidence, that the product is fault free

# Testing under test context of Quality Management

As described in **„International Software Testing Qualifications Board"** (ISTQB®), test activities also follows aspects of quality assurance, which is a part of QM. Following diagram shows the analytical dynamical test methods (see back background part). A test organization has to define their test activities against these test methods.

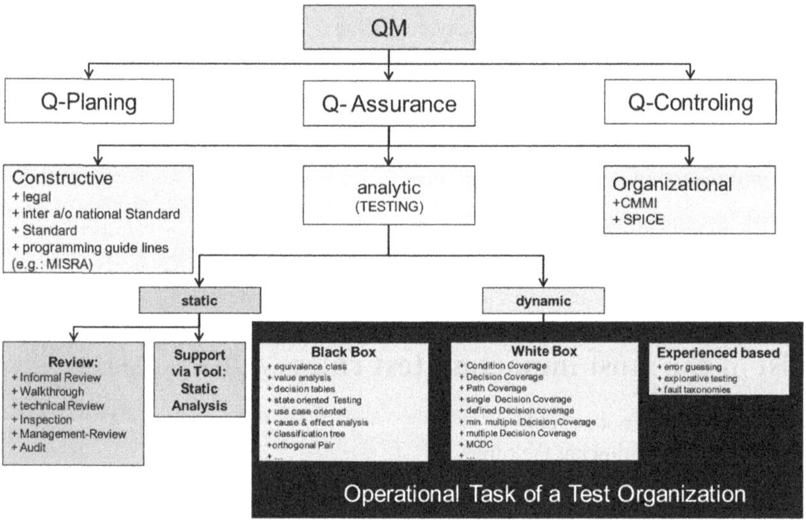


[1] Source Reference: ISTQB® – Certified Test Manager

The TRW EPS test organization is responsible for all test activities which are analytical dynamical test activities.

The "White Box Methods" of this diagram, refers to the part of SW Unit Testing only. The test method group „Black Box" & „Experienced based" is used for:

- ENG.7 SW Integration Testing against e.g. SW Architecture requirements
- ENG.8: SW Testing against SW Requirements
- ENG.9: System Integration against System Architecture Requirements
- ENG.10: System Testing against System Requirements

The test organization supports the analytic static testing, if requested.

# Test Process Structure & Documents

The structure and content of the test documents follows the IEEE-829 (SW Test Documentation) Standard.

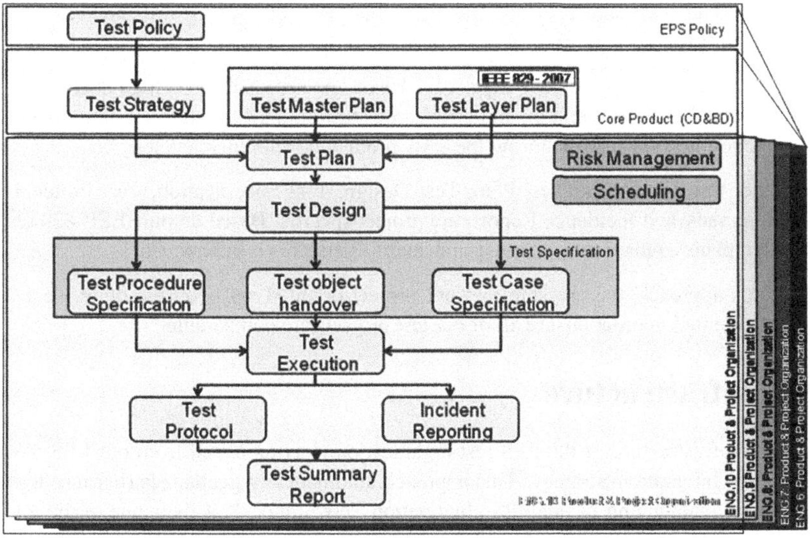


[2] Structure of Test documents against IEEE-829

„Test Policy“: Here you will find management commitment. In this policy our EPS product management engages them to install and maintain a test organization. Common expectations and tasks of a test organization are addressed (like Key Process Indicators a/o Measurement of performance …) in this paper. Based on IEEE-829, there is no clear content reference for the test policy.

For all other test documents, IEEE-829 has a clear structure and content recommendation.

Due to the fact, that IEEE-829 only was focussed on software; we adapted the test strategy to suit our EPS mechatronic system. That means, our test process covers all test activities from SW Unit Testing up to systematic integration of our software functionalities on vehicle level.

Therefore, our "system" definition matches at the end on vehicle level. That means Software itself is a sub-system or a system-element of your EPS system architecture.

Based on this, the breakdown activities Test Strategy, Master Test Plan and Test Layer Plan are product (see EPS Technology e.g. Colum & Belt Drive) and project (OEM) independent.

This approach guarantees a product release quality, which is comparable between different product types and projects. Based on this our reuse part of test methods, test tools & libraries…is very high.

This pre-condition supports our experience based test approach, with high quality and high amount of test automation and enables our test organization to release a product a/o function very efficiently during the early product development cycles.

All other test documents (Test Plan, Test Design, Test Specification, Test Protocol, Test Summary and Incidence Report) are project specific. Based on our IEEE-829 related templates, a maximum of reuse and quality standards is guaranteed.

With this approach, we are able to work project oriented and – on the other hand – quality oriented in order ensure all of our test projects are comparable.

## Test Infrastructure

Our test infrastructure of the EPS test organization is extrapolated from our EPS SW & System Integration Strategy. This approach automatically guarantees (because tool based) the maintaining of our EPS „Integration Test Strategy". This is one of the important base practices of ENG.7 (SW-Integration) and ENG.9 (SYS-Integration).

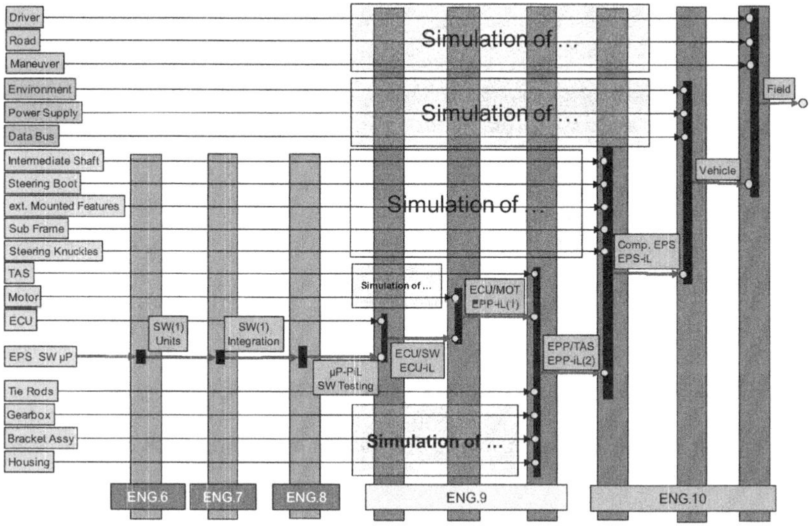


[3] Example TRW EPS Master Test Plan (MTP): "EPS Integration Tree"

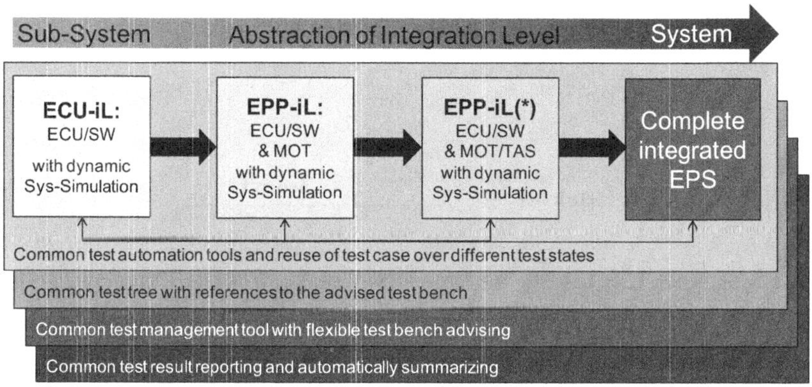


[4] Example: TRW EPS: Test Tool Infrastructure – System Integration (ENG.9)

Main properties of the test instance infrastructure:

- Simple comparison between deviations which are detected in various test states,
- High Reuse of test case generators and test control tools,
- Recording functionality of various test procedures based on various test bench types,
- Common test planning (tool & HR oriented) of all test activities,
- Embedded in one test management tool,
- Test design development and test result report covered in our requirement management tool (actually DOORS®),

Our slogan: "… Testing between „DOORS® & DOORS® " or „PTC® & PTC® " …"

## Fundamental Test Process (Test phases: P/P/P/P):

Our fundamental test process is split in four phases of testing, which has following dependencies (see picture).

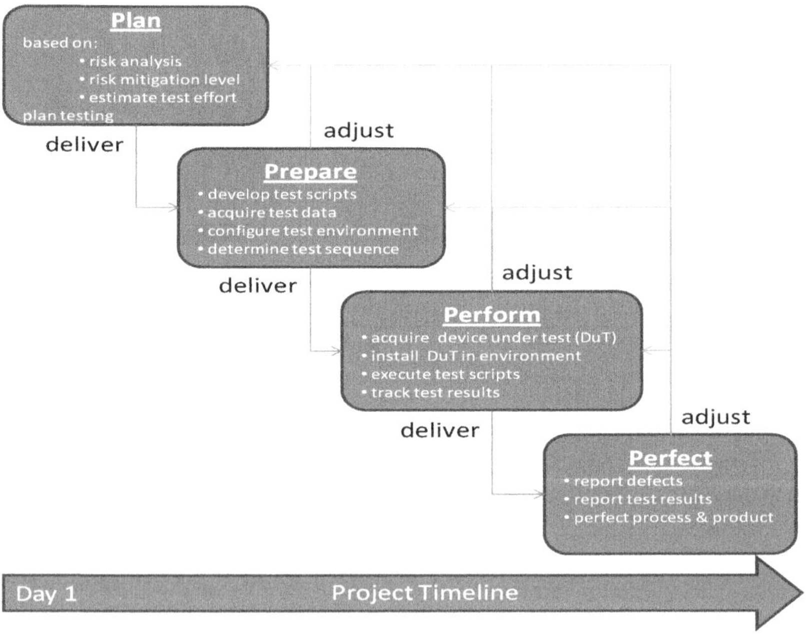


[5] Source: ISTQB® – Certified Test Manager

Plan: Test Manager [TM] takes care of these activities, planning split during project startup phase and during project deliveries.

Prepare: Test Manager setup this phase (fill in various information) and start a kickoff in order to hand over this phase to the test engineer (who is the owner of this phase).

Perform: Test Engineer is responsible and therefore the owner of this phase

Perfect: Test Engineer finalizes the active part of test activities. He supports the test manager with test results (test report, incident report, list of open point in order to start the problem resolution process) and his release recommendation. Archiving all test artifacts is also the responsibility of the test engineer.

Comment: In order to avoid unmanageable project risks, it is very important to involve a Test Manager very early (first day!) in the project planning a/o change request analysis phase.

# Testing of Requirements and/or Change Requests

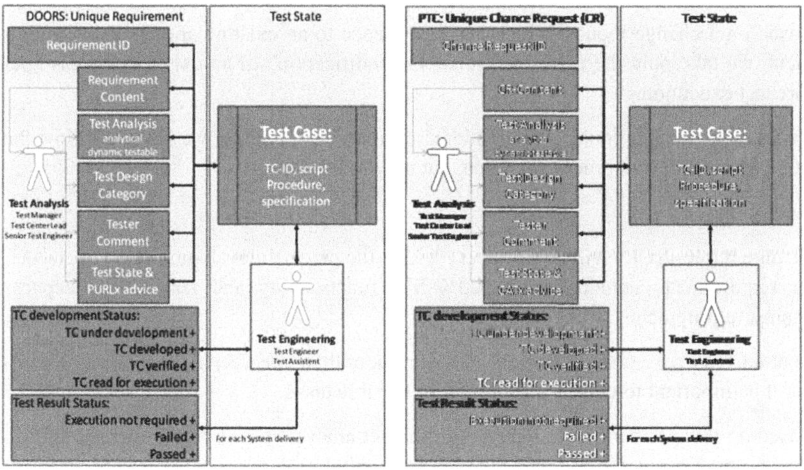


[6] TRW EPS: left side: requirements / right side: change request test analysis

Requirements based testing is established formally in the product life cycle process.

Test activities which are managed and planed in our requirement management tool (DOORS® a/o PTC®) pursue following targets:

- focusing of test case development against the unique requirement
- unmistakable allocation of requirement identification and related test case identification
- description of test design
- additional Information:
    - Definition of one or more test machines
    - relevance of test execution regarding release category level,
    - priority of execution (safety relevant, use case relevant, …),
    - …
- status report of test case development
- result reporting on test case versus requirement level

The picture [6] describes the difference between requirement based and change request test case development, progress reporting and test result reporting.

Based on the differences (we change the tool from DOORS® to PTC®) there is a high reuse potential.

Based on a change request which has a reference to an existing and testable requirement, we take only the decision: "reuse or modification" of an existing test case before test execution.

Based on a change request, which addresses a new SW feature, we have to follow the same test case development approach, but now in PTC®.

Definition:

Change request (CR) will address very late in the product development life cycle. After requirement content freeze, 100% SW functionality and 100% test deepness against requirements.

In order to cover a change request package (more than one CR) in a manageable manner it is important to cluster this request in feasible tasks.

Based on this pre-condition, we start an impact analysis based on the method Design Review by Failure Mode (DRbyFM is comparable with SW-FMEA®). The result of this analysis is the identification of the CR based test content plan, which covers the content of regression testing.

This type of regression testing focuses the test against the change request and possible side effects (see also bug fix confirmation).

# Assignment of test techniques and tools against test phases

Based on our fundamental test process (P/P/P/P), the following picture shows in which phase we are using which kind of tool.

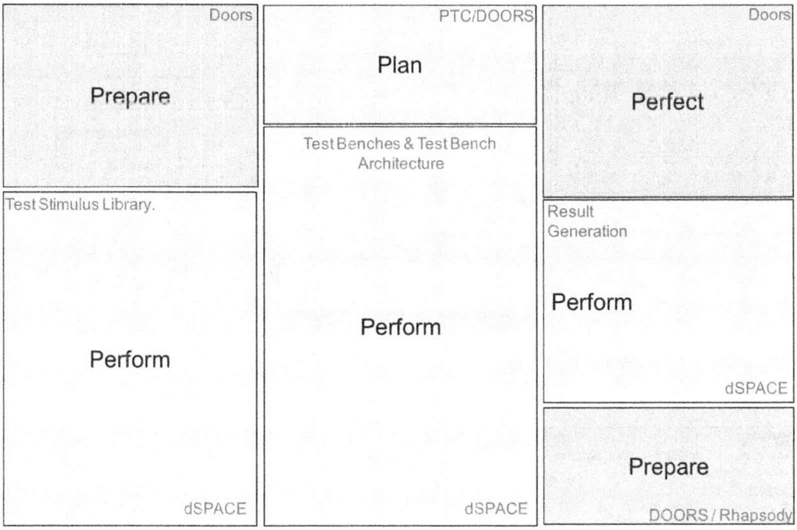


[7] Fundamental Test Process: 4P's-Phase cluster Tool Domains

Based on this cluster the next picture shows the P/P/P/P activities under the context of test tools, roles and work products.

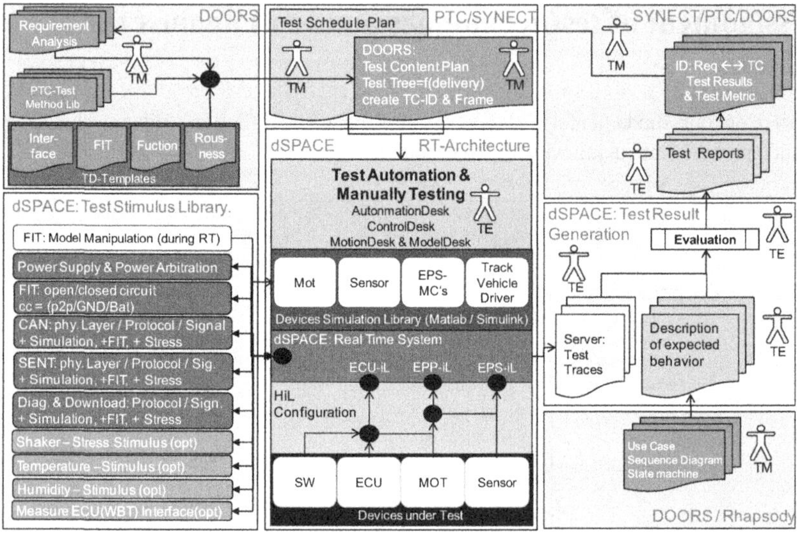

[8] Interaction between the P-phases

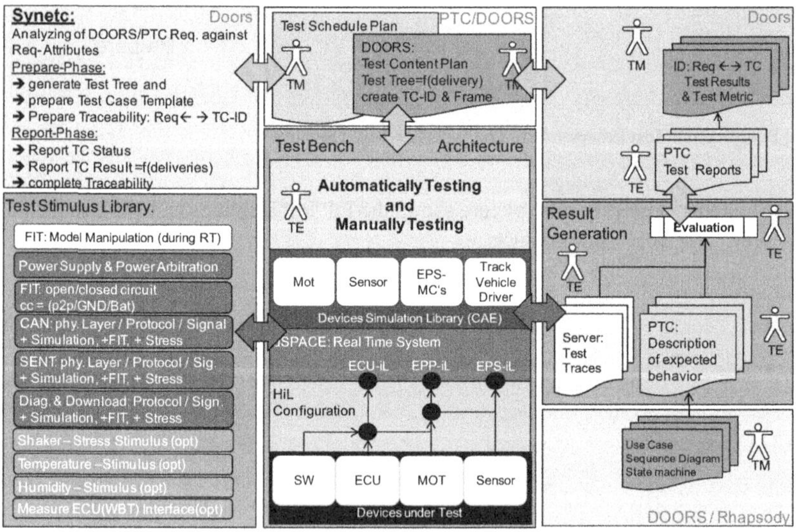

[9] Test Management Support (SYNECT®) / Test Automation (AutomationDesk®)

The yellow arrows identify the interactions which are supported via the dSPACE® tool SYNECT®.

The red arrow injunction with the dark-green area describes the functionality of test automation (dSPACE: AutomationDesk®).

Hint:
During my Chassis.Tech presentation this pictures will presented in animation mode

# Test Maturity Model:

A very important criterion is the difference between function and product release.

## Release of Functionality

In case of a function release we have the focus to identify a performance of a new feature under limited condition concerning time and budget (see also CV-phase = concept validation).

Typical example: Two days workshop with customer on a test track

In this phase many different engineering disciplines are involved, which follows an "agile" development approach.

These kinds of activities follow a horizontal communication approach.

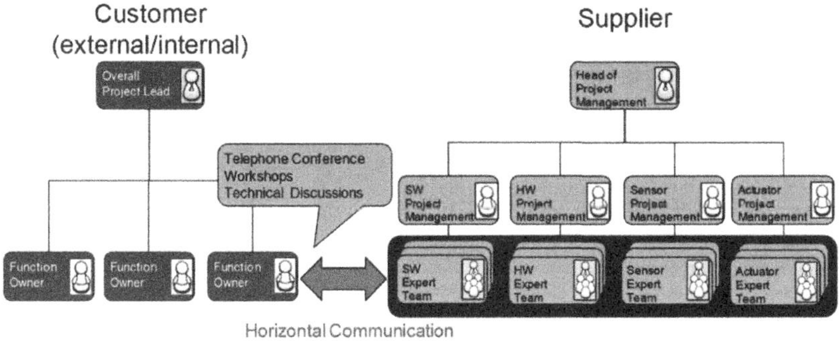


[10] Verbal f2f Communication on common level is very important to generate a common understanding of the performance of the product.

## Release of a Product

Regarding product releases, the test organization has the focus to solve all test activities which are necessary to support an unrestricted release recommendation for public roads and mass production.

Working on this kind product release support, it is very important to bring the evidence, that all determined test activities are covered completely.

There is only one test method available, which is able to cover the requested gapless test activities:

> **Requirement based testing with bidirectional traceability between unique requirement ID and test case ID's.**

In this product release phase we follow the "vertical " communication approach.

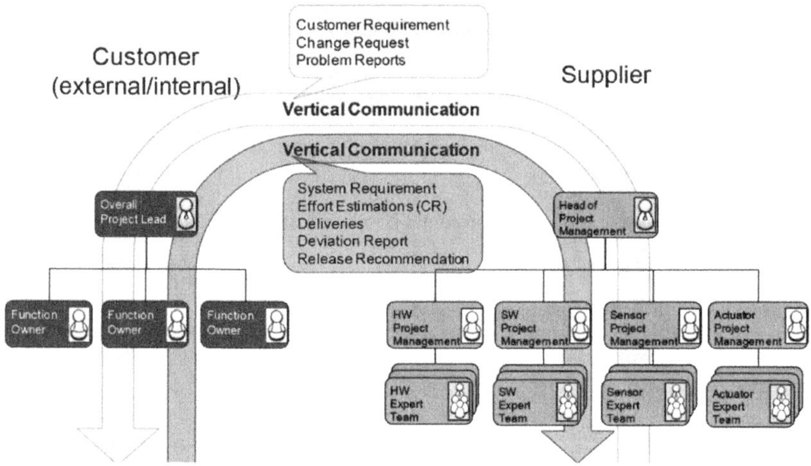


[11] The exchange of requirements has a high workload on documentation with a complex communication struchtire (formal)

Each single work packages have to follow the v-model.

Based on this v-model ten engineering processes have to be solved which are systematically synchronized via project management and supported by quality assurance.

During thc product development cycle requirement based testing takes place, which is also a pre-condition for our regression test strategy.

## Transition between function and product development

Following is important to handle the transition from function development to product development:

- Clear restriction of time and content of a function development phase,
- The quality aspects of working products which summaries the results of the function development phase (CV-phase), has to be assessed against product development quality
  - new safety goals,
  - new requirements
    - System & Sys-Architecture,
    - Sub-System & Sub-System Architecture
    - Test tools
  - test case descriptions,
  - ...
- All parallel executed function development projects have to be assessed against contradictions on:
  - functionality and
  - safety goals
- calculation of the complexity
  - estimation of non functional requirements like
    - processor load
    - stack load
    - program storage load
    - ...
- Influences on availability
  - self diagnostics routines
  - thresholds
  - ...
- Influences on reliability
  - damage-equivalent
  - ...

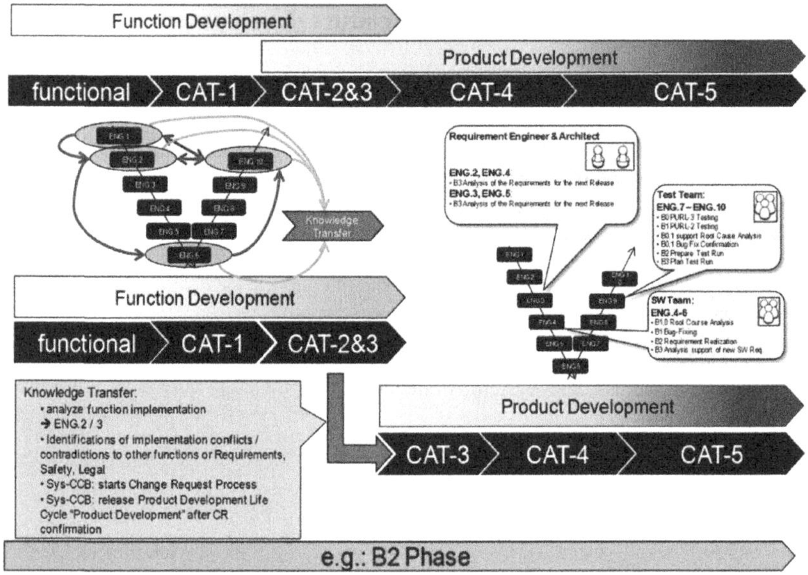


[12] Transition between function and product development

# Definition „free" Testing:

In the free testing mode the test engineer does not have fixed guidelines (test design templates, test checklists, written test specification …). During the first testing steps it is also not requested to generate test traces, which document the test run. He has a high freedom concerning the usage of existing test equipment, modification of test equipment or build-up of a new test environment.

It is recommended to document the results of this test session with the target of having documented information about the new knowledge.

This new recognition will lead to new:

- system & SW requirements,

and further formal:

- test designs, test specifications and test case

development.

## Advantages

- Short preparation phase
- Quick results
- Support the knowledge development
- Support function/feature development

## Disadvantages

- No statement against Completeness
- No clear statement against "Test End criterion is reached"
- No metric support
- No or bad reproducibility, because insufficiently documented
- Dependent on the experience level of the test engineer
- Not usable in a audit
- Not usable for product release
- Did not follow a reproducible analytical test principle, like:
    - Coincidence principle or
    - Common principle: „In case of a single failure detection the failure probability accumulates"
- For function development release usable with restrictions

## Typical area of application

- "Concept Verification" Phase (CV-Phase) with a-samples or modified product samples,
- Only useful for test & early acceptance investigations injunction with closed test track areas and instructed driver,
- Additional unplanned test session during the identification of a unclear deviation behavior in order to get clarification,
- Part of the „Route Cause Analysis"
- „Fire and Smoke" Tests: Stress testing with the target to destroy the Device under Test (DuT) in order to identify systemic reserves,
- incomplete or missing requirements or product descriptions, …,
- Testing under agile feature development conditions,
- Identification of requirement gaps, test gaps, warranty analysis,
- Testing under insufficient timing conditions
- „Testing in the Darkness"
- …

# Definition „experience based" Testing

In the case of „experience based" testing the responsible person of the test organization uses their product and/or test knowledge in order to "defend" their test content. In many cases this knowledge references to the common knowledge of a test organization.

If the test organization depends to the product development organization, which is specialized to develop and apply the same products (like: Airbags, EPS, EPS® ...) over a various number of product generations, this knowledge is enormous and very useful. In this case the test libraries (sum of all reusable test designs, test specifications, test cases ...) of this test organization represent the product knowhow, which is also quickly available for further applications.

The execution of the knowledge based test content is documented in test traces.

The test set-up uses test equipment which is still available and the confidence level of these test results is still high. For new product generations the test machines are simply modified and improvements based on the experiences of the last test session are automatically included.

Test result confidence level is in this case the result of the relationship between the previous and actual product design and functionality.

Based on common pedigree attributes and sufficient test depth of the reference project, many "base practices" and "outcomes" have a high potential to go into a review process in order to identify the reuse of existing work products.

## Advantages

- Short preparation time for test execution
- Quick test results with high trust level
- Work products capable of being offered in e.g. A-SPICE® audit injunction with a recovery plan which close the bidirectional traceability gap
- Follows a reproducible analytical test principle, like:
    - „experienced based",
    - Reference to predecessor project(s),
    - genetic qualities based on the capability of the predecessor
    - ...
- Useful on projects with a high "core" functionality portion
- Risk assessment based on the comparison between "core" and application related approach
- ...

### Disadvantages

- Incomplete statement in case of completion of the “test end criterion”
- Test metric reporting may lead to wrong statements
- Test quality depended on the experience of a test engineer a/o test organization
- Dependency against the history of previous test project
- No traceability between actual customer requirements and existing test cases
- Test process is incomplete
- Limited use for an Audit
- Not usable for a final product release recommendation
- Usable for a release of a functionality
- ...

### Typical Application

- “Concept Verification” Phase (CV-Phase) with a-samples or modified product samples,
- Design Verification Phase with b/c-samples
- Restricted release recommendation
- Important development step in case of product release life cycle (SPL.2)
- Testing against own and common product requirements
- Additional action against incomplete or not final released requirements
- Reuse of „old“ test libraries based on predecessor projects
- Bench Marking: Assessment of competitor products
- ...

## Definition „requirement based“ Testing

In case of „requirement based“ testing the test engineer has to focus his work against the structure and detail gradient of the supported requirements.

These guidelines are available on different levels:

1. Legal, Safety, OEM& TRW EPS Core Requirements
2. Project System Requirements = f(conclusion of 1.)
3. Project System Architecture Requirements
4. Sub-System Requirements
5. Sub-System Architecture Requirements

The verification method defines – for each requirement – additionally the recommended release concept.

In order to keep the test costs under control, it is useful to develop the content of the verification method in a common way, between requirement engineer and test engineer.

In many cases a full picture of a feature is given on both:

- Definition of the requirement content and
- Definition of the verification method

This makes sense, if the test organization belongs to the product development group and injunction with the test knowledge over product generations.

It is strongly recommended to develop test cases via test design and test specification description.

The execution to each test case has to be documented e.g. via test trace recording.

Test equipment has to be adapted to the "Device under Test" (product) and capable to the test target (right test tool for the right test task). This is very important for architecture (system & SW) related test activities.

Requirement based testing has the focus to cover all "base practices" and "outcomes" of the product development life cycle model (CMM® / A-SPICE® / ISO 26262).

## Advantages

- covers all ISO26262 / A-SPICE® & CMMI® related test aspects
- fulfills the evidence of test completeness: "Test End Criterion"
- support test metric report
- ...

## Disadvantages

- Requires accurate planning (time and content)
- Inflexible against interruptions and disturbances
- High costs
- Test deepness is a function of the requirement granularity
- ...

## Mitigation

- Introduction of professional test management
- Introduction of Test Automation (7/24)
- Opportunity of high reuse
- Capability of regression testing

## Typical Application

- "Concept Verification" Phase (CV-phase) with a-samples or modified product samples,
- Design Verification Phase (DV-phase) with b/c-samples
- Product Verification Phase (PV-phase) with d-samples or SOP parts
- Unrestricted release recommendation
- Product release for mass production
- ...

# Results

Requirement based, experience based and free testing are all test methods of a method tool box. In comparison to other method tool boxes, these test methods have – in various project situations – advantages and disadvantages.

Therefore for a test organization to be successful it is very important that they follow a dynamic test approach in order to react in the right way under various situations of project related circumstances and not just simply follow established routine testing methods.

This paper tries to support a product test & release life cycle model, which reacts to following parameters:

- Function or product development
- Different release categories based on use cases
- Sample phase of the system components
- History of the project

The described freedom of test methods supports a test organization, which also identifies gaps of requirements in order to improve their product description.

The test organization has to scrutinize themselves and optimize their capabilities and terms of working.

This is an important part of becoming a self-learning organization, which has to be able to understand and react on new product and technology challenges.

For us it is very important to be a capable test organization, which matches the challenges for autonomous driving:

- Test process support in a networking situation
- Test tools & methods which support this test process
- Test Tools based on a common standard
- Linkage information of test management attributes (Planning/Prepare & Perfect Phase) will exchange in a common tool infrastructure: e.g. DOORS®

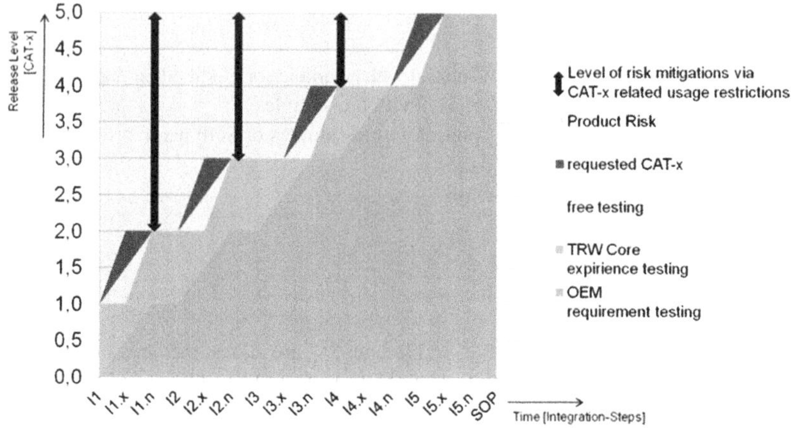


| | |
|---|---|
| Red Area: | drives the test case development |
| Yellow Area: | Test Case development (man. → auto.) in order to close gap=f(red / yellow) |
| Light green Area: | Test Cases are developed and a part of TRW Core test library |
| Green Area: | verified TRW Core test cases are referenced to project related requirements |

[13] Example: Areas of free, experienced & requirement based Testing during the product development cycle

# Realistic dynamic testing of EPS motors and ECUs by means of a hardware-in-the-loop test bench

Hermann Briese, Emad Farshizadeh, Stefan Oedekoven

DMecS Development of Mechatronic Systems GmbH & Co. KG, Köln;

Thomas Schubert, Hermann Henrichfreise,

CLM – Cologne Laboratory of Mechatronics, FH Köln

# Abstract

Conventional test benches with real steering mechanisms often have non-negligible parasitic properties and multiple control systems to apply stimuli. As a consequence the EPS motor and ECU interact with a system that has a different dynamic behavior than the steering mechanism in the vehicle. Depending on the actual test scenario, this might reduce quality of testing results significantly.

That was the motivation for the development of a test bench for realistic testing of EPS motors and ECUs as a complement for conventional test benches. In the test bench described here, the real steering mechanism is replaced by a test bench actuator that is mechanically connected to the EPS motor as device under test (DUT) by means of a measuring unit with a specifically tailored sensor concept. The cutting torque, which would act on the EPS motor in a real steering system, is computed by real-time simulation of a detailed model of the steering system that includes all relevant properties. Depending on the development stage, the EPS motor can be driven either by the full EPS ECU or a power stage only with e.g. a steering control implemented on a rapid control prototyping system. For HiL tests, the test bench might be used in combination with any suitable commercial vehicle model.

The realistic testing environment for the EPS motor and ECU is achieved by a highly dynamically controlled test bench actuator. The actuator control consists of an observer based state space control with extensions for compensation of nonlinear characteristics of the test bench. It is designed accounting for all dynamic components of the test bench setup: test bench actuator, measuring unit, and DUT. The actuator control adjusts the cutting torque in the test bench which would act on the EPS motor in the real steering system. This results in the EPS motor being operated in a realistic environment. The simple mechanical setup in combination with a sophisticated control leads to a high bandwidth and a testing quality that can hardly be achieved with conventional test benches. Additionally, it enables testing in early phases of development, even when the real steering mechanism might not be available yet. The model based approach allows for detailed investigation of the impact of special properties of the mechanism like backlash or altered friction on the EPS behavior. Furthermore, the influence of torque ripple of the EPS motor on the steering feel can be evaluated.


# 1 Introduction

For more than a decade, DMecS is working in the field of control development for EPS systems. During this period DMecS gained considerable experience with many types and performances of test benches for steering systems. The rather high effort for development and operation of test benches for the full steering system, as well as their sometimes limited performance plus the increased requirements for HiL tests induced the motivation to develop this new test bench approach. It has been developed in co-operation with the Cologne Laboratory of Mechatronics (CLM).

The core development target was to provide a test bench for realistic testing of EPS motors and their ECUs. The test bench was designed to apply that load torque (cutting torque) to the EPS motor that would act on it when operated in the real steering system. This cutting torque is highly dynamically controlled so that the dynamics of the test bench, in the relevant range of frequencies, is not visible in the results of the HiL tests. The cutting torque acting on the EPS motor is computed by real time simulation of the steering system. According to that, the quality of tests primarily depends on the used simulation model.

To achieve a sufficiently high performance it is necessary to account for the dynamics of the entire system (test bench and DUT) for controller design. This again means that all relevant properties and parameters of the DUT must be known. For the identification of the sometimes unknown parameters of the DUT, automated parameter identification is introduced [2]. For that, additional types of controllers for the test bench have been developed. This offers additional options for parameter identification and model validation like the determination of the torque-speed characteristics without disturbing vibrations and the investigation of the torque ripple of the DUT under different load conditions.

# 2 HiL Test Bench

The test bench shown in figure 1 consists of the test bench actuator, a proprietary measuring unit, and the EPS motor with ECU (DUT). Furthermore the figure shows a dSPACE system used for test bench control and real-time simulation of the steering system mechanics.

Instead of a real EPS motor and ECU, a high quality permanent-magnet synchronous machine is used to emulate different types of EPS motors during test bench development. A dSPACE MicroAutoBox is used to emulate the ECU.

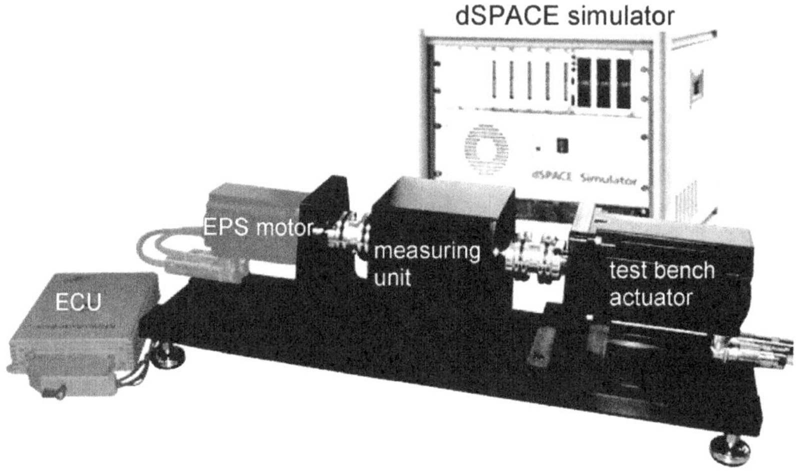


Figure 1: Test bench with DUT (EPS motor and ECU replacement) and dSPACE system

To achieve the high control dynamics required for realistic HiL testing the individual components were selected to satisfy special requirements. This includes, among other properties, low moments of inertia, low friction and low torque ripple at the test bench actuator. The latter is further reduced by specific control measures. In addition the sensors provide high accuracy and the actuator and sensor interfaces introduce only negligible signal delays. The actually used test bench actuator allows a maximum torque of 30 Nm. Thereby it is applicable for a high range of tests and various types of DUTs, even motor racing components.

Figure 2 shows the structure of the HiL test bench with the components described above and their interaction.

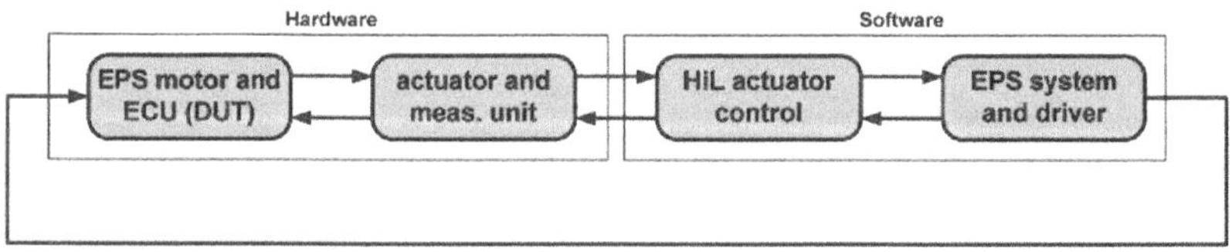


Figure 2: Structure of the HiL test bench

Besides the simulated test environment, in this case a simulation model of steering mechanics and driver, the test bench actuator control is implemented on the dSPACE system. It generates the control signal for the test bench actuator and provides additional signals for the simulation of the steering mechanism. The latter generates the reference signals for the test bench actuator control and the input signals for the EPS ECU.

# 3 HiL Test Bench Control

The realism of the testing environment for the EPS motor and ECU is achieved by a highly dynamically controlled test bench actuator which adjusts the cutting torque at the EPS motor. In addition to a high controller bandwidth, high control accuracy is requested. This must be realized despite disturbances like friction and torque ripple at the test bench actuator and measurement noise. Natural oscillations of the system should be damped actively by the control.

To achieve these goals, the cutting torque control consists of an observer based state space control. It is extended by a nonlinear model of the test bench actuator torque ripple. Figure 3 shows the corresponding control structure.

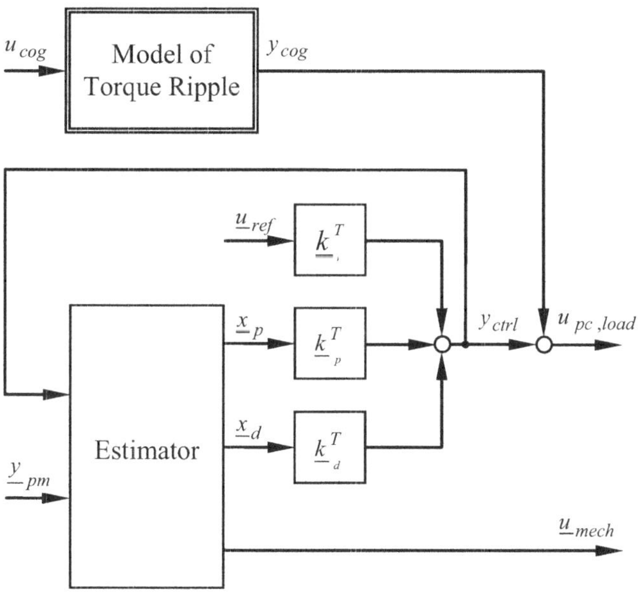


Figure 3: Structure of the test bench actuator control

The actual controller is built by a static feedback of the system state variables in the vector $\underline{x}_p$ as well as feedforward of the reference variables $\underline{u}_{ref}$ and of disturbance variables in the vector $\underline{x}_d$. Since not all system state and disturbance variables can be measured, an observer is required to estimate unknown variables by using the measurement variables in the vector $\underline{y}_{pm}$ and the control variable $y_{ctrl}$. Furthermore, the estimator (Kalman filter) calculates additional variables in the vector $u_{mech}$ that can be used for the simulation of the DUT test environment, e.g. a steering mechanism.

The controller and estimator gains are determined by the LQG/LTR design approach of an optimal robust control [3], [4].

To compensate the torque ripple of the Test bench actuator, the output signal of the torque ripple model $y_{cog}$ is added to the control output $y_{ctrl}$.

For purposes in connection with parameter identification, a position and velocity control was developed with the same design approach. With these controllers the control variable is adjusted at the DUT (not at the test bench actuator), analogue to the cutting torque control.

# 4 Results

This chapter describes the results achieved with the test bench. At first the results of the cutting torque control are shown to demonstrate the achievable performance. After that the results achieved for a HiL scenario with a simulated steering system are demonstrated.

## 4.1 Cutting Torque Control

A crucial requirement for realistic testing is to apply the requested cutting torque computed by HiL simulation exactly to the DUT under all conditions.

The following figure 4 shows the frequency responses of the reference transfer path of the closed control loop computed with linear models of the system (gray, —) and determined by measurement at the real test bench(black, ··).

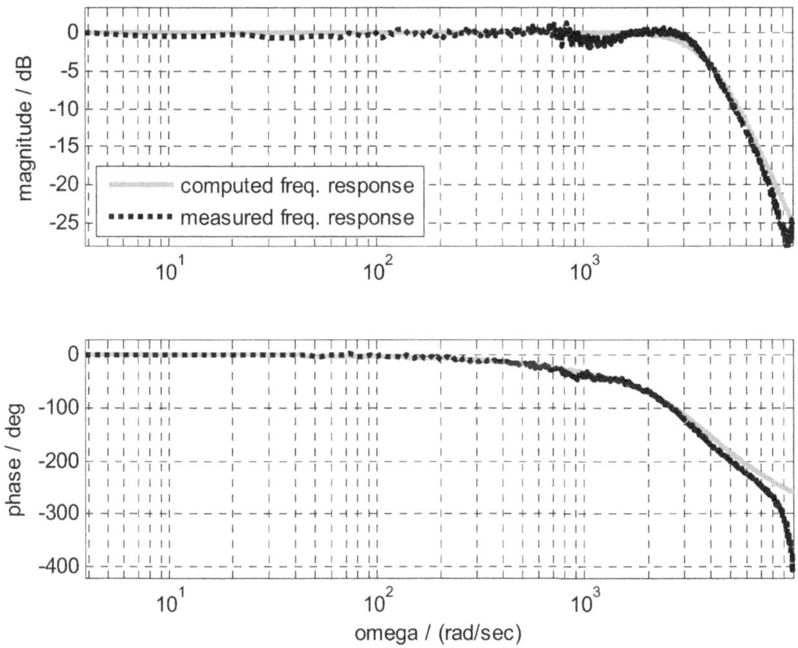


Figure 4: Closed-loop frequency responses from analysis of the linear system (gray, —) and measured at the real system (black, ··)

The bandwidth of appr. 500 Hz demonstrates the high performance of the test bench.

To demonstrate the reference transfer behavior in the time domain, figure 5 shows experimental results with passive DUT for a sinus sweep reference with a frequency up to 50 Hz and amplitude of 1 Nm.

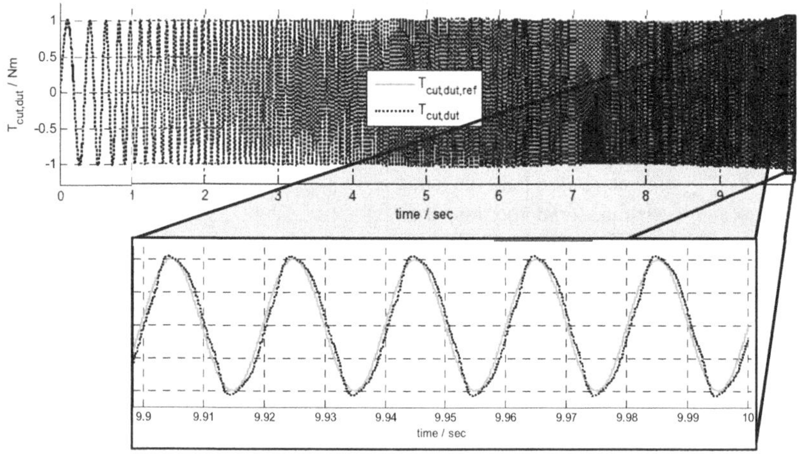


Figure 5: Time history with a sinus sweep reference signal for the cutting torque at the DUT (gray, —) and the measured one (black, ··)

The actual cutting torque (black, ··) corresponds very well with the reference (gray, —) even for higher frequencies.

In addition to the reference behavior, the disturbance behavior is important regarding the achievable HiL test quality. The disturbance behavior is tested by an experiment were the DUT is excited by a torque sinus sweep with a frequency up to 50 Hz and a amplitude of 1 Nm while the reference signal for the cutting torque at the DUT is zero.

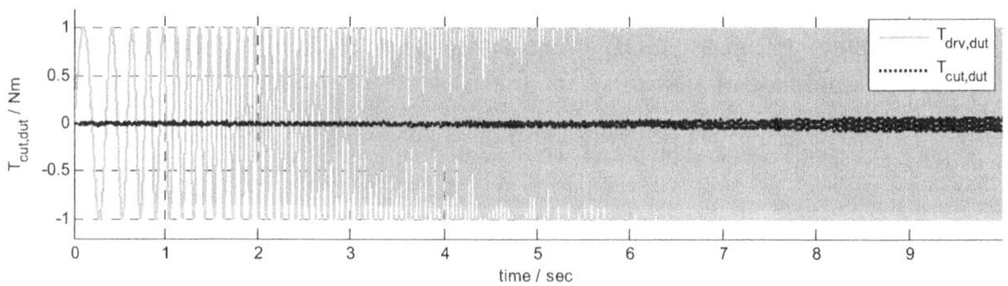


Figure 6: Time history with a sinus sweep disturbance induced by the DUT (gray, —) and the measured cutting torque (black, ··)

The impact of the disturbance excitation (gray, —) is suppressed very well. The maximal amplitude of the resulting cutting torque (black, ··) is appr. 0.07 Nm at a frequency of 50 Hz. The experiment whose results are shown above was conducted assuming that the torque requested from the DUT is roughly known. If not, the resulting cutting torque is less than twice as high.

## 4.2 HiL Test

Figure 7 shows a setup of an EPS system containing the EPS motor with its ECU and the steering mechanism. In the shown system (a dual pinion configuration), the torque generated by the EPS motor is applied to the steering rack by a transmission gear.

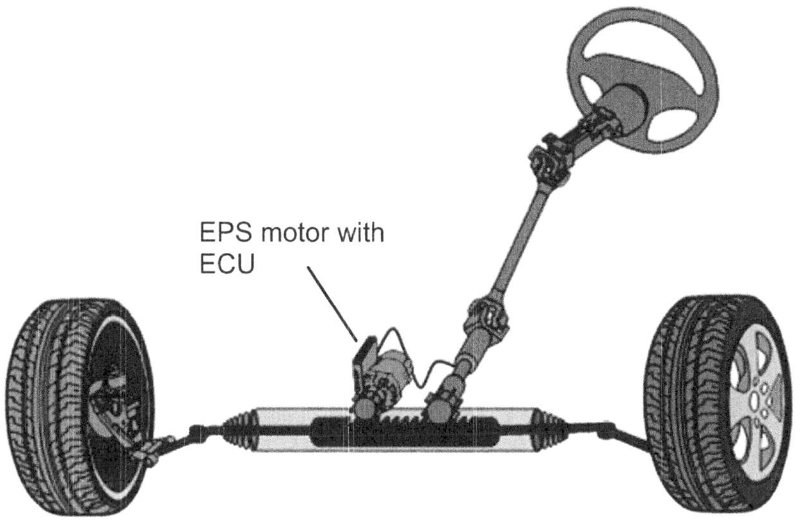


Figure 7: EPS system

For HiL testing a simulated test environment is created, that reproduces the main components of the environment of the EPS motor and ECU in a vehicle. Besides the steering system mechanics, in the setup discussed in this chapter it contains tire models and a driver model. The implemented EPS controller is based on a common boost curve orientated approach.

Figure 8 shows the time histories from simulation of the complete system (simulated test environment in combination with EPS motor and ECU) (gray, —) as well as the experiment with the real HiL test bench (black, ··). The simulated steering system is excited by the virtual driver following a triangular shaped steering wheel angle reference with maximum amplitude of 90° and a maximum angular speed of 600°/s.

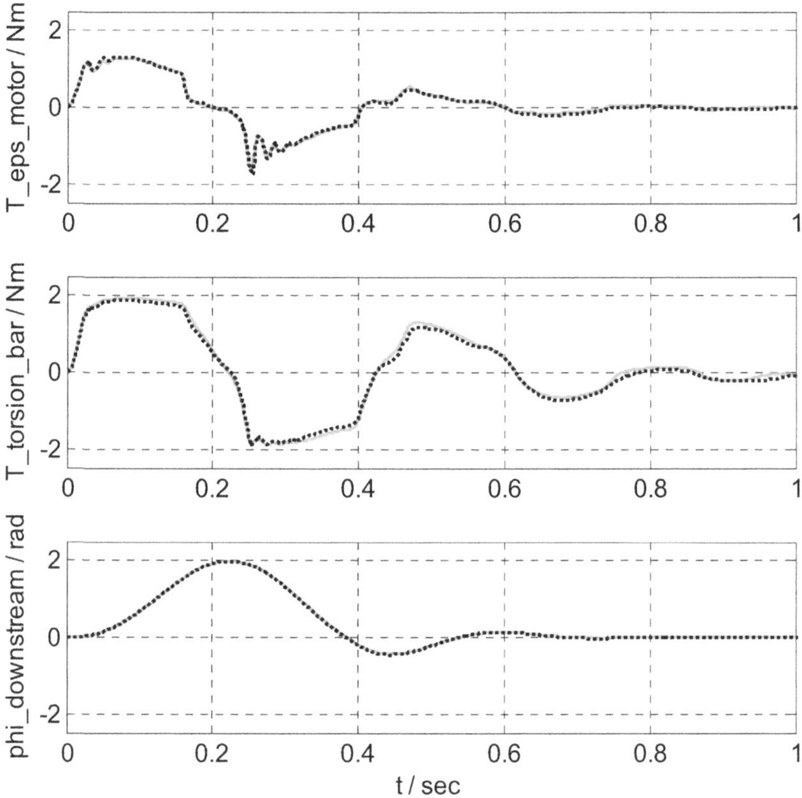


Figure 8: Time histories for triangular shaped reference steering wheel angle excitation, simulation (gray, —), HiL test bench (black, ··)

The diagram shows the EPS motor torque, the torsion bar torque and the lower torsion bar angle (pinion angle). The time histories from simulation are the reference for the results with the HiL test bench. The results measured at the test bench correspond very well with the reference.

In an additional experiment the simulated steering rack is exited with a force while the steering wheel is held at a constant reference position. This disturbance force is a block shaped impulse with amplitude of 15 kN and a duration of 0.04 s emulating the sideways approach of the front tire of a vehicle into a curb. Figure 9 shows the corresponding time histories.

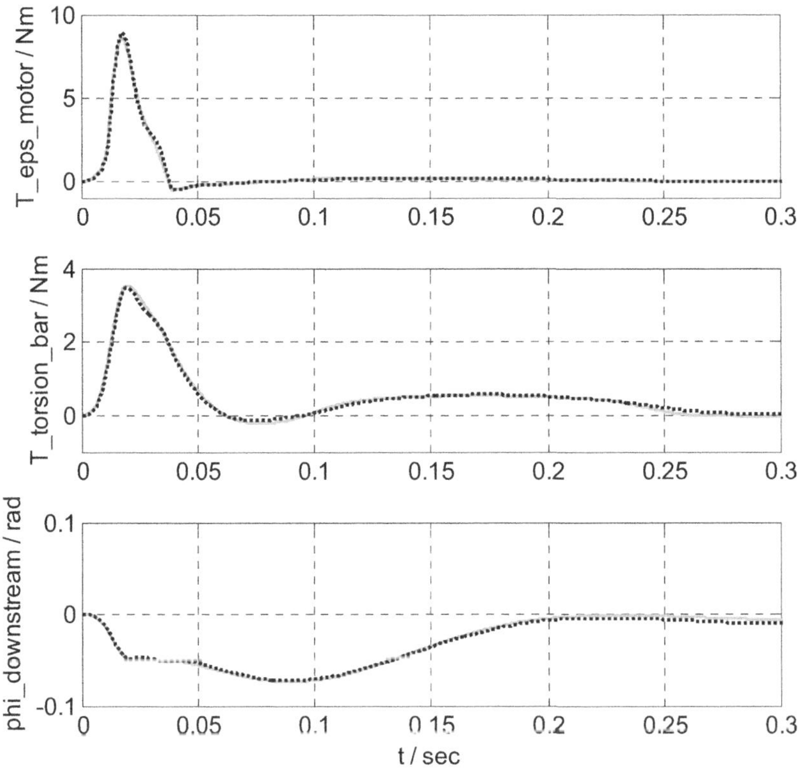


Figure 9: Time histories for block shaped disturbance excitation at the steering rack, simulation (gray, —), HiL test bench (black, ··)

As for the EPS system reference excitation, the results achieved with the HiL test bench for EPS system disturbance excitation correspond very well with the reference.

Another important disturbance excitation for the EPS system is formed by stochastic steering rack forces. Such rack forces occur when driving on rough driving surface (e.g. on a cobbled street). To simulate this excitation, a Gaussian distributed band-limited white noise with a standard deviation of 5 kN and a bandwidth of 50 Hz is applied to the simulated steering rack as in the experiment before. Figure 10 shows the corresponding time histories.

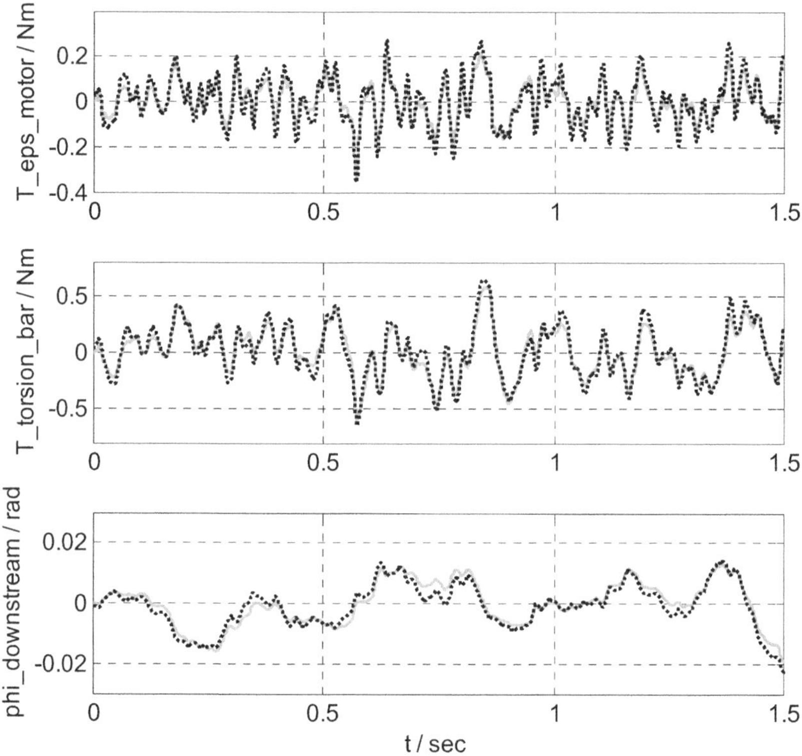


Figure 10: Time histories for stochastic disturbance excitation at the rack, simulation (gray, —), HiL test bench (black, ··)

Even for this disturbance excitation with high-frequency signal components the results correspond very well.

The results with the presented experiments show a realistic testing environment for the EPS motor and ECU regardless of the system excitation. The behavior of the EPS system is not distorted by the test bench. Thus the HiL test bench allows for a detailed analysis of the properties of motors and ECUs (or prototypical controllers) and of resulting effects.

This shall be demonstrated by the next HiL experiment. The reference is a linear increasing steering wheel angle with an angular speed of 12°/s. The EPS motor is driven with (gray, —) and without (black, ··) torque ripple compensation. The results are shown in Figure 11.

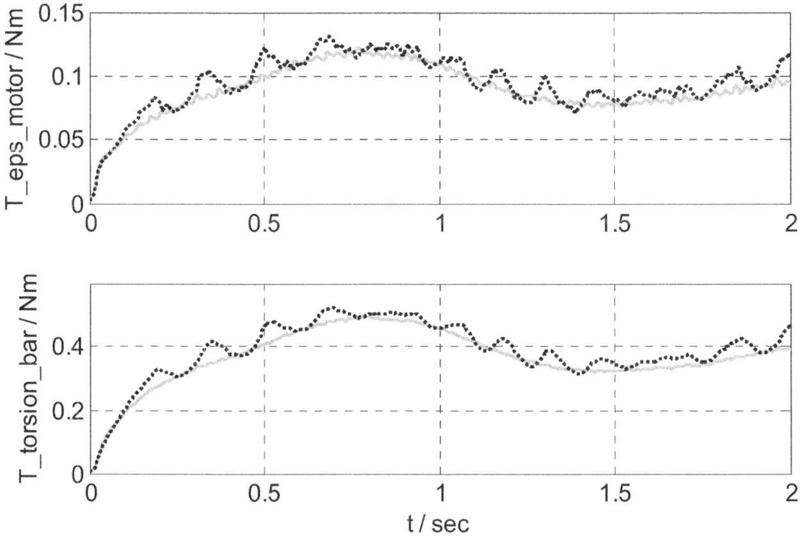


Figure 11: Measured time histories with (gray, —) and without (black, ··) torque ripple compensation at the EPS motor

Even the impact of the comparatively low torque ripple of the EPS motor with an amplitude of 0.03 Nm on the EPS system is visible in the time histories of the torsion bar torque. So the test bench is suitable to analyze the influence of the torque ripple on the steering feel.

# 5 Parameter Identification

Another field of application of the test bench is the parameter identification and model validation of the DUT. The identification of parameters is essential to ensure high quality and robustness of the test bench control. The implemented large extent of automation reduces the required effort essentially to the installation and start-up of the DUT in the test bench [2].

### *Mechanical Parameters*

In many cases the moment of inertia and viscous friction of the DUT are not well known. For optimal test bench performance however they must be known with adequate accuracy since the test bench controller accounts for the dynamics of the entire system (test bench with DUT). In addition, for a proper simulation of the DUT its complete friction characteristics must be known. For identification of these parameters, different experiments with special reference profiles are conducted. The determination of the parameters is carried out using a simulation model of the test bench. The stimuli from the test bench experiments are used for simulation of the system. Using a numerical optimization process the parameter values are determined ensuring congruence of measured and simulated system responses.

### *Torque Generation Dynamics*

Another important property is the transfer behavior of the DUT motor torque generation dynamics. In many cases it is strongly dependent on the operating point. The test bench provides the possibility to investigate the torque generation in a large area of operating points. To investigate the torque generation dynamics, a Kalman filter is used to estimate the effective air gap torque of the DUT. With accurate parameters values obtained by identification of the mechanical parameters, the used Kalman filter is fine-tuned to the specific DUT for this investigation.

### *Torque-Speed Characteristic*

A standard application of motor test benches is the measurement of the torque-speed characteristics (power curve). Vibrations often have a negative impact on the results. The test bench described in this paper enables the determination of the torque-speed characteristic without disturbing test bench vibrations. This is guaranteed by using a state-space controller as described in chapter 3 which enables active damping of vibrations during measurement. Again, a fine-tuned Kalman filter estimates the air gap torque of the DUT.

To show the quality of a determined power curve, the DUT is driven with an artificial preset speed dependent torque limiting characteristic with maximum torque demand. The test bench velocity control applies a linearly increasing rotational speed to the DUT. Figure 12 shows the measured torque limiting curve together with the preset one.

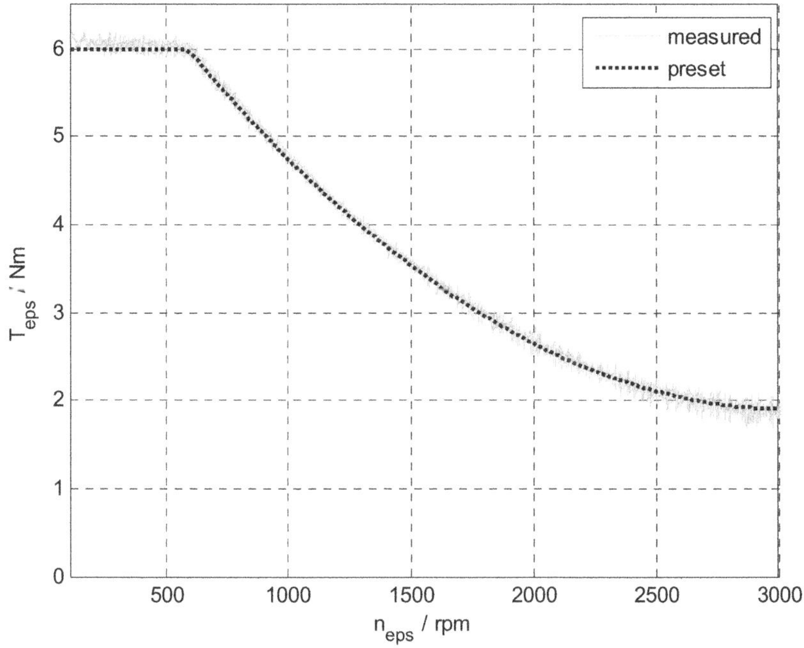


Figure 12: Torque limiting curve of the DUT determined with the test bench (gray, —) and preset (black, ··) for positive speed

The preset and the measured torque limiting curves are identical except small deviations caused by the torque ripple of the DUT.

## *Torque Ripple*

The investigation of torque ripple with conventional test benches is typically conducted using a torque sensor between the DUT and the test bench actuator and a control for this actuator (load motor) only. As described above, disturbing vibrations can have a negative impact on the measurement results. Furthermore the transfer behavior of the torque sensor becomes visible, especially at high-frequency DUT torque ripple. With the presented test bench, the torque ripple can be determined from the estimated air gap torque of the DUT using a fine-tuned Kalman filter. With that, it is possible to investigate the torque ripple in a large area of operating points, i.e. at different load states and at different rotational speeds. Figure 12 shows the torque ripple characteristic determined for the DUT mentioned in chapter 2 for load states between -6 Nm and 6 Nm at 100 rpm motor speed.

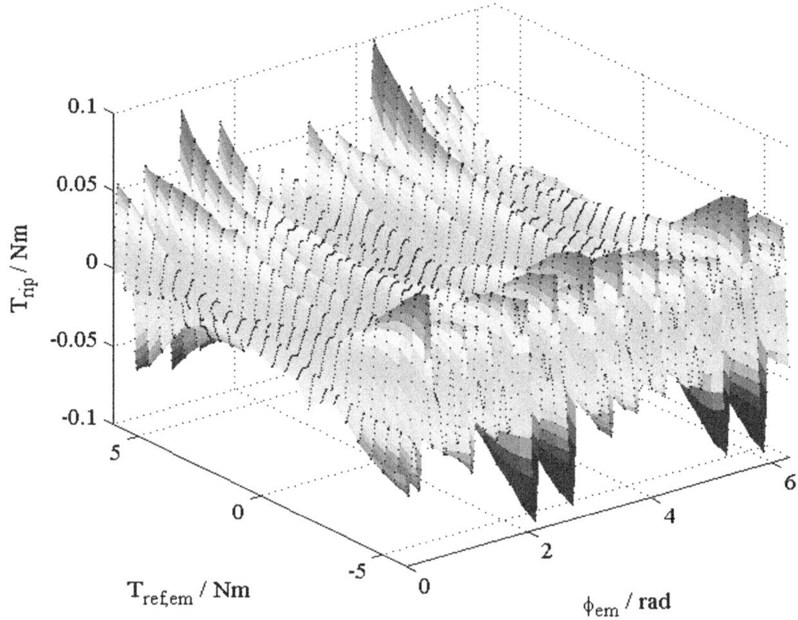


Figure 13: Determined torque ripple of the DUT for different reference torques at 100 rpm motor speed

In addition to an often used compensation of the torque ripple by means of a quasi-static measured characteristic with a feedforward control it is possible to develop and test a load-state dependent compensation. Figure 14 shows the results of the same measurement experiment as figure 13 but with load-state dependent torque ripple compensation.

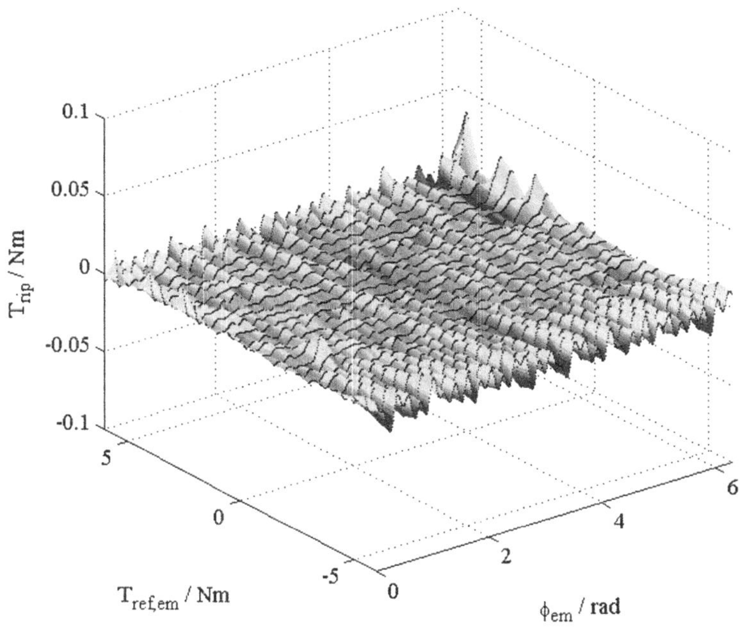


Figure 14: Determined torque ripple of the DUT with torque ripple compensation for different reference torques at 100 rpm motor speed

The results with the torque ripple compensation algorithm shown above are achieved with a high quality permanent-magnet synchronous machine and demonstrate the measurement and resulting development capabilities provided by the test bench.

# 6 Conclusion

This paper presents realistic dynamical testing of EPS motors and ECUs by means of a hardware-in-the-loop test bench. The real steering mechanism is replaced by a simulation model and the cutting torque is applied by a highly dynamically controlled test bench actuator. With this test bench the quality of tests primarily depends on the engaged simulation model of the steering mechanics and test environment. The behavior of the components to be tested is not influenced by the test bench itself. This enables the investigation of the effects of comparatively weak phenomena like the torque ripple of EPS motors. Testing is possible even when real steering mechanisms are not available (yet).

Furthermore the test bench is qualified for parameter identification and model validation. Besides the identification of the mechanical parameters of the motor, the torque generation transfer behavior, and the torque limiting curve, it provides the means to determine the torque ripple at different load conditions and rotational speeds.

The developed test bench concept can be applied to both prototypical components in the development phase as well as series production components. For steering systems, with appropriate extensions (e.g. sophisticated vehicle model or torque feedback steering wheel) advanced means for development and assessment of superposed control algorithms as well as the steering feel are provided.

The test bench concept presented here can be used for system identification and highly dynamic HiL test of any kind of (controlled) electric drives such as the actuator of an electrohydraulic brake system, the seat adjustment, or wheel hub drive. The test bench and the test bench concept, respectively, are applied by DMecS in several industrial development projects.

# References

[1] E. Farshizadeh, H. Henrichfreise, „Hardware-in-the-Loop-Prüfstand für den realitätsnahen dynamischen Test von EPS-Motoren und –Steuergeräten“, VDI-Tagung Mechatronik 2011, Dresden

[2] T. Schubert, H. Henrichfreise, E. Farshizadeh, “Automatisierte Parameteridentifikation für EPS-Motoren”, VDI-Tagung Mechatronik 2013, Aachen

[3] F.L. Lewis, “Applied Optimal Control and Estimation”, Prentice Hall, Englewood Cliffs, New Jersey, 1992

[4] H. Henrichfreise, “Prototyping of a LQG Compensator for a Compliant Positioining System with Friction”, TRANSMECHATRONIK – Entwicklung und Transfer von Entwicklungssystemen der Mechatronik, HNI Verlagsschriftenreihe, Vol. 23, 1. Edition, Paderborn, 1997

# A simulator study on the controllability of steering systems with reduced maximum steering wheel angle

Christian Dreßler
Research, TAKATA AG

Stefanie Eßers
Director EMEA Core Engineering & Research, TAKATA AG

# Abstract

Active Front Steering shows that with a variable gear ratio, it is possible to reduce the maximum steering wheel angle from zero position to lock to approximately +- 360°. Compared to a conventional power steering system, this is a reduction of about 40%. With technologies like reliable electric power steering or steer-by-wire coming to the market, it is possible to reduce the maximum steering angle even further. To assess the controllability of further reduced maximum steering angles, a driving simulator study with two different steering systems with reduced total steering wheel angle is conducted with 36 experienced normal drivers. The steering systems are modelled within Matlab/Simulink and are integrated into the simulator on a real time target. The steering wheel is controlled by a synchronous AC motor providing up to 24 Nm of steering wheel torque. One system is an adapted conventional steering system without variable gear ratio and a maximum steering wheel angle of +- 180°. The other system employs the advantages of a steer-by-wire system and uses a PID-controller to regulate the vehicles yaw rate proportional to the steering angle. The steering wheel torque is proportional to yaw rate, lateral acceleration and vehicle velocity. The maximum steering angle is set to +- 90°. Both systems are benchmarked with a conventional steering system with a steering wheel travel from zero to lock of +- 600°. For all three systems subjective measurements to evaluate the steering feel and objective measurements, e.g. lane keeping quality to evaluate the controllability, are taken. The study shows that there is a limitation regarding the reduction of the steering wheel stroke between +-180 and +-90°, as the system with a maximum steering angle of +-90° performs worst regarding overall controllability. Furthermore it showed that the system with a limited maximum steering angle of +-180° performed best with respect to subjective steering feel.


# Introduction

With active front steering, it is possible to reduce the steering angle to approximately two turns from lock to lock [1]. As these systems are already introduced in series production, they fulfill all requirements regarding safe and easy handling of the vehicle. With a Steer-by-Wire system designed as full-power steering system according to ECE r-79 [2], there is no mechanical or hydraulic linkage between the steering wheel and the front wheels [3]. As there is no torque applied by the driver to the steering wheel which is used to actually steer the front wheels, it is possible to reduce the range from steering lock to steering lock further. When reducing the steering angle furthermore, the requirements according to ECE r-79 have to be met. To "ensure easy and safe handling of the vehicle up to its maximum design speed" [2], it is necessary to provide the driver a certain range of steering angle stroke. In the following sections, a simulator study with two steering systems reducing the available steering angle stroke to +-180 deg and +-90 deg is presented.

# Tested Steering Systems

In total three different steering systems (two newly designed ones and one conventional system as benchmark) were tested. These systems were modelled within Matlab Simulink and executed on a real-time-target (see Figure 1). The models are provided with vehicle dynamics parameters, such as vehicle tyre forces or yaw rate by a bus connection to the simulation software where a double-track vehicle model is executed with a frequency of 240 Hz. The steering models themselves are running on 1 kHz and provide a torque set point for the motor controller which then controls the current of a synchronous motor. The motor is able to provide up to 22 Nm's of steering torque. This high torque is used to simulate the mechanical steering lock at different rotary positions. The driver's steering input is acquired by a multi turn encoder processed by the model and the desired steering angle for the front wheels is then provided to the vehicle dynamics model.

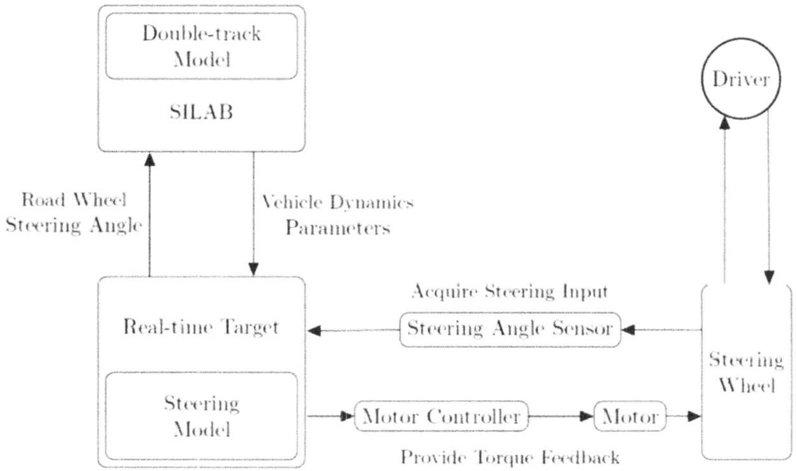


Figure 1: Implementation of Steering System Models in Driving Simulator

In the following sections the models for all three tested steering systems will be presented.

## Conventional Steering Model

The conventional steering model is a validated two mass oscillator according to Reisig [4]. Figure 2 shows an overview of the model used to calculate the steering wheel torque ($T_{StW}$). As inputs, the tie rod forces estimated from the double-track model and the currently measured steering wheel angle ($\delta_{StW}$) are used. Both masses which are used by the model use a simplified friction model. The only deformable element is the shown parallel spring damper element which represents the torsion bar for torque measurement and the stiffness of all other components in the steering system. The assisting force ($F_{SA}$) is applied to the steering rack and modelled as a quadratic function of the deflection of the spring and damper system. The whole model was validated by a real car test. Therefor tie rod forces and steering torque as well as steering wheel angle where measured. The steering angle of the front wheels is calculated with a fixed gear ratio. The total steering angle ranged from -600 to + 600 deg from lock to lock.

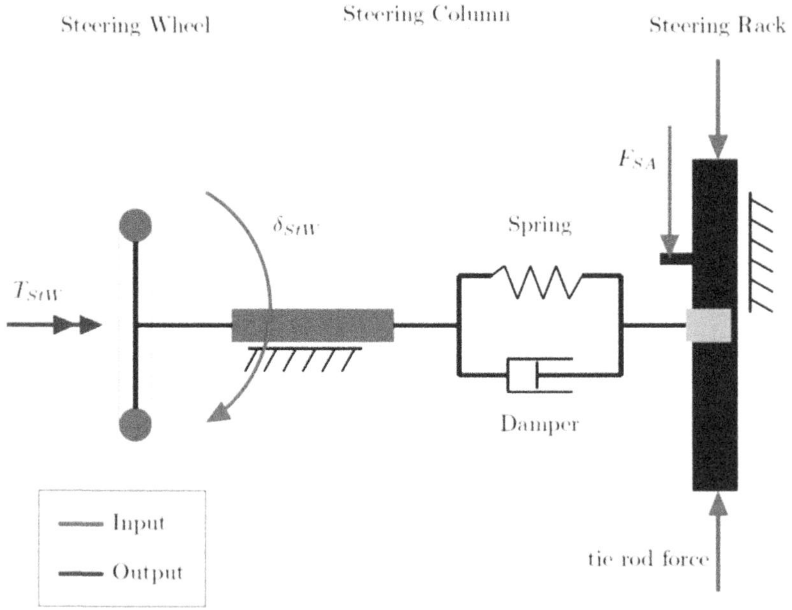


Figure 2: Model to calculate Steering Wheel Torque Feedback (according to [1])

### Conventional System with reduced Steering Angle

To follow up on the question of minimal steering angles required to control a vehicle, the tested systems which are compared with the conventional steering system should have a lower steering wheel angle range than an active front steering system. Thus the second model used has a reduced steering angle of 180 deg from zero to lock. Compared to the conventional system, it differs only in the fact that it has a total range of one turn. Torque feedback and front wheel angle are calculated as above mentioned. Only the assistance torque is adapted to provide a similar steering wheel torque than the conventional one.

### Steer-by-Wire System

In comparison to the two aforementioned systems, it is possible to design the torque feedback and steering ratio completely independent from any mechanical linkage by using a steer-by-wire system. To access the potential of such systems a control method proposed by Huang [5] is used. In this case, the steering ratio is no longer fixed and the steering wheel angle is used as input for a PID yaw rate controller. Dependent on speed, the sensed steering input is used to calculate the desired yaw rate for the vehicle. The speed dependent yaw rate gain is designed to provide a course control and a direct handling for vehicle speeds smaller than 25 km/h. For speeds above 45 km/h, it is designed to provide a more indirect behaviour. The necessary steering angle of the front wheel is then controlled by a PID controller. As mentioned before, it is also possible to provide a steering wheel torque independently from any tire forces. To provide a direct feedback of vehicle movement for the driver, the steering wheel torque is proportional to the lateral acceleration for high speed driving. In low speed driving conditions, it is proportional to the yaw rate of the vehicle. This complete system is designed to have a total steering angle range of -90 to 90 deg.

## Study Layout

To access the controllability of the two steering systems with reduced total steering wheel stroke, a study in a driving simulator is conducted. As reference the validated conventional steering system is used.

## Driving Simulator

A fixed base driving simulator equipped with a generic car interior is used. The position of steering wheel and driver's seat is similar to the position in a mid-size sedan car. The simulator has three projection walls with a size of 1875 x 1075 mm each. The projectors resolution is 1920 x 1080 Pixels resulting in a resolution of more than 26 ppi. This results in a 117 deg horizontal and 47 deg vertical driver's field of view. The used steering wheel is a leather-wrapped three spoke wheel and has a diameter of approximately 360 mm. The maximum time latency between driver's input and projection is 136 ms. As simulation software SILAB developed by the Würzburg Institute for Traffic Sciences GmbH is used.

## Measures

During the experiment, different subjective and objective measures are taken. On the subjective side, several questionnaires regarding steering feel and controllability [6] are used for the different manoeuvres described in the following section.

As objective measures, the lane keeping quality (see Figure 3), as well as over- and undershoot when entering new lanes are evaluated. The lane keeping quality can be calculated as the area between the driven path and the subjective lane centre. As the perception of the lane centre differs from subject to subject and is generally disturbed in driving simulators, the geometrical lane centre cannot be used to calculate the lane keeping quality. Instead the mean value of lane deviation is calculated as subjective perceived lane centre. The lane keeping quality can then be calculated as the integral of lane deviation over lane distance. As this value is dependent on the length of the track, the mean value of segments with a length of 100 m is used. The larger the value, the more difficulties a driver had to control the car to follow the lane.

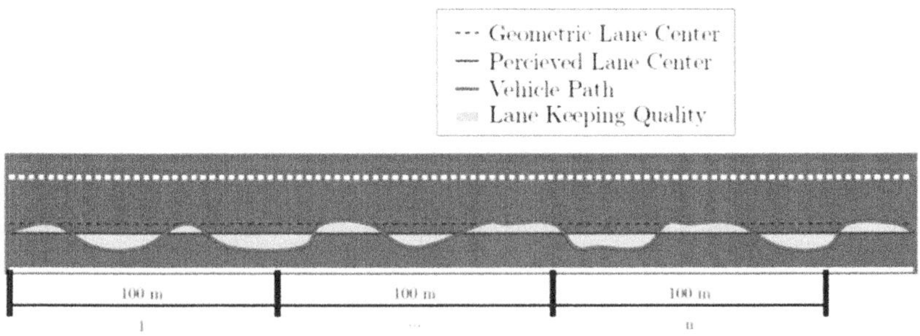


Figure 3: Definition of Lane Keeping Quality (according to [4])

For the lane change manoeuvre, the absolute over- or undershoot value, defined as the maximum distance between perceived lane centre and travelled path, is used. The smaller the deviation, the easier it is to steer the vehicle into the new lane. For the slalom manoeuvre, the lane keeping quality is calculated based upon the geometric lane centre, as the subjective perceived lane centre and curvature may differ from the real one and cannot be estimated in opposition to the straight line driving manoeuvre. For the cornering manoeuvres to the left and right, the over and undershoot values, like in the lane change manoeuvre, are used as well as the mean lane deviation from the geometric lane centre.

## Driving manoeuvres

After a familiarization track of approximately 5.9 kms, each subject drove a more or less 5 km track to get familiarized with the steering system. Afterwards, they filled out a questionnaire regarding their general impression of the steering system. The third track consisted of the following manoeuvres:

- Straight line driving on highway at 120 kph
- Lane change on highway at 120 kph
- Slalom on rural road with radius of 100 m's at 70 kph
- Left turn on intersection in urban area
- Right turn on intersection in urban area

After each manoeuvre was driven twice by the subjects, they were asked to stop the car on the road and to fill in a questionnaire with specific questions regarding controllability and subjective steering feel.

The second and third track were repeated for each steering system. The order of steering systems was randomized in order to avoid any sequence effects. The familiarization track at the beginning was driven with the conventional steering system by each participant.

## Subjects

In total 68 subjects were tested. Due to simulator sickness and technical failures only 37 subjects were analysed. Not all of the analysed subjects were able to test all three steering system due to technical failures. Among the subjects were 15 female drivers with an average age of 36.13 (SD = 7.03) years and 22 men with an average age of 39.95 (SD = 14.04). The age of all subjects ranged from 23 to 67 years. Only experienced drivers of compact and sedan cars were tested. The average driving experience was 17.92 (SD = 11.31) with an average yearly mile age of 19 622 (SD = 12 500).

# Analysis of results

According to Jones [6], a steering feel and performance sum score based on standardized z-values of different steering performance and steering feel parameters is calculated. This index allows a first analysis of the three steering systems. Regarding steering performance, there were significant differences between the Steer-by-Wire system with +-90 deg maximum steering angle in comparison to the conventional as well as to the Conventional system with reduced steering angle (+-180 deg). It turned out that the Steer-by-Wire system with the lowest maximum steering angle performed worst. Whereas the conventional reduced (+-180 deg) system perform a little worse than the conventional reference system, which turned out to not be significant.

Regarding steering feel there was a statistical difference between the conventional steering system with +-180 deg in comparison to both other systems indicating that the +-180 deg system performed best. There were no significant differences between the Steer-by-Wire and the reference system.

Table 1: Mean Lane Keeping Quality in m² for different manoeuvres

| Steering System | Straight Line Driving | Slalom |
|---|---|---|
| +-600 deg | 15,6 SD 9,9 | 47,5 SD 17,7 |
| +-180 deg | 14,7 SD 9,8 | 46,9 SD 15,4 |
| +-90 deg | 17,3 SD 11,4 | 45,0 SD 18,1 |

Having a more detailed look at the performance data, this result becomes more inherent. For the lane keeping quality (see Table 1) there is a difference of the systems ranking between straight line and slalom driving. As indicated from the overall score, the +-180 deg system performs best in the straight line driving manoeuvre. In the slalom manoeuvre at 70 kph it performs worse than the Steer-by-Wire system (+-90 deg), but still better than the conventional system.

Table 2: Mean Over- or Undershoot in m for different Manoeuvres

| Steering System | Lane Change | Left Turn | Right Turn |
|---|---|---|---|
| +-600 deg | -0,50 SD 0,37 | 0,02 SD 0,62 | 0,01 SD 0,51 |
| +-180 deg | -0,46 SD 0,35 | 0,15 SD 0,48 | 0,37 SD 0,47 |
| +-90 deg | -0,37 SD 0,34 | 0,12 SD 0,61 | 0,81 SD 0,35 |

For the over- and undershoot values (see Table 2), there is a similar picture. The conventional steering system performed best in all three manoeuvres. But for the left turn manoeuvre, the Steer-by-Wire system with +-90 deg steering angle is slightly better than the reduced conventional system with +-180 deg. During the other manoeuvres it performed worse.

# Conclusion

Summarizing, the conventional steering system like installed in current series production cars performed best with regards to steering performance, which is not surprising as this is the system to which all subjects are more or less used to. Obviously the subjects were able to adapt to a reduced steering angle within a short time. The conventional reduced system (+-180 deg) performed similar like the conventional system with regards to steering performance, and the subjects rated the system significantly better than the other ones. The Steer-by-Wire system with a total steering wheel stroke of +-90 deg performed worst in regard to steering performance. This can be an indicator that its steering angle might be too small and therefore the steering behaviour of the steering system to direct. Nevertheless, the system ratings showed that it is possible to achieve a similar subjective steering feel rating alike to the conventional steering system. [6]

However, it remains unclear whether the differences occur due to the reduced steering angle or the completely different control strategy of the +-180 deg and the +-90 deg Steer-by-Wire system. To distinguish this effect it is recommended to proceed with further tests comparing the two control strategies with each other when using the same maximum steering angle. In this case, it seems to be helpful to continue with a range of +-180 deg, as this system performed best. Regarding packaging and system complexity, there is no advantage for a further reduced steering angle as long as the total steering angle range is below 360 deg, because this makes it much easier to design a mechanical steering lock system for the steering wheel actuator in a Steer-by-Wire system. Furthermore, this can result in advantages for the clockspring and steering angle sensor design as the requirement for multi-turn capabilities will be obsolete. Additionally, it is recommended to proceed with in-vehicle tests to evaluation the steering performance and steering feel for systems with reduced maximum steering angles in a further stage of the development process, as there is a lack of fidelity in a driving simulator and the steering behaviour and subjective perception can only be modelled and simulated to a limited point.

# Summary

In this paper, a simulator study is presented to determine the minimal steering stroke required by normal drivers to be able to control a car safe and easily. 37 normal but experienced drivers without any special training participated in the study. They were presented three different steering systems with different maximum steering angles for each system. The first system was a model of a current series production steering system with a maximum steering angle of +-600 deg. The second one was the same model with an adopted steering ratio to provide the same level of feedback forces and a reduced steering angle of +-180 deg. The third system is a totally new approach to directly control the vehicles curvature with a total steering range of +-90 deg. The study shows that the +-180 deg system performed best with regards to subjective steering feel whereas the +-90 deg system performed worst in case of steering performance. The +-180 deg system seems to be the most promising system for further research. A direct comparison of both control methods is recommended, as there was no significant difference of steering feel between the new control method and the state of the art system. Reducing the steering range to less or equal than 360 deg total stroke enables easier designs for the steering wheel actuator of a full power Steer-by-Wire system.

# References


[1] Grimm, Roland: Active Steering Wheel, a new solution to implement active steering functionality into a vehicle, In: Pfeffer, Peter E. (Hrsg.) 5th International Munich Chassis Symposium 2014 – chassis.tech plus, Springer Fachmedien, Wiesbaden , ISBN: 978-3-658-05978-1, 2014

[2] United Nations: ECE-r79 – UNIFORM PROVISIONS CONCERNING THE APPROVAL OF VEHICLES WITH REGARD TO STEERING EQUIPMENT, Revision 2, 21. May 2005

[3] Pfeffer, Peter; Harrer, Manfred (Hrsg.): Lenkungshandbuch – Lenksysteme, Lenkgefühl, Fahrdynamik von Kraftfahrzeugen, Springer Fachmedien Wiesbaden, Wiesbaden, ISBN: 978-3-8348-0751-9, 2011

[4] Reisig, Richard: A real-time capable numerical Simulation model of torque Feedback in conventional power-assisted steering systems, Technische Universität Berlin, Institut für Energie und Automatisierungstechnik, Fachgebiet Elektronische Mess- und Diagnosetechnik, Master Thesis, 26. September 2012

[5] Huang, Pei-shih: Regelkonzepte zur Fahrzeugführung unter Einbeziehung der Bedienelementeigenschaften, Technische Universität München, Lehrstuhl für Ergonomie, Dissertation, 16. October 2003

[6] Jones, Penninah: Development and application of a questionnaire for evaluating the subjective Steering Feel of Steer-by-Wire steering systems, Technische Universität Berlin, Department of Psychology and Ergonomics, Master Thesis, 30. May 2015

# Implementation and testing of different control strategies on a steer-by-wire research platform

M. Sigilló, M. Dold, C. Delmarco, K. Polmans
ThyssenKrupp Presta AG, Liechtenstein

Author:

MSc. Michele Sigilló
Senior Development Engineer Steer-by-Wire

Co-Authors:

Dipl.-Ing. Markus Dold,
Project leader Steer-by-Wire

Ing. Claudio Delmarco, BSc
Project Engineer Mechanics Steer-by-Wire

MSc. (mult.) Kristof Polmans
Head of Vehicle Dynamics and Vehicle Testing

# Abstract

The progresses in computational speed and robustness made over the last decade by the automotive industry with respect to the automatization of road vehicles, reanimates the discussions about autonomous driving and X-by-Wire systems. ThyssenKrupp Presta AG has built up a modular research vehicle and equipped it with a Steer-by-Wire system in order to investigate the requirements and challenges that come along when the mechanical connection between steering wheel and steering rack is missing. In this paper, the architecture and concepts of the vehicle are presented and an overview about possible steering feedback strategies is given. Focus of the paper, however, is the adopted nonlinear position control of the steering rack, that is able to handle parameter variations, disturbances and sudden road condition changes while still guaranteeing an asymptotic tracking error convergence in an improved manner when compared to sliding mode compensators.

# 1 Introduction

Depriving the driver of a steered vehicle of the mechanical and thus physical link between the steering wheel and the steering gear, arises mainly two questions concerning the driver and the drivability of the vehicle, i.e. how can be ensured that the vehicle follows the desired path and what feedback may be needed to be given to the driver in order to safely drive the vehicle? To answer these questions, ThyssenKrupp Presta AG equipped a research vehicle with a Steer-by-Wire (SbW) system that is composed of an electrically actuated steering gear and a steering feedback actuator as main actors, both controlled through the use of dSPACE®'s Autobox that also allows to access various sensors and provides the platform for a communication network between the components of the system.

From a driver's point of view, the biggest change to a conventional steering system with mechanical linkage is, that in SbW no forces related to the lateral movement of the vehicle is fed back through the steering system to the steering wheel and thus to the driver. One could ask if that feedback is needed at all, since human perception of vehicle movement isn't limited to haptic feedback from the steering system only. This valid question is addressed in a more general way in [3], where haptic feedback is shown to improve the user's efficiency and reduce the error rate and stress level. A lot of work has therefore been done on the question of how the haptic feedback should be generated and what dependencies it should show. The most common approach is to use an electrical motor connected to the steering wheel and generate a defined torque that simulates a realistic driver experience such as in [1],[2] and [4]-[11]. In [4] an adaptive control approach that combines the control of the steering gear and the feedback actuator is presented, assuming the target dynamics and torque for the driver feedback exists and is known, [5] on the other hand tries to address the question of how the target dynamics can be described mathematically for the use in a control system.

It's natural to think that the target dynamics should match the real dynamics acting on the steering wheel when a conventional steering system is used, therefore many researchers make use of the vehicle's dynamics and kinematics to estimate the forces acting on the vehicle and its steering system and use this information to generate a target for the haptic feedback at the driver's input, i.e. the steering feel at the steering wheel, as done in [2], [9] and [10]. Another common approach is to identify the parameters that characterize the steering feel based on objective measurements of the steering system and to so define a target steering torque for the SbW system for different driving situations. [6],[7] and [11] found the characteristic components of steering feel to be essentially the systems base friction, the damping, a nonlinear base toque curve and a torque hysteresis. It was shown that these few and intuitive parameters were not only easy to tune and adjustable to the various driving situations, but also provided a satisfactory result when applied on real vehicles equipped with SbW systems.

While steering feel obeys a driver's subjective judgment and therefore can largely vary, the requirements for the steering gear's rack position are much more stringent, in the sense that independent from the steering feel, the car has to follow the driver's desired path. This calls for a fast and precise position control. Just like with many other applications, a linear control approach, as provided by [2], [8] and [10], shows a good tracking behavior of the proposed control systems, however, their performance is negatively affected by parameter variations, changes in road conditions, impact of external disturbances or dominating nonlinear system behavior when operating outside of the linearly described system range. [4] and [1] on the other hand developed a nonlinear control approach. Although [4] uses nonlinear control tools and remarkably avoids the measurement of torque values, the presence of external disturbances in the systems dynamic is not addressed and the torque estimation is successfully proved assuming exact model knowledge, which can't be assumed for a real system. The sliding mode compensator (SMC) approach chosen in [1] is in theory able to handle parameter variations, road condition changes and disturbances, but requires a single track vehicle model and assumes that the control torque is able to exert a discontinuous torque as typical for SMC to achieve asymptotic error convergence. This last assumption can't be met by any real actuator and leads to an ultimately bounded tracking error result as shown as well in [1].

In this paper, section 2 is aimed at introducing the chosen concepts and architecture built in the mentioned research vehicle, while section 3 gives an overview about the different challenges and opportunities that come along when trying to create a prope steering feedback. Section 4 focuses on the rack position control, where a Lyapunov based nonlinear controller able to handle parameter variations, external disturbances and road condition changes was deployed, while overcoming the disadvantages encountered with pure SMC as in [1]. At the end of this section a comparison to a conventional state space controller is made and field test results are discussed. Finally, in section 5 conclusions are drawn and further research work is presented.

## 2 System Concepts and Architecture

Given its advantageous available space underneath the dashboard and in the front end compartment of the vehicle, a "Roding Roadster R1" was chosen by ThyssenKrupp Presta AG as research and development platform.

Figure 1: Roading Roadster R1

The goal of the research project is to investigate the requirements for a SbW system not only with respect to the hard- and software, but also with respect to redundancy as required by the existing laws. In cases like this, where the applied technology additionally has to satisfy also the requirements of human perception, it is clear that the path leading to the final result is an iterative one. These iterations include also the combination of different hardware components, therefore the concept followed by ThyssenKrupp Presta AG is the one of modularity, meaning that with relatively small effort it is possible to interchange components and make adaptations to the hardware.

Since components and hardware inherently include sensors, the concept of modularity is also found in the data acquisition system with dSPACE®'s Autobox. In case of a change of sensor, the Autobox allows to change and use interface boards accordingly to the new sensor within few minutes. The rest of this section is dedicated to the description of the concepts applied for the feedback actuator and the steering gear.

## 2.1 Feedback actuator

As in [1],[2] and [4]-[12], the main actor in the feedback actuator is represented by an electric motor. The majority of the motors used in industry are brushless DC motors that show the effect of cogging and ripple. Since human haptic perception with respect to changes in torque is quite sensible, one of the requirements for SbW is the maximum allowable torque ripple on the steering wheel. Assuming a cogging free motor is used, torque ripple can be simulated by requesting such a torque exertion from the motor via software. Following this concept, a brushless DC motor without iron stator and a brushed DC motor where included onto the feedback actuator, given their cogging free properties.

Based on the concepts and results made by [6],[7] and [11], also in this project the steering feel is tuned by means of base torque, damping, hysteresis and friction. The feedback actuator is therefore equipped with an electrically adjustable friction element that is able to generate mechanical friction in a desired quantity and even simulate a mechanical end lock. Steering feel generated by the feedback actuator and trajectory control of the steering gear need information about angular velocity and angular position of the steering wheel. Due to their importance for the proper functioning and safety of the complete SbW system, several angular sensors are mounted (absolute and relative angle sensors) and available on the feedback actuator to ensure the systems redundancy. In [2],[4],[8],[10] and [12] the steering feel, independent from how it's calculated or defined, is a target value and becomes a controlled variable requiring a closed-loop control system and therefore also a torque sensor. On the other hand [11] reached good results without implementing any control algorithms. The feedback actuator presented here is equipped with a torque sensor that is used not only to measure the torque applied on the steering column, but also to serve as sensing unit for a closed-loop and open-loop control system in order to collect objective measurement results and be able to compare the performances of the two control approaches.

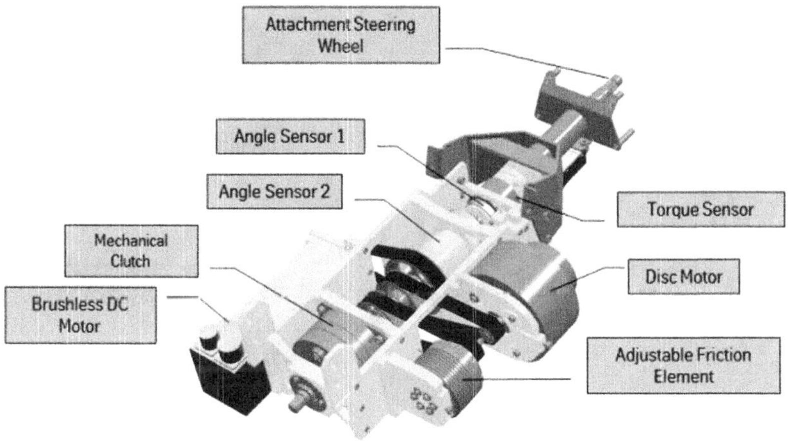


Figure 2: Feedback actuator SbW

With the elements describe above, the feedback actuator concept is completely described and ready for the usage in a SbW environment. For operational safety however, a mechanical clutch was mounted at the end of the actuator, such that in case of failure of the SbW system the clutch closes and establishes a mechanical connection from the steering wheel to the steering gear and the vehicle can so safely be driven aside.

## 2.2 Steering gear

On the basis of ThyssenKrupp Presta AG's serial product of electric power assisted steering (EPAS), the steering gear created for this project is based on such an in-house EPAS system.

The "Ball-Nut-Assembly" (BNA), pinion, rack and electric motor build together the EPAS system. For a SbW system the same components and the same build up would be enough and the properly controlled motor would be able to perform the required position control. The concept of redundancy however, calls for a backup solution in case of failure of the electric motor. The solution adopted in this project is a double motor solution in the sense that two electric motors , that together with their attached ECU form the so called power-pack (PP), are connected through a belt drive to one BNA mounted in the middle of the steering rack, as shown in Figure3.

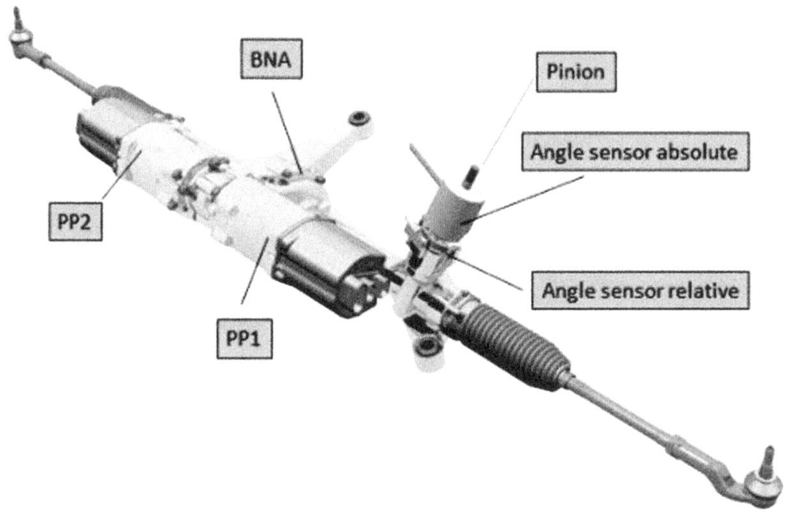


Figure 3: Steering gear SbW

Having two PP available gives rise to a series of questions and opportunities that are also part of the investigation of this research project. For instance, two PP could be actively used at the same time, each exerting half of the power of an EPAS PP. This approach would then open up the discussion of how the motors should be controlled in order not to run into stability problems. If on the other hand just one motor would be made responsible of moving the rack to the desired position, the question of how the other motor should be handled could lead to solutions where it stays passive and so becomes a generator, or it is actively controlled in a way that its presence doesn't affect the systems dynamics. As first approach for the present project the solution of an active and a passive motor are applied, meaning that in case of failure the active motor is shut down and the passive motor gets activated and takes over the positioning of the rack.

## 2.3 Complete SbW System and Architecture

The complete SbW system as built into the research vehicle is depicted in Figure4.

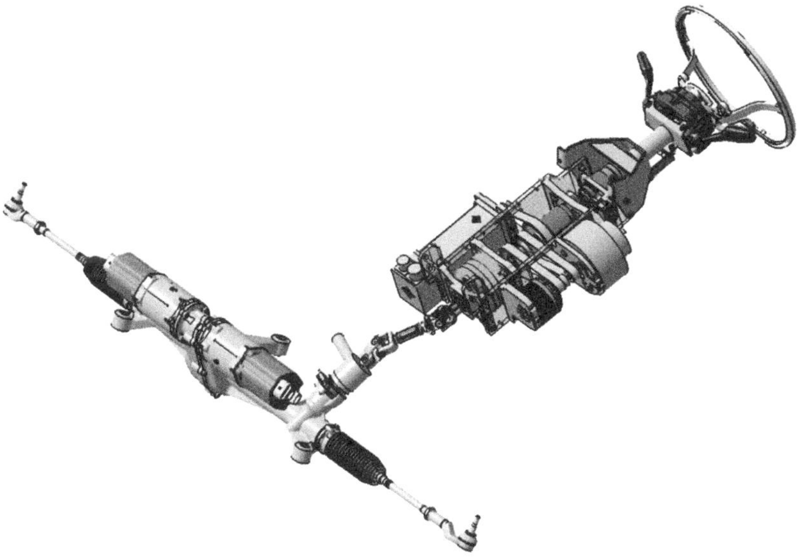

Figure 4: Complete SbW system

As mentioned in section 2.1, the research vehicle is equipped with a mechanical fall back solution composed of a clutch (green element in Figure4) and a short shaft connected to the pinion of the steering gear. In case of failure of the SbW system the clutch is closed and the steering wheel connected to the steering gear, allowing so the manoeuvrability of the vehicle in passive mode.

With the described concepts and components, together with dSpace®'s Autobox as computational unit, the whole system architecture can be set up as shown in Figure5.

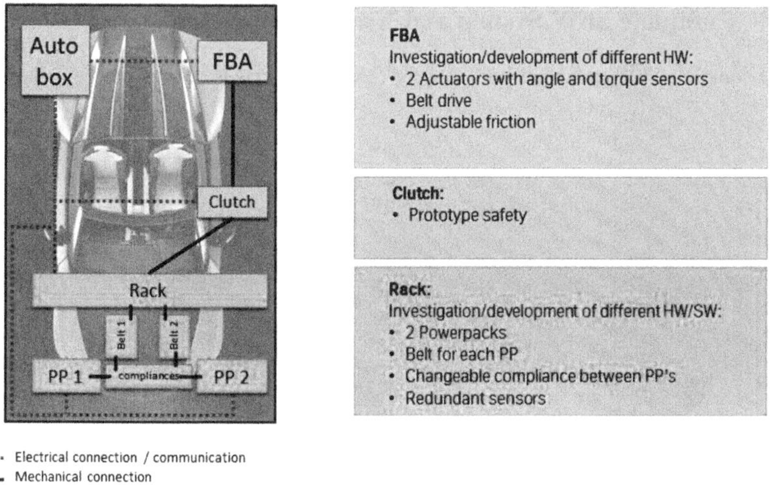


Figure 5: System architecture

dSpace®'s Autobox completes the system architecture and gives life to the SbW system. Its tasks are to wake up and establish the FlexRay® communication with the PP's, establish the CAN communication with the absolute angle sensors, supply the relative angle sensors and collect their measured values, read in the measured torque and serve as command interface for the clutch and control interface for the motors on the feedback actuator and the PP's on the steering gear respectively.

# 3 Steering Feedback

This section is dedicated to the challenges and opportunities that come along when the desired target haptic torque at the steering wheel, i.e. the steering feedback, is implemented. As mentioned in section 2.1, the steering feedback in this project is based on the concepts and results made by [6],[7] and [11].

While the open-loop approach in [11] is quite straight forward, the challenges and opportunities arise when a closed-loop approach is pursued. Main reason for a closed-loop control of the steering feel is, that a good controller can react on small parameter variations and disturbances like friction in the feedback actuator system, guaranteeing so always the targeted steering feel for the driver.

Throughout literature two approaches are used and described, one being a steering feel control by means of position control at the steering wheel like in [2],[4] and [8], the other one being a steering feel control by means of torque control at the steering wheel like in [11] and [12]. While the first approach relies on the equation of motion and its internal states (angular position and angular velocity) for the state-space control, the second approach, as used in [12] and applied on the feedback actuator used in this project, uses a transfer function where the input is represented by the motor torque and the output of the system is the torque measured at the steering wheel. For the resulting state-space representation the two states to be used are the measured torque at the steering wheel and its first time derivative. The respective experimental results, show that both approaches deliver a very good performance but as described in [12] the system properties represent the main criterion whether one or the other or both type of control can be applied. This becomes clearer with a closer look at the equation of motion of the steering wheel system like in [1]-[2] and [4], where it can be seen that damping and torsional stiffness are main contributors to the resulting torque. If the system shows a high torsional stiffness, like the one used for ThyssenKrupp Presta AG's SbW project, the complete system can be defined as a stiff system and thus poses higher requirements in terms of reaction time to the control algorithm and actuator. Roughly speaking, if the system is very stiff, it can be interpreted as a loss of degree of freedom that leads to a smaller system time constant and the need of faster actuators and algorithms.

From this observation it is clear that the feedback system stiffness and in general its dynamic behavior are part of the requirements this research project is meant to investigate.

# 4 Steering Rack Position Control

SbW systems get more driver acceptance, the less a driver can tell whether he's mechanically connected to the wheels or not. One indicator is the feedback at the steering wheel as presented earlier and the other one the tracking performance of the steering gear. While the steering feedback obeys subjective performance judgment and can therefore vary, the requirements for tracking are more stringent, i.e. the steering rack position has to track precisely the desired position at any given point in time. This calls for a fast and precise position control of the steering rack as independent as possible of parameter variations, road condition changes and disturbances on to the system.

The steering gear and its load can mathematically be approximated quite well by a linear system equation and as long as this system operates in the linear range, good tracking results can be achieved with linear control approaches like in [2], [8] and [10]. Unfortunately, system parameter variations, unpredictable varying road conditions and disturbances acting on the system, pose a very challenging task to linear control approaches. Under these conditions the operating point of the system diverges from the one assumed for the design and tuning of the controller and makes it therefore difficult to achieve good robustness and tracking performance. As shown in [1] and [4], Lyapunov based nonlinear control approaches successfully handle the problem of parameter uncertainty. The nonlinear adaptive control approach in [4] however, doesn't include external unknown disturbance forces in its dynamics and assumes that nonlinear effects of friction and damping are linearly parameterizable through a known regression matrix and a matrix of unknown but constant parameters. Although in [1] external disturbances and unknown system dynamics are compensated successfully through the use of a SMC, its theoretical asymptotic error convergence to zero can't be transferred to a real system, since the SMC part of the controller requires the motor torque to change proportional to the discontinuous signum-function, which in practice can't be achieved by any actuator.

The Lyapunov based nonlinear controller applied in this paper will be shown to be able to handle parameter variations, external disturbances and road condition changes while overcoming the disadvantages and concerns encountered in [1] and [4].

## 4.1 Model Description and Design Assumptions

The steering gear dynamics may be written in a simplified form as

$$m_{eq}\ddot{x}(t) + b_{eq}\dot{x}(t) + c(x_\Delta, t)x_\Delta(t) + F_{d1}(t) = pT(t) \qquad (1)$$

$\dot{x}(t), \ddot{x}(t)$ represent in (1) the rack velocity and rack acceleration respectively, $F_{d1}(t)$ describes the disturbances and not modelled forces, $m_{eq}$ is the equivalent lumped mass at the rack calculated from all in the system involved masses and mass moments of inertia, $b_{eq}$ represents the equivalent lumped damping friction coefficient calculated from the damping coefficients acting at different locations in the gear, and the expression $pT(t)$ describes the force applied to the rack by the motor torque $T(t)$ as control input through the ration $p$. The expression $c(x_\Delta, t)x_\Delta(t)$ represents elastic reaction forces acting on the system and deserves a separate discussion.

The biggest contribution to the reaction forces with elastic behaviour in a steering system is delivered by the so called self-alignment forces, which are mainly a function of the cornering stiffness of the front tire, the tire slip angle and to a minor extend also of time [1]. Further, cornering stiffness and slip angle are dependent on the nonlinear behaviour of the tire characteristic, which is very difficult to model and lets so the term $c(x_\Delta, t)x_\Delta(t)$ become a non-modelled nonlinear term that can be moved into the disturbance term of (1). Continuing this path of reasoning, it can be argued that the damping coefficient $b_{eq}$ is not a constant value as well, as it was done in [1], but similar to the reaction forces the part diverging from the constant value can be moved and included into the new disturbance term $F_d(t)$. The new system dynamics can therefore be rewritten as

$$m_{eq}\ddot{x}(t) + b_{eq}\dot{x}(t) + F_d(t) = pT(t) \quad (2)$$

For the controller development it's further assumed that $x, \dot{x}$ are measurable and that $x_d, \dot{x}_d, \ddot{x}_d \in \mathcal{L}_\infty$, where the three variables denote the desired rack position, desired rack velocity and desired rack acceleration respectively. Rack position and speed can easily be measured through the mounted sensors on the steering gear and the stated assumption of boundedness is indeed realistic, since all three signals derive from the steering angle input given by the driver that is physically limited and therefore also puts a limit to the desired speed and acceleration.

## 4.2 Open-Loop Error System

Just like any other controller also for the development of the Lyapunov based controller an error needs to be quantified. Let the tracking error be

$$e_1 = x_d - x \quad (3)$$

then two filtered tracking errors can be defined as

$$e_2 = \dot{e_1} + \alpha_1 e_1 \quad (4)$$

$$r = \dot{e_2} + \alpha_2 e_2 \quad (5)$$

where $e_1(t), e_2(t), r(t) \in \mathbb{R}^1$ and $\alpha_1, \alpha_2 \in \mathbb{R}^1$ represent positive control gains.

Multiplying (5) with the equivalent mass $m_{eq}$ allows to inject the dynamic system (2) in the open-loop error system as follows

$$m_{eq} r = m_{eq}(\dot{e_2} + \alpha_2 e_2) \quad (6)$$

Using (4) and (3) in (6) the error dynamics become

$$m_{eq} r = m_{eq}(\ddot{x}_d - \ddot{x} + \alpha_1 \dot{e}_1 + \alpha_2 e_2) \quad (7)$$

the second term on the right hand side of (7) can be substituted through the dynamic system in (2), resulting in

$$m_{eq} r = m_{eq}\ddot{x}_d + b_{eq}\dot{x} + F_d - Tp + m_{eq}\alpha_1 \dot{e}_1 + m_{eq}\alpha_2 e_2 \quad (8)$$

For a quicker stability analysis and easier bounding of the terms during the analysis it's convenient to make use of the mean-value theorem, which can be adopted through the use of the DECAL algorithm described in [15]. Applying DECAL on (8) leads to the open-loop error system

$$m_{eq} r = Y_d \theta + S_1 + F_d - Tp \quad where, \quad (9)$$

$$Y_d = [(\dot{x}_d - \dot{e}_1) \quad \ddot{x}_d] \qquad \theta = \begin{bmatrix} b_{eq} \\ m_{eq} \end{bmatrix} \qquad S_1 = m_{eq}\alpha_1 \dot{e}_1 + m_{eq}\alpha_2 e_2$$

Similar to [4], $Y_d$ is a known regression matrix and $\theta$ a matrix of constant coefficients that can be estimated if needed. It's easy to see that the term $S_1$ can now be bounded as a function of the states, i.e. the tracking error and filtered tracking errors, but it also can be seen that the assumption $F_d \in \mathcal{L}_\infty$ must be valid.

## 4.3 Control Law and Closed-Loop Error System

Given the open-loop error system in (9) a controller based on [14] of the form

$$T(t) = \frac{1}{p}\left(Y_d\hat{\theta}(t) + \mu(e_2, t)\right) \tag{10}$$

can be deployed and used in (9) to result in the following closed-loop error system

$$\begin{aligned} m_{eq}r &= Y_d\theta + S_1 + F_d - Y_d\hat{\theta} - \mu \\ \Leftrightarrow m_{eq}r &= Y_d\tilde{\theta} + S_1 + F_d - \mu \end{aligned} \tag{11}$$

where $\tilde{\theta} \in \mathbb{R}^2$ denotes the difference between the constant matrix $\theta$ and its estimate $\hat{\theta}$. In (10) and (11) the term $\mu(e_2, t)$ represents the key difference to the SMC in [1], where the signum-function directly acts on the calculation of the motor torque. Why this is not the case with the control law of (10) becomes clearer with a closer look at the closed-loop error dynamics

$$m_{eq}\dot{r} = \dot{Y}_d\tilde{\theta} - Y_d\dot{\hat{\theta}} + \dot{S}_1 + \dot{F}_d \underbrace{-(k_s + 1)r - \beta_1 sgn(e_2)}_{-\dot{\mu}} \tag{12}$$

In contrast to the SMC approach, the signum-function appears on a further time derivative level of the closed-loop error dynamics, such that its contribution to the control objective appears in the system dynamics, and thus in the calculation of the control input (10), in an integral, time continuous form. Since the control law for the control input is smooth in this form, it is indeed physically possible for an actuator to exert the requested torque output.

Equation (12) shows some more interesting facts and lays out the base for further needed assumptions and observations. The above describe open-loop error dynamics are going to be part of the stability analysis, thus it's worth looking at the bounding properties of (12). The first term to attract its attention is $\dot{F}_d$, because it's a term which structure can't be influenced and thus puts a further direct requirement and assumption to the disturbance, i.e. $F_d, \dot{F}_d \in \mathcal{L}_\infty$. For these assumptions to be valid upper bounds on the disturbance and its time derivative have to be known, in this case the upper bounds are represented by constants, so that $F_d \leq c_1$ and $\dot{F}_d \leq c_2$. Furthermore the time derivatives on $Y_d$,$\hat{\theta}$ and $S_1$ imply that some terms can be bounded as a function of the states and some simply by constants, but this influences at the same time the design of $\hat{\theta}$ accordingly to [14]. This means, that the algorithm for $\hat{\theta}$ and its time derivative have to be designed, such that this terms will be going to bound the new terms arising from the additional time derivative in (12) in order to guarantee stability convergence during the Lyapunov stability analysis. Nevertheless its key to recognize that the closed-loop error dynamics can be rewritten as

$$m_{eq}\dot{r} = N_B + \widetilde{N} - (k_s + 1)r - \beta_1 sgn(e_2) - e_2 \tag{13}$$

in which $N_B$ represents the terms that can be upper bound by a constant and $\widetilde{N}$ the terms that can be upper bound by a function of the states including the design of $\hat{\theta}$ and its time derivative.

## 4.4 Stability Analysis

For the Lyapunov stability analysis of the chosen controller and SbW system description, a Lyapunov candidate that satisfies the requirements listed in [13] is found in

$$V(t) = \frac{1}{2}e_1(t)^2 + \frac{1}{2}e_2(t)^2 + \frac{1}{2}m_{eq}r(t)^2 + P(t) \tag{14}$$

where $P(t)$ is defined as

$$P(t) = \beta_1|e_2(t_0)| - e_2 N_B(t_0) - \int_{t_0}^{t} L(t)dt \tag{15}$$

and the term under the integral is defined as

$$L(t) = r\big(N_B - \beta_1 sgn(e_2)\big) - \beta_2|e_1||e_2| - \beta_3|e_2|^2 - \beta_4|r||e_2| \tag{16}$$

Taking the time derivative of (14) results in

$$\dot{V} = e_1\dot{e_1} + e_2\dot{e_2} + rm\dot{r} + \dot{P} \tag{17}$$

in which the closed-loop error dynamics of (13) shows up. Using (4),(5),(13),(15) and (16) in (17), and after a series of mathematical manipulations, the time derivative of the Lyapunov candidate can be bounded as follows

$$\dot{V} \leq -||z(t)||^2\left(\lambda - \frac{\rho^2(||z(t)||)||z(t)||_1{}^2}{4\left(k_s - \frac{1}{2}\beta_4\right)}\right) \tag{18}$$

with the state vector $z = [e_1 \quad e_2 \quad r]$ and $\rho^2(||z(t)||)$ denoting a function of the states, the condition on $\lambda$ has to be

$$\lambda > \frac{\rho^2(||z(t_0)||)||z(t_0)||_1^{\ 2}}{4\left(k_s - \frac{1}{2}\beta_4\right)} \tag{19}$$

which essentially means that the initial conditions have to be known. At this point, if the conditions on the control gains $k_s, \alpha_1, \alpha_1$ and $\beta_1$ are met accordingly to [14], by invoking Barbalat's lemma as described and explained in [13], it is proven that the system is asymptotically stable and the tracking error $e_1$ as well as the filtered tracking errors asymptotically converge to zero and that the control input $T(t)$ as described in (10) is stable, bounded and real.

## 4.5 System Bounds

A side note has to made with regard to the upper bounds that are assumed to be known. As correctly addressed by [1], knowing the upper bounds is not always easy or even possible like in the case of external disturbances. There are several approaches, for instance the calculation of possible disturbances using extensive vehicle and system models as in [1] and [10]. Another method is the use of objective measurements and full utilization of the available control authority in the system. ThyssenKrupp Presta AG performed several objective measurements with the research vehicle and its conventional steering system and it was seen, that the peak disturbances acting on the steering system for this vehicle are coped with margin by the available control authority, therefore it can be stated that upper bounds exist and a stable controller tuning can be achieved also experimentally.

## 4.6 Test Results and Comparison to State-Space Control

Under the assumption of equal system knowledge a matrix representation for (2) was used and based on it a state-space controller designed and tuned in order to compare the tracking performances of the two controllers on the real system. The system equation in matrix form are then written as

$$\begin{bmatrix} \dot{x} \\ \ddot{x} \\ \dot{F}_d \end{bmatrix} = \underbrace{\begin{bmatrix} 0 & 1 & 0 \\ 0 & \frac{-b_{eq}}{m_{eq}} & \frac{-1}{m_{eq}} \\ 0 & 0 & \frac{-1}{\tau} \end{bmatrix}}_{A} \begin{bmatrix} x \\ \dot{x} \\ F_d \end{bmatrix} + \underbrace{\begin{bmatrix} 0 \\ \frac{p}{m_{eq}} \\ 0 \end{bmatrix}}_{B} T \tag{20}$$

$$x = \underbrace{\begin{bmatrix} 1 & 0 & 0 \end{bmatrix}}_{C} \begin{bmatrix} x \\ \dot{x} \\ F_d \end{bmatrix}$$

and the control gain vector $K$ is calculated through the pole placement method described in [16].

For the test bench test, the desired trajectory was defined as a sinusoidal signal with amplitude of 16mm and increasing frequency over time, ranging from 0.1Hz up to 1Hz. As expected the linear state-space controller struggles when nonlinear effects start to become more dominant and a phase delay between desired and measured position can be detected already at low frequencies. It is worth noting though, that around this operational point, where the load was simulated using springs, the overall behavior of the state-space controller is still very acceptable.

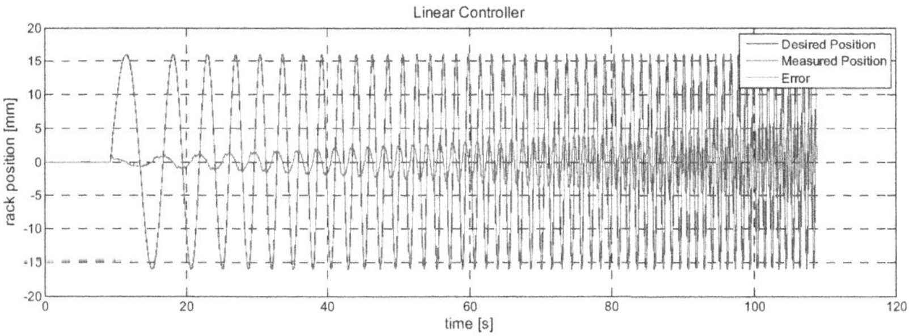


Figure 6: Tracking performance linear state-space controller

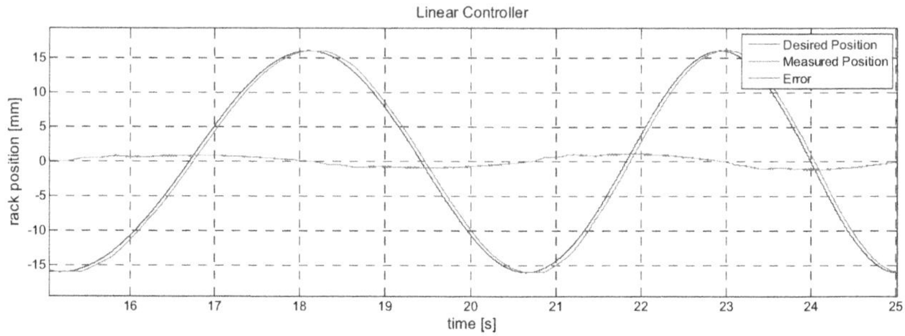


Figure 7: Tracking performance linear state-space controller zoomed

The same system and desired trajectory, but with the control law as defined by (10), gives a better tracking performance over the whole frequency range. The tracking error stays small compared to the one of the linear controller even though in both cases it increases with increasing frequency of the desired trajectory. This phenomena has a two folded explanation. On the one hand, the control theory guaranties for both system asymptotic error convergence, meaning that $e_1 \to 0$ as $t \to \infty$, therefore the continuously increasing frequency interferes with the concept of asymptotic convergence if the convergence time is big relative to the system's response time. On the other hand, the used PP's have their own integrated motor control, therefore if the internal motor controller isn't able to guarantee that at every point in time the exerted motor torque corresponds to the requested motor torque, tracking errors can and will arise. This last described phenomena was observed during the measurements depicted in Figure11, where the tracking error corresponded to a mismatch between requested and exerted motor torque.

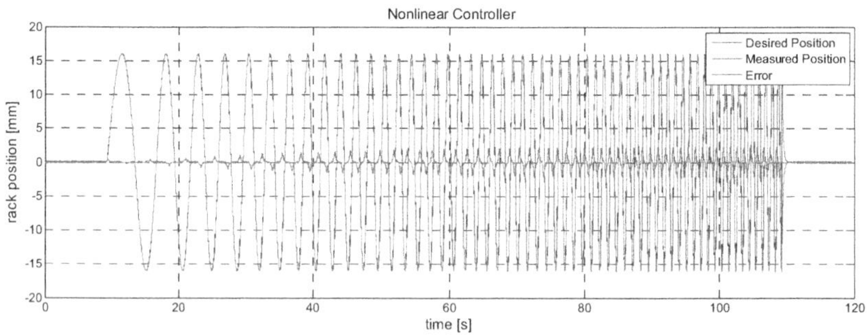


Figure 8: Tracking performance nonlinear control

Given the ability to handle the nonlinearities in the system better than the linear state-space controller and given its higher weight with respect to the measured velocity and desired velocity as well as acceleration, the nonlinear controller exhibits almost no phase shift in the lower frequency range as depicted by Figure9

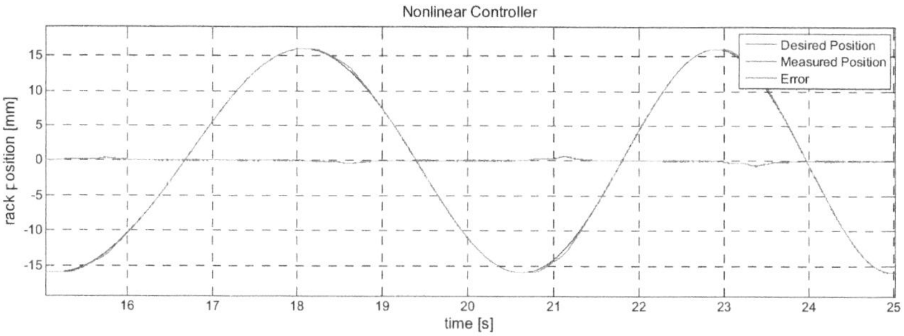


Figure 9: Tracking performance nonlinear control zoomed

Also on tests performed during driving, the tracking performance of the presented nonlinear controller shows very good and satisfying results as shown in Figure10.

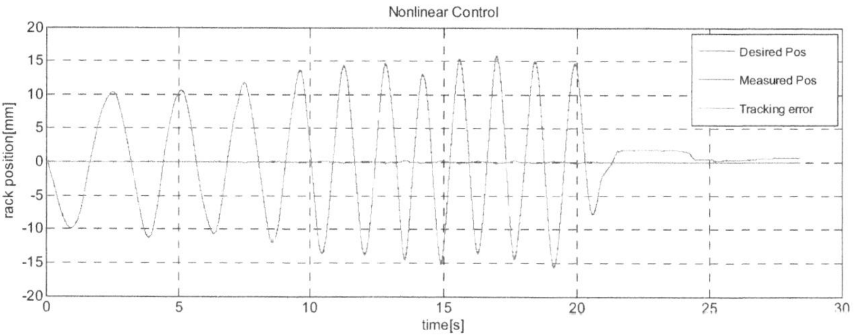


Figure 10: Tracking performance nonlinear control on test drive

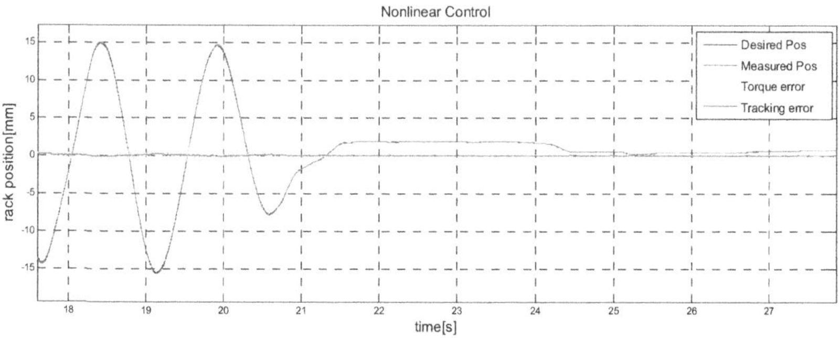


Figure 11: Tracking performance nonlinear control on test drive with Torque error

# 5 Summary and Further Work

The concepts and architecture of ThyssenKrupp Presta AG's SbW system as implemented in the company's research vehicle were presented and the different challenges and base requirements stated for the systems "feedback actuator" and "steering gear". Further, an introduction to the different strategies for the calculation of a target steering feel has been presented and the various control strategies for a close-loop steering feel control explained, in which the importance of the feedback system's mechanical characteristics has been emphasized since it influences directly the choice of the control approach for steering feel. The last part of this document focuses on the application and development of a nonlinear Lyapunov based control law for the steering rack position control able to handle and cope for parameter variation and unknown disturbances, hence guaranteeing an asymptotic error convergence and good tracking performance, especially when compared to linear state-space controllers.

Obviously still a lot of work has to be done with regard to the investigation of requirements for SbW systems, in the area of sensor precision, feedback actuator system properties and steering feel control as well as in the area of different control strategies for the rack position control. Being now in possess of the described research vehicle however, allows ThyssenKrupp Presta AG to consequently and efficiently continue the research and investigations needed in order to find and define those requirements that would allow to provide a good steering system to the customer at a fair price.

# References


[1] H. Wang et al. "Robust Control for Steer-by-Wire Systems With Partially Known Dynamics", IEEE Transaction on industrial informatics, 2014

[2] F. Claveau, Ph. Chevrel "A Multivariable Centralized Controller Design Methodology for a Steer-by-Wire System", European Control Conference, 2009

[3] Immersion "The Value of Haptics", Immersion.com, 2010

[4] P. Setlur et al. "A Nonlinear Tracking Controller for a Haptic Interface Steer-by-Wire Systems", IEEE Conference on Decision and Control, 2002

[5] U.B. Mandhata et al. "Evaluation of Customizable Haptic Feedback System for Ground Vehicle Steer-by-Wire Interfaces", American Control Conference, 2012

[6] T. Barthenheier "Potenzial einer fahrertyp- und fahrsituationsaghängignen Lenkradmomentgestaltung", Dissertation at the "TU Darmstadt", 2004

[7] S. Fankem, S. Müller „Modular concept for the calculation of the desired steering torque in steer-by-wire systems“, 3rd International Munich Chassis Symposium, 2012

[8] Á. Lörincz "Model reference control of a steer-by-wire steering system", Master Thesis at ThyssenKrupp Presta AG, 2004

[9] T. Koch "Untersuchungen zum Lenkgefühl von Steer-by-Wire Lenksystemen“, Dissertation at the „TU München“, 2010

[10] J.S. IM „ Control of Steer-by-Wire Vehicles with Passivity Approach“, International Federation of Automatic Control, 2008

[11] K. Polmans "Modelling and Implementation of Steering System Feedback for Steer by Wire Systems", Master Thesis at " Loughborough University", 2004

[12] M. Dähler "Steer-by-Wire Lenkgefühl-Regelung", Master Thesis at ThyssenKrupp Presta AG, 2015

[13] H.K. Khalil "Nonlinear Systems", ISBN: 978-0130673893, 2001

[14] P. Patre „Lyapunov-Based Robust and Adaptive Control of Nonlinear Systems using a Novel Feedback Structure“, Dissertation at the "University of Florida", 2009

[15] W.E. Dixon et. al. "Nonlinear Control of Engineering Systems", ISBN: 978-1-4612-0031-4, 2003

[16] J. Hespanha "Linear Systems Theory", ISBN: 978-0-691-14021-6, 2009

# Driving quality optimization based on cross-linked cause and effect chain models using the example of energy-efficient steering assistance

**Iwanicki, Marinette***
New Concepts „Semantic Validation Platform“
Calibration & Virtual Testing Solutions
AVL Deutschland GmbH
Peter-Sander-Straße 32
55252 Mainz-Kastel, Germany

**El-Haji, Mohanad**
SVP Product Manager
Institute of Vehicle Systems Technology (FAST)
Karlsruhe Institute of Technology (KIT)
Gotthard-Franz-Str. 9, 76131 Karlsruhe
Bld. 50.33, Room 202-203

**Freudenmann, Thomas**
SVP Project Manager
IPEK – Institute of Product Engineering
Karlsruhe Institute of Technology (KIT)
Gotthard-Franz-Str. 9, 76131 Karlsruhe
Bld. 50.33, Room 202-203

***Presenting Author**

# Abstract

This paper addresses the time- and information-intensive task of calibrating an electric power steering assistance system, which is optimised for energy efficiency and can be adapted to different vehicle variants. Hereby, the focus is on determining the customer-perceived properties steering feeling, energy consumption, system price and market availability date. It will be demonstrated how this aim can be achieved with a reduced testing scope by automated generation of behaviour models that predict the system behaviour in the defined parameter range.

In order to fulfil this task, the Semantic Validation Platform (SVP) is used. The SVP is an ontology-grounded expert collaboration system, designed to manage and support automotive development processes, in particular the validation and testing tasks. In combination with the intelligent test automation software AVL CAMEO and the virtual testing environment IPG CarMaker, the applied tool chain provides an environment that instructs the engineer step-by-step to plan the necessary tests, design and execute the tests automatically and generate behaviour models within the relevant context in order to make the right decisions. In each step, all relevant generated knowledge will be automatically documented in a sustainable way by the SVP, making it easily accessible and usable for future projects.


# Changes in Automobile Development

Modern vehicles feature a high complexity due to the interaction of numerous systems, especially mechatronic systems. Furthermore, increasing customer demands and legal restrictions must be considered when laying out vehicle properties. In doing so, information about the different vehicle variants as well as the characteristics of the vehicle's area of application, i.e. the external conditions such as weather and traffic of the designated market region but also characteristics of the different customers have to be taken into account (Seiffert, et al., 2008) (Wallentowitz, et al., 2009).

In the past decade, energy consumption has become one of the most important vehicle properties as the need for efficiency has been augmenting steadily caused by increasing gasoline prices and recent governmental regulations (Dannenberg, et al., 2007). As a result, manufacturers are urged to optimise an increasing number of vehicle systems as well as their interaction in the vehicle regarding energy efficiency. This optimisation has become particularly important for the development of electric vehicles due to the limited energy capacity of vehicle batteries (BMBF, 2013). However, it is essential that the increase in energy efficiency does not negatively influence other customer-relevant vehicle properties such as comfort or safety.

As a result, more and more mechatronical components are integrated into the vehicle in order to fulfil the required broad functional range and to control the vehicle behaviour accordingly (Valldorf, et al., 2004). Figure 1 shows and forecasts the increasing average number of ECUs per vehicle in the past and subsequent years.

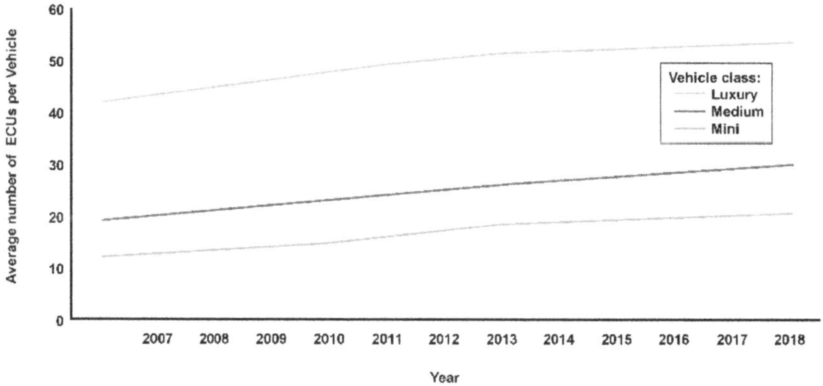


*Figure 1: Average number of ECUs in different vehicle classes (Riches, 2015)*

This evolution in automobile technology presents great opportunities as well as risks. On the one hand, it is possible to develop vehicles with sophisticated and adaptive system behaviour. On the other hand, increased complexity as a result of the high number of interacting systems and interdepending functions in combination with shorter product cycles represent a challenge for automobile manufacturers (Legner, et al., 2009). Facing these challenges can be supported significantly by methods and tools of knowledge management, in particular regarding the coordination of the activities of interdisciplinary development teams (Wallentowitz, et al., 2009).

Conventional system calibration and validation is a highly iterative and testing-intensive process that offers a great potential for reducing time and costs. As illustrated in Figure 2, the numerous different vehicle variants and optional extras lead to such a high number of possible adjustment options for each vehicle[1] that it is not possible anymore to validate each combination by testing (Günthner, 2007) (AVL LIST GMBH, 2014).

1 The BMW Group offers about 350 model variants with each up to 500 optional extras. For each single vehicle type this adds up to $10^{31}$ variation possibilities (Günthner, 2007).

This paper introduces a knowledge management system, the Semantic Validation Platform (SVP). The application of the SVP is illustrated by the example of the calibration of an electronic control unit (ECU) for an electromechanical steering system which has to be integrated into several vehicle variants. The SVP instructs expert teams to generate relevant and sustainable knowledge in form of behaviour models that reflect the system behaviour in its application area. Based on these behaviour models a calibration of the ECUs across different vehicle variants without further testing is possible. The SVP supports this task by a systematical workflow and intuitive user interface that guides engineers through the different activities. As a result, the efficiency of the calibration process is increased by avoiding the generation of useless or unused data based on inappropriate or incomprehensible simulation and testing (see Figure 3) (El-Haji, et al., 2014) (El-Haji, et al., 2012) (Freudenmann, et al., 2010).

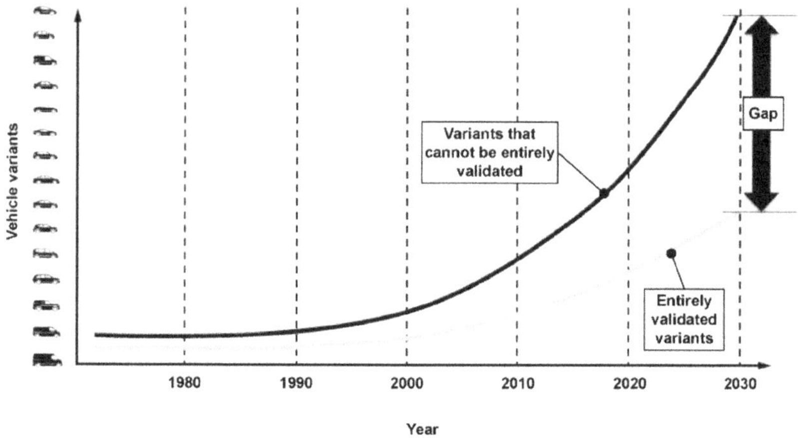


*Figure 2: Number of vehicle variants and part of variants that can be validated (AVL LIST GMBH, 2014)*

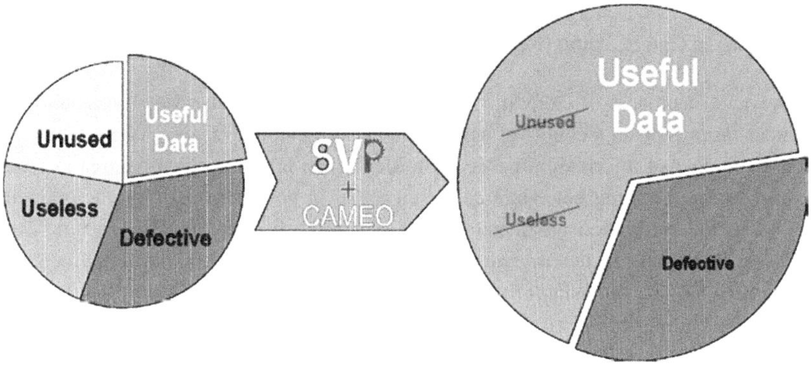


*Figure 3: Avoiding the production of unused or useless data with SVP and CAMEO (Freudenmann, et al., 2015)*

## Potentials and Challenges of Electromechanical Steering Systems

The steering of a vehicle, as direct interface to the driver, is a system that influences several customer-relevant, essential properties of the vehicle: apart from the precise realisation of the desired change of course, the most important property considering safety and comfort is the steering feeling, which is required to give feedback to the driver about the current driving state as well as to guarantee a stable directional stability. In addition, the steering system must damp shocks and vibrations from the track surface and is required to feature stability programmes such as the electronic stability control (ESC) (Proff, et al., 2012) (Harrer, et al., 2015).

Different customer segments will expect different characteristics of steering feeling, in particular regarding the steering wheel torque (see Figure 4). Therefore, it is an essential task of development to adapt the steering system characteristics to each targeted customer segment. In comparison to conventional steering systems, which possibly require complicated hardware adjustments to realise different system characteristics, this task can be processed very effectively with an electromechanical steering system by calibrating different parameter variants in the corresponding ECU. Another major advantage of electromechanical steering systems is the possibility to contribute to a low energy consumption of the vehicle[2,3] (Weißenborn, 2013) (ZF Lenksysteme GmbH, 2009) (Nexteer Automotive, 2015).

However, the process of calibrating electromechanical steering systems for multiple vehicle variants poses a couple of new challenges for automobile developers: There is a large amount of information needed considering the future application area as well as the expected driving behaviour of the customer in order to determine the optimal values of the calibration parameters of the electromechanical steering system. The process of parameter determination is a multi-disciplinary task, which requires experience in the domains of mechanical and electrical engineering as well as information technology. Experts in these different fields must work together in order to generate the desired system behaviour (Wallentowitz, et al., 2009). Figure 5 shows that only about a quarter of the working time can be used for actual value creation, whereas the rest of the time is spent for the search of information, communication and documentation (Forrester Research, 2000).

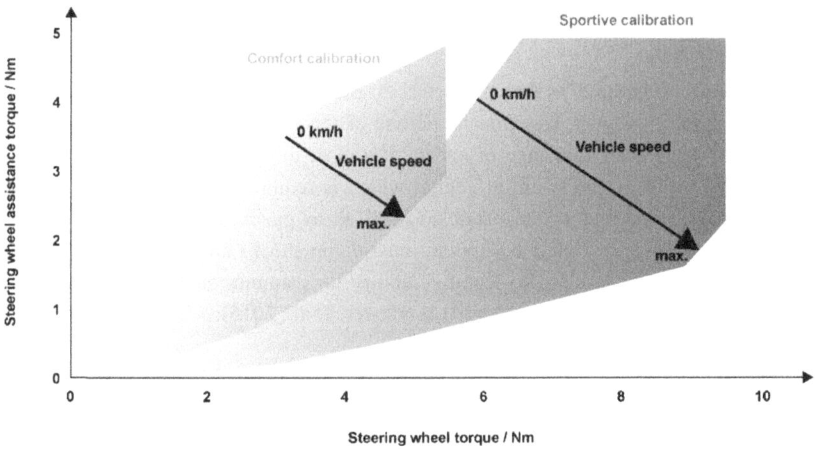


*Figure 4: Different possibilities of steering system calibration (ZF Lenksysteme GmbH, 2009)*

[2]The automotive supplier Nexteer Automotive has approximated the fuel savings induced by its over 28 million sold electromechanical steering systems to reach up to three billions of gallons since 1999 (Nexteer Automotive, 2015).

[3]The electromechanical steering system *ZF-Servoelectric* can save up to 90% of energy (approximately 0.4 litre/100 km less in the New European Driving Cycle) compared to hydraulic steering systems (ZF Lenksysteme GmbH, 2009)

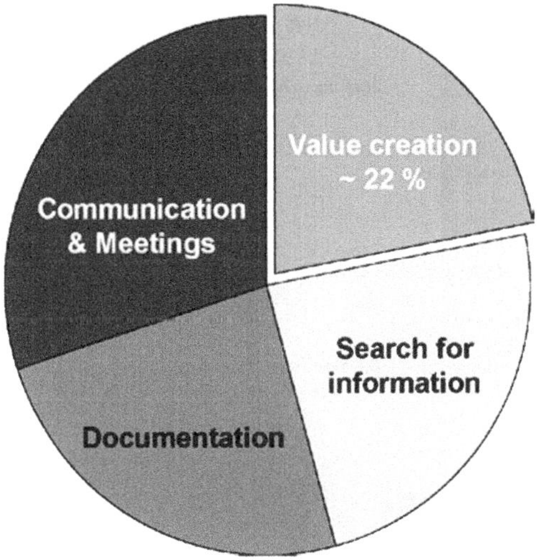


*Figure 5: Percentage of working time spend for actual value creation (Report)*

A well-coordinated calibration process, which is supported by suitable information technology, opens up great potential for the customer-specific adaptation of the steering system behaviour as well as for high process automation. In doing so, this cooperative und multi-disciplinary task can be accelerated and conducted more accurately compared to manual processing (see Figure 6). An essential objective is to avoid unnecessary and costly iterations in calibration activities, in particular in simulation and testing activities. This will be achieved by a context-sensitive management of all relevant information, which can be accessed and shared at any time by the development team members (Maune, 2001) (Seiffert, et al., 2008).

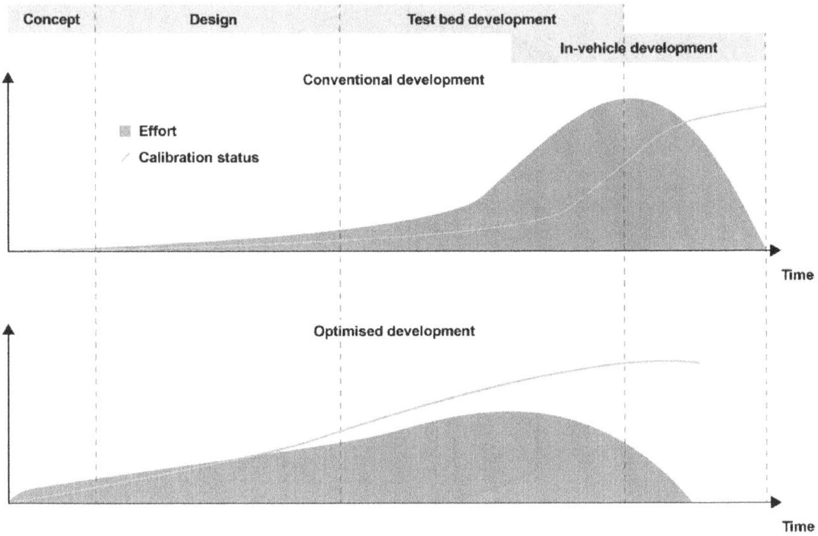


*Figure 6: Accelerated development by optimised and automated calibration (Seiffert, et al., 2008)*

# Handling Complexity and Efficient Calibration

As mentioned in the last chapter, the customer-specific calibration of an electromechanical steering system is a highly information-intensive and multi-disciplinary task. Diverse types of interrelated information from various domains have to be generated, accessed, combined and evaluated regarding the different variants of desired system behaviour. In order to manage this high complexity during the calibration process, methods and tools that allow for a holistic view on the vehicle properties as well as the properties of its subsystems are necessary. This includes the possibility of conducting simulations and tests in early stages of product development in order to achieve the satisfaction of the future customers (Maune, 2001) (Seiffert, et al., 2008). The SVP is a knowledge management system that was particularly developed to support automobile engineers handling these challenges (see Figure 7).

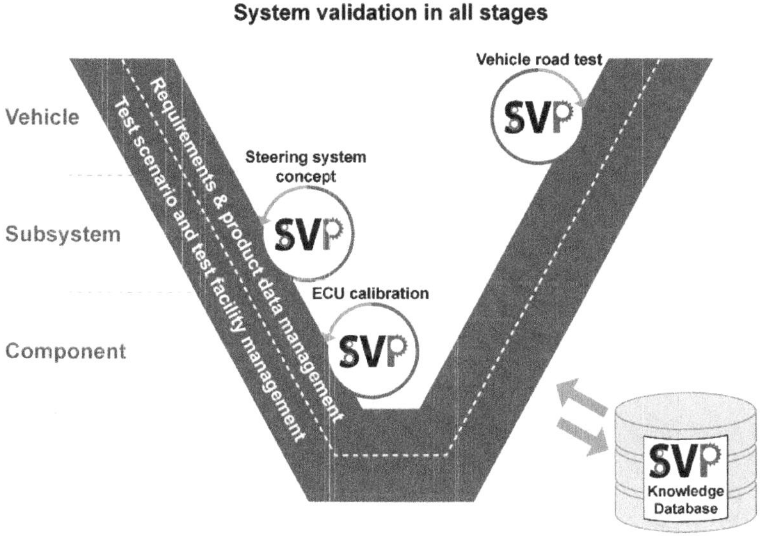


*Figure 7: Application of the SVP in different stages of system development allows for a holistic prediction of the characteristics of the vehicle (Freudenmann, et al., 2015)*

Meeting those challenges adequately is a difficult task as influencing parameters on the steering system behaviour and their interactions are not fully defined or unknown. In most cases, automobile manufacturers are not able to complete the huge amount of necessary simulation and testing for every single vehicle variant. This can lead to suboptimal system parameterisation, which results in customer dissatisfaction or in the worst case to unpredicted system malfunction. On the other hand, testing every vehicle variant increases development time and costs, which directs to higher purchase prices and later market entry.

The SVP is applied in combination with an intelligent virtual testing tool chain, AVL CAMEO and IPG CarMaker in order to handle this high complexity and to effectively support interdisciplinary teams in efficiently calibrating vehicle systems, i.e. an electromechanical steering system. The SVP is an ontology-grounded expert collaboration system, designed to manage and support automobile development processes, particularly the activities validation and testing. The underlying methodology of the SVP enables engineers to sustainably externalise, access and easily comprehend existing organisational knowledge as well as to generate new product-relevant knowledge by conducting statistical testing designs. The SVP supports a system calibration without any media breaks by interfacing with CAMEO (see Figure 8). The seamless workflow begins with the modelling of uncertain dependencies between the system parameters and the resulting system behaviour in the vehicle in form of Cause & Effect Chains and goes to the test design and test run through to data interpretation and the statistically confirmed decision making based on the gained knowledge. In doing so, the context of decisions is automatically documented and the lessons learned from one project are available for future development projects (Freudenmann, et al., 2010) (El-Haji, et al., 2012) (El-Haji, et al., 2014) (Freudenmann, et al., 2015).

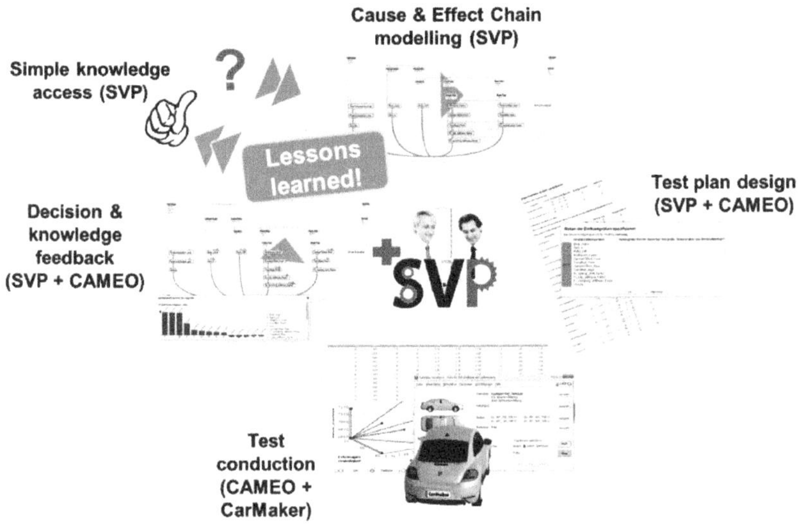


*Figure 8: Central activities of the SVP methodology and interfaces to CAMEO and CarMaker (Freudenmann, et al., 2015)*

The assumptions concerning the optimal system parameters represented in Cause & Effect Chains modelled by the members of the development team are validated by virtual testing (i.e. simulation) using IPG CarMaker. IPG CarMaker is a full vehicle simulation tool, including sophisticated driver and environment models. The application fields are the testing and development of chassis control systems as well as driver assistance systems (Holzmann, et al., 2008) (Schmidt, et al., 2015).

IPG CarMaker provides an interface to AVL CAMEO for test automation and data interpretation. Using this interface, the overall calibration process can be significantly accelerated by automatically conducting the relevant driving manoeuvres with different parameter value sets of the electromechanical steering system in the specified value range. After the generation of the simulation data, the corresponding behaviour model that represents the effects of different parameter value sets on the vehicle system behaviour is generated automatically. This behaviour model is based on statistical methods and enables engineers to make statements about parameter value combinations that have not been simulated; a behaviour model can reduce the amount of testing needed (Kuder, et al., 2003) (Seiffert, et al., 2008) (Martin, et al., 2012). Figure 9 illustrates the interaction of the SVP, CAMEO and CarMaker.

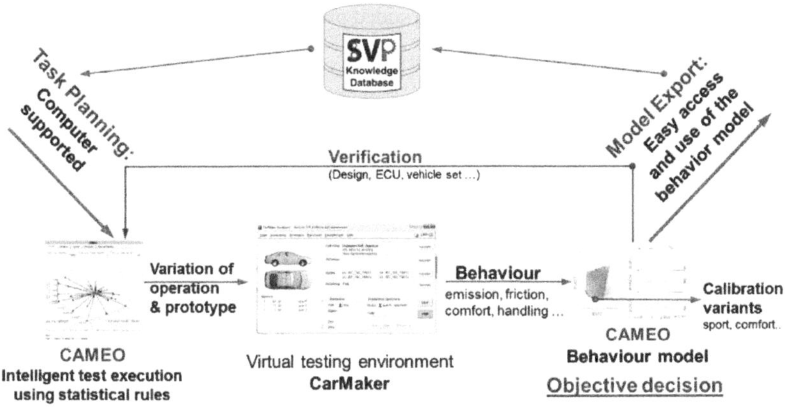


*Figure 9: Media-disruption free behaviour model generation with SVP in combination with CAMEO and CarMaker*

# Intelligent Steering System Calibration

This chapter will demonstrate the application of the knowledge management system SVP in combination with the tools for automated virtual testing, CAMEO and CarMaker for the calibration of an electromechanical steering system. The aim is to calibrate the steering system for a broad variety of vehicle variants under the boundary condition of energy efficiency. Hence, the steering system should feature the desired properties regarding the steering feeling, but at the same time consume as little energy as possible. By applying the mentioned tool chain this task is to be processed in an intelligent way, meaning that the optimal parameters for each vehicle variant should be determined by virtual testing with as few test runs as possible.

However, the resulting behaviour models should allow statistically secure statements considering the entire parameter range.

The following calibration process is based on the research of PFEFFER and HARRER who introduced an optimal steering wheel torque course depending on the lateral acceleration of the vehicle for steady state cornering. The steering assistance torque is defined as the difference of the steering torque at the steering axle and the steering wheel torque forced by the driver (Harrer, et al., 2015). The corresponding target area of steering assistance is displayed in Figure 10. It will be demonstrated in the following chapters how this ideal steering system behaviour can be generated by following the systematic and computer-aided process of the SVP.

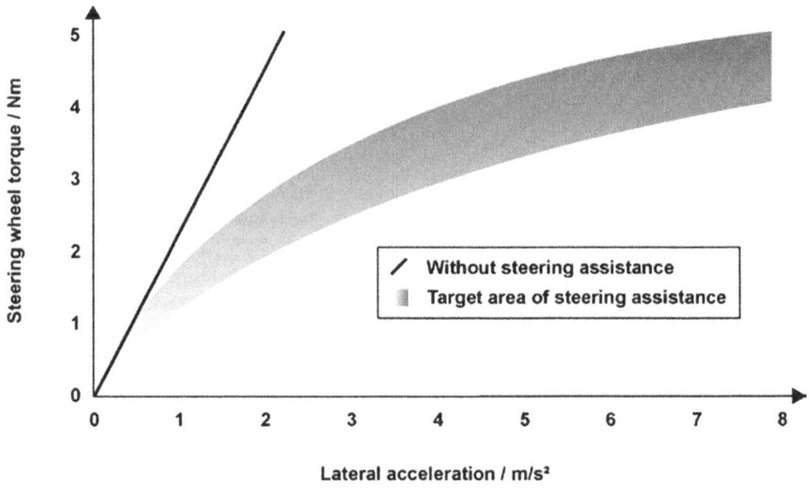


*Figure 10: Steering system assistance torque target area of steering wheel torque depending on the lateral acceleration of the vehicle (Harrer, et al., 2015)*

## Cross-linked Cause and Effect Chain Modelling

As displayed in Figure 8 the first step of the SVP methodology is to model the underlying assumption of all further activities in Cause & Effect Chains using the corresponding SVP functionalities. Figure 11 gives an overview of the interrelated Cause & Effect Chain Models for the described project and the steps up to the determination of the final calibration parameters and their respective values. Each Cause & Effect Chain represents the assumed interdependency of calibration parameters and depending parameters of other systems as well as the effect on specific vehicle behaviour.

Figure 12 displays the top level Cause & Effect Chain of the following calibration process: the influence of the steering assistance torque on the current consumption of the servo motor as well as the relation between steering wheel torque, steering torque at the steering axle and the steering assistance torque. In this functional model systems and their subsystems are represented by green boxes. Connections between green boxes represent subsystem-relations. Quantitative properties of systems are represented by yellow boxes and the connecting thick curved lines indicate the relation between quantitative properties, i.e. the causal connections. The unknown relation between quantitative properties that need to be verified by simulation or testing is represented as thick red line, the so called *test relation*. Thick green lines represent context information that is relevant for specifying the system boundary of the unit under test and the test plan as well as data interpretation and decision making. Thus, the Cause & Effect Chain Model includes all information needed to be able to conclude the vehicle behaviour from the simulated or measured behaviour of the unit under test.

Cause & Effect Chain Models are the central representation of all systems of the vehicle and the environment that are affected by the behaviour of the system under examination. Therefore, the generation of Cause & Effect Chain Models should be coordinated as a collaborative team work of experts of all relevant domains for the respective function under development. During the modelling process the experts are supported by various logical functions of the SVP for checking the model for completeness and reasonability. In addition, information represented in Cause & Effect Chain Models of other projects is filtered by the SVP and automatically made available for the users.

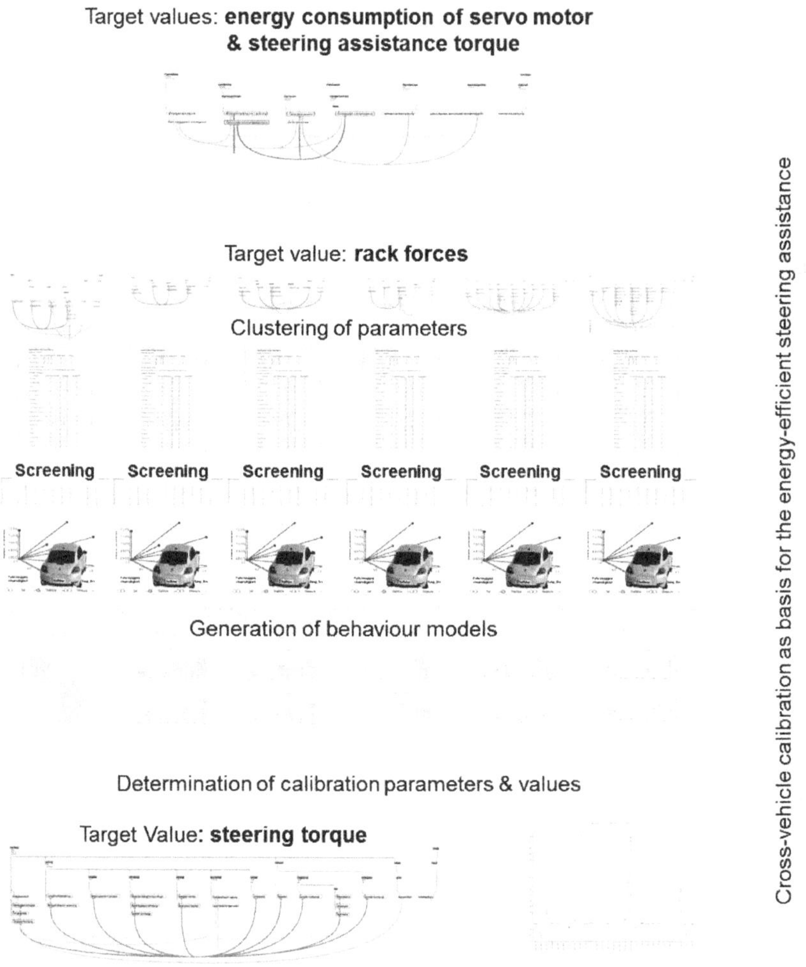


*Figure 11: Interrelated Cause & Effect Chain Models generated with the SVP and the processes of determining the final calibration parameters as well as their values*

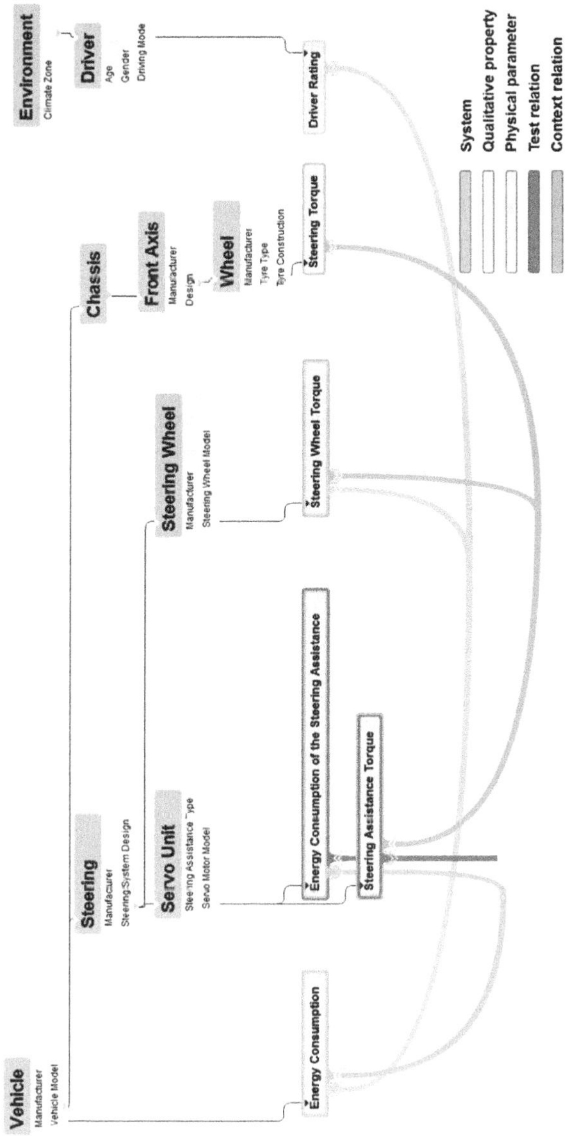


*Figure 12: Top-level Cause & Effect Chain of the steering system calibration process*

As the value range of the steering wheel torque course shall be given based on the research of PFEFFER and HARRER, the focus of testing is on the steering torque on the steering axle at the front tyres and shall be examined in this example as function of the right and left steering rack forces[4]. In order to gather all relevant parameters on the rack forces and not miss any significant influencing parameter or unknown interactions of parameters, all variable parameters within the steering subsystem as well as relevant vehicle parameters shall be examined to generate behaviour models valid for different vehicle variants.

The whole calibration process contains the completion of all SVP processes displayed in Figure 11. In the following, the functionality of the SVP and the automated virtual testing tool chain will be demonstrated using the example processing the assumption modelled in the Cause & Effect Chain Model displayed in Figure 13. The other processes are handled accordingly with different test runs, i.e. control and target variables as well as value ranges.

[4] This definition is a result of constraints of the applied testing tool chain.

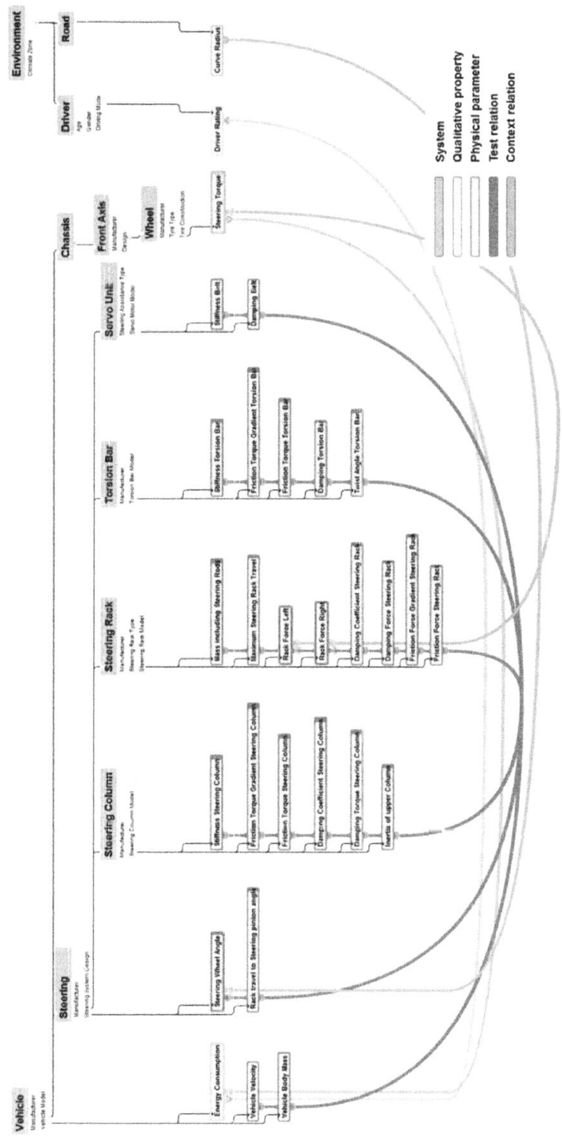


*Figure 13: Cause & Effect Chain Model focusing on the influences of the steering system properties on the steering rack forces*

## Specification of the Unit Under Test and the Test Plan

After all relevant parameters were identified by collaborative Cause & Effect Chain modelling with all experts the unit under test is identified by the SVP according to its logical system boundaries in the Cause & Effect Chain Model. In the case of the Cause and Effect Chain of Figure 13 the unit under test is the complete vehicle itself in order to vary all control parameters included within the test relation.

In a next step, the quantitative properties that are input to the test relation are assigned the role of either *control parameters* that can be set, *indirectly controllable parameters* or *disturbance parameters* that cannot be set but should be measured during a test run. The quantitative properties that are outputs of the test relation are assigned the role of *target parameters*. The change of the values of the target parameters depending on the value variation of the control parameters is the data input needed for the generation of the behaviour model. The SVP supports its users in this step by offering only the roles for quantitative properties that can be derived from the respective Cause & Effect Chain Model.

Subsequently, the value types and ranges are defined as a preparation of the test design Figure 14. The SVP offers multiple options of value types such as *constant*, *variable* or *profile* in order to be able to map any kind of possible testing manoeuvre. Furthermore, relevant characteristics of qualitative properties of systems can be defined on this screen. This information will be used for generating the behaviour model as well as documenting its limits of validity and testing context. Furthermore, the SVP will automatically make this information available for other projects and display relevant information in case a cross-link is detected.

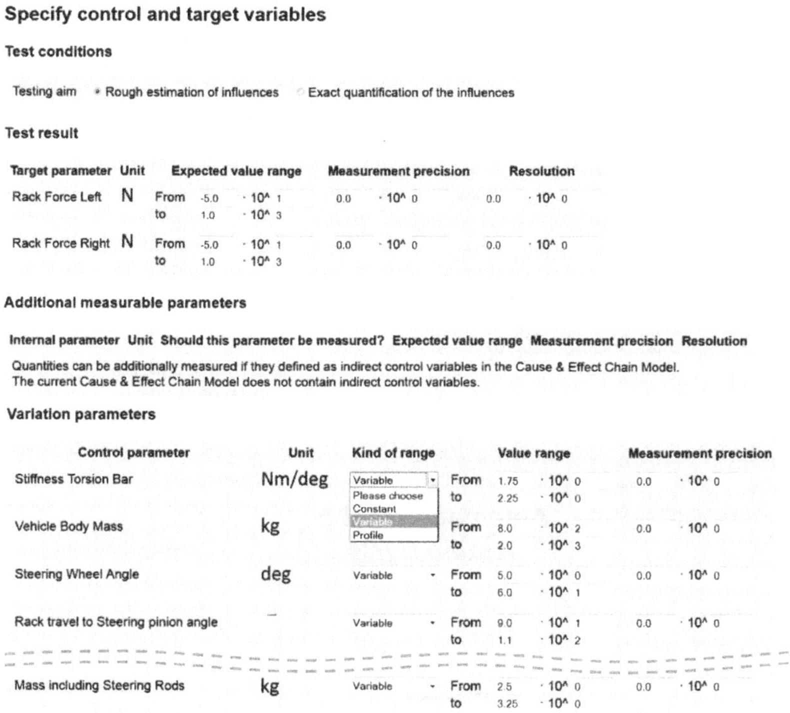


*Figure 14: Definition of value types and ranges for the test parameters identified in the Cause & Effect Chain Model as preparation for the test plan design*

## Automated Test Execution with AVL CAMEO and IPG CarMaker

In order to initiate the activity of intelligent testing, the information generated with the SVP has to be passed over to CAMEO. This is achieved by using the XML-interface between SVP and CAMEO. Based on the delivered data CAMEO automatically generates a statistical test plan according to the principles of the Design of Experiments without any further user input (see Figure 15).

Afterwards the test runs are executed by CAMEO that is able to remote control CarMaker via a test automation interface. In doing so, CarMaker simulates the different variants of vehicles by automatically parameterising its generic vehicle model as defined in the SVP and the subsequently generated test plan. The different parameterisations of the steering system model (based on PFEFFER and HARRER) according to the test relation in the respective Cause & Effect Chain (see Figure 13) are varied automatically in the defined parameter value range. The simulation data is recorded and fed back into the CAMEO database. After the whole test plan has been conducted, the corresponding behaviour model will be generated automatically.

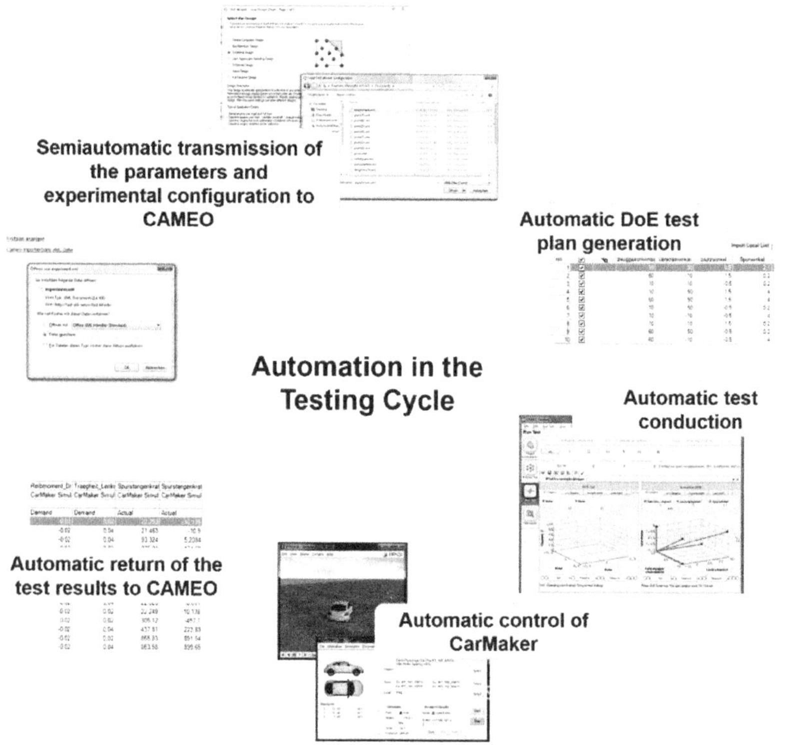


*Figure 15: Automated virtual testing with CAMEO and CarMaker based on Cause & Effect Chain Models generated with SVP*

## Decision Making

Using the XML-interface, the generated behaviour model is imported to the SVP and the included information is brought back into its original context, i.e. the respective Cause & Effect Chain Model (see Figure 16). Quantitative properties that have proven to have significant influence on the target variables are marked with a star symbol, where the size of the star represents the extent of the respective influence. Interdependencies of influencing parameters are represented by a chain symbol.

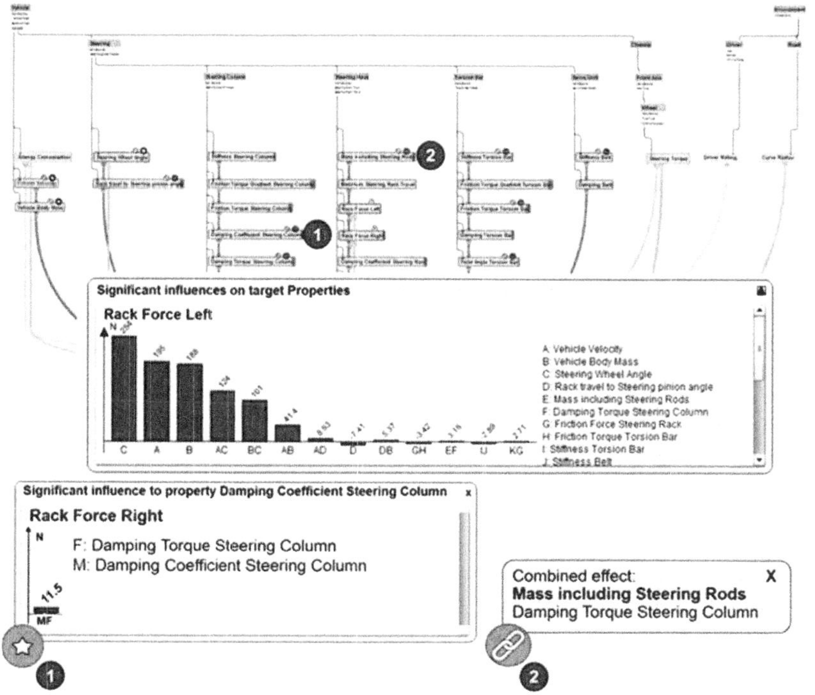


*Figure 16: Knowledge included in behaviour models fed back into its relevant context in the Cause & Effect Chain and made accessible and comprehensible for other users*

As a result, the complex information of the behaviour model is made comprehensible in a simple way and accessible for future projects through its additional context information provided by the Cause & Effect Chain. Decisions about the subsequent activities, i.e. either the final determination of the calibration parameter values or, in case the targeted system behaviour is not achieved yet, the further examination of influencing parameters on the system behaviour, are automatically documented and can be traced in a comprehensible way by other members of the organisation.

In the case of this calibration process, several cycles of SVP processes were conducted in succession until the final calibration parameters were determined. These cycles were built upon each other systematically and the number of unnecessary iterations could be minimised by the knowledge management functionalities of the SVP. The generated knowledge of each step is made available in the SVP in order to support engineers in the handling of similar calibration challenges in the future.

# Conclusion and Outlook

This paper introduced the SVP, a knowledge management system for increasing the efficiency and effectivity of testing and validation. This is achieved by enabling automobile engineers from different fields of expertise to exchange and combine their knowledge as well as to access already existing organisational knowledge with regard to a specific development task. The result of this multi-disciplinary team work, i.e. the assumption how required system behaviour can be achieved is represented in computer-understandable Cause & Effect Chain Models that are the basis of the subsequent behaviour model generation. Each behaviour model is generated using simulation or empirical data and allows engineers to predict system behaviour as effect of the variation of influencing parameters in the defined value range. By offering a data exchange interface to CAMEO behaviour models can be imported into the SVP for further processing. The knowledge gained from this project is accumulated within the organisation and represented in a simple way embedded in its relevant context. Hence, it is easily accessible and usable for other users that face similar challenges. Finally, the decisions made with the SVP are documented automatically and sustainably in order to allow the transparent tracing and under-standing of decisions within the organisation.

The practical application of the SVP and the connected tool chain for intelligent automated virtual testing was demonstrated using the example of the calibration of an electromechanical steering system. The task was to ensure a good steering feeling under the condition of a low energy consumption of the steering system. In addition, the steering system had to be calibrated for different vehicle variants. Therefore, behaviour models were generated that predict the influence of parameters of the steering system as well as vehicle and environmental parameters on parameters of the steering system. By following the SVP process, the initial number of 45 assumed influencing parameters could be reduced to a number of seven most significant parameters with a minimised number of necessary iterations. As a result, it is now possible to determine the optimal calibration parameter values as well as to predict the energy consumption of the steering system for different vehicle variants, without having to conduct further testing.

The gained expertise of this steering system calibration process is now available as objectified and computer accessible knowledge in the SVP database. Future activities will include the enhancement of the applied models by conducting road tests with test drivers as well as broadening the parameter value ranges in which the behaviour models are valid. The generated knowledge can be transformed on other application areas, e.g. the calibration of truck steering system, reducing the necessary testing and validation effort.

Furthermore, the presented tool chain will be applied to further vehicle subsystems that require calibration, e.g. combustion engines. This is possible due to the fact that the SVP is based on a generic process of testing and validation that can be applied independently from the system under development. In doing so, more and more knowledge will be accumulated in the SVP database, which reflects the growing organisational knowledge base. Apart from calibration tasks, this knowledge can also be used for enhancing innovation activities by supporting engineers to manage system complexity and interdepending functions. To achieve this aim, the SVP will be interfaced with requirement management systems supporting product development from the very beginning.

**References**

**AVL LIST GMBH. 2014.** *Whitepaper: Increasing Demands of Costumbers.* Graz : AVL LIST GMBH, 2014.

**BMBF, Bundesministerium für Bildung und Forschung. 2013.** *Energieeffizienz im Fokus: Forschung für Elektroniksysteme der Zukunft.* Bonn : Thiel Gruppe, 2013.

**Braess, H.H. and Seiffert, U. 2007.** *Handbuch der Kraftfahrzeugtechnik.* Wiesbaden : Vieweg+Teubner, 2007. Vol. 5.

**Dannenberg, J. and Burgard, J. 2007.** Innovationsmanagement in der Automobilindustrie. *Oliver Wyman: Car Innovation 2015.* 2007.

**El-Haji, M., et al. 2012.** Ontology-Grounded Test Facility Requirements Definition. *Tools and Methods of Competitive Engineering (TMCE).* 9, 2012.

**El-Haji, M., Freudenmann, T. and Albers, A. and Gauterin, F. 2014.** Ontology-Grounded Validation Methodology for Innovative Automobile Development Projects. *International Journal of Innovation and Technology Management (IJITM).* 2014.

**Forrester Research. 2000.** Development Portals Emerge. *The Forrester Report.* 2000.

**Freudenmann, T. and El-Haji, M. 2015.** *Media Disruption Free Product Development – Semantic Validation Platform (SVP).* Karlsruhe : Verein für deutsche Ingenieure (VDI), 2015.

**Freudenmann, T., El-Haji, M. and Albers, A. 2010.** An Approach to Increase the Efficiency of Experiments Applied for the Analysis of a Gear Unit. *International Federation of Automotive Engineering Societies (FISITA).* 205, 2010.

**Günthner, W. 2007.** *Neue Wege in der Automobillogistik: Die Vision der Supra-Adaptivität.* Heidelberg : Springer, 2007. Vol. 1.

**Harrer, M. and Pfeffer, P. 2015.** *Steering Handbook.* Heidelberg : Springer, 2015.

**Holzmann, H., et al. 2008.** Virtual Test Driving at General Motors Europe - Performance Simulation of Modern Chassis Control Systems. *ATZelektronik.* 1, 2008.

**Kuder, J., et al. 2003.** Calibrating Bosch Motronic Systems Efficiently Using Bosch/AVL iProcedures with AVL Cameo. *MTZ.* 12, 2003.

**Legner, C., et al. 2009.** Wandel in den Wertschöpfungsstrukturen der Automobilindustrie - Konsequenzen für Prozesse und Informationssysteme. *EBS European Business School: Whitepaper.* 2009.

**Martin, E. and Coghlan, N. 2012.** Emissions calibration yesterday, today and tomorrow. *Journal Combustion Engines (PTNSS).* 109, 2012.

**Maune, G. 2001.** *Möglichkeiten des Komplexitätsmanagements für Automobilhersteller auf Basis IT-gestützter durchgängiger Systeme.* Paderborn : Universität-GH-Paderborn, 2001.

**Nexteer Automotive. 2015.** *Nexteer: Corporate Profile.* 2015.

**Proff, H., et al. 2012.** *Zukünftige Entwicklungen in der Mobilität: Betriebswirtschaftliche und technische.* Heidelberg : Springer, 2012. Vol. 1.

**Renner, I. 2007.** *Methodische Unterstützung funktionsorientierter Baukastenentwicklung am Beispiel Automobil.* München : Technische Universität München, 2007.

**Riches, Ian. 2015.** Automotive Electronics System Demand Forecast 2013 to 2022. *Strategy Analytics.* 2015.

**Schmidt, S., et al. 2015.** Early PC-based Validation of ECU Software Using Virtual Test Driving. *ATZelektronik.* 1, 2015.

**Seiffert, U. and Rainer, G. 2008.** *Virtuelle Produktentstehung für Fahrzeug und Antrieb im Kfz.* Wiesbaden : Vieweg+Teubner, 2008. Vol. 1.

**Valldorf, J. and Gessner, W. 2004.** *Advanced Microsystems for Automotive Applications.* Berlin : Springer, 2004. Vol. 1.

**Wallentowitz, H., Freialdenhoven, A. and Olschewski, I. 2009.** *Strategien in der Automobilindustrie: Technologietrends und Marktentwicklungen.* Wiesbaden : Vieweg+Teubner, 2009. Vol. 1.

**Weißenborn, S. 2013.** Vollelektrische Lenkung: Der direkte Draht zur Straße. *Handelsblatt.* 2013.

**ZF Lenksysteme GmbH. 2009.** Elektrische Servolenkung. *Produktbroschüre.* 2009.

# The consequences of a closed rim design for the brakes of a high-efficiency vehicle

Dr. Ralf Stroph, BMW Group Research and Technology

Dipl.-Ing. Sebastian Gielisch, BMW Group Research and Technology

Dr. Alfred Pruckner, BMW Group Research and Technology

# Abstract

Highly efficient and especially electrified passenger vehicles labor to expand the $CO_2$ reduction or driving range through many measures. One significant factor is the optimization of the aerodynamics. The wheel rim design of a vehicle has a noticeable impact on the drag coefficient ($c_d$) of a passenger vehicle.

Especially the very challenging requirements of good aerodynamics gave the impulse to have closed wheels, i.e. a rim with an inner cover plate and an outer one. The effectiveness has been verified by aerodynamic tests. This closed wheel solution results in a great change to the cooling strategy of the brake disc, while having not a great impact on weight and no impact on the dimensions of the brake disc and caliper. A passive cooling channel with an actively closable intake, again for reasons of aerodynamics, has been sought and designed.

The inner cover also results in preventing brake dust settling on the rim, even if the outer cover might be transparent for design reasons. One additional advantage of electrified vehicles is the capability of regenerative braking which can be partly considered in the thermal design of the brakes and its cooling.

Simulation models of the temperature behavior were run, with an uncertain estimation of the level of cooling by closed wheels, giving a rough indication of the constraints within the thermal design of the wheel brakes. A prototype was built in which it was possible to verify the simulated cooling behavior of the brake discs.

A fading brake test has been checked and matched to the simulation results, leading to the awareness of whether the closed wheels and additional cooling is sufficient for this vehicle.

This contribution shows how it was achieved to use closed wheels to improve aerodynamics on this prototype and still comply with the requirements for brake disc temperature.

# 1 Introduction

Due to CO2 restrictions and efficiency goals, passenger vehicles are engineered to be optimized in that respect, whereas aerodynamics is one major impact. Especially the wheels have a noticeable contribution to the aerodynamics. The OEMs have worked for years on the improvement of the aerodynamics of the vehicles and also the wheels. The BMW research and technology department had the assignment to build a highly efficient vehicle with all possible measures to improve its fuel consumption.

For that vehicle, it was sought to design very light wheel rims, but also optimize them aerodynamically and take additional measures to keep the brake discs at a reasonable temperature in all conditions.

This paper will start with the purpose of wheel rims, look at their aerodynamical behavior and show the impact of a closed rim design. Simulation results of the brake disc temperature and the details of the rim lead to vehicle testing which is outlined at the end.

# 2 Purpose of a wheel rim

The purposes of a wheel rim for passenger vehicles are multifaceted. The main aspect is, of course, to connect the vehicle to the road. More specifically, to connect the wheel bearing of the vehicle chassis to the tire. Therefore, the tire needs to fit perfectly on the rim, no matter in which condition the vehicle is. Especially in high driving dynamic demands the wheel will be deformed but still needs to be kept on the rim so no air can escape and the tire forces can be forwarded into the vehicle chassis. Figure 1 shows a sectional view through a rim, where the area for the tire fit can clearly be seen.

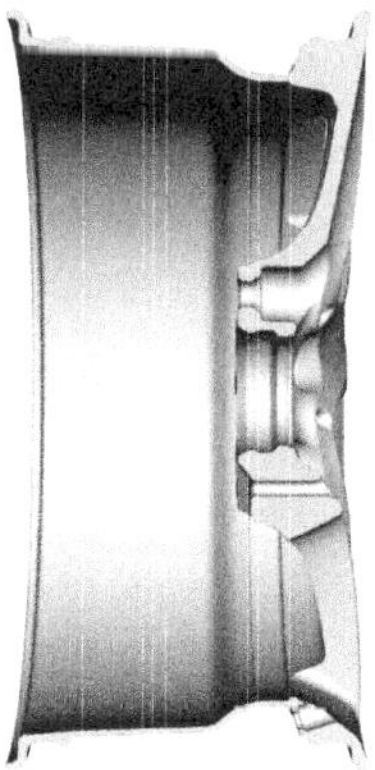

Figure 1: Sectional view through a rim.

Mentioning the forces which act on the rim, the other main requirement is the rigidity. The rim needs to be developed in such a manner, that the forces, which act on the rim, are well transported through the rim. However, lightweight and cost of material is always in opposition to rigidity. Hence, the right trade off has to be found to develop a rim which satisfies the previous requirements. An example of a finite element analysis is shown in figure 2, which is commonly used to determine the rigidity.

Figure 2: Finite element analysis of a rim with five spokes.

The most difficult requirement for a rim development is its styling, since no clear restraints or technical figures can be given for a good appearance, matching the vehicle for which it is used. Figure 3 shows four different designs which have a great variety [1].

Figure 3: Different wheel rim styles [1].

Last but not least, especially for the brake system one major requirement is the brake cooling. For a rim of 17 inch, for example, the axial ventilation opening is typically in the range of 300 – 350 $cm^2$ when no additional measures are in place.

In general, a rim with a very effective brake disc cooling feature has often fairly poor aerodynamics quality. As mentioned in the introduction, nowadays efficiency is ranked much higher than it used to be. Hence, the requirements of optimized aerodynamic characteristics of wheel rim design have been prioritized. A lot of engineering capacity is used to optimize the effects of aerodynamics of a vehicle.

Dominant for wheel rim design is the aerodynamic drag of the rim as well as the ventilation losses, see figure 5. Ventilation losses include all aerodynamic induced losses which result in turning moment working against the rotational direction of the wheel of a moving vehicle. The whole range of the ventilation losses include the turning moments which are induced by air pressure differences primarily at the tire tread, the rim spokes and the circulating air volume inside the tire. Both drag and ventilation losses should be minimized. In order to measure these losses, a special test bench is needed.

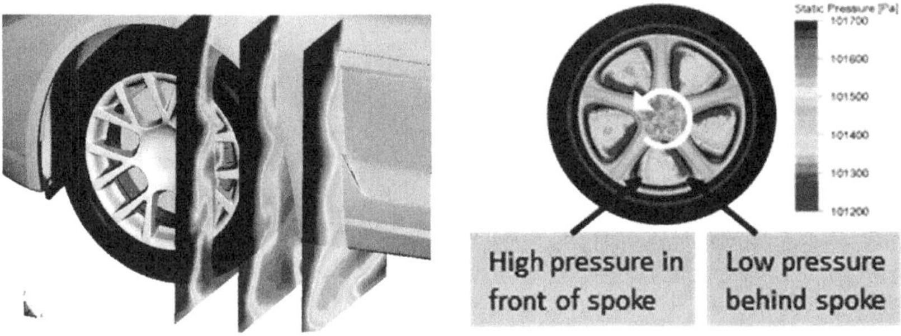


Figure 5: Aerodynamic drag and ventilation losses [2].

Requirements towards aerodynamics and brake cooling are often in conflict. Hence, in very efficient vehicles a good tradeoff between the brake disc cooling and the aerodynamics, still with an adequate styling is sought. Examples are the aero-rims of the BMW high efficiency and electrified vehicles [2] as shown in figure 6.

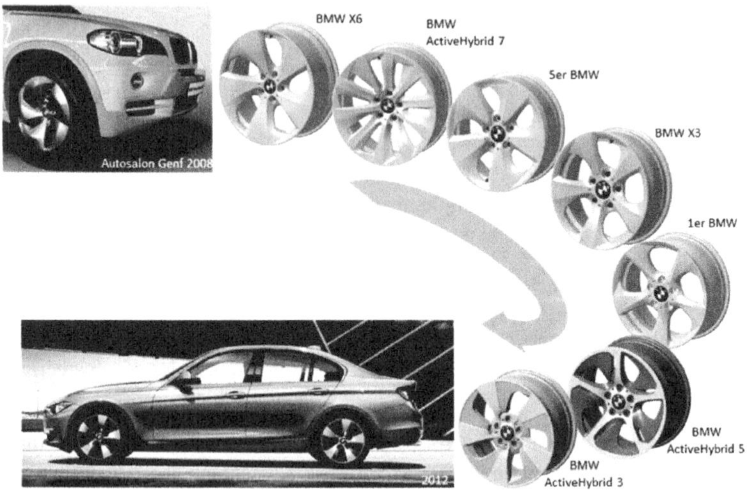


Figure 6: Development of aero wheel rims [2].

# 3 Consequences of closed rim design

The dimensioning of brake discs depend on the desired deceleration as well as the energy of the vehicle it has to absorb. The temperature characteristics while braking depends on the vehicle mass and velocity, combined with the mass of the brake disc, its temperature coefficient and the cooling of the disc.

For a typical highly efficient vehicle, with a rather low mass (m) of 1500 kg the maximum power for a deceleration of 1g from a speed (v) of 100 kph is 408 kW. The energy (E) of this vehicle at 100 kph would be:

$$E = \frac{1}{2}\, m\, v^2 \qquad (1)$$

hence, ½ * 1500 kg * (27,78 m/s)², being 579 kWs. With a brake force distribution of about 67% to 33% towards the front axle, the energy for one front axle brake disc ($E_B$) is 194 kWs. Taking a small massive front axle brake disc of a mass ($m_B$) of under 5 kg with a specific temperature coefficient (C) of about 500 J/(kg*K) [3], the temperature difference ($\Delta$ T) of one braking maneuver from 100 kph to stop would be:

$$\Delta T = E_B / (m_B * C) \qquad (2)$$

resulting in roughly 80 Kelvin.

This is of course a very rough calculation, as the brake disc is not entirely heated directly by the brake pads and is not the only part which is getting warmer. Also the cooling of the brake parts through convection and radiation is relevant for the temperature rise. A very common test to verify the thermal dimensioning of the brakes is a fading test, where ten braking maneuvers from 100 kph to stop are performed [4].

However, for normal customers, brake disc cooling is far less important than efficiency of the vehicle, as in normal driving cycles the brakes are not excessively used and their temperature are always well under a critical level. On the other hand, vehicle fuel consumption or electrical driving range is immediately visible for the customer. Consequently, it is more intelligent to engineer the vehicle setup that the basic layout is based on the efficiency requirements, and measures are provided to fulfill the cooling requirements of the brake discs only when needed. Even though in such a circumstance the aerodynamic characteristics might be far worse.

One possibility of optimizing the aerodynamics is to close the rim. Several publications [5, 6] have already suggested such a measure. For example, [5] came to the conclusion that a complete cover on the outside gives the best performance towards aerodynamic drag. We came to the same conclusion with even identical figures, which resulted in an improvement of about 9 percentage points for the prototype. Concerning the ventilation losses, [5] comes to the conclusion that a completely closed rim has disadvantages over partly open rims. This results from the effect, that the spokes are still moving within the air, but have no way to exhaust on the outer side where it is closed, hence the losses rise. If, however, the inner side of the rim is also completely closed, this disadvantage is eliminated and the ventilation losses are at its minimum. Furthermore, the aerodynamical characteristic is also slightly improved by this measure. The reduction in terms of aerodynamic drag ($c_d$) is 9 points for the outer cover and one additional point for the inner one for the complete vehicle. Figure 7 shows the results of measurements in terms of power consumption for aerodynamic drag only of the closed rim design with and without outer and inner cover of the rim. Taking the closed rim design as a reference with zero losses, the additional power consumption for the outer cover only and the open rim design is plotted. The figure represents the complete vehicle with four wheels. For a vehicle speed of 130 kph the difference is 500Watts, which rises over speed up to 2000 Watts at a speed of 210 kph.

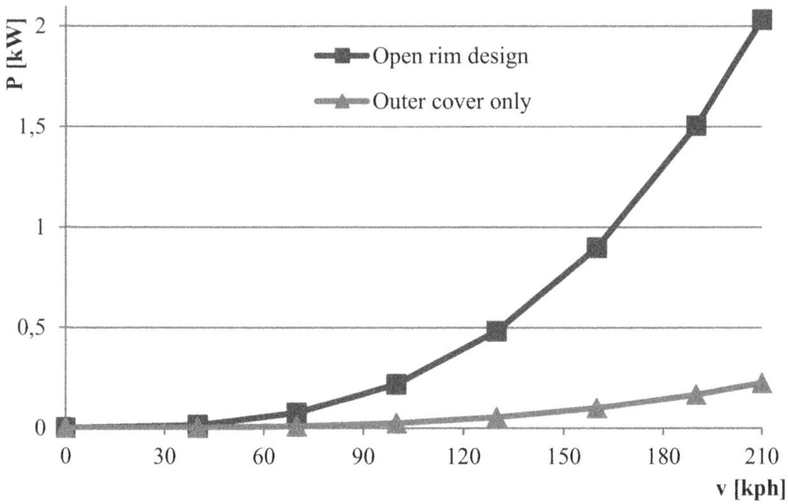


Figure 7: Power consumption of aerodynamic drag only.

The measurement results of the ventilation losses only are shown in figure 8. Here, it is differentiated between the open rim design, the outer cover only, the inner cover only and the outer and inner cover as reference (zero losses). The results are once more comparable with [5] for the outer cover only, which leads to a difference of about 700 Watts at a speed of 210 kph for the vehicle. At a speed of 130 kph the difference between the open rim and the fully closed rim is 100 Watts, which is still noticeable. The inner cover only has a greater impact on the ventilation losses, as it is directly fixed to the spokes and the air cannot escape between the spokes and the cover, which is not the case for the outer cover.

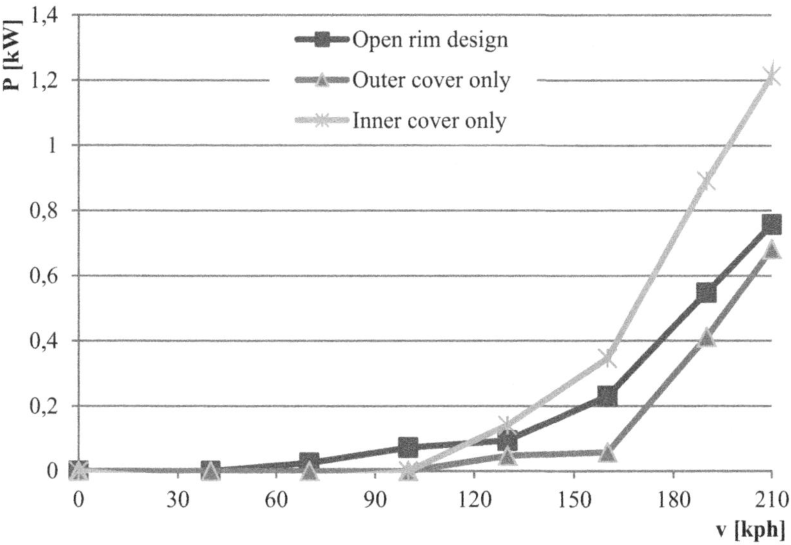


Figure 8: Power consumption of ventilation losses only.

## 3.1 Design option

In case of a rim with a full closed cover the carrying wheel can be designed and optimized for stiffness, durability and manufacturing process without consideration to rim and vehicle appearance. Usually the wheel is a styling driven part of the vehicle. The styling requirements often lead to heavier wheels which are more expensive due to the use of more raw material than it is necessary for stiffness and durability. Also the draft angles of the spokes might prevent a cost optimized manufacturing process.

Figure 9: Individually colored closed rim design.

Closed rims can be designed individually colored (compare figure 9) or even transparent. For such a transparent cover, the spoke styling can be seen with the enormous advantage that no dirt or brake dust gets onto them. Hence, one of the reasons for complaint, the brake dust on the wheels, will be eliminated and customer satisfaction is improved. An example in form of a photo composition is given in figure 10.

Figure 10: Transparent closed rim design (photo composition).

# 4 Simulation results

In order to get a feeling for the influences of the closed wheel rim design on the brake disc temperature a simulation has been run. However, rather than using a very complex and powerful simulation for this prototype test vehicle, a more simple simulation has been used.

The used simulation has, among others, basic input parameters like vehicle mass, velocity scheme, brake details and aerodynamical values, as well as parameters for the heat convection and the amount of air flowing past the brakes. This tool has given an idea of the difference in brake disc temperature between the open rim and the closed rim design. An additional cooling channel for the front brakes, which is explained in more detail in Chapter 5, was considered in the simulation.

Results of the simulation were analyzed for the mountain decent of Großglockner, a high-speed fading test as well as a special high speed fading test for a vehicle with a lower top speed and most important for us the fading brake test, which is commonly used [4]. The latter test consists of ten ABS braking maneuvers from a speed of 100 kph down to standstill with adjacent full throttle accelerations. This test was chosen for the vehicle, as for battery vehicles the acceleration from standstill is quite impressive (about 8.7 seconds 0 to 100 kph) and no degradation of the power management was likely to occur during this test. It must be mentioned that regenerative braking has not been considered in these simulations of the fading test, as the critical brake temperature was on the front axle and the vehicle was rear wheel driven.

The results are shown in figure 11. It can clearly be seen that the temperature of the front axle brakes (FA) is about 50 Kelvin higher for the closed rim design (dashed line) than for the conventional open rim (dotted line). However, with the estimation of the additional cooling through the cooling channels, it is not as much as one could have anticipated. The rear axle brakes (RA) do not have much difference in temperature between the open and the closed rim design in the simulation and are far away from the critical temperature.

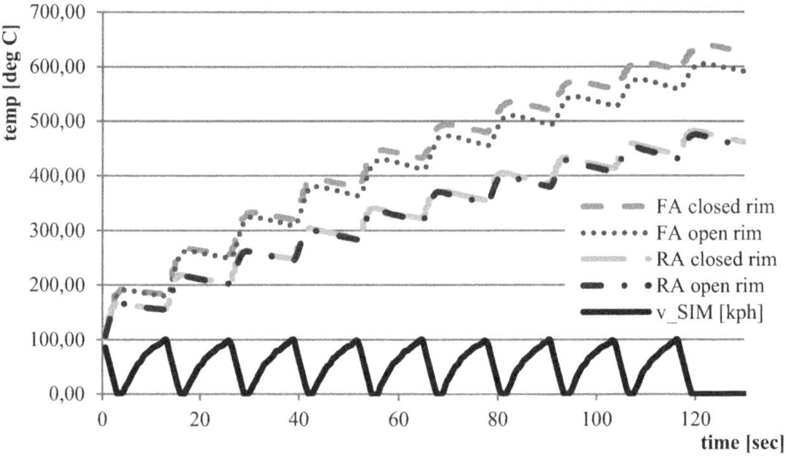


Figure 11: Simulation results of open and closed rim design.

# 5 Realization of the closed rim and cooling channel

In order to verify our considerations the wheel rims have been specifically designed to integrate the cover on the inner and outer side with a maximum in aerodynamic improvement and a reasonable solution in usability. The rim and its spokes have been designed to use as little raw material as possible for maximum weight savings. The inner side of the spokes have an even and smoothly shaped surface to easily facilitate the fixing of the inner cover with a special adhesive. On the outer side of the rim, where the cover is attached, a notch is integrated in the rim to hold the cover. Such a notch is provided on the outer and inner perimeter of the rim as well as the outer perimeter of the cover of the wheel bolts (compare figure 12). Additionally, for this prototype a lid for the valve access is provided [7]. However, it is also possible to integrate the valve where the wheel bolts are attached. It can clearly be seen in figure 12 that the outer cover is not directly attached to the spokes, which has an influence on the ventilation losses with the outer cover only, as outlined in Chapter 3.

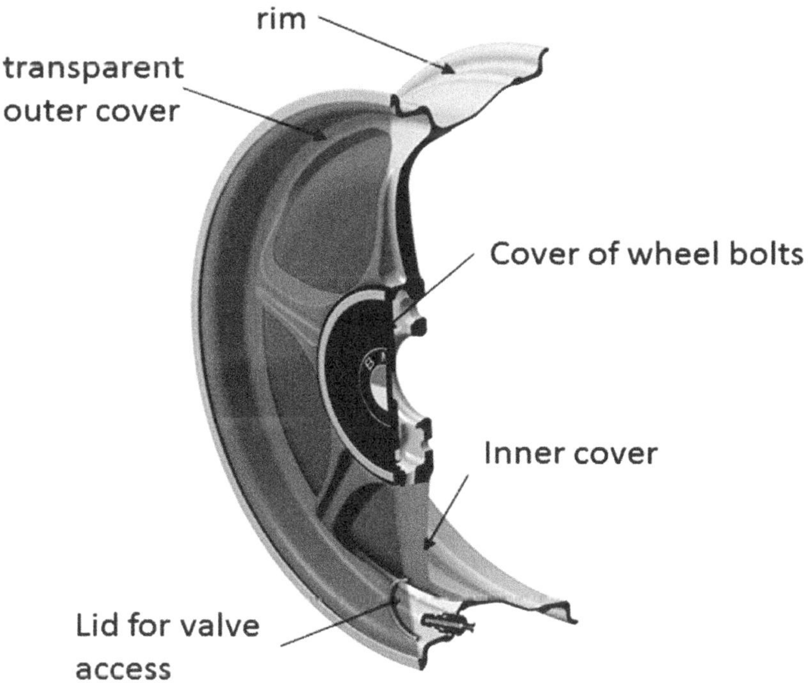


Figure 12: Closed rim design.

The design of the additional cooling channels for this prototype is shown in figures 13 and 14. As mentioned before, it is advantageous to be able to close the cooling channels when not needed for extra cooling of the brake discs. Especially in a battery electric vehicle, the usage of the friction brake is very rare as a majority of decelerations is performed through regenerative braking. Therefore, an electrically operated inlet door has been installed on the front of the vehicle to close the cooling channel and improve the aerodynamics of the car.

Figure 13: Design of the additional cooling channels.

Figure 14: Air flow through additional cooling channels.

# 6 Test vehicle results

In order to verify the thesis that the designed cooling channels are sufficient to compensate the missing cooling through the open rim design, the prototype has been used for real brake tests. The tests have been done on the BMW test track near Munich. One straight section on the high speed test track was sufficient to perform all ten accelerating and braking maneuvers. The tests started with the open rim design, for reference. The results are shown in figure 15. It can clearly be seen that the temperature of the front brake discs are close to 600 degrees Celsius after the ten braking maneuvers, very similar to the simulation as shown in figure 11. Since the rear axle might be partly decelerated regenerative, even during ABS situations (compare [8]), the more critical brake temperature is at the front axle which is shown.

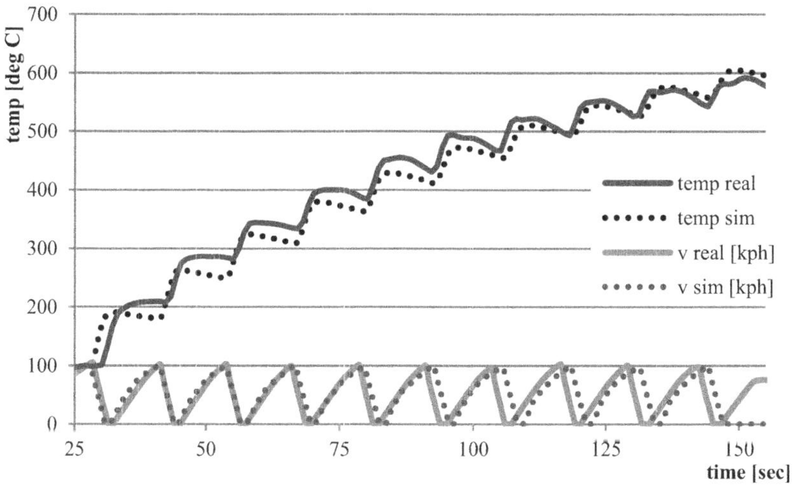


Figure 15: Comparison of Simulation and vehicle test with open rim design.

After changing the brake discs and brake pads, the brakes have been conditioned again and the tests with closed rim wheels were performed. The results are shown in figure 16. It shows that the temperature rises similar but the top temperature is about 650 degrees, which is slightly better than assumed by the simulation results.

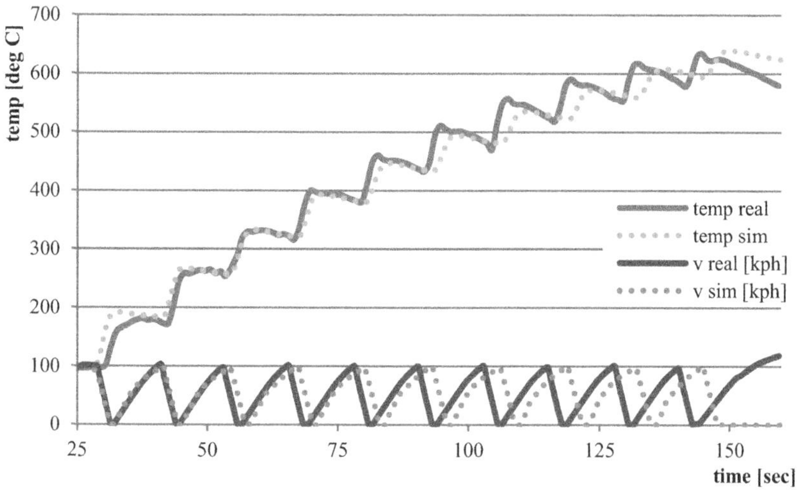


Figure 16: Comparison of Simulation and vehicle test with closed rim design.

The rim has been examined after the tests and no damage to the inner cover or the adhesive has been found. However, more extensive testing will be carried out, including the mountain decent and a high speed fading test.

# 7 Conclusion

This contribution has demonstrated that it is reasonable for the development of a highly efficient vehicle to put emphasis on weight and aerodynamics for the cost of brake cooling. It is outlined that additional cooling channels, and well designed paths for the air flow can almost compensate for complete closed wheel rim designs, which have a considerable impact on the aerodynamic drag coefficient of a vehicle and the additional power losses through the ventilation of the spokes.

Especially electrified vehicles do have the capability to decelerate regenerative and help to reduce the mass of the brake discs. However, for the use cases when regenerative braking is not possible, the capability of the brakes should still remain within the requirements as it was shown for this prototype for the front axle.

# 8 References

1. **BMW / Rolls-Royce:** BMW Media Pool Intranet, © Copyright BMW AG and Rolls-Royce Internet, http://www.rolls-roycemotorcars.com/, 2015.

2. **Tesch, G.; Modlinger, F.:** Die Aerodynamikfelge von BMW – Einfluss und Gestaltung von Rädern zur Minimierung von Fahrwiderständen, Tagung: Fahrzeug-Aerodynamik, Haus der Technik, 2012, Munich.

3. **Hasse S.:** Taschenbuch der Gießerei-Praxis 2010, Schiele & Schön. ISBN 978-3794908011.

4. **Breuer, B.; Bill, K-H. (Hrsg.):** Bremsenhandbuch.: Vieweg Verlag, Wiesbaden 2003. ISBN 3-528-03952-3.

5. **Vdovin, A.:** Investigation of Aerodynamic Resistance of Rotating Wheels on Passenger Cars, Thesis, Chalmers University of Technology, 2013, Gothenburg, Sweden.

6. **Wittmeier, F.:** Ein Beitrag zur aerodynamischen Optimierung von Pkw Reifen, Dissertation, 2014, Springer Vieweg, Stuttgart, ISBN 978-3-658-08806-4.

7. **Gielisch, S.:** (Bayrische MotorenWerke) Rad eines Krafffahrzeugs mit einer Verkleidung des Speichen-Raumes, Offenlegungsschrift DE 10 2013 211 216 A1, 18.12.2014, Munich.

8. **Teitzer, M.; Stroph, R.; Pruckner, A.:** Simulation of an anti-lock braking system with electric motors during regenerative braking in powerful BEVs, Chassis.Tech plus, June 8 – 9, 2010, Munich

# The contribution of brake emissions to the total vehicle emissions

R. Steege, D. Welp, J. Lange, TMD Friction

# Introduction

Over the last decades, the automotive industry has made huge efforts to lower vehicle emissions which have especially contributed to improve emission level in inner city areas. In Europe, limitations for particle emissions from diesel engines have dropped by 97% during the last two decades. In addition, friction materials have undergone many changes and functional components with an associated health hazard potential have been substituted step by step. Currently, the substitution of copper and its compounds have caused huge amounts of development work and resources in the friction industry dedicated to solve this challenge. However, with regards to a continuously further concentration of traffic in urban city areas, further measures and restrictions are to be expected. And although the exhaust emission values from vehicles is at an extremely low level, fine particle measurement devices in the inner cities still show there is room for improvement.

Vehicle homologation registration and release testing could in the longer term go beyond just emissions from engines to address total vehicle emissions. This would make the contribution from tires and road surfaces together with brake friction pads and discs a part of the overall balance. Next to the internal combustion engine, tire and brake wear are the main sources of fine dust.

Our investigation focuses on influences to and potential reductions of primary brake emissions, not taking into account any secondary means to collect already released brake particles and prevent them from entering the environment. A multitude of tests are run on a brake dynamometer replicating a typical endurance test widely used amongst OEMs and suppliers within the brake industry. Although this kind of test is far away from today's vehicle homologation tests braking parameters, it is able to deliver results in a known environment of realistic driving. Measurements are taken for brake pad and disc wear as well as the use of in-situ brake dust sampling devices. The results show a variety of dependencies for brake wear and emissions from the friction pads' materials concepts and the related brake rotors. They also open the view to relations of emissions under various braking conditions.

# Brake emissions contribution to total vehicles emission

Since the end of the last century EU legislation focused on solid emissions in the form of particulate matter (PM) besides such in gaseous form like carbon dioxide. The contribution of road transport (including resuspension) to PM is identified as the most important, according to the German Umweltbundesamt (figure 1). As a consequence, the automotive industry has undertaken significant measures to reduce emissions from combustion engines, whereas emissions related to brake and tyre wear are remaining on a constant level (figure 2).

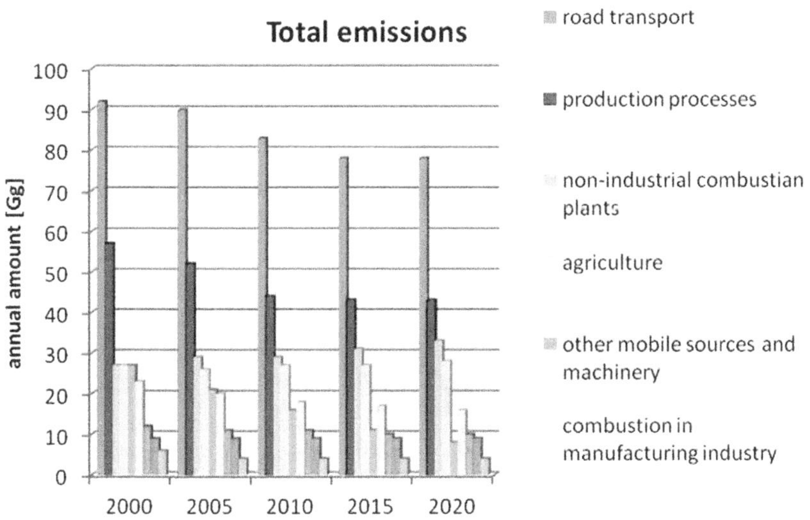


Figure 1: PM Emissions in Germany and related sources; Umweltbundesamt 2013 [1]

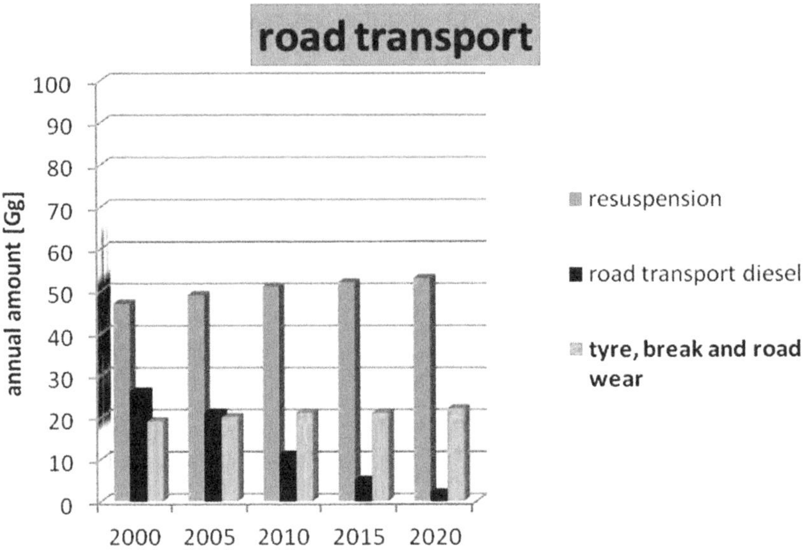


Figure 2: PM emissions from road transport; Umweltbundesamt 2013 [1]

A preliminary report of the German Environmental Agency presents an estimation for 2015 naming emissions in PM10 (particles with an aerodynamic diameter smaller than 10 μm) of 21 Gg from road transport in case of tire and brake wear per anno, compared to a contribution of 5 Gg from diesel engine emissions and 50 Gg caused by resuspension from roads. With the ongoing process to reduce fine and ultrafine dust emissions "wear-born" PM will be in the focus over the next years.

# Emissions from friction brakes

"Wear-born" PM is a result of at least two different kinds of processes including corrosion and mechanical wear. Vehicle deceleration with friction brakes builds on an interaction of friction pad and disc resulting in a dissipation of energy through a so called 3rd body or tribolayer. Coarse wear particles are produced in a rough mechanical way resulting with a tribochemical dissipation of energy in the 3rd body layer going alongside. The tribochemical conversions result in finer particles in a great variety of chemical substances, related to the wide range of conditions and the friction materials constituents. Such wear processes and the influences from friction material components as described by Österle [2] can even lead to new compounds formed in the friction process after several brake applications as reported by Kukutschova [3].

As brake wear particles are released in various sizes, a focus of prior investigations has also been with regards to ultrafine and nanoparticles. To assess the potential hazardous effect of nanoparticles, it is important to clearly specify their physical and chemical properties and their intended uses, the amount that will be produced, the likely exposure scenarios and the potential for accumulation in the bio-organisms or the environment. In studies performed with respect to such emissions from friction brakes, several scenarios are evaluated with up to now no distinct statement regarding the hazardous impact of PM from brake wear to human health. Gasser [4] reported a formation of reactive oxygen species (ROS) and a release of pro-inflammatory mediator, when directly contaminating brake wear debris, generated in cases of high friction loads, to lung cells. Kukutschova [3] compared the effect of wear debris and milled pad material in their study, with differing results for mutagenicity (AMES and SOS Chromotest) varying for the tested materials.

# The role of brake pad formulations

Friction material development has so far mostly been driven by the demand for rising safety and brake performance and at the same time been influenced by the availability or restriction of raw materials. Over all these years, three main material categories have been shaped to cover the full range of global requirements.

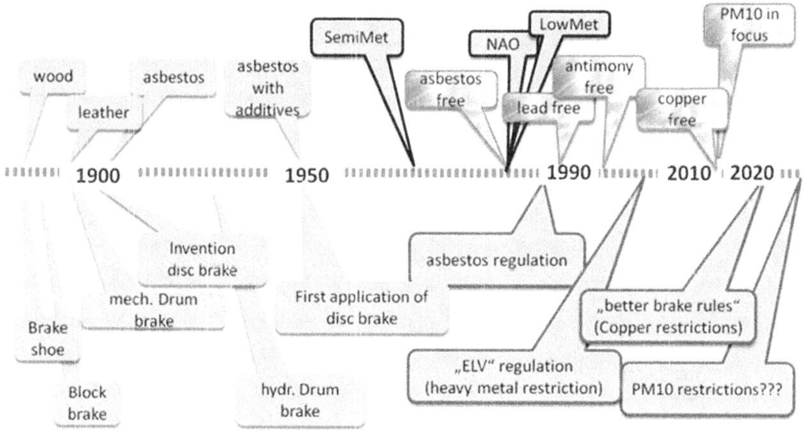


Figure 3: Material development over 135 years of friction braking (Source: TMD Friction)

Semi-Metallic and NAO-type products are typically designed for higher comfort and longer lifetimes compromising on lower friction levels. So called Low-Metallic (ECE) materials target much higher brake power and higher speed.

Any such friction material contains at least six distinct groups of raw materials. Fibres are used for reinforcement, abrasives to control friction level, lubricants to control wear with rubbers influencing noise and harshness behaviour. Fillers are needed to disperse functional additives. At least one binder is used, in most cases a phenol resin. Semi-Metallic friction materials contain more than 60% of steel and next to coke only a few other friction modifiers. NAO materials are based on a relatively large amount of copper and a multitude of different friction modifiers. ECE materials contain steel and copper in lower amounts to form part of the formulation. Banning heavy metals and especially lead in the 80´s and 90´s resulted in a big spread in kind and concentration of raw materials contained in the recipes of friction formulations [5].

The next step to fully eliminate copper to cope with US regulations will again diversify the friction material components as no single key substitute exists. All this complexity of the friction material composition is expected to be found in friction wear particles. As Thorpe and Harrison [6] showed, a similar huge variety of metal concentration can be found in friction materials linings and emitted brake dust (figure 4).

| Metal | Car brake linings (mg/kg, unless indicated otherwise) | Car brake dust (mg/kg, unless indicated otherwise) |
|---|---|---|
| Al | 3765 | 330-2500 |
| As | <2-18 | <2-11 |
| Ba | 2638 | 5900-74,400 |
| Ca | 14,300 | 920-8600 |
| Cd | <1-41.4 | <0.06-2.6 |
| Co | 6.4-45.8 | 12-42.4 |
| Cr | <10-411 | 135-1320 |
| Cu | 11-234,000 | 70-39,400 |
| Fe (%) | 1.2-63.7 | 1.1-53.7 |
| K | 857 | 190-5100 |
| Li | 55.6 | Not reported |
| Mg | 6140 | 83,000 |
| Mn | 181-3220 | 620-5640 |
| Mo | 0.4-215 | 5-740 |
| Na | 15,400 | 80 |
| Ni | 3.6-660 | 80-730 |
| Pb | 1.3-119,000 | 4-1290 |
| Sb | 0.07-201 | 4-16,900 |
| Se | <1-15 | 4.5-115 |
| Sr | 81.4 | 300-990 |
| Zn | 25-188,000 | 120-27,300 |

Figure 4: Metal concentration in brake linings and brake dust, Thorpe and Harrison [6]

# Methods of analysis with regards to particle emissions

Several methods were tested to assess the emissions of particulate matter. Based on different wear mechanisms all authors found next to coarse particles a varying portion of particles in the range of PM10 and PM2.5 which were named as "airborne". Sanders [7] tested different pad materials and found a unimodal mass-size distribution around 4-5 µm. Kukutschova [3] also tested an unimodal size distribution around 2-4µm. Particles smaller than 0.4-0.3µm were not detected due to technical limits of the measurement equipment. Focussing on particle numbers and using equipment which also includes particles smaller than 0.3 µm, the trend switches to lower particle size and sometimes to bimodal distribution.

A recent work from Sachse et al. [8] focused on particles smaller than 1µm and found a portion of brake dust with a mean diameter in particle numbers around 100nm. They also pointed out that there are around about 30 different influencing parameters on the measurements. Next to the location of sampling almost all environmental impact influenced their results. Strongly depending on the friction application sometimes NAO materials show lower emissions switching to completely the opposite result if the applications are forced into higher temperature regions.

Next to the above investigation on dynamometer, various authors investigated emissions directly from the vehicle. Riediker [9] used cars on lifting ramps with an exposure box around the brake for direct measuring. Sanders [7] researched the ideal position for measuring alongside the car. Augsburg [10] observed particle flow around the brake with high speed cameras and build up corresponding CFD models (computational fluid dynamics). Their findings point at a highly complex measuring situation. The smaller the particles, the more likely they are to follow the air flow and the more fundamental are the influences of the point of sampling and the measured results.

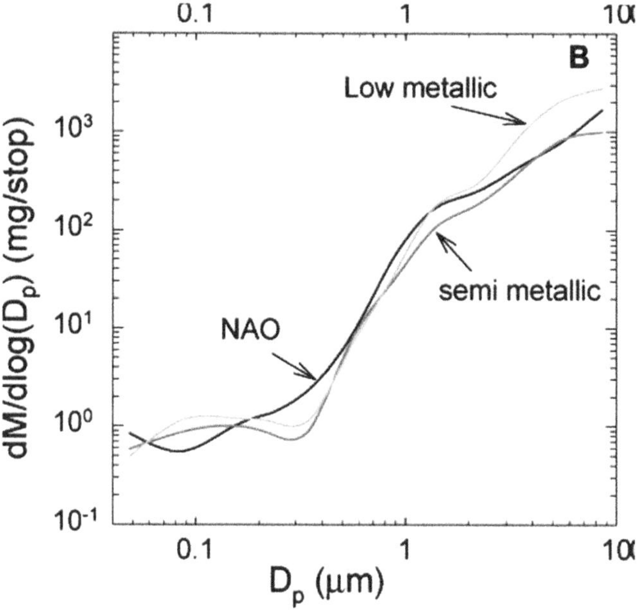


Figure 5: mass weighted wear debris size distribution over 3 AMS stops, [7, Sanders]

Beside direct particle measurement several papers focused on an emission factor which is useful for regulation purposes and estimation of dimensions in emissions. Typically about 50-70% of all worn mass become airborne under dynamometer conditions when including a wheel [7]. As typical value derived from published findings, the European Commission indicates 50% of all worn mass to become airborne of which 80% are regarded as PM10. An average emission is found at about 6.7 mg/km and vehicle [12]. Such values need to be reviewed and reconfirmed on a broader base of vehicles, brake systems, friction couples and test cycles but they can be taken as baseline reference values for this point in time.

Compared to the direct measurement of emitted particles there are some simple as well as pragmatic approaches quantifying brake emissions. Weighing gauges are commonly used for wear measurements of friction couples which are directly related to particulate matter. As wear figures can be measured very accurately, the application of the above mentioned "rules of thumb" can be used to relate such measurements to an expected airborne particle or PM10 outcome.

Some important aspects need to be considered in choosing an appropriate alternative for the simple weighing approach. Aggregated debris may become lost if components are dismantled from the test system for measurement purposes. Measurement gauges must be capable of quantifying the differences in weight within a given range. This can be achieved by extremely precise gauges and/or by attaining sufficient amounts of worn material. This becomes even more challenging with the scope (weight) of the test system, which limits the measuring range as well as the accuracy. Typically weighing scales in dynamometer test shops have a resolution 0.1 g for the measurements of a brake pad and/or rotor. Wear is calculated as the difference in the measurement before and again after the test. This accounts for a realistic measurement accuracy of some 0.3 g in the difference derived from a measurement system analysis covering the influences of the real measurement environment and operators.

One pragmatic weighing approach is to put a complete vehicle on a scale. Major advantages of this full scale vehicle measurement are; no need for any dismantling work, no loss of aggregated debris, the consideration of the real system and counting of airborne wear only. Disadvantages are one overall wear figure only comprehending brakes and tires at least. These provide no clue to the contribution to either front and rear brakes or even friction and rotor material. Another downside is the very high accuracy required. A vehicle requires a resolution of the scale of at least 3.6 g if the target of the same accuracy is to be achieved as for brake components above. This increases the full scale related accuracy requirement by a factor of 1000.

Another approach is to generate wear with a single-ended inertia type dynamometer. Wear measurements can be made on each dismantled component. Thus the contribution of friction and rotor material can be disclosed. Fewer influences are not under control which can add further advantages to this method, such as weather, traffic, load etc., these can significantly affect the reproducibility of any test procedures and the corresponding result. Cons are the dismantling effort which involves the danger to lose debris.

Similar to others, TMD Friction utilises aerosol monitoring systems as additional equipment for brake wear tests on dynamometers. The major added value is the information about the wear of each single brake application. However, the contribution of the either friction pads or rotors cannot be quantified in the overall measurements. Furthermore it counts airborne wear only. In contrary to this, weighing approaches allow such differentiation but need an significant amount of brake work to produce sufficient amounts of wear which can be weighed. Therefore numerous brake applications are obligatory between measurements.

# Appropriate test cycles

Depending on the considered areas Bukowiecki [13] stated differing shares for tyre, brake and resuspensions. Whereas brake emissions play an inferior role on a motorway with just with 3%, their share increases in city areas up to 21% of PM. Such dependencies are known and proven state through the friction industries' test cycles.

Sachse [8] pointed out, that numerous inputs and their interactions have influence on brake emissions (figure 6). Even more interactions must be considered in the development of a specific test cycle. The wheel brake layout and the driving parameters (e.g. speed, deceleration) interact in the definition of a load on the friction couple (e.g. friction speed, – force, – energy).

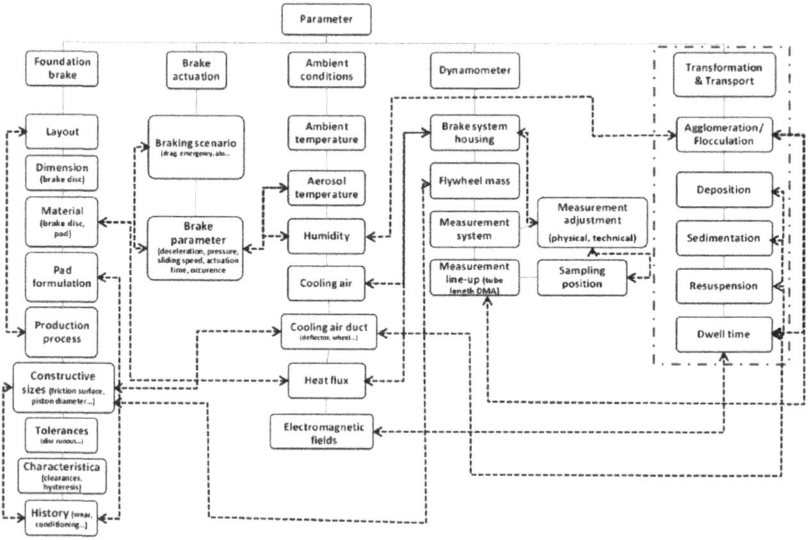


Figure 6: Influences on brake emission, Sachse [8]

The abrasive, adhesive and oxidative wear mechanism [Oehl, 14] and their interactions must be provoked by an appropriate test cycle in order to generate realistic figures for wear and/or particulate matter. Our previous investigations are showing, that the pre-condition of the friction couple before each brake apply has a dominant influence on wear besides the commonly cited brake application parameters leading to friction force, – work, – speed, – power and surface temperature [Steege, 15]. It therefore is as important to define the sequence of different brake applications as it is to define the range of the above mentioned parameters.

For our investigations, we concluded to focus on natural driving patterns to evaluate wear, particle emission and even apparent friction. A specified endurance run in southern Spain well known for vehicle release testing regarding brake comfort and life time was chosen for simulation on our brake dynamometers as criterion for the friction development. It consists of a driving cycle representing 200 km that is repeated some 25 times to simulate a driving distance of 5,000 km. Such a long distance is essential to either generate sufficient amounts of wear or describe the wear behaviour of friction couples versus the driven distance including bedding effects. Hundreds of tests generated robust "know-how" and some general conclusions can be stated, although numerous friction couples show an exception to these rules. To translate such wear measurements taken from weighing differences into comparable PM10 numbers, the rule of thumb is used, estimating 50 % of any brake wear particles as becoming airborne and 80 % of these being PM10 [12]. In this way all wear results can be expressed as PM10 per referred to the simulated mileage. Figure 7 shows the contribution of rotors and pads for various friction couples tested in this drive cycle. The variation is immense and can be quantified from 1–28 mg/km with a median of 15 mg/km. 38% contributed by the friction pads and 62% by the brake rotors. All this experience has been made so far with grey cast iron rotors.

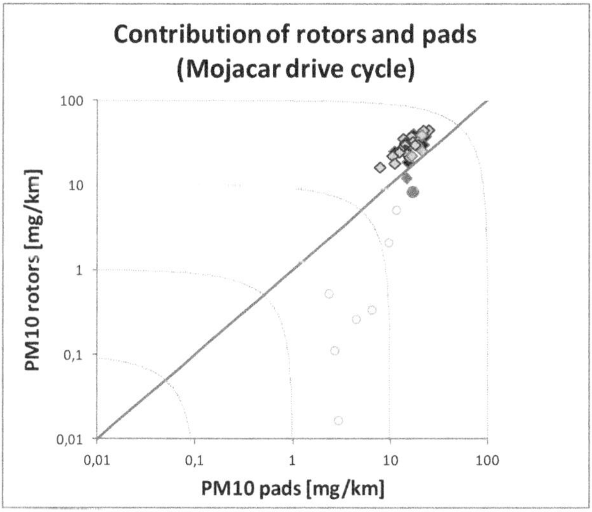


Figure 7: contribution of rotors and pads for various friction couples applying the Mojacar drive cycle in a dynamometer simulation (40% of wear are assumed to become PM10)

The lower the brakes' emissions are, the lower the contribution of the rotors become. TMD Friction test results disclose that brake rotors contribute 50–70% to the overall PM10 of friction brakes with an overall PM10 above 10 mg/km. Whereas brake rotors' contribution is only 0–20% for friction couples with an overall PM10 below 5 mg/km (figure 8).

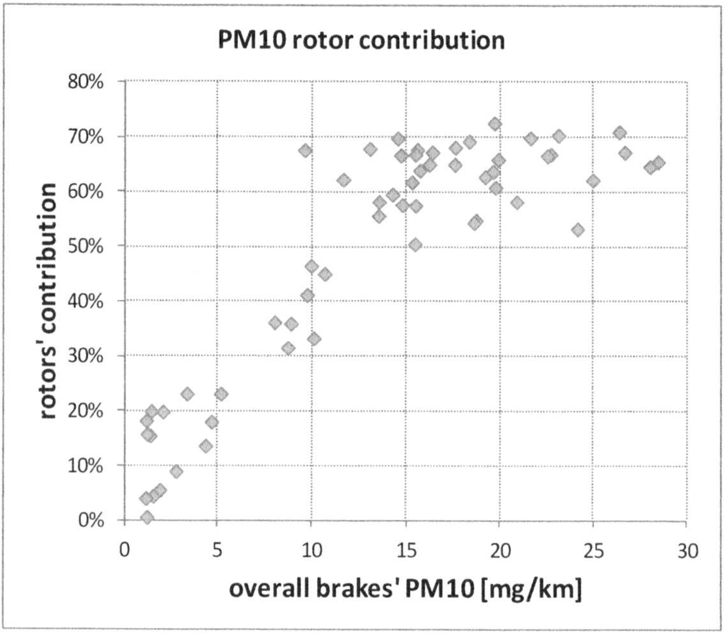


Figure 8: Brake rotor contribution to overall brake emissions from Mojacar dynamometer test (PM10 estimated as 40% of the wear measured by weight loss)

With road simulation tests it is possible to assign wear values measured after a sufficient number of brake applies under certain parameter sets of e.g. speed, temperature, deceleration. An aerosol monitor allows for the dedication of a cumulated particle concentration to a single brake apply (figure 9).

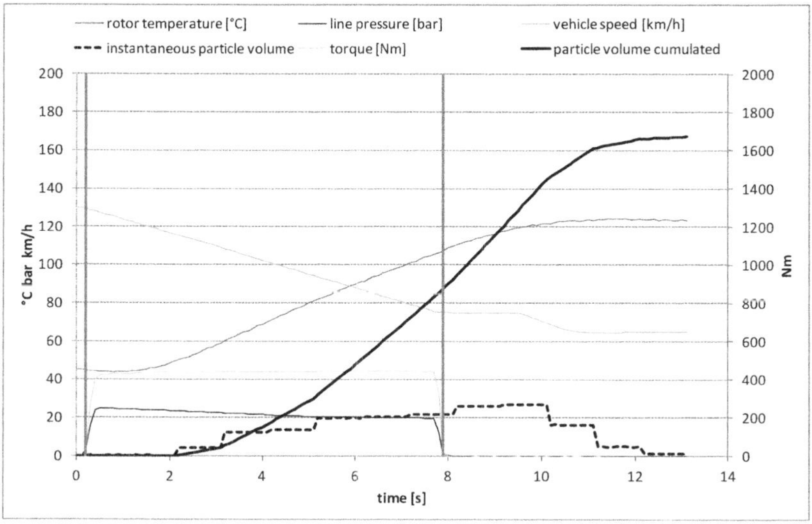


Figure 9: cumulative particle concentration for a single brake apply

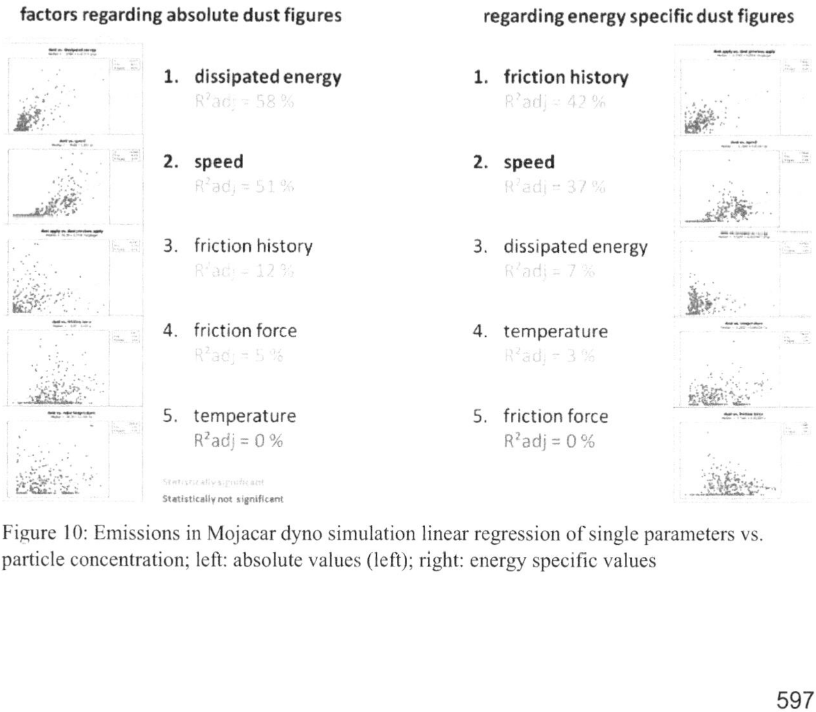


Figure 10: Emissions in Mojacar dyno simulation linear regression of single parameters vs. particle concentration; left: absolute values (left); right: energy specific values

This method was applied over a complete Mojacar test simulation. The driving brake application parameters were evaluated for an exemplary friction couple by singular, linear regression. The dissipated energy, friction speed, friction force, surface temperature and friction history were all considered. The friction history was defined as the apparent friction observed during the previous brake application

There is the common saying "no friction, no wear", which is confirmed by the fact, that the dissipated energy is the main driving factor for wear in this example. Therefore the specific wear was defined as wear related to the dissipated energy. This together with the friction history becomes the most important factor, followed by the friction speed and surface temperature. The friction force or surface pressure shows no statistically significant influence. A multiple regression with higher grades of parameters can disclose the interactions.

It is well understood that the brakes' PM10 figures for the Mojacar representing dynamometer test shown above are not comparable with those well known for engine exhaust derived from standardised fuel consumption release tests (e.g. NEDC (EU), JC08 (J), FTP US)). Thus the NEDC was implemented as a simulation for a dynamometer test to generate numbers comparable with those from the Mojacar simulation with equal energy dissipation (figure 11).

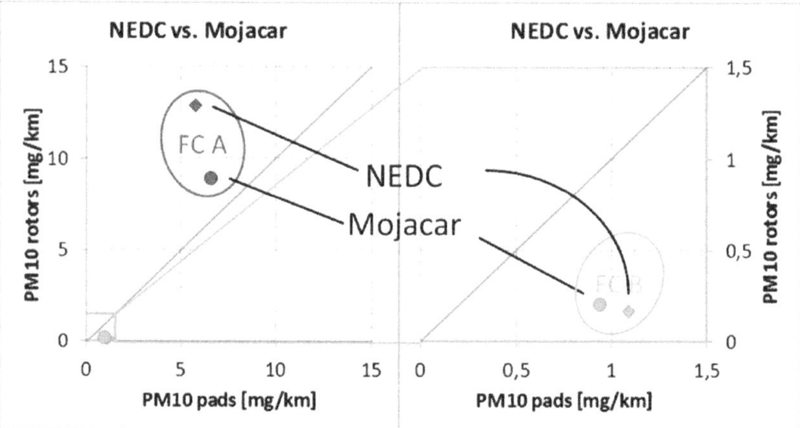


Figure 11: Emissions for two different friction couples (FC-A and FC-B) in either Mojacar and NEDC (PM10 estimated as 40% of the wear measured by weight loss)

The results for two exemplary friction couples can be directly compared for either NEDC or Mojacar. Both Friction couples tend to emit more on equal mileage under

NEDC load conditions, but the rotor contribution of both friction couples changes in different directions switching between the cycles. Friction couple A emitted 18.65 mg/km with a rotor contribution of 69 % in NEDC, but 15.49 mg/km with a rotor contribution of only 57 % in Mojacar. Friction couple B emitted 1.26 mg/km with a rotor contribution of 13 % in NEDC but 1.15 mg/km with a rotor contribution of 18 % in Mojacar.

A future task will be to determine a test procedure that will allow for the evaluation of brake and tyre emissions aside from exhaust emissions that are currently tested and limited with their well established procedures. Typical brake wear procedures considering natural driving cycles were compared with fuel consumptions cycles that are currently in use to evaluate exhaust emissions. Mojacar, Los Angeles city traffic and Taxi ville Paris were chosen as representatives for brake wear procedures and NEDC, FTP and WLTC (global) standard for fuel consumption cycles.

The most significant difference between fuel consumption cycles and brake wear tests is in their duration and mileage. Fuel consumption cycles take 19 to 30 minutes, whereas brake wear tests can last between 2 days to 1 week. The comparison of speed profiles discloses similarities between the driving patterns being considered. The deceleration profiles of fuel consumption cycles are much lower than those of brake wear cycles which were derived from "daily use" driving. Daily use is characterised by deceleration between 1 and 2.5m/s$^2$. More than the half of the deceleration phases in fuel consumption cycles don't even achieve 0.5 m/s$^2$, which is caused more or less by natural drag rather than any brake applications.

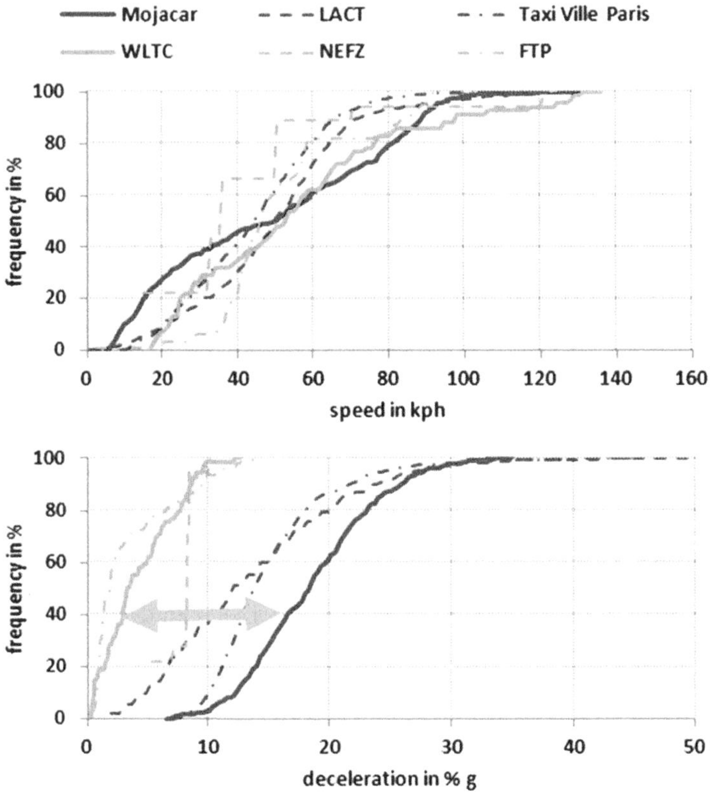


Figure 12: Distribution of speed and deceleration of brake test and fuel consumption cycles

There are several tasks that challenge the definition of a "vehicle overall emission test procedure". First of all the duration must be balanced between the low requirement of fuel consumption tests and the very high demand of brake wear procedures. If a solution can be found for this challenge the deceleration phases must be adjusted in the region of regular driving as in daily use. The worldwide harmonised Light vehicle driving Test Cycle developed by UN/ECE/WP.29/GRPE/WLTP-IG DHC subgroup is likely to be considered as a test in the future for fuel consumption limits. It was developed based on real driving data derived from various regions in the world. It would be an appropriate choice for the "vehicle overall emission test procedure", if the issues of duration and deceleration can be resolved.

# Summary

Brake emissions play a role in the overall emissions from road transportation. The German Environmental Agency indicates 21 Gg of PM10 emissions per year contributed from brake and tyres, compared to 76 Gg for the overall value related to road transport. The contribution from friction brakes is due to the wear of brake rotors or drums and friction pads or linings, as shown in this report.

Even though a hazardous impact to bio-organisms has not been proven so far for emissions from friction brakes, future regulations are expected to cover total vehicle emissions including those from friction brakes. With the continuous evolution of brake pad formulations, more and more raw materials with a potential hazardous impact had been taken out of the friction formulations and were replaced by new ingredients to compensate the functional effects and keep up with the demands to safety, performance and comfort. As a consequence the diversity of substances from the original friction material added by such generated during the friction process has grown over the years and led to a high complexity of emissions.

Brake emission particles are found in the micro range for a vast majority with a very small mass portion of particles going down to the submicron size. The quantitative measurement of brake emissions is difficult and complex as the observed system around the brakes has no defined boundaries and cannot be closed without interfering with the particle flow and influencing the measuring results.

Overall brake emissions are highly depending on the applied drive cycles and braking conditions. Real driving conditions as applied in this study led to a PM10 equivalent of 15mg/km for the median of over 60 tested friction couples. 68% of the measured emissions were contributed by the brake discs. The best cases with lowest PM10 equivalent emissions of 1-5mg/km were achieved through driving down the contribution from the rotors to below 20%.

Comparing emissions for various test cycles led to non transferable results from one cycle to another. A friction couples' emissions better in one cycle were not necessarily better in another cycle. Future definitions for measurement test cycles will have to carefully look at the impact of any influencing parameter or boundary condition.

# References


1 Jörß, W.; Kugler, U.; Theloke, J. (2013). Emissionen im PAREST Referenzszenario 2005-2020

2 Österle, W., Griepentrog, M., Gross, T. and Urban, I. (2001). Chemical and micro structural changes induced by friction and wear of brakes. Wear 251: 1469-1476

3 Kukutschová, J., Moravec, P., Tomášek, V., Matějka, V., Smolík, j., Schwarz, J., Seidlerová, J., Šafářová, K. and Filip, P. (2011). On airborne nano/micro-sized wear particles released from low-metallic automotive brakes. Environmental Pollution 159:998-1006

4 Gasser, M., Riediker, M., Mueller L., Perrenoud, A., Blank F. Gehr, P. and Rothen-Rutishauser, B. (2009). Toxic effects of brake wear particles on epithelial lung cells in vitro. Particle and fibre toxicology

5 Wiaterek Dr C., Bremsenhandbuch, Breuer/Bill, Springer Verlag

6 Thorpe, A. and Harrison, R.M. (2008). Sources and properties of non-exhaust particulate matter from road traffic: A review. Science of the Total Environment 400:270-282

7 Sanders, P.G., Xu, N., Dalka, T.M. and Maricq, M.M. (2003). Airborne brake wear debris: size distributions, composition, and a comparison of dynamometer and vehicle tests. Environmental Science and Technology 37:4060-4069

8 Sachse H., Augsburg K., Ivanov V., Trautmann C., Egenhofer F.; An approach to lower particle emissions of friction brakes on vehicles; 58th Scientific colloquium Ilmenau 2014

9 Riediker, M., Gasser, M., Perrenoud, A., Gehr, P. and Rothen-Rutishauser, B. (2008). A system to test the toxicity of brake wear particles. 12th Intern. ETH-Conference on Combustion Generated Nanoparticles, 23–25 June 2008, Zurich

10 Augsburg K. Horn R., Sachse H., Characterization of particulate emissions of vehicle wheel brakes; 56th International Scientific Colloquium 2011

11 Garg, B.D., Cadle, S.H., Mulawa, P.A. and Groblicki, P.J.; Brake wear particulate matter emissions. 2000, Environmental Science and Technology 34:4463-4469

12 Grigoratos T, Martini G. (2014) Non-exhaust traffic related emissions. Brake and tyre wear PM, JRC science and policy report for the EU commission

13 Bukowiecki N., Lienemann P., Hill M.,Figi R., Buchmann B., Furger M., Richard A., Mohr C., Weimer S. Prevot A. Baltensberger U. (2009). PM10 Emission factors of abrasion particles from road traffic (APART). Swiss Assoziation of Road and Transport Experts (VSS)

14 Oehl K.H., Paul H.G. „Bremsbeläge für Straßenfahrzeuge", Verlag Moderne Industrie (1991)

15 Steege R., Bauer G., Lange J., Holzapfel T., „Life time prediction for brake linings" SAE brake colloquium 09BC-0065/2009-01-3027 (2009)

# An innovative production method for a C/C-SiC brake disc, suitable for a large-scale production

Nestler, D.; Nier, N.; Roder, K.; Todt, A.; Wielage, B.; Wagner, G.
Institute of Materials Science and Engineering,
Technische Universität Chemnitz

Kroll, L.; Päßler, E.
Institute for Lightweight Structures, Technische Universität Chemnitz

Spange, S.; Weißhuhn, J.; Würfel, H.
Institute of Chemistry, Technische Universität Chemnitz

# Introduction

The industrial progress and the market-oriented management concepts result in increasing demands of the industry for improved performances while simultaneously improving on the energy efficiency. This results in the necessity for a progressive implementation of high-performance materials in areas such as the automotive industry, engineering, as well as applications in aerospace industry. Especially construction elements taking on large stresses call for excellent mechanical properties even at extremely high temperatures. In these high-temperature ranges only engineering ceramics, which unfortunately are very limited regarding the application due to the very brittle characteristics, showcase sufficient strength values.

Fibre-reinforced ceramics are characterised by an improved breaking elongation and damage tolerance compared to the conventional ceramics. This means, they can also be utilised as a structural materials in cases dealing with impacts straining. Additionally they are considerably lighter than metals, as well as being more resistant toward corrosion, wear and thermal shocks and more durable. However, the high production costs and limited feasibility of complex geometries currently limit the application of these materials to niche products of the high-price segment. One of the most commonly known examples are ceramic brake discs. These products showcase increased performances, a significant reduction of the weight and improved operation times compared to normal grey cast brakes. The overall weight reduction of the automobile results in the considerable improvement of the energy and $CO_2$ efficiency due to the reduced amount of required fuel and therefore decrease of the $CO_2$ emission.

The development and implementation of fully automated, reproducible production methods suitable for a large-scale production of fibre-reinforced ceramics, which offer a high flexibility regarding the geometry would lead to a considerable cost reduction and offer other great potentials for the economy. An innovative manufacturing technology is the key to open up these materials for various industrial applications such as friction clutches, gears, superchargers and prostheses.

# Motivation and objective

## Common production method of C/C-SiC break discs

Commercial long fibre-reinforced (1-100 mm) ceramic matrix composites (CMC) are produced by means of liquid silicon infiltration (LSI) (Fig. 1). The complex production method consists of three key steps. First a carbon fibre-reinforced plastic (CFRP) is created by utilising phenolic resin (moulding). This step is followed by the pyrolysis of the polymer, which results in carbon, creating a so-called C/C material. Finally, the silicon infiltration follows, during which the ceramic silicon carbide is produced by the reaction of silicon and carbon. The final product is a composite material, which can withstand immense thermal and mechanical straining (C/C-SiC composite).

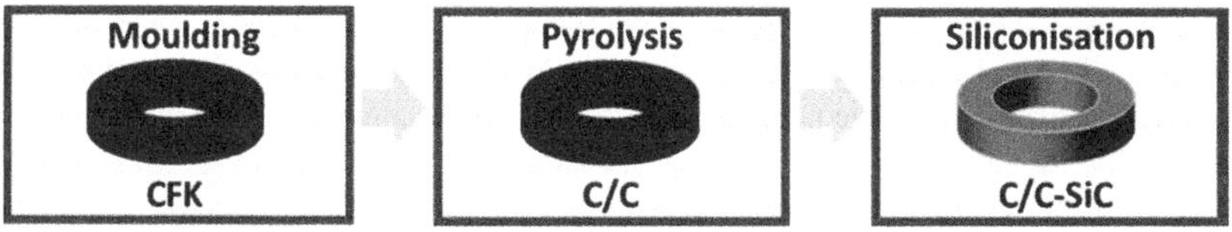


Figure 1: Production of ceramic C/C-SiC composite materials by means of LSI

Currently hot pressing is the state of the art production method to mould CFRPs as a preliminary step to produce CMCs [1]. This process is characterised by various manual steps, resulting in high production costs and limiting the reproducibility. The geometries, which can be obtained with this method, are limited by the process and only allow simple parts to be produced in this fashion.

## Innovative production method of C/C-SiC break discs

In order to create the framework for fibre-reinforced ceramics [2] to be utilised on a broad industrial scale, large-scale production methods and their positive impact on the processing costs are a mandatory requirement. The increased production volumes, which can be achieved with these technologies, enable the decrease in production costs.

The main objective of any research regarding the CMCs has to be the transfer of any advantages large-scale production methods to fibre-reinforced ceramic parts and components. None of the currently ongoing or already completed papers deal with creating C/C-SiC composites (fibre-reinforced ceramics) via LSI on a large-scale CFRP production.

The processing steps, known as compounding and injection moulding, which are utilised for the CMC production, have to be developed, adapted to the individual and tailored properties of CMCs and need to be optimised.

In this paper the development of innovative production methods of carbon fibre-reinforced high-performance ceramics (C/C-SiC materials) by means of thermosetting injection moulding as a large-scale production method are investigated in a holistic fashion.

# Experiment and results

## Selection of the precursor materials

As a result of the versatile requirements regarding the process a large number of experiments had to be conducted to find a suitable precursor material and appropriate apparatus settings. The large number of different starting compounds and the possible adjustments of the entire process chain are examples of that. The following paper deals with the report of an innovative procedure to create C/C-SiC composites on a large scale on the basis of one typical precursor.

The starting point of the LSI process is the polymer precursor. It should show a sufficient viscosity and wetting of the carbon fibres for an optimal processability during the compounding process. It is of special interest to apply a precursor containing a large amount of carbon and small amount of oxygen. Suitable materials are polyfurfuryl alcohol and phenolic resins, however the latter provide higher yields of carbon after pyrolysis [3]. The created carbon shows a specific pore-volume and morphology of the pores, which can be described by a function of the polymer introduced to the process and the temperature regime during the pyrolysis [4]. The dimensional stability during the pyrolysis is guaranteed by sufficient crosslinking of the curing agent. Any gaseous by-products created during the curing reaction have to be kept at a minimum in order to avoid the development of a closed porosity within the CFRP. The development of an open porosity and networks of cracks after pyrolysis is crucial for the infiltration with silicon.

Finding a precursor, which meets all of the requirements set by the process is challenging. Due to the demand for a high carbon-yield after the pyrolysis phenolic novolacs were utilised and a curing agent was added. These resins can be manufactured as thermoplastic polymers below the temperature required for any crosslinking reactions. After the heating during the transformation the dimensional stability is similar to the one of thermosetting resins. A novolac type phenolic resin (purchased from Dynea Erkner GmbH) was used together with hexamethylenetetramine (99%) as a curing agent (purchased from Carl Roth GmbH & Co. KG). The resin displays a glass transition at 60 °C (DSC 30 from Mettler Toledo) and has an average molar mass of 1124 g/mol (PL-GPC 50 Plus from Polymer Labories). Carbon fibres provided by Zoltek Companies, Inc. were used as roving. Zinc stearate, purchased from Bearlocher GmbH, was used as a lubricant in order to improve the processability during the injection moulding. The composition of the mixture is the following: 41.74 wt% novolac, 6.26 wt% curing agent, 50.00 wt% carbon fibre, 2.00 wt% lubricant.

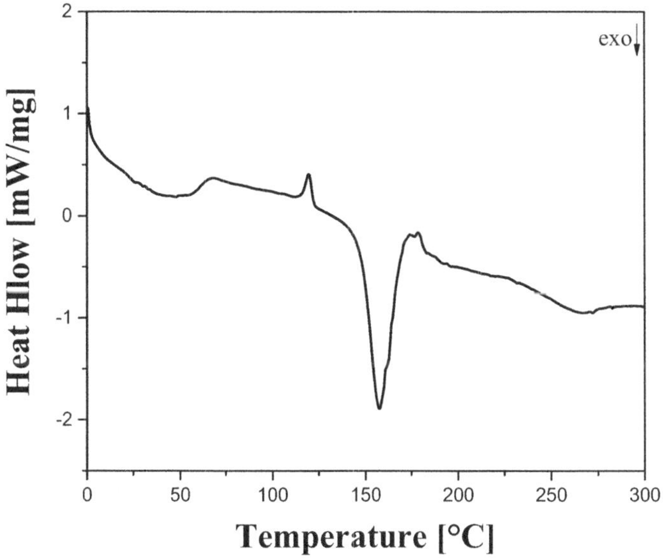


Figure 2: Differential scanning calorimetry (DSC) of the mixture containing novolac, curing agent and lubricant. The heating process (0-300 °C, 10 K/min) shows the glass transition at 60 °C and an exothermic signal indicating the crosslinking reaction at 148 °C.

A DSC measurement of the mixture without carbon fibre was conducted to determine the processing temperature for the preparation of the compound and the sample (Fig. 2). The mixture shows a glass transition of the novolac at 60 °C. The crosslinking reaction starts at 148 °C. As a result the compounding was performed below this temperature (vide infra). The amount of carbon fibre was set at 50 %. The lubricant was set at 2 % according to the results of a previous paper [5].

## Moulding process

### Compounding

The objectives of the compounding process are to combine all ingredients and to distribute them in a uniform fashion in order to obtain homogeneous properties of the material. Furthermore, the material has to be converted in a granular form. During the course of this investigation, the compounding tests were carried out utilising a co-rotating twin screw extruder (Noris ZSC 25) (Fig. 3). It consists of an engine incl. the controls, a transmission, a processing unit (cylinder, screws, heating system), feeding units (weighing system, hoppers, side feeders) and a nozzle. The screws are divided into several segments. The geometry of the screws can be adapted to an individual task by combining the kneading, mixing and conveying elements.

Figure 3: Twin-screw extruder Noris ZSC 25

## *Production Parameters*

The most suitable process parameters regarding the compounding process (temperatures of cylinder zones, location of feeding the ingredients, screw geometry) were determined during numerous trials. The novolac, curing agent and lubricant were stored in hoppers, which stood on weighing systems. They were introduced to the processing unit through screw feeders. This ensured the exact and continuous supply of the formulation components. The fibres (rovings) were pulled into the processing unit directly. The granulation process utilised a hot die face pelletizing system to obtain a uniform grain size of granulate. This is an important requirement to ensure a proper "feeding" during the injection moulding process. The subsequent transportation by air acted as a cooling step. This prevented the single particles from sticking to each other.

The novolac resin and lubricant were introduced toward the front region of the processing unit. There they were melted, as well as homogenised. In the mid-section, the fibres were pulled in and cut as a result of shear stresses within the plasticisation unit. The curing agent was added at the end of the cylinder. This was early enough to ensure the even distribution within the melt-fibre mixture and late enough to prevent any premature crosslinking

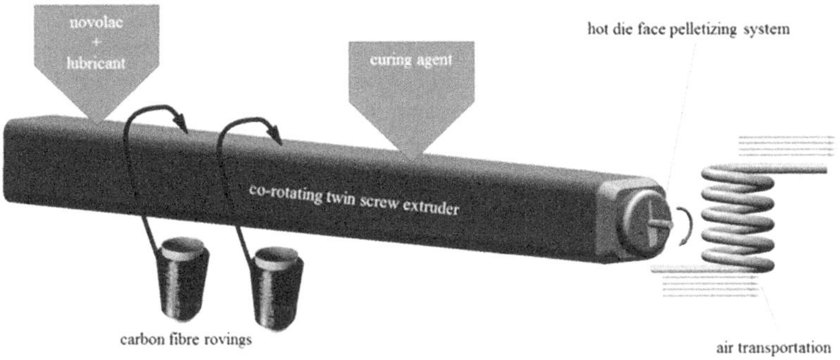


Figure 4: Experimental set-up for compounding (schematic)

The result of compounding process is a carbon fibre-reinforced phenolic resin in form of granular particles (Fig. 5). The granulate was utilised during the subsequent characterisation (fibre content, flow-curing behaviour) as well as the production of samples by means of injection moulding.

Figure 5: Carbon fibre-reinforced phenolic compound

## *Characterisation*

The actual fibre content of the precursor after compounding was determined by means of extraction experiments of the granulated material with acetone over a period of 120 h. This experiment allows the separation of the carbon fibres from the resin and the lubricant. The compound contained 48 wt % carbon fibres. The actual amount is slightly lower as the amount initially desired. This is a result of inevitable slipping effects during the pull-in of the carbon fibre rovings.

During the processing of thermoset moulding compounds softening (due to the melting of the resin) and the solidification (due to the chemical curing reaction) overlap. In addition to other parameters, the so-called flow-curing behaviour influences the cycle time and the maximum dwell time of the material in the cylinder. These are important parameters, which facilitate an optimal adjustment of the injection moulding machine. Long dwell times at plasticisation temperature and short curing times at mould temperature regarding the injection moulding process are desirable.

The flow-curing behaviour was determined using a Brabender Plasticorder. It consists of two counter-rotating kneading blades, driven by an electric motor, and an oil-tempered kneading chamber. The chamber is fitted with a sensor, which measures the temperature of the moulding material. The torque, which is proportional to the viscosity of the compound, is recorded too. The result of the test is a torque-temperature-time diagram in style of DIN 53764. During the initial phase the torque increases due to chamber being filled up. After this the torque drops, which can be attributed to the melting of the material. This trend continues until a minimum of viscosity is reached. After that, the viscosity rises due to the incipient crosslinking reaction. Fig. 6 shows a schematic plastogram with characteristic values.

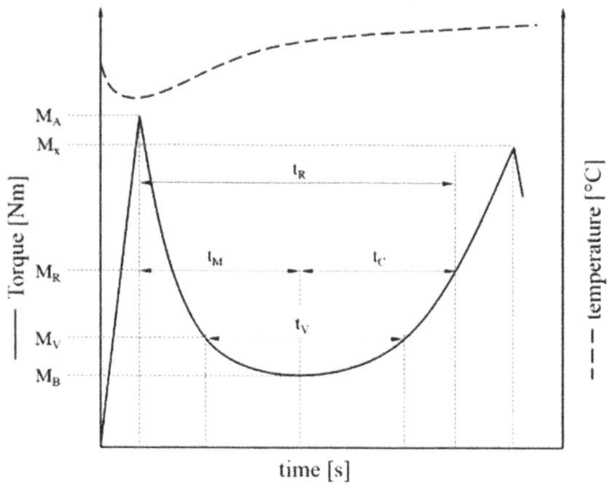


| | |
|---|---|
| $M_A$ | Filling torque |
| $M_B$ | Minimum of torque |
| $M_V$ | Torque $M_B + 3$ Nm |
| $M_R$ | Torque $M_B + 10$ Nm |
| $M_X$ | Maximum of torque |
| $t_M$ | Melting time |
| $t_V$ | Residence time |
| $t_C$ | Curing time |
| $t_R$ | Reaction time |

Figure 6: Plastogram (schematic)

The flow-curing behaviour at 90 °C was determined during previous tests. At this temperature, the initial reaction took place after only 12 minutes. This means the dwell time was long enough to avoid premature curing within the plasticisation unit. The flow-curing behaviour was also determined at temperatures of 140 °C, 155 °C and 170 °C. For each temperature two curves were recorded. These plastograms are shown in Fig. 7 left. For reasons of clarity, the temperature curves are not displayed. These figures showcase the shortening of the dwell, melting and curing time with increasing temperatures. The curing time at 170 °C is only about a quarter of the curing time at 140 °C (Fig. 7 right).

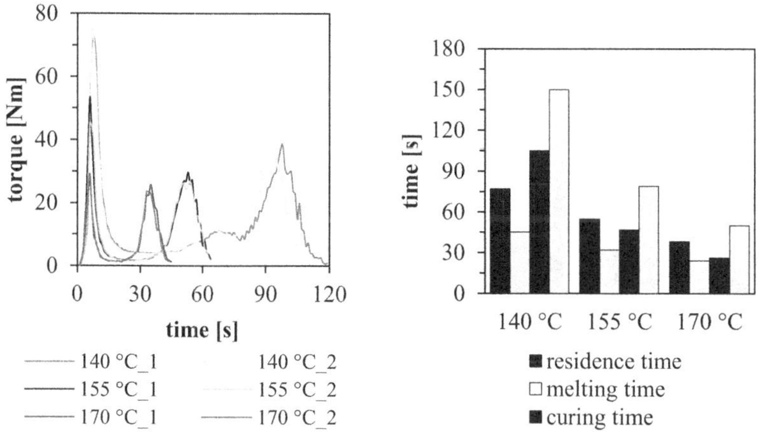


Figure 7: Plastograms of the investigated material for 140 °C, 155 °C and 170 °C (left) and resulting times of the flow-curing tests (right)

The results indicate the great influence of the temperature on the flowing and curing behaviour of the compound. This shows the necessity of a precise management of the temperature throughout the entire plasticisation and solidification process.

## Injection moulding (CFRP production)

The aim of the first step of the LSI preparation route is the production of a highly dense CFRP-preforms with a low porosity and a homogeneous distribution of the fibres. Due to the high degree of automation the injection moulding process is predetermined by high processing pressures and a good reproducibility.

### *Production Parameters*

#### Process cycle

The injection moulding tests were conducted on an Arburg injection moulding machine having a clamping force of 150 kN and a screw diameter of 22 mm (Fig. 8, left). The granular material is stored in the hopper above the filling zone of the plasticisation unit and pulled into the cylinder by the rotating screw. After the material has melted due to shear stresses and heat of the cylinder it is injected into the hot cavity with an axial motion of the screw. Once the material is cured the mould opens, the finished part can be removed and the process starts again.

A cuboid specimen geometry (50 x 10 x 3 $mm^3$) was chosen (Fig. 8, right). On one of the ends encompasses the sprue, while the opposite end is characterised by an overflow. The overflow was added to improve the venting of the cavity and to prevent any air pockets from forming at the end of the flow path. For the subsequent tests, the sprue and overflow were removed from the part.

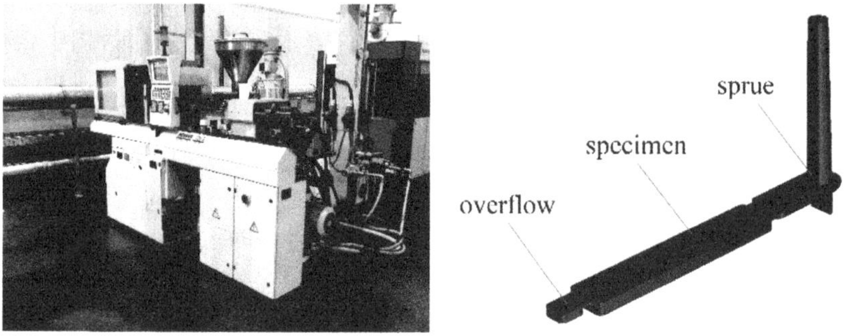


Figure 8: Injection moulding machine (left), moulded part (right)

**Process parameters**

The most suitable process parameters for the injection moulding process were determined iteratively too. The parameter configuration shown in Table 1 could be found for the production of samples.

Table 1: Parameter-set for injection moulding process

| **parameter** | | **unit** | **value** |
|---|---|---|---|
| temperatures | feeding zone | °C | 75 |
| | plasticisation zone | °C | 90 |
| | nozzle | °C | 90 |
| | mould | °C | 170 |
| clamping force | | kN | 150 |
| rotary speed of screw | | $min^{-1}$ | 175 |
| melt volume | | ccm | 5.50 |
| backpressure | | bar | 50 |
| mass flow | | ccm/s | 4 |
| injection pressure | | bar | 1000 – 1200 |
| switch over to holding pressure | | ccm | 1.00 |
| profile of holding pressure | step 1 | bar | 750 |
| | step 2 | bar | 500 |
| | step 3 | bar | 50 |
| curing time | | s | 70 |
| temperatures | feeding zone | °C | 75 |
| | plasticization zone | °C | 90 |
| | nozzle | °C | 90 |
| | mould | °C | 170 |
| clamping force | | kN | 150 |
| rotary speed of screw | | $min^{-1}$ | 175 |
| melt volume | | ccm | 5.50 |
| backpressure | | bar | 50 |
| mass flow | | ccm/s | 4 |
| injection pressure | | bar | 1000 – 1200 |
| switch over to holding pressure | | ccm | 1.00 |
| profile of holding pressure | step 1 | bar | 750 |
| | step 2 | bar | 500 |
| | step 3 | bar | 50 |
| curing time | | s | 70 |

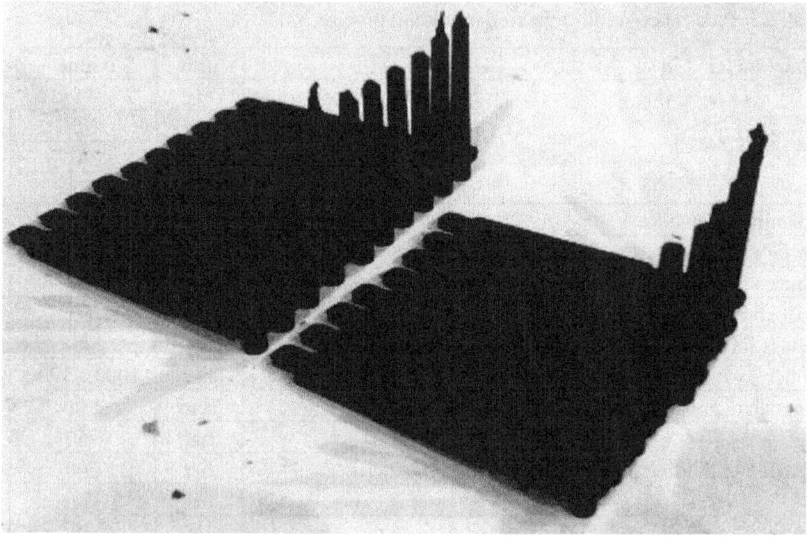

Figure 9: Injection moulded parts

## *Characterisation of the CFRP composites*

The structure of the CFRP composites was examined at the cross sections by means of a light microscope (LM) (reflected-light microscope GX51, Co. Olympus). In addition, the closed porosity was quantified by means of a picture analysis via LM. For this examination 20 different areas of the same sample were analysed. Furthermore the open porosity and density were determined utilising the buoyancy test given in DIN EN 1389.

## *Structure of the CFRP composites*

The fibre-matix-distribution is homogeneous, as displayed in the LM picture (Fig. 10). The only preferred fibre orientation detected was in direction of the flow and of low significance. This alludes to the fact that most of the fibres were cut in the length direction. No micro-cracks are visible. The material is characterised by a high fibre-matrix-binding. There are closed pores to be found within the sample, which are not distributed homogeneously throughout the volume of the sample. Due to the evaporation of the by-product (ammonia), the pores are created during the curing reaction between the phenolic resin and curing agent. If it does not disappear completely from the sample, closed pores arise. The development of the closed porosity cannot be interpreted as a thermal decomposition process of the stearate-lubricant because the decomposition temperature is not reached throughout the entire injection moulding process [5].

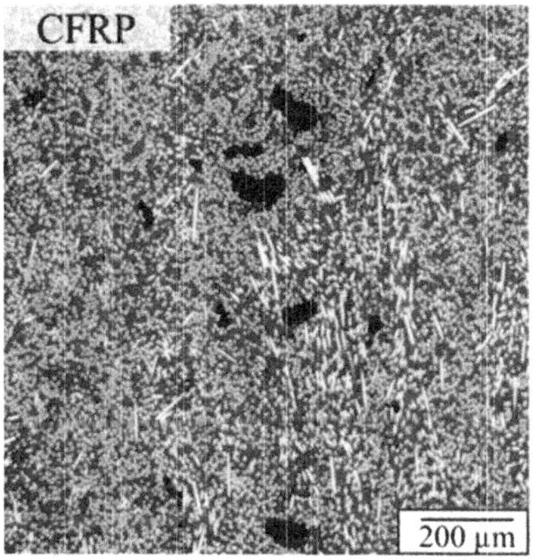


- good fibre-matrix bonding
- homogeneous distribution of the fibres and matrix
- low preferred fibre orientation
- inhomogeneous pore distribution
- visible pores (closed porosity) on a microscopic scale (up to approx. 10 µm)
- no visible open porosity
- no visible cracks

Figure 10: Cross-section image by LM of the microstructure of the CFRP samples. White – carbon fibres, gray – matrix, black – closed pores

The quantitative evaluation of the picture analysis (Fig. 11) shows a closed porosity between 2.46 %±1.70 %. The pores are distributed irregularly throughout the volume, which could be the result of an inhomogeneous distribution of the curing agent, followed by an irregular curing in the sample volume.

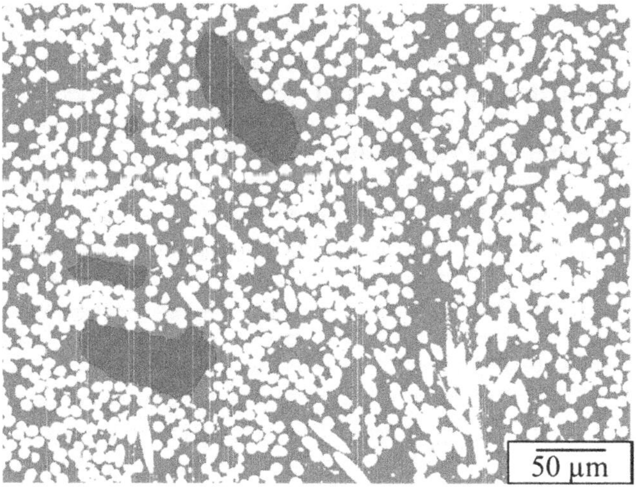


Figure 11: Picture-analysis of phases in CFRP samples.
Yellow – carbon fibres, blue – matrix, red – closed pores

*Open porosity and density of the CFRP composites*

The relative open porosity of CFRP is about 0.2 % (Fig. 12), which is a relatively low figure. This value was determined by an analysis of the microstructure, which showed no signs of cracks or open pores (Fig. 10). The low values regarding the open and closed porosity result in a high density of about 1.4 g/cm³.

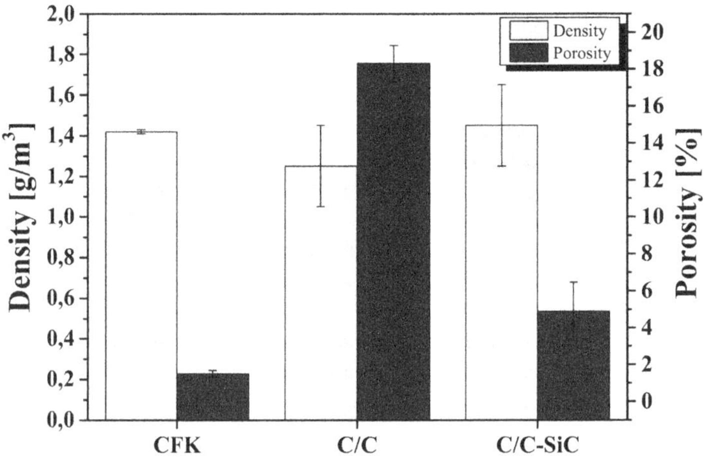


Figure 12: Open porosity and density of CFRP-, C/C- and C/C-SiC-composites.

# Ceramisation (Production of the C/C and C/C-SiC composites)

## Pyrolysis (Production of the C/C-composites)

*Process parameters for the production of the C/C-composites*

The CFRP composite underwent a pyrolysis process in an argon-filled tube furnace (CTF 12/65, Co. Carbolite), until reaching the final temperature of 960 °C. The result was a C/C composite with a crack-pattern, which is a result of the matrix shrinking.

*Characterisation of the C/C composites*

The structure and the open porosity of these samples were examined in a similar fashion as the CFRP samples were. Additionally the shrinkage behaviour of the composites after the pyrolysis was analysed. This process was conducted to provide evidence of the accurate size of components from this material. This required the comparison of geometric values of the samples before and after undergoing the pyrolysis. The form stability and relative remaining matter were determined by analysing the weight-loss from CFRP-state to C/C-state (named "C-yield")

### *Shrinkage and "C-yield" after the pyrolysis process*

The dimensional change after undergoing pyrolysis is shown in Fig. 13. The diagram shows the longitudinal shrinkage (about 1.5 %), as well as the shrinkage of the depth (about 1.7 %). Both these values are very low. In comparison, the reduction of the width is obviously higher (about 5.5 %). These numbers indicate the fact that fibres are not oriented in direction of the width, but rather follow the direction of the flow. This means the reduction of the width cannot be prevented, however the length is only reduced by a low percentage. The volumetric shrinkage is about 8.5 %.

After the pyrolysis of the CFRP samples a "carbon yield" of about 80 % can be detected. This value is quite high. This can be explained by the presence of carbon fibres in the sample (compared to resin-curing agent-samples without carbon fibres).

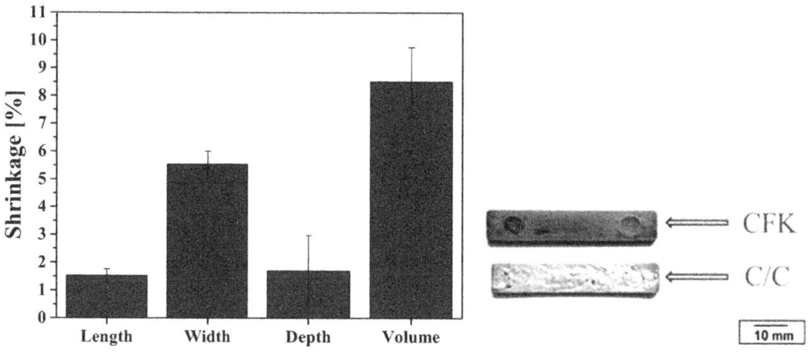


Figure 13: Left: shrinkage of the C/C samples after pyrolysis, right: CFRP (before the pyrolysis) and C/C (after the pyrolysis) materials

### *Structure of the C/C composites*

The microstructure of C/C composites is shown in Fig. 14. Closed pores can be found in the CFRP-state, which also remain in the C/C-state. The closed pores will not burst open after pyrolysis and are not connected to the open crack-network. Furthermore the cracks are orientated vertically to the image plane, which verifies the theory of the majority of the fibres being aligned in direction of the flow path (crossing the image plane). The shrinkage in this direction is prevented by these fibres. The flow path is crossed by cracks, which are distributed homogeneous throughout the volume of the sample. They form a network at several points.

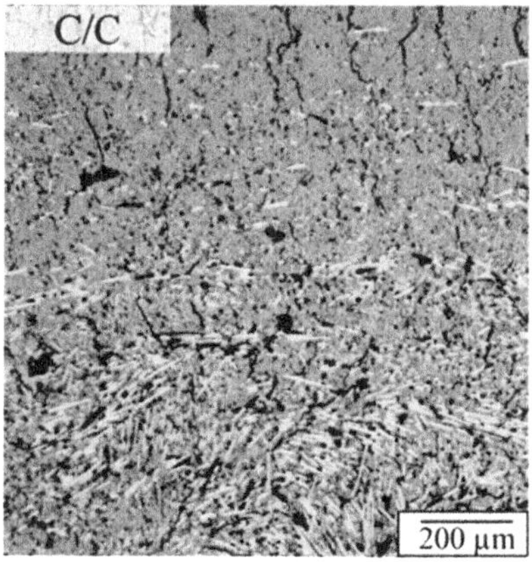


- visible pores (closed porosity) on a microscopic scale (up to approx. 10 µm
- inhomogeneous distribution of closed pores
- visible open pores at cracks crossing the float direction
- homogeneous distribution of cracks

Figure 14: Cross-section image by LM of the microstructure of the C/C samples. White – carbon fibres, grey – matrix, black – closed pores and cracks

### *Open porosity and density of the C/C composites*

According to the analysis of the microstructure, the open porosity of the C/C composites rose to about 18.2% after undergoing the pyrolysis (Fig. 12). Many cracks (Fig. 14) represent this open porosity.

## Siliconisation (C/C-SiC production)

### *Production parameters*

The C/C composite was infiltrated with liquid silicon in a vacuum at ≥ 1410 °C (gravity siliconisation). The amount of silicon necessary to ensure a complete infiltration was calculated for every sample individually, depending on the open porosity and actual mass [6]. The finished product is the C/C-SiC composite.

### *Characterisation*

The structure and the open porosity of these samples were examined similarly as was the case with the CFRP and C/C composites. In order to characterise the mechanical properties of the C/C-SiC composites a 3-point bending test (also in-situ via SEM) was preformed (micro bending module Co. Kammrath & Weiss). For this purpose 5 samples with the dimensions of 45 mm x 10 mm x 3.5 mm were used. The application spacing during every test was 40 mm and the testing speed was set at 5 µm/s (Fig. 15).

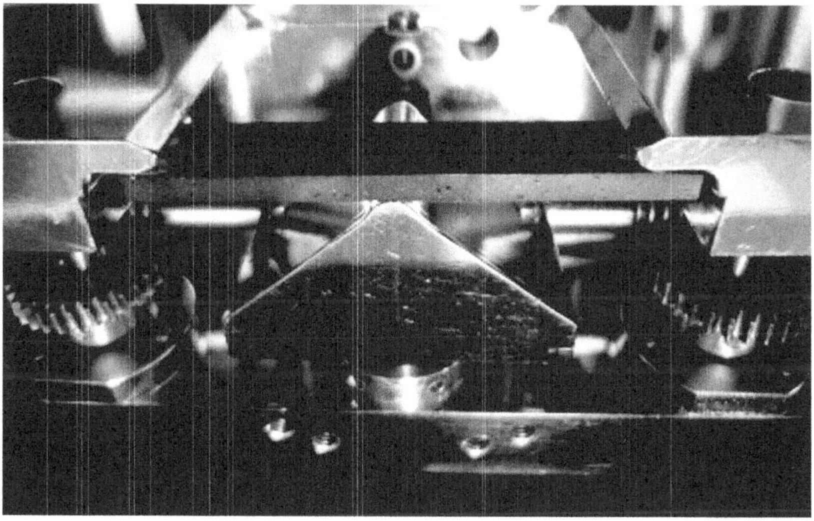

Figure 15: Test set-up for the in-situ 3-point bending test via SEM

*Structure of the C/C-SiC composites*

Fig. 16 shows the multiple cracks, which are a result of the pyrolysis, being concentrated after undergoing the infiltration process with liquid silicon. This is made possible by the reaction of the silicon with the carbon matrix and the transformation to SiC. A small part of the silicon did not have any access to the carbon, due to the small width of the cracks and the resulting fast sealing of the channels by the development of SiC. This means there is a residual amount of silicon to be found within the material. The closed pores are not filled with the silicon, as the melt has no access to those. As a result of this the finished products contain cavities, which influence the mechanical properties in a negative way.

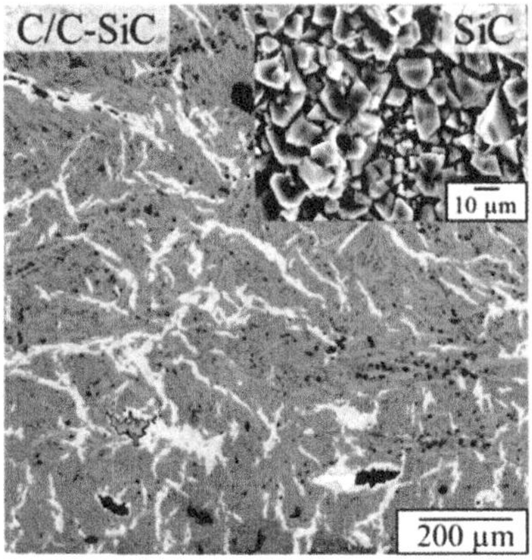


- infiltrated cracks
- Si and SiC areas
- non-infiltrated closed pores
- partially damage of the C-Fibre

Figure 16: Cross-section image by LM of the microstructure of the C/C-SiC samples. White – free Silicon, dark gray – matrix, light gray – silicon carbide, black – closed pores

## *Open Porosity of the C/C-SiC composites*

The open porosity reduces significantly after the infiltration process to approx. 4.9%. This means that 73.3% of the initial C/C porosity (18.3%) is filled with the siliconisation (Fig. 12). The residual porosity is a result of not all cracks being infiltrated. There are two main reasons for this happening. The one is the reduction of the capillary forces compared to gravity for wide cracks (>100 µm) with an increasing diameter. This leads to the silicon not being kept within the cracks. The other main reason is the silicon compressing the entrance to the crack during the transformation to SiC, which blocks any further silicon from entering the area. These types of cracks can be seen in Fig. 16. Due to the fact that there is no visible penetration of epoxy resin during the preparation of the cuts the entrance of the cracks has to be blocked.

The C/C-SiC composites are characterised by a density of approx. 1.4 g/cm$^3$, which is relatively low. This is a result of the presence of numerous closed pores, imperfectly filled cracks and pores, as can be seen in Fig. 19.

## *Mechanical properties of the C/C-SiC composites*

Fig. 17 shows the stress-strain diagrams. The individual bending strength values of the samples differ. The average value was determined to be 58.42 ±15.69 MPa. This suggests different structures of the samples. A reason for this could be the inhomogeneous character of the compound.

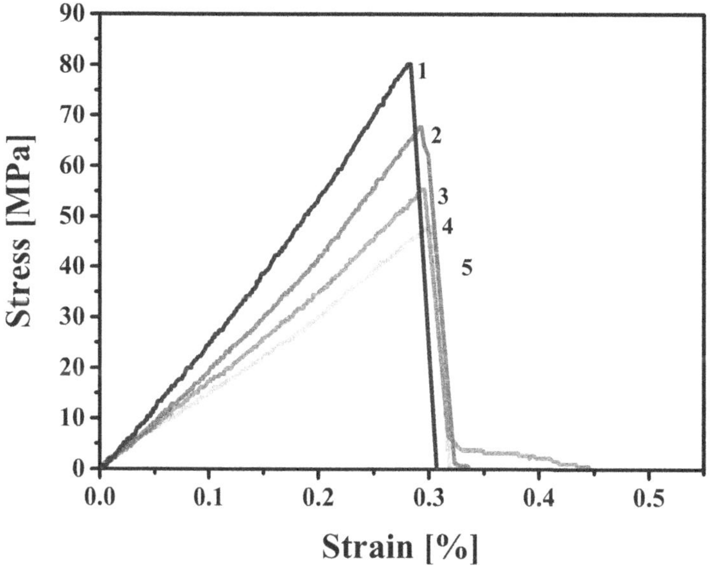


Figure 17: Stress-strain-diagrams of the C/C-SiC composites

Due to the small dimensions of the samples it is very likely that the composition (ratio of the fibre and matrix-content) varies for every individual sample. The breaking elongation values of these materials differ slightly as well. The average is 0.3 ±0.02 %. All of this indicates a brittle behaviour of the compounds. However, a more detailed examination of the curves 3 and 5 display the malfunction of the samples does not take place solely in a brittle fashion. The fractographic images display the course of the cracks during the highest measured strain as well as at the end of the tests (Fig. 18).

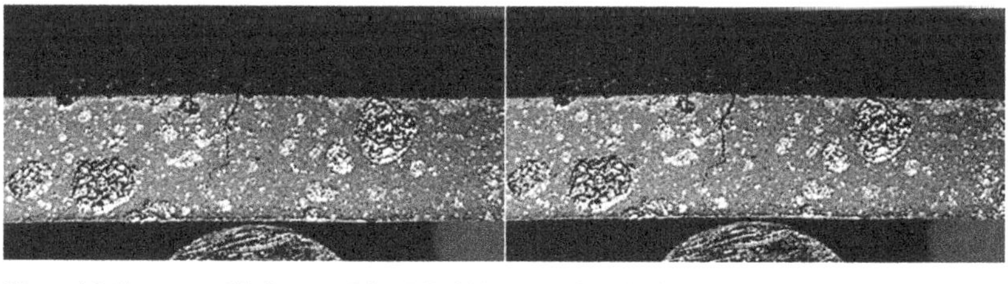

Figure 18: Fractographic image of the C/C-SiC composites during the 3-point bending test

Both images show the end of the crack being located slightly beneath the centre of the sample. This indicates the crack being deflected in direction of the depth of the sample, which is not shown in the image. The in-situ test by means of a scanning electron microscope revealed the deflection of the crack at the carbon fibres (Fig. 19).

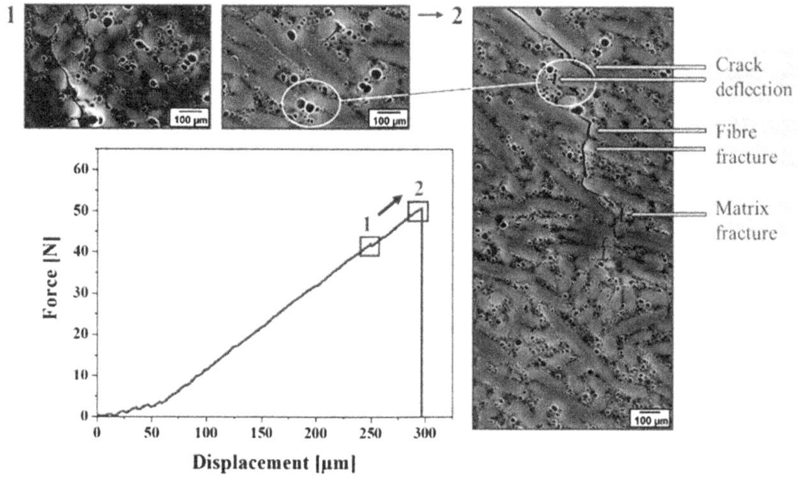


Figure 19: Microstructure image of the C/C-SiC by in-situ 3-point bending test via SEM

During follow-up examinations of the fracture point the pull-out effect of the fibres could be seen clearly (Fig. 20). This shows that energy dissipating effects are functional at the short carbon fibre-reinforced C-SiC composites.

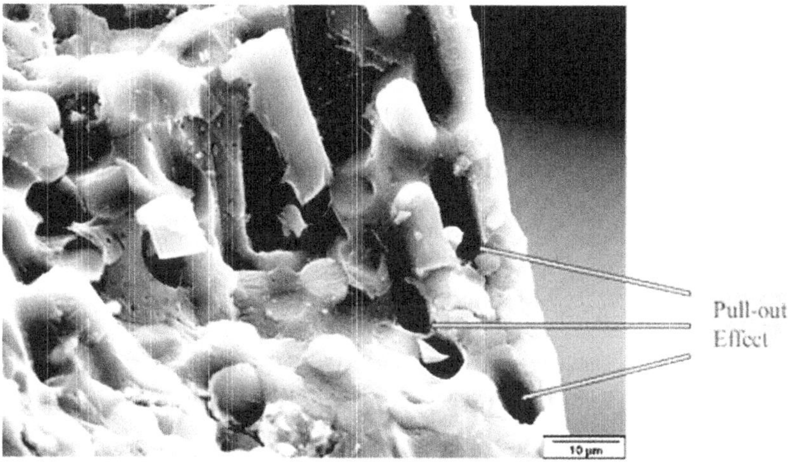


Figure 20: SEM image of the pull-out effect of the C/C-SiC composites

# Conclusions

The claim of this paper is the development of an innovative production method of C/C-SiC materials by means of thermosetting injection moulding as a large scale production method via LSI route. Therefore, a large number of preliminary test were done to define the right precursors. In cause of the demands like high C-Yield, cross-linking temperature and processability phenolic novolacs were preferred as precursors. Furthermore, 6.26 wt% hexamethylentetramine as a curing agent, 50 wt% carbon fibre as a reinforcing component and 2 wt% zinc stearate as a lubricant were chosen.

To get a uniformly distributed material with homogeneous properties, all ingredients were compounded with the help of a co-rotating twin screw extruder and the obtained compound got granulated. During the compounding process the carbon fibres were added directly as roving and chopped due to shear stresses inside the plasticization unit. After characterisation of the compound a suitable parameter-set for the injection moulding process was determined (Table 1). The injection moulded specimens had a cuboid geometry of 50 x 10 x 3 $mm^3$.

After the characterisation of the CFRP, a low preferred fibre orientation lengthwise to the flow direction, no micro-cracks, a high fibre-matrix-binding, an inhomogeneous distributed closed porosity (2.46 %±1.70 %), a very low open porosity (0.2 %) and a density of about 1.4 $g/cm^3$ were determined. The CFRP underwent a pyrolysis process in an argon-purged tube furnace up tp 960°C, to create a C/C composite. The resulting shrinkage in length was lower than the high shrinkage of the width. The necessary homogeneous crack-network for the following silicon infiltration process was formed and the open porosity raised up to 18.2 %. The C/C composite was infiltrated in a vacuum at a temperature of ≥ 1410 °C. This process aimed to obtain C/C-SiC and to condense the numerous cracks. However, the closed pores were not infiltrated. A remaining open porosity of 4.9 % and a relative low density of about 1.4 g/cm³ was determined. After the characterisation of the mechanical properties a bending strength of 58.42 ± 15.69 MPa by 0.3 % strain was measured. It can be observed, that some composites did not show any brittle fatigue behaviour. By means of in-situ bending tests via SEM examinations energy dissipating effects like crack deflection and fibre pull-out can be detected at the short carbon fibre-reinforced C-SiC composites.

The feasibility of a scalable process for production of C/C-SiC materials by means of thermosetting injection moulding method via LSI route was shown in this article.

# Outlook

## C/C-SiC-composites with C-filament-inserts

The integration of C-filaments by means of inserts enables the further improvement of physical properties of fibre-reinforced ceramics. Therefore, woven or non-woven fabrics have to be integrated to the injection moulding process.

In preliminary examinations first trials by back injection moulding of the textile inserts were performed. An unidirectional carbon fibre non-woven was inserted to mould and got back injection moulded with short C-fibre-reinforced compound. In Fig. 21 a good resin impregnation of the textile insert can be observed.

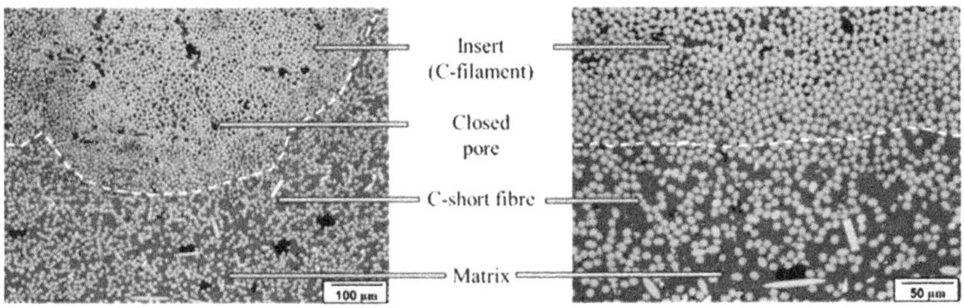


Figure 21: Cross-section image by LM of the microstructure of the C/C-SiC specimen with integrated C-filament as an insert

## Multicomponent injection moulding

A further aim is also the development of a multicomponent injection moulding process for the production of fibre-reinforced ceramics. By this technique material compounds with different compositions of the single components and different properties can be produced. An example for a commercial product would be the ceramic brake disc, which deposes a supporting body and a friction layer.

## Reference

[1] Rosenlöcher, J.; Deinzer, G.; Waninger, R.; Münchhoff, J.: Hochleistungsbremsscheiben aus Faserverbundkeramik, Materialwissenschaft und Werkstofftechnik 38 (2007) 11, 883-942

[2] Krenkel, W.: Keramische Verbundwerkstoffe, 1. Aufl. Weinheim: Wiley-VHC, 2003,ISBN 3-527-30529-7

[3] Fitzer, E.; Schaefer, W.; Yamada, S.: The formation of glasslike carbon by pyrolysis of polyfurfuryl alcohol and phenolic resin. In: Carbon 7 (1969), 643-648.

[4] Kuo, H. et al.: Effect of carbonization rate on the properties of a PAN/phenolic-based carbon/carbon composite. In: Carbon 43 (2005), 229-239.

[5] Nestler, D.; Nier, N.; Roder, K.; Päßler, E.; Weißhuhn, J.; Todt, A.; Würfel, H.; Kroll, L.; Spange, St.; Wielage, B.: Development and characterisation of phenolic resin moulding materials for the production of new short fibre-reinforced C/C-SiC composites, Proceedings 20th Symposium on Composites 2015, Wien, accepted.

[6] Krenkel, W.: Keramische Verbundwerkstoffe, WILEY-VCH, Weinheim, 2003.

# Combined control strategy for the combustion engine and brake system to enhance the driving dynamics and traction of front-wheel-drive vehicles

**Dipl.-Ing. Daniel Killian**

Stefan Fischer, M.Sc.

Prof. Dr.-Ing. Markus Lienkamp

Dipl.-Ing. Stephan Poltersdorf

Dr.-Ing. Ralf Schwarz

# 1 Motivation

In very dynamic situations vehicles with front wheel drive can have certain limitations in terms of traction and handling characteristics. Because of the shift in dynamic wheel load from the front axle to the rear axle and the additional reduction of contact force of the inner front wheel when accelerating out of bends, the traction and handling capabilities can be limited.

One approach to improve traction during cornering, which comes along without additional actuators is to apply brake torque preventively on the inner wheel. Such approaches exist in several production cars for some time already. Popular realization for example is the *XDS*-System by Volkswagen [1] or Ford's *eTVC* [2], which both use the front wheel brakes to enhance cornering and traction of their front driven vehicles.

During the years the brake systems of passenger cars have evolved to a complex mechatronic system. While the variety of driver assistance systems using automatic brake interventions are steadily increasing, the requirements to the brake system itself in terms of precision, dynamics and noise level during pressure generation are getting higher accordingly [3].

Due to that development it is imaginable that the brake system can be incorporated more and more as a device for influencing driving dynamics within a wider range of driving situations.

A beneficial range of using brake-torque-vectoring are driving situations on low-μ, due to the comparatively low brake torque that has to be applied for an effective influence of driving behaviour. However, one characteristic of torque vectoring systems using the front axle solely is pronounced especially on low-μ surface:

The active torque distribution is able to support the desired driving behaviour as long as the front axle is in a stable condition. Once the friction potential is used up, the vehicle cannot maintain its neutral driving behaviour anymore. As a consequence the understeering tendency is increasing rapidly and the electronic stability program (ESP) has to come into action. The ESP-interventions are perceptible by the driver and sometimes considered to be uncomfortable.

In this article a control strategy will be presented, which is able to mitigate this undesired behaviour by a combined control of the combustion engine and individual brake torque on all wheels with friction adaption. This will help to preserve a more natural driving behaviour at the limit of driving dynamics.

# 2 Initial situation

## 2.1 State of the art of combined engine and brake control

Besides ESP, enhanced brake-intervention systems that many vehicles are equipped with, are mostly bound to high-μ driving situations. On low-μ surface like snow, brake-interventions are usually constricted to those caused by the electronic stability program. In conjunction with an engine torque control based on the feedback of actual wheel slip, the handling characteristic is significantly reduced due to safety reasons.

Because of the less pronounced adaption to friction between tyre and road and the feedback-based manner of the ESP algorithm for directional stability, the interventions of engine and brake torque can therefore be very intense. The engine torque requested by the driver especially when accelerating out of bends is sometimes reduced in a very restrictive way.

## 2.2 Physical limitations of torque distribution of front driven vehicles

Limited traction and driving dynamics of front driven vehicles can be influenced positively especially when accelerating during cornering by dynamic torque distribution of the driven wheels. Firstly, the inner wheel will be prevented from spinning due to longitudinal force exceeding the friction limit. Secondly the understeering tendency which is caused by an increased demand of side slip angle on the front axle due to longitudinal force and shift in wheel load can be counteracted by the resulting yaw moment.

Irrespective of the way of generating this yaw moment at the front axle, whether through an active differential or selective brake intervention, a supporting effect of the driver input cannot be maintained arbitrarily to the limits of driving dynamics:

As soon as the front axle reaches its physical limit, which is mainly defined by the cornering grip of the outer wheel, the vehicle changes its driving characteristic from a neutral behavior to heavy understeering (figure 1).

In this regard vehicles with rear- or all-wheel drive and an active torque distribution at the rear axle show a consistent behavior considering the self-steering-gradient which is decreasing progressively. A pure torque distribution at the front axle always has a physically related “two-phase” self-steering characteristic as illustrated in figure 1.

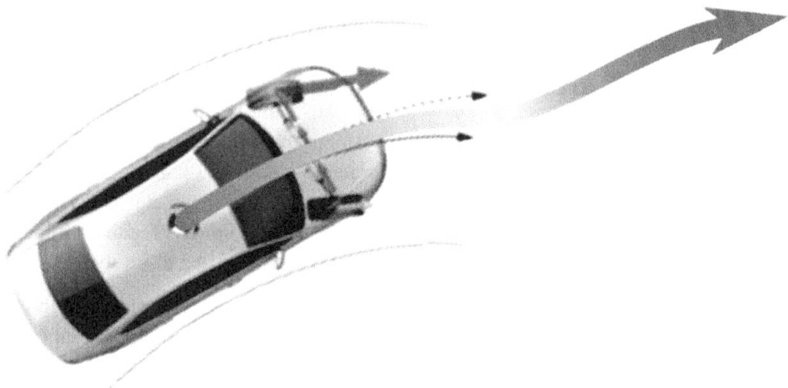

Figure 1: Two-phase vehicle behaviour by active torque distribution at the front axle: Strong contrast between linear self-steering characteristic to understeering when reaching the physical limit of the front axle during acceleration out of bends

The problem from the driver's point of view is, that he cannot rely on a uniform vehicle behavior, and is mostly surprised when it comes to such an unexpected change in self-steering characteristics.

## 2.3 Derivation of a control strategy for vehicles with front wheel drive by interaction of engine torque control and active brake control

Based on this insight, a control concept will be illustrated in the following, how driving experience and traction capabilities of vehicles with front wheel drive can be positively affected by a combined control strategy of combustion engine and wheel individual brake intervention.

A main advantage of wheel individual brake intervention is, that both axles can be actuated. The idea of this approach is, to preserve the lateral grip of the front axle by limiting the total engine torque and the applied yaw moment on that axle without completely overruling the drivers input. This is achieved by a torque control which adapts to friction connection and drivers throttle input. A concerted shift of yaw moment from the front to the rear wheel brakes depending on the driving situation should decrease the understeering effect to a more natural driving behaviour. The concept of this combined approach is explicated in the following.

# 3 Actuation concept of the brake system

## 3.1 Attributes of wheel-load based traction enhancement

The aim of this brake control approach is the equalization of used friction potential of the driven wheels under the influence of lateral acceleration to achieve optimal traction conditions during cornering. An open differential would limit the traction to the capability of the inner wheel. An assimilation of the tractive potential of both wheels can be accomplished in a good approximation by a proportional distribution of longitudinal forces according to the wheel loads (3.1) [4].

$$\frac{F_{\mathrm{long,i}}}{F_{\mathrm{N,i}}} = \frac{F_{\mathrm{long,o}}}{F_{\mathrm{N,o}}} \ . \tag{3.1}$$

This equation can be satisfied by an active brake intervention by the following torque request

$$M_{\mathrm{Br,i}} = M_{\mathrm{Dr,i}}\left(1 - \frac{F_{\mathrm{N,i}}}{F_{\mathrm{N,o}}}\right) \quad \text{with} \quad M_{\mathrm{Dr}} = M_{\mathrm{Dr}} = \frac{1}{2}\,\eta_{\mathrm{total}} \cdot i_{\mathrm{total}} \cdot M_{\mathrm{Eng}} \ . \tag{3.2}$$

With the brake torque request based on equation (3.2), the tractive potential of the inner wheel can be raised to the level of the outer wheel.

However this approach comes along with a number of drawbacks:

- The single sided brake torque generates a yaw moment which influences the driven trajectory but does not necessarily support the driver input and can be counterproductive while steering the car.
- The calculated torque is based on the wheel shift, which in turn depends on the measured lateral acceleration. As shown in figure 2, the sensor signal of the lateral acceleration is lagging behind the driver's input at the steering wheel. Under real driving conditions this effect is noticeable especially at the end of cornering, where there is still an active brake intervention at the inner wheel although the vehicle is moving straight already. The driver perceives this as an unwanted detaining during acceleration after curves. The additional time delay of the brake system itself supports this effect.
- Another unwanted side-effect is a certain self-amplification of the applied brake torque which can also be traced back to the lateral acceleration signal as a significant parameter for the brake request calculation. The generated yaw moment causes a further increase of lateral acceleration which in turn leads to a further increasing yaw moment.

In summary it can be stated, that this approach is not appropriate as a function to control the brake torque requests. Therefore, it has to be substituted by control strategies which reflect the drivers intent.

Nevertheless this function plays a main role when it comes to the brake torque distribution between front and rear axle, which is described in chapter 4.

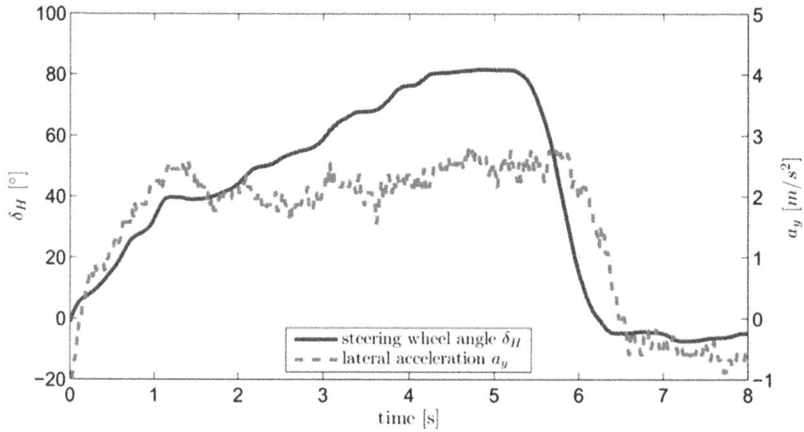


Figure 2: Time delay between steering wheel signal path and measured acceleration

## 3.2 Feed-forward control based on the linear single track model

For a better interpretation of the driver input a model-based feed-forward control is applied, which is built on an inverse single track model with additional terms for the active yaw moment. A full set of equations can be found in [4], where it is used for influencing the dynamic steering behavior. In this context, the algorithm is transferred in its steady-state version and will be enhanced by aspects of longitudinal dynamics to consider the needs of a neutral handling characteristic when accelerating out of bends. To improve the vehicle's behavior in these situations according to the driver's input, the side slip stiffness $c_{fa/ra}$ of front and rear axle are modified as a function of the acceleration pedal position in such a manner that the resulting yaw moment counteracts the understeering tendency. In a similar way the steady-state characteristics of the vehicle can be influenced when side slip stiffness of target and actual vehicle model are different in its original state already (3.3).

$$\dot{\psi}_{\text{desired}} = \frac{v}{l + \frac{m}{l}\left(\frac{l_r}{c_{fa}} - \frac{l_f}{c_{ra}}\right)v^2} \delta_{fa} \quad . \tag{3.3}$$

The extended single track model in [4] is equalized with the desired yaw rate in (3.3). The resulting equation (3.4) has to be solved for the yaw moment $M_z$, to get the basic term for the further brake torque control.

$$\dot{\psi}_{\text{desired}} = \dot{\psi}_{\text{actual}} = \frac{c^*_{\text{fa}} \cdot c^*_{\text{ra}} \cdot l_{\text{f}}}{c^*_{\text{fa}} \cdot c^*_{\text{ra}} \cdot l^2 + m \cdot v^2 (c^*_{\text{ra}} \cdot l_{\text{r}} - c^*_{\text{fa}} \cdot l_{\text{f}})} \delta_{\text{fa}} + \frac{(c^*_{\text{fa}} + c^*_{\text{ra}})\, v}{c^*_{\text{fa}} \cdot c^*_{\text{ra}} \cdot l^2 + m \cdot v^2 (c^*_{\text{ra}} \cdot l_{\text{r}} - c^*_{\text{fa}} \cdot l_{\text{f}})} M_z \tag{3.4}$$

with $c^*_{\text{fa}} = f_{\text{fa}}(\alpha_{\text{throttle}}, a_x, a_y)$ and $c^*_{\text{ra}} = f_{\text{ra}}(\alpha_{throttle}, a_x, a_y)$ .

Due to the most important driver inputs like steering wheel angle $\delta_H$ and acceleration pedal position $\alpha_{throttle}$ , the driver's intention can be considered much better than on a purely physical point of view.

## 3.3 Derivation of further control strategies for an improved driving experience

As mentioned before, the realization of an active yaw moment through brake actuation is possible on the front as well as on the rear axle as a combined actuation of both. The advantage of an actuation of the front axle is the more effective generation of yaw moment by a direct distribution of driving torque. Furthermore the front axle generally has the more powerful brake system, which can cope with a higher amount of energy and a higher torque. These are the main reasons why most of such systems in series production cars are only using the front wheel brakes.

However by an actuation of the front wheel brakes solely, the undesired behavior as described in chapter 2.2 cannot be avoided.

In order to maintain a trustable driving behavior without cutting the drivers input sharply, a continuous and defined blending of brake intervention from the front to the rear axle has to be realized. As a second part, the front wheels have to be protected from an exaggerated driving torque by limiting the engine torque according to the driving situation especially on low-µ-surface. At the same time the constriction of the driver's intention has to be as slight as possible.

Summing up, the control strategy is based on a balanced combination of

- adaptive limitation of engine torque to preserve the lateral grip of the front axle
- continuous shift of yaw moment generation to the rear axle for maintaining the handling characteristics applied by the brake torque vectoring of the vehicle.

# 4 Combined actuation concept of engine and brake system

## 4.1 Engine torque control

In order to realize a harmonic driving behavior, it is more convenient to limit the requested engine torque preemptively instead of reducing it on a feedback-based manner.

By restricting the engine torque to the physically transmittable limit with the help of a friction estimation algorithm, an adaptive torque control can be realized in an ordinary way (4.1)

$$M_{\text{Eng,lim}} = \frac{2}{\eta_{total} \cdot i_{total}} \mu \cdot r_{\text{Wheel}} \cdot \max\left(F_{\text{N,fl}}, F_{\text{N,fr}}\right) \quad . \tag{4.1}$$

The main drawback of this simple approach is, that the drivers intention is completely overruled when the maximum frictional connection is reached. This behavior is displeasing especially for drivers, who are able to adjust the correct drive torque better than the basic estimation used for this limitation.

In order to reach a more satisfying driving experience, a more sophisticated algorithm has been developed. The basic idea of this approach is not to limit the driver's input but to manipulate it in a certain direction. Thereby the driver does not have the impression of being overruled or restricted.

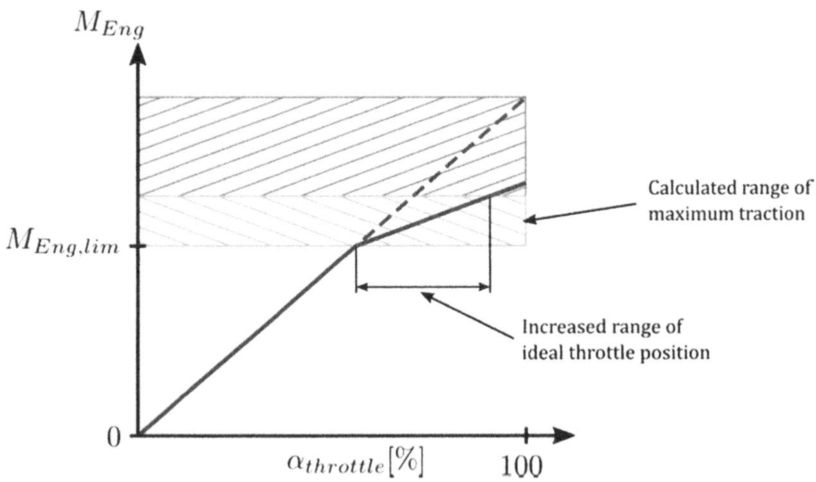


Figure 3: Enhanced torque control with adaptive throttle characteristic

Figure 3 shows the requested engine torque according to the position of the acceleration pedal. In the original application, there is a linear relationship between the engine torque request and the pedal position on the whole range (blue dashed line). The torque which the front axle is able to transmit can be calculated as shown in (4.1). This torque describes the point, where the gradient of the acceleration pedal characteristic is changed from the original to a modified one. Different forms like progressive or degressive development of this characteristic are conceivable. In the example above a static gradient is shown.

Due to several uncertainties, e.g. in calculating the actual wheel load $F_N$ or in estimating the frictional coefficient $\mu$, a range of optimal traction is defined (yellow area). By this adaptive manipulation of the pedal characteristic, the driver experiences a greater sensibility within the threshold. It is still possible to increase the driving torque, but only within reasonable limits.

Especially in winter driving situations this approach shows a major benefit:

The conventional traction control restricts the engine torque to a level, that a specific longitudinal slip according to the tire characteristic is adjusted, without incorporating the driver's excessed input. On snowy tracks it has turned out, that the typical tire force-slip-characteristic does not come into full effect so that the target slip of the traction control is actually too low.

With the new control algorithm, the driver is able to adjust his own target slip through the acceleration pedal very precisely inside the transient area of the tire-slip-curve, which adapts itself according to the current maximum of calculated friction.

The effectiveness can be noticed in figure 4, which shows a comparison between the maximum attainable longitudinal acceleration on the same snowy track. Under those driving conditions it can be shown, that a raised slip level is beneficial for longitudinal acceleration performance.

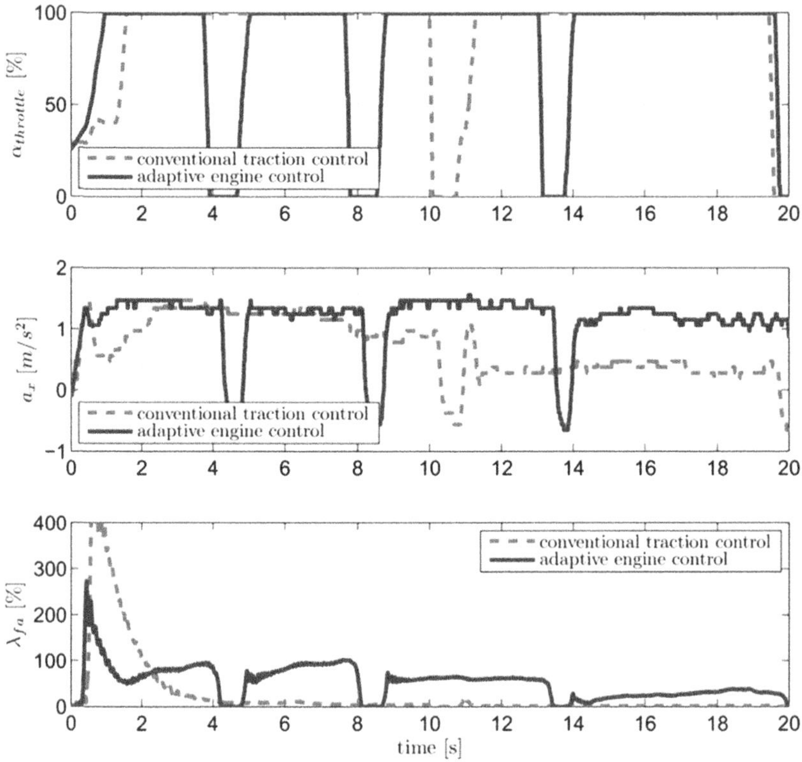


Figure 4: Comparison of longitudinal dynamics on a snow track

## 4.2 Brake torque distribution control

As mentioned in chapter 3.1, a brake control which is based on shift in wheel load induced by lateral acceleration is not constructive. However, the function comes into effect in terms of distributing the required yaw moment as one of two relevant parameter to relieve the front axle of additional torque distribution. The second distribution parameter is the actual factor $k_{M_z}$ that divides the total yaw moment into parts of front and rear axle.

As long as the front axle is in a stable driving state, the generation of yaw moment takes place at this axle. This is more beneficial due to the active distribution of drive torque and the higher achievable torque as mentioned in chapter 3.3. With the front axle reaching the limit of driving dynamics, a continuous relocation of yaw moment to the rear axle should be initiated. The driven axle is close to the limit when the demand of side slip angle is increasing disproportionately. The most conclusive way to determine this behavior is a feedback of the actual yaw rate which is internally compared to the calculated desired yaw rate. The difference between these two signals can be scaled to an inverse understeering ratio with values between 0 and 1. The limits are as follows:

- $k_{M_z} = 1$: no perceptible deviation of actual and desired yaw rate
- $k_{M_z} = 0$: maximum considered deviation of actual and desired yaw rate

In that way the factor $k_{M_z}$ ensures, that the front axle will not be weakened by the additional brake torque-vectoring task. At the same time however the brake torque at the front axle must not decrease below the required level for optimal traction. The wheel-load-based function ensures the last-mentioned fact (4.2).

$$M_{z,\text{fa}} = \min\left(k_{M_z} \cdot M_{z,\text{BTV}}, M_{z,\text{Diff}}\right) \quad \text{and} \quad M_{z,\text{ra}} = M_{z,\text{BTV}} - M_{z,\text{fa}} \quad . \tag{4.2}$$

The equivalent yaw rate of optimal traction is derived by the geometric properties of the two track model according (4.3) with the required brake torque $M_{\text{Br,i}}$ derived in (3.2)

$$M_{z,\text{Diff}} = M_{\text{Br,i}} \left(\frac{1}{2}\, b_{\text{f}} \cdot cos\delta_{\text{f,i}} - l_{\text{f}} \cdot sin\delta_{\text{f,i}}\right) \frac{1}{r_{\text{wheel}}} \quad . \tag{4.3}$$

The entire control strategy is summarized in figure 5.

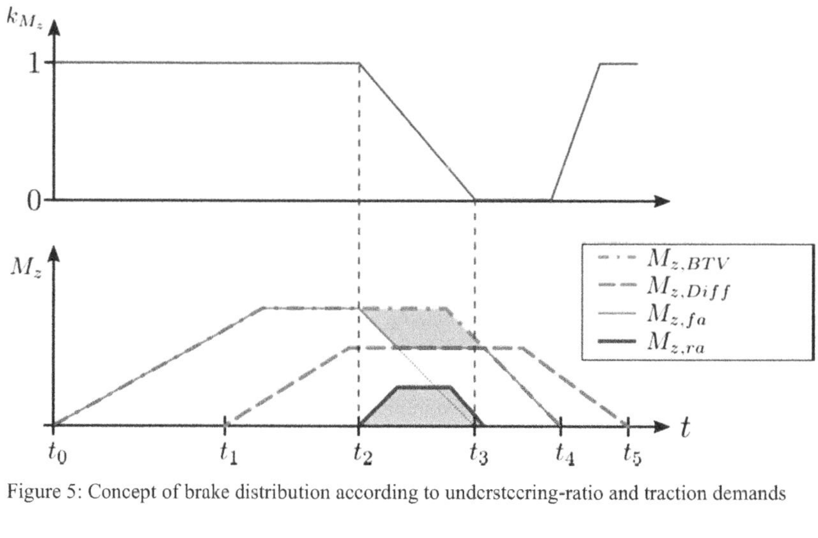


Figure 5: Concept of brake distribution according to understeering-ratio and traction demands

In this hypothetical driving manoeuvre, the driver is entering a curve at time $t_0$. According to the steering wheel angle, the system is requesting a driver supporting yaw torque $M_{z,\mathrm{BTV}}$ . Since the algorithm detects no instability regarding the front axle, the total torque is applied at this one ($M_{z,\mathrm{fa}} = M_{z,\mathrm{BTV}}$). During the steady state phase of cornering, the driver accelerates at $t_1$ which leads to an increasing value of required traction torque $M_{z,\mathrm{Diff}}$ . Due to additional load of longitudinal force, the front wheels start to reach their physical limit at $t_2$. The front axle gets partly discharged from its torque vectoring task until it is kept at a minimum level for optimal traction. The residual torque is applied to the rear wheels. The driver is leaving the corner at $t_4$. Since the traction-based function has no direct access to brake torque requests, the driver is not distracted from accelerating between $t_4$ and $t_5$ as described in chapter 3.1.

# 5 Implementation under real driving conditions

## 5.1 Test vehicle

The experimental vehicle for the real driving manoeuvers is an Audi sedan with front wheel drive and manual transmission. For the actuation of engine torque and brakes individually, a rapid prototyping interface is implemented to the electronic control unit of the brake system. This software interface transfers the incoming brake torque requests into equivalent hydraulic pressures.

The control algorithms are running on a dSPACE micro autobox which is connected via CAN to the rest of the vehicle. Furthermore the car is equipped with additional pressure sensors in each of the four hydraulic circuits from the ESP hydro-unit to the wheel brakes.

## 5.2 Measurement results

For the analysis of the effectiveness of the introduced control concept the driving manoeuver is a section from a winter test track where an acceleration during cornering can be seen. A selection of relevant signals is shown in figure 6.

The first plot shows the steering wheel angle $\delta_{\mathrm{H}}$ and the vehicle's yaw rate $\dot{\psi}$. At the entry of the curve the yaw rate matches the drivers intention quite good, which of course indicates a stable driving state. Since the friction potential is rather low and the driver requests additional engine torque (as noticed in plot 4), the front wheels come to their limit at around $t = 1.5s$. The shift factor $k_{\mathrm{M_z}}$ in plot 2 starts to raise between second 1 and second 3. Thus the yaw moment is relocated from front to rear simultaneously. In the further course, the front yaw moment adopts the minimum level for optimum traction. The remaining yaw moment necessary for the vehicle's handling characteristics is applied to the inner rear wheel. The actual pressures of both inner wheels are shown in plot 3.

During the second half of the manoeuver, the driver is requesting an exaggerated engine torque, which can be effectively controlled without harsh intervention. On the basis of the actual yaw rate it can be noticed, that the vehicle is following the further turning radius of the track well. The last plot shows the longitudinal wheel slip during the manoeuver. Is can be proved that the traction-enhancing function can distribute the brake torque correctly. Especially during the steady-state phase of the manoeuver between $t = 4$ ... $5.5s$ the inner and outer slip of the front wheels are similar to each other. At around $t = 6s$ the driving enhancing function suppresses a further torque intervention which ensures an unimpeded acceleration with evenly distributed friction utilization as indicated by similar wheel slip of left and right wheel also at the end of the manoeuver.

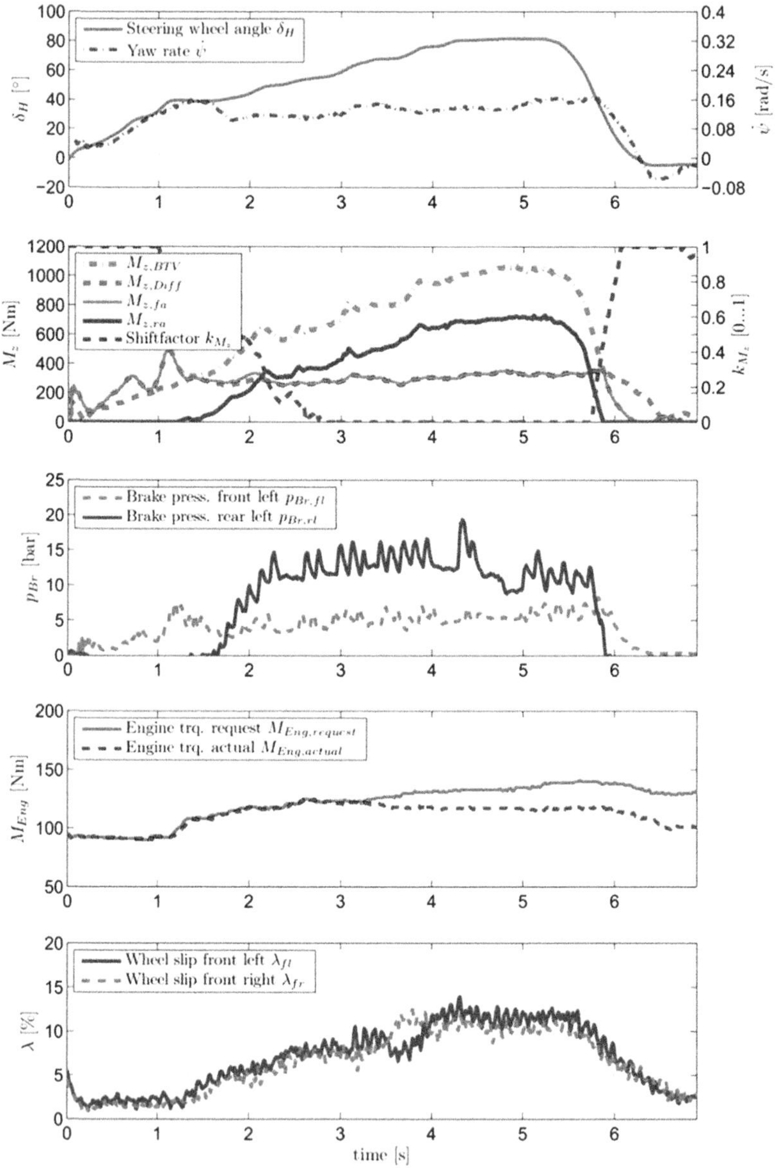


Figure 6: Intervention mechanism under real driving situation during accelerated cornering

# 6 Conclusion

With the introduced approach of a concerted control of engine torque and blending of brake torque vectoring between front and rear axle an effective algorithm has been developed to improve handling characteristics and traction of front wheel drive vehicles. Particularly on surface with low friction potential the driven front axle could be kept on track in a way that control intervention is not perceived as disturbing by the driver. The concept of this approach was first introduced and explained on a theoretical basis. After that the effectiveness was shown under real driving situations.

# Acknowledgements

The authors of this article would like to thank Audi AG for initiating and supporting this research-project.

# List of symbols

## Symbols

| | | |
|---|---|---|
| $F_{\text{long}}$ | [N] | longitudinal force at wheel contact patch |
| $F_{\text{N}}$ | [N] | wheel load |
| $M_{\text{Br}}$ | [Nm] | brake torque |
| $M_{\text{Dr}}$ | [Nm] | drive torque |
| $M_{\text{Eng}}$ | [Nm] | engine torque of the vehicle |
| $M_{\text{Eng,lim}}$ | [Nm] | engine torque limited by the control algorithm |
| $M_z$ | [Nm] | yaw moment of the vehicle |
| $a_x$ | [m/s²] | longitudinal acceleration |
| $a_y$ | [m/s²] | lateral acceleration |
| $b$ | [m] | wheel track |
| $c_{\text{fa}}$ | [N/rad] | cornering stiffness of the front axle |
| $c_{\text{ra}}$ | [N/rad] | cornering stiffness of the rear axle |
| $i_{\text{total}}$ | [-] | total transmission ratio according to the current gear |
| $k_{\text{fa}}$ | [-] | factor for yaw moment distribution |
| $l$ | [m] | wheel base |
| $l_{\text{f}}$ | [m] | distance from vehicle's center of gravity to front axle |
| $l_{\text{r}}$ | [m] | distance from vehicle's center of gravity to rear axle |
| $m$ | [kg] | vehicle mass |
| $p_{\text{Br}}$ | [bar] | brake pressure |
| $r_{\text{wheel}}$ | [m] | wheel radius |
| $v$ | [m/s] | velocity |
| $\alpha_{\text{throttle}}$ | [-] | throttle position |
| $\delta_{\text{H}}$ | [°] | steering wheel angle |
| $\delta_{\text{fa}}$ | [°] | steering angle of the front axle |
| $\eta_{\text{total}}$ | [-] | transmission efficiency |
| $\lambda$ | [-] | longitudinal wheel slip |
| $\mu$ | [-] | friction coefficient |
| $\dot{\psi}$ | [rad/s] | yaw rate |

### Indices

| | |
|---|---|
| BTV | brake torque vectoring |
| Diff | differential |
| f | front |
| r | rear |
| fa | front axle |
| ra | rear axle |
| fl | front reft |
| fr | front right |
| i | inner |
| o | outer |

## References

[1] Volkswagen AG. (2015, May 05). Technical glossary. Retrieved from http://www.volkswagen.de/content/de/brand/de/technologie/technik-lexikon/elektronische_differenzialsperrexds.html

[2] Ford Motor Company. (2015, May 05). Media Kits and Press Materials. Retrieved from http://fiestast.fordmedia.eu/factsheets/fiesta_ST_performance_EU.pdf

[3] Reif, K.: Brakes, Brake Control and Driver Assistance Systems - Function, Regulation and Components, Springer Vieweg, Wiesbaden, 2014

[4] Meißner, T.-C.: Verbesserung der Fahrzeugquerdynamik durch variable Antriebsmomentenverteilung, Dissertation, Cuvillier Verlag, Göttingen, 2008

# Real-time simulation of braking interventions in heavy commercial vehicles

Dr.-Ing. Philipp Wagner (1)
Development Engineer Driving Dynamics Simulation

Dipl.-Ing. Florian Bauer (2)
PhD Student

Dipl.-Ing. Thomas Ille (3)
Department Manager Vehicle Dynamics & Simulation Technology

Dr.-Ing. Christian Kohrs (4)
Central Division Manager Research

(1, 3, 4) MAN Truck & Bus AG, Engineering Research,
Dachauerstr. 667, 80995 München

(2) Institut für Mechanik, Universität der Bundeswehr München,
Werner-Heisenberg-Weg 39, 85577 Neubiberg

# Introduction and motivation

Increasing efficiency, safety, reliability and comfort of vehicles is a crucial task in the commercial vehicle industry. Considering the wide range of possible vehicle applications, it is obvious that commercial vehicle manufacturers deliver a wide range of vehicle variants to meet customer's needs. Offering a full range vehicle portfolio can lead to more than 50.000 chassis variants differing in axle configurations, wheel bases, suspensions, engine and gear specifications, cabin sizes, and fuel tank options, to name only a few [8]. Fulfilling high quality standards leads to an increasing development and testing effort that can be supported by virtual methods to optimise development process efficiency. Some examples of MAN and NEOPLAN products can be found in figure 1.

Figure 1: Some examples of MAN and NEOPLAN commercial vehicles.

In this paper, one specific example of this optimisation process considering brake systems is described. Heavy duty vehicle brake systems use pneumatic circuits to deliver braking power to the wheels. Advantages are the high density of energy, the availability of the medium, the easy and clean option to connect trailer brake systems, and the use of compressed air for further commercial vehicles subsystems such as driver seat suspension, air suspension and door actuation of buses. In any case, using a source for brake energy is necessary in heavy duty vehicles since human force is insufficient for braking [8]. Nowadays, electro-pneumatic brake systems can be regarded as state of the art. In addition to the redundant pure pneumatic fail safe brake function, an electronic control system operates the brakes "by wire". Pressing the brake pedal is interpreted as a deceleration request. The amount of deceleration controls the brake pressure and therefore the brake torque. Thus the driver does not influence the air pressure by means of a direct feedthrough. This configuration offers a wide range of additional opportunities such as the optimised cooperation between pneumatic brake system, engine brake and retarder. Furthermore, the interaction with other electronic control units is used to enhance the vehicle functions. One example is the activation of the brakes by the adaptive cruise control [8]. For more detailed information on brake systems in general see for example [3].

To optimise brake functions and driving dynamics for the huge variety of chassis variants it is necessary to support the testing department with simulation and virtual methods. Considering critical driving situations with interactions of the anti-lock braking system (commonly referred to as ABS) or the electronic stability control (ESC) interactions, a sufficient representation of the driving dynamics as well as the function of the electronic and pneumatic brake system have to be provided in a simulation model.

Furthermore, testing and homologation procedures can be efficiently supported by hardware-in-the-loop (HiL) simulation. This has been presented for passenger cars, for example by Holzmann et al. [4]. Wurster et al. presented a HiL simulation based approach for the homologation of brakes for trailers [12].

In this paper, an approach is presented that combines a system dynamic vehicle model with a simulation model of the pneumatic brake system. The electronic side of the brake system is represented by the real electronic control unit of the brake, which requires a HiL environment. Therefore, a crucial demand for the whole system is real-time capability. In this way models from different physical domains are connected virtually and with real hardware in a real-time environment. The structure of the simulation environment will be presented, as well as some details of the simulation models. First results will be shown simulating driving situations including ESC interactions.

# System requirements and structure

This section covers the description and explanation of the vehicle systems. The system requirements and structure of the hardware and software system will be derived. As mentioned before, real-time ability for all components is crucial. Although successful approaches have been performed using complex real-time multi-body simulation software for vehicle dynamics simulation [9], it is still very common to use system dynamic models for real-time simulation.

## System dynamic models

System dynamic models have been established as a good compromise for real-time simulation. The different software packages offer a basic vehicle model that can be parametrised in a wide range. The complexity of the model itself, considering the number of differential equations to solve, is smaller than given by a comparable multi-body model to achieve real-time capability. However, this advantage also leads to reduced precision in certain results compared to a highly detailed multi-body model. One example is the limited possibility to simulate the exact function of some suspension systems such as rigid axle configurations of commercial vehicles. Here the given interaction between left and right wheel is hardly describable in the independent wheel-carrier based elasto-kinematics templates. On the other hand, these software packages offer comfortable opportunities to generate tracks, roads, environments and drivers. In this work, a system dynamic model will be used.

To generate the characteristic maps for the axle suspensions, data has to be derived from kinematics and compliance test rigs, either in reality or in simulation. In this case, the detailed axle submodels from a multi-body vehicle model have been measured in a virtual test rig. A picture of the virtual test rig with a leaf spring front axle as an example is given in figure 2.

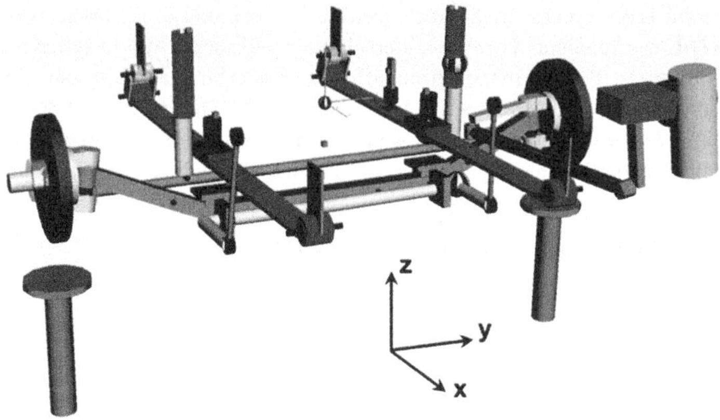


Figure 2: Kinematics and compliance test rig with an example of a leaf-spring front axle.

dSPACE ASM offers a template for rigid axles that requires characteristic maps showing the reaction motion on roll angle and spring deflection input as well as forces applied to the wheel centre points. Performing the kinematics and compliance analysis with the front axle given in figure 2, translations and rotations of the wheel center points are given depending on the roll angle α and the deflection z in the middle of the axle. These two motions have to be calculated from the deflection of the left and right wheel centre points, as these are the test rig inputs. Some examples of the characteristic maps are given in figure 3.

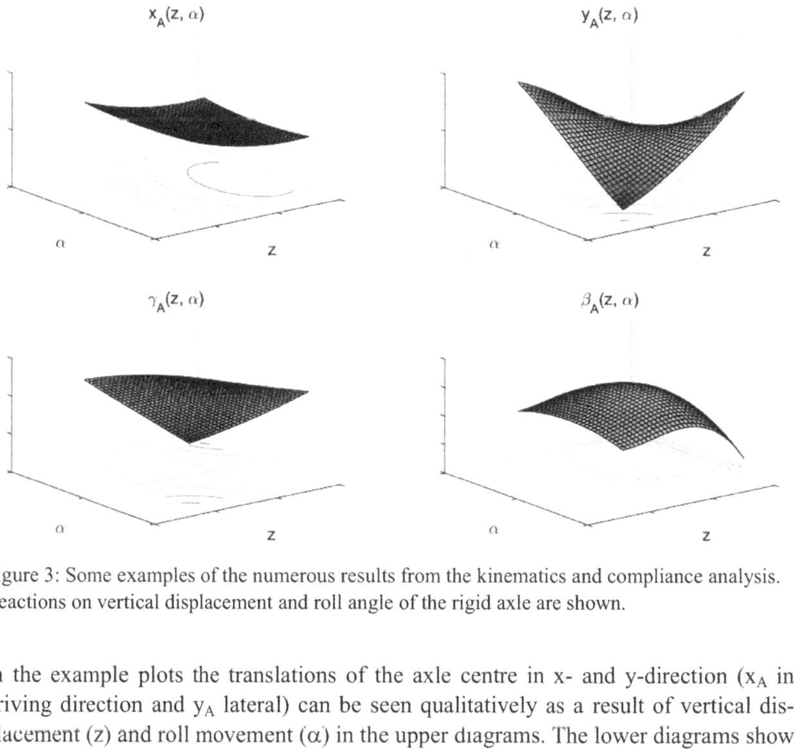


Figure 3: Some examples of the numerous results from the kinematics and compliance analysis. Reactions on vertical displacement and roll angle of the rigid axle are shown.

In the example plots the translations of the axle centre in x- and y-direction ($x_A$ in driving direction and $y_A$ lateral) can be seen qualitatively as a result of vertical displacement (z) and roll movement (α) in the upper diagrams. The lower diagrams show yaw ($\gamma_A$) and pitch ($\beta_A$) movements around the z- and y-axis of the axle as a function of z and α. Yaw movements, although they might seem unexpected at first glance, only occur for combinations of z and α. Having either vertical displacement or roll movement only, the projected straight lines under the surface show that yaw movements of the axle body are negligible in this case. In the same manner, the rear axle has been examined. The results of the kinematics and compliance analysis for the axles are processed for the use in the system dynamic vehicle model in appropriate look-up tables. As regarding all possible combinations of displacements, roll angles and steering angles for each wheel would lead to an enormous amount of data, only a certain set of independent look-up tables is generated. Under the assumption of linear superposition combined motions can be calculated. This is a major disadvantage compared to a multi-body model, where the differential equations of motion of the system are solved regardlessly of the complexity of the elasto-kinematic structure.

The performance of the system dynamic vehicle model, compared to a validated and highly sophisticated multi-body model, is shown in the following two figures. Figure 4 illustrates the steering wheel angle (upper diagram) and the roll angle (lower diagram) against the lateral acceleration for an accelerated circle at a constant radius. The vehicle models are operated closed-loop with appropriate driver models.

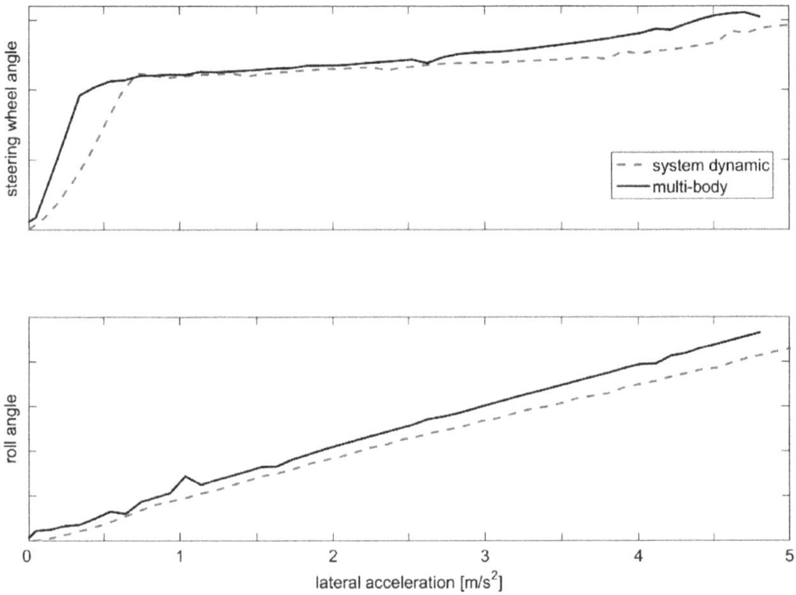


Figure 4: Comparison of the system dynamic and the multi-body model during driving an accelerated cirlce at a constant radius.

On a global scale, the system dynamic model shows satisfactory behaviour, however the steering angle demand is smaller compared to the multi-body model. The differences for small lateral accelerations are caused by the different driver models and their acceleration procedures and are therefore not relevant for the evaluation of vehicle dynamics. From figure 4 it can also be seen that the system dynamic model reaches smaller roll angles.

Figure 5 shows a more dynamic manoeuvre (sine-with-dwell). Here, instead of a driver model, a constant initial speed of 70 km/h and a kinematic steering input are given, so the vehicle is operated in an open-loop mode. The input steering angle for both models can be seen in the upper diagram.

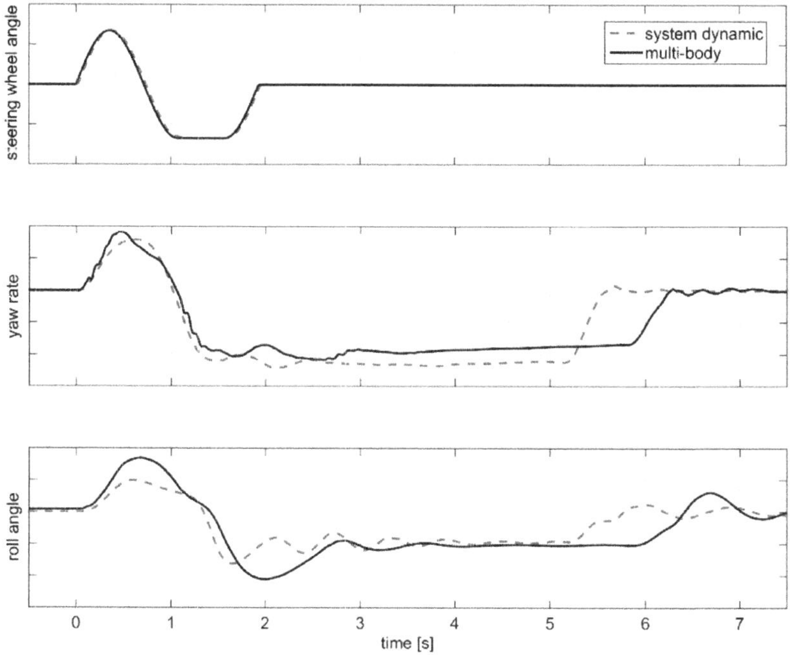


Figure 5: Comparison of the system dynamic and the multi-body model during a sine-with-dwell manoeuvre at an initial speed of 70 km/h.

Again, the global behaviour can be called sufficient, however there still are minor local differences in yaw rate and in roll angle. The main difference in this example shows in the time between 5 and 6 seconds. Here the system dynamic model ends the sliding phase earlier. During this extreme driving manoeuvre, even minor differences in modelling tires and axle kinematics can lead to these deviations. However, both models use the TMeasy tire model with the same parametrisation, so from this side no relevant differences should be expected. Detailed information on the tire model itself is given by Rill [10].

## Pneumatic submodel

It is a very common approach to combine a vehicle model with a real electro-pneumatic brake system in HiL simulators. However, the effort to rebuild a complete commercial vehicle brake system or even variants for different axle configurations in the laboratory is very high. For that reason a simulation model for the electro-pneumatic brake system is desirable to avoid expensive and complex hardware. Representing the pneumatic system in a model is not trivial, considering the required level of precision and obtaining real-time capability.

As the brake system is in the focus of interest in this work, after the general comments on commercial vehicle brake systems in the introduction a more detailed description will be given. A possible implementation of a commercial vehicle brake system is given in figure 6. The illustration only shows the relevant components for the current work and neglects any additional elements. The air tanks are connected to the supply pressure to guarantee sufficient air reserves for braking maneuvers at any time. Pneumatic connections lead from the air tanks to the electronic pressure modulators (EPM), where the brake pressure is generated for one wheel or one axle. In the latter case, having just one EPM per axle, to generate individual brake pressure for each wheel in case of ABS brake maneuvers, additional pressure control valves (PCV) can cut the brake pressure to avoid wheel lock-up. The pressure in the brake cylinders pushes the pistons and generates a brake torque at the brake discs. Both EPM and PCV are operated by the electronic brake system control unit (EBS) [11]. Based upon these different subsystems, also more complex configurations can be generated with limited effort.

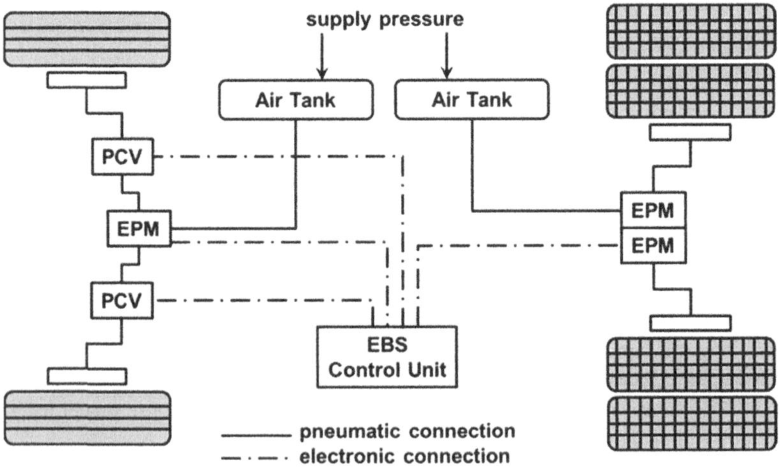


Figure 6: Simplified overview over a possible commercial vehicle brake configuration.

As mentioned before, for simplification reasons different parts of the brake system as for example the pneumatic backup system or the vehicle sensors are not shown.

The main modules of the pneumatic model shall be described briefly. First of all, the pneumatic medium air is modeled following the ideal gas law. To model the air tanks or vessels the first law of thermodynamics, the conservation of energy, and the conservation of mass can be used. The brake cylinders follow the same mathematical descriptions; however, their volume is not constant over time. To describe mass flow across the system borders the first derivation of the conservation laws is used. In this work, extended mathematical deductions that build the base of the model shall be avoided. An excellent overview on modelling pneumatic systems that also gave precious information for this work has been given by Beater [2]. More details on the described pneumatic model can be found in [1].

In the given approach the mass flow rate through the valves is modelled according to ISO 6358 [6]. The solenoids of the valves can be described very easily. They open above a certain voltage and close below that voltage. The delay of the solenoid valves is very small, however regarded as a first order delay.

Finally, the brake calipers transform the brake pressure into a braking torque at the wheel hub. The brake membrane cylinder generates a force at the piston which is amplified over a lever before moving the brake pads to the disc. Brake pad force and friction at the brake radius generate the desired brake torque. As identification of friction and damping within the brake mechanics is laborious and not reliable, a simpler approach has been chosen. The piston stroke was measured with respect to the pressure in the brake cylinder in a static experiment. As long as the piston is closing the clearance between brake disc and brake pad, small pressure changes result in comparably high piston displacement. When the pad contacts the disc, very small piston stroke comes along with a significant increase in pressure. Considering this extreme non-linear relation between stroke and pressure, a look up table rather than a mechanical model with the first order explicit solver is used here. A first order delay is used to account for effects of inertia and friction and to avoid numerical instabilities. All other assumptions for modeling the components have been validated through component measurements. Time delay in the pneumatic transition lines can be modelled by a first order delay.

To test the general function of the brake pneumatics, the real vehicle brake system was compared to the pneumatic model, both controlled by the vehicle model running on the HiL simulator. Typical results can be seen in figure 7.

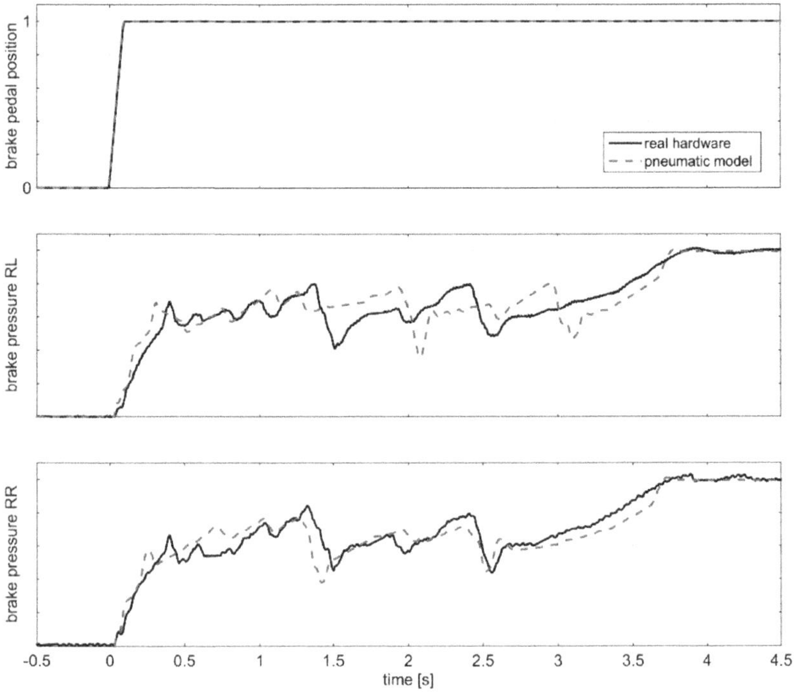


Figure 7: Braking manoeuvre from 70 km/h on high friction underground with ABS interaction, comparison real brake and pneumatic model.

A brake manoeuvre from 70 km/h to zero is used as an example. The friction of the road surface is high, however there is ABS activity. In the first diagram the brake input for both configurations is shown which is identical. The brake pressures of the rear left (RL) and rear right (RR) brakes are shown in the second and third diagram. In general, the simulated pressure corresponds very well with the real pressure for the left brake, only few ABS interactions do not occur at the same time. Having in mind that due to different reasons arising from the communication of several electronic components, HiL simulations are not totally repeatable, as can be found in [5], the shown results can be called satisfying. For the right brake the correlation is even better and has only minimal differences, the corresponding releases of local brake pressure are easy to find. All in all, this example, as well as several more results, render the pneumatic model applicable for the next steps towards the full simulation environment.

## Architecture of the simulation environment

After the description of the system dynamic and the thermodynamic submodels a short overview of the complete system will be given. Figure 8 shows the schematic architecture in a simplified way. The HiL simulator runs the system dynamic vehicle model as well as the electro-pneumatic brake model that also communicate with each other. If a brake request is generated by the driver model, the signal condition module of the HiL simulator communicates this request to the EBS control unit of the brake. The electronic control unit (ECU) calculates the required brake pressure and communicates it to the EPMs that are connected to the HiL simulator. Brake requests from the ECU, generated for example in case of ABS or ESC situations, are communicated in the same manner.

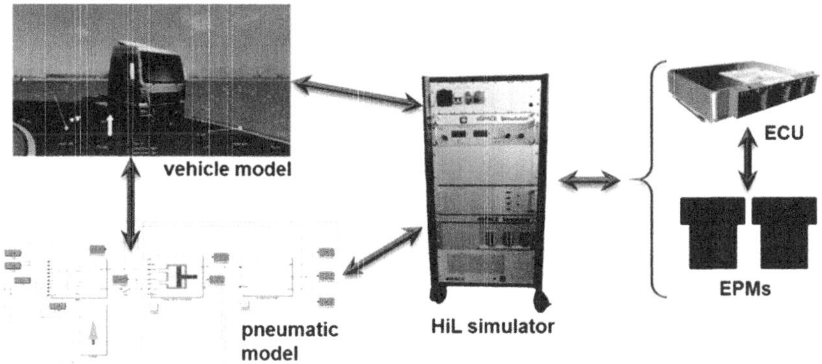


Figure 8: Schematic architecture of the simulation environment.

The HiL simulator transfers the signals to the pneumatic model where brake pressures and brake torques are generated to influence the vehicle model. The change of the motion of the vehicle model is continuously reported to the EBS control unit.

The modular structure of both the system dynamic and the pneumatic model allows a simulation environment where different vehicle variants can be generated automatically from given basic modules according to the specifications of the real vehicles. These model variants can then be subjected to different driving manoeuvres. Having predefined limit values for interesting result parameters, also automatic result interpretation is possible to a certain extent. Similar systems for development and homologation processes, however without the integration of hardware components, have already successfully been implemented by Litter et al. [7].

# First results

After the validation of the pneumatic and the vehicle dynamic submodels first experiments within the complete HiL environment have been performed. One example of a successful simulation result can be found in figure 9. The manoeuvre of interest is an open-loop sine-with-dwell at an initial velocity of 70 km/h. In this case, the system dynamic model is used in combination with the EBS control unit in the two parameter configurations ESC on and ESC off.

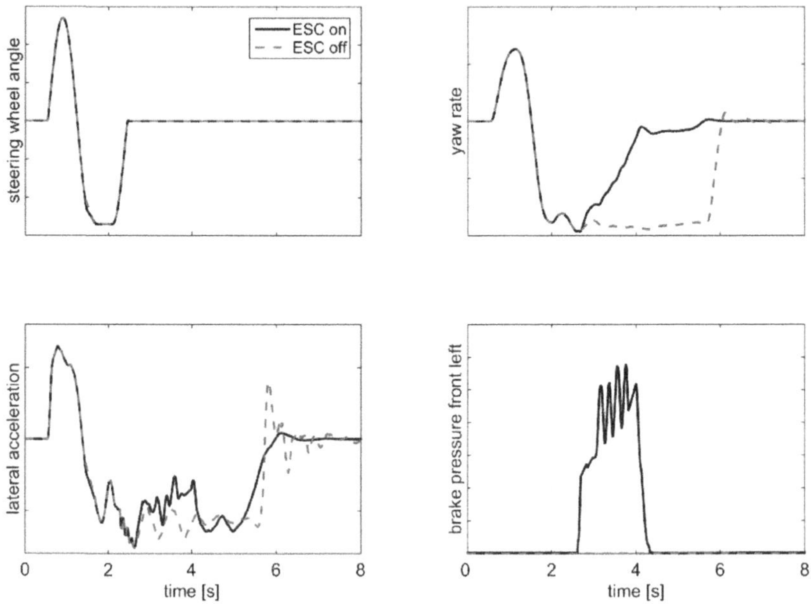


Figure 9: Sine-with-dwell manoeuvre at 70 km/h initial velocity with the system dynamic model in the HiL environment. Comparison of the vehicle reaction with ESC activated and deactivated.

The upper left diagram shows the steering angle imposed upon the vehicle model which is the same for both variants. In the upper right diagram the yaw rates of the two variants can be seen. After the end of the steering input at about 2.5 seconds, the vehicle model without ESC still shows a high yaw rate that does not decrease for several seconds, which means that the vehicle model is still sliding. In comparison to that, the vehicle model with activated ESC shows a rapidly decreasing yaw rate and a much more stable behaviour. In the lower left diagram, the lateral acceleration for both variants can be seen. Without ESC there is a significant overshoot at about 6 seconds that does not occur with ESC. The last diagram (lower right) finally shows the brake pressure caused by the EBS control unit. Whereas no braking occurs in the vehicle model with deactivated ESC, the model with ESC generates a brake pressure at the left front wheel at the end of the steering input. During the ESC interaction, the ABS is also activated. This can be seen in the brake pressure oscillations in the lower right diagram of figure 9.

An example of the user interface appearance during simulation is given in figure 10, showing the vehicle in motion from different views in the upper part. Vertical wheel forces are symbolised with yellow vertical arrows, longitudinal forces (in this case brake forces) are symbolized with red arrows. The dashboard in the lower part of the image contains both vehicle-look-alike instruments as well as driver input elements for manual driving or the initiation of predefined driving manoeuvres. As the shown manoeuvre leads to an ESC interaction, the warning light is enabled on the virtual dashboard to indicate the control system activity.

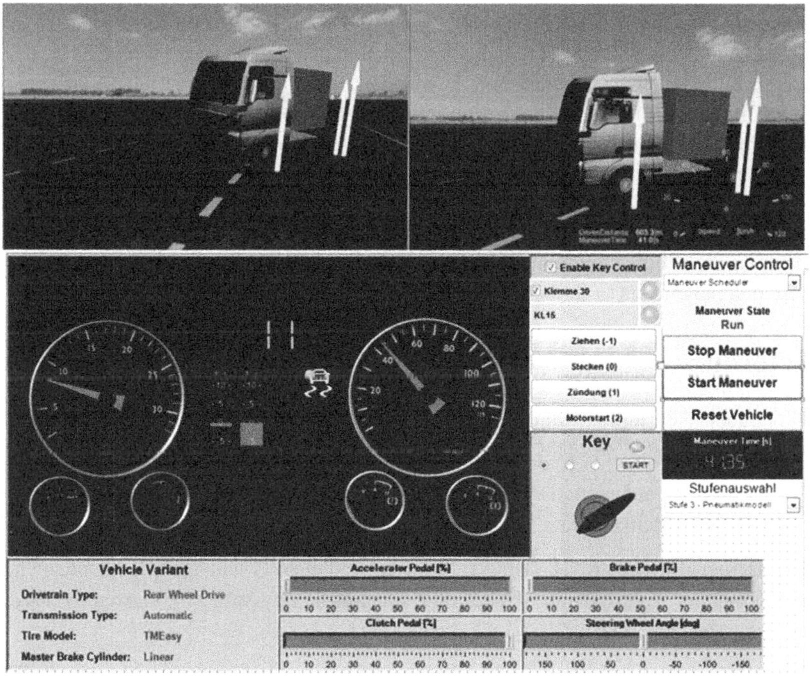


Figure 10: Animation of the sine-with-dwell manoeuvre with active ESC, vehicle from different perspectives and virtual dashboard.

## Conclusion and outlook

Increasing development efficiency is a crucial topic in the commercial vehicle industry, especially considering the high number of available chassis variants. In this paper, a HiL simulation environment for the simulation of brake interventions in heavy commercial vehicle applications has been presented. In this environment a system dynamic vehicle model, that has been accurately parametrised and compared to a validated multi-body vehicle model, has been combined with a physical model of an electro-pneumatic brake system of a commercial vehicle. Both models are real-time capable to enable the integration of a real EBS control unit within a HiL simulator. Some model properties and details have been described as well as the simulation architecture. Examples of the model accuracy have been given. First results show that ESC and ABS interactions occur and successfully stabilise the vehicle model. A significant advantage of the presented system is that it is possible to omit real pneumatic brake systems in the development environment and thus increase efficiency and flexibility.

Future work comprises tests with real-time capable multi-body vehicle models. For the presented case this means that the development and maintenance of models in different simulation environments could be avoided, leading to higher efficiency and less effort. In this way it is possible to concentrate on high sophisticated multi-body models with different levels of detail in one modeling environment. Furthermore, the simulation environment shall be extended to enable the automated simulation of different vehicle variants to cover the immense variety of available vehicle configurations. The simulation environment can then be used for homologation purposes, for the virtual optimisation of brake components and systems as well as the optimisation of control loops and parameters to further increase overall system performance.

## References

[1] Bauer, F.; Lion, A. Modeling of electro-pneumatic brake systems for real-time simulation. 85th GAMM Annual Meeting, Erlangen, 2014.

[2] Beater, P. Pneumatic Drives. System Design, Modelling and Control. Springer, Berlin Heidelberg New York, 2007.

[3] Breuer, B.; Bill, K. H. (Hrsg.). Bremsenhandbuch. Grundlagen, Komponenten, Systeme, Fahrdynamik. ATZ/MTZ-Fachbuch, 3. Auflage. Vieweg Verlag, Wiesbaden, 2006.

[4] Holzmann, H.; Hahn, K. M.; Webb, J.; Mies, O. Simulationsbasierte ESP-Homologation für PKW. ATZ 114 (2012), no. 9, p. 698 – 702, September 2012.

[5] Isermann, R. Fahrdynamik-Regelung. Modellbildung, Fahrerassistenzsysteme, Mechatronik. ATZ/MTZ-Fachbuch. Vieweg+Teubner Verlag, Wiesbaden, 2006.

[6] The International Organisation for Standardisation (ISO). Pneumatic fluid power – Determination of flow-rate characteristics of components using compressible fluids – Part 1: General rules and test methods for steady-state flow. ISO 6358-1, 2013-05.

[7] Litter, S.; Fleischhacker, J.; Ille, T.; Weinfurter, H. Model Based Homologation Process with SIMPACK. SIMPACK User Meeting 2014, Augsburg, October 2014.

[8] Grundlagen der Nutzfahrzeugtechnik. Basiswissen Lkw und Bus. MAN Nutzfahrzeuge Gruppe, München, 2008.

[9] Mansvelders, R.; Trautenberg, W. SIMPACK Realtime. SIMPACK News, SIMPACK AG, Gilching, July 2013.

[10] Rill, G. Simulation von Kraftfahrzeugen. Fundamentals and Advances in the Engineering Sciences. Vieweg, Braunschweig, Wiesbaden, 1994.

[11] Wallentowitz, H.; Reif, K. Handbuch Kraftfahrzeugelektronik. Grundlagen, Komponenten, Systeme, Anwendungen. ATZ/MTZ-Fachbuch. Wiesbaden, Vieweg & Sohn Verlag, Wiesbaden, 2006.

[12] Wurster, U.; Ortlechner, M.; Schick, B.; Drenth, E.; Crawley, J. Simulation-based homologation of brakes for trailers. ATZ worldwide 112 (2010), no. 9, p. 48 – 52, September 2010.

# Efficient digital development of brake components with multiple requirements

Konrad Meister, Dr. Tobias Rößler, Dr. Victor Fäßler,
Dr. Stefan Staudacher

# 1 Abstract

This paper is addressed on efficient digital development on the example of brake components using TOKEN, a tool developed at TWT GmbH in Stuttgart. The main features of TOKEN are explained and an insight on how to use the tool is given. To demonstrate the potential of the software it is shown how TOKEN can be integrated e.g. in the development of a brake dust shield. In doing so it is presented how TOKEN can be used to provide additional information about aero- and thermodynamics in a defined design envelope. We present how this knowledge is able to lead to a more mature concept of a brake dust shield in the early development process. The maturity of the concept is verified in a detailed three-dimensional coupled CFD based analysis of the design and a comparison to an already existing brake dust shield.

# 2 Introduction

Nowadays, the design process of systems and parts often consists of a succession of multiple design and validation phases. This leads to a time-consuming overall development process. In order to reduce turnaround times in the overall design process TWT GmbH proposes a tightly coupled design approach. This approach can decrease the number of necessary iterations between design and validation significantly. The development of a brake dust shield serves as an example to illustrate how the digital development of brake systems can be improved with this approach.

Like most components of a car, a brake dust shield has multiple requirements at the same time. On the one hand it is supposed to prevent the rims from brake dust pollution and on the other hand it shouldn't influence brake cooling and aerodynamic qualities of the car. Considering these requirements in combination with design envelope limitations is a challenging task, which has to be solved in a coupled process between design and validation. As geometry focused approaches are well known to the design engineer while physical processes are not, a transfer of knowledge from the simulation engineer to the designer is needed.

The transfer of knowledge can be achieved with the proposed TWT-software tool TOKEN (Mangelsdorf, 2015). TOKEN holds all relevant simulation results and provides them concentrated to the design engineer by using problem characteristic parameters and three-dimensional data. Hereby, the design engineer receives a better understanding of the technical impact of different designs without the need to understand the technical details of the simulation process.

# 3 Description of TOKEN

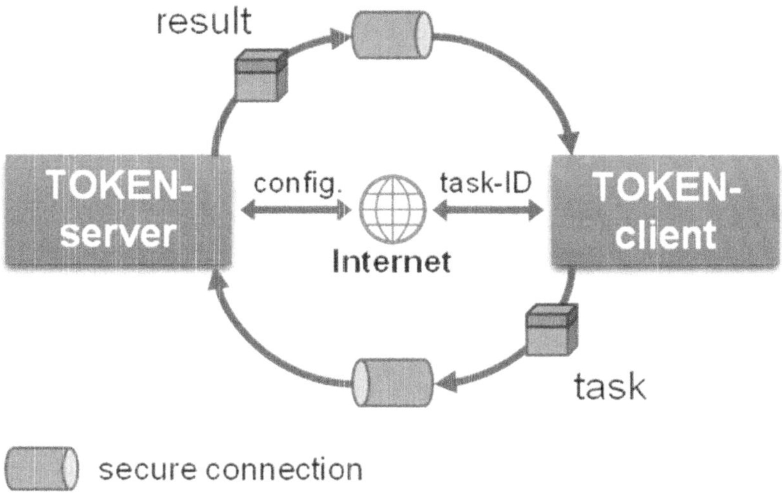


Figure 1: Server-client structure of TOKEN

TOKEN was developed at TWT GmbH Stuttgart and is used as an exchange platform between two or more development parties. TOKEN is realized as a server-client-based application environment consisting of two main components (see Figure 1). The server-side component provides a central database to store simulation job data and user information. Typically the server component can provide several clients with the requested information while limiting each client's data access to specific data. The client component is a windows application running on the user's local machine. Communication between client and server is performed by using a secure connection and an internet connection. While only limited, non-critical data for authorization and about minimal assignment information is transferred via the internet, major and confidential data is sent via a secure connection (see Figure 1).

## 3.1 The client-interface of TOKEN

To use the client-interface of TOKEN an assignment and an authorized user-ID is needed, which are both created by the administration-team of TOKEN. Afterwards the user is able to log in. The client serves as frontend to the services the web component provides and lets the user add, track and view simulation tasks. When the user logs in a brief overview of all tasks of the assignment is given containing the task-ID, status, a description of the task, a token-value[1], and the last modification date (Figure 2). To support the user changes are presented in bold letters and tasks with simulation results are highlighted by a green background color. Besides the task information the user is able to view details or results of a selected task. Moreover, the user is able to add a new task.

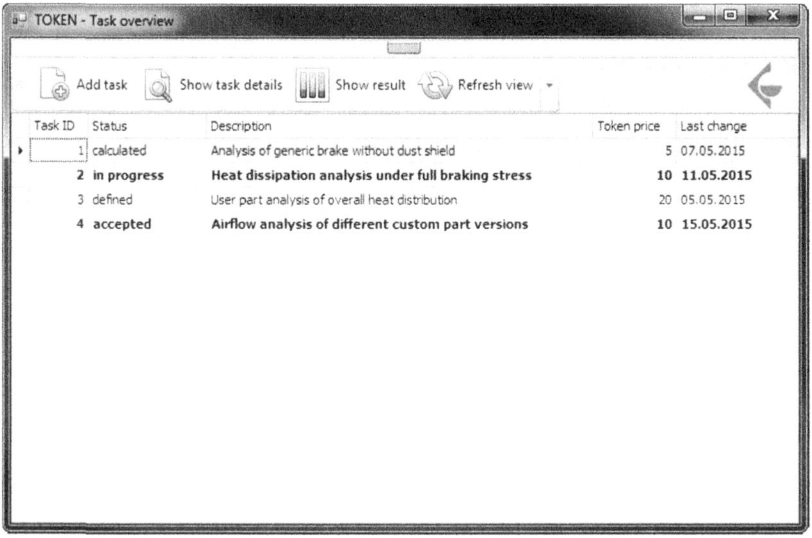


Figure 2: Overview of current simulation tasks after login

Adding a new task is done via the task editor (Figure 3), which allows the user to define a task in three steps. In the first step the user has to define the components (geometry containing files) for the simulations, while the second step is used to specify the boundary conditions. Therefore, the user can choose predefined boundary conditions which are commonly used in his assignment or he can define custom boundary conditions by filling in the predefined constraints panel. In the last step the user receives an overview of the newly defined task and is able to add a description. If the user confirms the task description the task is sent via the secure connection to the simulation engineer and is added to the task overview in Figure 2. After confirming a new task the simulation engineer has the possibility to contact the user to get additional information about the task or to clarify outstanding questions.

1 A token-value is a virtual currency which can be used to determine and bill the costs of a task.

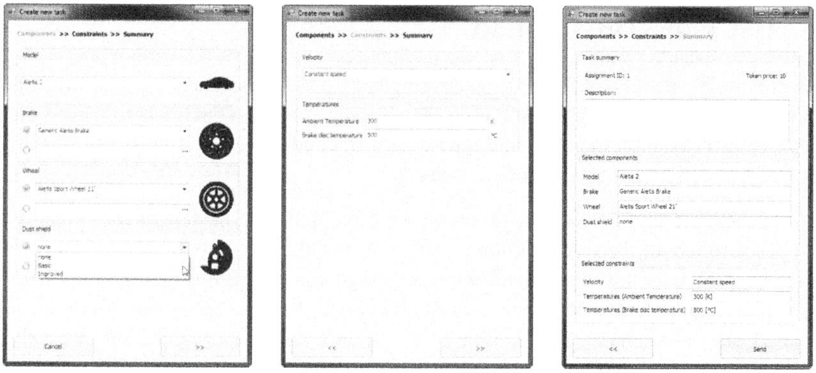

a) Component definition b) Constraints definition c) Summary

Figure 3: Three steps to define a new task using TOKEN

When a simulation result is available it can be opened through the task overview menu. This menu (Figure 4) is divided in three regions. On the upper left it features a display engine capable of depicting WebGL data to provide the user with multiple interactive 3D representations of the simulation results. Alongside a brief description of the simulation is given, while below textual representations of the main results (e.g. maximum temperatures, average pressures etc.) are listed. Thus, the result screen provides relevant data in concentrated form to the user.

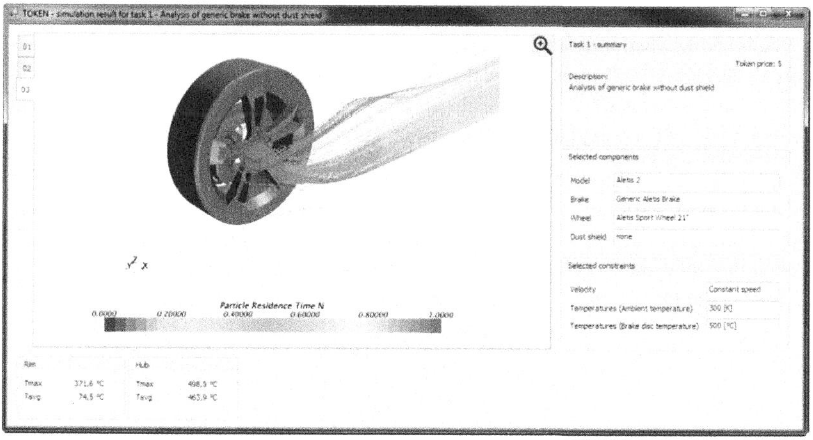

Figure 4: Result window with 3D representation of simulation results

# 4 Use Case Dust Shield

As previously mentioned, TOKEN is developed by TWT as an exchange platform between two or more development parties. The keynote behind TOKEN is to reduce the development iteration loops of a product by a structured documentation and transposition of simulation and development results.

To demonstrate the TOKEN process a brake dust shield of a passenger car, which is supposed to reduce the rim pollution by brake dust is chosen. As base of operations an already existing brake dust shield, which is plugged subsequently between the rims and the brake system is redesigned. Dust shields of this type are already offered for sale e.g. in online markets. This type of brake dust shield is designed subsequently and can be compared with the starting point of the iterative optimization of a car development process. For this reason it is an ideal comparative case to a development process improved by TOKEN.

In this paper it is assumed that the brake dust shield has been developed only with regard to its target function to prevent the rims from brake dust pollution and that there is no real thermal an aerodynamic optimization of the shield. This "stand-alone solution" is, concerning the optimization process of the whole car, a worst case example because it has been developed encapsulated.

In an iterative development process of a passenger car, the mentioned dust shield could be seen as a first draft of the design engineer, standing at the beginning of a development chain. Using TOKEN consequently can lead to a more mature version of this first draft.

Below it is described how a brake dust shield can be designed, also with respect to requirements of aerodynamics and heat protection, applying TOKEN in a fictive development process. But first of all we do an analytical consideration of the dust shield mentioned before and explain why there is capability for optimization.

Figure 5 shows a redesigned "stand-alone solution" of a dust shield as it is actually offered e.g. in online markets. If we have a closer look we can notice, that there are cooling openings in the dust shield. The arrangement of the openings leads to the conclusion that they are supposed to transport fresh cooling air from the car outer flow to the brake system.

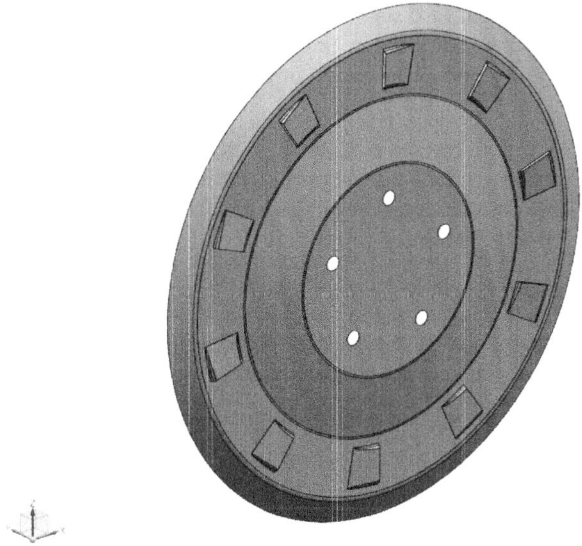

Figure 5: "Stand-alone solution" dust shield

It is obvious, that the brake dust shield is blocking the cooling air flow for the brake system because it is covering the whole rim opening. The shield openings which are arranged like rudimentary rotor blades are an attempt of the design engineer on the one hand to keep the brake dust particles inside the wheel and on the other hand to re-establish the cooling air flow by transporting air from the flow around the vehicle into the rims for a better cooling of the brake system.

This erroneous assumption is caused by a lack of information. The crucial information the design engineer has needed, is that there is a pressure gradient from the inside of the wheels to the outside and that therefore the air flows from the underbody and the engine compartment through the wheels to the vehicles outer flow. This effect can be explained by Bernoulli's law which indicates that an increase in the speed of the fluid occurs with a decrease in pressure. The car body causes a strong displacement of the air flow. As a result the air is accelerated and the local static pressure besides the vehicle body is falling. This means, that the shown dust shield is indeed very effective in preventing the outer rims from brake dust pollution (Figure 15) but is counterproductive concerning the brake cooling because of transporting the air against the natural flow direction into the inside of the rims.

To underline this statement, several CFD simulations are performed to show in a first step the pressure gradient and the particle flow next to the brake system of the base configuration and in a second step the overheating of the brake environment caused by the dust shield.

The results from the first step will be used as input for TOKEN even before the first draft is designed. These results are the basic information for the design engineer and provide the crucial indication that the natural direction of the cooling air flow has to be maintained to ensure the brake cooling.

The results of the second step describing the overheating caused by the dust shield can be provided in our fictive development process via TOKEN as results from a former iterative optimization process of a brake dust shield. That prevents, that development work is done twice and that mistakes will be repeated respectively.

In a third step design and CFD results from an improved dust shield are analysed and presented.

## 4.1 Numerical description

The simulation process and the approach to get the CFD results are not in the focus of this paper. The CFD results are only used to show the functionality of TOKEN. Anyway, the boundary conditions and model details of the CFD simulations are to be described in the following subchapters.

### 4.1.1 Code description

For the simulation of aerodynamics, heat transfer and brake dust particle flow the commercial CFD code StarCCM+ is used. StarCCM+ is a RANS based code which offers various physical models to deal with challenges arising from the application mentioned before. The models which are used in this investigation are specified below.

### 4.1.2 Setup and model description

As mentioned before we will have a look at three simulation cases. In the table below the three cases are listed. The essential case for the design engineer is CASE A, because it gives him fundamental information about the flow condition in the component environment.

Table 1: Model description of the analysed cases

| | CASE A | CASE B | CASE C |
|---|---|---|---|
| Model setup | Without dust shield | - With dust shield<br>- front openings<br>- short slots | - With dust shield<br>- back openings<br>- long slots |
| Case description | Basis | Stand-alone solution | Improved dust shield |

The brake dust shields are adapted to the rims of the car "ALETIS2". The chassis and the brake system of the model are simplified as well as the wheels and the engine compartment. This approach is chosen due to high computational effort arising from the simulations and is completely sufficient to demonstrate the flow conditions around the dust shield.

To simulate the brake heating a conjugated heat transfer approach is used. Therefore the brake disc, the brake caliper, the wheel hub and the suspension are modelled as solid, to represent the heat conduction in the components of the brake system and of its periphery. For the simulation of the brake dust particles, the Lagrangian Multiphase model is used. This model solves the equation of motion for representative parcels of the dispersed phase as they pass through the system. It is intended for systems that consist mainly of a single continuous phase carrying a relatively small volume of discrete particles, droplets, or bubbles. It is suited where the interaction of the discrete phase with physical boundaries is important.

For the investigation of the particle interaction with the surface of the rim the escape wall boundary is chosen. That means if there are any particles colliding with the wall they will stay there. For the simulation of the particle interaction with the flow the phase model "Two-Way-Coupling" is chosen. That means that the state of the continuous phase depends on the dispersed phases, through inter-phase mass, momentum, and energy transfer effects.

In the following table the boundary conditions of the three simulations are listed. The value for brake temperature is a typical value used in investigations for brake cooling as used e.g. from Schütz (Schütz, 2009). The properties of the brake dust particles are estimated on the basis of the work of Iijima et al. (Iijimaa A., 2007).

Table 2: Boundary conditions for the multiphase simulations

| Boundary conditions | CASE A, CASE B, CASE C |
|---|---|
| Flow velocity | 100 km/h |
| Number of volume cells | 12,5 million |
| Initial brake disc temperature | 500 °C |
| Density particles | 2950 kg/m$^3$ (only CASE B and C) |
| Particle diameter | 5 E-6 m (only CASE B and C) |
| Particle mass flow | 2,1 E-5 kg/s (only CASE B and C) |
| Wall boundary for Lagrange | escape (only CASE B and C) |

## 4.2 Providing information to the design engineer using TOKEN

The intention behind TOKEN is to give as much information as possible to the design engineer before he starts with the first design. In a conventional development process the design engineer would first design a concept and then give this concept to the CFD engineer to evaluate his proposal. Then normally a long iteration process starts and the dust shield is passed several times from one department to another. In this example we skip this first step and try to reduce the iteration loops by providing as much information as possible to the design engineer.

In this use case the result of a CFD based vehicle simulation without a dust shield has to be transferred and presented respectively. A typical development process with TOKEN could proceed the following way:

1. The design department is supposed to design a break dust shield to prevent the rims from brake dust pollution
2. The design engineer is sending a request for information concerning the development of a brake dust shield via TOKEN to the CFD engineer
3. The CFD engineer knows the flow characteristics in the wheel house of a passenger car and provides this information as a bundle via TOKEN back to the design engineer (Figure 6 – Figure 8).

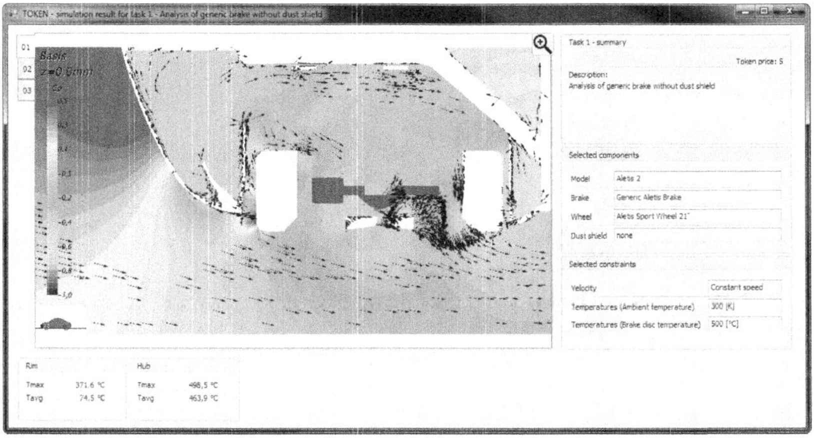


Figure 6: Pressure gradient in the front wheelhouse of the basis configuration without dust shield (CASE A)

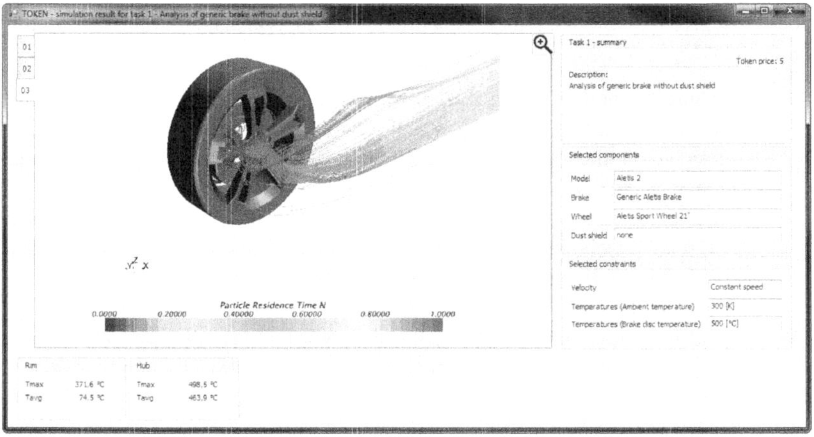


Figure 7: Flow of the brake dust particles of the basis configuration (CASE A)

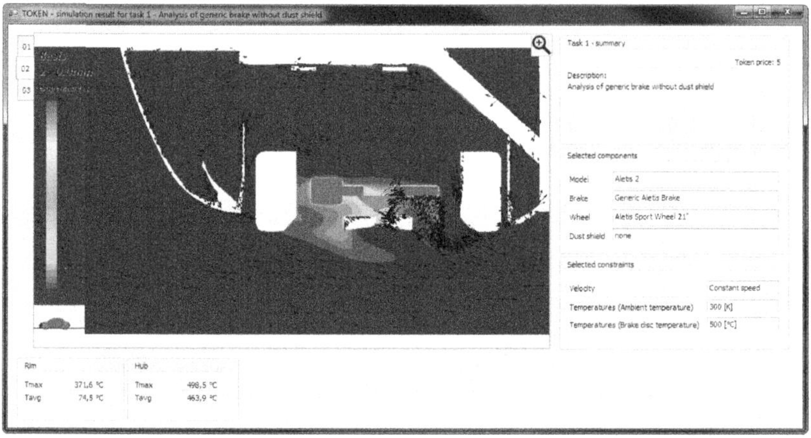


Figure 8: Temperature distribution in the brake periphery without dust shield (CASE A)

Usually it needs a little experience to interpret the simulation results. So the CFD engineer can give the design engineer support writing an explanation in the box on the right side of Figure 6 – Figure 8. This support text could look as follows:

- To ensure the brake cooling it is essential to keep the flow direction from the inside of the rims to the outer flow of the car as can be seen in the illustration of the pressure gradient on the picture on the left. So if you are designing cooling openings on the dust shield, they should support the natural flow direction
- Most of the brake dust particles in the basis configuration leave the inner of the wheel immediately in direction to the outer flow (as can be seen in the picture on the left side). If you are designing the openings like described before it is for sure, that a few particles will leave the inner of the wheels. So it will be an advantage to design the openings as slim slots which are pointing in opposition to the rotation direction of the dust shield (e.g. designed as small ramps curved out of the dust shield). The aim should be to transport the particles between the rim openings to the outer flow of the car.

4. As additional information the CFD engineer can provide the CFD results from a former investigation of a dust shield with the hint, that there have already been CFD simulations with a first draft of a dust shield in the past. This case would be our CASE B described in Table 2.

The analysis of the flow conditions of this draft show, that the arrangement of the openings, which are supposed to transport cooling air inwards of the rim, are counterproductive concerning the brake cooling (see comparative analysis in chapter 5). This effect is caused by the pressure gradient mentioned at the beginning of this chapter.

In the GUI of TOKEN the design engineer now has all relevant information he needs to design a first reasonable draft of the dust shield at a glance. Due to the CFD results of the basis configuration of the car and the hints from the CFD engineer the design engineer receives an impression how the dust shield should look like. He can now incorporate this information into his work and will get an appropriate draft of a dust shield in one single step.

In Figure 9 a draft of the dust shield designed with the information from TOKEN is shown. The design engineer has combined the advices given to him from the CFD engineer via TOKEN and the design requirements like e.g. the design envelope, structural requirements etc.

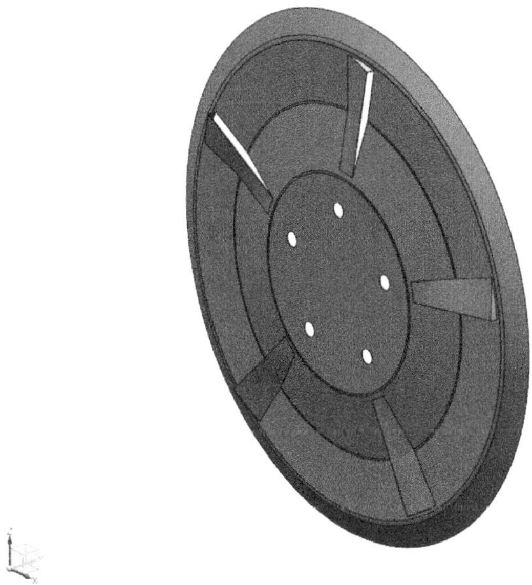

Figure 9: CASE C, dust shield developed with TOKEN

# 5 Results

To analyse the positive effect of the TOKEN-based brake dust shield development, CFD simulations are performed to illustrate the improvement of the design process. Therefore the simulation results of the “TOKEN-developed dust shield” (CASE C) are compared to the “stand-alone solution” (CASE B). In the following table a few relevant results are listed.

Table 3: Results from the CFD simulations

| | CASE A | CASE B | CASE C |
|---|---|---|---|
| drag coefficient | 0,360 | 0,373 | 0,373 |
| T avg hub [°C] | 463,9 | 480,6 | 472,4 |
| T max hub [°C] | 498,5 | 499 | 499 |
| T avg rim [°C] | 74,5 | 83,3 | 66,6 |
| T max rim [°C] | 371,6 | 267,5 | 228,7 |

The evaluation of the results in Table 3 shows that the drag of the car gets worse by placing a dust shield between the rims and the brake system. The installation of the “TOKEN-developed dust shield” doesn’t improve the drag coefficient in the first step. It remains on the same level as the “stand-alone solution”. Clear improvements can be seen by comparing the average temperatures of the rims and the wheel hub and the maximum temperatures of the rims. The arrangement of the cooling openings of CASE C cause that there is more convection through the openings concerned to CASE B as can be seen in Figure 10 and Figure 11.

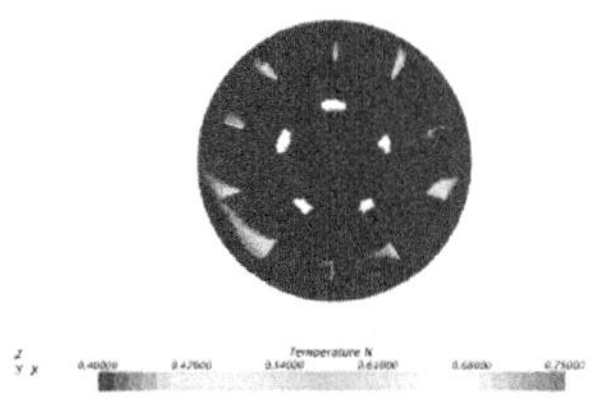

Figure 10: Surface temperature on the dust shield of CASE B

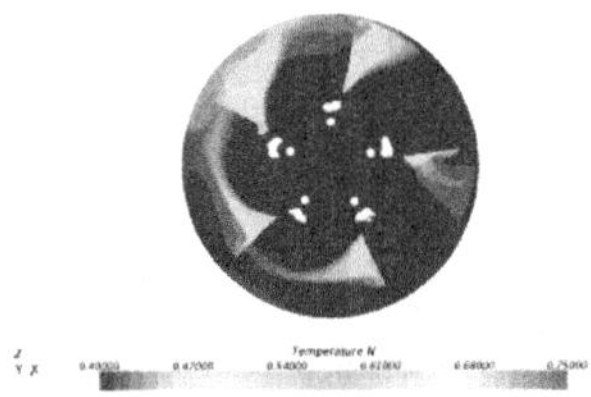

Figure 11: Surface temperature on the dust shield of CASE C

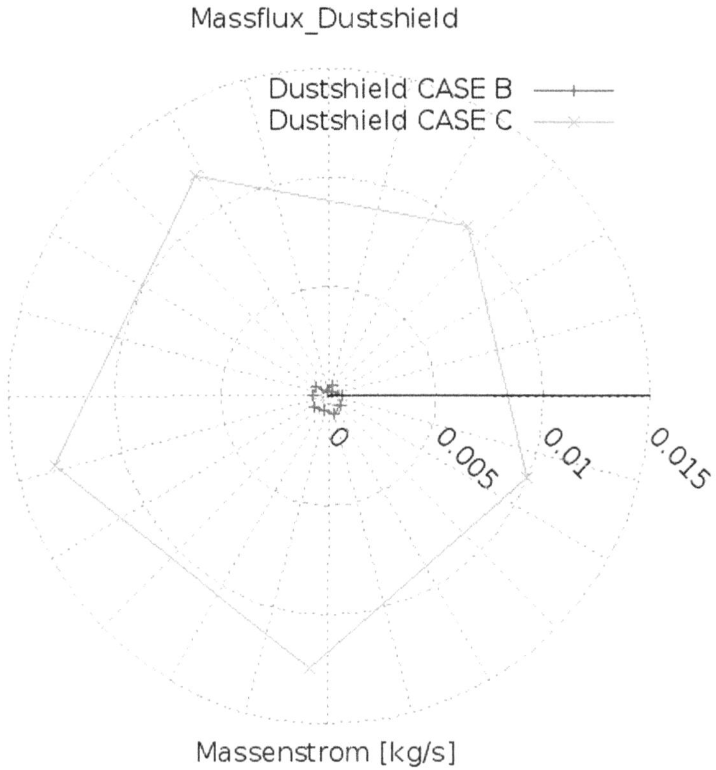


Figure 12: Mass fluxes through the cooling openings of CASE B and CASE C

The quantitative analysis of the mass fluxes through the cooling openings of the dust shields CASE A and CASE B (Figure 12) shows that there is hardly any mass flux through the openings of CASE B as suggested in the analytical analysis in chapter 4. Maintaining the natural flow direction through the rim openings leads to a significant higher mass flux in CASE C (Figure 12) and to a better cooling of the inside of the rim respectively.

This positive effect can also be seen by a qualitative consideration of the surface temperatures on the inside of the wheel and the rim respectively. If we have a closer look at the temperatures on the dust shield and the inside of the rim of CASE C (Figure 14) it is obvious compared to CASE B (Figure 13) that the surface temperatures are falling with the new dust shield.

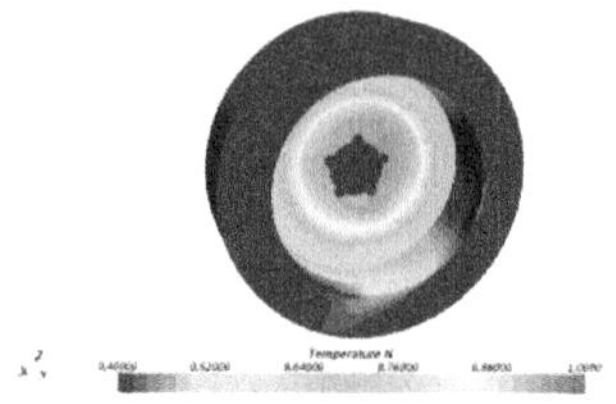

Figure 13: Surface Temperature on the inside of the wheel and the dust shield of CASE B

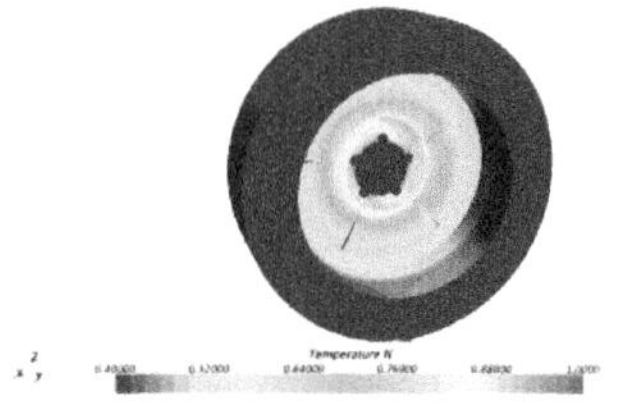

Figure 14: Surface Temperature on the inside of the wheel and the dust shield of CASE C

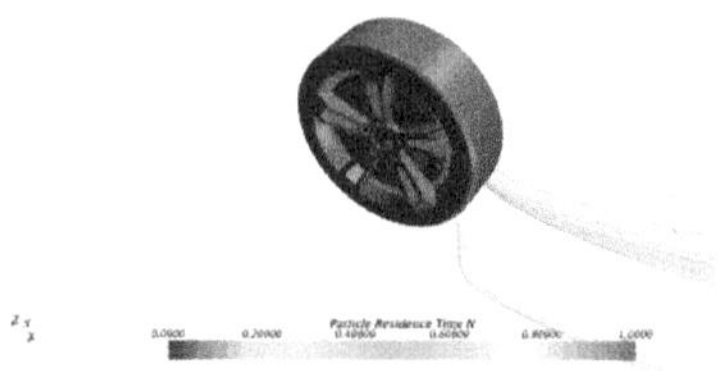

Figure 15: Brake dust particle flow CASE B

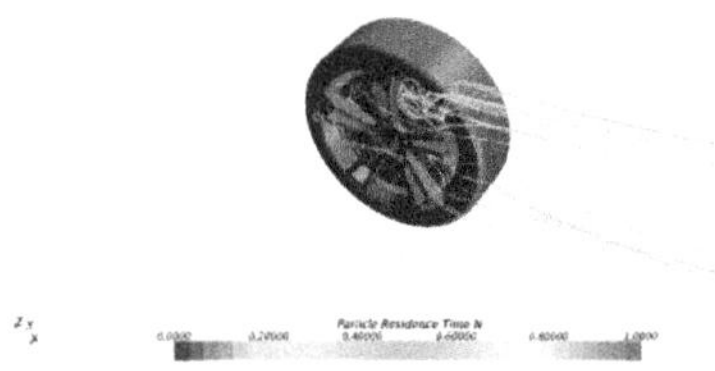

Figure 16: Brake dust particle flow CASE C

Another conspicuous detail is the outflow of hot air next to the ground as can be seen on the inside of the wheel in Figure 13. The hot spot disappears using the "TOKEN-developed dust shield" (Figure 14). This cooling effect is due to keeping the natural flow direction, from the inside to the outside of the wheel. It is obvious that the flow in this direction transports positive a few brake dust particle to the outer flow (Figure 17). The design engineer has tried to avoid using design measures that particles impacting on the rim surface. The openings of the dust shield are designed in a manner that the particles are conducted between the rim openings and are carried away with the outer flow. The visualisation of the brake dust particle flow in Figure 16 shows that indeed most of the particles are carried away. If we have a closer look at Figure 17 it can be seen that a few particles impact the rim in driving direction. But those particles are very rare. The majority of the brake dust particles don't impact on the surface of the rim. The brake dust shield developed with TOKEN is therefore a good compromise between preventing the rims from brake dust pollution and ensuring a better component cooling.

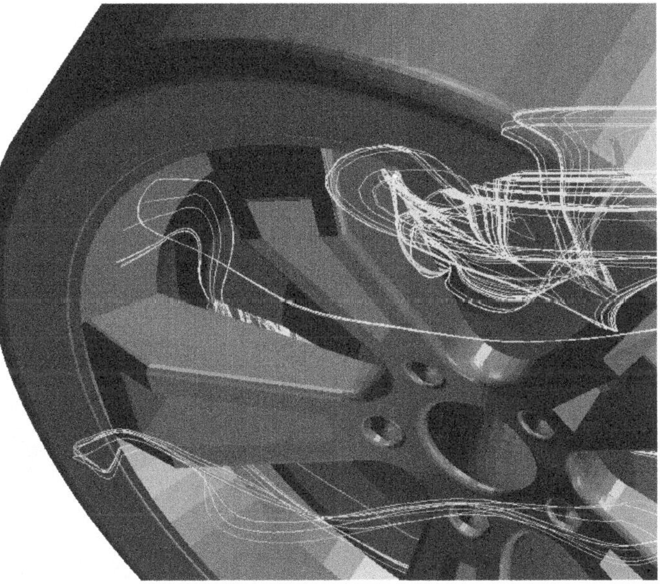

Figure 17: Impact of a few brake dust particles (CASE B)

# 6 Conclusion

The current paper proposes TOKEN, a tool for efficient digital development of components with multiple requirements, which was developed at TWT GmbH Stuttgart. The basic idea of TOKEN is to provide a communication platform between a design engineer and e.g. a simulation engineer in order to reduce iterations in the development process and attain a more mature concept in the early development process.

The paper gives an introduction to TOKEN and the main features of the tool were explained. In addition to this a TOKEN based process was presented on the example of a brake dust shield development, which couples aero- and thermodynamic questions with common design problems. It is demonstrated how knowledge transfer from an analysis to a design engineer by using TOKEN is possible and how this can affect and improve the development process. As a result of this coupled process an improved brake dust shield concept was obtained with a slender development process.

To evaluate this new concept a three-dimensional coupled CFD based comparison between an existing brake dust shield and the new design was performed. This indicated a reduction of the rim and hub temperature using the new design compared to the existing one by keeping the drag coefficient identical. The design process led to a good compromise between brake dust pollution and surface temperature on the rim.

# 7 Bibliography

**Iijimaa A., Satob K., Yanoc K., Tagoa H., Katoa M.,Kimurad H., Furut N. 2007.** *Particle size and composition distribution analysis of automotive brake abrasion dust for the evaluation of antimony sources of airborne particulate matter.* 2007.

**Mangelsdorf, Marcus. 2015.** *Konzipierung und Implementierung einer Plattform für die kooperative Produktentwicklung zwischen Entwicklungspartnern.* s.l. : Institut für Automatisierungs- und Softwaretechnik, Uniersität Stuttgart, 2015.

**Schütz, Thomas. 2009.** *Ein Beitrag zur Berechnung der Bremsenkühlung an Kraftfahrzeugen.* 2009.

# Electrically controlled parking brake (EPB) for heavy commercial vehicles

Dr. Falk Hecker, T. Weinhold,
Knorr-Bremse Systeme für Nutzfahrzeuge GmbH

**This manuscript is not available according to publishing restriction. Thank you for your understanding.**

# Electric parking brake meets drum brake – synergy or contradiction

Christian Breiten, B. Schmittner, Continental Teves AG & Co. oHG

**This manuscript is not available according to publishing restriction. Thank you for your understanding.**

# Large aircraft landing gears – a brief overview

H.-Martin Besch (Dipl. Ing.)
retired Airbus Loads Senior Expert

# Abstract

The presentation provides a simple view on large aircraft landing gears. Following items are touched:

Landing gear types, civil, military and special ones are shown with several examples of existing and past lay outs.

The landing gear design objectives along with its constraints are given as well and discussed.

Surprising is the large run time of aircraft on ground which is expressed by typical usage.

An important item worth to mention is the weight relation of landing gears in relation to the total structure weight of an aircraft.

A short view on the airworthiness requirements is made which also provides a view on the suspension modeling for design load determination which is briefly expressed.

Landing gear overall structure components with their specific parts e.g. main fitting and shock absorber, braking system and retraction system are shown for nose and main landing gears.

Tyre for landing gears and their characteristics are discussed. Since aircraft tyre data are insufficient supplied by the tyre manufacturer special test bench has been developed and are successfully applied for structure design data. Some data and results are discussed.

Some of the landing gear structure loads conditions as are landing, braking and handling on ground are given.

If time allows, some interesting videos will be shown, providing an impression of dynamic landing impacts with excessive deformation at wheels and structure which indicates the enormous load at landing gear and associated airframe structure.

# Large Aircraft Landing Gears

A Rough Overview

Presented by

**H.-Martin Besch (Dipl. Ing.)**

retired Airbus Loads Senior Expert

ATZ Munich, June 17th, 2015

## Large Aircraft Landing Gears

**Content**

- Characteristics of Landing (Ldg.) Gears and Classification
- Example of Ldg. Gear Types
- Airworthiness Requirements
- Ldg. Gear Structure Components
- Shock Absorber
- Ground Loads Model Composition
- Aircraft Ldg. Gear Tyres
- Ground Loads Design Conditions
- Test videos/Photos

2

ATZ Munich, June 17th, 2015

## Landing Gear Types

**Landing Gear Types - Miscellaneous**

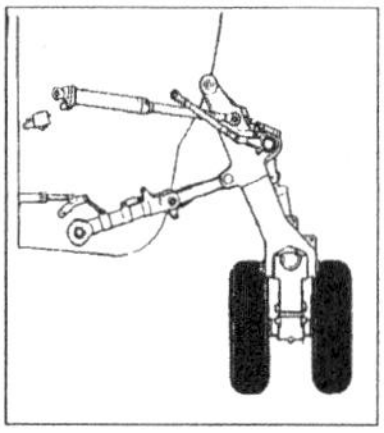

BAe 146

Lockheed C-5A

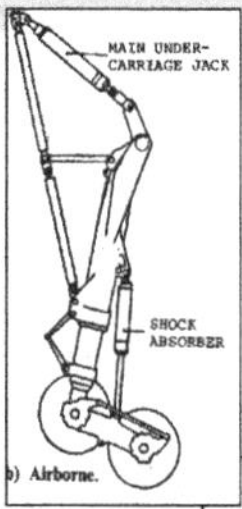


BAe Vulcan

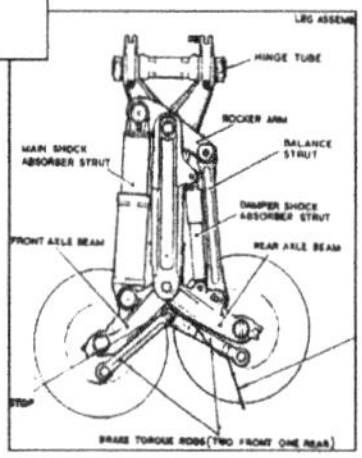

BAe Nimrod

Not shown here: Floaters, Skis, air cushion usw.

ATZ Munich, June 17th, 2015

3

## Landing Gear Types: Classification

**Classification of Landing Gear Types:**

Gen. according to the number of wheels and their arrangement

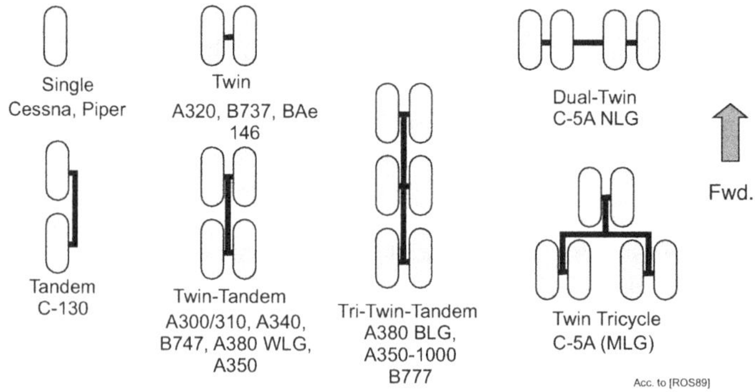


Acc. to [ROS89]

4

ATZ Munich, June 17th, 2015

Ancient Landing Gear
Wright Flyer
ATZ Munich, June 17th, 2015

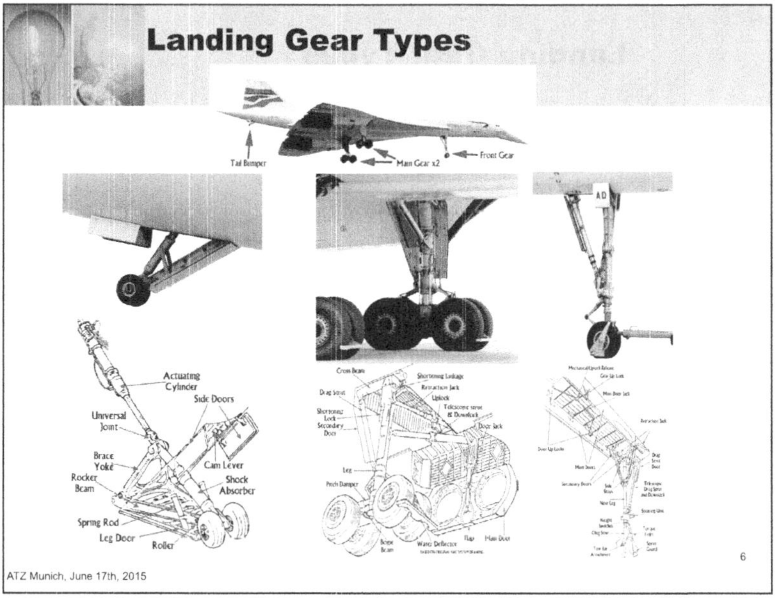
Landing Gear Types
Tail Bumper
Main Gear x2
Front Gear
Actuating Cylinder
Side Doors
Universal Joint
Brace Yoke
Cam Lever
Rocker Beam
Shock Absorber
Spring Rod
Leg Door
Roller
Cross Beam
Shortening Linkage
Retraction Jack
Drag Strut
Uplock
Telescopic strut & Downlock
Shortening Lock
Secondary Door
Door Jack
Leg
Pitch Damper
Bogie Beam
Water Deflector
Flap
Main Door
6
ATZ Munich, June 17th, 2015

## Landing Gear Types

ATZ Munich, June 17th, 2015 7

## Landing Gear Types

ATZ Munich, June 17th, 2015 8

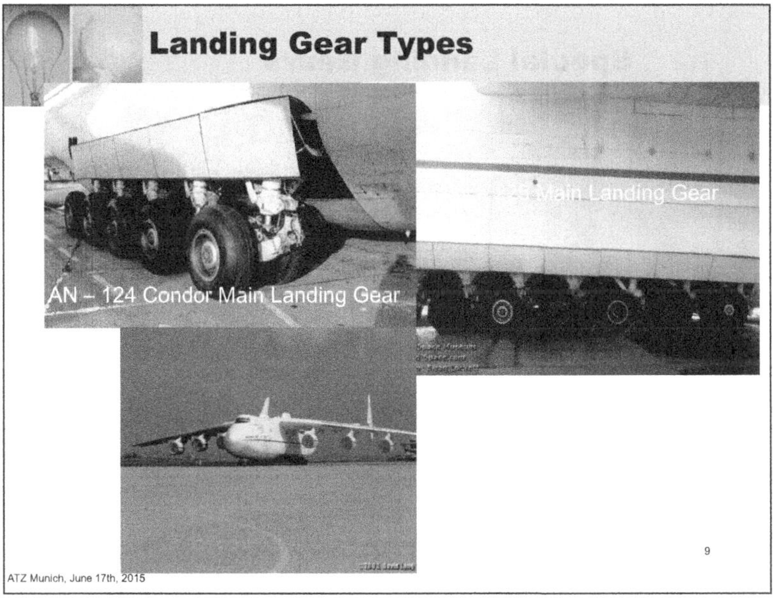
Landing Gear Types
AN – 124 Condor Main Landing Gear
ATZ Munich, June 17th, 2015
9

Landing Gear Types
Lockheed C5A Galaxy
Main Landing Gear
ATZ Munich, June 17th, 2015
10

## Special Landing Gears

11

ATZ Munich, June 17th, 2015

## Special Landing Gears

12

ATZ Munich, June 17th, 2015

## Special Landing Gears

Convair XB-36 „Peacemaker“
Main Landing Gear

ATZ Munich, June 17th, 2015

13

## Military Aircraft Landing Gear

Military Jet Main Landing Gear

ATZ Munich, June 17th, 2015

14

## Mil Landing Gear

Avro Arrow 201
Main Landing Gear

15

ATZ Munich, June 17th, 2015

## Landing Gear Types

A330/A340
Main Landing Gear

16

ATZ Munich, June 17th, 2015

## Design Objectives
## Aircraft Landing Gears

**Design Functions**

- Absorb landing shocks (vertical speed) and taxiing shocks
- Provide ability for ground maneuvering: taxi, take-off run, landing roll and steering on ground (commonly nose gear but main gear group as well)
- Retract in flight
- Provide braking capability
- Allow for aircraft towing
- Protect the ground surface

**Design Constraints**

- Limited storage area, incl. Minimum effect on aircraft drag
- Minimum weight
- Minimum maintenance effort
- Ensure design goal on ground
- Acceptable ground clearance
- Minimum costs

17

ATZ Munich, June 17th, 2015

## Aircraft on Ground
## Example Airport Frankfurt/Main

Note:
Distances and length of Taxi ways

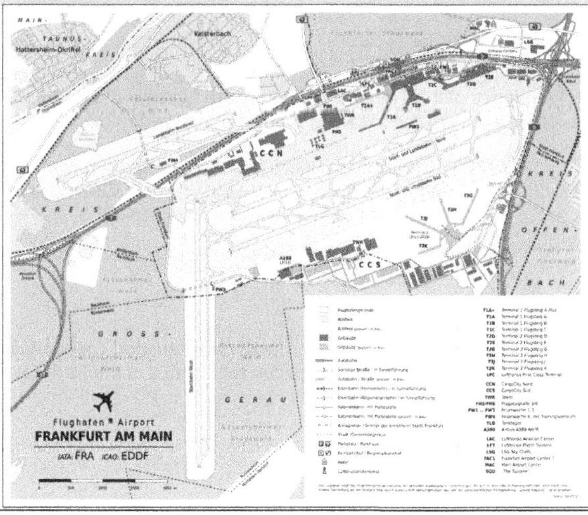


ATZ Munich, June 17th, 2015

## Aircraft on Ground
## Life Distance

- Taxi distance: 1.5km
- A/C on runway: 1.5km
- Short range aircraft: 3 flights/day
- A/C in service: 30 years

**Total distance on ground:**

⇒ 2 * 3km * 3 flights/day * 365 days/year * 30 years

**≈ 200,000km landing gear run on ground**

ATZ Munich, June 17th, 2015

19

## Airworthiness Requirements
## The rules for Structure Loads Assessment

Civil Airplanes of the Transport Category are typically designed and certified according to one of the following widely harmonized rules:

- **CS 25** - Certification Specification for Large Aeroplanes - representing the European regulations of the **European Aviation Safety Agency (EASA)**, in operation since 2003

  *Note: The CS25 rules correspond largely to the former JAR 25 (Joint Airworthiness Requirements) of the JAA (Joint Aviation Authorities)*

- **FAR 25** - Airworthiness Standards: Transport Category Airplanes – re-presenting the American regulations of the **Federal Aviation Administration (FAA)**

ATZ Munich, June 17th, 2015

20

## Airworthiness Requirements
### The rules for Structure Loads Assessment

**Safety rules :**
Several bodies of rules exist giving minimum requirements for structure design and qualification/certification according to the usage of the aeroplane (military or civil), the category of the aeroplane etc.

Civil aeroplanes have to comply with **FAR** (Federal Aviation Requirements) rules in the USA, or the almost equivalent **JAR** (Joint Aviation Requirements), now EASA **CS** rules in Europe (FAR/JAR 23 being applicable to normal, utility, acrobatic, and commuter aeroplanes, **FAR/JAR/CS 25** being applicable to transport category aeroplanes). Most of the loads requirements are in **Subpart C (Structure) of FAR/JAR/CS**, but other side requirements are also in the subparts B 'Flight', D (Design and Construction), E (Powerplant), F (Equipment) and G (Operating Limitations and Information)

Several Military Requirements also exist. We recall here the USA MIL Standards, the UK DEF-STAN, and the French AIR.

*As matter of principle, the **Civil Rules are only safety oriented**. The **Military Requirements are both safety and performance oriented** since they are issued by the customer of the product.*

**Commercial rules:**
e.g. LG robust design, Top Level Aircraft Industry Requirements for low cost maintenance

ATZ Munich, June 17th, 2015 21

## Ground Loads Airworthiness Requirements
### CS/FAR §§ 25.471 – 25.519 & 25.561 – 25.563

**Ground Loads**
**EASA Part 25 C**

**Dynamic calc./ Full flexible a/c**
- Dynamic landing
  - MLW-landing
  - MTOW-landing
  - Heavy landing
- Emergency landing
- Dynamic braking
- Dynamic taxi

**(Quasi-)static calc. Rigid a/c**
- Rigid ground loads, resp. so-called
- "book cases"

ATZ Munich, June 17th, 2015 22

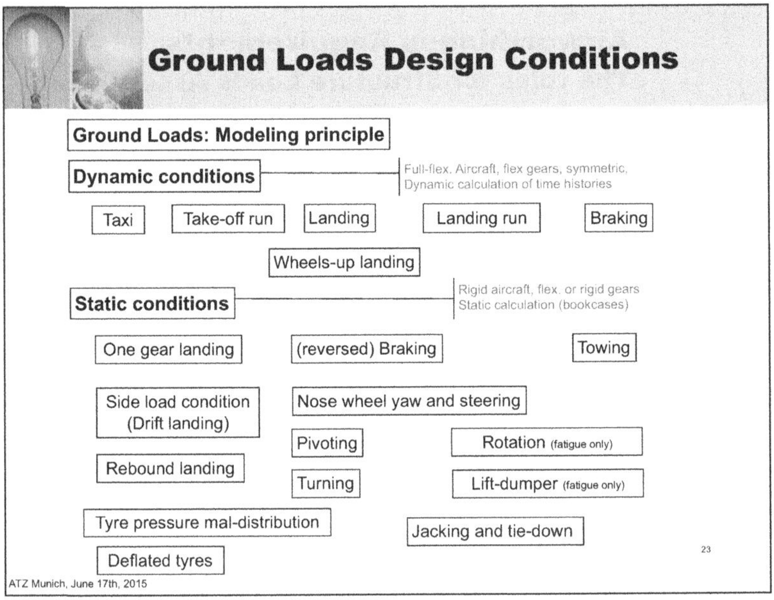
Ground Loads Design Conditions
Ground Loads: Modeling principle
Dynamic conditions
Full-flex. Aircraft, flex gears, symmetric,
Dynamic calculation of time histories
Taxi
Take-off run
Landing
Landing run
Braking
Wheels-up landing
Static conditions
Rigid aircraft, flex. or rigid gears
Static calculation (bookcases)
One gear landing
(reversed) Braking
Towing
Side load condition (Drift landing)
Nose wheel yaw and steering
Pivoting
Rotation (fatigue only)
Rebound landing
Turning
Lift-dumper (fatigue only)
Tyre pressure mal-distribution
Jacking and tie-down
Deflated tyres
23
ATZ Munich, June 17th, 2015

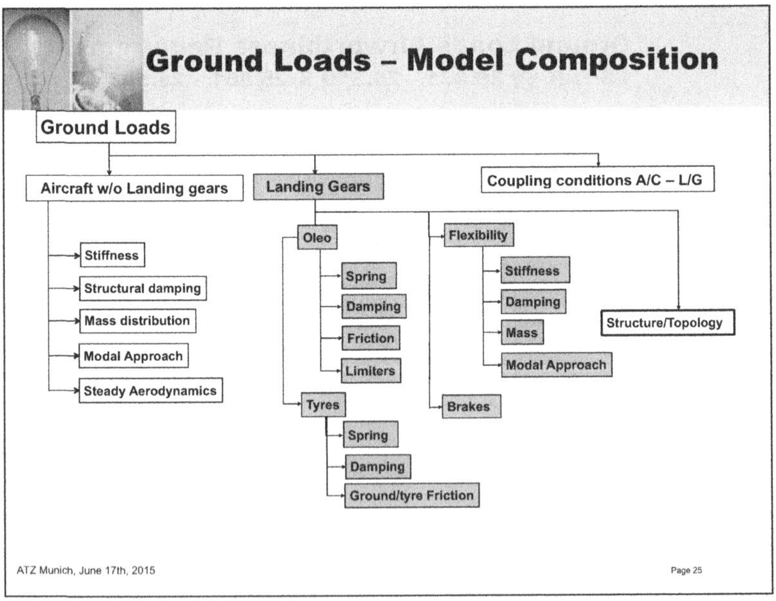
Ground Loads – Model Composition
Ground Loads
Aircraft w/o Landing gears
Landing Gears
Coupling conditions A/C – L/G
Stiffness
Structural damping
Mass distribution
Modal Approach
Steady Aerodynamics
Oleo
Spring
Damping
Friction
Limiters
Tyres
Spring
Damping
Ground/tyre Friction
Flexibility
Stiffness
Damping
Mass
Modal Approach
Brakes
Structure/Topology
ATZ Munich, June 17th, 2015
Page 25

## Nose Landing Gear

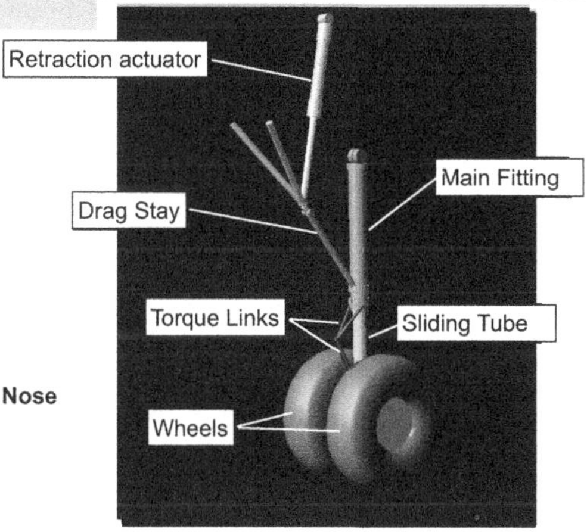


**Sketch of a typical Nose Landing Gear**

25

ATZ Munich, June 17th, 2015

## Nose Landing Gear (Components)

Twin Landing Gear

Nose Gear, front view

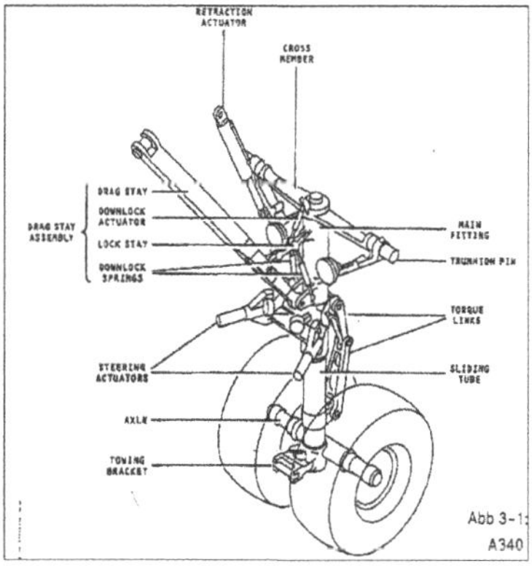


Nose Gear, left view

26

ATZ Munich, June 17th, 2015

## Main Landing Gear Components

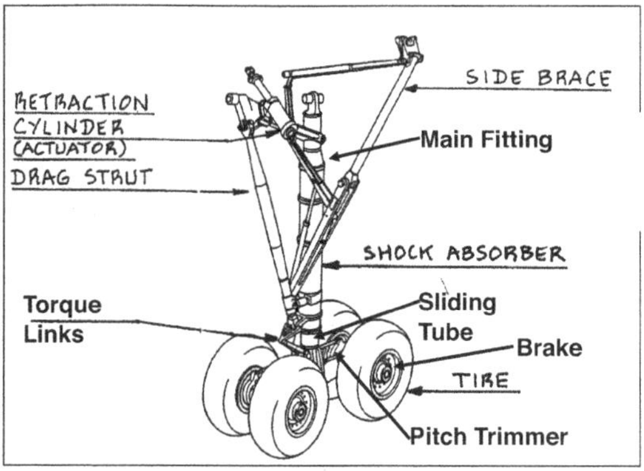


Sketch of a typical Nose Landing Gear

27

ATZ Munich, June 17th, 2015

## Main Landing Gear (MLG)

ATZ Munich, June 17th, 2015

28

## Landing Gear: Shock Absorber

Landing Gear Parts– Hydro-pneumatic Shock Absorber (Oleo)

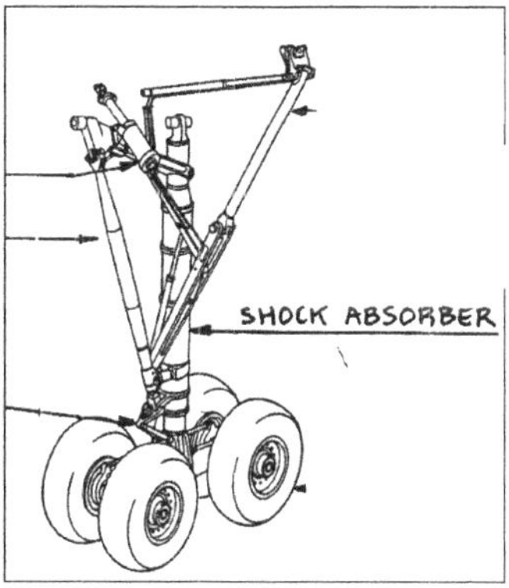


**Internal oleo forces:**

- **Gas spring (single-, double-stage)**
- **Hydraulic damping**
- **Stroke limiters**
- **Seal friction**
- **Bearing friction**

29

ATZ Munich, June 17th, 2015

## Landing Gear: Shock Absorber

The shock absorber itself contains two components

- one storing and then releasing mechanical energy (spring function),
- another one transforming mechanical energy in thermal energy (damping function). The shock absorber is usually designed as an

Oleo-pneumatic (Nitrogen and oil) system. The key roles of the shock absorber are:

- Absorb vertical speed of the aircraft at landing
- Provide smooth taxiing

30

## Shock Absorber Model

The shock absorber total force ($F_{SA}$) is modeled as an element capable of transmitting a ground reaction in the direction of its own axis to the airframe depending on the actual stroke and on the rate of change of the stroke versus the time for a given temperature and shock absorber setting:

$$F_{SA}=F_S+F_D+F_F$$

- Where $F_S$ is the spring force
- Where $F_D$ is the damping force
- Where $F_F$ is the friction force

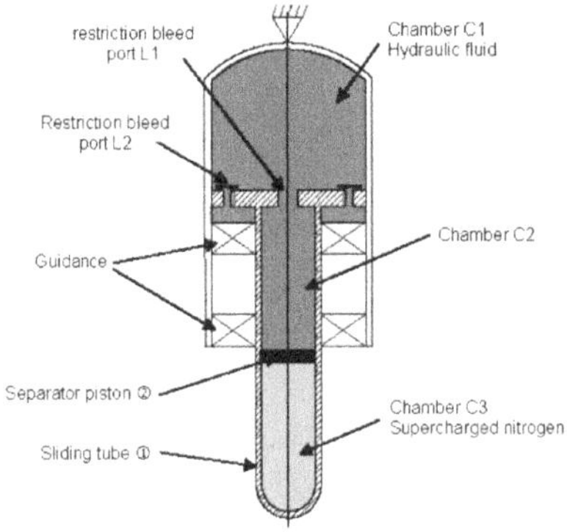


**Figure 2:** ***Elements of a shock absorber***

31

ATZ Munich, June 17th, 2015

## Tyres

32

ATZ Munich, June 17th, 2015

## Michelin Tyres Seize (A350)

33

ATZ Munich, June 17th, 2015

## Landing Gear: Tyre Vertical

Landing Gear Parts – Tyre

**Vertical force (Example)**

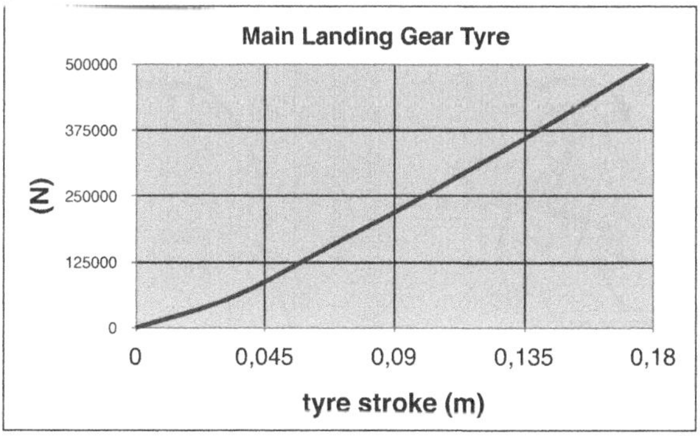


34

ATZ Munich, June 17th, 2015

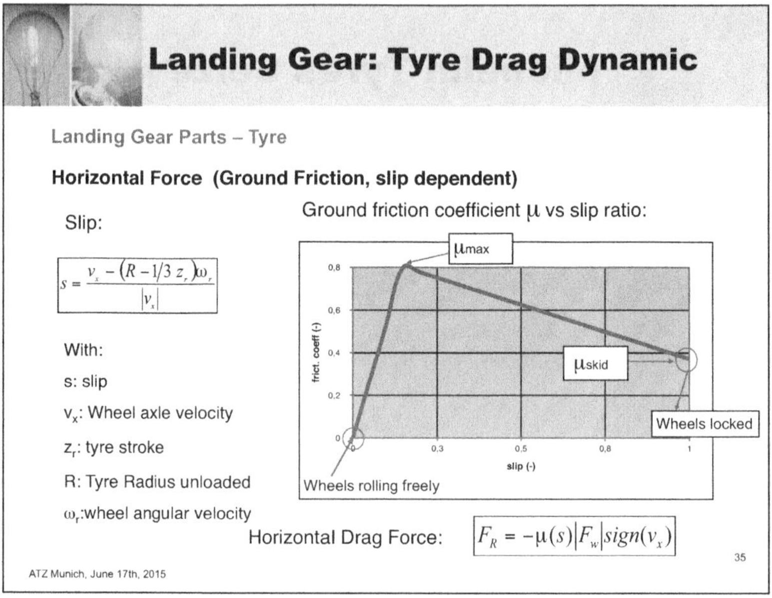
Landing Gear: Tyre Drag Dynamic
Landing Gear Parts – Tyre
Horizontal Force (Ground Friction, slip dependent)
Slip:
$s = \frac{v_x - (R - 1/3\, z_r)\omega_r}{|v_x|}$
With:
s: slip
$v_x$: Wheel axle velocity
$z_r$: tyre stroke
R: Tyre Radius unloaded
$\omega_r$:wheel angular velocity
Ground friction coefficient μ vs slip ratio:
μmax
μskid
Wheels locked
Wheels rolling freely
frict. coeff (-)
slip (-)
0,8
0,6
0,4
0,2
0
0,3
0,5
0,8
1
Horizontal Drag Force:
$F_R = -\mu(s)|F_w| sign(v_x)$
35
ATZ Munich, June 17th, 2015

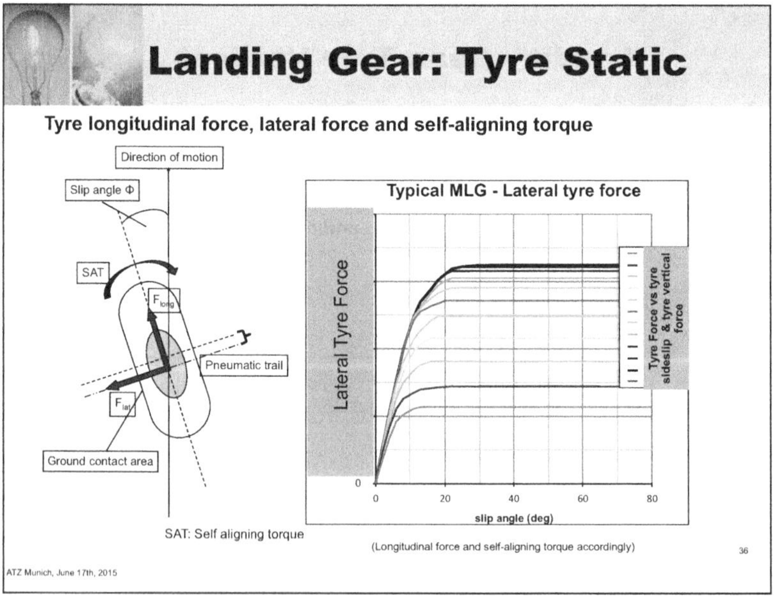
Landing Gear: Tyre Static
Tyre longitudinal force, lateral force and self-aligning torque
Direction of motion
Slip angle Φ
SAT
$F_{long}$
Pneumatic trail
$F_{lat}$
Ground contact area
SAT: Self aligning torque
Typical MLG - Lateral tyre force
Lateral Tyre Force
Tyre Force vs tyre sideslip & tyre vertical force
0
0
20
40
60
80
slip angle (deg)
(Longitudinal force and self-aligning torque accordingly)
36
ATZ Munich, June 17th, 2015

## Dynamic Conditions for Design Loads

37

ATZ Munich, June 17th, 2015

## Touch Down at Landing

Landing ( 25.473 )

- $V_Z$ = 10 fps (3.05 m/s) at MLW
- $V_Z$ = 6 fps ( 1.8 m/s) at MTOW

**Level Landing**

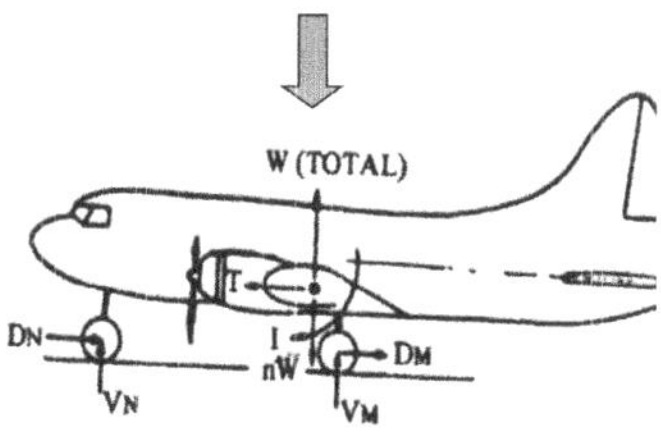


**Tail down Landing**

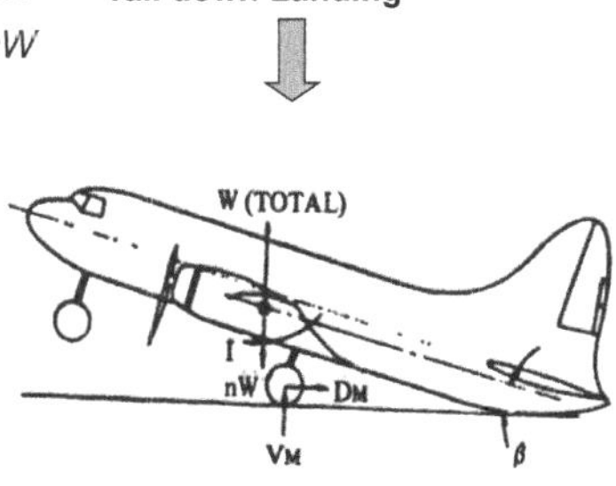


β = Angle for main gear and tail structure contacting ground except need not exceed stall angle.

38

ATZ Munich, June 17th, 2015

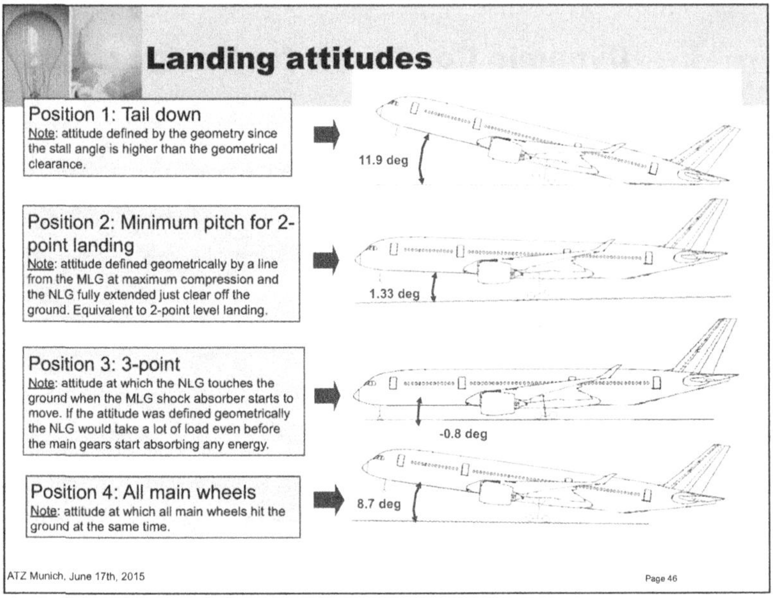
Landing attitudes
Position 1: Tail down
Note: attitude defined by the geometry since the stall angle is higher than the geometrical clearance.
11.9 deg
Position 2: Minimum pitch for 2-point landing
Note: attitude defined geometrically by a line from the MLG at maximum compression and the NLG fully extended just clear off the ground. Equivalent to 2-point level landing.
1.33 deg
Position 3: 3-point
Note: attitude at which the NLG touches the ground when the MLG shock absorber starts to move. If the attitude was defined geometrically the NLG would take a lot of load even before the main gears start absorbing any energy.
-0.8 deg
Position 4: All main wheels
Note: attitude at which all main wheels hit the ground at the same time.
8.7 deg
ATZ Munich, June 17th, 2015
Page 46

Dynamic Braking
A380
F-WWDD
(Dynamic) Braking
© 2006 Ulrich Grueschow - www.MilitaryAircraft.de
40
ATZ Munich, June 17th, 2015

## Main Landing Gear Brakes

41

## Brake System

BRAKES, WHEELS, AND SKID CONTROL

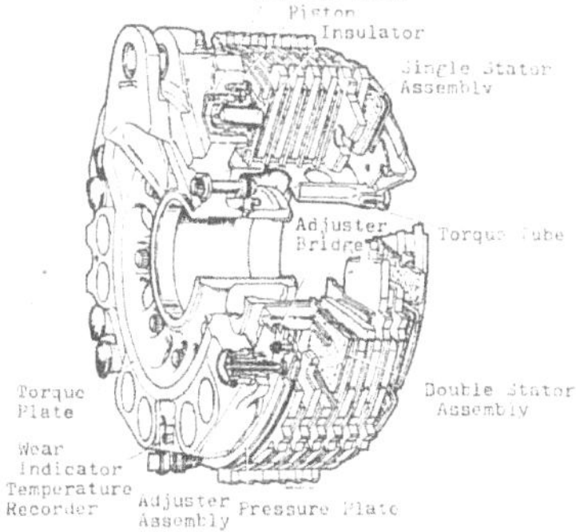


Fig. 7.3 Brake detail (source: Dunlop).

## Dynamic Braking

Moment due to brake drag increases NLG vertical load

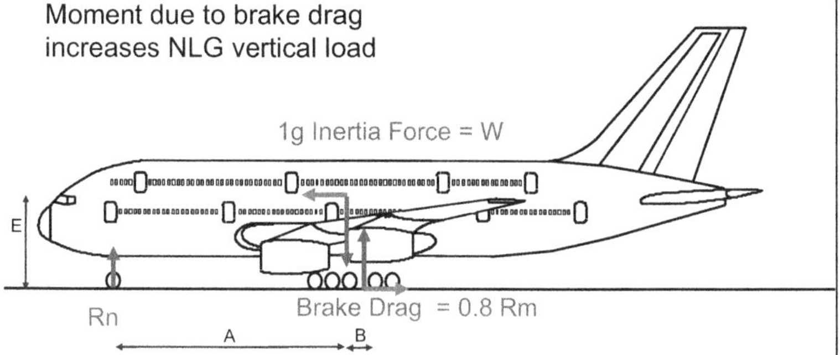


$$Rn = \frac{W}{A + B} \cdot \left( B + \frac{0.8\, f.A.E}{A + B + 0.8E} \right)$$

where 0.8 is the coefficient of friction (bookcase) and f is a dynamic over-swing factor

EGL 43

## Dynamic Braking

### Dynamic braking

- The torque rise rate is the main influential parameter on dynamic braking loads. The dynamic braking maneuver gives high vertical load on the Nose LG, and **the higher the braking torque rise rate, the higher the Nose LG load**
- The torque rise measured in flight tests during specific braking tests, with warm brakes, are commonly covered by the theoretical torque rise used in calculation. Therefore the model is conservative

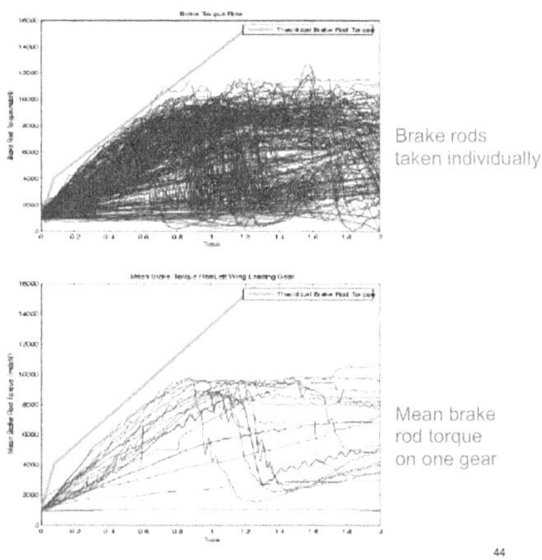


44

ATZ Munich, June 17th, 2015

**Ground Loads:**
**Dynamic Taxi/Dynamic Braking**

## Dynamic Braking - Brake Torque

Brake Torque, applied at each wheel equipped with brakes (acc. CS CC10):

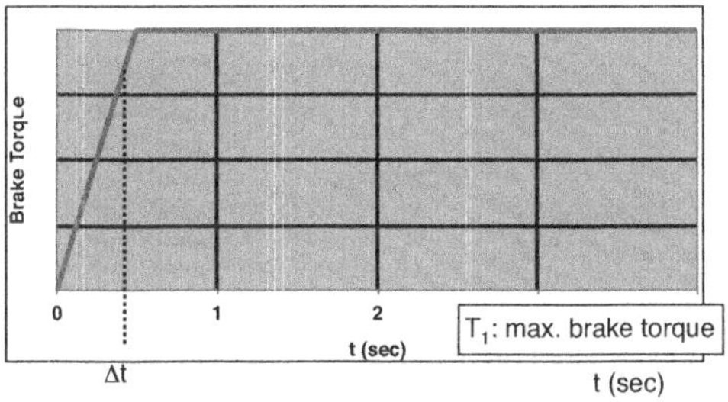


**Note:** During simulation brake torque must not exceed the torque arising from ground friction. Otherwise wheel would turn backwards!

45

ATZ Munich, June 17th, 2015

**Ground Loads:**
**Dynamic Taxi/Dynamic Braking**

## Dynamic Braking – Ground Friction Coeff.

Ground Friction at Low Speed

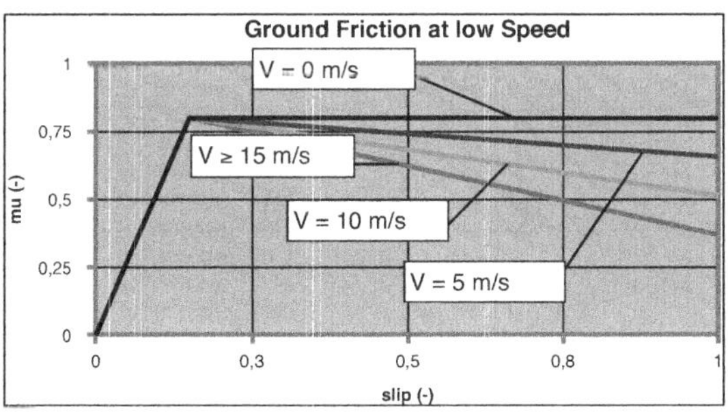


46

ATZ Munich, June 17th, 2015

Static Ground Loads (Bookcases)
Ground Handling
47
ATZ Munich, June 17th, 2015

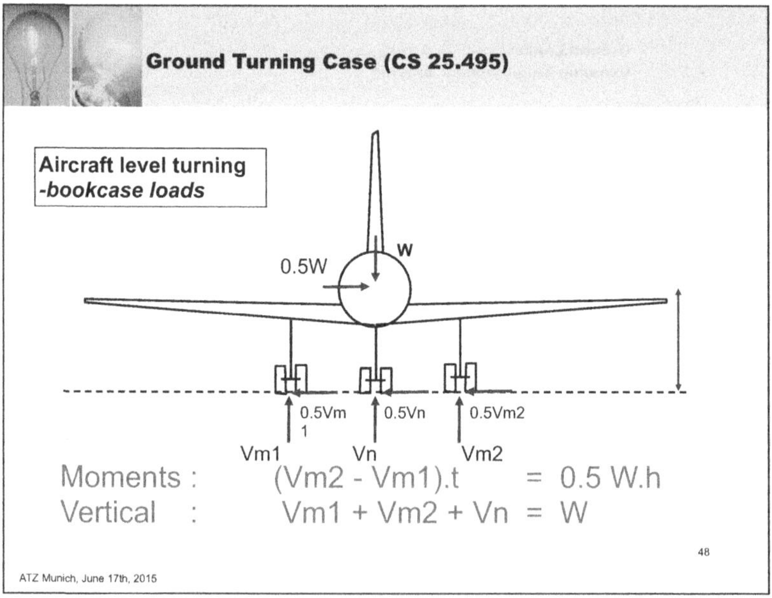
Ground Turning Case (CS 25.495)
Aircraft level turning
-bookcase loads
W
0.5W
0.5Vm1
0.5Vn
0.5Vm2
Vm1
Vn
Vm2
Moments : (Vm2 - Vm1).t = 0.5 W.h
Vertical : Vm1 + Vm2 + Vn = W
48
ATZ Munich, June 17th, 2015

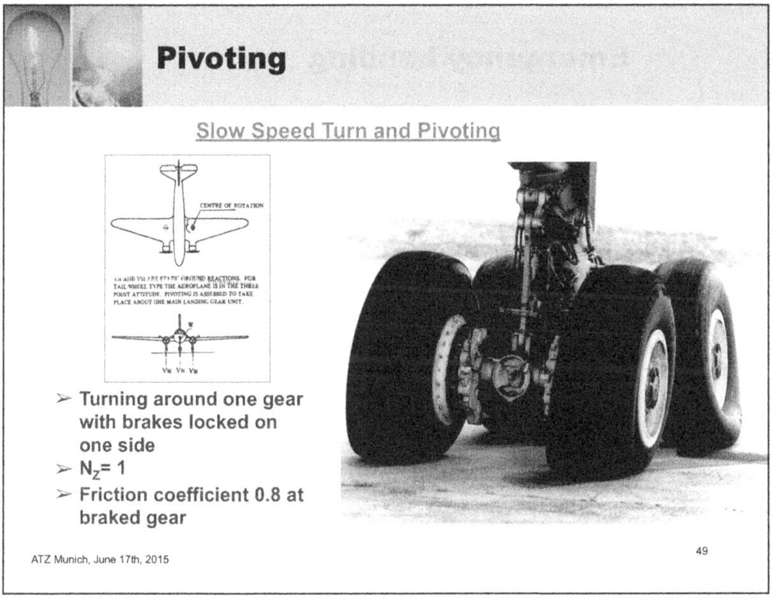
Pivoting
Slow Speed Turn and Pivoting
CENTRE OF ROTATION
Turning around one gear with brakes locked on one side
N_Z= 1
Friction coefficient 0.8 at braked gear
ATZ Munich, June 17th, 2015
49

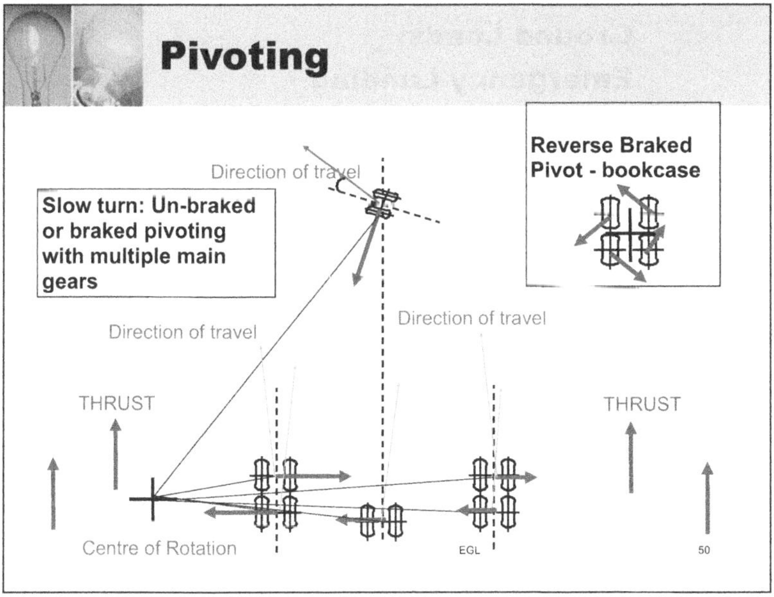
Pivoting
Reverse Braked Pivot - bookcase
Direction of travel
Slow turn: Un-braked or braked pivoting with multiple main gears
Direction of travel
Direction of travel
THRUST
THRUST
Centre of Rotation
EGL
50

## Emergency Landing

### Emergency Landing (1)

Landing with one ore more gears retracted

- Landing Impact (dynamic)
- Sliding after touch-down (rigid ground loads)

Request:

No rupture that could be catastrophic for the safety of the occupants including ruptures leading to fuel spillage

## Ground Loads: Emergency Landing

### Emergency Landing (2): Impact Cases

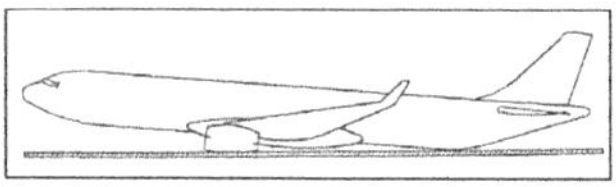

All gears up

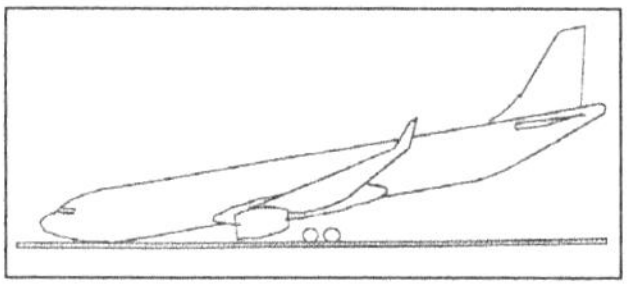

Nose gear up, main gear down

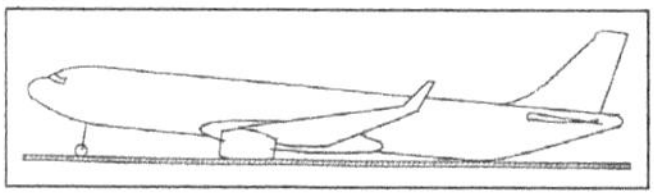

Nose gear down, main gear up

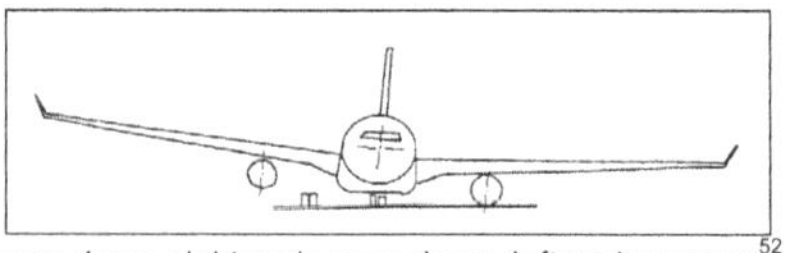

Nose gear down, right main gear down, left main gear up

## QUESTIONS

THANK YOU FOR YOUR ATTENTION
ANY QUESTION???

WHAT ARE THE DESIGN CRITERIA FOR FLYING CARS?

53

ATZ Munich, June 17th, 2015

## Acknowledgement & References

**The presentation is partly based on Airbus material, especially a presentation of Airbus Ground Loads Engineer Rainer Sonder's presentation in the frame of Friday Lectures. Many thanks to his contribution.**

CUR88 N. S. Currey: Aircraft Landing Gear Design: Principles and Practices, AIAA Education Series, 1988

ROS89 J. Roskam: Airplane Design Part IV: Layout Design of Landing Gear and Systems, Ottawa, Kansas 1989

DEF97 Thomas DeFiore/ Richard Micklos: Video Landing Parameter Survey - John F. Kennedy International Airport Final Report - July 1997, U.S. Department of Transport / Federal Aviation Administration, DOT / FAA / AR - 96 /125

BAR99 Terence Barnes/ Thomas DeFiore/ Richard Micklos: Video Landing Parameter Survey - Washington National Airport Final Report June 1999, U.S. Department of Transport / Federal Aviation Administration, DOT / FAA / AR - 97 /106

54

## Aircraft Landing Gears

### Backup Material

1. Tests
2. Landing Gear Types: Footprints

55

ATZ Munich, June 17th, 2015

## A Very Hard Landing

56

ATZ Munich, June 17th, 2015

Landing in Turbulence
57
ATZ Munich, June 17th, 2015

# Continuous wheel force measurement for passenger vehicles and commercial vehicles

Dr. Manfred Kraus, Dr. M. Bäuml,
Schaeffler Technologies AG & Co. KG

**This manuscript is not available according to publishing restriction. Thank you for your understanding.**

# Adaptive state observers for driving dynamics – online estimation of tire parameters under real conditions

J. Bechtloff, M.Sc.
Dipl.-Ing. C.Ackermann
Prof. Dr.-Ing. Dr. h.c. R. Isermann

Institut für Automatisierungstechnik und Mechatronik der
TU Darmstadt (Institute of Automatic Control and Mechatronics),
64283 Darmstadt

# Abstract

Model based control of lateral vehicle dynamics needs accurate tire models and their parameters to work efficiently. Tire properties are changing, due to wear, tire pressure, road conditions or a change in tire type or brand. Therefore the goal is to estimate the variable properties online by using the available ESC sensors only.

An extended Kalmanfilter is used to estimate the side slip angle, cornering stiffness and the maximum friction coefficient at the front and rear axle. The longitudinal velocity is assumed to be known from a preceding estimation algorithm. To model the vehicle dynamics an extended single track model with a nonlinear tire model is used. The tire model covers the lateral and longitudinal physical limits by considering longitudinal slip and shape parameters for the nonlinear characteristics.

Results are shown for a dynamic test drive on a plane test track and a public country road with a BMW 530d F11.


# 1 Introduction

Model based open and closed loop control of the lateral vehicle dynamics are among others strongly dependent on a tire model and its parameters. These parameters can change significantly due to wear or a change of the tire pressure, type or brand. This is a main reason why fully model based control is yet not used in production vehicles. One example for application is to generate a desired force at the rear axle with a rear steering, in order to influence the yaw dynamic of the vehicle. Then at least the cornering stiffnesses have to be known.

Algorithms that estimate cornering stiffnesses during test drives for highly dynamic excitation on plane test tracks were published in e.g. [1] [2]. The problem was the difficulty to differentiate the cornering stiffnesses at the front and rear axle during stationary cornering and normal drives. In this contribution an estimation of the cornering stiffnesses with normal excitation on public road is shown.

# 2 Modelling

The lateral tire force directly depends on the vertical load $F_z$, the side slip angle $\alpha$ and tire properties such as e.g. cornering stiffness $c_\alpha$. Other parameters like the maximum friction coefficient $\mu_{\max} = F_{\mathrm{ymax,f}}/F_{\mathrm{z,f}}$ and shaping parameters are used to cover situations with high lateral accelerations and to avoid the underestimation of cornering stiffness. The influence of possible longitudinal slip $\lambda$ is also included to avoid the underestimation of corning stiffness during situations with combined lateral and longitudinal slip conditions. The lateral forces can be calculated with eq. (1) and (2) for the front and rear axle [3]

$$F_{y,f} = f\left(F_{z,f}, \alpha_f, \lambda_f, c_{\alpha,f}, \mu_{max,f}, \alpha_{max,f}, \dots\right) \cdot \cos(\delta_f) \tag{1}$$

$$F_{y,r} = f(F_{z,r}, \alpha_r, \lambda_r, c_{\alpha,r}, \mu_{max,r}, \alpha_{max,r}, \dots) \tag{2}$$

When the nonlinear tire model (1) and (2) is linearized during driving straight ahead without longitudinal slip

$$F_{y,f} = \alpha_f \cdot c_{\alpha,f} \tag{3}$$

$$F_{y,r} = \alpha_r \cdot c_{\alpha,r} \tag{4}$$

the classic linear tire model is obtained. Fig. 1 shows the nonlinear tire model and the linearization for small side slip angles $\alpha$.

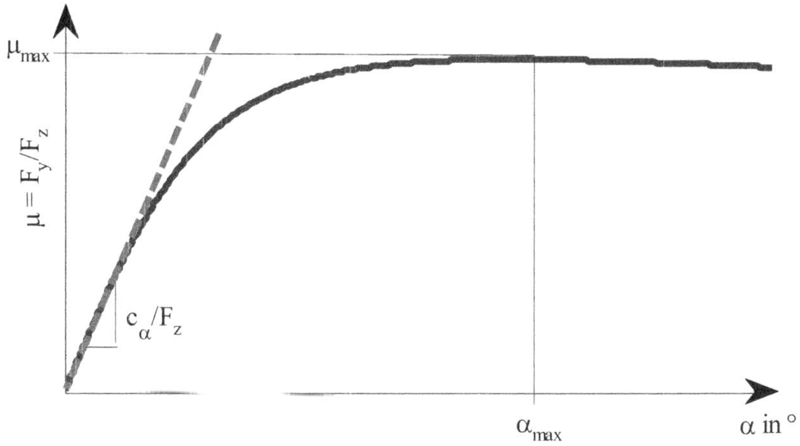


Fig. 1: Nonlinear tire model and the linearization for small side slip angles $\alpha$

The axle side slip angles depend from the lateral and longitudinal velocity $v_y$ and $v_x$, the yawrate $\dot{\psi}$ and on the front middle steering angle $\delta_f$. The lateral velocity or more common the sideslip angle

$$\beta = \arctan\frac{v_y}{v_x} \tag{5}$$

cannot be measured directly in production vehicles. Therefore it has to be estimated. With the force and moment balance equations of the single track model [4] it holds

$$\dot{v}_{\mathrm{y}} = \frac{1}{m}(F_{\mathrm{y,f}} + F_{\mathrm{y,r}}) - \sin(\phi_{\mathrm{road}}) \cdot g - v_{\mathrm{x}}\dot{\psi}. \tag{6}$$

The influence of the banked corner angle $\phi_{\mathrm{road}}$ is considered. Applying the output model

$$a_{\mathrm{y}}^{\mathrm{ECU}} = \frac{1}{m}(F_{\mathrm{y,f}} + F_{\mathrm{y,r}}) + \sin(\phi_{\mathrm{roll}}) \cdot g \tag{7}$$

the lateral acceleration $a_{\mathrm{y}}^{\mathrm{ECU}}$ is linked with the lateral velocity and the tire parameters. It is important to compensate the part of earth gravity $\sin(\phi_{\mathrm{roll}}) \cdot g$ in the lateral acceleration sensor which is caused by roll motion. The roll angle $\phi_{\mathrm{roll}}$ is calculated by the use of an simple roll model [5].

To increase the robustness of the filter algorithm it is useful to consider the yaw dynamics as a system state

$$\ddot{\psi} = \frac{1}{J_{\mathrm{z}}}(F_{\mathrm{y,f}} \cdot l_f - F_{\mathrm{y,r}} \cdot l_r + \Delta M_{\mathrm{z}}) \tag{8}$$

with the moment of inertia $J_{\mathrm{z}}$ and distances from the center of gravity to the front axle $l_{\mathrm{f}}$ and the rear axle $l_{\mathrm{r}}$.The additional yaw torque $\Delta M_{\mathrm{z}}$ that covers ESC intervention is calculated based on braking and driving torques.

# 3 State estimation

To estimate the dynamic states and parameters of the nonlinear vehicle model an extended Kalmanfilter is used [6]. More sophisticated estimation methods for nonlinear processes like the Unscented Kalmanfilter produced similar results.

The estimation of influencing quantities, such as the banked corner angle $\phi_{\mathrm{road}}$ and tire parameters, is realized by models based on stochastic differential equations

$$\dot{\mathbf{x}} = \mathbf{f}_c(\mathbf{x}, \mathbf{u}, \mathbf{w}) \tag{9}$$

that will be discussed in the following sections. The whole state vector

$$\mathbf{x} = \begin{bmatrix} v_{\mathrm{y}} & \dot{\psi} & \phi_{\mathrm{road}} & c_{\alpha,\mathrm{f}} & c_{\alpha,\mathrm{r}} & \mu_{\mathrm{max,f}} & \mu_{\mathrm{max,r}} \end{bmatrix}^{\mathrm{T}} \tag{10}$$

involves the relevant motion states, influencing quantities and tire parameters.

The inputs

$$\mathbf{u} = \left[\delta_{\mathrm{f}}, \omega_{\mathrm{fl}}^{\mathrm{ECU}}, \omega_{\mathrm{fr}}^{\mathrm{ECU}}, \omega_{\mathrm{rl}}^{\mathrm{ECU}}, \omega_{\mathrm{rr}}^{\mathrm{ECU}}, F_{\mathrm{z,fl}}, F_{\mathrm{z,fr}}, F_{\mathrm{z,rl}}, F_{\mathrm{z,rr}}, \phi_{\mathrm{roll}}, \Delta M_{\mathrm{z}}\right]^{\mathrm{T}} \tag{11}$$

are the mean steering angle at the front wheels $\delta_{\mathrm{f}}$ from a steering model. The wheel speeds $\omega_{ij}^{\mathrm{ECU}}$ are needed for the longitudinal slip calculation. The inputs wheel loads

$F_{z,ij}$ and roll angle $\phi_{roll}$ are calculated by a suspension model. The additional moment $\Delta M_z$ is calculated by brake and drive torques.

The plant noise **w** is white and has a normal probability density function.

$$p(\mathbf{w}) \sim N(0, \mathbf{Q}) \tag{12}$$

Its covariance **Q** describes the magnitude of the disturbances and their correlations.

## 3.1 Tire parameters as state variables

The tire parameters $c_{\alpha,f}$, $c_{\alpha,r}$, $\mu_{max,f}$, and $\mu_{max,r}$ are modeled as very slow changing states. Therefore they are modeled as random walk processes

$$\dot{x}(t) = w\,(t). \tag{13}$$

A random walk is an integrated white noise [7]. Figure 2 shows the signal flow and figure 3 a simulation of this process.

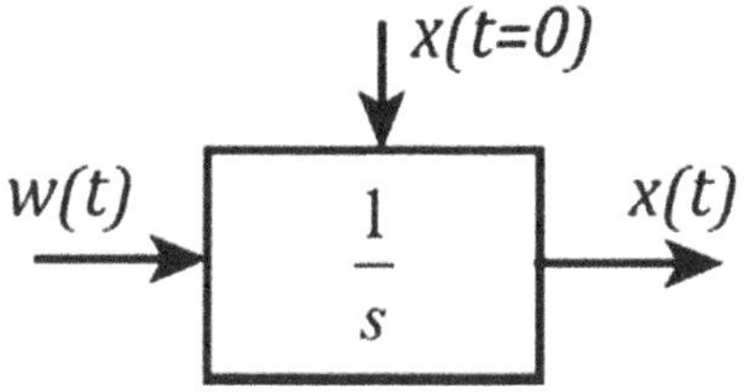


Fig. 2: Block Diagram of the Random Walk Process

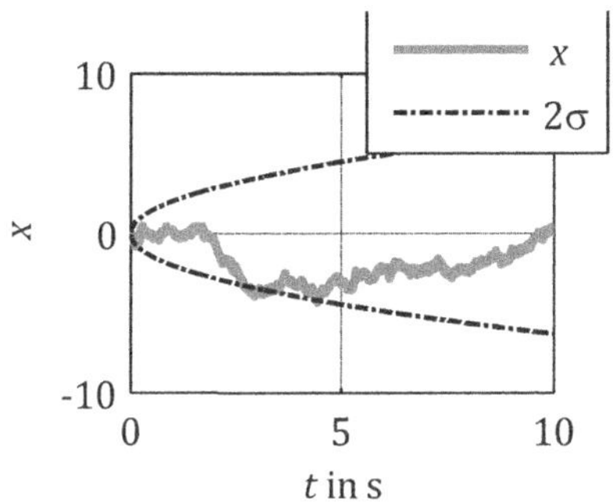


Fig. 3: Random Walk Process $\sigma(w(t)) = 1$

## 3.2 Banked corner angle as sate variable

In order to model the banked corner angle $\phi_{road}$ it is assumed, that the road has in an average a zero banked corner angle. In other words the mean of this stochastic process is known. With a stochastic process of first order

$$\dot{x}(t) = -\frac{1}{T} \cdot x(t) + \sigma\sqrt{\frac{2}{T}}w(t) \tag{14}$$

this property can be fulfilled. The application parameter $\sigma$ determines in which borders the stochastic process will be bounded. This way of modeling will cause the banked corner angle to converge to zero (like a PT1 element) with the time constant $T$ in case it is unobservable. Figure 4 shows the signal flow and figure 5 the simulation of this process.

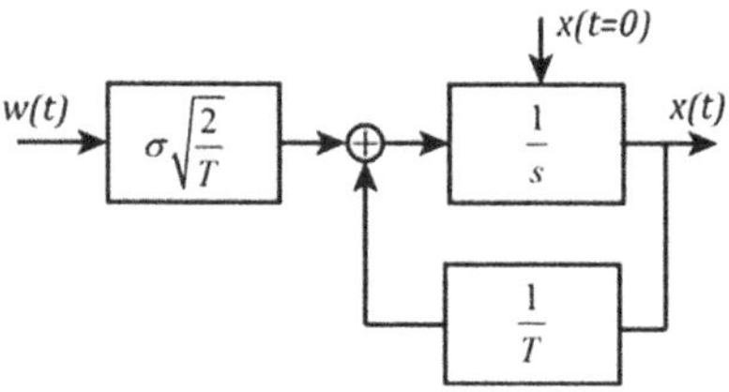


Fig. 4: Block Diagram of the Exponential Correlated Process

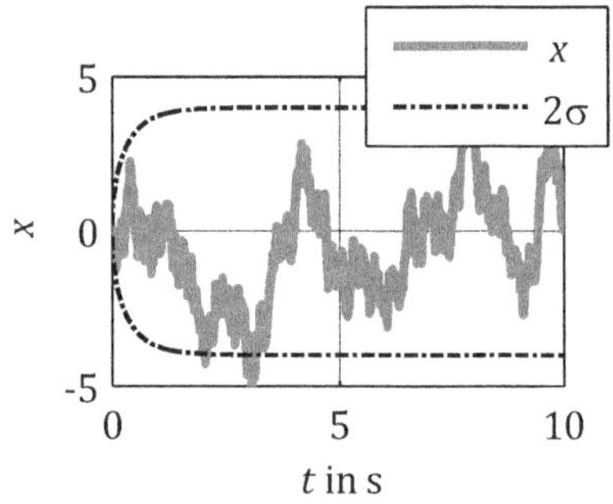


Fig. 5: Exponential Correlated Process and double of standart deviation calculated with a EKF, $x(t=0)=0,\ 2\sigma=4,\ T=1$

## 3.4Overview

The nonlinear state dynamics

$$\dot{\mathbf{x}} = \mathbf{f}_c(\mathbf{x}, \mathbf{u}, \mathbf{w})$$

$$\dot{v}_{\mathrm{y}} = \frac{1}{m}(F_{\mathrm{y,f}} + F_{\mathrm{y,r}}) - \sin(\phi_{\mathrm{road}}) \cdot g - v_{\mathrm{x}}\dot{\psi} + w_{\mathrm{vy}} \tag{15}$$

$$\ddot{\psi} = \frac{1}{J_{\mathrm{z}}}(F_{\mathrm{y,f}} \cdot l_f - F_{\mathrm{y,r}} \cdot l_r + \Delta M_{\mathrm{z}}) + w_{\dot{\psi}} \tag{16}$$

$$\dot{\phi}_{road} = -\frac{1}{T} \cdot \phi_{road} + \sigma\sqrt{\frac{2}{T}} w_{\phi} \tag{17}$$

$$\dot{c}_{\alpha,\mathrm{f}} = w_{c_{\alpha,\mathrm{f}}} \tag{18}$$

$$\dot{c}_{\alpha,\mathrm{r}} = w_{c_{\alpha,\mathrm{r}}} \tag{19}$$

$$\dot{\mu}_{\mathrm{max,f}} = w_{\mu_{\mathrm{max,f}}} \tag{20}$$

$$\dot{\mu}_{\mathrm{max,r}} = w_{\mu_{\mathrm{max,r}}} \tag{21}$$

follow from the equations (1) – (14). The nonlinear output model

$$\mathbf{z} = \mathbf{h}(\mathbf{x}, \mathbf{u}, \mathbf{v}) \tag{22}$$

$$a_{\mathrm{y}}^{\mathrm{ECU}} = \frac{1}{m}(F_{\mathrm{y,f}} + F_{\mathrm{y,r}}) + \sin(\phi_{\mathrm{roll}}) \cdot g + v_{\mathrm{ay}} \tag{23}$$

$$\dot{\psi}^{\mathrm{ECU}} = \dot{\psi} + v_{\dot{\psi}} \tag{24}$$

describes the linkage of measureable quantities with the system states. The vector $\mathbf{v} = [v_{\mathrm{ay}}\ v_{\dot{\psi}}]$ includes the white output noise

$$p(\mathbf{v}) \sim N(0, \mathbf{R}) \tag{25}$$

and has the covariance **R**.

Tab. 1: Initial model parameters

| Parameter | Description | Value | Unit |
|---|---|---|---|
| $m$ | overall vehicle mass | 1970 | kg |
| $J_{\mathrm{z}}$ | yaw moment of inertia | 3403 | $\mathrm{kgm^2}$ |
| $g$ | gravity acceleration | 9.81 | $\mathrm{m/s^2}$ |
| $l_f$ | distance front axle center of gravity | 1.52 | m |
| $l_r$ | distance rear axle center of gravity | 1.45 | m |
| b | wheel base | 1.62 | m |
| $c_{\alpha 0,\mathrm{f}}$ | front axle cornering stiffness | 120 | kN/rad |
| $c_{\alpha 0,\mathrm{r}}$ | rear axle cornering stiffness | 205 | kN/rad |
| $\mu_{\mathrm{max0,f}}$ | maximum friction coefficient of the front axle | 0.95 | - |
| $\mu_{\mathrm{max0,r}}$ | maximum friction coefficient of the rear axle | 1.05 | - |
| $\alpha_{\mathrm{max0,f}}$ | side slip angle with max. friction coefficient of the front axle | 8 | ° |
| $\alpha_{\mathrm{max0,r}}$ | side slip angle with max. friction coefficient of the rear axle | 10 | ° |

Fig. 6 illustrates the whole signal flow of the Kalmanfilter based estimator.

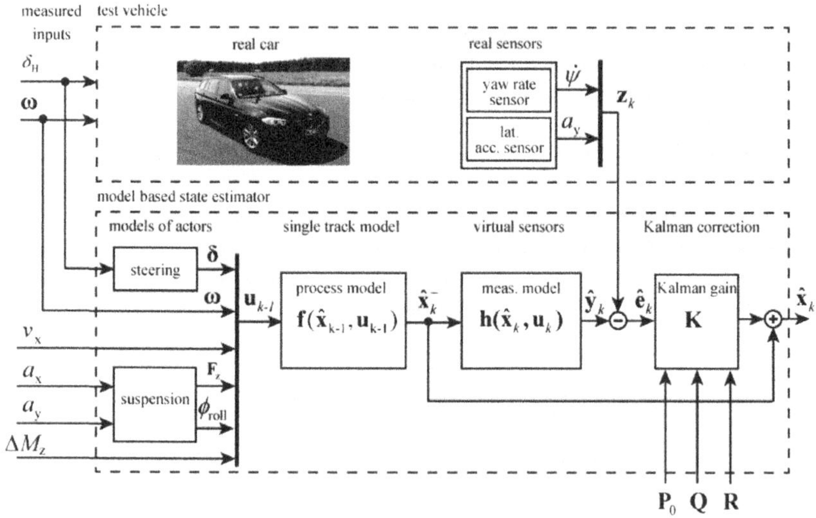


Fig. 6: Signal flow of the estimator

# 4 Sensitivity of cornering stiffnesses

The main difficulty of estimating cornering stiffnesses is the unknown side slip angle. Only the lateral acceleration and yaw rate can be measured. Another difficulty is the fact that different combinations of cornering stiffnesses at the front and rear axle can produce equal self-steering coefficients. In other words different combinations of cornering stiffnesses can result in the same stationary lateral acceleration and yaw velocity.

Fig. 7 should illustrate that phenomenon. The described single track model is simulated with input quantities of a real test drive. The signal STM 1 is the lateral acceleration with perfect identified cornering stiffnesses $c_{\alpha 1,f} = c_{\alpha 0,f}$ and $c_{\alpha 1,r} = c_{\alpha 0,r}$ from tab. 1. STM 2 shows the lateral acceleration with different cornering stiffnesses ($c_{\alpha 2,f} = 1.5 \cdot c_{\alpha 0,f}$) but equal self-steering coefficients $EG_1 = EG_2$. It is obvious that the outputs of the two models are only differentiable during the dynamic parts of the test drive.

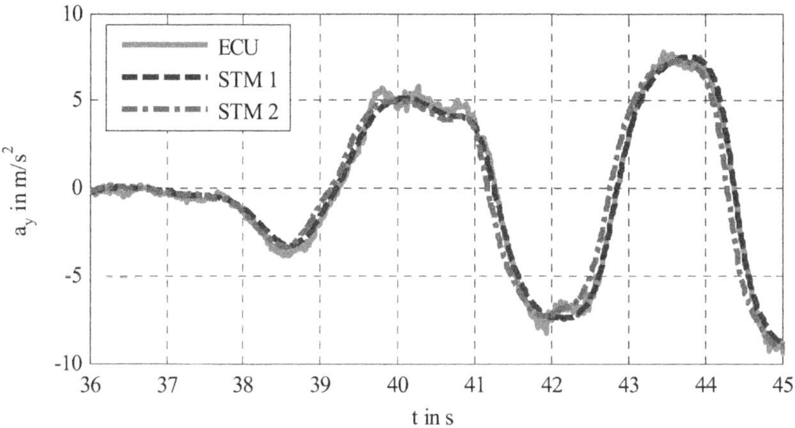


Fig. 7: Lateral acceleration during slalom maneuver from two single track models with equal self-steering coefficients and different cornering stiffnesses

Another method to show the dynamics depending sensitivity is a Bode diagram of the transfer function

$$G_{\text{ay}}(s) = \frac{a_y(s)}{\delta_f(s)} \tag{26}$$

with the same example of Fig. 7. Fig. 8 illustrates the differences of the lateral acceleration depending on the input frequencies of the steering angle. The amplitude at 0.5 Hz input frequency differs by approx. 5% with a cornering stiffness deviation of 150%. This example illustrates the challenge of estimating the cornering stiffnesses under real conditions with disturbances of banked corner angles regarding the lateral acceleration signal.

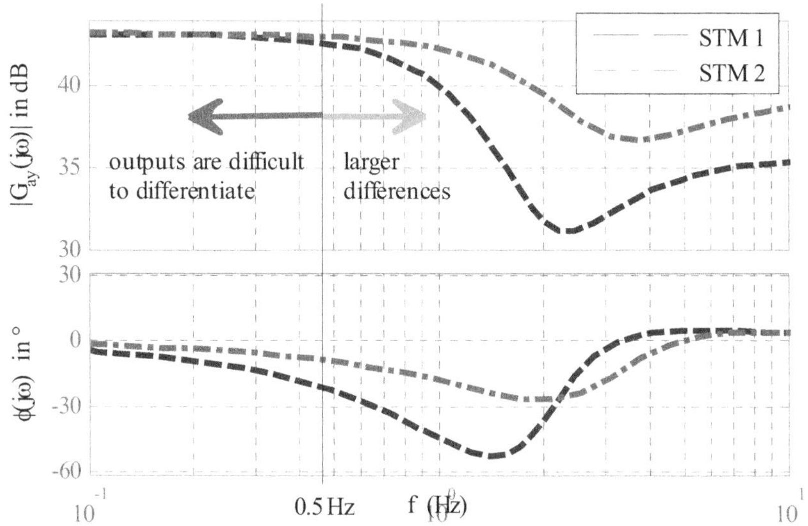


Fig. 8: Bode diagram of transfer function from steering angle to lateral acceleration. The linear single track models STM 1 and STM 2 have the same self steering coefficient but different cornering stiffnesses

Fig. 9 shows the simulation and the estimation of the random walk process of the cornering stiffness at the rear axle during stationary cornering (t<5s) and dynamic slalom (t>5s). One important detail of Kalman filtering is that the filter estimates its own estimation error. In Fig. 9 (right side) the estimation error and the estimated error in form of the 2σ-borderlines are displayed. The 2σ-borderlines are growing with time during the stationary maneuver and shrinks during the dynamic slalom. Thus the Kalmanfilter "knows" the unobservability of cornering stiffnesses during stationary maneuvers. Therefore no heuristic modifications are needed.

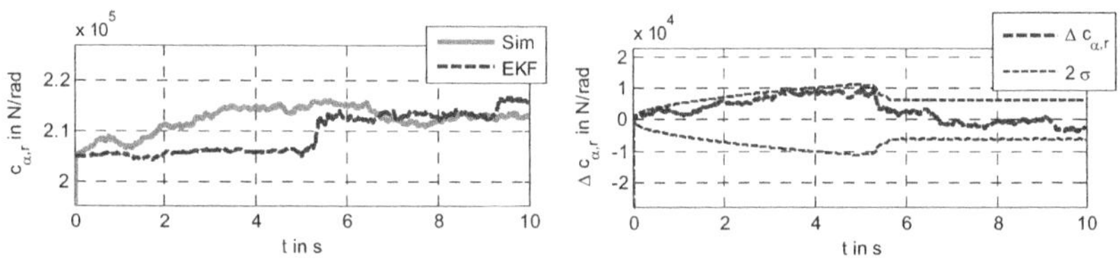


Fig. 9: Simulation and estimation of changing rear cornering stiffnesses during a stationary cornering (t<5s) and dynamic slalom (t>5s)

# 5 Results

First the estimation results of a test drive with large excitations on a plane test track are shown. Then a normal test drive on a public country road is examined.

## 5.1 Reference test drive

The reference test drive is a repeated slalom maneuver with a slight drift at the end followed by stationary cornering with an overdrawn steering angle. Thus the side slip angles on both axles are in highly nonlinear areas of the tire characteristic curve. The assumed initial cornering stiffnesses are different compared to the ones identified offline but result in an equal self-steering coefficient. This should challenge the estimation algorithm much more than starting with cornering stiffness that are obviously wrong even during stationary maneuvers.

Fig. 10 shows the measured and estimated states. The reference velocity $v$ is measured by a D-GPS, the side slip angle by an optical Correvit Datron sensor. To gain more insight to the data, the side slip angle at the rear axle is also plotted. The reference cornering stiffnesses and maximum friction coefficients (Tab. 1) were identified by an offline optimization method from ECU, Correvit and GPS data [8].

After approx. 60 seconds the tire parameters converge close to the reference values. The estimated side slip angle $\beta$ follows the Correvit signal well. Even during a drift with side slip angles about 20° at 53 seconds the side slip angle estimation fits the reference data accurately.

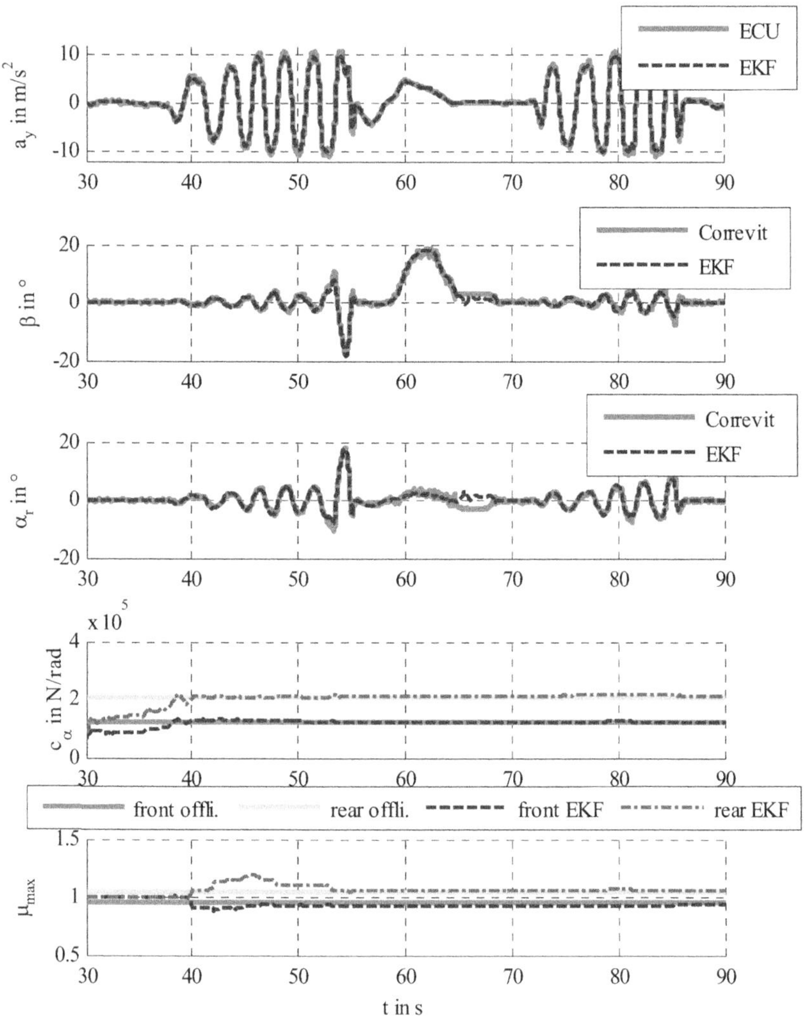


Fig. 10: Estimation of side slip angle, cornering stiffness and maximum friction coefficient during a test drive on an even test track, $v \approx 80$ km/h

To illustrate the final accuracy of the tire parameter estimation, the lateral axle forces

$$F_{y,f} = \frac{1}{cos(\delta_f)} \cdot \frac{J_z \cdot \ddot{\psi} + a_y \cdot l_r}{l_f + l_r} \qquad F_{y,r} = \frac{-J_z \cdot \ddot{\psi} + a_y \cdot l_f}{l_f + l_r} \tag{27}$$

are estimated with ECU signals. Fig. 11 shows the estimated tire forces (27) plotted against the axle side slip angle $\alpha_f$ and $\alpha_r$ measured via the Correvit sensor. The final estimated tire curve fits well to the measured data.

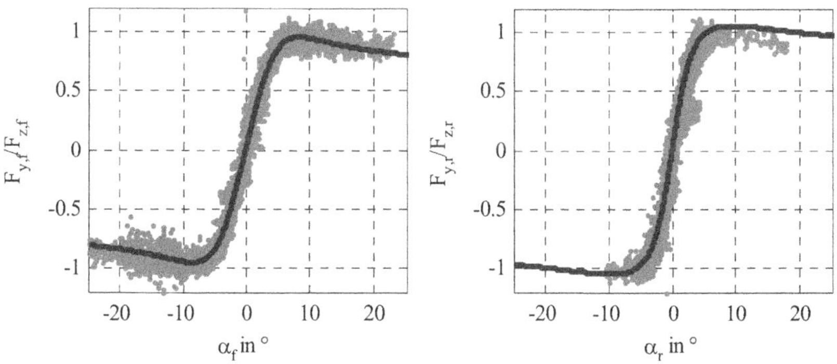


Fig. 11: The resulting estimation of cornering stiffness and maximum friction by the EKF and data measured with ECU and Correvit sensor.

## 5.2 Public Country Road

Compared to test drives on test tracks such large, dynamic excitations are not given during a normal drive on public roads. The lateral accelerations are lower and less dynamic. Anyway when e.g. the tires are changed the new cornering stiffnesses have to be recognized even with smaller excitations and more disturbances from the road to allow a model based control of lateral dynamics.

To take the real disturbances into account it is necessary to tune the Kalmanfilter correctly. Thus the tradeoff between fast and robust identification can be adjusted.

The 11.4 km long route on public country roads from “Großer Feldberg” to “Hohemark” near Frankfurt is shown in Fig. 12.

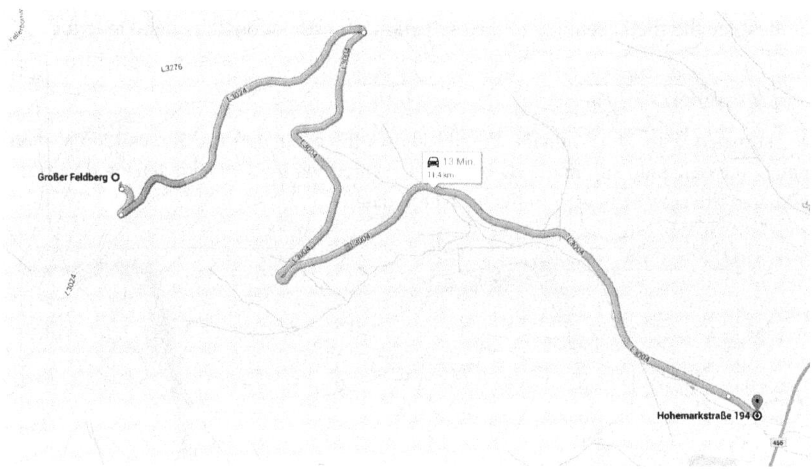


Fig. 12: 11.4 km long route of a normal ride on public country road from Großer Feldberg to Hohemark near Frankfurt

Fig. 13 shows the measured and estimated states. The maximum lateral acceleration is about 6 m/s². It takes more time for the convergence of cornering stiffness estimation (about 4 minutes). The resulting cornering stiffnesses are underestimated a slightly by 15%. $\mu_{\max}$ stays equal to the initial value because of the lower excitations.

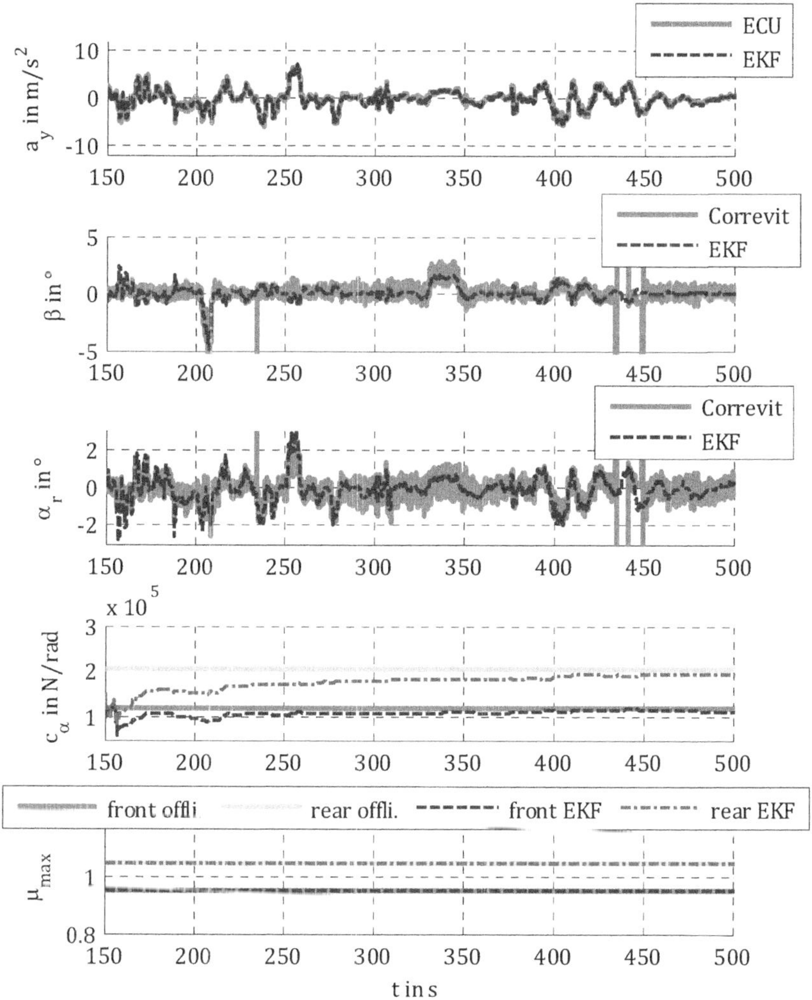


Fig. 13: Estimation of side slip angle, cornering stiffness and maximum friction coefficient during a normal ride on a country road

# 6 Conclusion

By studying the sensitivity of cornering stiffness, it is obvious that only dynamic maneuvers ($f > 0.5$ Hz) are useful to estimate cornering stiffnesses.

The estimation method presented in this publication is able to show good estimation results even under less extreme and more stationary maneuvers on public roads. The relative error of the cornering stiffnesses was about 15%. To take the real disturbances into account it was necessary to tune the Kalmanfilter covariance matrix $\mathbf{Q}$ and $\mathbf{R}$ correctly. During high accelerations up to the physical limits, it was also possible to estimate the maximum friction coefficient accurately.

Hence, adaptive observers of the side slip angle $\beta$ or adaptive control of the lateral vehicle dynamics in productions vehicles becomes possible by knowing of changing tire parameters.

# References

[1] D. Wesemeier, "Modellbasierte Methoden zur Schätzung nicht messbarer Größen der Fahrzeugquerdynamik und des Reifenluftdrucks," Dissertation, Institut für Automatisierungstechnik und Mechatronik, TU Darmstadt.

[2] Matthias Schorn, "Modelle zur Beschreibung des Fahrzeugverhaltens," in *Fahrdynamik-Regelung: Modellbildung, Fahrerassistenzsysteme, Mechatronik*, R. Isermann, Ed, Wiesbaden: Friedr. Vieweg & Sohn Verlag, 2006, pp. 27–46.

[3] H. B. Pacejka, *Tyre and vehicle dynamics,* 2nd ed. Oxford: Butterworth-Heinemann, 2006.

[4] R. Isermann, Ed, *Fahrdynamik-Regelung: Modellbildung, Fahrerassistenzsysteme, Mechatronik*. Wiesbaden: Friedr. Vieweg & Sohn Verlag, 2006.

[5] C. Ackermann, M. Bauer, and R. Isermann, "Identification of roll dynamic behaviour of vehicles using a gyro-box and GPS," in *13. Stuttgarter Symposium*

[6] Greg Welch and Gary Bishop, "An Introduction to the Kalman Filter," *UNC-Chapel Hill*, no. TR 95-041, 2006.

[7] M. S. Grewal and A. P. Andrews, *Kalman filtering: Theory and practice using MATLAB,* 3rd ed. Hoboken NJ: Wiley, 2008.

[8] J. Bechtloff, M. Bauer, C. Ackermann, and R. Isermann, "Fast Identification of a detailed Two-Track Model with Onboard Sensors and GPS," in *chassis.tech*: ATZlive, Springer Vieweg, 2014.

# Damage on alloy wheels for motor vehicles – limits of rework

Thomas Kollmeier, TÜV SÜD Product Service GmbH

## Damage on alloy wheels for motor vehicles – limits of rework

Thomas Kollmeier, TÜV SÜD Product Service GmbH

**17th of June 2015 @ chassis.tech plus**

TÜV SÜD Product Service GmbH 17th of June slide 1 TÜV®

## Structure

**Part 1 Fundamentals and initial situation**

Chapter 1: Representation of existing international and national regulations, standards and guidelines for rework

Chapter 2: Presentation of damage

Chapter 3: Situation in the market and rework companies

**Part 2 Realized investigations**

Chapter 1: Clasification of damage

Chapter 2: Studies on the individual damage areas

Chapter 3: Derivation and summary of considered admissible rework at wheels

TÜV SÜD Product Service GmbH 17th of June slide 2 TÜV®

## Part 1 Fundamentals and initial situation

### Chapter 1: Representation of existing international and national regulations, standards and guidelines for rework

**EUWA (Association of European Wheel Manufacturers):**
„The repair of a damaged rim or disc by heating, by welding, by addition or removal of material is absolutely forbidden".

**ETRTO (Technical organization of the European tire and wheel manufacturers):**
"Damaged or deformed wheels and wheels with cracked or deformed bolt holes can not be repaired or put into operation."

**ISO 14400, section 5.**
"After removing the wheels and their fasteners, this in-depth review with respect to their safe state. Namely those parts without breakage, deformation, corrosion, significant wear or other imperfections are. In addition, no technical changes to the wheel that should be made. Weld repair in the sense of material gifts in addition rim or hub area, due to fractures, cracks, tears or significant erosion are not allowed because they can lead to additional stresses in the critical areas."



## Part 1 Fundamentals and initial situation

### Chapter 2: Presentation of damage

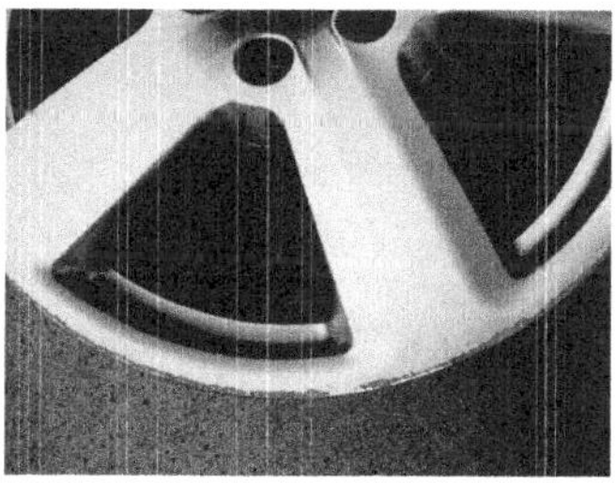

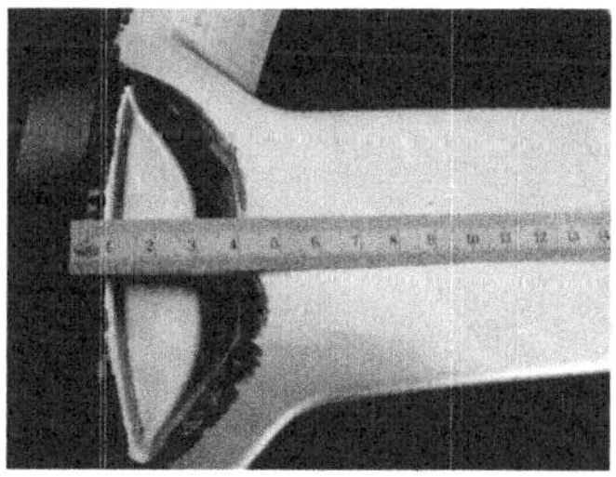

Pictures: Guideline for retreatment process of wheels (Audi AG)

- left: Only superficial damage on the varnish
- right: No deeper damage than 1mm in the basic material and distance less than 50 mm from the outer rim

## Part 1 Fundamentals and initial situation

### Chapter 2: Presentation of damage

Pictures: Guideline for retreatment process of wheels (Audi AG)

- left: Deformation inner rim flange
- right: Damage deeper than 1mm in the basic material

TÜV SÜD Product Service GmbH 17th of June slide 5 TÜV®

## Part 1 Fundamentals and initial situation

### Chapter 2: Presentation of damage

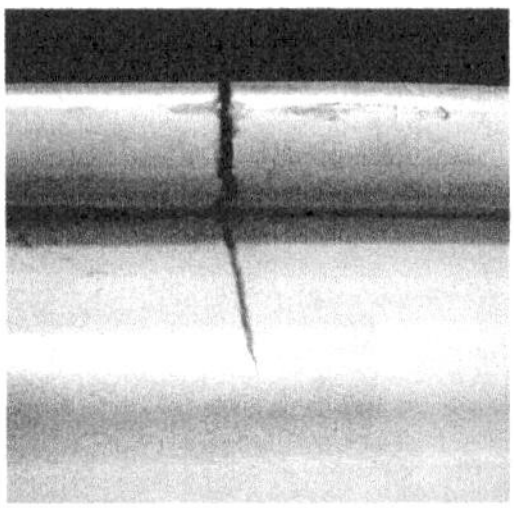

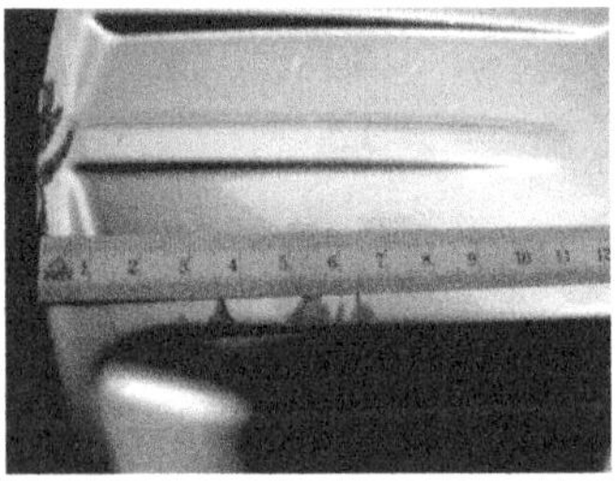

Pictures: Guideline for retreatment process of wheels (Audi AG)

- left: Crack in the rim (area inner rim flange)
- right: Damage more than 50 mm distance from the outer rim flange

TÜV SÜD Product Service GmbH 17th of June slide 6 TÜV®

## Part 1 Fundamentals and initial situation

### Chapter 3: Situation in the market and rework companies

Example:

CARTEC Autotechnik Fuchs „Wheel Doctor"

Die Aluklinik Mertens Mönchengladbach

Felgenretter at Oberhausen

Felgendoktor.de Munich

- Repair of serious deformations
- Heating up of the wheel
- recovery at the wheel
- Welding work
- Mechanical post processing

TÜV SÜD Product Service GmbH 17th of June slide 7 TÜV®

## Part 2 realized investigations

### Chapter 1: Clasification of damage

**Abrasive wear and corrosion**

- Influencing factors:
  - Growing wheel load
  - Growing engine power
  - Material pairing bolt/wheel bearing
  - Condition of wheel attachement face

**Cracks, fracturess**

- In presence of:
  - Pothole
  - Road curbs
  - barriers (mostly with low section tires)
  - overload

**Deformations**

- Resulting from:
  - Road condition
  - Driving style
  - Wheel stiffness

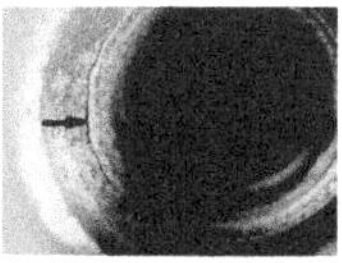
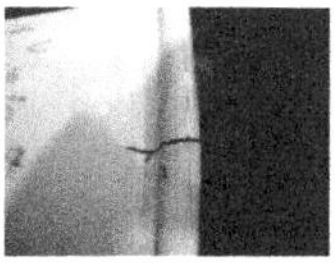

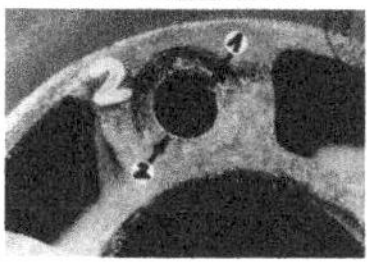

TÜV SÜD Product Service GmbH 17th of June slide 8 TÜV®

## Part 2 realized investigations

### Chapter 2: Studies on the individual damage areas

**Limitations and self-understanding :**

- Considering the well-known recommendations, guidelines and standards
- At Rework it is not allowed in any case to change material structure and -behaviour as well as reduce the life time of a wheel
- Mechanical post precessing is just on non critical areas possible

**The following is deriving:**

Recovery of (serious) deformaions and welcding work is not permissible!

TÜV SÜD Product Service GmbH 17th of June slide 9 TÜV®

## Part 2 realized investigations

### Chapter 2: Studies on the individual damage areas

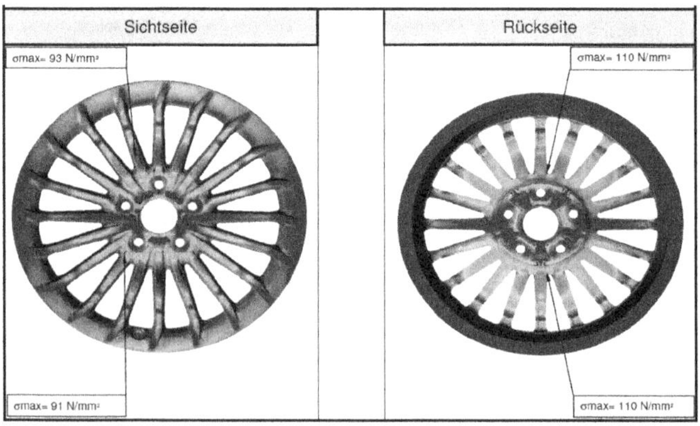


Picture Borbet Austria

TÜV SÜD Product Service GmbH 17th of June slide 10 TÜV®

## Part 2 realized investigations

### Chapter 2: Studies on the individual damage areas

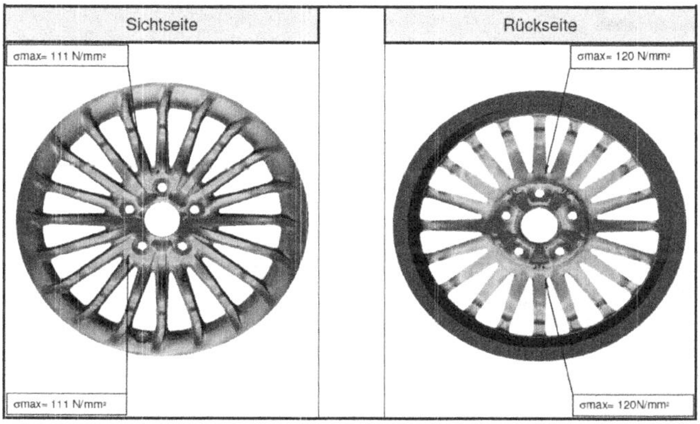


Picture Borbet Austria



## Part 2 realized investigations

### Chapter 2: Studies on the individual damage areas

**Result from the FEA:**

Change of stress due to aditional machining of the design side of about 1 mm

- Outside: + 20 %
- Back: + 9 %

⇨ Reduction of life time round about 50%

## Part 2 realized investigations

### Chapter 2: Studies on the individual damage areas

**Damage on wheels**
2 mm deep notches as well as damaged areas on outer rim flange and spokes within and beyond the 50 mm of the outer rim flange

**The following tests were conducted**
- Bending fatigue test under dry conditions
- Bending fatigue test under corrosion
- Energy absorption test

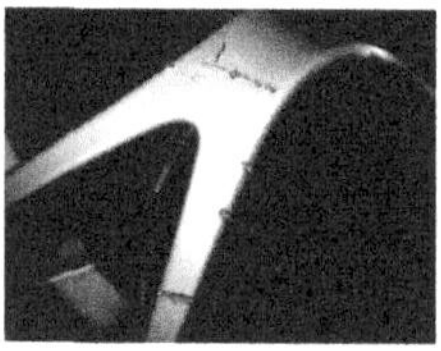
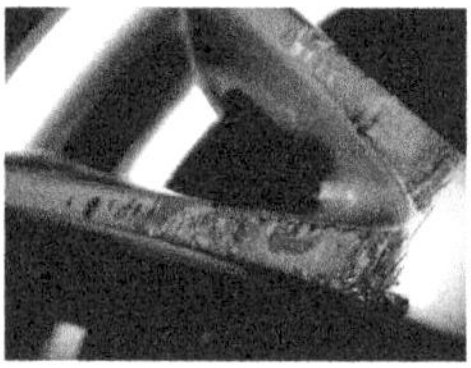
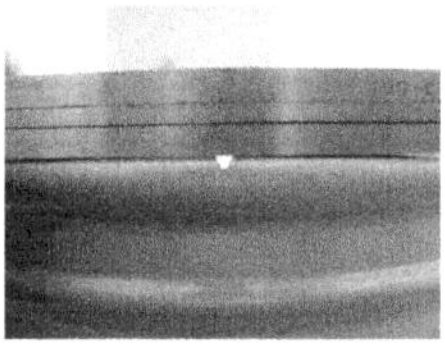

Pictures: guideline for rework on wheels (Audi AG)

TÜV SÜD Product Service GmbH 17th of June slide 13 TÜV®

## Part 2 realized investigations

### Chapter 2: Studies on the individual damage areas

**Result of bending fatigue test under dry conditions and corrosion:**

| | new wheel | | damaged wheel | | |
|---|---|---|---|---|---|
| | reached Lc till failure | failure area | reached Lc till failure | failure area | remarks |
| BFT dry Design 1 | 4.031.843 | spoke at bolt hole | 4.740.262 | spoke at bolt hole | no cracks at reworked area within 50 mm cracks at rework area beyond 50 mm visible |
| BFT corrosion Design 1 | 517.900 | area of bolt holes | 730.574 | spoke at reworked area beyond the 50 mm | no cracks at reworked area within 50 mm |
| BFT dry Design 2 | 5.981.111 | middle of spoke | 3.471.890 | middle of spoke | no cracks at reworked area within 50 mm cracks at rework area beyond 50 mm visible |
| BFT corrosion Design 2 | 492.453 | area of bolt holes | 802.430 | spoke at reworked area beyond the 50 mm | no cracks at reworked area within 50 mm |

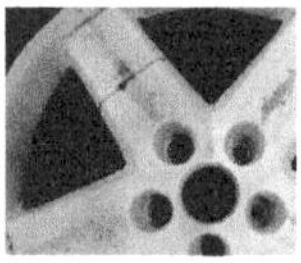
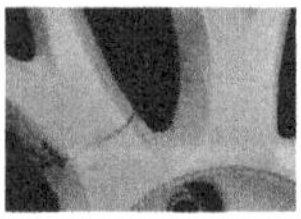

Pictures: guideline for rework on wheels (Audi AG)

TÜV SÜD Product Service GmbH 17th of June slide 14 TÜV®

## Part 2 realized investigations

### Chapter 2: Studies on the individual damage areas

**Result of energy absorption test:**

- The energy absorption at notches with sharp edges (should represent the damaged areas) are very low cause of notching effect
- The energy absorption is much higher at the rounded notches (should represent the damaged areas after reworking) due to lower notch effect

| | Notch with sharp edges | Rounded notch | New area |
|---|---|---|---|
| Energy absorption till breack Design 1 | 107 Joule | 154 Joule | 373 Joule |
| Energy absorption till breack Design 2 | 159 Joule | 198 Joule | 294 Joule |
| Energy absorption till breack Design 3 | 239 Joule | 373 Joule | 414 Joule |

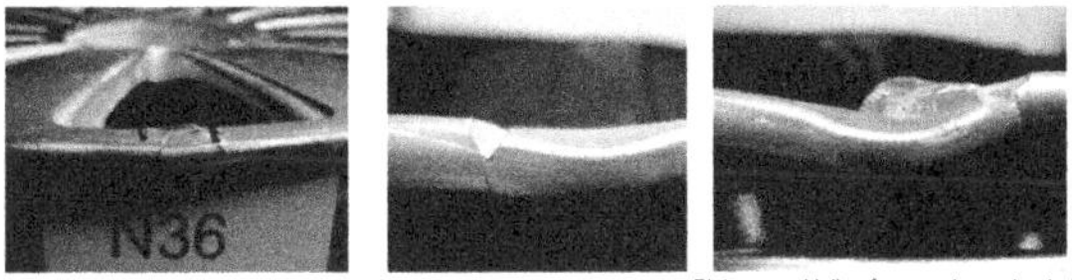


Pictures: guideline for rework on wheels (Audi AG)



## Part 2 realized investigations

### Chapter 2: Studies on the individual damage areas

**Thermal paint removal for repainting**

Due to heat nearly 40 percent loss of material hardness and up to 90 percent of their lifetime, tests were oriented on in-market thermal processes :

- Heating of wheels for paint stripping and applying new granules to over 200°C during several hours
- The test wheels were several light-alloy wheels from OE quality GK-Al-Mg Si7 T6
- After heating, bending fatigue tests were conducted

## Part 2 realized investigations

### Chapter 2: Studies on the individual damage areas

**Decrease of hardness after a temperature influence of 3 hours**

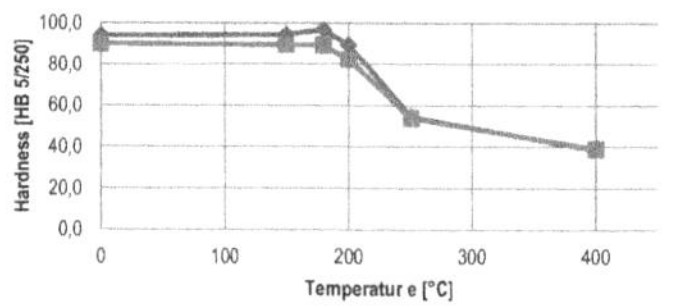


**Decrease of hardness after a temperature influence of 0,5 hours**

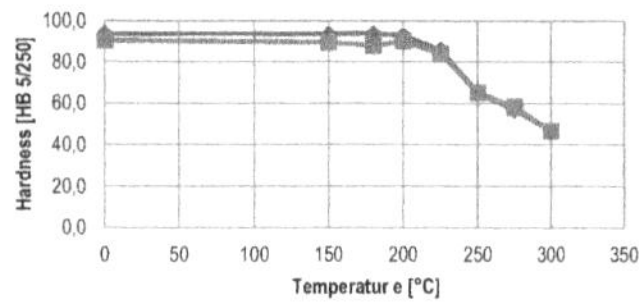


Result of hardness tests:

- After a temperature influence of 3 hours was already at 200°C a noticeable decrease of hardness (approx. 5%) realized
- After 3h at 250°C a decrease of hardness of about 40% is already existing. At higher temperatures the trend is increasing even further.
- Within 30 minutes of temperature influence, beginning from 225°C a noticeable decrease of hardness were determine (approx. 8%)
- After 0,5h at 250°C there is a decrease in hardness of about 25%. At higher temperatures the trend is increasing even further.

TÜV SÜD Product Service GmbH 17th of June slide 17 TÜV®

## Part 2 realized investigations

### Chapter 2: Studies on the individual damage areas

**Result comparative bending fatigue test:**

| | Bending moment | reached Ls till failure | failure area |
|---|---|---|---|
| new wheel | 6.200 Nm | 1.626.762 | crack at spoke |
| wheel after temperature influence (3h at 250°C) | 6.200 Nm | 173.023 | crack at spoke |
| new wheel | 5.000 Nm | 15.000.000 | no cracks |
| wheel after temperature influence (3h at 250°C) | 5.000 Nm | 1.564.178 | crack at spoke |

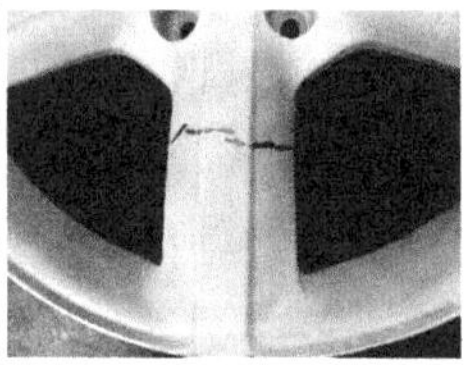

- ➢ Wheel durability minimized by about 90 %
- ➢ Legal guidelines (§30 StVZO/ECE-R 124) no longer fulfilled

TÜV SÜD Product Service GmbH 17th of June slide 18 TÜV®

## Part 2 realized investigations

### Chapter 2: Studies on the individual damage areas

**Damage on wheels**
Deformation at the inner rim flange
Material removal at outer rim flange

**Rework**
Recovery of the deformation
Welding on the outer rim

**The following tests were conducted after rework and finishing of the wheels**

- Hardness tests
- Material testing
- Micrograph

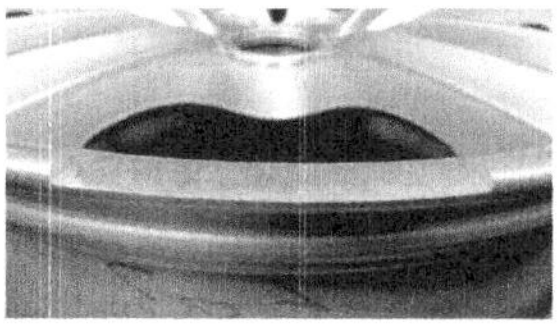

TÜV SÜD Product Service GmbH 17th of June slide 19 TÜV®

## Part 2 realized investigations

### Chapter 2: Studies on the individual damage areas

**Result material testing:**
max. permissible deviation of results to new wheel ± 3% (expected variation at production)

**After recovery of deformation:**

| Rim Well | $R_{p0,2}$ [MPa] | $R_m$ [MPa] | A [%] |
|---|---|---|---|
| New wheel | 204,9 | 240,2 | 1,7 |
| Reworked wheel | 209,6 | 243,6 | 1,9 |
| Variation [%] | 2,3 | 1,4 | 11,8 |

| Inner rim flange | $R_{p0,2}$ [MPa] | $R_m$ [MPa] | A [%] |
|---|---|---|---|
| New wheel | 230,2 | 297,3 | 9,3 |
| Reworked wheel | 247,3 | 301,5 | 6,5 |
| Variation [%] | 7,4 | 1,4 | -30,1 |

- At rim well the ultimate strain is rising up over 3% so the wheel is a bit softer after rework
- At inner rim flange the ultimate strain is heavily decreasing after recovery, so the trend to a crack in the becoming more brittle rim flange is much higher while driving, even if the wheel gets a new deformation at the inner rim flange

**After welding on outer rim flange:**

| Outer rim flange | $R_{p0,2}$ [MPa] | $R_m$ [MPa] | A [%] |
|---|---|---|---|
| New wheel | 235,1 | 290,6 | 8,3 |
| Reworked wheel | 110 | 221 | 15,5 |
| Variation [%] | -53,2 | -24,0 | 86,7 |

- Due to heating input during the welding, the strength of the outer rim flange is heavily decreasing, so an impact test 13° at that area could be negative and legal guidelines (§30 StVZO/ECE-R 124) are no longer fulfilled

TÜV SÜD Product Service GmbH 17th of June slide 20 TÜV®

## Part 2 realized investigations

### Chapter 2: Studies on the individual damage areas

**Result micrograph**

- No changes at the reworked areas inner rim flange and rim well
- higher porosity at the welding zone than in the wheel material
- high porosity at the transition zone welding / wheel

**Result brinell hardness test (HB)**

| | Outer rim flange at welding [HB] | Rim well at recovery deformation [HB] | inner rim flange at recovery deformation [HB] |
|---|---|---|---|
| New wheel | 92,5 | 86,2 | 94,3 |
| wheel | 62,3 | 85,9 | 96,2 |
| Variation [%] | -32,6 | -0,3 | 2,0 |

- The hardness test confirmed the tensile testing with the softer outer rim, due to heating input, the little changes at the rim well and the becoming more brittle structure at the inner rim flange



## Part 2 realized investigations

### Chapter 2: Studies on the individual damage areas

**Summary of the realisation:**

- Recovery the primitive state of deformations on wheels not possible till now
- Every material changes its properties by deformation processes
- Reduction of cross-sections in critical areas leads to the reduction of life time
- Negative influences on material upon heating
- Changing the material composition by separation and crystallization processes
- Embrittlement by rapid cooling, inhomogeneity by diffusion processes and incorporation of additives
- Altered properties by re-heat treatment

## Part 2 realized investigations

### Chapter 3: Derivation and summary of considered admissible rework at wheels

- Only casted and forged alloy wheels can be reworked
- To be sure that no deformation occurred, a concentric- and axial run out must be done (target max. 0,5 mm)
- The depth of the damage in the base material of max. 10% of the cross section (width rim flange), but not more than 1 mm, may be exceeded under any circumstances
- Proper preparation to the maximum damage depth in the base material of 1 mm is only in the range of 50 mm in the radial direction from the outer rim flange allowed
- Beyond the reworkable areas only a preparation of the coating is permitted
- Max. exposure temperature / time of $90^{\circ}C$ / 40 min. must not be exceeded during the painting
- Heat input and welding processes of any kind are not permitted
- Wheels with cracks may not be reworked and must be replaced immediately
- Material recovery is not allowed
- The attachment face of the wheel, bolt holes, center hole, valve seat, and the inner surface of the spokes and the rim must not be reworked or painted.
- The tire seat surfaces on the rim must be considerate of the danger of slipping not be reworked or painted and must be covered with suitable materials

⇨ ***So only optical rework, which means basically the professional technical restoration of the wheel in terms of optical defects by polishing, local grinding, rounding of notches, possible filling, priming and painting, is permitted!***

TÜV SÜD Product Service GmbH 17th of June slide 23 TÜV®

# ....thank you for your attention

TÜV SÜD Product Service GmbH 17th of June slide 24 TÜV®

# Evaluation of different modeling approaches for the tire handling simulations – analysis and results

Author: Francesco Calabrese *

Co-authors: Manfred Baecker *,Axel Gallrein *

*Fraunhofer Institut fuer Techno- und Wirtschaftsmathematik (ITWM)
Fraunhofer-Platz 1 D-67663 Kaiserslautern, Tel: +49 631 31600-4749

# Abstract

In the last years the need of more accurate vehicle handling simulation is increased. This is due to the wider possibilities that the new CAE methods offers to the vehicle dynamic engineers. In the past the semi-empirical tire models have been the most used, but today it seems to be not enough. In vehicle handling simulation more component become to be important like 3D road, thermal effect on rubber, asphalt conditions and electronic safety systems interaction with the tire dynamic. The cited semi-empirical tire model seems to be not adequate to simulate such complex scenario also if they remain still attractive for a group of simulations scenario due to their simplicity and efficiency.

This paper shows that in order to increase the reliability of the tire models also in the described extreme conditions, it is necessary to couple a thermo-dynamical model with a mechanical one. The thermal model is important to estimate the temperature propagation inside the tire structure and the temperature evolution over time. The mechanical model is responsible for the heat and temperature creation and to understand how the temperature has an influence on the tangential force that the tire applies on the road.

The described thermal model is designed to be used as a module and is applicable to tire models of various modeling details. In this paper, the coupling with different type of tire models, semi-empirical or fully physical structural model, is described and the weaknesses and strengths of this different solutions are analysed in order to full fill all the possible simulation scenario.

At the end of the paper, the capabilities of the overall models are demonstrated and qualified in some illustrative tire and vehicle simulation scenario.

# 1 Introduction

For many years, in the past, semi-empirical tire models like magic formula [1] have been the only one used in multi-body handling vehicle simulations. This model is a full fitting scale scheme with the big advantageous of the fast calculation time given the analytical formula. Meanwhile the full 3D structural tire models [6] have been used only in the field of comfort, ride and sound propagation.

In the last 5 years, with the increasing of the complexity of the CAE vehicle simulation software, engineers are trying to simulate even more complex scenario for handling, than full physical structural tire model became fundamental also in this kind of applications. A list of possible simulation studies is showed in the Table 1. Many of these experiment, like steady state manoeuvre, can be fully simulated by using Magic Formula with a good accuracy but others cannot be performed for the following reason:

- Temperature effect
- 3D Road Surface
- Contact patch accurate behavior
- Friction coefficient temperature and asphalt dependent
- Tire lag
- Longitudinal, lateral and vertical dynamic up to 30 Hz
- Tread pattern influence on handling

In addition it has to be considered also the temperature effect [3] [4] [12]. When the tire works in large slips values (longitudinal slip major than 8% and slip angle major than 3°) the dynamic friction has a preponderant role and it is strongly influenced from the temperature level, as shown in many tribological studies [5] [13] [14]. In the past in tire modelling rarely the effect of the temperature on the tire behaviour has been taken into account. In all the applications in which the tire works in the linear slip stiffness zone, to neglect the temperature effect on the tire characteristic can be also considered a good compromise to simplify and optimize the modelling approach without losing quality.

Table 1. Possible Handling simulation scenario and

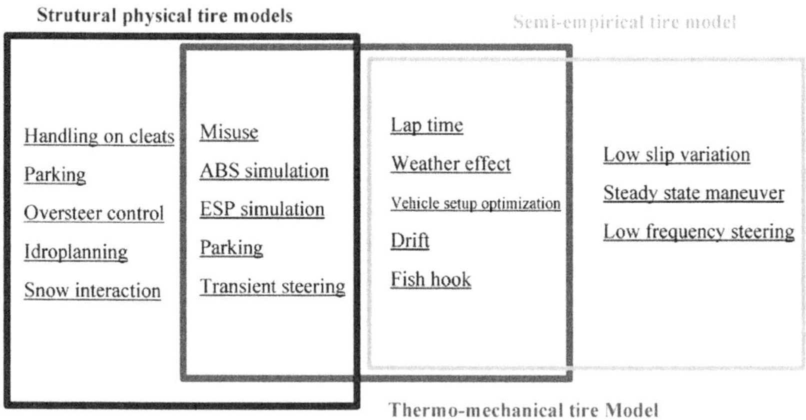


Looking again at the Table 1 basically all the experiment can be full filled by using a structural physical tire model in combination with a thermal model. The Magic Formula results still attractive or necessary when:

- The calculation time is important. Real Time.
- In all the cases in which the experiment are very long.
- For some operation of optimization in combination with regressive algorithm.
- When 3D road, structural effect, asphalt property, tread pattern, predictive capability are not important.

In this paper the two strategy are presented and compared trough analysis and results:

- CDTire/MF++, a MF based model sensible to thermal effect
- CDTire/3D combined with CDTire/Thermal, a fully structural model able to perform hard handling simulation scenario with thermal effect.

  In the following paragraphs the Thermal model, the MF++ and CDTire/3D will be explained and some application scenario will be analysed in order to understand the differences.

# 2 Thermal tire model

Thermal modeling is relatively new in tire simulations [2] [3] [4]. For this reason the new Thermal model [12] will be analysed in details. In the following paragraphs details regarding importance on handling characteristic, equations, and some results of simulations will be shown.

## 2.1 Importance of the thermal effect on tires and the mechanism of Temperature generation

Tires have a very complex composite structure made of different layers such as carcass, bandage and steel cords kept together with the filled rubber. The rubber compound has a prominent role for the tread behaviour. Indeed due to its viscoelasticity properties the rubber is capable of dissipating energy in order to create friction and absorb impact.

Usually the heat generation due to the rubber strains is the most important source for temperature creation. At the same time as temperature arises, also the rubber loss and storage modulus decrease.

The rubber-asphalt interaction mechanisms are influenced as result of these phenomena because the rubber penetration into the asphalt asperities depends on the rubber compliance. Therefore the resulting friction coefficient is very sensitive to the temperature variation. This effect can be seen really clearly during a sweep angle experiment (See figure 1). In the following figures the temperature-force correlation during this kind of standard test is shown [2]. If the test is performed over a relatively long time (dark gray line – quasi-static) the temperature can also reach variations from 60 ° C up to 80° C respect to the initial temperature (See figure 1). In this case it is possible to observe that the peak of the force characteristic curve changes when the temperature rises. Obviously these phenomena have an impact on the tire performance given that the friction is not always at its optimal level but changes with the temperature. If the test is performed in a relatively short time (grey line) the heat has not enough time to penetrate into the tread and warm up the tire, so in this case the temperature results are quite constant during the experiment, but the measurement is not suitable for parameter identification purposes because of the presence of the relaxation length effect in the slip stiffness.

Furthermore because of temperature creation and propagation there are also some secondary effects like the internal pressure and rolling resistance change. However in this article only the friction and inflation pressure dependency will be analyzed.

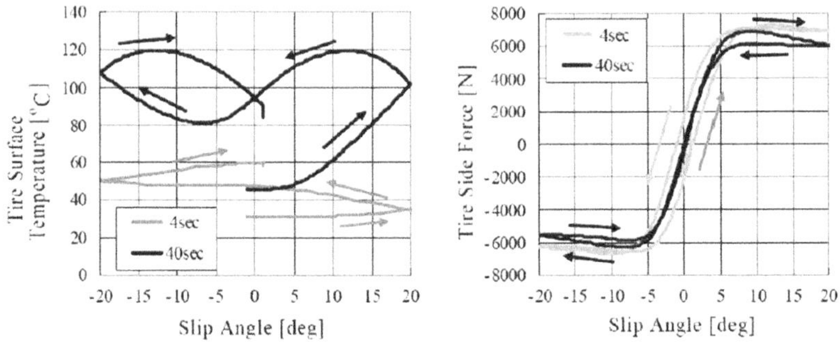


Figure 1. Lateral force during a side slip angle variation for different time length of the experiment, right figure. Temperature during a side slip angle variation for different experiment durations, left figure.

The phenomena responsible for the temperature variation of the rubber are basically the following [2] [3] [4]:

- Energy dissipated by the rubber strains
- Convection with the external air
- Friction in the sliding zone of the contact patch
- Conduction with the asphalt through the contact patch
- Propagation trough the tire volume

## 2.2 Model equation description

The heat conduction phenomena can be described by the Fourier diffusion equation (1). The equation can be integrated on the discretized volume of the tire (figure 2) using the so-called finite volume approach and applying the Navier-Stokes theorem (2)

$$\frac{\partial}{\partial x}\left(k\frac{\partial T}{\partial x}\right)+\frac{\partial}{\partial y}\left(k\frac{\partial T}{\partial y}\right)+\frac{\partial}{\partial z}\left(k\frac{\partial T}{\partial z}\right)+\dot{q}_g=\rho c_p\frac{\partial T}{\partial t} \tag{1}$$

$$\int_V(\nabla\cdot k\nabla T+\dot{q}_g)dV=\oint_{s_i}k\frac{\partial T}{\partial n}ds_i+\int_V\dot{q}_g\cdot dV=\int_V\rho\cdot c_p\dot{T}\cdot dV \tag{2}$$

The equation (2) obtained after the integration represents the energy equilibrium of a single finite volume element: In this equation the sum of the thermal fluxes $q_{tot}$ that enter in the element through the surfaces with the internal energy generation $E_g$ gives the variation of internal energy of the thermal cell $E_{st}$ for an infinitesimal instant of time.

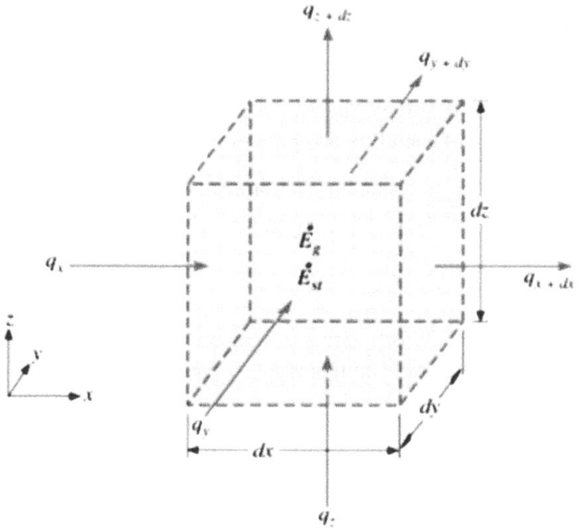

Figure 2.Representation of the thermal energy equilibrium of a single finite volume element.

The finite volume approach has the advantage over finite differences in that it allows using volume elements of arbitrary shape. This aspect is important because it allows having a non-uniform discretization without error in the energy balance calculation of the domain.

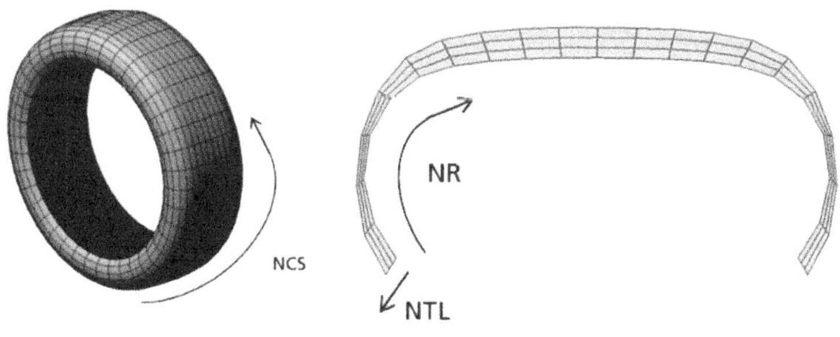


Figure 3.Example of discretization. NR, NTL, NCS are the three integer number that define how fine is the discretization along the thickness, the section and the circumferential direction.

The full volume of the tire is discretized along the circumferential, sectional and thickness direction. Such a fine discretization is necessary in order to also describe asymmetric scenarios of generation and propagation (e.g. large camber or full sliding condition during braking) and the various material properties along the thickness direction.

All the possible thermal phenomena in the tire thermodynamics can be included as flux generation at the boundary of the single thermal cell without changing the form of the equation (2). The basic form of each flux is represented as follow:

- Friction power density: it can be described as the product of the tangential force multiplied for the sliding velocity.

$$\vec{q}_{fp-i} = \tau_i \cdot v_s \tag{3}$$

- Road Conduction: it depends on the difference in temperature between road and asphalt and on the thermal resistance. The thermal resistance depends on rubber-asphalt conductivities.

$$\vec{q}_{rc} = R_{asphalt-rubber}(T_{road} - T_{rubber}) \tag{4}$$

- Convection with the external gas: It depends on the difference of temperature between air and tire and on the convection coefficient. The convection coefficient H is calculated by using the Nusselt number (Nu). For this calculation the basic assumption is that the tire has the shape of a cylinder. The convection coefficient will be a function of the tire geometry, the relative velocity between tire and external air and of the air conductivity

$$\vec{q}_{ig} = H_{fc}(T_{ext-air} - T_{rubber}) \tag{5}$$

- Heat generated by the rubber strains: It can be locally described as integral on the volume of the stress that multiplies the rubber strains. During the normal operational condition it will depend on the amplitude and frequency of the global forces Fx, Fy, Fz applied at the contact patch.

$$\dot{q}_{g-i} = \int_V \sigma \frac{\partial \varepsilon}{\partial t} dV \tag{6}$$

Another important property of the model is the modularity. Readily available tire models as described in literature can be of various nature, semi-empirical (as the well-known Magic Formula), empirical or totally physical. The proposed thermal model can be applied like a module in combination with different kinds of tire model. In the scheme in figure 4 the general approach of the coupling concept is described:

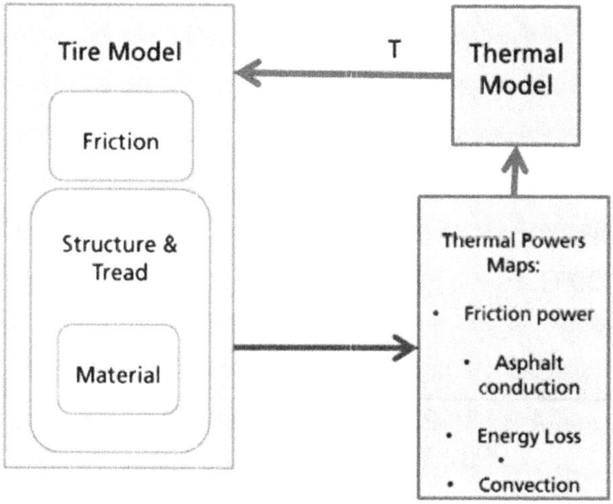


Figure 4. Scheme of coupling between the thermal and the mechanical (tire) model

From the information coming from the tire model the thermal model creates several maps of energy inputs for the thermal cells. The input energies are used to solve the energy equilibrium described in the paragraph 2 and to calculate the temperature. Then the temperature is used to modify some of the tire model properties like the internal pressure and friction coefficient (but also stiffness and damping if the tire model is a fully physical structural model).

## 2.3 Friction-Temperature dependency

A very important part of the coupling concept regards the friction-temperature modification. As described in the previous paragraph the friction varies with temperature. As outlined in various article (Persson [5], figure 5) the William Landel Ferry law can be used to describe this kind of variation. The adapted WLF principle says that the master curve of friction can shift in the friction-frequency plane when the temperature changes. The shift value $a_t$ can be calculated with the WLF formula.

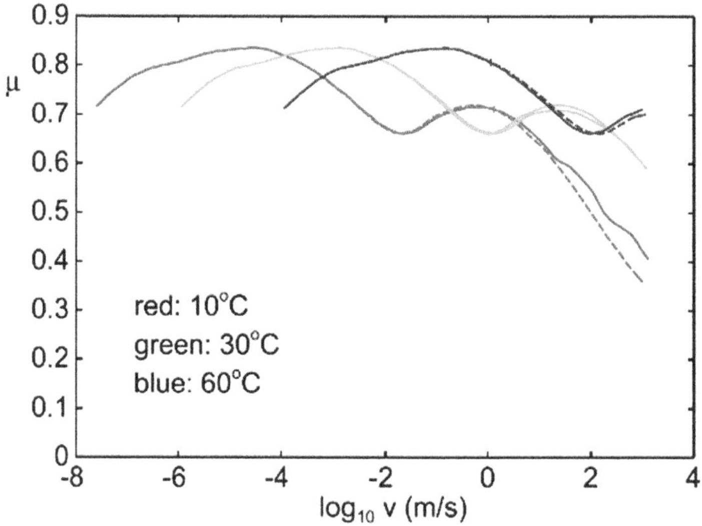


Figure 5. Example of friction coefficient as function of sliding velocity for three different temperature[6].

In the figure 6 is shown the effect of this coupling mechanism. A sweep slip angle experiment is performed. In this case the thermal model is coupled with the Pacejka MF 2002. The lateral force divided for the vertical load is plotted against the slip angle. In the figure 6 it is very clear that the obtained effect of grip variation can be reproduced in quite an accurate way with this approach. Anyway this approach can be used in the same way at a local level with structural tire model or global in the case of MF.

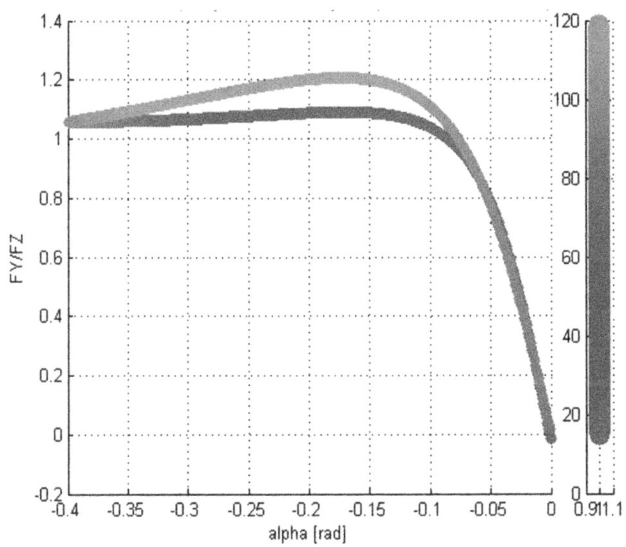


Figure 6.Slip angle sweep test by using the thermo-mechanical model described in the article. On the right side the temperature color-band. On the left side the FY/FZ versus slip angle plot. Simulation time 40s.It's clear how the temperature variation can have an influence on each point of the characteristic shape due to the friction change

## 2.4 An application example of the Thermal model

In the next paragraph the results of a simulation scenario will be shown and explained in detail.

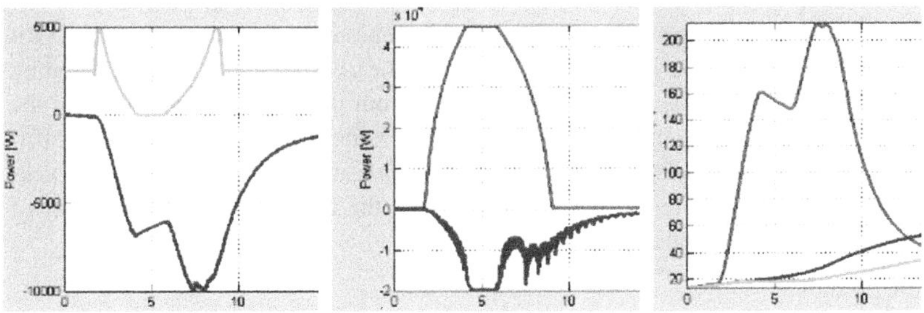

Figure 7.Braking results: left picture strain energy (green) and external air cooling(blue); center picture power due to friction (red) and asphalt cooling (blue), right picture surface temperature(red) and internal liner temperature (blue)

In the first ten seconds the tire passes from pure rolling condition to the full braking. It is possible to observe how during this operation the energy generated from the friction power increases because of the longitudinal force. Meanwhile the energy generated by the rubber strains decay dramatically, because without rolling (the tire is skidding) the bending deformations of the belt are zero. On the other hand, because of the tire temperature increase, also the differences between air temperature-rubber temperature and asphalt temperature- rubber temperature became larger and as consequence, the same is also true for the relative energies.

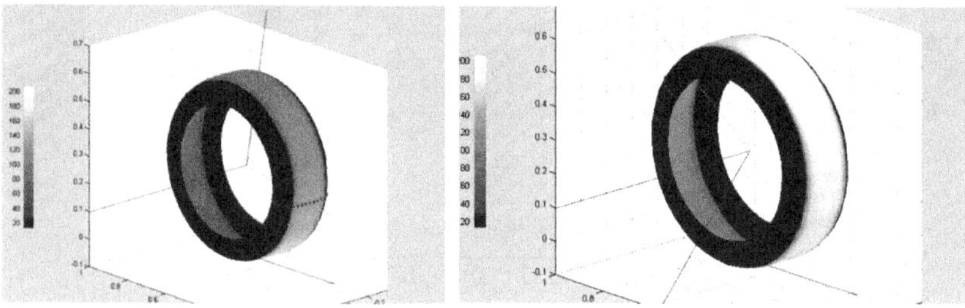

Figure 8.Right:Tire model representation during the braking (instant 3s of the simulation). It's possible to observe a gradient of temperature along the ring direction. This is because each section receive a different quantity of heat while the breaking is increasing. Instant 7s of the simulation. Left: presence of an hot spot on the surf due to the skidding

After the full braking in skidding condition it is possible to note a hot spot on the surface of the tire. The hotspot is created exactly in the zone that was sliding on the asphalt. The fact that this hotspot travels on the road periodically, with the same frequency of rotation, creates the oscillation of the asphalt absorbed energy that is visible around the second number ten in figure 8.

The last observation regards the temperature evolution during the simulation. As it will be clear in the experimental data shown in the next paragraph the temperature seen on the tire surface is usually very different from the internal temperature of the tread. This happens because of the high thermal insulation properties of the rubber, which create a certain delay in the temperature propagation through the section thickness. It is indeed possible to observe this effect in the figure 9 (where the history of the propagation in a certain section is shown).

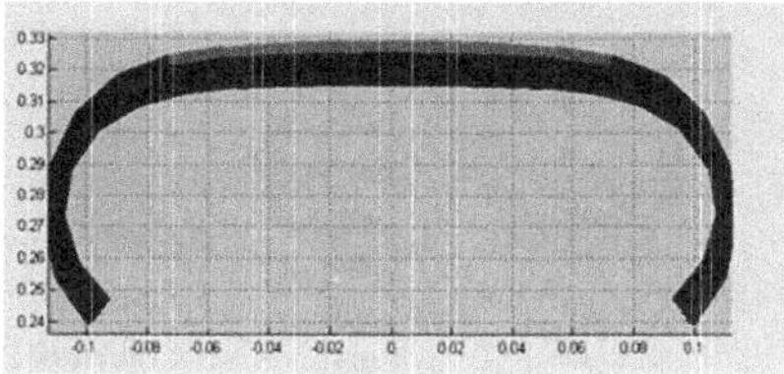

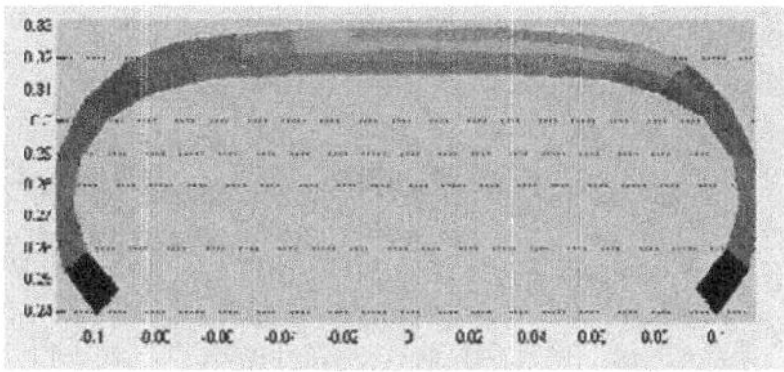

Figure 9.History of the propagation in a certain section at instant 2s (left picture) and 25s (right picture)

# 3 Magic Formula with thermal effects : the MF++

The combination of the thermal model with the semi-empirical tire model 'Magic Formula' it's not straight forward. The reason is that in this case no structural local information is available and it is necessary to create them empirically. The two additional empirical functions used are a procedure for the contact patch shape evaluation and a model for the estimation of the global dissipated energy during tire deformation. Basically, the information about contact patch shape and position in the thermal volume domain are necessary to understand how to divide the finite volumes in elements in contact with the asphalt and elements cooled by the external air flux in order to distribute in a correct way the friction power, the heat absorbed by the asphalt and the heat taken by the air. Given that the tangential stress and the sliding velocity are not available locally in the MF, the original formulation (3) of the Friction Power has to be modified in the following way:

$$\vec{q}_{fp-i} = \tau_i \cdot v_s = R_{fp} \frac{(F_x v_{sx} + F_y v_{sy})}{A_{cp}} \tag{7}$$

Where:

Fx Global longitudinal force

Fy Global lateral force

$R_{fp}$ Contact patch thermal resistance

$V_{sx}, V_{sy}$ Sliding velocities

$A_{cp}$ Contact patch area

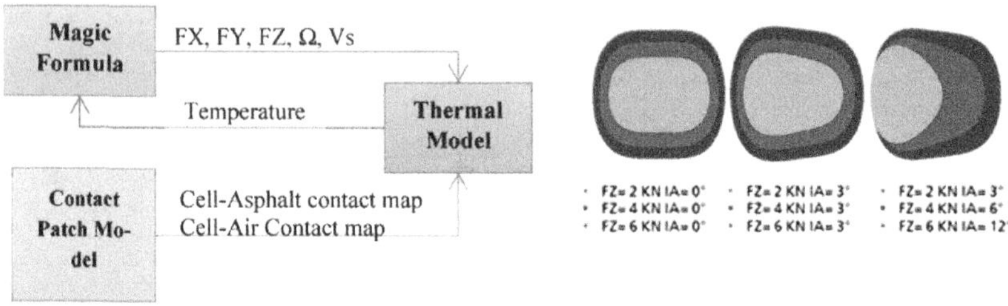


Figure 10. Contact patch shape empirical model for the MF expansion. The shape evaluated by the model depends on vertical load FZ, internal pressure, camber angle, lateral force FY (upper figure). The coupling scheme: The contact patch model create the map of thermal cell in contact and cell in contact with external air. The MF model passes forces and velocities values to the thermal model. The thermal model passes the temperature to the MF for the Grip modification.

Furthermore in this case is not possible to have the energy dissipated by the structure strains locally (as expressed in (6)) and another formulation is necessary to fill this other gap. As various authors suggest the whole energy dissipated by the structural deformation can be expressed as a function of the loads' magnitude and frequency. The frequency can be assumed equal to the frequency of rotation. So the suggested local original formulation (6) becomes:

$$\dot{q}_{g-i} = \int_V \sigma \frac{\partial \varepsilon}{\partial t} dV = P(F_x, F_y, F_z, \omega) \tag{8}$$

The formula (8) it's an empirical formula and its parameters can be estimated with two strategies. One is by using rolling resistance measurements for various loads and rolling velocities at free rolling conditions .Another, by using a structural tire model (validated on a local basis) to get this information in a virtual way.

The MF is modified simultaneously using the scaling coefficient relative to the friction coefficient and the braking/cornering coefficients (Lmu and Lk). The scaling coefficients are used as function of the temperature, but the structure of the whole formula has not been changed.

After this adaptation, also the MF model can be used without losing so much accuracy in the thermal model simulation, as will be shown in the next paragraphs.

## 3.1 MF++ Application results

In the following paragraph will be shown two application in which the MF++ is used. The first results regard the so called sweep of the slip angle.

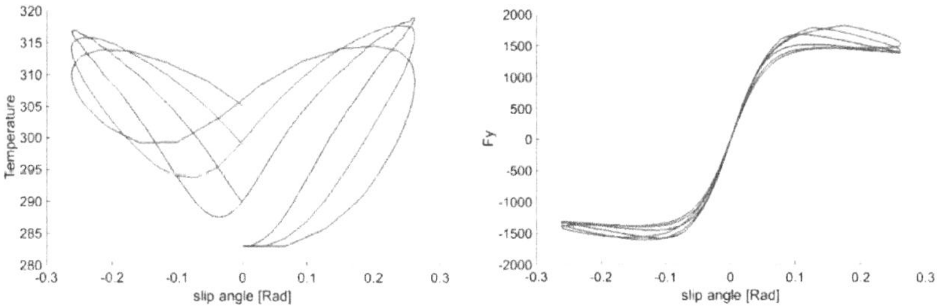


Figure 11. Sweep of the slip angle experiment. Experiment length 20s (blue line), 25s (red line), 30s (purple line)

In figure 11 on the left side the variation of temperature is plotted for various time length of the experiment. It is clear that shorter is the experiment duration lower are the temperature. On the left side the relative lateral forces are plotted. It is possible to observe the softening of the treads block for higher temperature, in fact the braking stiffness decay for higher temperature. At the same time also the grip change when the temperature move from its optimal value.

In the next example is shown how the effect of the temperature can be founded also on car measurement [12]. The full lap of a racing competition is simulated and compared with the relative measurement. It is clear how the temperature effect on grip, braking and cornering is important and how well can be reproduced with the model (figure 12).

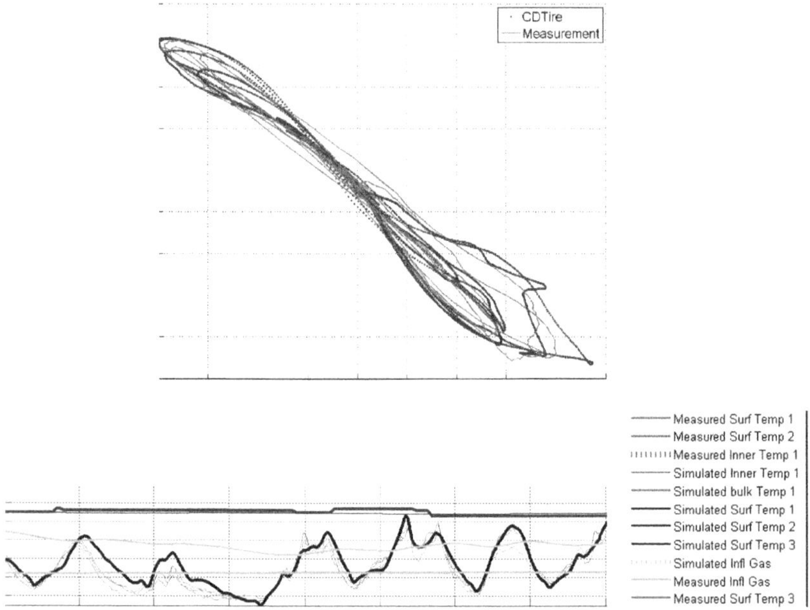


Figure 12. top: lateral force vs slip angle during a motorsport application, red line measurement, blue simulations. Bottom: temperature evolution during the same event. Red line measurement, black simulation [12].

In the temperature comparison plot (figure 12) it is possible to appreciate how fast is the dynamic of the temperature of the surface that is excited from the external air cooling, asphalt conduction and friction. Basically it is possible to notice that when the tire is accelerating the tire cools down and the dominant effect are the forced convection and the asphalt conduction. Usually the power used by the braking is more than the accelerating one. Indeed the surface temperature increase especially during the braking event, in this condition the friction power is very high and dominant respect the other effect. It is important to remark that the energies fluxes cannot be measured, but they can be estimated by using the model.

# 4 CDTire/3D, a structural 3D tire model with thermal effect

The coupling with a fully 3d structural tire model is quite straight forward. In this case the CDTire/3D model is used [6].

Experimental and FEM studies [7] show that the tire energy dissipation are due to bending of the sidewall (20 – 30 %) and due to the belt and tread compression, bending and shearing (60 – 70%). Thanks to the 3D structural tire model [6] of all this behavior can be properly simulated. In this full 3D structural model the tire structure is described by using the shell Kirchhoff-Love formulation.

The entire anisotropic layers in the thickness are modeled. The coupling is straightforward because each energy loss due to strain of carcass, steel cords, deformation of the treads, sidewall and belt is available locally in the tire model and it can be passed to the thermal one for the energy balance calculation in the exact location in the thickness. The coupling scheme is described in the figure 4. The Temperature will be available for each location and can be used to modify the visco-elasto-plastic material properties of the structural tire model. In this way the whole model could also be used to make rolling resistance predictions.

In the figure 13 are shown the simulation results for a particular cornering condition with camber. It is possible to appreciate the deformation of the structure due to camber, lateral force (applied at the contact patch) and vertical load. The contact patch shape for example result conic for this particular extreme condition. Due to the particular deformed condition only the left side of the tire is affected from large temperature, this is because the large slide occurs in the left part of the belt (figure 13). It is clear from this example that also extremely complex non-symmetric thermo-mechanical condition can be described with this approach.

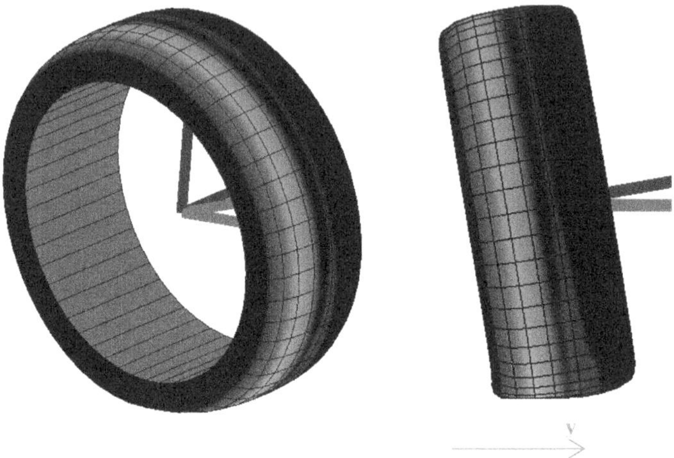

Figure 13 .CDTire/3D with thermal coupling .On the top a snapshot of a cornering scenario (Slip angle 0.3) for an Ortho and longitudinal view. In the bottom picture the contact pressure distribution and the local used friction coefficient contours for the dynamic contact patch.

## 4.1 CDTire/3D applications: Dynamic Contact Patch Simulations

The contact patch it's a really important aspect in tire mechanics. This aspect of tire dynamic is influencing:

- The wear, that it's higher where there is a higher pressure.
- The global grip that locally depends on local pressure, sliding velocity and temperature.
- The force lag or relaxation length.
- The braking and cornering stiffness strongly depends on the longitudinal and lateral dimensions of the contact patch area.

As consequence to reproduce in a very accurate way the contact patch dynamic has a great importance for the vehicle simulations in hard handling applications. There are several case in which a steady state punctual tire model like the magic formula is not enough, for example in hard braking simulation or hard cornering. In this cases the contact patch dynamic has an influence on the transient dynamics (tire lag and vibrations) and grip. If the contact patch area, pressure distribution, and temperature are not well reproduced from the model in this case also the vehicle response will be not in agreement with the reality because the grip could be over/under estimated from the as well as the forces oscillation. A 3D structural tire results fundamental than also for a correct simulation of the suspension behaviour. Usually the tire during suspension analysis is represented by a 3D spring, but this element is not accurate enough to reproduce the torque and moment that are transmitted from the tire to the rim given that these are strongly non-linear dependent from displacement and rotation.

Many expert know that actual FEM models are not able to represent the high dynamic contact patch status at very large slip level. In this case, due to the fact that the model is MBS based, due to the proper description of the tire structure and the high stability of the implicit solver, the model show good behaviours also at large slip and transient status. In the following paragraph is shown the comparison between the measurement realized by I. Kuwayama [11] and a simulation of a similar tire with CDTire/3D.

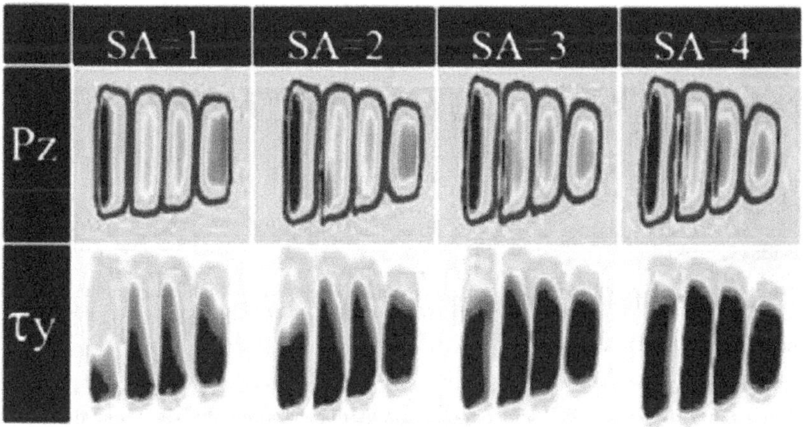


Figure 14: Contact patch dynamic for an eco-tire: 175/65R15. V=100 Km/h, lateral slip experiment on a drum that allows to measure the CP by using piezoelectric sensor.

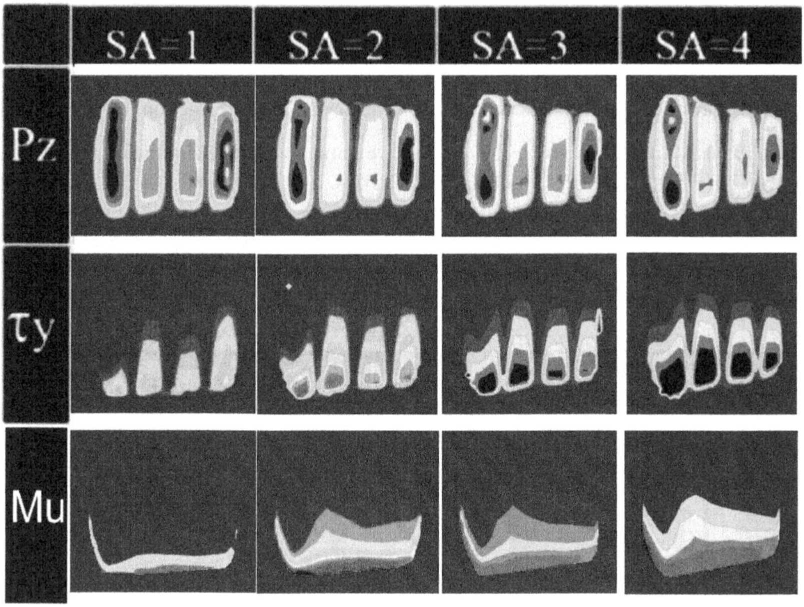


Figure 15: Contact patch dynamic for a tire 175/65R15. V=100 Km/h, Simulation results. Pressure contour band 0-0.9 MPa, tangential stress contour band 0-0.5 MPa, Mu (grip) band 0-1.2

In the figure 14 is showed the comparison between some measurement done by Bridgestone and presented in the article [11]. In this article I. Kuwayama shows how by using a drum with embedded quartz sensors allocated on the surface is possible to measure all the component of the stress due rubber-drum contact. The tire used during the experiment is a 175/65R15, this experimental tire has 4 continuous rib as tread pattern. In the figure 15 the same tire is modelled in the CDTire/3D environment by using the geometrical details that can be found on the paper. The results, for this tire rolling on a drum as well, show quite good agreement with the experiments. It is necessary to notice a couple of phenomena that are well reproduced:

- In the contact patch the contact pressure in bigger on the sidewall than in the centre of the contact patch. This effect is caused basically from the sidewall that make the structure stiffer in that zone. Tire designer try usually to avoid this effect in order to have a uniform pressure that brings to uniform wear. However, also if a tire is well optimized, because of the centrifugal forces at high rotational speed the pressure will ever be slightly higher on the side.

- When the slip angle increase also the shape of the lateral force change and become like a trapezium. This append basically because the lateral tire force 'pull' the contact patch laterally. The effect increase with force as showed in the results. Another important aspect is the tire lag; it is well known that is slower at lower velocities. In fact at slow velocities the 'trapezium shape' has a bigger evidence as shown in the following figure, this because maybe the tire cannot recover the deformation fast:

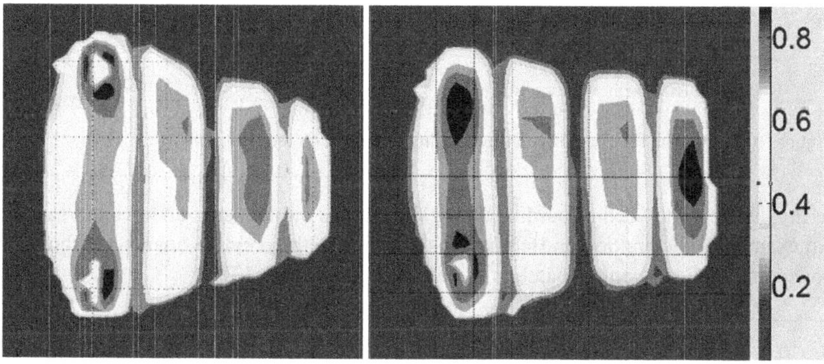


Figure 16: Contact patch dynamic for a slip angle of 3 Deg. Effect of the velocity on the contact patch

- A consequence of the pressure distribution is that the tangential stress is bigger where the pressure is lower. Basically the treads with less normal pressure can slide more easily. This is why the two central ribs has major deflection that the two on the side. Another observation is that regarding the tangential stress it is not easy to do a one to one comparison with the measured tire, this is because the colour map of the measured one is not known. In the simulated tire the dark black colour is associated with the treads that are in total slide, by the other side in the measured case it has to be supposed that for 1 Degree of slip angle the slippage area cannot be so extended, so it is believed from the author of this paper that the contour colour map has a low upper limit.
- Last considerations regards the slippage area. This is showed in the last row of the figure 15 third row . The various colour represent how much of the actual available friction is used. When the colour is red it means that the sliding limit is reached, so the red are is in sliding. It has to be noticed that the sliding limit is not the same for each point because the friction is function of pressure, temperature and sliding velocity (figure 16). However the zone with higher vertical pressure, like the two external ribs, slide less than the zone with lower pressure, the two central ribs.

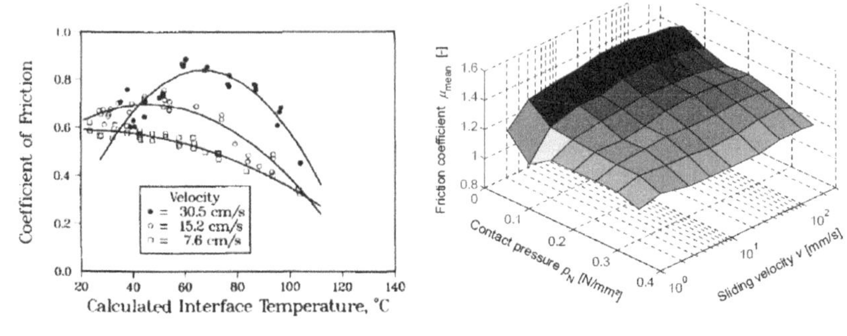


Figure 16: Friction as function of pressure, sliding velocity and temperature [13].

In the last year the need of hard handling simulations on uneven road have increased. An example is the braking on Belgian cleat. Similar to the previous analysis is showed how complex is the contact patch in these conditions (figure 17).

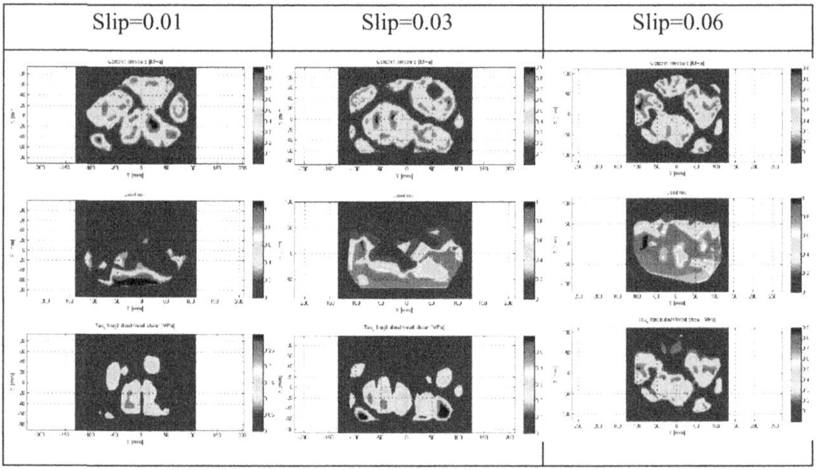


Figure 17: Breaking on Belgian block for 3 slip conditions. First row contact pressure, second row slip/stick condition and third row tangential stress.

From the figure 17 it is clear that under braking the contact patch area in contact is less respect to the case with flat road. Of course if the area of contact is smaller the global grip will be also less. The consequence of this is that the braking operation on a Belgian cleat road will require more time as showed in the figure 18. In this case is showed the force on the rear tire under full braking in a vehicle simulation. The initial vehicle velocity is 50 Km/h. The braking is performed without ABS. In the case in which the surface of the road is flat the braking is done in 3 s. On the Belgian cleat, same conditions of braking, 5 s. It is clear the importance of the 3D surface of the simulation of the braking operation that cannot be characterized with a MF.

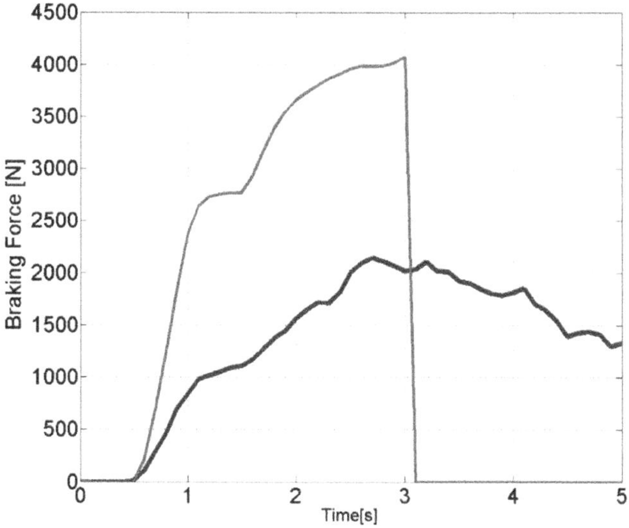


Figure 18: Breaking on Belgian block (blue line) and on flat surface (red line).Simulation results. Due to the lower grip on the Belgian block the braking take more time.

After this analysis is clear that in order to increase the physicality in vehicle simulation and to close the gap with the physical inconsistency a structural tire model is needed. Only with this kind of model several additional information about the tire status are available. In the next paragraph the final consideration regarding the two approaches will be analysed.

# 5 Conclusion

In the article two different approach for tire handling simulation have been analysed. The two approach consist of a MF tire model and of a structural 3D fully physical tire model. Both model are sensible to the thermal variation of the rubber property.

In the first part the article shows how fundamental to use a thermomechanical model in handling application, given the high sensitivity of cornering performance and grip of a tire from the temperature.

In the second part the two approach semi-physical versus fully physical are analysed.

Both approach presented are valid, but each one have his advantageous and limitation:

The MF++ approach has the limitation that it cannot be used when:

- The road has a 3D geometry and the main wavelength of the obstacle is shorter than the contact patch length
- The information regarding rubber/asphalt interaction are important. In this case no contact patch information like contact pressure, temperature, sliding velocity are available
- The vibration of the belt respect to rim is important for the evaluation of the real slip (f.e. ABS braking simulations)
- The road is deformable (snow, ice, water).

The main MF++ advantageous are:

- Real time factor close to 0.1
- It is the only model that can be used for automatic regression technique in optimization procedure (car / lap optimizations).
- It can be used without losing quality in slow slip steady state manoeuvre

Regarding the CDTire/3D temperature sensible version, the main limitations are:

- The real time factor is to large and it cannot be used with regression technique for automatic optimization study
- Real time application are not possible

The main advantageous are:

- It is physical, it has predictive capabilities
- High transient structural effect are represented, so ABS simulation are possible
- 3D road is supported
- All local information are available, like contact patch contact pressure, temperature or sliding velocity
- In combination with a soft soil also snow/water tire interaction can be performed

To conclude, both solution results very attractive for handling simulations. The sensation is that with the increasing of the computational efficiency and performance the physical models could have a chance to overcome the semi-physical approaches.

# Citation


[1] H. Pacejka,Tire and Vehicle Dynamics, 3rd Edition

[2] Masahiko Mizuno, "Development of Tire Side Force Model Based on "MagicFormula" with the Influence of Tire Surface Temperature ", Toyota research report

[3] Giordano D.,"Temperature prediction of high performance racing tyres." PhD thesis

[4] Février P. (2008): 'Thermal and Mechanical Tire Force & Moment Model presentation'. 4thIntelligent Tire Technology Automotive Conference, Wiesbaden October 2003.

[5] Lorenz B1, Persson BN, Fortunato G, Giustiniano M, Baldoni F.Rubber friction for tire tread compound on road surfaces. J Phys Condens Matter. 2013 Mar 6;25(9):095007. doi: 10.1088/0953-8984/25/9/095007. Epub 2013 Jan 18.

[6] Gallrein, A., Baecker, M., and Gizatullin, A., "Structural MBD Tire Models: Closing the Gap to Structural Analysis SAE Technical Paper doi:10.4271/2013-01-0630.

[7] Société de Technologie Michelin (2001): 'Rolling Resistance and fuel saving'

[8] G. Leister (1997) New Procedures For Tyre Characteristic Measurement, 27:S1, 22-36, DOI:10.1080/00423119708969642

[9] Theodore L. Bergman, Adrienne S. Lavine, Frank P. Incropera, David P. DeWitt, Fundamentals of Heat and Mass Transfer, 7th Edition

[10] Gerald W. Recktenwald , Finite-Difference Approximations to the Heat Equation, lessons notes

[11] Development of Dynamic Contact Patch Measurement System, Ph.D. Isao Kuwayama, M.Eng. Hiroyuki Matsumoto, Ph.D. Hisashi Heguri ,22nd Aachen Colloquium Automobile and Engine Technology 2013

[12] Calabrese, F., Baecker, M., Galbally, C., and Gallrein, A., "A Detailed Thermo-Mechanical Tire Model for Advanced Handling Applications," *SAE Int. J. Passeng. Cars – Mech. Syst.* 8(2):2015, doi:10.4271/2015-01-0655.

[13] Coefficients of dynamic friction as a function of temperature, pressure, and velocity for several polyethylene resins, Mark A. Spalding and Kun S. Hyun, Article first published online: 26 AUG 2004, DOI: 10.1002/pen.760350702

[14] Investigations of local contact phenomena between tyre and road, SILENCE Final Training Conference, Bergisch Gladbach,Gunnar Gäbel, Prof. Dr. Matthias Kröger

# Rolling resistance modeling for electric vehicle consumption

Dipl.-Ing. Andrea Ficht, Prof. Dr.-Ing. Markus Lienkamp

Institute for Automotive Technology, TU Munich

# 1 Introduction

Due to increasing oil shortage and ever stricter $CO_2$-Emissions regulations vehicle energy consumption simulations are nowadays indispensable for the car industry. Especially the limited range of today's electric vehicles makes it particularly necessary to model all energy loss sources as precisely as possible. As the drive train efficiency of electric vehicles is higher than those of conventional vehicles, the percentage of the energy loss caused by driving resistance forces is higher, too. These are composed of the aerodynamic drag, inertial drag, friction in the power train, rolling resistance force and slope resistance.

For standard vehicle consumption simulations as presented in [1] the rolling resistance is assumed to be constant, in some cases, if needed it changes as a function of speed. This is most relevant for high velocities from 120 km/h upwards. However, these rolling resistance coefficients correspond to steady state values which were determined according to standards such as the ISO 28580 [2] at one operating point of the tyre. Thus measured at a given pressure, load, ambient temperature and velocity after a warming-up procedure of 30 minutes. At the beginning of the test, the rolling can be roughly 20 % higher than the steady state value and it then drops quickly until it reaches its minimum after 20-30 minutes. This phenomena results in a false calculated energy consumption of the vehicle for a specific cycle. A similar study on the transient rolling resistance was presented in [3] suggesting that the tyre label should take account of this phenomena.

This paper deals with the quantification of the rolling resistance variance at driven operation points in the case of compact electric vehicles, such as the electric vehicle Visio.M developed by the "Technische Universität München", which is equipped with a 115/70R 16 tyre.

# 2 Contribution of Rolling Resistance to Vehicle Energy Consumption

The rolling resistance is defined as the energy consumed by a tyre per unit of distance. The main cause of the energy dissipation is due to the viscoelastic material characteristics of rubber. Viscoelastic materials lose energy in form of heat whenever they are deformed. This energy loss results in a force that opposes the tyre's direction of movement and therefore the vehicle's movement. [4]

## 2.1 Contribution to Energy Consumption

The contributions of each resistance to the driving resistance and hence the total energy consumption is strongly dependent on the operational profile. As presented in [4], figure 1 shows the distribution for a conventional vehicle with a combustion engine, a mass of 1100 kg and a frontal vehicle surface of 0.65 m², an internal friction of 50 N and a maximum engine power of 51 kW. The rolling resistance coefficient deployed to calculate these contributions is 12 ‰. The rolling resistance coefficient is the value that was determined according to the standard ISO 28580 [2].

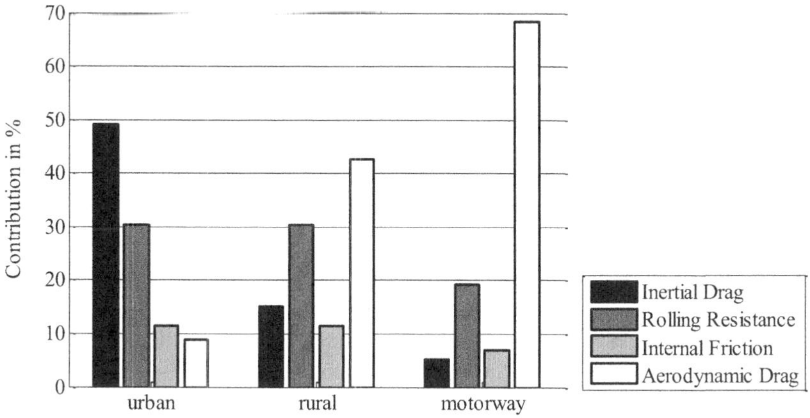


Figure 1: Contribution to the driving resistance forces [4]

Due to the limited charging infrastructure and the limited range, compact electric vehicles appear to be most suitable for city rides. Regarding the contributions for the urban areas in figure 1, these may slightly differ for compact electric vehicles as brake energy regeneration is possible. Therefore the contribution of inertia loss is presumably smaller for electric vehicles. Hence the energy consumption contribution of all other resistance forces will be higher. This also concerns the rolling resistance.

## 2.2 Measuring Rolling Resistance of Tyres

According to the standard ISO 28580 [2], the rolling resistance is measured under precisely defined conditions. The ambient temperature and tyre temperature at the beginning of the test are 25 °C, the tyre load is usually 80 % of the maximal load. The pressure is 2.1 and 2.5 bar for a standard tyre and respectively an extra load tyre. During a warming-up procedure, the tyre runs on the test bench for 30 minutes at a speed of 80 km/h. The procedure can be carried out for different velocities if a velocity dependent rolling resistance is needed. Once the tyre is at its steady state, the rolling resistance is measured. The determined rolling resistance coefficient is therefore valid for the mentioned conditions with the intention to easily compare different tyres with one another. The standard ISO 28580 suggests empirical formulas to calculate the steady state rolling resistance at different conditions of ambient temperature, pressure and load.

Primarily the rolling resistance is used to compare the energy efficiency of different tyres. But note that the determined rolling resistance coefficient is also used to simulate vehicle energy consumptions for cycles such as the NEDC without warming-up procedure and for different velocities, although velocities from 0-120 km/h are driven.

## 2.3 Influencing Parameters on Rolling Resistance

Under real driving conditions, which would mean ambient temperature that are most of the time below 25 °C in Germany, without warming-up procedures and for varying velocities, the rolling resistance can differ considerably from the standard ISO 28580 rolling resistance value.

The left plot of the following figure 2 shows the evolution of the rolling resistance in the case of the Visio.M tyre for a constant speed of 80 km/h without warming-up procedure at 25 °C and 15 °C ambient temperatures. At the beginning of the test, the rolling resistance is about 20 % higher. After about 10 minutes at a constant speed of 80 km/h the rolling resistance reaches its minimum steady state value. The instantaneous rolling resistance force was measured on a test drum using measures of the resistive force at the tyre spindle. The presented rolling resistance coefficients represent measurements that were conducted on a test drum with a diameter of 1.7 m under the conditions shown in table 1. The correction of the test drum curvature was taken into account in order to obtain a rolling resistance value for a flat surface. A 25 % increase of rolling resistance of was added to attain realistic values for rough road surfaces [4].

For different velocities, the tyres reach different operating states. The right plot of figure 2 shows the same measurements of rolling resistance as above for a constant velocity of 40 km/h. At the beginning of the tests at 40 km/h for ambient temperatures of 25 °C and 15 °C, the rolling resistance is 12.2 % respectively 19.5 % higher. Compared to the constant velocity test at 80 km/h, the warming-up procedure at 40 km/h is a lot slower and the steady state rolling resistance is merely reached after 30 minutes.

Table 1: Rolling resistance testing conditions of Visio.M tyre

| Tyre dimension | 115/70R 16 |
|---|---|
| Load | 2000 N |
| Pressure | 3.0 bar |
| Ambient temperature | 25 °C |
| | 15 °C |
| Constant speed | 80 km/h |
| | 40 km/h |

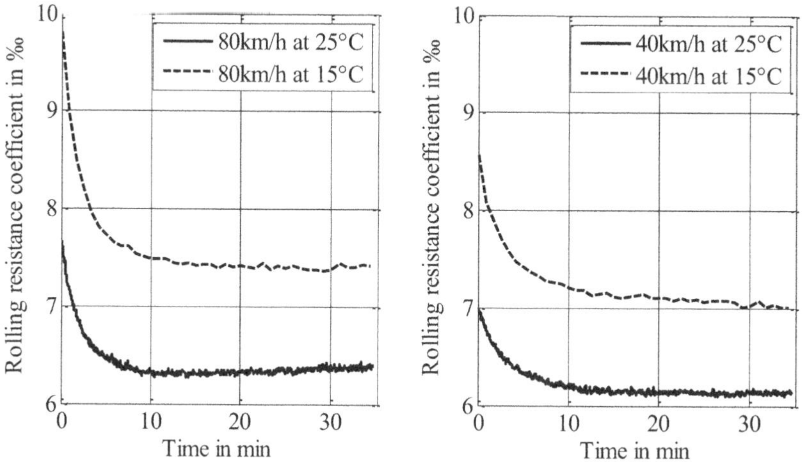


Figure 2: Measured rolling resistance coefficient for 80 km/h and 40 km/h at ambient temperatures of 25 °C and 15 °C

# 3 Driving Scenarios for Compact Electric Vehicles

Having noticed and understood the factors influencing the energy loss of the tyre, one has to take a closer look on the real driving conditions of compact electric vehicles to understand and determine the instant tyre energy dissipation. The driven velocities as well as other environmental conditions can have an effect on the tyre's dissipated energy. This following paragraphs explain how an operational profile for compact electric vehicle can be characterized. Thanks to the results of field trials with MINI E's that have been conducted from July 2012 to February 2014 and recorded anonymously throughout a project at the "Technische Universität München", the driven trips were evaluated and analysed.

## 3.1 MINI E Field Test

For this field trials 10 vehicles were used by two companies as company cars and 33 vehicles by private test persons in Munich and its surroundings. The MINI E is a MINI Cooper converted to an electric vehicle. It has an electric motor with a maximum power output of 150 kW, the maximum velocity is electronically limited to 150 km/h and the battery allows ranges up to 160 km. Unlike conventional vehicles, regenerative braking is possible and can be applied by releasing the acceleration pedal in a controlled way. Only high decelerations above 2.35 $m/s^2$ require the use of the mechanical brake. [5]

The following analysis includes over 9 000 trips that have a total added up distance of more than 100 000 km. In order to log the GPS-velocities, the GPS-positions, accelerations and angular rates of the driving vehicles, they were equipped with Smartphones (iPhone 4). These were fixed and maintained fastened in the boot with the exclusive purpose of logging driving data. The GPS-velocities and positions were logged at a frequency of 1 Hz, the (iPhone) sensor allowing a logging frequency of 25 Hz for the accelerations and angular rates. All recorded data are divided into tracks using an algorithm which consider certain criteria e.g. standstill phases, tunnel sections and are registered in a database. Further details of the MINI E field test are presented in [5].

Throughout this paper, the term "user" corresponds to the vehicle and not to the driving person. In the case of private use of the vehicle, it can be assumed that only one person drove the car whereas for the company car use, several persons might have driven the same vehicle.

Table 3: Data of MINI E fleet

| | |
|---|---|
| number of vehicles | 43 |
| number of trips | 9 327 |
| total covered distance | 102 664 km |

## 3.2 Evaluation of Driven Distances and Velocities

In the following, driven trip distances, velocities and lateral accelerations are evaluated statistically. The evolution of speed and the corresponding distance covered define how intensively the tyre heats up and cools down during a trip and hence how much energy is dissipated by the tyre. Figure 3 shows the cumulative percentage of all driven distances per trip. The average driving distance is about 11 km. 70 % of all driven distances are under 11.5 km, 90 % under 30 km and 95 % under 39.2 km. The lowest mean distances per user is 4.24 km and the highest is 26.4 km.

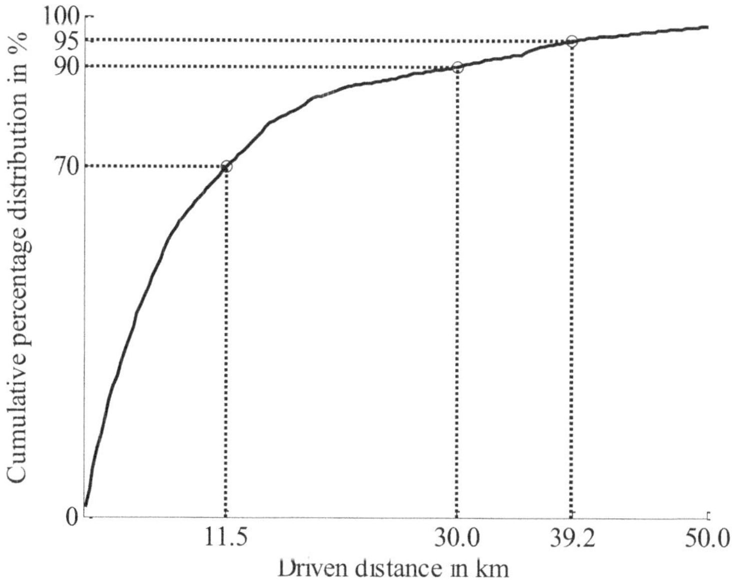


Figure 3: Covered distance of trips

Table 2: Driven distances of the MINI E fleet

| | |
|---|---|
| Average driving distance | 11.0 km |
| Lowest mean distance per user | 4.24 km |
| Highest mean distance per user | 26.4 km |

Following figure 4 shows the cumulative velocities driven by the MINI E fleet. The standstill phases thus velocities equal zero are not included.

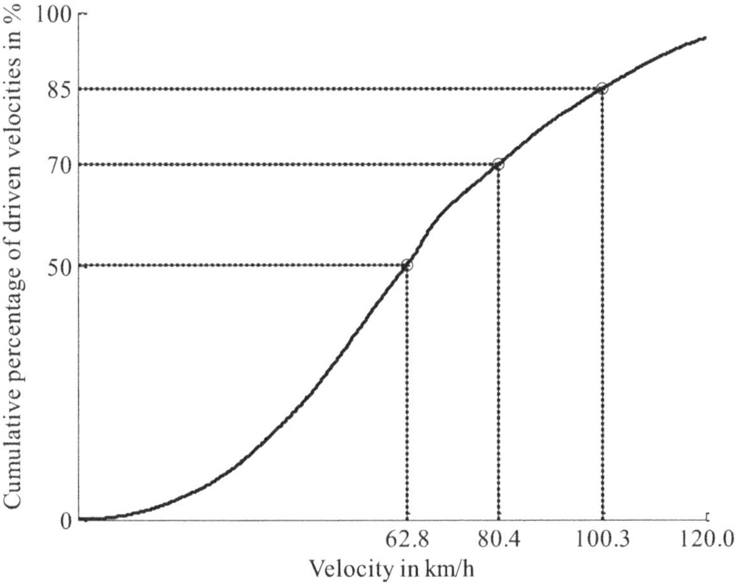


Figure 4: Driven velocities

Table 3: Mean speed of driven velocities throughout the MINI E fleet test

| | |
|---|---|
| Mean speed excluding standstill | 39.8 km/h |
| Mean speed excluding standstill of the slowest user | 24.9 km/h |
| Mean speed excluding standstill of the fastest user | 55.3 km/h |
| Mean speed including standstill | 33.0 km/h |
| Mean speed including standstill of the slowest user | 19.8 km/h |
| Mean speed including standstill of the fastest user | 50.4 km/h |

The above figure shows that 50 % percent of the driven velocities are below 62.8 km/h, 70 % are below 80.4 km/h. The mean speed of all users excluding standstills is about 40 km/h and including the standstills it is about 33 km/h.

Using the mean driven distance and the mean driven velocities including the standstills, the mean travelling time would be 20 minutes. Recalling the test results of the transient rolling resistance at a constant speed of 40 km/h (see paragraph 2.3), the tyre would have just reached its steady state.

The projection of the lateral forces on the tyre's direction of movement also increases the rolling resistance and heats up the tyre additionally. Figure 5 shows which lateral accelerations were driven by the MINI E fleet excluding the accelerations equal zero. It shows that the lateral accelerations are beneath 0.6 m/s² in 80 % of the cases and rarely exceed 4 m/s².

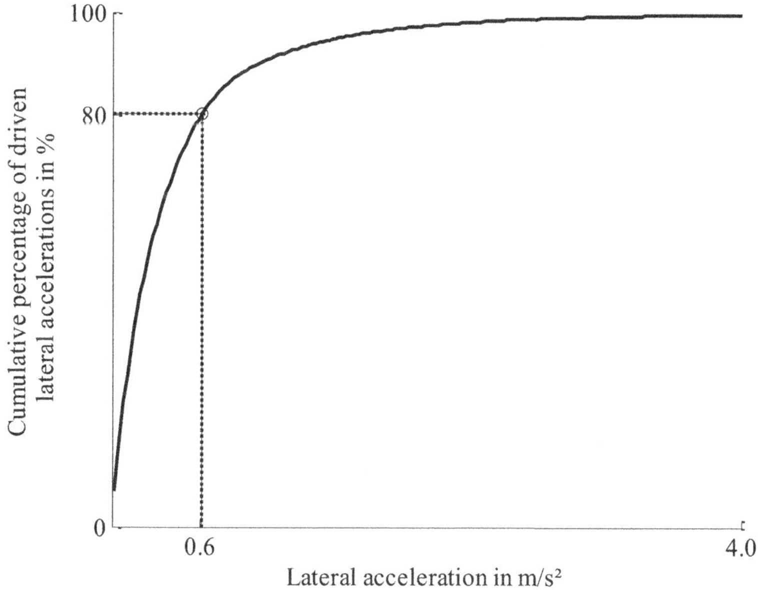


Figure 5: Driven lateral accelerations

## 3.3 Conclusion of MINI E Fleet Evaluation

The driven distances and the driven velocities strongly depend on the user. The data show a rather high use of vehicles in city areas, as the users of the vehicles work and live in the surroundings or the city centre of Munich. Test persons who live in the suburbs of Munich tend to have a higher mean speed and longer driven distances. The range of the vehicle limits the driving distances to short trips to places that provide enough battery charging possibilities e.g. the city centre of Munich or to longer trips where a known battery charging possibility is available e.g. a user that lives 40 km from his place of work and charges his vehicle at home. Analysis in paragraph 4.6 point out the relevance of the effects of lateral accelerations on the instantaneous rolling resistance value.

## 3.4 Example City Trip

Nowadays compact electric cars are mainly used for short trips in urban areas as the preceding study has shown. Therefore an example of a city trip from the MINI E database has been chosen for further analysis. The trip was recorded in Munich city, the average speed was 26 km/h, and the distance is 6.65 km. Figure 6 shows the driven stretch on a map.

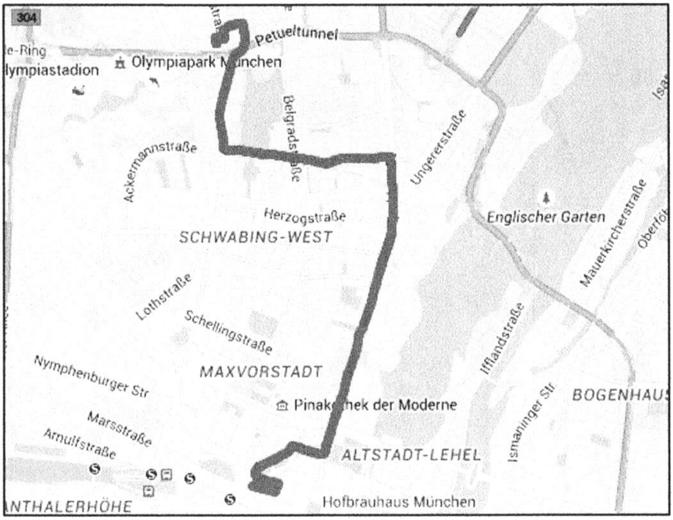


Figure 6: Example trip on map

Table 4: Data of MINI E example trip

| | |
|---|---|
| Driven distance | 6.65 km |
| Mean velocity | 25.8 km/h |
| Maximum velocity | 61.3 km/h |

The top plot of figure 7 shows the corresponding speed evolution of the trip. As we can see on the map, the driven stretch comprises several turnings. Therefore the evolution of the lateral acceleration shown in the lower plot of figure 7. The lateral accelerations do not exceed 4 $m/s^2$. According to [6], this corresponds to normal driving behaviour.

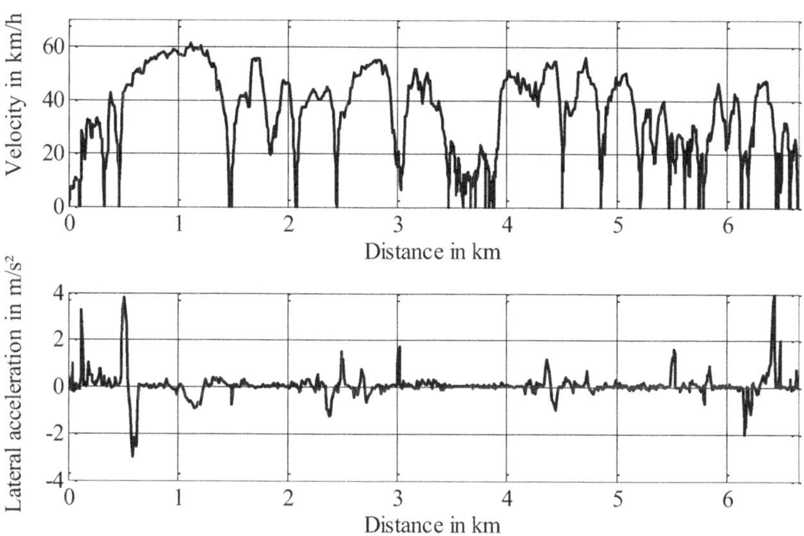


Figure 7: Evolution of velocity and lateral acceleration for the example trip

# 4 Rolling Resistance Model

Experimental studies of the phenomena of the transient rolling resistance have already conducted as presented in [7]. Apart from these, it can be helpful to have a model which describes the transient rolling resistance behaviour, so further optimization can be carried out on tyre energy efficiency. A physical model of the transient rolling resistance already exists and was developed by Mars and Luchini presented in [8] and [9]. This analytical model is based on the heat transfer of the tyre with the environment. It is assumed to be only a function of the steady state rolling resistance at a given operating condition and the instantaneous tyre temperature. Following assumptions have been used to develop the model:

1. The tyre has a single, uniform temperature
2. The work done by the instantaneous rolling resistance is entirely converted to heat in the tyre.
3. At a given set of conditions, it is assumed that a linear proportionality approximates the dependence of the rolling resistance on the instantaneous tyre temperature, which is also used in standards to correct the rolling resistance for ambient temperature differences.

For the Visio.M tyre an extended model based on [8] and [9] is developed as follows. One of the inputs of the model is a steady state rolling resistance value according to the standard ISO 28580. Based on this reference, other steady state values are calculated for other operating conditions according to empirical formulas proposed by the standard ISO. The second input is the instantaneous tyre temperature that can be determined using the first law of thermodynamics. This approach is detailed below for the steady state and transient of the given tyre.

## 4.1 Steady State Rolling Resistance

The steady state rolling resistance corresponds to the minimal stabilized value that the tyre attains after a certain period for a given condition, corresponding to its warming-up time. This value can be determined according to the standard ISO 28580. For other ambient temperatures following formula according to the mentioned standard is used to calculate the corresponding steady state rolling resistance values.

$$C_{rr,iso} = C_{rr,amb}[1 + \alpha(T_{amb} - T_{iso})] \tag{1}$$

For varying passenger car tyre pressure and load following formula is used according to the same standard and are also mentioned in [4]:

$$C_{rr} = C_{rr,iso}\left(\frac{p}{p_{iso}}\right)^{-0.4}\left(\frac{Z}{Z_{iso}}\right)^{-0.15} \tag{2}$$

Having had a closer look at the test bench results (figure 2), one can observe that there is a velocity dependency of the rolling resistance of the Visio.M tyre. At steady state, there is a difference of the rolling resistance coefficient of about 0.2-0.3 ‰. Therefore the presented rolling resistance model includes the dependency of the velocity which is assumed to be linear as presented in equation (3).

$$C_{rr} = C_{rr,iso} + k_1 v \tag{3}$$

Due to its viscoelastic properties the tyre dissipates heat energy when rolling at a certain speed. In the tyre's steady state, the generated internal energy $\dot{Q}_{int}$ balances exactly the heat energy dissipated $\dot{Q}_{diss}$ to the surroundings. According to the first law of thermodynamics following equation is obtained:

$$\dot{Q}_{int} - \dot{Q}_{diss} = 0 \tag{4}$$

The generated internal heat is equal to the work that maintains the tyre at a constant speed under the influence of the rolling resistance:

$$\dot{Q}_{int} = vF_R^* \tag{5}$$

The energy dissipated to the surroundings corresponds to the convective heat transfer of the tyre according to Newton's cooling law as equation (6) presents. A is the heat exchange surface of the tyre.

$$\dot{Q}_{diss} = hA(T^* - T_{amb}) \tag{6}$$

The heat exchange coefficient is dependent on the driven velocity according to (7), while $h_0$ is the reference heat exchange coefficient for rubber and air. The parameter k is the sensibility exponent and is tyre dependent.

$$h = h_0 \left(\frac{v}{v_0}\right)^k \tag{7}$$

By substituting the equations (4)-(6) in (3), the steady state temperature of the tyre is obtained as follows:

$$T^* = \frac{F_R^* v_0^k v^{1-k}}{Ah_0} + T_{amb} \tag{8}$$

## 4.2 Transient Rolling Resistance

When the tyre's speed increases or decreases, there is a change of internal generated heat. The tyre has a thermal inertia which means that the tyre can store energy. In this case following equation corresponds to the first law of thermodynamics:

$$\dot{Q}_{int} - \dot{Q}_{diss} = \dot{Q}_{stored} \tag{9}$$

The stored energy corresponds to the specific heat density of the tyre and the instantaneous time rate of the change of temperature

$$\dot{Q}_{stored} = mc_p \frac{dT}{dt} \tag{10}$$

The internal heat generation is calculated according to equation (5), except that the rolling resistance is not constant, but changes according to the correction formula to ambient temperature differences resulting in:

$$\dot{Q}_{int} = vF_R(t) \tag{11}$$

For the dissipated heat energy to the surroundings the equations (5) and (6) are still valid for the transient behaviour. The equations (5), (6), (9) and (10) are substituted in equation (8), obtaining equation (12).

$$vF_R^*[1+\alpha(T^*-T(t))]-h_0\left(\frac{v}{v_0}\right)^k A(T(t)-T_{amb}) = mc_p\frac{dT}{dt} \qquad (12)$$

Finally the steady state temperature is still replaced by equation (8) in equation (11) which results in the following differential equation:

$$mc_p\frac{dT}{dt} = (T_{amb}-T(t))\left[\alpha vF_R^* + h_0\left(\frac{v}{v_0}\right)^k A\right] + vF_R^* + v^{2-k}F_R^{*2} \qquad (13)$$

Once this differential equation is solved by applying the same method as in [8], the instantaneous tyre temperature is used to calculate the instantaneous rolling resistance:

$$F_R(t) = F_R^*[1+\alpha(T^*-T(t))] \qquad (14)$$

The parameter values in table 5 retrieved from [8] have been used for the model.

Table 5: Heat transfer coefficient from [8]

| | | |
|---|---|---|
| Velocity sensitive exponent | k | 0.5 |
| Specific heat capacity rubber | $c_p$ | 1670 kg/K |

Other heat transfer parameters, which are tyre dependent, were determined using the test results (as presented in paragraph 3) and are presented in table 5. They differ from the parameter values in [8] as the used tyre is fairly small and light with other heat transfer properties.

Table 6: Determined coefficients for Visio.M tyre

| | | |
|---|---|---|
| Reference heat exchange coefficient at 80 km/h | $h_0$ | 55 W/mK |
| Rolling resistance sensitivity to temperature | $\alpha$ | 0.014 1/K |

## 4.3 Comparing Test Bench Results with the Model

For the same conditions as for the test bench measurements the model is used to calculate the instantaneous rolling resistance. Figure 8 shows the test bench results compared to the model for the velocities 80 km/h on the left plot and 40 km/h on the right plot at the ambient temperatures 25 °C and 15 °C. For different conditions the fit of the model varies from 65-87 % (table 7). The fit is calculated according to following formula where $x_{mes}$ is the measured value and $x_{sim}$ is the simulated value.

$$fit = 1 - \sqrt{\frac{\sum_i (x_{sim,i} - x_{mes,i})^2}{\sum_i (x_{sim,i} - \overline{x_{mes}})^2}} \tag{15}$$

Further actions still have to be taken to improve and finally validate the model.

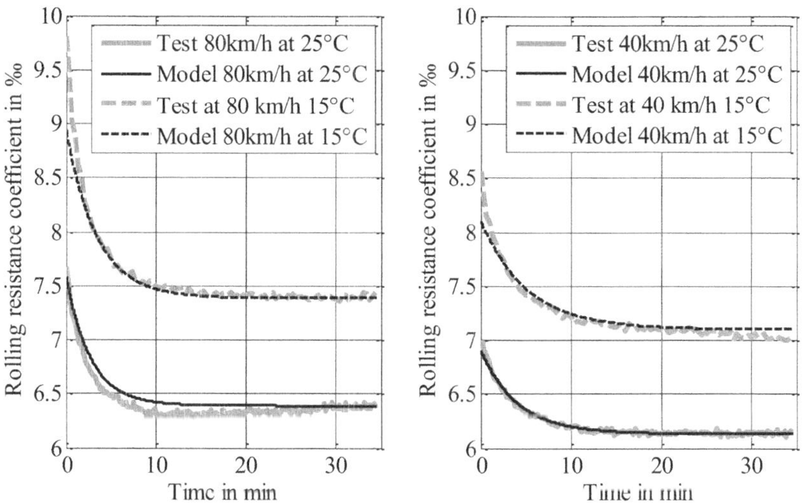


Figure 8: Measured and modelled rolling resistance

Table 7: Fit of the model compared to the test bench results at different operating conditions

| | 25 °C | 15°C |
|---|---|---|
| 80 km/h | 65.4 % | 65.5 % |
| 40 km/h | 87.1 % | 71.9 % |

## 4.4 Rolling Loss for the New European Driving Cycle (NEDC)

The rolling resistance model is used to analyse the error if the transient rolling resistance is not taken into account when driving the New European Driving Cycle (NEDC). By using equation (3) to calculate the steady state rolling resistance values for the given velocities, the mean is at 6.28 ‰ for a load of 2000 N and a tyre pressure of 3 bar at 25 °C. With the developed model, that includes the transient effects, the average rolling resistance is at 6.62 ‰, thus 5.30 % higher. Figure 9 illustrates the driven velocity in the top plot. The bottom plot shows the evolution of the rolling resistance using the model for two different ambient temperatures 25 °C and 15 °C.

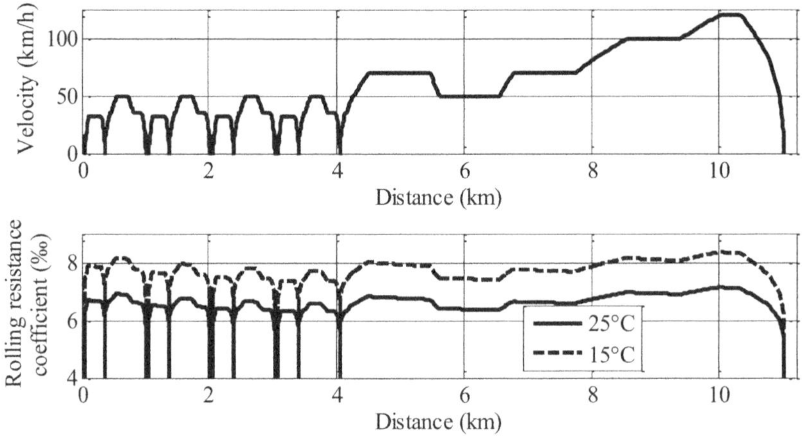


Figure 9: Evolution of velocity and rolling resistance at 25 °C and 15 °C

Table 8: Mean rolling resistances coefficient for the NEDC

| Ambient temperature in °C | 25 | 15 |
|---|---|---|
| Mean steady state rolling resistance coefficient in ‰ | 6.28 | 7.30 |
| Mean transient rolling resistance coefficient in ‰ | 6.62 | 7.75 |

Table 8 shows the mean values of the rolling resistance coefficients for the two different ambient temperatures at steady state and by using the presented model. The steady state values are calculated according to equation (3) for the driven velocities.

The corresponding rolling resistance energy loss for all four tyres on the vehicle is presented in figure 10 as a bar plot. “SR 25°C” corresponds to the rolling loss of the tyres using the steady state rolling resistance coefficient at 25°C and respectively for “SR 15°C” at 15°C ambient temperature. For the same conditions “Model 25°C” and

"Model 15°C" represent the rolling loss with the transient rolling resistance model. The difference of the model compared to the steady state values can rise up to over 20 % rolling loss.

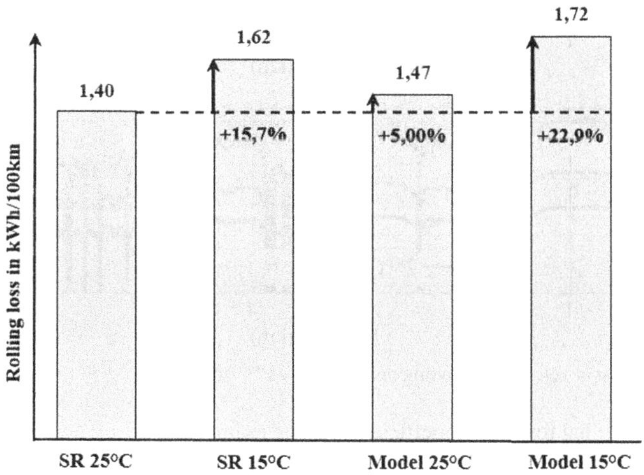


Figure 10: Rolling loss for four tyres with the steady state values and the transient model at 25 °C and 15 °C for the NEDC

## 4.5 Rolling Loss for the Example Trip

The rolling resistance model is used for the example trip as described in paragraph 3.4. It represents a more realistic driving stretch than the NEDC. Neither load transfer variations due to turnings nor other effects caused by lateral dynamics have been taken into account for the time being.

In figure 11 the top plot illustrates the evolution of the velocity, the bottom plot shows the evolution of the rolling resistance using the model for different ambient temperatures at 25 °C, 15 °C and 5 °C. Winter days in Germany can drop below 5 °C. This causes a higher tyre rolling loss. The rolling resistance coefficients for the different temperature conditions for the steady state values and calculated with the model are presented in table 9. The total rolling loss for the different scenarios are illustrated in figure 12.

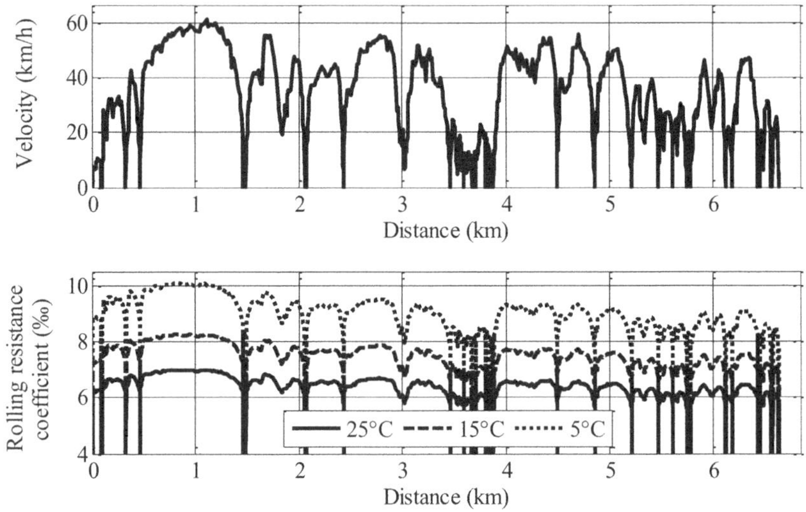


Figure 11: Evolution of velocity and rolling resistance at 25 °C and 15 °C

Table 9: Mean rolling resistance coefficients for the example trip

| Ambient temperature in °C | 25 | 15 | 5 |
|---|---|---|---|
| Mean steady state rolling resistance coefficient in ‰ | 6.12 | 7.11 | 8.48 |
| Mean transient rolling resistance coefficient in ‰ | 6.36 | 7.44 | 8.95 |

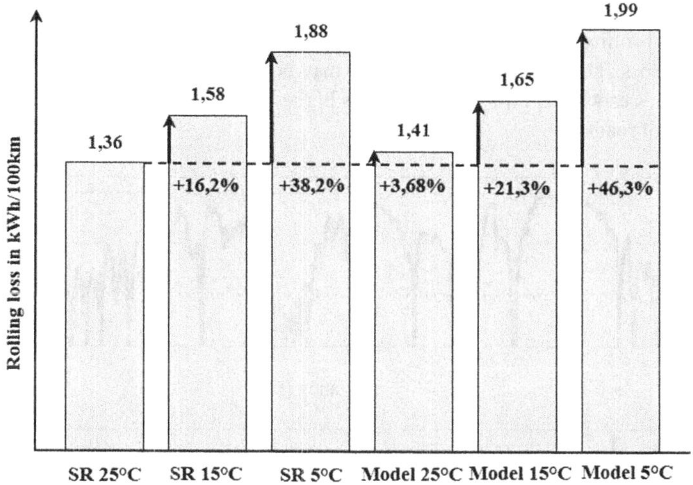


Figure 12: Rolling loss for four tyres with steady state values and the transient model at 25 °C, 15 °C and 5 °C for the example trip

## 4.6 Rolling Loss for Example Trip with Turnings

A verified two track model of the Visio.M allows firstly to simulate the load transfers due to turnings, traction and braking. Bearing in mind equation (3), a varying tyre load also influences the steady state rolling resistance, thus the transient value. Secondly, the two track model simulates the instantaneous slip angles and lateral forces on all four wheels so the opposing force to the wheels' directions can be calculated. Studies [10] show that this force can be determined as follows:

$$F_{R_{lat}} = F_y \sin\alpha = C_\alpha \alpha \sin\alpha \tag{16}$$

This additional force on each tyre leads to an additional internal heat. This phenomena is simulated with the presented model with an additional force in turnings that is calculated with the cornering stiffness of the tyre and the instantaneous slip angle.

Figure 13 presents the evolution of velocity (top plot), lateral acceleration (middle plot) and the evolution of the rolling resistance in only longitudinal direction compared to the rolling resistance in longitudinal and lateral direction (bottom plot) for the left rear tyre of the vehicle.

For the three scenarios, the rolling resistance coefficient values are listed in table 10 and the total rolling loss is illustrated in figure 14. The turnings do have an impact on the rolling loss. The rolling loss in turnings may be slightly higher for trips on curvy country roads as well as for sporty drivers as higher lateral accelerations are driven on those kind of roads.

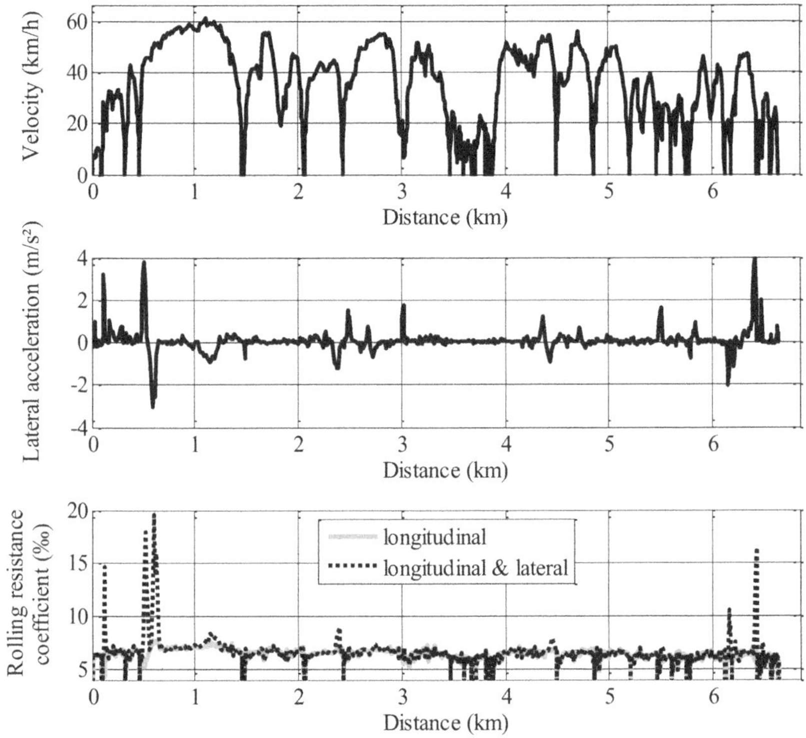


Figure 13: Evolution of velocity, lateral acceleration and rolling resistance excluding and including lateral dynamics at 25°C for the example trip

Table 10: Mean rolling resistance coefficients for the example trip excluding and including lateral dynamics

| Ambient temperature in °C | 25 |
|---|---|
| Steady state rolling resistance coefficient in ‰ | 6.12 |
| Mean transient rolling resistance coefficient long. in ‰ | 6.36 |
| Mean transient rolling resistance coefficient long.&lat. in ‰ | 6.61 |

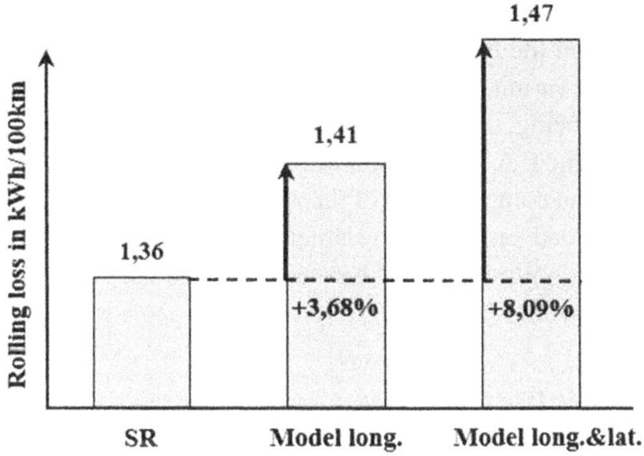


Figure 14: Rolling loss for four tyres with steady state rolling resistance values and the transient model with and without turnings for the example trip at 25 °C

# 5 Summary

This paper presents how the rolling resistance influences the consumption of vehicles, especially for compact electric vehicles. During the tyre's warming up procedure, test bench results show that the rolling resistance is higher than its steady state value and is dependent on the driven velocity, the rolling duration, the ambient temperature, the load and the tyre pressure. Commonly, the rolling resistance coefficient used to calculate vehicles energy consumption is the standard ISO 28580 value that is determined at one operating point of the tyre. This results in inaccurate calculated energy consumption.

In order to know for which operating points of the tyre the rolling resistance has to be analysed, realistic driving scenarios have to be defined. Thus, recorded trips from an electric vehicle fleet with MINI E's have been evaluated regarding the driven distances, the driven velocities and the lateral accelerations. The trips were mainly city trips, which means rather short rides not exceeding 30 minutes duration. During this period, the tyre has barely reached its steady state.

A rolling resistance model for the VisioM tyre is developed and presented to better estimate the evolution of the rolling loss for a given trip. For the NEDC as well as for a realistic compact electric vehicle driven trip, the rolling loss calculated with the model is clearly higher than the standard steady state value. Furthermore, the model can take account of turnings which causes an increase of the rolling resistance and heats up the tyre additionally.

This approach represents the next step to a more realistic estimated rolling resistance value in particular when the tyre is not in its steady state. This is mainly the case for driving scenarios with various accelerations and braking manoeuvres, standstills and turnings. For the time being, the model has been compared to rolling resistance measurements on the test bench. A comparison to realistic measurements on track might be interesting, however, the comparability of the results is not given due to exterior, interfering factors like the road surface, the weather and the driver. The presented model is the next step to energy optimization of tyres and will be used for analysis of vehicle consumption simulations.

# 6 Acknowledgement

The experiments for this work were conducted by Michelin. The authors would like to thank them for the cooperation and support.

# List of Symbols

| | |
|---|---|
| $\alpha$ | Sensitivity of rolling resistance to temperature (1/K) |
| $A$ | Outside tyre surface area (m²) |
| $C_\alpha$ | Cornering stiffness of tyre (°/N) |
| $c_p$ | Specific heat transfer factor (J/kg/K) |
| $C_{rr,iso}$ | Rolling resistance coefficient according to ISO standard (‰) |
| $C_{rr,amb}$ | Rolling resistance coefficient for another ambient temperature (‰) |
| $C_{rr}$ | Rolling resistance coefficient (‰) |
| $F_R(t)$ | Instantaneous rolling resistance force (N) |
| $F_R^*$ | Steady state rolling resistance force (N) |
| $F_{R_{lat}}$ | Additional rolling resistance force in turnings (N) |
| $F_y$ | Lateral force (N) |
| $h$ | Heat transfer coefficient (W/m²/K) |
| $h_0$ | Reference heat transfer coefficient (W/m²/K) |
| $k$ | Heat transfer exponent, dimensionless |
| $k_1$ | Velocity dependent rolling resistance coefficient (‰.s²/m) |

| | |
|---|---|
| $m$ | Mass of tyre (kg) |
| $p$ | Tyre inflation pressure (bar) |
| $p_{iso}$ | Tyre inflation pressure according to ISO standard (bar) |
| $\dot{Q}_{diss}$ | Rate of dissipated heat to surroundings (W) |
| $\dot{Q}_{int}$ | Rate of internal heat generation (W) |
| $\dot{Q}_{stored}$ | Rate of internal heat storage (W) |
| $t$ | Time (s) |
| $T_{amb}$ | Ambient Temperature (°C) |
| $T_{iso}$ | Reference temperature according to ISO Standard (°C) |
| $T^*$ | Steady state operating temperature (°C) |
| $T(t)$ | Instantaneous operating temperature (°C) |
| $v_0$ | Reference tyre velocity (m/s) |
| $v$ | Tyre velocity (m/s) |
| $x_{mes}$ | Measured value |
| $x_{sim}$ | Simulated value |
| $Z$ | Tyre load (N) |
| $Z_{iso}$ | Tyre load reference according to ISO standard (N) |

# References


[1] Liebl, J., Lederer, M., Rohde-Brandenburger, K., Biermann, J.-W., Roth, M., and Schäfer, H.: *Energiemanagement im Kraftfahrzeug. Optimierung von CO2-Emissionen und Verbrauch konventioneller und elektrifizierter Automobile*. pp. 163 - 241. Springer Vieweg, Wiesbaden, 2014.

[2] ISO 28580:2009: *Passenger car, truck and bus tyres - Methods of measuring rolling resistance - Single point test and correlation of measurement results*.

[3] Greiner, M., Unrau, H.-J., Pfriem, M., and Gauterin, F.: *Bewertung des Rollwiderstands auf Basis transienter Rollwiderstandsverläufe im Hinblick auf elektromobilitätstypische Fahrerprofile*. 14. Internationale VDI-Tagung „Reifen-Fahrwerk-Fahrbahn", Hannover, 2013.

[4] Michelin: *Der Reifen - Rollwiderstand und Kraftsstoffersparnis*. Clermont-Ferrand, 2005.

[5] Helmbrecht, M., Bengler, K., and Vilimek, R.: *Adaptations in Driving Efficiency with Electric Vehicles*. 15th International Conference, HCI International, Las Vegas, NV, USA, 2013.

[6] Bachmann, T., Bielaczek, C., and Breuer, B.: *Der Reibwert zwischen Reifen und Fahrbahn und dessen Inanspruchnahme durch den Fahrer*. *ATZ* 97, 10., 1995.

[7] Schuring, D. J., Siegfried, J. F., and Hall, G. L.:*Transient Speed and Temperature Effects on Rolling Loss of Passenger Car Tires*. SAE Technical Paper Series, Detroit, Michigan,USA, 1985.

[8] Mars, W. V. and Luchini, J. R.:*An Analytical Model for the Transient Rolling Resistance Behavior of Tires*.Tire Science and Technology, Akron, Ohio, USA, 1999.

[9] Luchini, J. R. and Popio, J. A.: *Modeling Transient Rolling Resistance of Tires*. Tire Science and Technology, Akron, Ohio, USA, 2007.

[10] LaClair, T.: *"Rolling Resistance"* in *"The Pneumatic Tire"*, U.S. Department of Transportation, National Highway Traffic Safety Administration (NHTSA), 2006.

# The non-steady-state tire model as a set of physical submodels for driver assistance systems analysis

Dipl.-Ing. Pavel Sarkisov, Dresden University of Technology

Prof. Dr.-Ing. Günther Prokop, Dresden University of Technology

Dr.-Ing. Sergey Popov, Bauman Moscow State Technical University

# Abstract

The challenges of advanced driver assistance systems development bring a necessity to describe a complex tire motion, which incorporates non-steady-state rolling, combined slip and a whole range of slip up to 100%.

Empirical description of such a complex mode requires much more experimental work than for convenient steady-state rolling with sideslip only. FE-models are time-consuming and usually not reasonable for handling analysis in case of flat rigid road surface. These issues motivate an investigation of physical background of tire behaviour, as long as physical model is able to respond to any correct excitation it receives.

The paper describes a modular combination of several simplified physical blocks of tire model: model of carcass deflection, model of tread shear, friction model and later thermal model. These models are selected and parametrized with help of separate measurements of tire carcass and tread properties.

Due to two-dimensional contact patch deformation distribution inside of its area is provided avoiding FE-method. It refines the simulation quality of lateral force and aligning torque, makes it possible to estimate remaining grip potential in any contact point during the non-steady-state tire excitation, what is necessary nowadays for further development of advanced driver assistance systems control algorithms.


# Motivation

Nowadays improvement of road safety has a legislatively defined priority. One of the most efficient ways to do it is automated control of the vehicle in addition to the ordinary actions of a human: advanced driver assistance systems. There is large group of systems, which correct the forces, acting on the vehicle, by means of individual kinematic or force impact on its wheels.

Calculation of forces on the four wheel hubs, which are required, for example, to ensure vehicle stability, is relatively simple. But determination of such an impact on the wheel, which causes a desired force response of the tire, is still one of the most complicated issues of Newtonian mechanics. That is why tire is usually simulated in particular cases of its motion, which represent lower complexity, like steady-state handling dynamics, parking maneuvers, braking dynamics or ride comfort.

The research and development of driver assistance systems put forward new requirements to the tire modelling.

1. Such electronic systems are able to adjust longitudinal force on the tire by means of drive and brake torque management and lateral force through individual steering and camber control. That means the tire model has to be able to consider combined load in both directions and consequently combined slip.
2. The highest efficiency of vehicle control is provided, when whole grip potential of the tire is used. This requires a tire model that describes a whole range of slip values – from 0 % to the 100 %.
3. Finally, electronic control systems operate with frequency from 10 to 30 Hz during vehicle transient motion. That is why the non-steady-state tire model is required.

Applicability of existing methods of tire modelling to these requirements is limited.

# State of the art

There are three general approaches of tire modelling for purposes of vehicle dynamics simulation [1].

- **Empirical models** simulate specific dependencies rather accurate, but they provide no understanding of the tire background processes and are not able to derive further tire properties without a specific experiment.
- **Physical models** represent a flexible body of a tire as certain mechanism, whose behaviour roughly corresponds to the tire behaviour. Such models have usually simple structure with low number of parameters, but they are not accurate enough.
- **Finite Element methods** are a high-accuracy tool of tire modelling, but because of its complexity it is not always reasonable to apply the method to simulate tire handling properties in case of rigid and flat road surface.

That is why the most suitable approach for driver assistance systems development, analysis, and testing is a simplified physical modelling, as soon as physical system is able to response to any correct excitation it receives.

The limitation of existing physical tire models is low accuracy, caused by simplified structure: the majority of models are oriented to one or few specific motion modes, such as steady-state rolling with a slip angle, parking maneuvers, slip-free rolling and so on. These models are as a general rule intentionally developed in such a manner in order to provide an ability of realtime operation.

Consequently present requirements to tire model highlight a necessity of rational increase of model level of detail, in order to take into account more relevant physical properties of this system.

The analysis of tire structure shows, that the tire can be considered as a composition of two main components – anisotropic tire carcass and layer of isotropic rubber of tire tread. That is why the reasonable approach to describe such a system is its representation as two physical models – carcass model as a core and tread model as a periphery. As long as tread is responsible for friction and heat development due to slip, the friction and thermal sub-models are coupled only with the tread model.

This paper is focused on investigation of different combinations of carcass and tread models. Literature analysis shows a number of such combined models.

**Brush-and-Ring-Tire** [1]. According to its structure tire carcass and breaker are considered as a number of flexible strings, and a tread of the tire is represented as brush model. These strings are describing the tire carcass body, still they are independent from each other. Such a description cannot completely consider different carcass deformation on the left and right side, even though the difference of tread deformation is taken into account. Furthermore, the contact patch is represented as several curved lines instead of limited area. These two features lead to an error of aligning torque simulation for different slip angles, but the lateral force is described rather accurate.

**Treadsim** [1]. Such an approach combines brush-based tread model with the approximated carcass deflection using several single rows of brush elements. The carcass model considers its physical properties such as lateral, torsional and bending stiffness values, still the approximation is made via second-order function, which represents one-dimensional curve. Consequently a deformation of a carcass as a whole coherent body is described neglecting this feature.

**TameTire** [2-4]. This approach is one of the most advanced physical models, combining mechanical tire deformation description with thermal model. Tire tread is described as a brush model, tire carcass deflection is simulated with help of second-order approximation function, based on lateral force, tire lateral flexibility and kinematic constrains. The system is solved with help of iterative numerical method. In case of high-frequency excitation the parabolic approximation function may limit the carcass description. The detailed understanding of carcass deformation behaviour may significantly improve the simulation quality and contribute to avoiding scaling functions, used in TameTire structure.

To conclude: present-day challenges of driver assistance systems development require a description of two-dimensional tire contact patch and carcass deformation in order to be able to consider combined slip from 0 to 100% in a non-steady-state.

The global goal of this research is a refinement of description of non-steady-state tire handling characteristics by understanding the physical processes in two-dimensional contact patch. General strategy can be summarized in a following way:

1. Model concept development considering only handling-relevant tire properties.
2. Analysis of the physical processes in the rolling tire in order to derive from them as many constraining equations as possible.
3. Development a stable numerical solving method for the set of these equations.
4. Model parametrisation and validation.

# Developed tire modelling method

As a result of tire construction and deformation analysis there was developed following structure of a physical model: The carcass and the breaker are described as a flexible belt, connected with the rim through an endless number of elastic elements. Tread is considered as brush elements, which are constructed on the flexible belt.

Then the structure is expanded to three-dimensional form: the belt becomes wide, and the brush model turns into two-dimensional array of brush elements (Fig. 1).

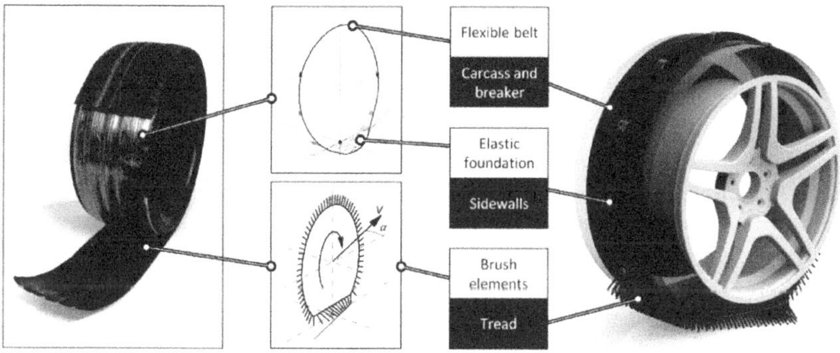


Figure 1: Concept of a combined tire model.

The main advantage of such a structure compared to "brush-and-ring-tire" is a consideration of common wide carcass instead of several independent lines and a description of two-dimensional contact patch.

But as a consequence of this advanced design the mathematical representation of the model becomes much more complicated, than for conventional models [5, 6]. Same to the real tire, deformation of the model consist of deflection of the belt (representing carcass) and shear of the brush (representing tread). If to unfold the endless circular belt on the road surface, it may be considered as a flat beam, based on elastic foundation [7] (Fig. 2).

The deflection of such a beam $y(x)$ as a function of its longitudinal coordinate $x$ is defined through fourth-order differential equation, including belt bending stiffness $EI$, tensile force $T$ in a tangential direction, elasticity of the foundation $k$ and load, induced by tread on the belt (Eq. 1). This load function contains distributed lateral force on the belt $q_Y(x)$ and a derivation of distributed bending torque $\mu'_Z(x)$ that cannot be considered by string model [7, 8].

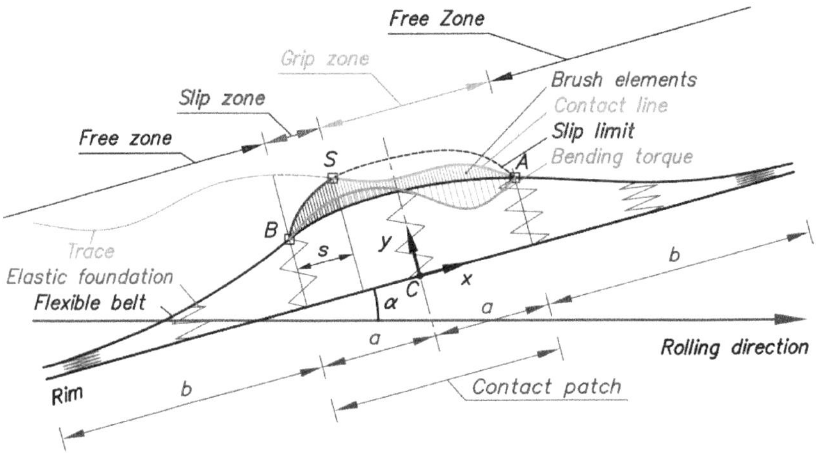


Figure 2: Tire representation as a "flat beam on the elastic foundation".

$$EI\, y^{IV}(x) - T\, y^{II}(x) + k\, y(x) = q_Y(x) + \mu'_Z(x) \tag{1}$$

Fourier transformation of the load function makes it possible to find the analytical form of belt deflection, so the carcass deformation formula can be defined purely analytically (Eq. 2). The constant parameters $C_i$, $\lambda_i$ as well as Fourier series parameters $a_0$, $a_n$, $b_n$ of this expression are defined numerically.

$$y(x) = \sum_{i=1}^{4} C_i e^{\lambda_i x} + \frac{a_0}{2} + \sum_{n=1}^{\infty} \left( a_n \cos\frac{\pi n x}{L} + b_n \sin\frac{\pi n x}{L} \right) \tag{2}$$

As soon as the form of the belt is found, it is possible to calculate position of any of its points based on plane-sections-hypothesis. In such a manner the position of root points of brush elements are determined. If to connect these root points with the road surface and consider kinematic connection between them, a connecting vector will describe a deformed brush element. In case of steady-state rolling with a non-zero slip angle the tread shear increases from the front to the rear of the contact patch, but only up to certain limit, due to the slip.

Tread deformation calculation determines a load, which is applied from the tread to the belt. This function goes to the belt deflection calculation, what makes the whole set of equations consistent, but there are significant difficulties in the set. From the one hand, there are several partial differential equations. From the other hand, the slip limitation is implemented with help of Heaviside step function, which makes it complicated to solve the equations through operational calculus. Consequently the set of model equations cannot be solved analytically, represents high complexity of direct numerical solving and requires a specific mathematical tool to solve the model equations. For that purpose was an iterative solving process developed (Fig. 3). Within the each moment of time several successive approximations to the accurate solution are performed.

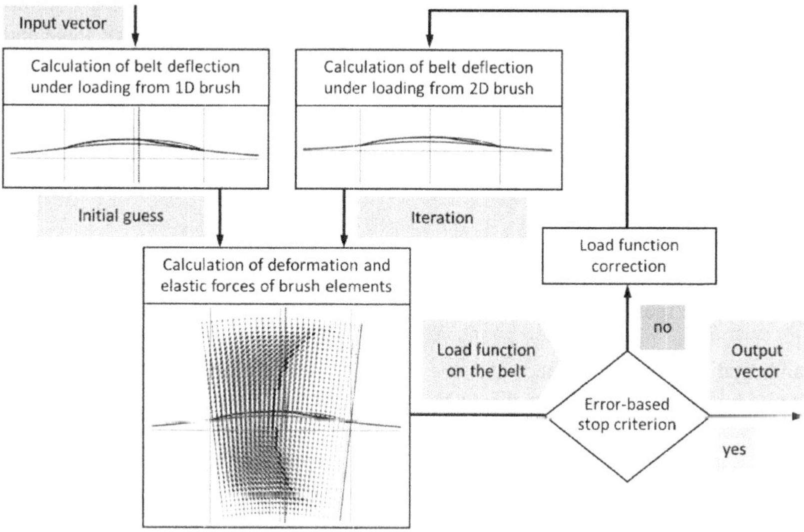


Figure 3: Operation scheme of iterative solving method.

- As an initial guess there is used a solution of one-dimensional model, such as string-based system, neglecting bending torque.
- On the deformed belt there is built a two-dimensional array of brush elements.
- With help of kinematic connection between road and tire the deformations and elastic forces of brush elements are determined.
- These forces are recalculated to the distributed lateral force and distributed bending torque on the belt, which compose the load function.
- This function is transmitted once again to the belt deflection calculation, so the cycle becomes closed.

In order to provide process convergence the load function difference between two iteration is reduced in the correction block [9]: it helps to avoid too fast approximation to the solution and possible throw-over [10].

Every point of two-dimensional contact patch creates elementary longitudinal and lateral force, but also an elementary bending torque for the tire carcass. So the load, induced from the tread to the belt, is calculated with both lateral distributed force and bending torque. It is visible, that the difference between load function of two iterations is significant, so it is necessary to make the approximation slower. As soon as iteration step is reduced the curves of load functions of current and previous iteration are coming closer to each other and finally coincide, what means an exact solution.

# Testing tools

For both parametrisation and validation of the model is experimental data required. As a primary test tool there was selected the drum tire test bench of Dresden University of Technology: tire is rolling on the drum 2 m in diameter, there is one frame for camber adjustment and the second frame for slip angle variation. Horizontal forces are carried by the bearing 1.5 m in diameter. The vertical load is applied by hydraulic cylinder, which is connected to the framework of the test bench (Tab. 1). As long as overall dimensions of the frame are about 2 m, a metrological evaluation was conducted: the error caused by frame flexibility is below 1.4% [11].

As a secondary testing tool was chosen hydraulic pulsing machine of Dresden University of Technology. The device consist of hydraulic cylinder, table and frame (Tab. 2). This facility makes it possible to investigate material properties of rubber block of tire tread.

Table 1: Tire test rig technical data.

| Max. velocity | 320 km/h |
|---|---|
| Max. wheel load | 30 kN |
| Max. slip angle | 90° |
| Max. lateral force | 20 kN |
| Drum diameter | 2000 mm |
| Drum width | 500 mm |
| Max. tire diameter | 900 mm |
| Slip angle variation | up to 35 °/s |
| Camber range | +10°.. –45° |
| Camber variation | up to 35 °/s |

Table 2: Hydraulic pulsing machine technical data.

| Force range | ± 50 kN |
|---|---|
| Displacement range | ± 125 mm |
| Accuracy | 0,01 mm |
| Frequency range | 0 .. 100 Hz<br>(ca. 0.5 mm) |

# Model parametrisation

In order to define model parameters it is necessary to measure the physical properties of tire carcass and tread separately.

According to ordinary production process, the rubber belt of a tread layer is put or wound up on the carcass, then this item is vulcanized and it is not possible anymore to separate the tread from the carcass. The problem may be solved in a following way:

The carcass is firstly covered with heat-resistant film, consequently tread belt is put on this film. The vulcanization is made as usual, but due to the film tread and carcass do not stick to each other and can be easily separated. This approach makes it possible to measure two components of tire body deformation separately – the bending of its carcass and a shear of the tread elements (Fig. 4).

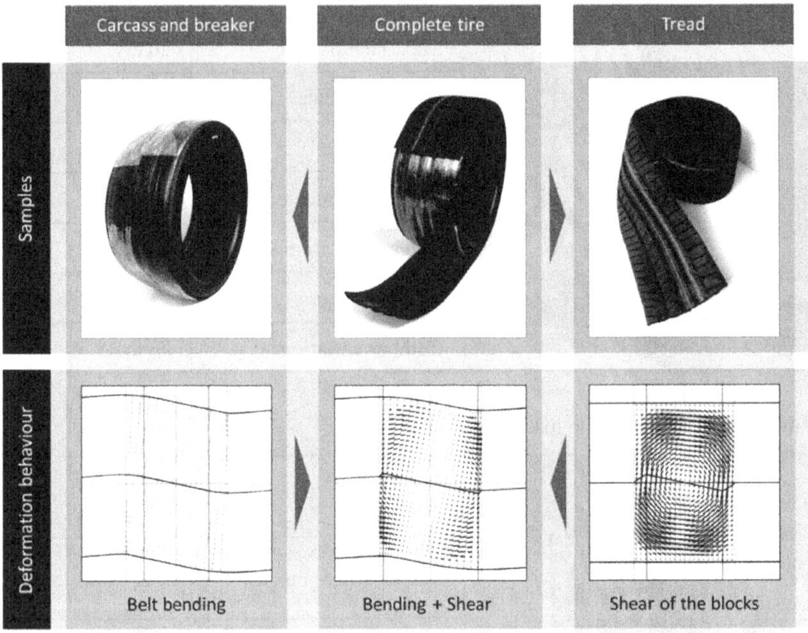


Figure 4: Separation of different components of tire deformation with help of separated carcass and tread samples.

To investigate this issue there were measured tire lateral and torsional static stiffness-values. In first case the support platform of a wheel was turning around a vertical axis in order to determine the connection between turning angle and tire resistance torque. In the second case the platform was moving slowly in the lateral direction (Fig. 5).

All measurements were conducted for three different wheel load values and for two items: for carcass without tread and for complete tire. The linear parts of the curves corresponds to the little displacement, where there is practically no slip yet. Based on these linear stiffness values the parameters of tire carcass and tread may be determined.

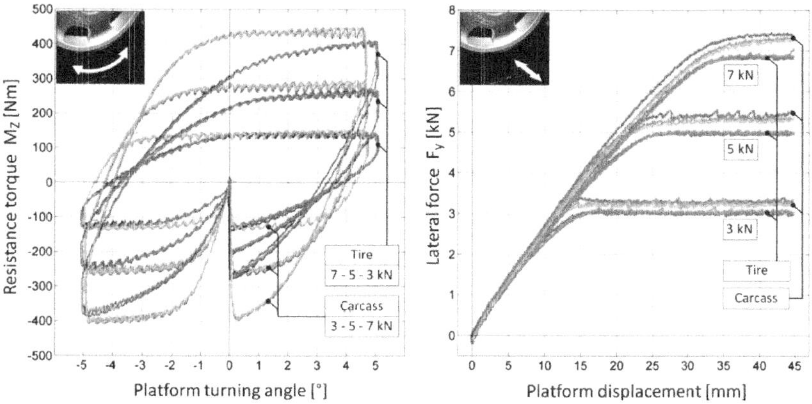


Figure 5: Measurement of static tire and carcass stiffness values.

# Model validation

As soon as model parameters are defined, it is possible to simulate tire transient behaviour. The main feature of tire non-steady state concerning handling is not inertia, damping or gyroscopic issues, but so called relaxation length, which defines the delay between kinematic excitation of the wheel and force response of the tire.

In order to investigate the influence of carcass deformation on the dynamic tire properties the "slip angle step"-maneuver was measured and simulated: tire is rolling straight at 3 km/h, the slip angle is changed stepwise from zero to three degrees and after 4 s is returned back to zero.

Four different approaches to carcass bending simulation in case of two-dimensional contact patch were investigated:

1. Rigid carcass [1, 5, 6] (Fig. 6).
2. Rigid carcass with only lateral flexibility [1] (Fig. 7).
3. Beam-based carcass bending calculation, where each its section stays perpendicular to the belt center line. This case corresponds to the Euler-Bernoulli hypothesis of a beam bending (Fig. 8).
4. Beam-based carcass bending calculation, where each its section stays parallel to the wheel rotation axis, that means no section is turned due to carcass deformation. This approach corresponds to the methods applied by Brush-and-Ring-Tire and Treadsim models, but as distinct from them it takes into account distributed bending torque on the belt (Fig. 9).

Due to the description of two-dimensional contact patch it is possible to investigate the deformation of every point of contact patch traveling from its front to its end.

To investigate the deformed condition equilibrium state of the tire is compared on figures 6-9, which is achieved already in 2 s after excitation. Lateral force and aligning torque responses to the slip angle step are depicted for each approach as well.

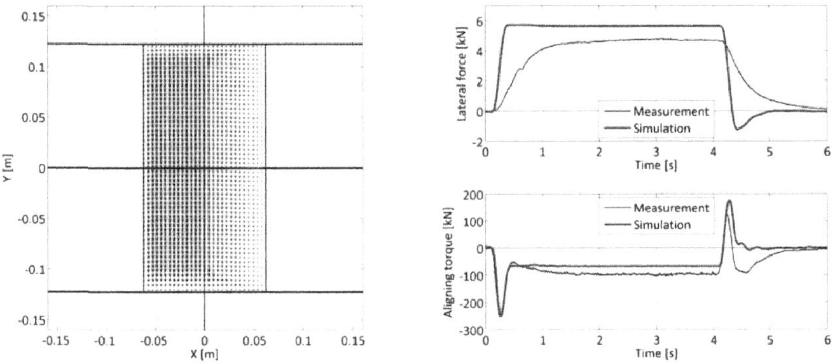


Figure 6: Simulation results for the model with rigid carcass.

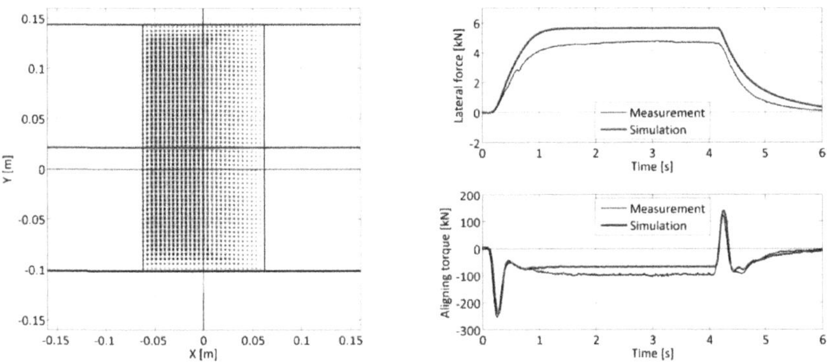


Figure 7: Simulation results for the model with a lateral carcass flexibility.

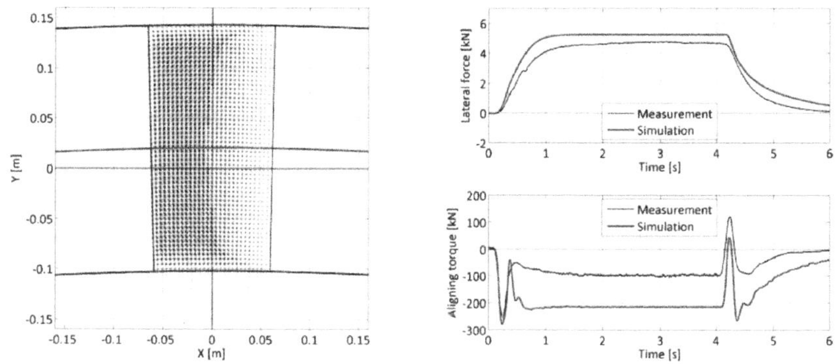


Figure 8: Simulation results for the model with beam-based carcass, where each its section stays perpendicular to the center line (Euler–Bernoulli hypothesis).

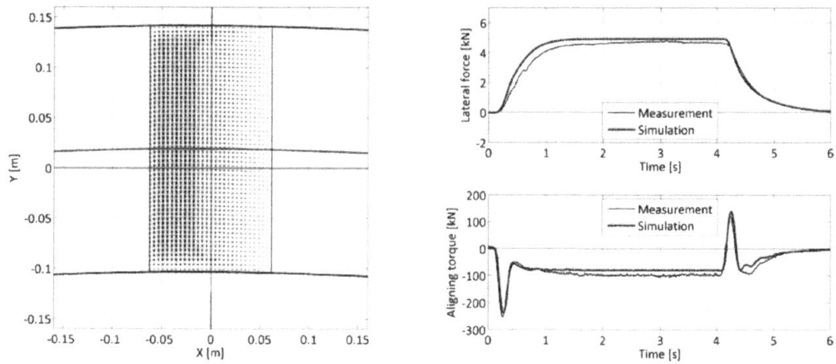


Figure 9: Simulation results for the model with beam-based carcass, where none of its section turns.

Following features should be noted based on this comparison:

1. Model with rigid carcass shows too short delay of force response because it is not able to describe big lateral deformation of a tire (Fig. 6). It motivates to consider a rigid carcass, which can be still displaced as a solid body in a lateral direction.
2. Such a model with lateral flexibility improves response delay partly, but still depicts a large deviation of steady-state lateral force (Fig. 7). The reason is a straight position of the carcass, which forces brush elements to compensate kinematic difference and consequently deform excessively.

3. Implementation of a flexible carcass, described with help of beam theory, improves the simulation of lateral force, but causes a bigger error of aligning torque, especially in a steady state (Fig. 8). The background of this effect originates from end part of the contact patch, where tire carcass approaches rim plane back, but brush elements still have to follow the road considering sliding.
4. The rejection of application of Euler-Bernoulli hypothesis for wide body of tire carcass results in the case, where its cross sections are still flat, but not perpendicular to the belt center line anymore. Consequently the shear stress between them appear. Nevertheless such an assumption delivers a better results of both lateral force and aligning torque simulation (Fig. 9).

# Conclusion and outlook

In the paper there were considered existing tire models in sense of suitability to the new requirements of advanced driver assistance systems development. The necessity of refinement of the modelling by means of understanding of the physical background was detected.

A combined method of tire modelling was proposed, which incorporates brush model of the tire tread, defined with help of kinematic constrains, and a model of carcass deflection. Several different approaches to carcass bending description were investigated. The outcome emphasizes a limited applicability of classic beam theory based on Euler-Bernoulli hypothesis of plane sections to the simulation of carcass deflection. The best simulation results were achieved with a beam-based carcass with distributed lateral force and bending torque on it and enabled shear between its cross sections.

Further development should be focused on dynamic variation of vertical load distribution in a contact patch and improvement of solving routine in order to reduce operation time.

The description of physical processes inside the two-dimensional contact patch provides a complete strain figure for non-steady-state combined loading with any slip value and direction. The practical utility of the method is refined description of non-steady-state tire characteristics for vehicle handling and stability analysis; a possibility to investigate deformation of any point inside the contact patch avoiding finite-elements-method and provided understanding of the physical background of the tire deformation.

In such a manner the developed method contributes to non-steady-state tire modeling refinement, what is necessary nowadays for further development of the advanced driver assistance systems and improvement of a road safety.

# References


[1] Pacejka, Hans B.: Tyre and Vehicle Dynamics. 3. Auflage. Elsevier BH, 2002. ISBN: 978-0-08-097016-5.

[2] US patent 2010/0010795 A1, Fevrier et al.: Method for simulating the thermo-mechanical behavior of a tire rolling on the ground, issued 2010-01-14.

[3] Février, Pierre; Fandard, Gérard: A new thermal and mechanical tire model for handling simulation. VDI-Berichte Nr. 2014, 2007.

[4] Février, Pierre; Fandard, Gérard.: Thermal and mechanical tire modelling to simulate handling. Automobiltechnische Zeitschrift (ATZ) 05/2008.

[5] Popp, Karl; Schiehlen, Werner: Fahrzeugdynamik, Stuttgart, Teubner Verlag, 1993.

[6] Van Putten, Sebastiaan: Tire Bore Torque. A semi-mechanical approach to measurement and modeling of standstill and rolling conditions. Intelligent Tire Technology Conference, Darmstadt, 2012.

[7] Ellis, John R.: Vehicle Dynamics. Advanced School of Automobile Engineering, Cranfield, London Business Books Limited, 1969.

[8] Sarkisov, Pavel; Prokop, Günther; Popov, Sergej: Reifenmodellierung für den instationären Betrieb durch Kombination der Funktionsansätze. 1. WKM Symposium, RWTH Aachen, 2014.

[9] Korn, Granino A.; Korn, Theresa M.: Mathematical Handbook for Scientists and Engineers, New York, McGraw Hill Book Company, 1968.

[10] Pacejka, Hans B.: A hybrid computer model of tire shear force generation. Document 3, Automobile Manufacturers Association, Highway Safety Research Institute, 1971.

[11] Sarkisov, Pavel; Kubenz, Jan; van Putten, Sebastiaan; Prokop, Günther: Steifer Reifenprüfstand für höhere Messgenauigkeit. Automobiltechnische Zeitschrift (ATZ) 04/2015, p. 46-51.

## Acknowledgements

Research project is supported by "Erasmus Mundus Action 2 Programme of the European Union".

Authors would like to thank Mrs. Denise Beitelschmidt, Mr. Axel Wildgrube, Mr. Thomas Thüringer, Mr. Stefan Eckert and Mr. Dirk Schlimper for their contribution to the project development.

# Evolution of the requirements on vehicle tires and insights from 15 years of test operation

Lars Netsch, TÜV SÜD Product Service GmbH
Reifen & Räder / Tires & Wheels
Daimlerstrasse 11, 85748 Garching

Co author:

Michael Staude, TÜV SÜD Product Service GmbH
Reifen & Räder / Tires & Wheels

# 1 Introduction „The tire for the car of tomorrow“

40 years ago some scientists and engineers worked on ideas with cars that are riding on an air cushion and they assumed the function of wheels mounted on cars in the "future" (i.e. year 2000) would maybe just consist in supporting shunting and parking manoeuvres /1/. The future technology of tires was not considered at all.

But today between the beginning and the specified end of their service lives, vehicle tires must satisfy a large number of requirements. For all specified vehicle/tire combinations and under all possible operating conditions, they must be able to fulfil their challenging functions reliably under the most varied environmental stresses (e.g. solar radiation, temperature, ozone, depending on the region in which they are used) and on an immense range of different grounds, extending from optimum road surfaces to the most demanding off-road terrain. Apart from all the influencing factors due primarily to users and how careful they are, the interaction between vehicle and tire is equally decisive for tire stress. For example, the load profile of a tire is crucially influenced by geometrical, kinematics, elasto-kinematics, and thermodynamic correlations as well as the drive concept. Following the evolution of road vehicles and change of several requirements the tires are subject to a continuous improvement process, driven by the development progress of the vehicles, events in the field and increasing requirements in the legislation /2/.

In this lecture some important milestones will be displayed, which had in the history of the last 15 years more or less significant impact on the development of passenger car tires in Europe and that built up the trends for design of safe and economic tires to be fitted on the car of tomorrow. Today one of the most safety relevant performance parameters of tires consists in the braking performance on wet road surfaces; the paper shows the improvement of the parameter "wet braking performance" over a wide range of products and also under different test conditions. It demonstrates how successful the optimization of this parameter could be realized in the area of conflict between rolling resistance and rolling noise.

# 2 Drivers of tire development during the last 15 years in Europe

## 2.1 General

Target of tire development for the European market is to deliver efficient and reliable tires and to provide the consumer with maximum levels of safety.

The development is mainly influenced by the following impact factors:

- innovations and evolution in vehicle technologies
- growing preferences and expectations in driving comfort and visual appearance
- permanent increase in vehicle dynamics performance
- demands for prevention of fatalities due to tire failures or poor tire performance
- requirements for environmental and health protection
- regulatory environment, revision of statutes
- economy, commercial crises and fuel availability / – cost
- energy policy
- necessity of customer information and transparency.

## 2.2. Development progress of the vehicles and requirements of vehicle owners

Trends in tire development driven by vehicle evolution are mainly increasing car performance with rising vehicle maximum speed at higher vehicle mass at the same time. This requests more robust tires with higher structural stability (speed- and load-index). Improving vehicle dynamics performance demands better driving stability, provided by wider tires with lower aspect ratio for optimized transfer of the forces acting in the contact patch; the consequence is then application of larger rim diameters, as this helps to increase the size of the contact patch. High vehicle speed requests more space for bigger brake systems, especially on the front axle of the vehicle (resulting in larger rim diameter, too). As an example for the evolution of tire dimensions the size of the standard tire fitment for a Mercedes S-Class 15 years ago was 225/60R16 (type W 220) /3/ and today for the current model it is 245/55R17 (type W222). Briefly summarized, the nominal section width was increased by 20mm and the rim size by one inch (25mm). As one advantage a bigger tire diameter supports optimized rolling resistance, lower noise levels and better wear behaviour; but: drivers´comfort requirements are limiting the use of too low aspect ratios, leading to higher vertical stiffness, especially as application of higher inflation pressure is necessary in this case.

Every vehicle manufacturer sets its own priorities in the technical characteristics: safety, comfort, driving performance (including braking capabilities) or economy might be emphasized. Tires for original equipment of the cars of a certain manufacturer will then be customized individually according to their performance profile. The total number of performance criteria at a car manufacturer counts to about 50 or more. However, since all these properties are not without conflict, an optimal compromise is the goal of the whole development. The determination of the optimal compromise is the task of the vehicle manufacturer; the technical application is under responsibility of the tire manufacturer /3/. Another challenge for the tire- and vehicle manufacturer is to ensure proper interaction of the tires with the sophisticated control systems of the vehicles, such as ABS or ESP, VSA...

Of course, individual tire characteristics are to be maintained, allowing tire manufacturers to distinguish themselves from their competitors and to place a stronger focus on certain features than on others.

The customer of a vehicle manufacturer expects to have the opportunity to select a tire according to his/her individual preferences and driving habits. This, in fact, is the reason why the majority of tires as original equipment or in the replacement market is deliberately designed to offer certain properties, be it riding comfort, low tire noise or rolling resistance, high mileage, sporty driving characteristics or excellent wet or snow-performance.

In the last 15 years the trends in tire development driven by vehicle drivers were especially:

- attractive styling: wider tires with lower aspect ratio and larger rim diameters provide a sporty impression and put emphasis on the design of wheels
- wide tires are associated with high safety (wide is „strong“)
- new popular wheel designs and geometries can have impact on the dimensions of the tires to be fitted.

## 2.3. Events and milestones in the Automotive field

Triggered by a series of fatal accidents in the US, involving certain tires of one brand and mounted on a certain vehicle model in the years 2000 and 2001, tire related accident statistics worldwide showed a clear potential for improvement. Prevention of future similar failures became an important public health goal /4/, /5/.

In 2002, the USA passed the Tread Act to require an installation of a Tire Pressure Monitoring System (TPMS) to warn the driver if the tire is significantly underinflated. This legislation could be considered as the first strong step for the intelligent tire development. With the Tread Act a rule was published to make Tire Pressure Monitoring Systems a mandatory feature on new vehicles in 2005, with a phase-in period until 2007 for all new vehicles. Since latest the year 2014 TPMS are state of the art, specified in European technical standards and mandatory for new vehicles by law within the European Union, too. And this measure is intended not only to improve vehicle safety, but also to reduce $CO_2$ emissions. A further consequence was that also application of run-flat tire systems, guaranteeing extended mobility in case of tire failure became an important part of tire/vehicle systems development.

Today there is a huge variety of such systems in the market, for example (the following list does not claim to be comprehensive):

- Bridgestone: RFT (Run-Flat-Tire)
- Continental: SST (Self Supporting Tire)
- Dunlop: DSST (Dunlop Self Supporting Technology) or ROF (RunOnFlat)
- Goodyear: ROF (RunOnFlat)
- Hankook: HRS (Hankook Runflat System)
- Michelin: ZP (Zero Pressure) / SST (Self Supporting Tire) / PAX System
- Pirelli: Run Flat, or Eufori@.

The world wide attention to above mentioned series of severe accidents also induced a kind of rebirth of the, already since the early 1990s, existing concepts for an „Intelligent Tire“, including monitoring of the inflation pressure. In 1996 a sensor system developed by the researchers at Darmstadt University of Technology, Germany, the so called „Darmstadt Tire Sensor“ had been presented as a research project /6/.

As one of the representative projects of this new era the project „Apollo“ (full name: Intelligent Tire for Accident Free Traffic ), starting in 2002 shall be mentioned /7/.

According to the target description of the project the future intelligent tire:

- knows tire pressure, temperature and wheel load
- can monitor tire wear
- can warn the driver about a concealed tire damage
- can warn the driver of aquaplaning or low friction
- is always active, even when the car is stationary
- can measure tire slip and forces, knows its position under the car
- doesn't need a battery
- doesn't cause weight, vibration or tire assembly/ balancing problems

and is a significant part of the car's active control systems for safety and comfort.

Additionally an integrated intelligent tire system shall include a self harvesting power system, wireless data transmission system and vehicle safety assistant system. Today, in 2015 we can observe really promising progress in achieving the targets described above in the area of research and development; but systems with a high technical maturity are not yet available in the market.

Later on the attention was moved from the concept of the intelligent tire to the new generation of the "**Green Tire**": EU leaders set the targets of the European climate and energy package in March 2007, when they committed Europe to become a highly energy-efficient, low carbon economy.

The climate and energy package is a set of binding legislation which aims to ensure the European Union meets its ambitious climate and energy targets for 2020.

These targets, known as the "20-20-20" targets, set three key objectives for 2020:

- 20% reduction in EU greenhouse gas emissions from 1990 levels
- raising the share of EU energy consumption produced from renewable resources to 20%
- a 20% improvement in the EU's energy efficiency.

The targets were enacted in the year 2009 /8/.

Finally the potential role of the new generation of Green Tire to support the goal 20% cut in $CO_2$ in by 2020 was in mind.

National development plans to implement E-Mobility (e.g. in Germany in 2009) reinforced the sourcing of $CO_2$ reduction potential by tire technology improvement.

One new trend coming up at the same time was triggered by the increasing activities in the field of electric vehicles′ development and the need for special tires equipping those cars: the concept of the **large&narrow** tire technology was presented to the market; the concept consists in usage of relatively big outer diameters and slim section widths compared to conventional tire sizes. The target is to realize very low rolling resistance along with a high level of dry and wet grip performance by fully utilizing size and inflation pressure effects /9/. The key factor for realizing very low rolling resistance is the suppression of tread deformation. Such tires are already in the market with electric (hybrid) vehicles today. But, however, we have to keep in mind that, even though this is clearly representing a new trend in the tire market the idea and findings are not completely new: already in the year 2001 the fuel efficient Audi A2 1.2 TDI 3L (a so-called "3-litre car") was making benefit from a tire dimension having exceptional proportion of outer diameter and section width, with size designation 145/80R14; only with this dimension the excellent fuel efficiency and low emission performance could be achieved. And up-to-date fuel efficient tire types with this dimension still are available in the market today.

After the economic crises in Europe with considerable impact on the tire industry, reaching beyond the year 2010 the pressure to curb $CO_2$ emissions continued and the tires were identified as the key to meet the $CO_2$- and safety challenges of road transport. Today the tire is an integrant of the energy efficiency program of the European Commission.

## 2.4 Legislation on tires

The rapid development in the field of automotive technology over the past decades imposes great demands on legislators to adequately address this technological progress. Ideally, legislation should be easy to enforce, ensuring the highest possible levels of safety for road users. Any regulation or directive must be liberal enough not to obstruct innovation and technical progress. At the same time, it should not restrict, but rather promote, technological progress. In the field of tire technology considerable achievements have been made, aimed at increasing traffic safety and enhancing reliable functional performance and including economic, social and environmental issues.

Under the legislation of the year 2000, 15 years ago a passenger car tire merely had to pass a high speed test (which approves the speed and load indices applied) and, as of 2003, adhere to certain limits of tire noise emissions to gain type approval. Within this statutory framework, a tire manufacturer was allowed to design and build the tire according to its own and the vehicle manufacturer´s (OEM´s), respectively the customer's desired performance characteristics.

With regard to driving safety and stability, it was obvious that a tire has to provide much more than merely the guaranteed durability under given speed and load conditions. This made the importance of tires as an essentially safety-relevant component appear to be underestimated in the past, considering the fact that there were no limits established so far.

In the first years of the new millennium within the scope of the UNECE GRRF Ad-Hoc Group "Tire Wet Grip", the UNECE developed a new regulation addressing the wet braking capability of tires, while Article 3 of the European directive 2001/43/EC, amending Council directive 92/23/EEC already demanded that safety aspects of tires be investigated, namely "...to what extent technical progress would, without compromising safety..." allow the introduction of stricter noise emission limits. In improving the directives to integrate environmental aspects, it was highly important also to consider the aforementioned environmental/safety performance interactions (e.g. noise and wet braking capability characteristics) and the service life (anti-wear and abrasion performance) in tire design. Due to the necessity of curbing global $CO_2$ emissions, the automotive industry put high effort into minimizing the fuel consumption of vehicles. This was also taken into account when tire directives and regulations, respectively were further refined by increasingly defining limits on tire rolling resistance. One example showing that such limits were necessary is the "3-litre car" (as mentioned in section 2.3 above), which, for design reasons, was only able to achieve its high fuel economy in combination with certain tire types. Since, however, current directives do not allow the specification of unequivocal vehicle/tire make and tire type combinations, optimised environmental characteristics of vehicles may be adversely influenced, if consumers, as they are entitled to do, replace tires with optimised rolling resistance by a free choice of aftermarket tires.

With statistics showing that noise disturbs two third of the population, mainly road traffic noise (Germany /10/) with a significant contribution by tire-road noise and posing a risk to human health the type approval procedure for car tires was amended by a test of the noise emissions of the tire and corresponding limit values.

A report, commissioned by the European Commission („FEHRL") /11/ recommended to label tires according to their noise emissions. Such a label was necessary for a regulation according to which cars must be operated with tires no noisier than those used during type approval, to empower every car owner to take environmental aspects into consideration when purchasing replacement tires and for the introduction of incentive schemes.

The EU Directive 2001/43/EC on the limitation of tire rolling noise had been effective since 4 August 2003 for approval of new types of tires. However, the limit values it set were too high to stimulate an advance in the state of the art. Thus the Directive 2001/43/EC needed to be revised: The limit values had to be lowered to reflect the state of the art.

For this reason, within Europe there were calls for markedly lowered limit values. The proposed reduction versus the limit values of the directive amounted to 1 to 5 dB(A) for passenger-car tires (71dB(A) for all section width categories) and 4 to 6 dB(A) for tires for commercial vehicles, depending on category of use. Furthermore, it was proposed that ambitious limit values for rolling resistance shall be introduced as soon as possible. For better consumer information, it was proposed that all tires should be labelled with the type approval values for tire rolling noise and rolling resistance /12/.

In the year 2009 with Regulation (EC) No 661/2009 of the European Parliament and of the council "concerning type-approval requirements for the general safety of motor vehicles, their trailers and systems, components and separate technical units intended therefore" (General Safety Regulation) as an amendment to Directive 2007/46/EC an extended framework for the type-approval has been established, stricter requirements for tires have been implemented and Tire Pressure Monitoring Systems (TPMS) have become mandatory for a certain vehicle category. **The new regulation amended the limits for tire rolling noise (change of groups and lower values, see table 1) and introduced new minimum requirements for wet grip (braking performance on wet surface) and rolling resistance (as an indicator for fuel efficiency).**

The regulation applied from November 2011 and the first impact on tire type approval was effective in November 2012.

Table 1: Class C1 tires (passenger car) rolling sound emission limit values

| Nominal section width [mm] | Limit dB(A) before 2012 |
|---|---|
| 145 and lower | 72 |
| Over 145 up to 165 | 73 |
| Over 165 up to 185 | 74 |
| Over 185 up to 215 | 75 |
| Over 215 | 76 |
| | **from 2012** |
| 185 and lower | 70 |
| Over 185 up to 245 | 71 |
| Over 245 up to 275 | 72 |
| Over 275 | 74 |

From table 1 it can be read that e.g. for tires with dimensions having a section width of 195mm or 205mm, representing sizes with a high volume of sales in Europe, the rolling sound limits were lowered by 4 dB(A).

The preparation towards conformity of tires to the new minimum requirements set out by the European Commission was a quite big challenge to be solved by the tire industry inside and outside of Europe. The complexity of fulfilling all the requirements at the same time will be further described in section 3.

Beyond fulfilling the above requirements the European Union decided that tire manufacturers should be encouraged to optimise all parameters beyond the standards already achieved. It was found that technological developments make it possible to significantly decrease energy losses due to tire rolling resistance, to significantly reduce external rolling noise and to significantly improve wet grip and thus to reduce wet braking distances beyond those minimum requirements. It was therefore appropriate to lay down provisions to encourage end-users to purchase tires with performance characteristics beyond the minimum standards by providing harmonised information on the above parameters.

With regulation (EC) No 1222/2009 of of the European Parliament and of the council of 25th November 2009 "on the labelling of tires with respect to fuel efficiency and other essential parameters" the European Union made the provision of harmonised information on external rolling noise, rolling resistance and wet grip performance mandatory with effect from 1st November 2012.

The tire labelling was expected by the tire industry to have an important role in promoting the adoption of tires that go beyond the minimum requirements. Now, three years after the implementation of the labelling system we can confirm that tire manufacturers put high effort into development in order to provide tires with labels that show performance levels for rolling resistance and wet grip significantly higher than the minimum requirements of the General Safety Regulation, typically corresponding to the penultimate grading of the scale from A to G (aiming on gradings C, B or even A), while maintaining the balance of performances at the same time. In addition we are observing that the Member States of the European Union are taking measures against products without labels or against non-compliant products in the market.

At least in case of the rolling resistance the efforts for improvements are obligatory, because starting already next year, in November 2016, the limit values for type approval of passenger car tires according to the General Safety Regulation will be tightened: the rolling resistance threshold value for passenger car tires (measured according to a world wide standardized test procedure) will be reduced by 13%!

For the sake of completeness, but without detailed dedication within this paper following aspects of legislation shall be mentioned:

Regulation (EC) No 1907/2006 of the European Parliament and of the council of 18 December 2006 "concerning the Registration, Evaluation, Authorisation and Restriction of Chemicals (REACH), establishing a European Chemicals Agency"; certain harmful chemicals shall not be placed on the market and used for the production of tires or parts of tires; furthermore, the tires and treads for retreading manufactured after 1st January 2010 must no more be placed on the market if they contain those substances exceeding the indicated limits.

Legislation on winter tires: The use of winter tires is affected by different national legislation in EU member states and the occurrence of winter conditions; member states with high occurrence of winter conditions typically mandate the use of Winter tires for a fixed period; member states with incidental winter conditions, or occurrence in specific regions of member states use legislation for conditional fitment, mandated during occurrence of winter conditions. The technical performance requirements on winter tires are currently subject to amendment and to harmonization, respectively.

# 3 The conflict between rolling resistance, wet braking behaviour and wear

It is relatively difficult for a tire manufacturer to deliver top performance in all relevant criteria, as it is often necessary to compromise one characteristic in order to improve another. In the past usually the tires displayed below-average performance in one aspect, while performing above average in another. In fact, the admittedly highly difficult balancing act of delivering top performance in all criteria at a reasonable price was only managed by very few manufacturers, and only with a few, selected tire specifications.

During the progress of implementation of new legislation, especially tire labelling, the „magic triangle", built up by the three key performance parameters of a tire: safety (grip / traction) – wear / durability and fuel economy was one of the most frequently used terms for describing the technical conflict of tire development.

"Magic triangle" means in this case that improving one of the three characteristics above meant worsening another or even the other two.

One of the challenges is that grip on wet roads can be improved by making the rubber softer. However, soft grades of rubber usually wear more quickly, leading to more frequent tire replacement and higher fuel consumption. There are similar conflicts between the other corners of the triangle. Consequently, anyone wanting to improve one of these three properties in the past inevitably had to compromise on the other two. With the advancement of suitable synthetic rubbers the suppliers of the tire industry are helping tire manufacturers to overcome this dilemma to a certain extent and thus offering tire developers greater scope /13/.

From the description above we may expect that thanks to the evolution in the development of new compounds for tire tread and sidewall it is possible to completely improve rolling resistance and wet traction without having a negative effect on wear and, therefore, the life of the tires.

In addition we have to be aware of the impact on rolling noise of a tire when the key performance parameters above shall be improved: in order to reduce tire rolling sound levels it is advantageous to use a thicker tread base and lower groove width; but this can be in conflict with the other properties, especially rolling resistance. That means that in fact we should rather describe this multidimensional conflict more specifically as the "magic square" of modern tire development.

From a safety standpoint, the obvious concern from the "magic square" is a loss of tire grip in order to achieve lower rolling resistance (better vehicle fuel economy) or lower rolling noise.

The following section of the paper shows the evolution of the parameter "wet braking performance" over a wide range of products and also under different test conditions. It demonstrates how successful the optimization of this parameter could be realized in the area of conflict between rolling resistance, rolling noise and wear.

# 4 Insights from 15 years of test operation

## 4.1 General information

In several larger projects since 1999, TÜV SÜD has tested the wet braking capability of a total of more than 2000 passenger car tire sets on the market; also tests for assessment of the exterior rolling noise were an important part of this testing (picture 1). For these activities different commonly used tire dimensions were tested. The tests included the tires of well-established tire manufacturers from Europe, USA and Asia, all of them approved for the European market without any restrictions.

Picture 1: Wet braking performance and exterior rolling noise testing 1999 – today

Within many of the tire groupings tested (a grouping in terms of this project is made up by commercially available tires of the same dimension and range of use), the spread of average deceleration between the best and the worst tire was at more than 20%. Assuming, for example, an average deceleration of the best tire by 7 $m/s^2$ on a wet surface with low friction level, the actual consequences become apparent when applied to real-world road conditions. A difference of 25%, for example, as experienced within these tests, when suddenly braking down from 100 km/h on a wet road surface, would increase the braking distance to standstill by 18 metres. A vehicle with poorly performing tires was still travelling at 50 km/h, where the same vehicle equipped with the best tires of the test had already stopped.

Another reason for these strongly differing deceleration capabilities of a vehicle with different tires might be found in the interaction between the tire and the vehicle's ABS system. It is important to know that any tire has more or less differing μ-slip characteristics, caused by its specific tread compounds, carcass stiffness and tread design. The response behaviour of an ABS brake, i.e. the permitted limits of slip that are allowed before brake pressure is released or applied again, varies from car to car. For improved deceleration ability, ABS can be harmonised with tire characteristics (respectively the μ-slip-characteristic), which holds considerable potential of reducing stopping distance in case this interaction is optimally harmonised. On the other hand, an unnecessary loss of safety performance might be caused by using tires with characteristics strongly differing from the pre-assumed conditions.

**Therefore for selecting testing data for the evaluation presented in this paper we tried to limit the variety of vehicle types used for testing to a minimum.**

The quality and texture of a road surface, resulting in the friction (or grip) level has considerable influence on the road performance of a tire and, along with it, on the driveability of a vehicle. Depending on the surface friction values, at wet braking, differences in the absolute deceleration values of up to 36% for groups of the same tire dimension and more than 40% for individual test samples, tested on different proving grounds on an artificially wetted asphalt track surface (conform to the current technical specifications), were noted. The limitation of surface characteristics of test tracks on which wet braking tests according to European technical specifications are permitted was introduced relatively late, quite in the middle of the observation period: in the year 2007 the first revision of UNECE Regulation 117 entered into force, limiting the micro and macro roughness of the track surface first time.

**For this reason the results of testing for further consideration in the statistical analysis had to be carefully pre-selected and put into clusters of test sessions on surfaces with comparable friction level.**

The database for the following considerations has been established by in-house results and insights gained in numerous tests in the tires&wheels department of TÜV SÜD within the past 15 years. Part of those tests providing data in the very beginning of the observation period (1999-2002) were commissioned by the public sector and have been published (references to the relevant publications are found in the references list of /2/) while others may only be quoted anonymously with respect to customers and our obligations under pertinent confidentiality agreements.

For better understanding of the wet braking performance tests the key parameters at a glance are shown below:

- initial test speed min. 80 km/h
- ABS supported full-braking 80 to 0 km/h (evaluated in range down to 20 km/h)
- test vehicle load according to 60-90% of tire max. load index (LI)
- uniform tire inflation pressure 2,2 bar
- water depth on the artificially wetted surface between 0,5 and 1,5 mm
- measurement of braking distance and evaluation of mean deceleration.

For assessing the balance of performance of the tires tested, the third edge of the "magic triangle", represented by the rolling resistance of the selected tires is included into the statistics.

Fortunately TÜV SÜD can take advantage of a long experience with the assessment of tire rolling resistance, especially using the same equipment all time, which was modernized and adapted to the current state of standards and technology on a regular base since 2002 (picture 2).

Picture 2: Rolling resistance testing at TÜV SÜD from 2002 – today

The standardized test of rolling resistance for a passenger car tire is briefly described as follows:

- indoor test on a roller drum machine; steel drum with diameter 2.0m
- test speed and load: 80 km/h at 80% of maximum tire load capacity (LI)
- inflation pressure: 2.1 bar (or 2.5 bar for reinforced or extra load tires)
- tire warm up before actual measurement: 30 min (no further run in of new tires)
- the final result is the measured rolling resistance force (RRF) or calculated from that the rolling resistance coefficient (RRC) of the tire.

In order to get the most reliable information of technological progress over the complete, 15 years covering observation period, tire types of pioneers in the field of economic and ecologic tires in Europe, well managing the "magic square" of performance were included into the analysis; those pioneers of tire technology already had put several generations of fuel efficient tires with overall good balance of performance into the market around the year 2000.

Regarding the tire rolling noise the TÜV SÜD database does not include a satisfying amount of test results for the tire types selected continuously over the observation range; in addition the conditions under which the results were gained significantly changed during the last 15 years: it was found that the existing technical specifications for the surface of test facilities did not ensure sufficiently comparable results and consequently the standards were amended soundly; this in turn subsequently had additional significant impact on the test results where test track surfaces were re-built according to the new standards. Therefore the result on improvement of the parameter rolling sound emission within this paper is limited to the statement that all of the selected tire types fully comply with the declining limit values for rolling sound emission.

## 4.2 Statistical analysis – introduction

As described in section 3 the evolution of the parameter "wet braking performance" in the area of conflict between rolling resistance, rolling noise and wear is a demanding challenge for the development of tires. By statistical analyses it shall be demonstrated how successful the optimization of this key tire performance parameter could be realized in the past.

In order to assess the evolution of braking on wet surface (represented by the braking distance between 80 and 20 km/h in a standardized test) a sample of test results was selected according to the following criteria:

- tire dimensions of types representing the population with data available over a period of at least 7 years; these are 195/65R15, 205/55R16 and 225/45R17
- tires, state of the art and provided to the market at the date of testing
- the results of braking distance are are split into two clusters depending on the provided friction level of the wetted surface (high or low)
- test vehicles used within the selected data: VW Golf, generations IV to VII.

The sample population consists in a total of **300 sets** of test specimens providing results for braking on wet surface (at least 8 valid cycle repetitions each) and / or results for tire rolling resistance. The wet braking test results were gathered on a total of **10 test tracks in Europe**. The rolling resistance was always measured using the same machine.

Within the sample population subgroups with single dimensions or two dimensions expected to show similar performances are extracted, provided the number of data points (years) does not fall below 5. If not possible a wider spread of results including the data of up to three tire dimensions is accepted to support reliability of the findings. It was tried in all cases to find the best compromise under consideration of in the rule: the larger the sample size, the more reliable it´s mean; he larger the variation of data values, the less reliable the mean.

The analysis shall demonstrate if and to what extend the key tire performance characteristics are changing (improving) in respect to time, from year to year. As indicator the time dependent mean value of results derived by a simple linear regression model, such that the effect of time can be tested, is chosen. The mean is a particularly informative measure of the "central tendency" of the variables if it is reported along with its confidence intervals, thus all graphs are showing the linear trendline of the mean values over time plus the confidence interval of the mean value on a level of 95%. The confidence intervals (limits) for the mean give us a range of values around the regression line where we expect the "true" (population) value of the mean is located (with a given level of certainty). For performing the statistics we request confidence intervals for the p-level 0.05 (representing the error probability); se we can conclude that there is a 95% probability that the real population mean is located between the two limits of the confidence interval.

All trends represented by the mean values shown in the analysis charts below are verified by assessing the statistical significance versus the "null hypothesis" that the observation of a time-based trend could have been caused by effects not considered in the analysis or that the observed effect is random. The statistical tool used is the

ANOVA (Analysis of Variance); the purpose of ANOVA is to test for significant differences between means by comparing (i.e., analyzing) variances.

Within the ANOVA the t-test (the most commonly used method to evaluate the differences in means) is applied to assess if the effect of time, represented by the slope of the trendlines is statistically significant. For this test also an error probability of 0.05 and thus a confidence level of 95%, respectively, is selected.

## 4.3. Statistical analysis – results

Braking distance of full sample population in tire dimensions 195/65R15, 205/55R16 and 225/45R17 (the data points are represented by the mean values of observed results for each of the three tire dimensions in the particular year)

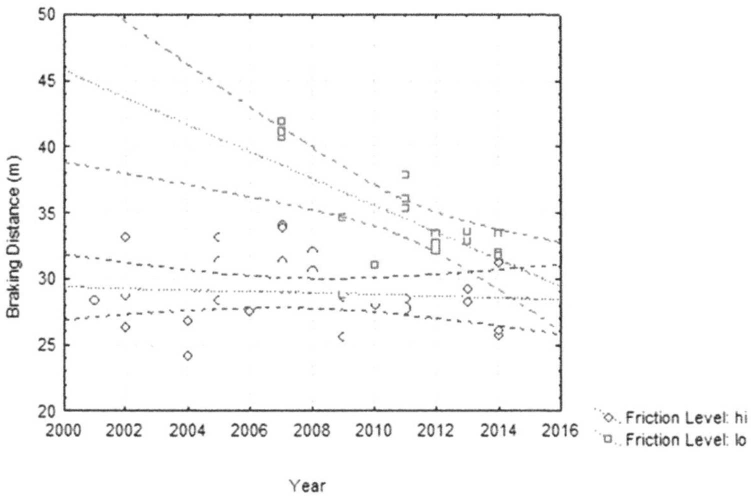


Result: significant decreased braking distances between years 2007 and 2014 on low friction level; average slope 1,0 m/year. On high friction level slope of 0,1 m/year observed, but due to high variation of the values statistical significance is not proven.

<u>Braking distance of subgroup with 2 brands and specifications of efficient tires in dimension 205/55R16</u>

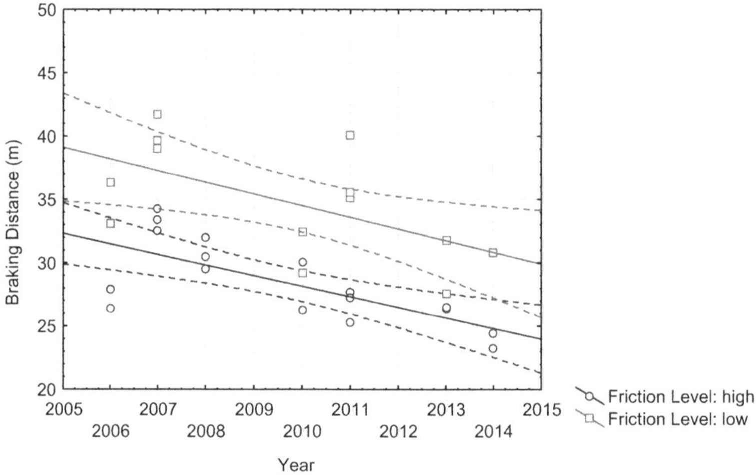


Result: significant decreased braking distances between years 2006 and 2014; average slope 0,8m/year on high friction level and 0,9m/year on low friction level.

Braking distance of subgroup with 2 brands and specifications of efficient tires in dimension 195/65R15

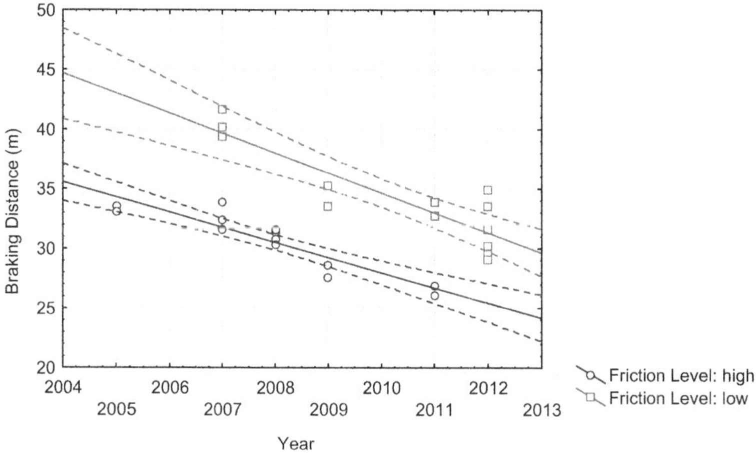


Result: significant decreased braking distances between years 2005 and 2012; average slope 1,3m/year on high friction level and 1,7m/year on low friction level.

**The following statistics show the evolution of tire rolling resistance (represented by the rolling resistance force "RRF", measured according to a standardized test procedure):**

<u>Rolling resistance (RRF) of full sample with tire dimensions 195/65R15, 205/55R16 and 225/45R17 (the data points are represented by the mean values of observed results for each of the three tire dimension in the particular year)</u>

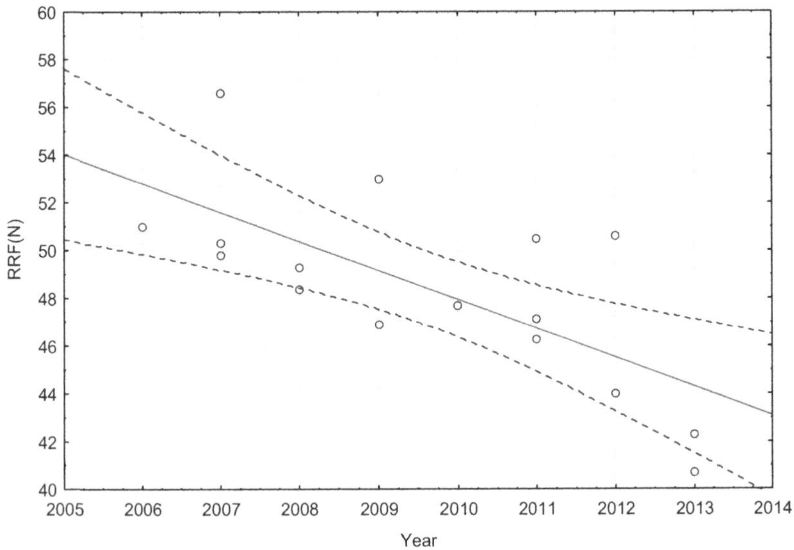


Result: significant decreased RRF between years 2006 and 2013; average slope 1,2N/year.

Rolling resistance (RRF) of sample sub group with 2 brands and specifications of efficient tires in dimensions 195/65R15 and 205/55R16

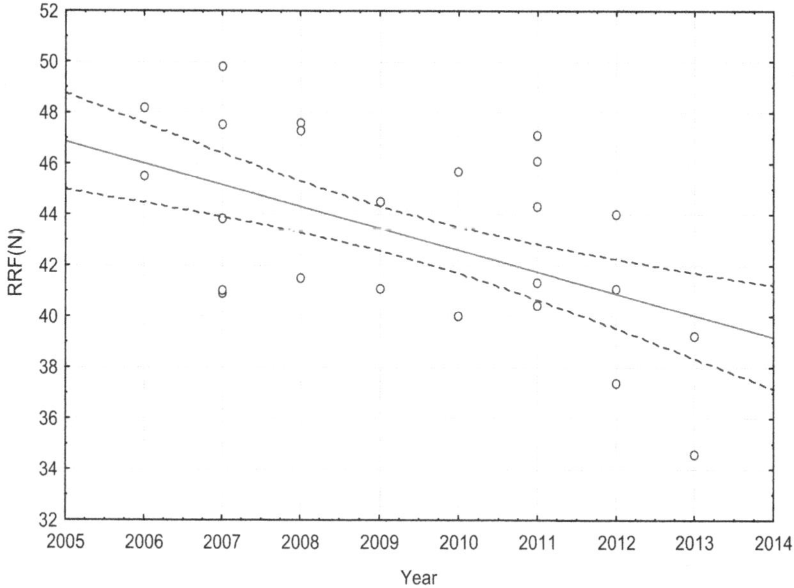


Result: significant decreased RRF between years 2006 and 2013; average slope 0,9N/year.

Comparison of the evolution of rolling resistance (RRF) and braking distance (on all friction levels) for sample sub group with 2 brands and specifications of efficient tires in dimensions 195/65R15 and 205/55R16

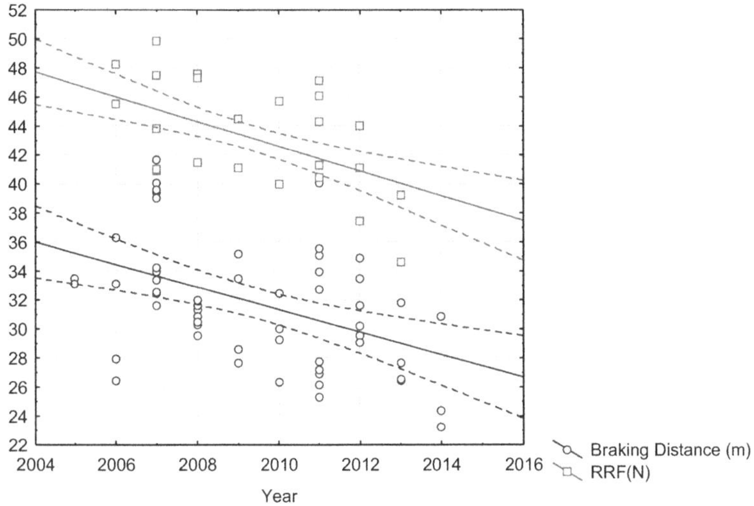


Result: significant decreased average braking distances on both friction levels between years 2005 and 2014; average slope 0,8m/year; within the same time period significant decreased RRF between years 2006 and 2013; average slope 0,9 N/year.

## 4.4 Statistical analysis – results summary

The analysis of the test data including the complete sample population shows that both key performance parameters, rolling resistance (rolling resistance force "RRF") and braking distance on wet surface were improved significantly during the observation period.

In case of the sub group with energy efficient specifications included in the population, it is obvious that the improvement could be achieved even though the starting position of evolution was already located at low levels of braking distance and rolling resistance, representing the high degree of technology and development over the whole observation period.

# References


/1/ Was ist Was – Das Auto, Neuer Tessloff Verlag, 1974

/2/ SURVEY ON MOTOR VEHICLE TIRES & RELATED ASPECTS, TÜV SÜD, commissioned by the European Commission, 2003

/3/ Günter Leister, Fahrzeugreifen und Fahrwerkentwicklung, ISBN 978-3-8348-0671-0, 1st edition, Vieweg+Teubner / GWV Fachverlage GmbH, Wiesbaden, 2009

/4/ Quality Control Systems Corporation, R. A. Whitfield, 2006

/5/ Siemens VDO, Philippe Fournet-Fayat, tire.wheel.tech, Germany, Munich, 2006

/6/ Darmstädter Reifenkolloquium, Fortschritt-Berichte VDI, Reihe 12, Nr. 285, Volker Bachmann et.al., TH Darmstadt, Germany , 1996

/7/ APOLLO, IST-2001-34372, Intelligent Tire for Accident-free Traffic, Project funded by the European Community under the "Information Society Technology" Programme (1998-2002), Report 2003

/8/ The 2020 climate and energy package, European Commission, online (http://ec.europa.eu/clima/policies/package/index_en.htm)

/9/ Bridgestone Corporation, I. Kuwayama et.al, chassis.tech 2013, Munich, Germany, 2013

/10/ Federal Environmental Agency (UBA), Dessau , Dr. Lars Schade , tire technology EXPO, Cologne, Germany, 2007

/11/ Forum of European National Highway Research Laboratories (FEHRL), "Study SI2.408210 Tire/Road Noise", 2006

/12/ Federal Environmental Agency (UBA), Berlin, Reiner Stenschke, tire technology EXPO, Cologne, Germany, 2005

/13/ Lanxess AG, Background Information "Better tires thanks to LANXESS", Leverkusen, 2012

# Concurrent design of vehicle tires and axles

Dipl.-Ing. (FH) Jens Wimmler, BMW Group

Prof. Dr.-Ing. Dieter Schramm, University of Duisburg-Essen

Dr.-Ing. Martin Wahle, BMW Group

Dr. Markus Zimmermann, BMW Group

# Abstract

Concept design of both axles and tires is difficult, because they simultaneously affect many different objective quantities in vehicle dynamics related to, e.g., self-steering behavior, transient behavior, maximum lateral acceleration, etc. Typically, the tire performance is evaluated using a specified axle design (or vice versa) and then optimized. This way, the optimal tire will depend on the axle that was chosen. If the axle design changes during the development process, the vehicle performance will suffer. Similarly, the optimal tire or axle for one vehicle may not be sufficient for another vehicle from the same platform or vehicle architecture.

This paper proposes a new method that computes permissible ranges of tire and axle properties, so-called solution spaces. Within these solution spaces, all possible combinations of tires and axles are guaranteed to deliver the required vehicle performance in all relevant disciplines. During concept design, solution spaces are maximized for flexibility in the development process. Large solution spaces can relax conflicts of goals when they are large enough to allow for optimization with respect to additional requirements, for example the optimization of tire properties with respect to fuel consumption and CO2 emissions. Solution spaces of many vehicles may be overlaid in order to identify parameter settings for modules or vehicle platforms. Developing axles and tires to lie in the center of solution spaces provides maximum robustness with respect to unintended variations or uncertainties.

The method is illustrated using examples from the vehicle development process at BMW. Fast computing simplified vehicle and tire models are linked with appropriate numerical methods for solution space analysis. Requirements on tires and axles are derived and expressed as permissible ranges for their functional properties, such as the cornering stiffness. Examples of tires and axles are provided that may be realized and will make the vehicle reach the performance goals.

# 1 Introduction

The development of vehicle properties on a component level is very time consuming and expensive. One reason is the large number of objective quantities that depend on many design variables. Design work is a complex task and requires typically several iteration loops. In order to increase efficiency, objectives are often cascaded downwards along several abstraction levels (e.g., subjective vehicle level, objective vehicle level, subsystem level and component level), see Fig. 1.

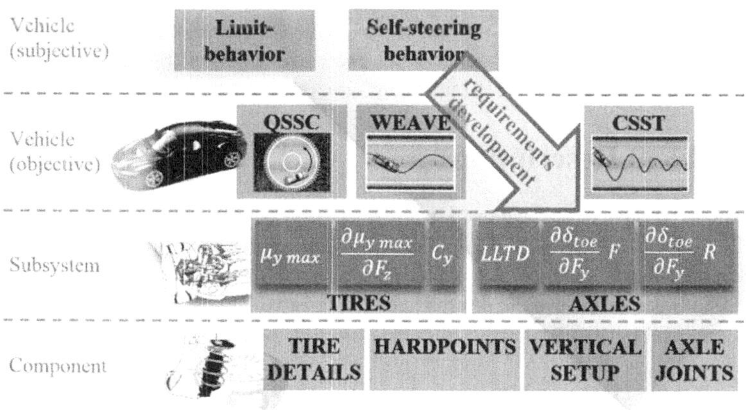


Fig. 1: Abstraction levels for tire and axle design in the V-Model.

In the classical development process for vehicle dynamics, the basic design variables of the vehicle (e.g., mass, wheelbase, etc.) are prescribed. After this, tires and then axles are designed. Classical optimization may yield a design which comes as close as possible to given performance targets. This is desirable if unintended variations and uncertainties of design variables can be neglected. The major disadvantage of this method is the lack of flexibility with respect to changes, which may occur, e.g., during the manufacturing process or may be necessary due to new requirements from other departments during the development process. An alternative approach which produces flexibility is the so-called solution space method. The inputs for this method are the design objectives formulated as inequality constraints for objective quantities, and a list of design variables that may be adjusted so as to reach these design objectives. From this input, the method calculates a range of solutions, a so-called solution space, which will be explained in Chapter 3. For detailed information regarding the method see [1] and [2].

This paper describes how the solution space method is adapted to the simultaneous design of axles and tires along the V-Model. Vehicle performance is taken into account as well as feasibility on the component level. The concurrent design of tire and axle is a complex design problem, because the design variables associated with the axle will affect the appropriate choice or the design variables associated with the tire and vice versa. In addition, the values of these design variables are uncertain due to missing knowledge about feasibility or additional requirements that cannot be considered explicitly. Solution spaces represent all solutions with respect to the objectives within which tires and axles may be realized. They may be used to provide maximum transparency by offering the designer a choice among several solutions.
The article is organized as follows. Chapter 2 describes the design problem. In Chapter 3, solution spaces and the design method are presented. In Chapter 4, the new method is applied and results are shown. Conclusions are drawn in Chapter 5.

# 2 Problem Description

Tires and axles have to be developed in a way that all relevant design objectives on the vehicle level can be reached. The different abstraction levels for tire and axle design are shown in Fig. 1. In Table 1, the key design variables on the subsystem level are defined. In order to achieve the desired vehicle performance, the design ranges for the subsystem level have to be derived from the objective quantities of the vehicle.

Table 1: List of design variables.

| Design variable | | Subsystem | Cluster |
|---|---|---|---|
| Maximum lateral friction coefficient [-] | $\mu_{y\,max}$ | Tire | A |
| Lateral load transfer distribution [-] | $LLTD$ | Axle | A |
| Variation of lateral friction coefficient with normal load $[kN^{-1}]$ | $\partial\mu_{y\,max}/\partial F_z$ | Tire | A |
| Toe variation under cornering force on the front axle $[min/kN]$ | $(\partial\delta_{toe}/\partial F_y)_F$ | Axle | B |
| Toe variation under cornering force on the rear axle $[min/kN]$ | $(\partial\delta_{toe}/\partial F_y)_R$ | Axle | B |
| Cornering stiffness $[°/N]$ | $C_y$ | Tire | B |

As the vehicle performance on the system level depends on the interaction of tire and axle, both have to be designed simultaneously to reach all development goals. This may lead to a seemingly inferior component performance, e.g., a tire with a lower lateral

friction coefficient. Yet, the overall setup will result in a better performance on the system level. Even if a satisfying solution on the subsystem level can be identified, it is uncertain that proper solutions on the component level exist.
The objective quantities which are used to evaluate the performance of the vehicle are defined in Table 2. Design goals are specified as

$$\text{subject to } {f_c}^l \leq f(\boldsymbol{x}) \leq {f_c}^u \text{ for all } \boldsymbol{x} \in \Omega\,. \tag{1}$$

Where $f(\boldsymbol{x})$ is the system response, $f_c$ are upper and lower boundaries for the permissible region of the objective quantity and $\boldsymbol{x}$ is the vector of design variables of a vehicle in the design space $\Omega$.

Table 2: List of objective quantities.

| Objective Quantity | Definition |
| --- | --- |
| Maximum lateral acceleration (m/s²) | $a_{y,max} = \max(a_y(t))$ |
| Stability reserve = difference of the maximum lateral acceleration between rear and front axle (m/s²) | $a_{y,reserve} = a_{y,max,R} - a_{y,max,F}$ |
| Lowest vertical tire load at maximum lateral acceleration (N) | $F_{z,min} = \min(F_z(a_{y,max}))$ |
| Change in steering angle between three different lateral acceleration ranges (°) | $\Delta\delta_{i-j} = \delta_i - \delta_j$ |
| Change in body slip angle between three different lateral acceleration ranges (°) | $\Delta\beta_{i-j} = \beta_i - \beta_j$ |
| Yaw velocity response at three different velocities (s$^{-1}$) | $\left(\frac{\dot{\psi}}{\delta}\right)_i$ |
| Phase difference angle between yaw velocity and lateral acceleration at 1Hz (°) | $\angle\varphi°$ |
| Time delay between steering input and yaw velocity (s) as explained in [3] [4] [5] | $T_{eq}$ |
| Maximum amplification of yaw velocity response normalized w.r.t. its static value (-) | $\frac{\dot{\psi}}{\delta}_{norm} = \max\left(\frac{\dot{\psi}}{\delta}_{norm}(f)\right)$ |
| Maximum amplification of the body slip angle normalized w.r.t. its static value (-) | $\frac{\dot{\psi}}{\delta}_{norm} = \max\left(\frac{\beta}{\delta}_{norm}(f)\right)$ |

# 3 Methodology

The development method proposed in this paper relies on cascading of the system into different abstraction levels and the subsequent balancing of targets on multiple levels. Due to the high complexity in vehicle development, it is an advantage to break down the vehicle in different abstraction levels. This helps the engineer to handle the complexity and focus on the key variables first. In this paper, we distinguish between the subjective vehicle level, objectified vehicle level, subsystem level and component level, see Fig. 1. Based on experience and objectification, an expert transforms the subjective targets into objective targets of the vehicle. The solution space for the underlying subsystem level is calculated using a particular algorithm as presented in [1], [2] and [6]. For simplicity, the objective quantities and design variables on the subsystem level are assigned to two separate clusters A & B. Cluster A includes objective quantities and design variables that are strongly associated with the static limit behavior of the vehicle. Cluster B includes the objectives and design variables for self-steering behavior. In Fig. 2 the dependency structure of design variables and objectives is shown within each cluster.

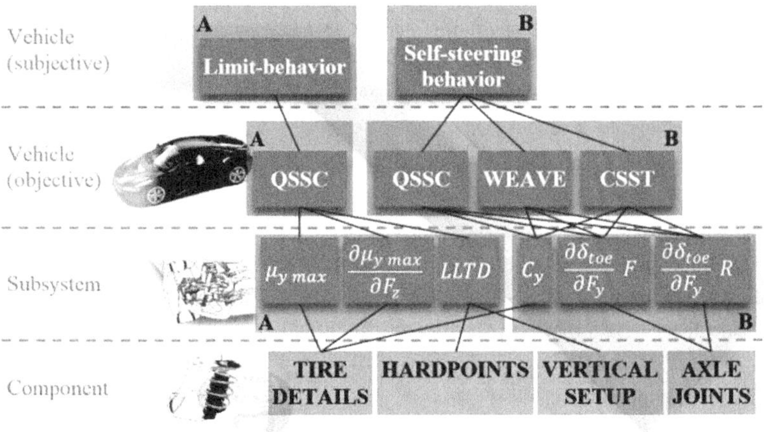


Fig. 2: Dependency graph of tire and axle design along the V-Model.

For Cluster A, a two dimensional solution space can be constructed, when $\partial\mu_{y\,max}/\partial F_z$ is fixed, see Fig. 3. Within this solution space the required vehicle performance is guaranteed and all configurations are equal in that they all reach their design objectives. Within the solution, space a so-called solution box can be constructed. The benefit of the solution box is that the design variables can be varied independently from each other

without violating the objective targets. Developers of vertical components as well as tire developers can use their associated parameter intervals resulting from the solution box as target region for their component design.
As presented in Table 1, Table 2 and Fig. 2 the static limit behavior depends on more than two design variables. Taking the degression into account the solution space of the friction coefficient and LLTD varies. Fig. 4 shows the slices of the two dimensional solution space for different tire degressions and gives an idea of its three dimensional shape. In this particular case, the solution space decreases with a lower tire degression which is a result of the static rear axle load of more than fifty percent.
In order to prove the possibility of realizing tires in the determined solution box, the solution box is compared with tire performance values that were measured for existing tires. This data may used to identify ranges for tire detail variables such as width or diameter, as shown Fig. 5.

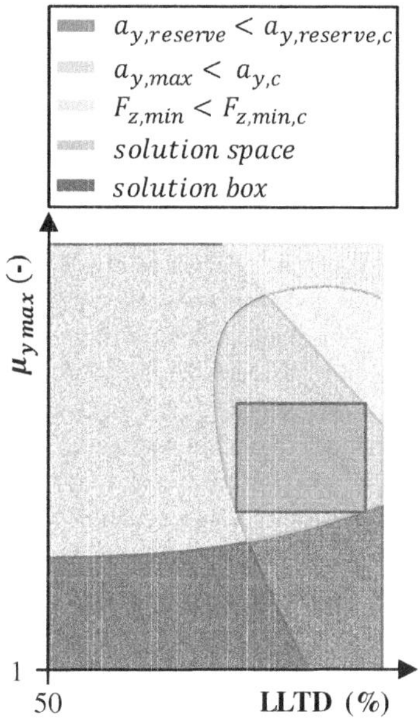


Fig. 3 Solution space for static limit behavior (degression of friction coefficient is fixed).

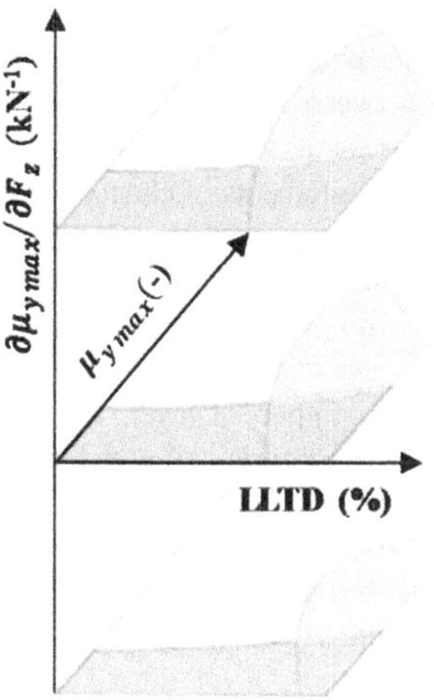


Fig. 4 Solution space for static limit behavior for varying degression of friction coefficient.

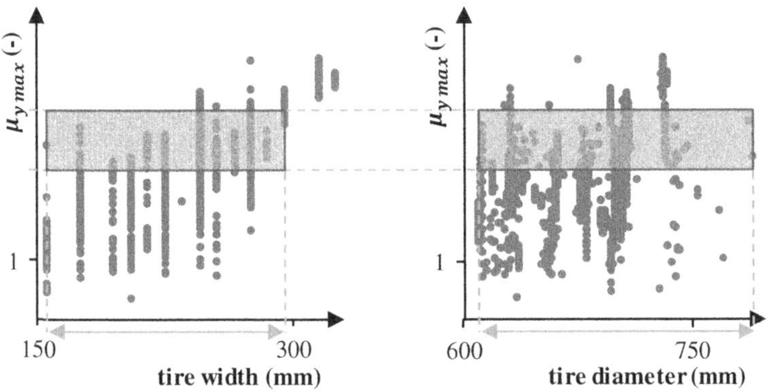


Fig. 5 Hardware test data from the subsystem and component level. The green box shows the requirement on $\mu_{y\,max}$ derived from objectives of cluster A.

For cluster B, the same approach is suitable. In a first step, the solution space for the cornering properties of the axle is calculated (see Fig. 6). In a second step, the combined behavior between axle and tire can be analyzed. The chosen solution box is verified with the tire databases again. In contrast to Fig. 5, in Fig. 8 only tires which fulfill the functional requirements of cluster A and B are considered as suitable tires. The chosen solution box is verified with the tire databases again. In contrast to Fig. 5 in Fig. 8 only tires which fulfill the functional requirements of cluster A and B are considered as suitable tires. If no existing tire can be identified that fulfils the requirements of both clusters, this does not necessarily mean that there is no possibility to reach the overall design goals on the component level. The tire developer may investigate if and under what conditions an appropriate tire can be developed. The shape of the solution spaces in Fig. 4 and Fig. 7 show that axles and tires affect the relevant objective quantities simultaneously, because the solution space boundary is diagonal. If only design variables from either axles or tires were relevant, boundaries would vertical or horizontal lines. Due to this dependency, tires and axles have to be designed simultaneously.

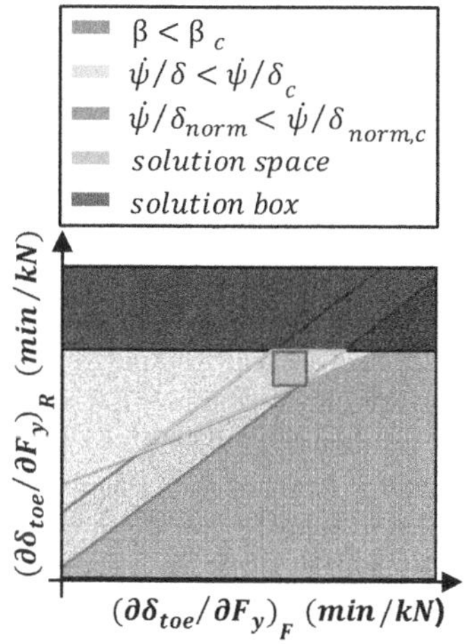


Fig. 6 Solution space for toe variation and constant cornering stiffness.

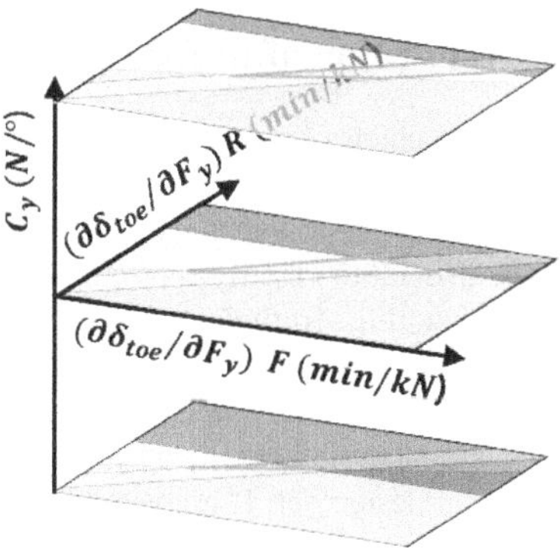


Fig. 7 Quasi 3D solution space for self-steering behavior.

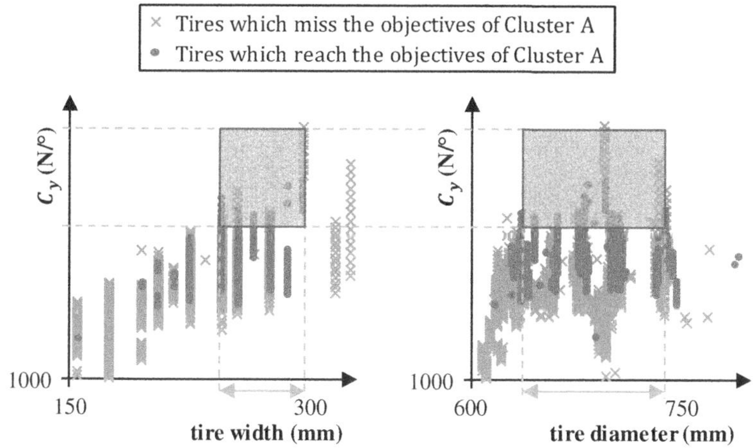


Fig. 8 Hardware test data from the subsystem and component level. The green box shows the requirement on the cornering stiffness derived from objectives of cluster B.

# 4 Application: How to reach design goals?

## 4.1 Description of initial situation

When new vehicles are developed, some design goals may not be reached. In particular, new types of vehicles, for example Battery Electric Vehicles (BEV), lead to major changes of the basic vehicle variables such as mass or rotational inertia, which may make it difficult to identify appropriate designs.
In Fig. 9 and Fig. 10 the solution spaces for two vehicles are shown. The grey color represents an existing mid-size vehicle, while the blue color is used for a BEV which is currently in the concept stage. The boundaries of the solution spaces are shown as lines, while nominal designs are depicted as dots. As the BEV is in the concept stage, there is no nominal design yet. If a given concept has a solution space in both clusters, static limit behavior and self-steering behavior, the vehicle may be realized. In case the adaption of tire and axle do not deliver the required performance, other more extensive measures are necessary. As the target vehicle performance in cluster A can be reached easily we will focus on cluster B in the following example.

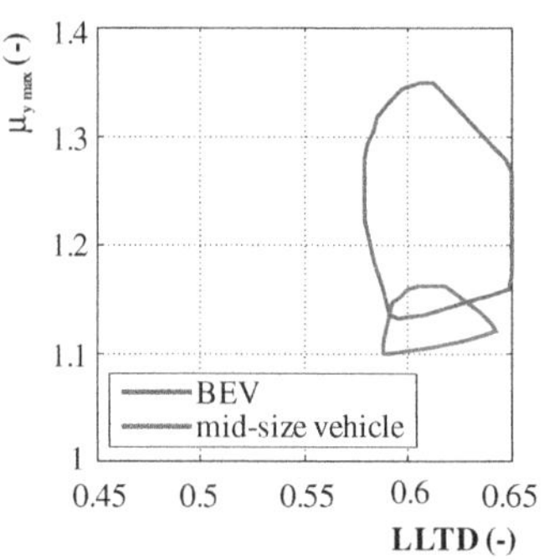


Fig. 9 Solution space for the limit behavior (cluster A).

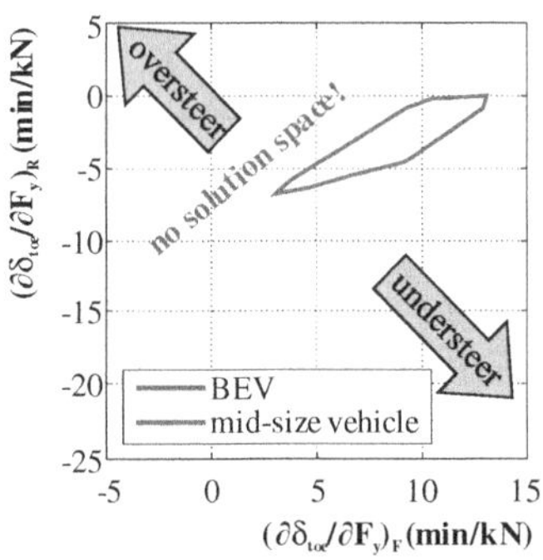


Fig. 10 Solution space for the self-steering behavior (cluster B).

## 4.2 Relaxation of the objectives

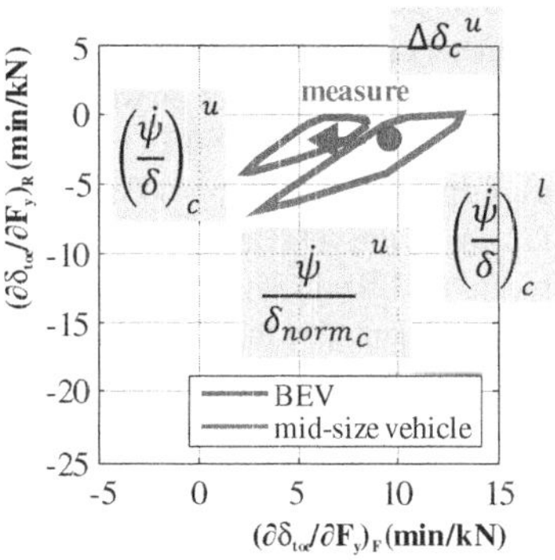


Fig. 11 Solution space for axle design variables that is created by relaxing the design objectives (blue line).

If there is no possibility to reach the required vehicle performance due to adaption of tire and/or axle, one may relax the design goals. The four critical objective quantities which restrict the size of the solution space are shown in Fig. 11. This enables the decision makers to overthink the objectives they have derived from the subjective specifications. Here, the upper threshold value for the maximum amplification of the yaw velocity response was relaxed in order to get a suitable solution space.

## 4.3 Changing of rear tire cornering stiffness

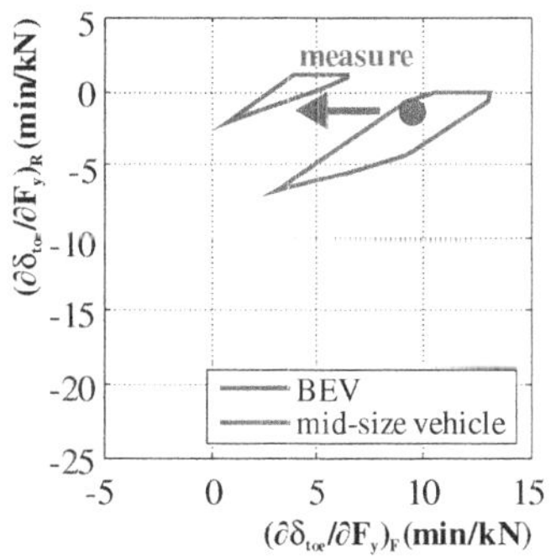


Fig. 12 Solution space for axle design variables that is created by changing the cornering stiffness of the rear tire (blue line).

Another possibility to reach the desired vehicle performance is the adaption of the rear tire cornering stiffness and subsequent change of the axle design. Changing the tire concept from all-same-tires to individual-tires creates a solution-space (see Fig. 12). Here, the cornering stiffness of the rear tires was increased by 10 percent compared to the value of the nominal design. Having done this, the axle design parameters need to be adjusted in order to lie within the solution space.

## 4.4 Changing basic layout parameters

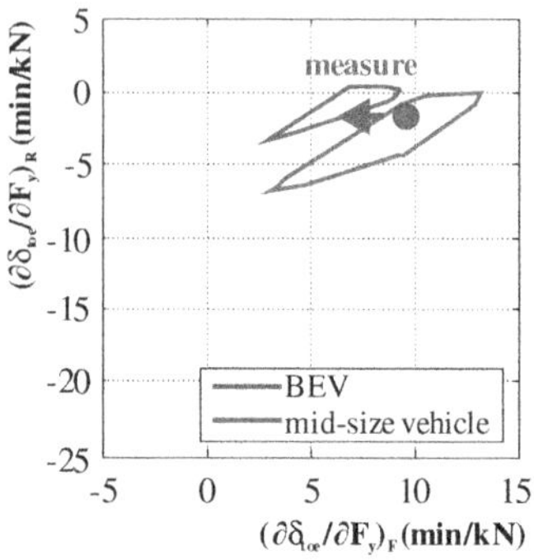


Fig. 13 Solution space for axle design variables that is created by reducing the vehicle mass (blue line).

The vehicle mass or other basic layout parameters such as wheelbase etc. are influential design variables as well. In Fig. 13 a mass reduction of 60 kg opens a solution space. With an appropriate change of the axle, the vehicle satisfies all design objectives. Like in Fig. 12, the front axle has to be adapted to the set of vehicle variables to lie in the solution space.

# 5 Conclusion and Outlook

Solution spaces for tires and axles were derived from objectives on the objectified vehicle level. The classification into cluster A (limit behavior) and B (self-steering behavior) leads to only three design variables for each cluster. This way, the shape of the complete solution space may be visualized. Solution boxes rather than arbitrarily shaped solution spaces help to define target intervals for each design parameters. These target intervals are independent from each other and may be used as design goals for component developers. Intervals rather than point targets allow for component development that is restricted by other requirements related, e.g., to cost or CO2 emissions. Solution spaces may be used to identify solutions for systems that consist of components with strong interaction, such as a vehicle with tires and axles as subsystems. This paper presents three scenarios in which solutions may be generated in two steps: first, a solution space for axle design variables is created by either appropriately changing the design goals, the tire cornering stiffness or basic layout parameters. Second, the solution is realized by adapting the axle design variables to lie within their solution space.

For simplicity, the approach presented here relies on separately designing cluster A & B which is permissible in an early design phase. For accuracy, however, all design variables and design goals should be considered simultaneously. This poses no conceptual difficulty to the design method presented and will be further explored in future research.

# 6 Literature


[1] Zimmermann, M., and von Hoessle, J.: Computing solution spaces for robust design. International Journal for Numerical Methods in Engineering, 2012

[2] Eichstetter, M., Redeker, C., Müller, S., Kvasnicka, P., Zimmermann, M..: Solution spaces for damper design in vehicle dynamics. chassis.tech, 2014

[3] Harrer, M.: Characterisation of Steering Feel. Thesis, University of Bath, Department of Mechanical Engineering. Bath UK, 2007

[4] Decker, M.: Zur Beurteilung der Querdynamik von Personenkraftwagen. Thesis, Technische Universität München. München, 2009

[5] Weir, D., DiMarco, R.: Correlation and Evaluation of Driver/Vehicle Directional Handling Data. SAE Technical Paper 780010, 1978

[6] Avigad G., Moshaiov A.: Interactive Evolutionary Multiobjective Search and Optimization of Set-Based Concepts. IEEE Transactions on Systems, Man, and Cybernetics, No. 4, Vol. 39, 2009.

[7] Milliken, W. F., Milliken, D. L.: Race Car Vehicle Dynamics, Warrendale: Society of Automotive Engineers, 1995.

# New testing technology to evaluate lateral ice performance of tires

Klaus Wiese, Jens Heyne, Burkhard Wies, Thomas Neddenriep

Continental Reifen Deutschland GmbH

Hannover

# Abstract

To drive with a vehicle on an icy road surface with standard winter tires (non studded) is due to the low friction potential one of the most critical driving situations appearing in the winter season. To lower the risk for accidents it is necessary to develop tires with increased ice performance. To do so reproducible test methods to evaluate tires potential to transmit forces on ice are necessary. Driving in an ice circle to evaluate the lateral force transmission of tires on ice is a difficult test procedure and its reproducibility can be improved. The newly developed test method is coupling a normal vehicle to a guiding rail and enables the measurement of lateral forces in dependency of the side slip angle of the tires. The new method is highly reproducible and correlating to the ice circle measurement procedure. Not just getting an average lap time to drive in the ice circle, but complete lateral force characteristics is a big benefit of the new test method enabling the tire developers to build up further knowledge about the force transmission of tires on ice, but in future to increase the friction potential even more.

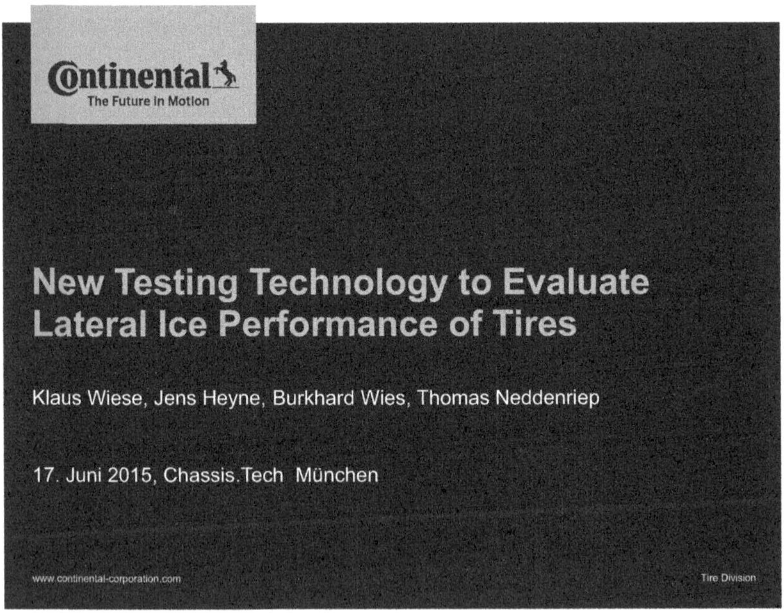

## Agenda

1 Motivation and Requirements for New Test Method
2 Development of New Test Method
3 Lateral Tire Force Characteristics on Ice
4 Correlation of New Test Method with Ice Circle Measurements
5 Correlation of New Test Method with Outdoor Measurements
6 Summary and Conclusions





## Motivation and Requirements for New Test Method

- Driving on an ice surface is the most slippery driving condition. Especially wet ice.
- High relative probability for accidents due to low friction coefficients.

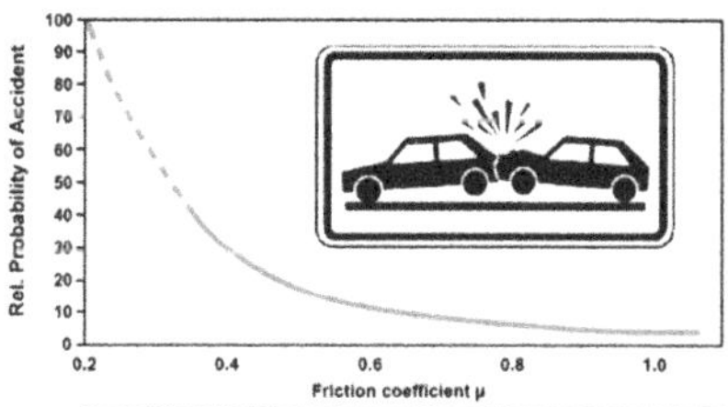


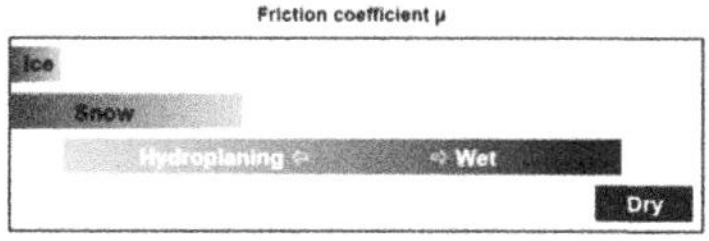

New Testing Technology to Evaluate Lateral Ice Performance of Tires

## Motivation and Requirements for New Test Method

- Well known tire testing procedure to evaluate the lateral tire performance on ice:
  - Evaluate the lap times driving in an ice circle.

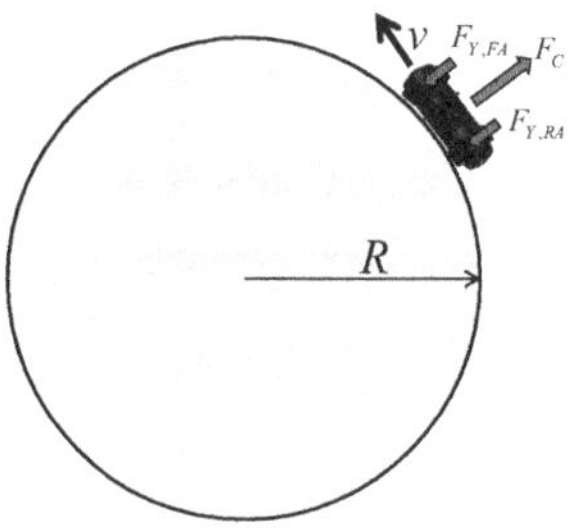


$$F_C = m \cdot \frac{v^2}{R} = F_{Y,FA} + F_{Y,RA}$$

Typical ice skating rink

Continental

17. Juni 2015
K. Wiese © Continental AG
4

New Testing Technology to Evaluate Lateral Ice Performance of Tires

## Motivation and Requirements for New Test Method

- Tire Characteristics and Force Transmission on Ice

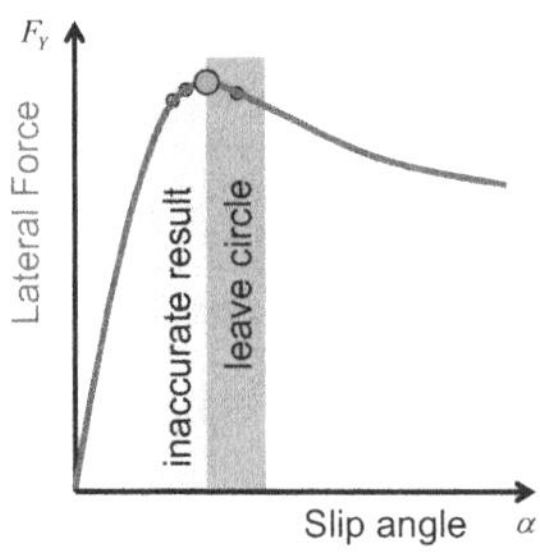


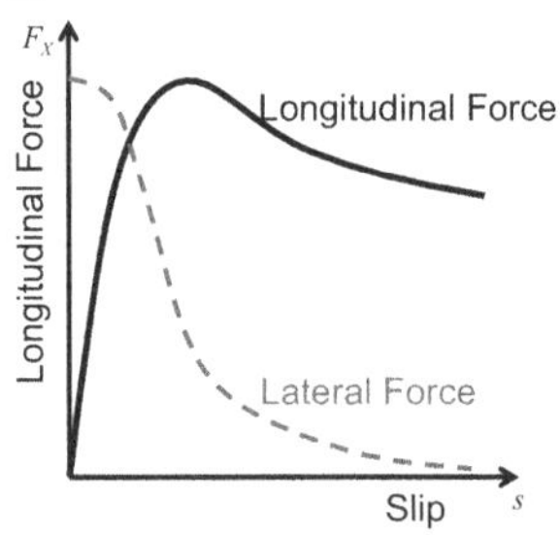


Goal: Maximum lateral Force → fastest lap time
Increasing slip angle leads to increasing lateral force up to maximum
Additional steering wheel input reduces lateral force → unstable situation
Increasing speed needs longitudinal force, that means longitudinal slip
→ lateral force is decreasing immediately → leave the circle → unstable situation

Continental

17. Juni 2015
K. Wiese © Continental AG
5

New Testing Technology to Evaluate Lateral Ice Performance of Tires

## Motivation and Requirements for New Test Method

- The unstable driving situation during ice circle measurements leads to high fluctuations in test results.
- High testing effort for several repetitions (average lap time of several laps and different drivers) to enable tire development.
- Additionally the **ice surface** itself is close to its melting point and therefore also changing its friction potential significantly depending on **surrounding conditions** (e.g. ice and air temperature and humidity of air)
- Several runs on different days are needed to determine an average tire performance, which can be used for tire development.

- **Requirements for new test method:**
  - Good climate control for stable surrounding conditions
  - Stable and repeatable driving situation, reduced influence of test driver

Continental

17. Juni 2015
K. Wiese © Continental AG 6

New Testing Technology to Evaluate Lateral Ice Performance of Tires

## Development of New Test Method

- **Idea:**
  - **Measure the lateral force characteristics vs. slip angle** and
  - **determine the maximum lateral force** to judge the tire performance for lateral ice performance.
  - Additional benefit: complete tire characteristics available for better tire optimization (not just a lap time)

Automated Indoor Braking Analyzer (AIBA) Design

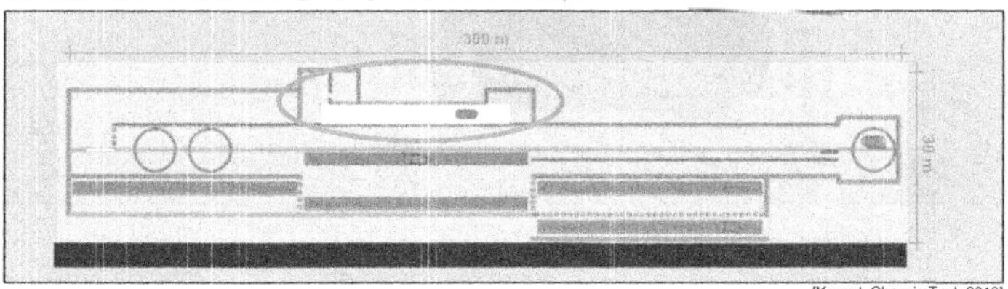


[Kessel, Chassis.Tech 2013]

Continental

17. Juni 2015
K. Wiese © Continental AG 7

New Testing Technology to Evaluate Lateral Ice Performance of Tires

## Development of New Test Method

- **Idea:**
  - **Measure the lateral force characteristics vs. slip angle** and
  - **determine the maximum lateral force** to judge the tire performance for lateral ice performance.
  - Additional benefit: complete tire characteristics available for better tire optimization (not just a lap time)

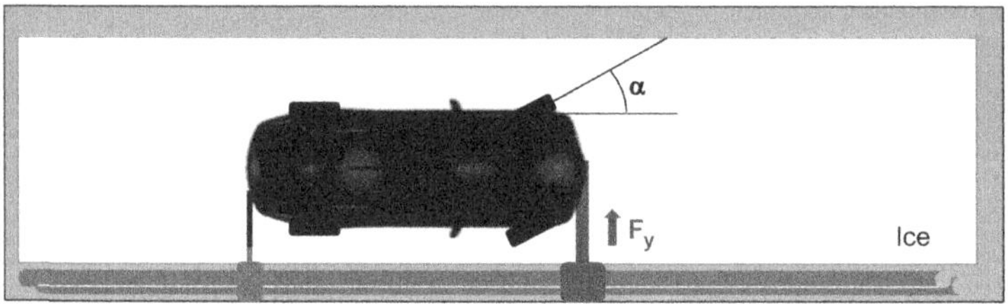


Continental

17. Juni 2015
K. Wiese © Continental AG 8

New Testing Technology to Evaluate Lateral Ice Performance of Tires

## Development of New Test Method

**Equilibrium of Moments around rear axle:**

$l_1 * F_{y\_Front} + l_3 * F_{y\_FA\_Calc} - l_2 * F_{y\_Rear} = 0$

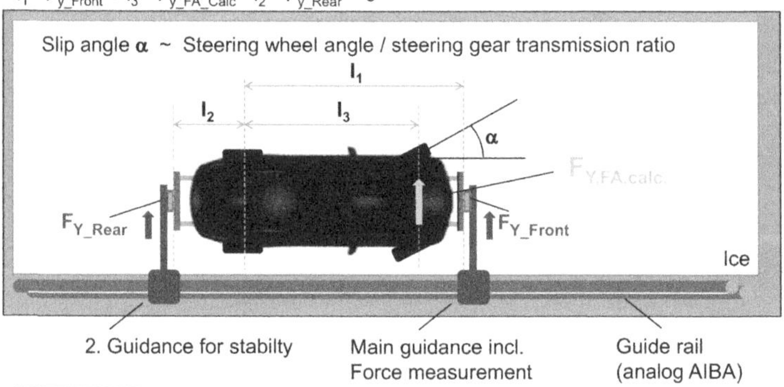


Continental

17. Juni 2015
K. Wiese © Continental AG 9

New Testing Technology to Evaluate Lateral Ice Performance of Tires

## New Test Method

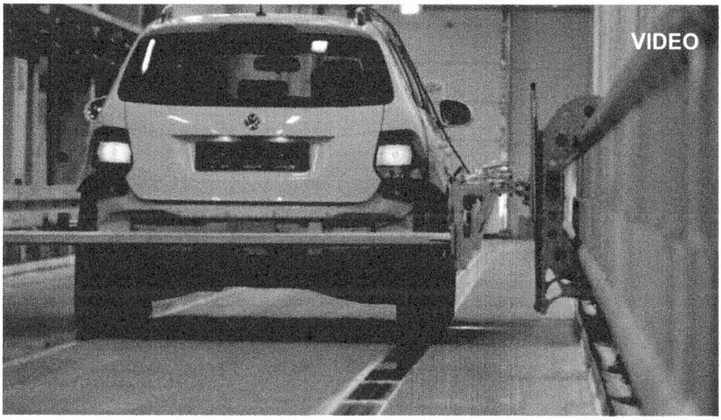


Continental

17. Juni 2015
K. Wiese © Continental AG
10

New Testing Technology to Evaluate Lateral Ice Performance of Tires

## New Test Method – Testing Parameters

- Vehicle velocity: constant
- Slip angle 0° to 10°
  (steering wheel angle ca. 0° to 180°)
- Rail guided vehicle
- Measured variables:
  Lateral forces, steering wheel angle
- Data processing with DigiTalker
- Evaluation:
  Lateral Force (or $\mu_y$) versus slip angle

Continental

17. Juni 2015
K. Wiese © Continental AG
11

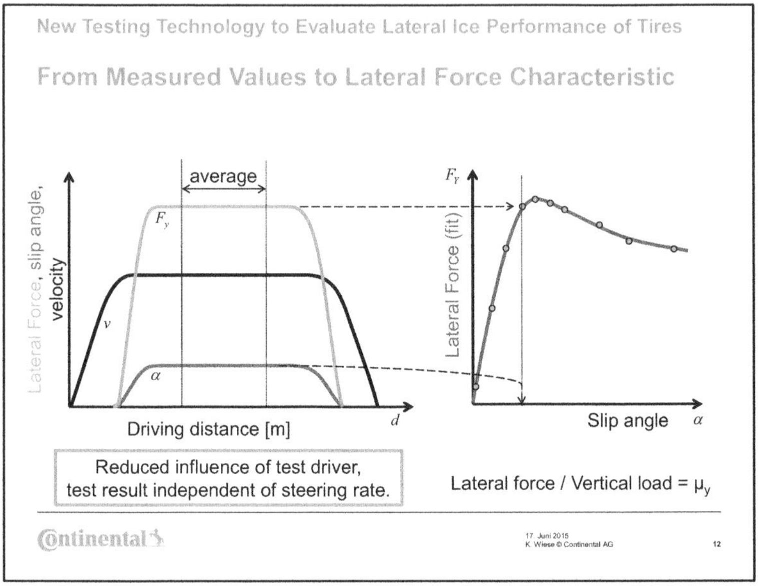
New Testing Technology to Evaluate Lateral Ice Performance of Tires
From Measured Values to Lateral Force Characteristic
average
Lateral Force, slip angle, velocity
$F_y$
$v$
$\alpha$
Driving distance [m]
$d$
$F_Y$
Lateral Force (fit)
Slip angle
$\alpha$
Reduced influence of test driver, test result independent of steering rate.
Lateral force / Vertical load = $\mu_y$
Continental
17. Juni 2015
K. Wiese © Continental AG
12

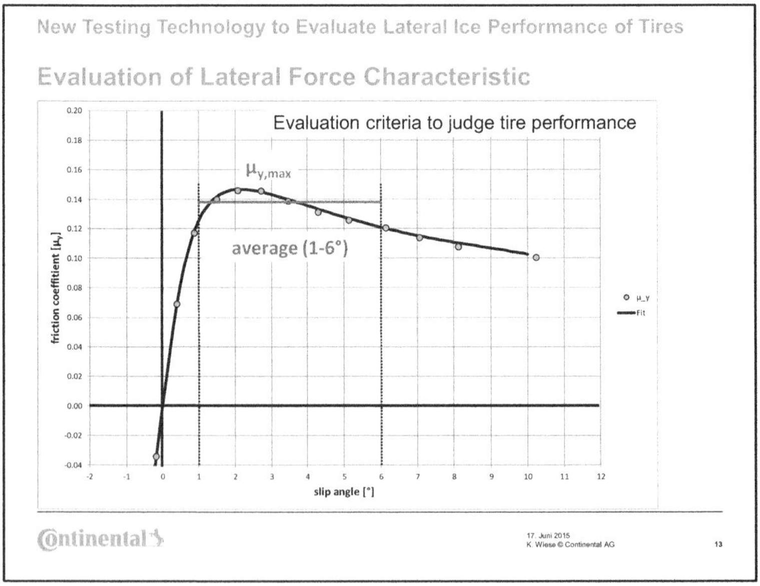
New Testing Technology to Evaluate Lateral Ice Performance of Tires
Evaluation of Lateral Force Characteristic
Evaluation criteria to judge tire performance
$\mu_{y,max}$
average (1-6°)
friction coefficient [$\mu_y$]
slip angle [°]
μ_y
Fit
Continental
17. Juni 2015
K. Wiese © Continental AG
13

New Testing Technology to Evaluate Lateral Ice Performance of Tires

## Comparison of Front vs. Rear Wheel Driven Vehicle

Continental

17. Juni 2015
K. Wiese © Continental AG

14

New Testing Technology to Evaluate Lateral Ice Performance of Tires

## Comparison of Front vs. Rear Wheel Driven Vehicle

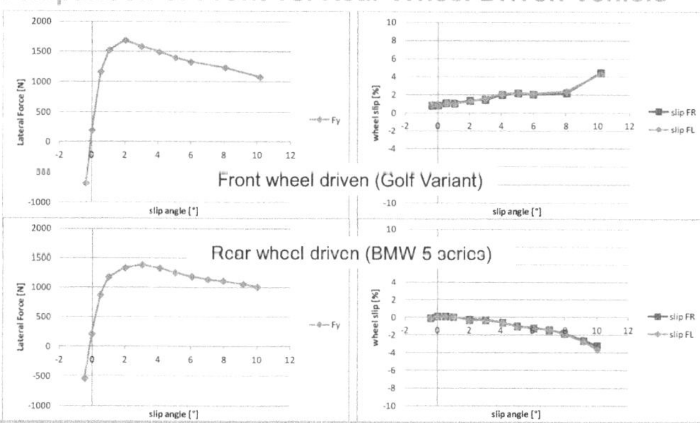


Less influence of driver for BMW due to decoupling of traction force to keep vehicle speed and application of lateral force at the front axle.

Continental

17. Juni 2015
K. Wiese © Continental AG

15

New Testing Technology to Evaluate Lateral Ice Performance of Tires

## Validation of New Test Method

**Comparison of 8 different tire sets tested with 3 different test methods in several runs on different days**

- **Test Methods:**
  - **Ice circle -** Ice skating rink
  - Lateral Characteristics Ice, New Test Method – Contidrom
  - **Lateral acceleration ice outdoor** – Sweden
- **Tire Sets:**
  - Nordic 1
  - Nordic 2
  - **Wi EU 1 – Test reference = 100%**
  - Wi EU 2
  - AS EU
  - US AS
  - SU 1
  - SU 2

Note: values > 100% → better performance

Continental

17. Juni 2015
K. Wiese © Continental AG
16

New Testing Technology to Evaluate Lateral Ice Performance of Tires

## Ice Circle Results – Single Run, 1 Driver

High basis trends possible, due to **changing surrounding conditions**
(e.g. switching off/on air conditioning or ice surface cooling, ...)

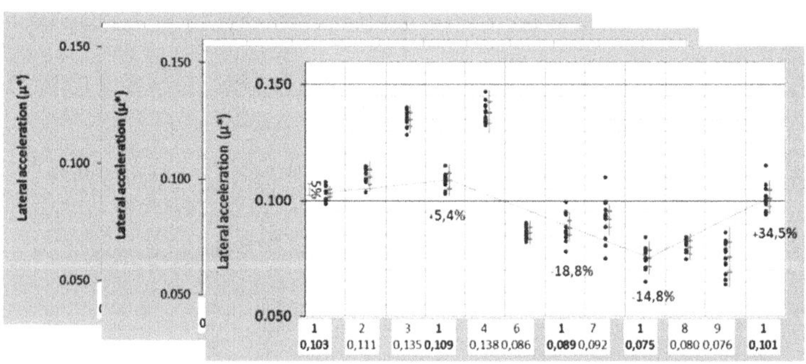


Continental

17. Juni 2015
K. Wiese © Continental AG
17

New Testing Technology to Evaluate Lateral Ice Performance of Tires

## Ice Circle Results from Ice Skating Rink

| Tire Set | Day 1 Driver 1 | Day 2 Driver 1 | Day 3 Driver 1 | Day 4 Driver 1 | Day 5 Driver 1 | Average Driver 1 | Std. Dev. |
|---|---|---|---|---|---|---|---|
| Nordic 1 | 134.8 | 131.3 | 128.2 | 137.9 | 138.4 | **134.1** | 4.4 |
| Nordic 2 | 126.2 | 127.9 | 120.8 | 133.8 | 131.8 | **128.1** | 5.1 |
| Wi EU 1 | 100.0 | 100.0 | 100.0 | 100.0 | 100.0 | **100.0** | |
| Wi EU 2 | 105.0 | 108.0 | 110.7 | 107.2 | 109.7 | **108.1** | 2.2 |
| AS EU | 95.5 | 103.7 | 101.2 | 103.3 | 97.0 | **100.1** | 3.7 |
| US AS | 90.4 | 105.2 | 116.4 | 115.2 | 112.0 | **107.8** | 10.7 |
| SU 1 | 113.0 | 121.3 | 136.9 | 110.5 | 112.1 | **118.8** | 11.0 |
| SU 2 | 82.7 | 108.3 | 117.7 | 103.5 | 109.3 | **104.3** | 13.1 |
| $R^2$ Driver 1 - 2 | 0.88 | 0.91 | 0.93 | 0.95 | 0.89 | 0.98 | |

| Tire Set | Day 1 Driver 2 | Day 2 Driver 2 | Day 3 Driver 2 | Day 4 Driver 2 | Day 5 Driver 2 | Average Driver 2 | Std. Dev. |
|---|---|---|---|---|---|---|---|
| Nordic 1 | 121.5 | 117.9 | 125.9 | 129.5 | 126.0 | **124.2** | 4.5 |
| Nordic 2 | 120.1 | 117.0 | 119.0 | 124.4 | 125.2 | **121.1** | 3.5 |
| Wi EU 1 | 100.0 | 100.0 | 100.0 | 100.0 | 100.0 | **100.0** | |
| Wi EU 2 | 109.2 | 102.5 | 107.7 | 101.8 | 106.0 | **105.4** | 3.2 |
| AS EU | 102.8 | 99.2 | 98.2 | 97.8 | 98.9 | **99.4** | 2.0 |
| US AS | 89.6 | 104.0 | 113.2 | 108.4 | 102.7 | **103.6** | 8.8 |
| SU 1 | 104.8 | 117.0 | 135.8 | 103.3 | 107.4 | **113.7** | 13.5 |
| SU 2 | 79.3 | 107.8 | 125.1 | 105.2 | 112.8 | **106.0** | 16.8 |

| Correlation Driver 1 | $R^2$ |
|---|---|
| Day 1 - Day 2 | 0.77 |
| Day 1 - Day 3 | 0.26 |
| Day 1 - Day 4 | 0.70 |
| Day 1 - Day 5 | 0.66 |
| Day 2 - Day 3 | 0.62 |
| Day 2 - Day 4 | 0.79 |
| Day 2 - Day 5 | 0.84 |
| Day 3- Day 4 | 0.32 |
| Day 3 - Day 5 | 0.43 |
| Day 4 - Day 5 | 0.93 |

| Correlation Driver 2 | $R^2$ |
|---|---|
| Day 1 - Day 2 | 0.25 |
| Day 1 - Day 3 | 0.00 |
| Day 1 - Day 4 | 0.37 |
| Day 1 - Day 5 | 0.31 |
| Day 2 - Day 3 | 0.75 |
| Day 2 - Day 4 | 0.59 |
| Day 2 - Day 5 | 0.70 |
| Day 3- Day 4 | 0.22 |
| Day 3 - Day 5 | 0.35 |
| Day 4 - Day 5 | 0.87 |

Continental

**High standard deviation for SU and US AS tires.**
**Bad reproducibility of results on different days.**

18

New Testing Technology to Evaluate Lateral Ice Performance of Tires

## Ice Traction Result from Ice Skating Rink

Why do the SU and US AS tire sets show a bad reproducibility?

Quick check with available equipment during one testing night:
→ Measurement of μ-slip curves of Nordic 1, Wi EU 1 and SU 1 sets

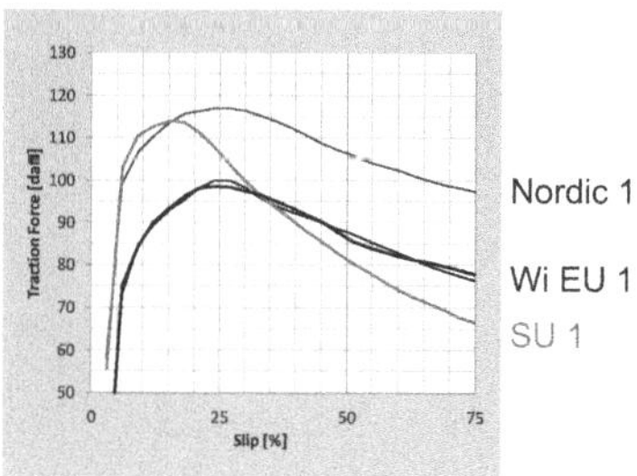


More peaky μ-slip characteristics of SU 1 leads to more unstable results than for Winter Tires. High influence of driver and very sensitive to testing conditions.

Continental

17. Juni 2015
K. Wiese © Continental AG

19

New Testing Technology to Evaluate Lateral Ice Performance of Tires

## Test Results from New Test Method - Contidrom

| Tire Set | Evaluation - Peak [%] Day 1 | Day 2 | Day 3 | Average | Std. Dev. |
|---|---|---|---|---|---|
| Nordic 1 | 133.3 | 157.1 | 151.0 | **147.1** | 12.4 |
| Nordic 2 | 126.5 | 132.8 | 137.3 | **132.2** | 5.4 |
| Wi EU 1 | 100.0 | 100.0 | 100.0 | **100.0** | |
| Wi EU 2 | 109.4 | 116.6 | 113.9 | **113.3** | 3.6 |
| AS EU | 100.1 | 100.4 | 101.4 | **100.6** | 0.7 |
| US AS | 104.5 | 99.2 | 95.6 | **99.8** | 4.5 |
| SU 1 | 84.4 | 74.1 | 72.9 | **77.1** | 6.3 |
| SU 2 | 69.6 | 65.4 | 64.9 | **66.6** | 2.6 |

| Correlation | $R^2$ |
|---|---|
| Day 1 - Day 2 | 0.96 |
| Day 1 - Day 3 | 0.96 |
| Day 2 - Day 3 | 0.99 |

| Tire Set | Evaluation - Slip Angle 1-6° [%] Day 1 | Day 2 | Day 3 | Average | Std. Dev. |
|---|---|---|---|---|---|
| Nordic 1 | 124.6 | 137.7 | 134.2 | **132.2** | 6.8 |
| Nordic 2 | 123.6 | 126.0 | 134.7 | **128.1** | 5.8 |
| Wi EU 1 | 100.0 | 100.0 | 100.0 | **100.0** | |
| Wi EU 2 | 106.7 | 113.6 | 113.2 | **111.2** | 3.9 |
| AS EU | 96.4 | 99.9 | 101.5 | **99.3** | 2.6 |
| US AS | 102.5 | 98.3 | 94.8 | **98.5** | 3.9 |
| SU 1 | 84.1 | 74.6 | 74.1 | **77.6** | 5.6 |
| SU 2 | 67.7 | 66.5 | 65.8 | **66.7** | 1.0 |

| Correlation | $R^2$ |
|---|---|
| Day 1 - Day 2 | 0.95 |
| Day 1 - Day 3 | 0.95 |
| Day 2 - Day 3 | 0.98 |

**Good reproducibility of results on different days.**

Continental 17. Juni 2015 K. Wiese © Continental AG 20

New Testing Technology to Evaluate Lateral Ice Performance of Tires

## Correlation of Ice Circle vs. New Test Method

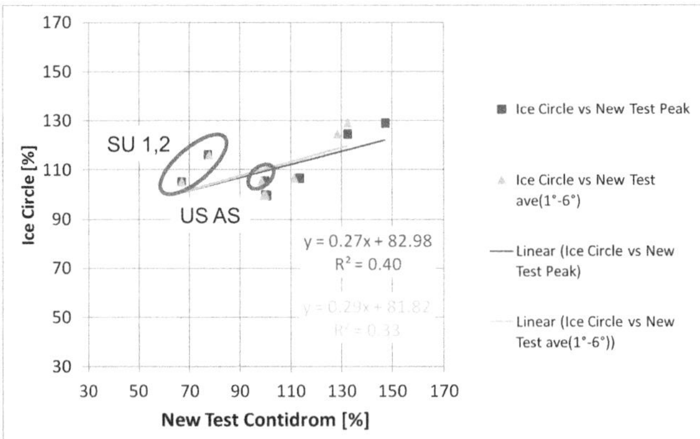


Continental 17. Juni 2015 K. Wiese © Continental AG 21

New Testing Technology to Evaluate Lateral Ice Performance of Tires

## Correlation of Ice Circle vs. New Test Method

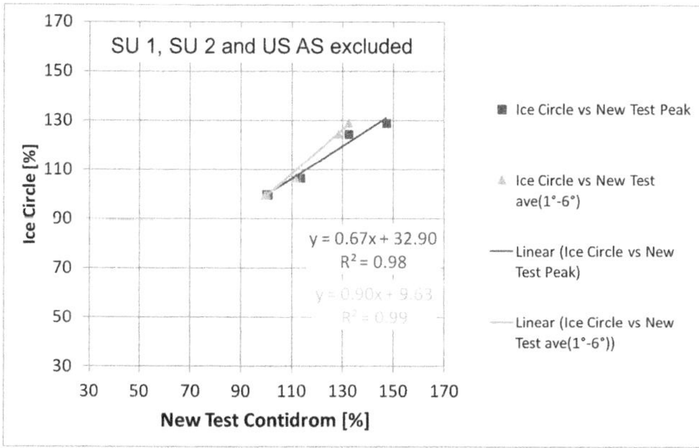


Continental

17. Juni 2015
K. Wiese © Continental AG

22

New Testing Technology to Evaluate Lateral Ice Performance of Tires

## Correlation of Outdoor Measurements vs. New Test Method and Ice Circle

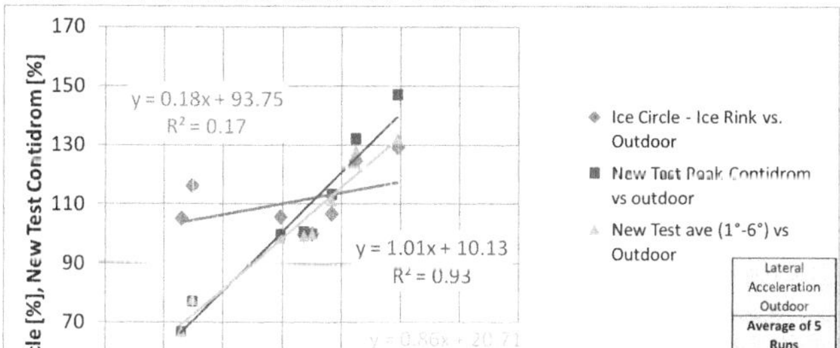


| | Lateral Acceleration Outdoor |
|---|---|
| | Average of 5 Runs |
| Tire | $\mu_y$* |
| Nordic 1 | 128.8 |
| Nordic 2 | 114.7 |
| Wi EU 1 | 100.0 |
| Wi EU 2 | 106.6 |
| AS EU | 97.3 |
| US AS | 89.2 |
| SU 1 | 59.5 |
| SU 2 | 55.8 |

Continental

17. Juni 2015
K. Wiese © Continental AG

23

New Testing Technology to Evaluate Lateral Ice Performance of Tires

## Summary and Conclusions

- ✓ New Test Method to measure **lateral force characteristics of tires on ice** is developed.
- ✓ The measurement of lateral force characteristics enables more detailed **tire performance evaluation and optimization.**
- ✓ **Sensitivity** of test results on **ice and testing conditions** is obvious.
- ✓ Test conditions in simple ice skating rinks can lead to opposite rankings compared to average outdoor ice conditions.
- ✓ The new test method in the **for tire testing designed ice hall** enables optimization of winter tires **to improve the safety on the most slippery road conditions**.

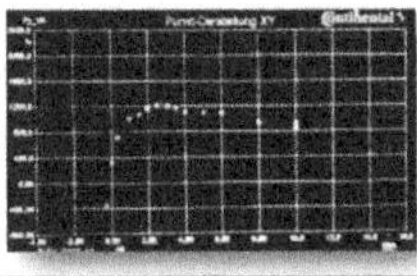

17. Juni 2015
K. Wiese © Continental AG 24

New Testing Technology to Evaluate Lateral Ice Performance of Tires

## Literature

[1] Manfred Mitschke und Henning Wallentowitz, *Dynamik der Kraftfahrzeuge*, 5., überarb. u. erg. Aufl. 2014, DOI: 10.1007/978-3-658-05068-9, © Springer Fachmedien Wiesbaden 2014

[2] P. Pfeffer, M. Harrer (Hrsg.), *Lenkungshandbuch,* DOI 10.1007/978-3-8348-8167-0_5, © Vieweg + Teubner Verlag | Springer Fachmedien Wiesbaden GmbH 2011

[3] Günter Leister, *Fahrzeugreifen und Fahrwerkentwicklung*, Strategie, Methoden, Tools, DOI: 10.1007/978-3-8348-9312-3, © Vieweg+Teubner Verlag | GWV Fachverlage GmbH, Wiesbaden 2009

[4] T. M. Kessel, R. Mundl, B. Wies and K. Wiese, "An Analytical Thermodynamical Approach to Friction of Rubber on Ice," *Tire Science and Technology,* p. 124, 2012.

[5] M. Gießler, F. Gauterin, K. Wiese and B. Wies, "Influence of Friction Heat on Tire Traction on Ice and Snow," Tire Science and Technology, TSTCA, Vol. 38, No. 1, January – March 2010, pp. 4-23.

[6] M. Gießler, "Mechanismen der Kraftübertragung des Reifens auf Schnee und Eis", Dissertation, Karlsruher Institut für Technologie Fakultät für Maschinenbau, 2011, Karlsruher Schriftenreihe Fahrzeugsystemtechnik Band 11

17. Juni 2015
K. Wiese © Continental AG 25

# Sense-Plan-Act – the role of chassis systems

Ralph Lauxmann, Continental Teves AG & Co. oHG

**This manuscript is not available according to publishing restriction. Thank you for your understanding.**

# Consumer acceptance ensured through safety

Klemens Schmiederer, TÜV SÜD AG

# Abstract

As automotive technology has advanced rapidly, safety has joined comfort and performance as one of the key requirements. Analysis and functionality testing of all safety components throughout the vehicle service life play a central role in this context.

Working together, OEMs and third-party testing and inspection organisations make sure that the above requirements are fulfilled. This safety partnership also plays a prominent role for consumer acceptance. People will only accept technical innovations to take over control in cars if they continue to feel safe.

Innovations in active safety have particularly improved vehicle safety in recent years. Passive safety, too, has advanced by leaps and bounds. The safety of adults and children in cars has been enhanced further. Passenger cells and car bodies have been developed to a level of maturity at which the focus of interest can now move towards the protection of other road users outside the car. Today, pedestrian protection and accident prevention are at the top of the agenda of automotive developers.

To reach these goals, we need to further improve the interoperability of active and passive safety systems in cars. In situations where an accident can no longer be avoided, these systems need to interact to ensure higher safety. In line with the above, Integrated Safety has become a modern byword – all the more as cars take over more and more of the driver's responsibilities.

Innovative chassis systems, which also extend to electric cars, play an outstanding role in this context. Electronic chassis components, active steering, drive-by-wire, automatic levelling and tyre pressure monitoring systems are some of the keywords representing the interoperability of vehicle safety components.

Preventing accidents and ensuring the functionality of all safety systems throughout a vehicle's service life is only possible with the combined efforts of all stakeholders throughout the vehicle lifespan. TÜV SÜD accompanies the design and development activities of experts in industry. Our multi-disciplinary professionals support customers throughout the design and development process. TÜV SÜD's experts and vehicle inspectors hold designations and accreditations for more than 400 international regulations.

The portfolio of services extends from 360-degree safety concepts and risk analysis for whole vehicles and vehicle systems to the measurement of rolling resistance aimed at improving fuel consumption. The service package also includes consulting during the development of the system, from the definition of suitable services to employee training. Also included in the services is the integration of advanced driver assistance systems (ADAS), both active and passive such as airbags and seat belt tensioners. Other issues range from quality assurance for failure modes and effects analysis (FMEA) to driving tests. Safety management, hazard and risk analysis and safety certificates for hybrid and electric vehicle projects round off the range.

The manufacturer's authorised representative in the EU is Springer Nature Customer Service Centre GmbH, Europaplatz 3, 69115 Heidelberg, Germany. If you have any concerns regarding our products, please contact ProductSafety@springernature.com

Printed and bound by CPI Group (UK) Ltd, Croydon, CR0 4YY

07/01/2026

02030924-0007